MICROWAVE ENGINEERING

Third Edition

ABOUT THE AUTHORS

Annapurna Das received her BSc (Hons.) and MSc degrees in Physics from Calcutta University, India; MTech in Microwave Electronics and PhD degrees in Electrical Engineering from University of Delhi, India. As a senior research fellow of CSIR, New Delhi, she was engaged in doctoral research, and teaching undergraduate courses in the Electrical Engineering Department, Delhi College of Engineering, during 1980 to 1982, and doctoral research in Indian Institute of Technology (IIT) Madras, Chennai, during 1983 to mid-1985.

She was faculty in the Department of Electronics and Communication Engineering, Anna University, Chennai, since 1985 and became Professor in 1998. She was in charge of the microwave laboratory and was engaged in a number of training, research and teaching programmes of the Anna University in the areas of microwaves, EMI/EMC, electromagnetics and antennas. Presently, she is Principal of Guru Nanak Institute of Technology, Kolkata, under West Bengal University of Technology. She has nearly 60 research publications.

Prof Das worked at the Interference Control Technology, USA, for two months (June–August '91) as Research Fellow under the United Nations Development Programme. She visited several R&D Institutions and Industries in USA and was Chief Investigator of R&D projects of DST, AICTE, and the Indian Navy.

She is the co-author of *Microwave Engineering,* published by McGraw Hill, USA, Singapore and India. She is a life member of Society of EMC Engineers (India) and Indian Society of Technical Education (ISTE). Her current and major areas of research interest are microwaves, EMI/EMC, Micro-strip Antennas and Electromagnetics.

Sisir K Das obtained BTech, MTech and PhD degrees from Calcutta University, IIT Kharagpur and Anna University, respectively. He started his carreer as teaching and research faculty at the Delhi University in 1977, and joined the Ministry of Commerce & IT, Govt. of India, as a scientist in 1980. As head of EMC, Dr Das led EMC evaluation and design of electronics products manufactured by the industry, meeting international standards, and electromagnetics research projects in the country and abroad for 30 years. He continues to conduct numerous EMC and high-speed/frequency training programmes for engineers in the industry and educational institutions. He has conducted numerous classes at a large number of engineering colleges as visiting faculty. Presently, he is Professor and Dean—Research & Administration, Guru Nanak Institute of Technology, Kolkata.

Dr Das worked at National Institute of Standards and Technology, US Department of Commerce, USA, for three months (July–October '91) as Research Fellow under the United Nations Development Programme. He visited several R&D institutions and industries in USA, Japan and South Korea, for technical interaction/ collaboration programmes.

He is the co-author of *Microwave Engineering,* published by McGraw Hill, USA, Singapore and India. He has written three chapters in *Engineering EMC*, published by IEEE Press. He has nearly 110 research publications in journals and conference proceedings.

Dr Das is a Senior Member of IEEE, Life member of Society of EMC Engineers (India), and Chief Editor of *EMC Journal of Society of EMC Engineers*. His earlier professional involvements are Associate Editor for *IEEE EMC Journal*, USA (1994–2000), and Manager of ISO 9001 & NABL 17025 Quality Systems (1998–2004).

Dr Das received Society of EMC Engineers' (India) highest award in 2002 in recognition of his contribution to the EMI/EMC solutions for Indian Industrial Products, and received a silver medal for his contribution to the workshop "SAMEER 2001 and Beyond", organized by DoE/SAMEER in 1995.

MICROWAVE ENGINEERING

Third Edition

Annapurna Das

Principal
Guru Nanak Institute of Technology
JIS Group Education Initiatives
Kolkata, West Bengal

Sisir K Das

Dean
Guru Nanak Institute of Technology
JIS Group Education Initiatives
Kolkata, West Bengal

McGraw Hill Education (India) Private Limited

Published by McGraw Hill Education (India) Private Limited,
444/1, Sri Ekambara Naicker Industrial Estate, Alapakkam, Porur, Chennai 600116, Tamil Nadu, India.

Microwave Engineering, 3e

Eleventh reprint 2019
RZCYCRBTDRRBB

Print Edition
ISBN (13 digits): 978-93-329-0287-9
ISBN (10 digits): 93-329-0287-9

E-book Edition
ISBN (13 digits): 978-93-329-0288-6
ISBN (10 digits): 93-329-0288-7

12 13 14 15 16 AVO 23 22 21 20 19

Managing Director: *Lalit Singh*
Senior Manager—Content Development & Production Services: *Shalini Jha*
Head—Production Services: *Satinder S Baveja*
Assistant Manager—Production Services: *Suhaib Ali*
General Manager—Production: *Rajender P Ghansela*
Manager—Production: *Reji Kumar*

Typeset at Text-O-Graphics, B-1/56, Aravali Apartments, Sector 34, Noida 201301, and,
Printed and Bound in India at Avon Printers, Plot No. 16, Main Loni Road, Shahdara, Delhi 110094.

Visit us at: www.mheducation.co.in ; (Toll free in India): **1800 103 5875**
Write to us at: info.india@mheducation.com

CIN: U80302TN2010PTC111532

Dedicated

to

Our Parents

CONTENTS

PREFACE

The importance of RF and Microwave Engineering is increasing rapidly with progress in the applications of radar, wireless Communications, astronomical research, domestic appliances, security systems, etc. The first edition of this book was published in 2000 in India, in 2004 in Singapore and in 2007 in USA. We are gratified with its high acceptability and use by university communities as a textbook.

The second edition was published in 2009 in India to enhance the educational contents of the book. In addition, corrections of some errors noticed in the first edition, introduction of a new Chapter 4 on Microwave Integrated Circuit Design and Manufacturing, and coverage of new topics on Microwave Passive Devices, Microwave Amplifiers and Oscillators were some of the changes that reflected in the second edition.

The authors have been fortunate enough to interact with Professor H Brian Sequeira, Johns Hopkins University, Laurel, MD, USA, who is referring to the second edition to teach his students a course on Electromagnetic Transmission Systems in the Engineering Program for part-time professionals (EPP) at The Johns Hopkins University (JHU), USA. He has appreciated the authors for producing good coverage of topics in lucid language. At the same time, he has suggested changes to some inconsistent notations and pointed out typographical errors in the book. He has extended tremendous support to the authors by developing a table of some erroneous contents and corresponding corrective measures for the entire book almost line-by-line. This has motivated the authors to bring out the third edition of the book incorporating his suggestions and some additional topics which the authors thought appropriate for further advancement of the book. Professor Brian's comments are given separately at the end of the preface.

In the third revised edition, fundamental principles are stressed and complete derivations are provided for all significant formulas and relationships. All important fundamental concepts and principles of electromagnetic and transmission line theory, which are useful for understanding microwave engineering, are covered to the extent possible within a text of reasonable size.

Salient Features

- Application-oriented approach: expanded coverage on Radar
- Updated with newest technologies: Gunn diodes, IMPATT, etc.
- Detailed derivations help in conceptual understanding
- Inclusion of newer technologies such as MESFET, HMT, etc.
- Rich pedagogy:
  - ✧ Diagrams: *417*
  - ✧ Examples: *132*
  - ✧ Review Questions: *185*
  - ✧ Also includes *MATLAB* results

Chapter Organization

The following is the summary of the chapters and changes made in the third edition from the second one.

Chapter 1 is an updated introductory chapter. **Chapter 2** covers a comprehensive summary of basic electromagnetic theory needed as background for proper understanding of microwave engineering. It serves as concise reference or review of familiar topics to those who have prior knowledge of these.

Chapter 3 provides a more basic knowledge on RF and microwave transmission lines using distributed circuit models. A more detailed coverage is included in this edition. This chapter includes transmission line theory and characteristic parameters, propagation of pulse signals, characteristics of coaxial and planar transmission lines such as strip lines, microstrip lines, coplanar lines, and slot lines. The chapter covers the theories of rectangular and circular waveguides. Power-handling capability of microwave transmission lines is also covered.

Chapter 4 describes microwave integrated circuit design and manufacturing with some modification of the content of the second edition. **Chapter 5** covers impedance transformation for matching with more clarity and modifications to the second edition.

Chapter 6 develops the basic microwave circuit theory and passive devices. The chapter includes detailed impedance, admittance, ABCD, and scattering matrix descriptions of microwave junctions. Properties of scattering parameters for shift in reference planes, matched network, reciprocal network and lossless networks are described with proof. Derivation of *S*-matrix for a large number of network configurations is given for clear understanding. Some additional microstrip discontinuities are covered for design guidelines.

Chapter 7 includes microwave resonators where some corrections and modifications in the write-up are given in the theory of coupled cavities. **Chapter 8** covers microwave filters with correction and modification to the material given in the second edition. No major modifications are carried out in **Chapter 9**, which describes describing microwave vacuum tube devices.

Chapter 10 on microwave solid-state devices and circuits now includes a different treatment of the Gunn diode with more lucid explanation of its operation and applications. A little more theory is added to IMPATT diodes. The High Electron Mobility Transistor (HEMT) is added in this chapter in addition to some theoretical formulae for MESFET.

Chapter 11 on applications of microwaves is modified with the microwave radar system covered in more detail such as tracking radars and phased array radar. **Chapter 12** on microwave radiation hazards has only a minor change.

Finally, **Chapter 13** on microwave measurements includes additional diagrams and some modifications in scattering parameter measurements. A new measurement technique for cavity *Q* using network analyzer with brief theory of coupled cavity has been added. This will help establish the experimental set-up in laboratories for microwave measurements.

The authors have tried to provide a broad, comprehensive treatment of the fundamental theory, principles, and designs that are the foundation for studying Microwave Engineering. Many references have been included for the benefit of the reader who wishes to pursue a given topic in greater depth.

Almost all chapters include many solved numerical examples for better understanding and also given exercise problems for practice.

Online Learning Centre

The book is accompanied by an online learning center, available at *http://www.mhhe.com/das/me3* that offers PowerPoint lecture slides.

Acknowledgements

The authors gratefully acknowledge Professor H Brian Sequeira, Johns Hopkins University, Laurel, MD, USA, for his enormous help, criticisms, and constructive suggestions for the improvement of this edition. Special thanks goes to their scientist friend Dr U C Ray, Scientist G (Retd.), SSPL, DRDO, for his interest and suggestive write-up which helped in modifying chapters 4, 10 and 11. The authors express hearty thanks to their colleagues at GNIT, Kolkata: Antara Ghosal, Anurima Majumdar, Aditi Sengupta, Avali Ghosh and Somalina Chowdhury for proofreading and some computational assistance for preparing the manuscripts. The authors thank all the teachers in the past for their contributions during their student lives.

The authors are also grateful to the editorial team of McGraw Hill Education (India) for their interest in the production of the book. They acknowledge with pleasure the invaluable suggestions and constructive criticisms of the reviewers of the book. Their names are given below.

Ramnath Datta	St. Thomas College of Engineering and Technology, Kolkata
Milind Shah	Government Engineering College, Modasa
A Ponraj	Easwari Engineering College, Chennai
Arun Kumar BV	APS College of Engineering, Belgaum
Nrimala Priya	Sona College of Technology, Salem

The authors extend their special thanks and love to their son, Amlan Das, daughter-in-law, Sanghamitra, and granddaughter, Saavarna, for inspiring them to complete the book.

Annapurna Das
Sisir K Das

Publisher's Note

McGraw-Hill Education (India) invites suggestions and comments from you, all of which can be sent to *info.india@mheducation.com* (kindly mention the title and author name in the subject line).

Piracy-related issues may also be reported.

REVIEW SUMMARY OF SECOND EDITION

H Brian Sequeira
Johns Hopkins University
Laurel, MD, USA

A review of the second edition of the above titled text revealed excellent content. Coverage spans a good breadth of topics which have been treated to the appropriate depth. Discussions and explanations are lucid, and easy to comprehend in almost all cases.

I recommend that attention be focused on the form of presentation. Top priority should be given to elimination of typographical errors that pervade the text. Next, attention should be given to notational inconsistencies, which I have described in additional detail in this document. Third, some sentences are semantically ambiguous, and should be altered to eliminate confusion for the reader. Attention to these factors should improve readability significantly.

I have prepared chapter-unique attachments that identify typographical errors, notational inconsistencies, ambiguous phrasing, and sequencing to undertake some alterations and corrections.

This documents confined to notation and the use of fonts and layout of equations to improve readability. It is intended to cover recurring instances that were too numerous to flag at every occurrence.

Since 1990 onwards, this reviewer has been teaching a course on Electromagnetic Transmission Systems in the Engineering Program for Part-time professionals (EPP) at The Johns Hopkins University (JHU), USA. A typical student has had an undergraduate background in electromagnetics about 10 years on the average prior to attending this course. Nevertheless, inconsistent notation has confused the student and motivated this instructor to develop a "convention" that has greatly improved comprehension of his class notes distributed to students. This system of notation has been recommended to avoid ambiguity for the beginning or returning student.

It is suggested to represent instantaneous quantities in script font, and vector quantities in bold script font. Independent variables and material parameters such as space coordinates, and time by Times Roman italics font thus: x, y, z, t. Similar treatment can be given to other independent chosen variables such as frequency, f. Also use Times Roman italics to represent scalar quantities such as the amplitude of a scalar or vector field or the components of a vector. Thus, J_1 represents the amplitude of a current density field, and E_x, E_y, E_z, represent the components of a vector. Similarly, use italics symbol font for permittivity, ε; permeability, μ; conductivity, σ; wavelength, λ; radian frequency, ω; skin depth, δ; and other variables like these.

The functions may be represented by Times Roman plain (regular) font such as sin, cos, sinh, cosh, exp, or its abbreviation, e, Bessel function, J_1 (compare with current density J, etc. Also use italics for the subscript/superscript such as E_x, δ_c, or Z^m, respectively.

If, however, the subscript/superscript represents a numerical index, use plain font, such as H_{mn}, for the mnth mode in a waveguide. The representation of scalar phasor quantities follows the prescription described under "Independent Variables and Material Parameters." The representation of vector phasor quantities is in Bold Times Roman font.

It is important to adhere to the convention that **E** and **B** and phasor quantities in general are NOT dependent on time. Rather, the time dependence resides with their instantaneous counterparts through the relation E = Re[$\mathbf{E}ej\omega t$], where Re denotes "real part of".

The unit vectors along rectangular coordinate axes may be represented by $\hat{\mathbf{x}}$, $\hat{\mathbf{y}}$, $\hat{\mathbf{z}}$. Likewise, a unit vector perpendicular to an interface is $\hat{\mathbf{n}}$.

A Bit About H Brian Sequeira

H Brian Sequeira, born in India, obtained a BSc and MSc in Physics from University of Bombay and Indian Institute of Technology, Bombay (IIT-B), respectively, and MTech in Physical Engineering and PhD in Electrical Engineering from Indian Institute of Science, Bangalore, and University of Delaware, respectively. Prior to his PhD degree, he developed receiver hardware for a command-guided system for DRDL, Hyderabad, and worked as Instrumentation Officer at IIT-B.

Following completion of his PhD, he worked for Martin-Marietta (now Lockheed-Martin) Corporation where he developed millimeter-wave integrated circuits based on 0.1 μm pseudomorphic High-Electron Mobility Transistors (p-HEMTs) at frequencies up to 94 GHz. At that time, Dr Sequeira developed techniques which were not then available for characterizing materials, devices, and circuits at these frequencies. He was then appointed Technical Director for development of active homing missile radars which has been successfully completed and deployed.

Dr Sequeira next turned his attention to architecting low-cost terminals for fixed satellite services at C- and Ku-bands. These were deployed in rural areas where the terrain is inhospitable for cost-effective roll-out of land-based telephony infrastructure. Dr Sequeira is now with Applied Physics Laboratory of The Johns Hopkins University where he participated in development of spacecraft assets for Deep-Space communications, and radar instruments for Synthetic Aperture Radar (SAR) imaging of the Moon.

While at Martin-Marietta, Dr Sequeira began teaching Microwave Engineering in the Part-Time Program for Engineering Professionals at Johns Hopkins University, and has maintained that association to the present day. He is a senior member of the IEEE, and was Program Planning Chair of the recently concluded 2011 International Microwave Symposium held in Baltimore.

1 INTRODUCTION

INTRODUCTION 1.1

Microwave is the region in the electromagnetic (EM) wave spectrum in the frequency range of about 1 GHz (= 10^9 Hz) to 100 GHz (= 10^{11} Hz). This corresponds to a range of wavelengths from 30 cm to 0.3 cm in free space ($\lambda = c/f$). The free space is characterized by the electrical medium parameters: permittivity $\varepsilon_o = \frac{10^{-9}}{36\pi}$ F/m, permeability $\mu_0 = 4\pi \times 10^{-7}$ H/m, conductivity $\sigma_0 = 10^{-14}$ S/m, intrinsic impedance $\eta_o = \sqrt{\frac{\mu_o}{\varepsilon_o}} = 120\pi \approx 377\ \Omega$, and velocity of wave propagation (light) $c = 2.9979 \times 10^8$ m/S. During World War II, microwave engineering became a very essential consideration for the development of high-resolution radars capable of detecting and locating enemy planes and ships through a narrow beam of EM energy. Such a beam could be achieved by means of a paraboloid antenna of large diameter compared to the wavelength of radiation. Therefore, size of the antenna to be carried by an airplane on board could be reduced by using such short wavelengths of microwaves.

The microwave band is capable of making economic transmission of a large number of communication channels or TV programmes by modulating all these channels or programmes into a single microwave carrier and transmitting them over one communication link. This is possible because the modulation side-bands are just a few per cent of the microwave carrier frequency.

Because of very short wavelengths, microwaves are capable of almost freely propagating through the ionized layers in the atmosphere. This facilitates radio-astronomic research of space, and communication between the ground stations and space vehicles.

According to the quantum theory, the quantum of energy at microwave frequencies becomes comparable to the difference in energies between adjacent energy levels of atoms and molecules. Moreover, molecular, atomic and nuclear systems exhibit resonances in the microwave range. These lead microwaves to become a powerful tool in microwave radio spectroscopy for material analysis.

In more recent years, microwaves are being widely used in domestic microwave ovens for rapid cooking, and also in industrial and medical uses for microwave heating.

Microwave engineering has a marked difference from conventional electronics engineering because of the short wavelengths involved. Conventional low-frequency circuit analysis based on Kirchhoff's laws and voltage-current concepts of the distributed transmission line theory no longer apply here. This is because the propagation time of electrical effects from one point in a circuit to another point at microwave frequencies is comparable with the period of the oscillation of currents and charges. Moreover, conventional circuits or lines radiate out the microwave energy as electromagnetic waves resulting in high loss in signal transmission. Thus, microwave transmission involves propagation of EM waves consisting of changing electric and magnetic fields in a medium. These specific properties of microwaves steered engineers to develop techniques for microwave-guided structures such as waveguides, coaxial lines, striplines, microstrip lines, slot lines and microwave sources like klystrons, magnetrons, travelling-wave tubes, backward wave oscillators, microwave solid-state devices, and other microwave circuit components like attenuators, phase-shifters, isolators, circulators, directional couplers, detectors, mixers, etc.

Microwave frequencies are grouped into several smaller bands, which are designated and listed in Table 1.1.

Table 1.1 Microwave bands (IEEE and commercial designation)

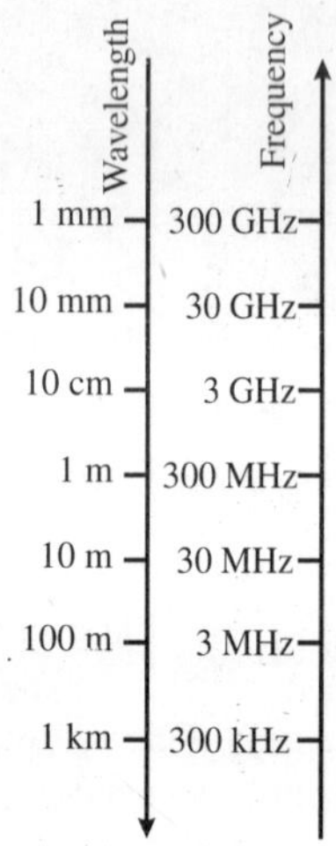

Frequency range	Band designation
110–170 GHz	D-band
75–110 GHz	W-band
60–90 GHz	E-band
50–70 GHz	V-band
40–60 GHz	U-band
33–50 GHz	Q-band
26.5–40 GHz	Ka-band
18–26.5 GHz	K-band
12.4–18GHz	Ku-hand
8–12.4 Ghz	X-band
4–8 GHz	C-band
2–4 Ghz	S-band
1–2 GHz	L-band
300–3000 MHz	UHF-band
30–300 MHz	VHF-band

High-frequency signals are used in wireless communication systems for the efficient transmission of information because an antenna radiates efficiently if its size is comparable to the signal wavelength. Since the signal frequency is inversely related to its wavelength, smaller size antennas operating at RFs and microwaves have higher radiation efficiencies. Also the transmission of broadband signals requires a high-frequency carrier signal to accommodate large number of channels. In addition to the traditional applications in communication, such as radio and television, RF and microwave signals are now used in cordless phones, cellular communication, local area networks, wide area networks, and metropolitan area networks, personal communication service, radio-frequency identification (RFID), cordless mouse or keyboards for computers. RF is superior to infrared technology since any reasonable obstruction does not affect line-of-sight operation.

It is known that the electrical characteristics of the atmosphere affect the propagation of electromagnetic signals. Unlike RF signals the ionosphere does not reflect microwaves, and the signals propagate line of sight. Hence, curvature of the Earth limits the range of a microwave communication link to less than 50 km. One way to increase the range is to place a human-made *satellite* in the space to received LOS microwave signals through ionosphere and retransmit to the earth through the transponders. This called *satellite communication system*. Another way to increase the range of a microwave link is to place repeaters at periodic intervals of about 30-50 km. This is known as a terrestrial communication system. Tables 1.2–1.4 list selected devices used at RF and microwave frequencies and applications of microwaves.

Table 1.2 Microwave devices

Microwave Devices
- Active
  - Solid-State
  - Vacuum-Tube
    - Linear Beam
      - Klystrons
      - Traveling wave tubes
    - Magneton
- Passive
  - Directional couplers
  - Attenuator
  - Phase shifters
  - Resonators
  - Filters
  - Circulators/Isolators
  - Antennas

Table 1.3 Solid-state devices used at RF and Microwave Frequencies

Solid-State devices
- Transistors
- Transferred Electron
  - Gunn diode
- Avalanche Transit Time
  - BARITT diode
  - IMPATT diode
  - TRAPATT diode
  - Parametric devices

Table 1.4 Some applications of microwaves

Microwave Applications
- Communication
  - Terrestrial
  - Satellite
  - Wireless/Mobile
- Radar
  - Civilian
    - Air traffic control
    - Aircraft navigation
    - Ship safety
    - Space vehicles
    - Remote sensing
    - Law enforcement
  - Military
    - Surveillance
    - Navigation
    - Guidance of weapons
    - Electronic warfare
- Industrial and Biomedical
  - Process control
  - Drying
  - Curing
  - Treatment of elastomers
  - Waste treatment
    - Nuclear
    - Cellulosic
  - Monitoring
  - Imaging
  - Hyperthermia
- Heating
  - Industrial
  - Household

HISTORICAL RESUME 1.2

There are some pioneers who have laid the foundations of microwave engineering. Some of them are worth mentioning and are described as follows:

James C Maxwell, the founder of the electromagnetic theory of radiation, presented to the Royal Society in 1864 a paper titled *A Dynamical Theory of the Electromagnetic Field*, which described the properties of electromagnetic fields in terms of 20 equations. The equations constitute the well-known Maxwell's equations. He had predicted theoretically the existence of electric and magnetic fields associated with electromagnetic wave propagation and made the foundation of modern electromagnetic theory in 1873. **Oliver Heaviside** gave a further modern form to Maxwell's formulation for practical applications of guided waves.

In 1893, **Heinrich Hertz** first conducted an experiment to show that a parabolic antenna fed by a dipole, on excitation by a spark discharge, sends a signal by wave motion to a similar receiving arrangement at a distance. He gave strong experimental support for the theoretical conclusions drawn by Maxwell for electromagnetic fields.

William Thompson (1893) developed the waveguide theory for propagation of microwaves in a guided structure. Later, **Lodge** (1897–1899) established the mode properties of propagation of EM waves in free space and in a hollow metallic tube known as the waveguide.

Sir J C Bose (1895, 1897, 1898) generated millimetre waves using a circuit developed in his laboratory and used these waves for communication, and developed microwave spectrometers, diffraction gratings, polarimeters and detectors for conducting microwave experiments. He also developed microwave horn antennas, which are still considered useful feeds for reflector antennas.

In 1937, the microwave vacuum-tube Klystron was developed by **Russel** and **Varian Bross**. **J D Kraus** (1938) developed the corner reflector antenna for electromagnetic wave transmission. **J T Randall** and **H A H Boot** first made the cavity magnetron as high-power microwave oscillator in 1940. **Kompfner** (1944) developed the microwave travelling wave tube. **Percy Spencer** (1946) built the microwave oven for domestic cooking. Some modern devices were developed after 1950. **Deschamps** (1953) developed the microstrip antenna. **S B Cohn** (1955), **H A Wheeler** (1964) and followers advanced the microwave integrated circuits and applications. **J B Gunn** (1963) developed the Gunn diode for microwave generation using solid-state material such as GaAs not mentioned here. Interested readers may refer textbooks, journals and scientific magazines for further information.

MICROWAVE INTEGRATED CIRCUITS 1.3

Microwave integrated circuits can be made by hybrid or monolithic techniques on a variety of substrate materials as described in Table 1.5.

Table 1.5 Properties of microwave substrates

Type of Substrate	*Relative dielectric constant (Î$_r$) at 25°C*	*10^4 at tan δ at 10 GHz*	*Thermal conductivity (W/cm°C)*
SEMICONDUCTOR			
Si	11.7–12	40–150	0.9 (high resistivity)
GaAs	12.8–13	16	0.3 (high resistivity)
InP	12.5–14	–	0.68
CERAMIC			
Alumina	9.6–9.9	1–2	0.2–0.3
Beryllia	6.6–6.8	3	2.5
Fused Silica	3.8	<1	0.013
Sapphire	9.3–11.7	<1	0.38–0.46
POLYMERIC			
Woven PTFF/glass	2.2	9	0.0026
Non–woven PTFE/glass	2.1	4.5	–0.003
Ceramic filled PTFE/glass	10.5	15	0.004
Polyolefin	2.3	1	0.001
Ferrite/Granite	1316	2	0.03

Source: Frey J and Bhasin K [1.5]

Hybrid and Monolithic Integrated Circuits 1.3.1

In a hybrid integrated circuit, the circuit interconnections are made by metal lines deposited on an insulating substrate. The active devices in the circuit are bonded to the substrate and these devices may be made of different materials (e.g. Si, GaAs, InP); hence, the term *hybrid*. Hybrid circuits can be thin-film or thick-film type. In thin-film hybrid circuits, the metallization is deposited by evaporation and the desired circuit pattern is made by photolithographic techniques. In thick-film hybrid circuits, the metallization is usually deposited by silk-screening, in which the desired pattern is incorporated into the screen itself. In hybrid circuits, both active and passive components are separately fabricated and bonded to the substrate by soldering, thermocompression bonding, or by using a conductive adhesive. Sometimes capacitors, resistors and inductors may be fabricated using the substrate metallization itself, along with deposited insulating films. But the component values achievable by the latter method are usually somewhat limited.

In a monolithic integrated circuit, all components are formed together by various processes, such as diffusion or ion implantation, which changes the physical properties of the same block of substrate material. Inductors are not available at all in low-frequency monolithic ICs. Non-integrable external components, such as piezoelectric crystals, can be used to establish circuit resonances. Circuit interconnections are established by metallization in the same fashion as in thin-film hybrid ICs.

In microwave hybrid integrated circuits, interconnections are usually established by Quasi-TEM-mode transmission lines such as microstrip lines. Table 1.5 shows the characteristics of the various insulating substrates suitable for microwave integrated circuits. To minimize loss, thin-film is preferred than thick-film techniques at microwave frequencies. Active devices, such as IMPATT, Gunn diodes or transistors, are now commonly used in hybrid microwave integrated circuits.

In a monolithic microwave IC, active devices are fabricated in a single block of semiconductor material and connected by metal lines. The metal lines are segments of microstrip lines and must be designed with care.

For various reasons, gallium arsenide is preferred over silicon for microwave monolithic ICs. Many of the microwave solid-state active devices achieve their optimum performance in GaAs. In addition, the higher dielectric constant of GaAs permits narrower transmission lines, leading to compactness in the design.

The two key elements of microwave circuits are transmission lines and active solid-state devices. Both are described in this book in appropriate chapters.

REFERENCES

1. *Historical Perspectives or Microwave Technology*, Special Centennial Issuc, IEEE Trans., Vol. MTT-32, September, 1984
2. Montgomery, C G: *Technique of Microwave Measurements*, McGraw-Hill Book Company, New York, 1947
3. Ginzton, E L: *Microwave Measurements*, McGraw-Hill Book Company, New York, 1957
4. Bailey, A E (Ed.): *Microwave Power Engineering*, Academic Press, New York, 1961
5. Frey, J and Bhasin, K B: *Microwave Integrated Circuits*, Artech House, Inc. 1985

2 PROPAGATION CHARACTERISTICS OF ELECTROMAGNETIC WAVES

INTRODUCTION 2.1

Electrical signals are transmitted from the sending end to the receiving end in the form of currents and voltages through conductors of transmission lines, in the form of electromagnetic waves through guiding structures such as waveguides and optical fibres, or in the form of electromagnetic waves by radiation in space. Some fundamental theories of propagation of EM waves is described in this chapter. Electromagnetic waves are generated and radiated by accelerated electric charges or by time-varying electric current in a conductor.

In electromagnetic wave propagation in a medium, a changing magnetic field will induce an electric field and vice versa. The waves propagate with two oscillatory fields, one electric (**E**) and other magnetic (**H**). The regions close to the radiating sources are most likely to have high intensity of fields, having both longitudinal and transverse components with respect to the direction of propagation. In general, such locations are characterized by complicated field structure, including reactive (stored) and real (propagated) energies, irregular phase surfaces, and unknown field polarization.

The propagation characteristics of electromagnetic waves, in general, depend on the electrical parameters (ε, μ, σ) of the medium and the presence of the boundaries or interfaces between two media. In free space, at sufficiently large distances from the source, wave propagation takes place in Transverse Electromagnetic Mode (TEM) from one point to another one. In this mode, **E** and **H** fields are orthogonal to each other and both are perpendicular to the direction of propagation as shown in Fig. 2.1 for time harmonic fields. This is also called **plane waves**.

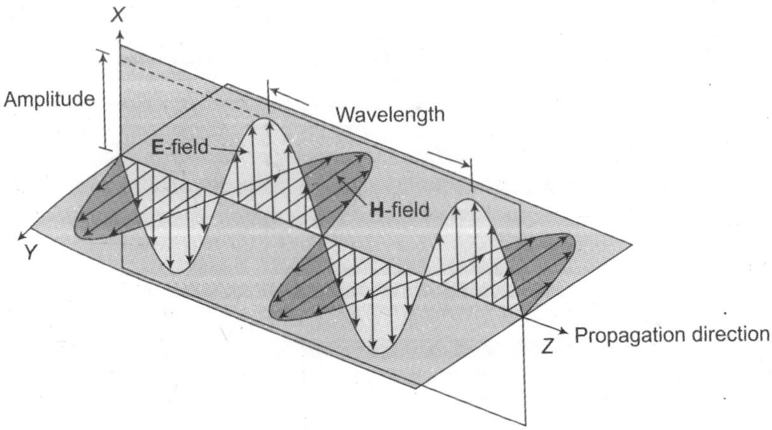

Fig. 2.1 Plane-wave propagation

Wave Propagation 2.1.1

The electromagnetic signal may be propagated from the transmitter to a distant receiver as a direct wave, ground reflected wave, surface wave or as a sky wave depending on the frequencies. These propagation paths are explained in Fig. 2.2.

The ionospheric region (50 km above the earth) contains gases which are ionized due to solar radiation. All the frequencies which are lower than a critical frequency f_c, given by

$$f_c = 9\,(N_{max})^{1/2}\ \text{Hz} \tag{2.1}$$

are reflected back to the receiver, on the earth, where N_{max} is the maximum electron density in the ionosphere. Ionospheric-reflected waves are used for long-distant radio broadcast. Since microwave frequencies are greater than f_c, microwave signals are not reflected back to the earth, but penetrate the ionosphere to travel in space. The component of the microwaves, diffracted around the surface of the earth, is guided by the ground–air interface as a surface wave which gets attenuated very fast. Therefore, only propagation of surface-to-surface microwave takes place through direct and ground-reflected waves in the atmosphere. Because of the curvature of the earth and the presence of other obstacles like buildings and trees, microwave signals are attenuated and the range of microwave direct transmission is limited to about 50 km between a transmitter and a receiver.

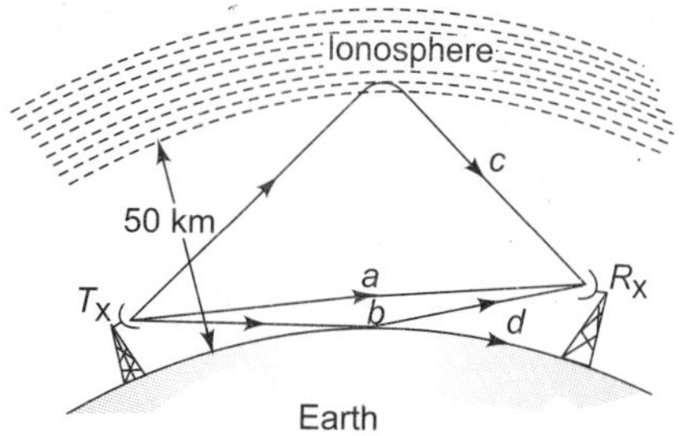

a. Direct wave
b. Ground-reflected wave
c. Ionospheric-reflected wave
d. Surface wave

Fig. 2.2 Electromagnetic wave propagation above the earth

The following symbols as given in Table 2.1 will be used in this book for further description:

Table 2.1 Symbols Used with MKS units

E = Electric field strength (V/m)	ρ = electric charge density (C/sq.m)	I = electric current (A)
H = Magnetic field (A/m)	ε = permittivity (F/m)	**J** = electric current density (A/sq.m)
D = Electric displacement density (C/sq.m) = ε**E**	μ = permeability (H/m)	c = speed of light (m/S)
B = Magnetic flux density (T)	**M** = Magnetization (A/m)	**P** = Polarization (C/sq.m)
ϕ_B = Magnetic flux (weber)	e = Electric charge = 1.602×10^{-19} (Coul)	σ = Conductivity (S)

The fields of electromagnetic waves are functions of both space coordinates and time. For the time harmonic fields ($e^{j\omega t}$), the instantaneous quantities are described by

Electric field $\mathcal{E}(r, t) = \mathbf{E}(r)e^{j\omega t}$ (2.2a)

Magnetic field $\mathcal{H}(r, t) = \mathbf{H}(r)e^{j\omega t}$ (2.2b)

Electric flux density $\mathcal{D}(r, t) = \mathbf{D}(r)e^{j\omega t}$ (2.2c)

Magnetic flux density $\mathcal{B}(r, t) = \mathbf{B}(r)e^{j\omega t}$ (2.2d)

Electric current density $\mathcal{J}(r, t) = \mathbf{J}(r)e^{j\omega t}$ (2.2e)

Since the time-dependent factor or solution is uniquely defined by $e^{j\omega t}$, for most of the cases only space dependent solution **E**(r), **H**(r), **D**(r), **B**(r), **J**(r) will be used in mathematical expressions.

Basic Parameters of Electromagnetic Waves 2.1.2

Propagation of EM waves is significantly influenced by the characteristics of the medium. The electrical characteristics of a medium are described by three parameters ε, μ, σ, called permittivity, permeability and conductivity, of the medium, respectively. Suffix 'o' will be added to these electrical parameters if the medium is free space. Accordingly, we define the following universal constants in electromagnetic:

1. Velocity in free space: $c = 3 \times 10^8$ m/s
2. Permittivity in free space: $\varepsilon_0 = \dfrac{10^{-9}}{36\pi}$ farad/m (or F/m)
3. Permeability in free space: $\mu_0 = 4\pi \times 10^{-7}$ henry/m (or H/m)
4. Conductivity of free space: $\sigma_0 = 0$ mho/m (or S/m)
5. Intrinsic impedance of free space: Ratio $E/H = \eta_0 = 120\pi$ ohm (or Ω)

In air, these constants are same, except the conductivity is taken as $\sigma_0 = 10^{-14}$ mho/m due to losses in upper atmosphere in presence of varieties of particles and water vapour, etc.

Classification of Media 2.1.3

Media of EM wave propagation can be classified as follows:

1. **Linear**: Electrical parameters (ε, μ, σ) are independent of field strength or source current density.
2. **Homogeneous**: Electrical parameters are independent of position within the medium.
3. **Isotropic**: Electrical parameters are independent of the direction of **E**.
4. **Time-invariant**: Electrical parameters are independent of time.
5. **Non-dispersive**: Electrical parameters are independent of frequency.

Simple materials or media: Materials or media that are homogeneous, linear, and isotropic is called **simple materials or media**.

In EM theory, three different types of media are of much use: conductors, dielectric materials, and magnetic materials.

Conductors

A good conductor is defined as a material having electrical parameters ε_0, μ_0, and $\sigma >> \omega\varepsilon_0$ at operating frequency $f = \dfrac{\omega}{2\pi}$ below optical region. In this, some of the electrons can moves freely from molecule to molecule under the influence of an applied electric potential gradient or electric field. These free charges can be electrons, holes or ions.

Dielectrics

In most dielectrics, electrons are tightly bound to individual nuclei of the nonpolar molecules in equilibrium (Fig. 2.3a). These bound charges can move only a fraction of an atomic distance away from their equilibrium positions. When an external electric field $\mathbf{E}_e$ is applied, in most dielectrics, the nuclei are unable to move because of molecular lattice forces, but the electron orbits can be distorted as shown in a simplified model (Fig. 2.3b). Here, the +ve and –ve charge centres q are offset by an effective distance **d** to form an electric dipole moment $\boldsymbol{p} = q\boldsymbol{d}$. The induced dipole generates an electric field $\mathbf{E}_i$ in opposite direction of $\mathbf{E}_e$, which in combination with the applied electric field alters the total field $\boldsymbol{E}$ both inside and outside the dielectric.

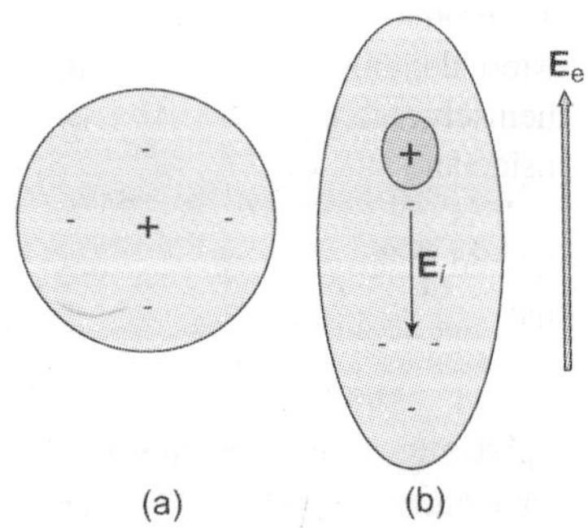

Fig. 2.3 Dielectric polarization

The electric dipoles formed throughout a dielectric are discrete and distributed continuously. Accordingly, a **dipole moment per unit volume** of the dielectric or **polarization vector P** is defined as

$$\mathbf{P} = \varepsilon_0(\varepsilon_r - 1)\mathbf{E} \text{ C/sq.m} \tag{2.3a}$$

where ε_r is the dielectric constant of the dielectric medium. The quantity $\varepsilon_0\, \varepsilon_r\, \mathbf{E} = \varepsilon\, \mathbf{E}$ is called the electric flux density **D** [C/sq.m]

$$\mathbf{D} = \varepsilon\, \mathbf{E} \tag{2.3b}$$

Magnetic Materials

Unlike an electric field, magnetic fields interact only with moving charges or electric current. The magnetic materials are said to interact with an externally applied magnetic field. There are two types of electron motion within atoms and molecules: orbital motion and spin motion.

As shown in Fig. 2.4, an electron orbits a nucleus and results in **orbital current**

$$I_0 = \frac{\text{Charge/Orbit}}{\text{Time/Orbit}} = \frac{e}{2\pi/\omega_0} = \frac{e\omega_0}{2\pi} \text{ A} \tag{2.4a}$$

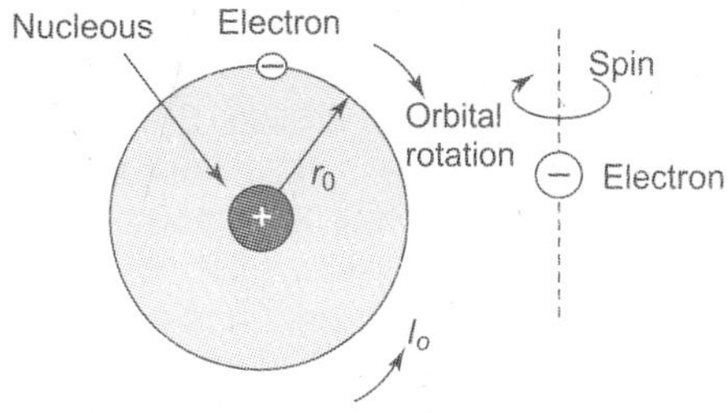

Fig. 2.4 Atomic current

Here, ω_0 is the angular velocity of the electron and e is the electron charge. Thus, this loop current produces N-pole on the side of the loop where from the current is seen anticlockwise and S-pole on the other side. Therefore, a magnetic dipole is formed with the **magnetic dipole moment**

$$\mathbf{m}_0 = I_0 S_0 \hat{\mathbf{z}} = \frac{e\omega_0}{2\pi}\pi r^2 \hat{\mathbf{z}} = \frac{e\omega_0 r_0^2}{2}\hat{\mathbf{z}} \text{ A sq.m} \tag{2.4b}$$

Here, S_0 = Surface area of the loop or orbit, r_0 = Radius of the orbit.

It is assumed that electrons are solid particles of charge that spin about their own axes. The spin of electrons can be modelled as circulation of a charge about an axis of rotation which constitutes an infinite number of current loops. The sum of magnetic moments due to these loops create a resultant **spin dipole moment**

$$m_s = \frac{eh}{4\pi m} \tag{2.4c}$$

where h is Planck's constant (6.6262×10^{-34} JS) and m is the mass of the electron. In most materials, in an absence of any externally applied magnetic field, the orbit and spin magnetic moments are randomly oriented and produce no resultant internal magnetic field and net zero magnetic moment for each atom. When an external magnetic field **B** is applied to the material, the magnetic moments of the electrons more or less align themselves with the **B** so that net magnetic moment is not zero. This results in change in the net **B** field both inside and outside the material.

It is convenient to model the properties of magnetic materials in terms of **magnetization vector M**, defined as the net magnetic dipole moment per unit volume at each point throughout a magnetic material so that

$$\mathbf{M} = \mathbf{B}/\mu_0 - \mathbf{H} = (\mu_r - 1)\mathbf{H} \text{ A/m}$$

Or

$$\mathbf{B} = \mu_0\, \mu_r \mathbf{H} = \mu\mathbf{H} \text{ A/sq.m} \tag{2.4d}$$

WAVE IMPEDANCE 2.2

The two fields, **E** and **H**, are related by impedance, called the wave impedance, given by

$$\eta = \frac{\mathbf{E}}{\mathbf{H}} \text{ ohms} \tag{2.5}$$

One experiences variable wave impedance as one move away from the source of the electromagnetic radiation. It is seen from Eq. (2.5), that a very high impedance is encountered in the region where the **E** field is predominant. This happens near a high-impedance voltage-driven source like a straight wire. On the other hand, if the waves are generated by a low-impedance current-driven source like a loop or coil, the wave impedance is very low in the vicinity of the source where the **H** field predominates. In these predominant field regions, most of the energy is stored in the dominant field components **E** or **H**. As one moves away from the field source, correspondingly, fields **H** or **E** are induced more in the predominant **E** or **H** field environment, respectively. Finally, after a distance r_0 the fields are stabilized to give a constant wave impedance in the region beyond the distance r_0, called the ***far field region***. This region is characterized by the TEM mode of propagation. The variation of wave impedance with the distance from the source is shown in Fig. 2.5 where it is seen that **E**-field impedance is very high and the **H**-field impedance is very low. Both are merged to plane wave field impedance (120π ohm in free space). For a small sized ($D << \lambda_o$) source

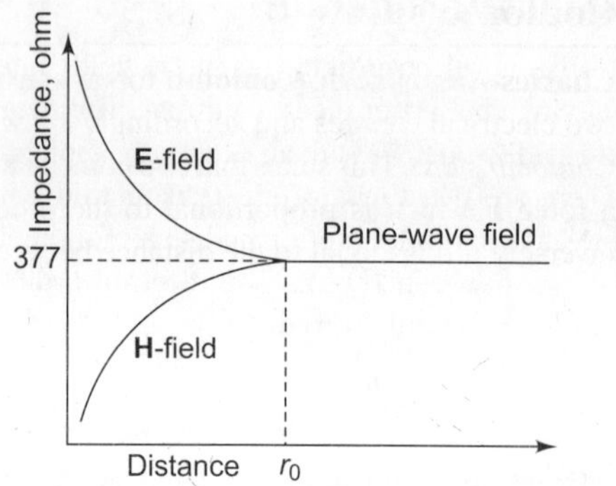

Fig. 2.5 Variation of wave impedance with distance

$$r_o \approx \frac{\lambda_o}{2\pi} \tag{2.6a}$$

When the current source is larger than the wavelength,

$$r_o \approx \frac{2D^2}{\lambda_o} \tag{2.6b}$$

MAXWELL'S EQUATIONS 2.3

Development of a complete theory of electromagnetism started with **James C Maxwell (British, 13 June 1831–05 Nov. 1879)**. He proposed that just as time-varying magnetic fields can produce electric fields, the opposite is also true. He had predicted theoretically the existence of electric and magnetic fields associated with electromagnetic wave propagation and made the foundation of modern electromagnetic theory in 1873. In his theory, many prior works of other scientists are considered as the base of electromagnetic which are now described first.

Ohm's Law 2.3.1

In 1826, **George Simon Ohm** established the **Ohm's law**, of circuit theory. When free electrons in a material move from molecules to molecules, conduction current ***I*** will flow due to the influence of an applied **E**-field. It may be noted that the **E** field may be generated by an impressed current density $\mathbf{J}_i$, which is different from medium conduction current density **J**. For most materials, the conduction current density **J** and the **E** field are related by an equation:

$$\mathbf{J} = \sigma \mathbf{E} \text{ [Amp per unit length or area or volume]} \quad (2.7)$$

Here, σ [S/m] is called the conductivity of the material. Equation (2.7) is called the point form of Ohm's law used in electromagnetic. In a perfect conductor, $\sigma = \infty$, so that $\mathbf{E} = \mathbf{J}/\sigma = 0$ inside the perfect conductor.

Coulomb's Law 2.3.2

Charles-Augustin de Coulomb found existence of a force between two electrical charges and accordingly a law was established, called *Coulomb's law*. This states that two point charges q_1 and q_2 experience a force **F** which is proportional to the product of the charges and inversely proportional to the distance between them (Fig. 2.6):

$$\mathbf{F} = \frac{q_1 q_2 \hat{\mathbf{r}}}{4\pi\varepsilon_0 r^2} \text{ (N)} \quad (2.8a)$$

The force is attractive for charges of opposite polarity and repulsive for charges of same polarity and dependent on the medium (ε).

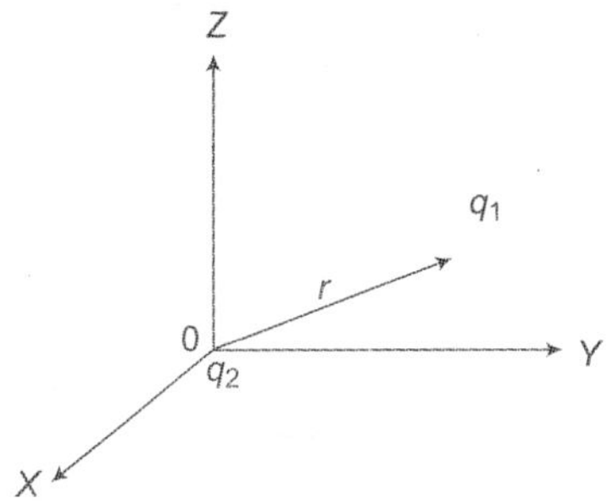

Fig. 2.6 Coulomb's law

Electric Field and Potential 2.3.3

Field

If a unit charge is placed at a point r, the **E** field at r is defined as the force on it by a source charge q_2 located at the origin:

$$\mathbf{E} = \frac{\mathbf{F}}{q_1} = \frac{q_2}{4\pi\ \varepsilon_0}\frac{\hat{\mathbf{r}}}{r^2}$$

$$= -\frac{q_2}{4\pi\varepsilon_0}\nabla\left(\frac{1}{r}\right) \text{ (N/C or V/m)} \quad (2.8b)$$

Thus, electric field is also dependent on the medium and can be expressed in terms of gradient of a scalar $\nabla\left(\frac{1}{r}\right)$. Since curl of the gradient of a scalar is zero, above equation results in

$$\nabla \times E = \nabla \times \left[-\frac{q_2}{4\pi\varepsilon_o}\nabla\left(\frac{1}{r}\right)\right]$$

$$= -\frac{q_2}{4\pi\varepsilon_o}\nabla x \nabla\left(\frac{1}{r}\right) = 0 \quad (2.8c)$$

The electrostatic field is thus, irrotational ($\nabla \times \mathbf{E} = 0$) and is completely defined by a charge distribution.

Potential

Electric potential at a point is defined with respect to a reference value at some other point in a medium and is expressed in terms of the electric field in the medium. Let a point charge q be moved from a point A to another point *B* in an existing **E**-field as shown in Fig. 2.7. From Coulomb's law, the charge is moved when a force on *q* is $\mathbf{F} = q\mathbf{E}$. Therefore, the ***work done*** in moving the charge by an infinitesimal distance *dl* is

$$d\mathbf{W} = \pm \mathbf{F} \cdot d\mathbf{l} = \pm q\mathbf{E} \cdot d\mathbf{l} \qquad (2.9a)$$

Here, the +ve sign indicates that work is being done by an external source which establishes the **E**-field. The –ve sign indicates that work is done by the **E**-field. The total potential energy is the total work done in moving *q* from *A* to *B* and is given by

$$W = \pm q\int_A^B \mathbf{E} \cdot dl \qquad (2.9b)$$

The **potential difference** between points *A* and *B* is **the potential energy** per unit charge and is defined as

$$V_{AB} = \frac{W}{q} = \pm\int_A^B \mathbf{E} \cdot dl \text{ J/C or V} \qquad (2.9c)$$

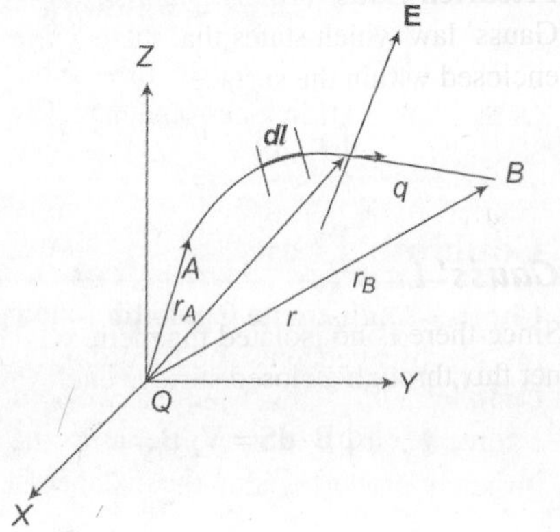

Fig. 2.7 Electric potentials at points *A* and *B*

(i) The –ve value of V_{AB} indicates that the work is being done by the **E**-field and thus, there is a loss in potential energy in moving *q* from *A* to *B*.

(ii) The +ve value of V_{AB} indicates that the work is being done by an external source and thus, there is a gain in potential energy in moving *q* from *A* to *B*.

(iii) V_{AB} is independent of the path of movement of the charge.

Potential at a Point

If it is assumed that the **E** field is generated by a point charge *Q* placed at the origin then

$$\mathbf{E} = \frac{Q}{4\pi\varepsilon r^2}\hat{\mathbf{r}} \qquad (2.9d)$$

From Eqs. (2.89c) and (2.89d),

$$V_{AB} = -\int_{r_A}^{r_B} \frac{Q}{4\pi\varepsilon r^2}\hat{\mathbf{r}} \cdot \hat{\mathbf{r}}dr$$

$$= \frac{Q}{4\pi\varepsilon}\left[\frac{1}{r_B} - \frac{1}{r_A}\right]$$

$$= V_B - V_A \text{ Volt} \qquad (2.9e)$$

Here, V_B and V_A = Absolutes potentials at *B* and *A*, respectively. Since electric potential at infinity is considered to be zero, the absolute potential at any point *r* is given by

$$V = -\int_{\infty}^{r} \mathbf{E} \cdot dl = \frac{Q}{4\pi\varepsilon r} \text{ Volt} \qquad (2.9f)$$

where it is considered $r_A \to \infty, V_A = 0, r_B = r, V_B = V$

From Eq. (2.9f), we can write

$$\mathbf{E} = -\frac{\partial V}{\partial l} \text{ V/m} \qquad (2.9g)$$

Thus, the electric field is the –ve gradient of potential.

Gauss' Law for Electricity 2.3.4

Friedrich Gauss (1835) established that an electric charge produces an electric field **E**(r, t). He formulated Gauss' law which states that the electric flux out of any closed surface S is proportional to the total charge enclosed within the surface:

$$\oint_S \mathbf{E}\cdot d\mathbf{S} = \frac{q}{\varepsilon_0}\nabla\cdot\mathbf{E} = \frac{\rho}{\varepsilon_0} \tag{2.10}$$

Gauss' Law for Magnetism

Since there is no isolated magnetic pole exists, but two opposite poles can be present to form a dipole, the net flux through a closed surface enclosing a dipole is zero:

$$\oint_S \mathbf{B}\cdot d\mathbf{S} = \nabla\cdot\mathbf{B} = 0 \tag{2.11}$$

Gauss's Divergence Theorem 2.3.5

In 3D medium, Gauss's divergence theorem states that the total electric flux (surface integral) of the vector **F** through a closed surface S enclosing a volume V is equal to the volume integral of divergence of the vector **F**. Here, the convention is that, the positive orientation of the elementary surface vector **d**S on S is outward normal, that is, $\hat{\mathbf{n}}dS$, where $\hat{\mathbf{n}}$ is the unit normal vector in the direction outward from V.

$$\int_S \mathbf{F}\cdot d\mathbf{S} = \int_V \nabla\cdot\mathbf{F}dV \tag{2.12}$$

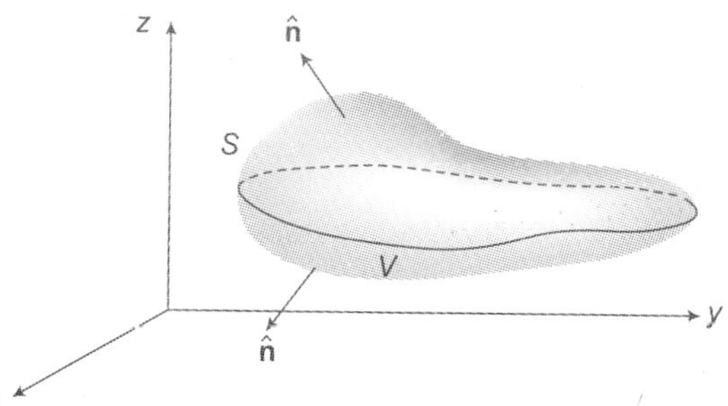

Fig. 2.8 Representation of Gauss's divergence theorem

Stokes' Theorem 2.3.6

Stokes' theorem states that a line integral of a vector **F** is equal to surface integral of curl of the vector **F** in 3D vector fields:

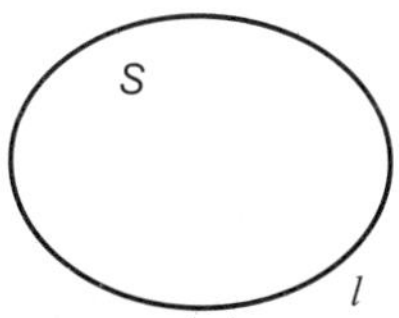

Fig. 2.9 Representation of Stokes' theorem

$$\int_l \mathbf{F}\cdot dl = \int_S (\nabla x \mathbf{F})\cdot d\mathbf{S} \tag{2.13}$$

and
$$\mathbf{F} = q(E + v \times B) \tag{2.14}$$

Conservation of Charge and Equation of Continuity 2.3.7

The law of conservation of charge states that electric charges can neither be created nor destroyed. Since the flow of charges results in electric current and charge is conserved, there is a relationship between the current flowing out through the surface S enclosing a homogenous region of volume V as shown in Fig. 2.10, and the rate of change of the charge contained within the region. If the medium permittivity is ε and conductivity σ containing charge density ρ, the total current I leaving V through S is equal to the time rate of decrease of the total charge Q within V. This conservation formula can be expressed mathematically as

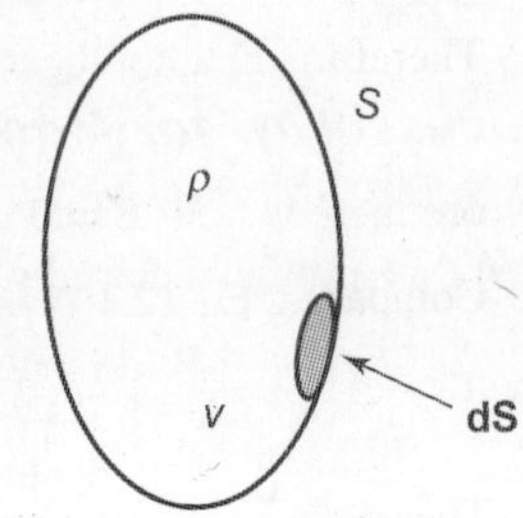

Fig. 2.10 Conservation of charge

$$I = -\frac{dQ}{dt} = \oint_S \mathbf{J}\cdot\mathbf{dS} \tag{2.15}$$

where

$$Q = \int_V \rho\, dv \tag{2.16}$$

From Eqs. (2.15) and (2.16), we have total outward flow of current

$$I = \oint_S \mathbf{J}\cdot\mathbf{dS} = -\frac{d}{dt}\int_V \rho dV \tag{2.17}$$

= Total rate of decrease of charge

Since ρ is a function of position and time,

$$\frac{d}{dt}\int_V \rho\, dv = \int_V \frac{\partial \rho}{\partial t}\, dv \tag{2.18}$$

Applying the **divergence theorem**, we can write

$$\oint_S \mathbf{J}\cdot\mathbf{dS} = \int_V \nabla\cdot\mathbf{J}\, dv \tag{2.19}$$

Thus, from Eqs. (2.18) and (2.19),

$$\int_V \nabla\cdot\mathbf{J}\, dv + \int_V \frac{\partial \rho}{\partial t}\, dv = 0 \tag{2.20}$$

The relation (2.20) is true for all homogeneous regions. Therefore, we can write

$$\nabla\cdot\mathbf{J} + \frac{\partial \rho}{\partial t} = 0 \tag{2.21}$$

This last equation is called the **equation of continuity**.

For steady or dc currents,

$$\frac{\partial \rho}{\partial t} = 0 \tag{2.22}$$

In this case, the equation of continuity becomes

$$\nabla \cdot \mathbf{J} = 0 \tag{2.23}$$

Therefore, for a dc source **J** is called a ***solenoidal*** vector field.

In terms of **E**-field, the continuity equation for steady current leads to Ohm's law in a conducting medium, which states that

$$\mathbf{J} = \sigma \mathbf{E} \tag{2.24a}$$

Therefore, in a homogeneous medium,

$$\nabla \cdot \mathbf{J} = \nabla \cdot \sigma \mathbf{E} = \sigma \nabla \cdot \mathbf{E} = 0$$

or

$$\nabla \cdot \mathbf{E} = 0 \tag{2.24b}$$

Comparing Eq. (2.19) with Gauss's law:

$$\nabla \cdot \mathbf{E} = \frac{\rho}{\varepsilon} \tag{2.25}$$

Therefore, in a homogeneous conducting medium, the volume charge density, ρ, must be zero.

Ampere's Law 2.3.8

Ampere's law, known historically as **Ampere's circuital law**, relates the magnetic field **B** to its electric current source I and states that a magnetic field is produced in space due to an electric current source and this field is proportional to this electric current. The direction of the magnetic field is given by Right Hand Rule as shown in Fig. 2.11.

The integral form of Ampere's law states that for any closed-loop path, the sum of the length elements times the magnetic field in the direction of the length element is equal to the permeability times the electric current enclosed in the loop (Fig. 2.11).

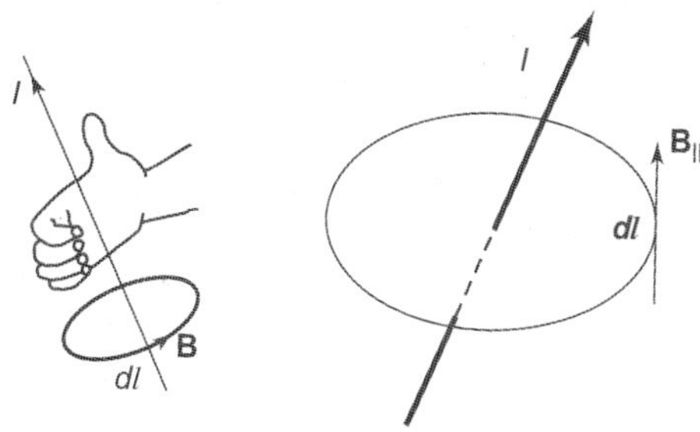

Fig. 2.11 Right-hand rule for **B** due to current I

Thus,

$$\sum \mathbf{B}_{\parallel}\, dl = \mu_0 I; \; dl \to 0 \tag{2.26}$$

where $B_{\parallel}$ is the component of **B** tangential to the loop. In the limit $dl \to 0$, Ampere's law becomes

$$\oint_l \mathbf{B} \cdot d\mathbf{l} = \mu_0 \int_S \mathbf{J} \cdot d\mathbf{S} \tag{2.27}$$

Here, l is the total closed line contour, **B** is in web/m, **J** is the electric current density through the surface S enclosed by the curve l; $d\mathbf{S}$ is the vector area of an infinitesimal element of surface S.

The directions of all these vectors are governed by right-hand rule: *With the palm of the right-hand toward the area of integration, and the index finger pointing along the direction of line integration, the outstretched thumb points in the direction that must be chosen for the vector area* ***dS***.

Displacement Current and Inconsistency of Ampère's Circuital Law

In free space, the time rate of change of electric flux density $\mathbf{D}(= \varepsilon_o \mathbf{E})$ is defined as the displacement current $\left(\frac{d\mathbf{D}}{dt}\right)$. In a dielectric, the above contribution to displacement current is present too, but a major contribution to the displacement current is related to the polarization of the individual molecules of the dielectric material. Even though charges cannot flow freely in a dielectric, the charges in molecules can move a little under the influence of an electric field **E**. The positive and negative charges in molecules separate under the applied electric field, causing an increase in the state of polarization, expressed as the polarization density **P** (Fig. 2.12). A changing state of polarization is equivalent to a current $\partial \mathbf{P}/\partial t$. Both contributions to the displacement current are combined by defining the displacement current as $\partial \mathbf{D}/\partial t$, where $\mathbf{D} = \varepsilon_o \mathbf{E} + \mathbf{P}$. Comparing with Eq. (2.3), it is apparent that **D** is independent of medium. The displacement current was originally conceived by Maxwell.

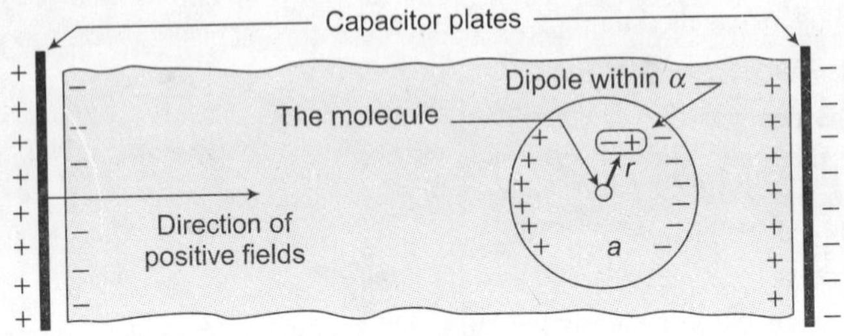

Fig. 2.12 Contribution of a dielectric to the **E**-field
[Source: Panofsky & Phillips Classical Electricity and Magnetism, Addison-Wesley Pub. Co., Inc., 2nd Edn.1962]

In the Ampère's circuital law, two important inconsistencies exist: first, there is an issue regarding the continuity equation for electrical charge. A theorem in vector calculus states that the divergence of a curl must always be zero. Hence, $\nabla\cdot(\nabla\times\mathbf{B}) = 0$ and so the original Ampère's law implies that $\nabla\cdot\mathbf{J} = 0$. But, in general, according to the continuity equation, $\nabla\cdot\mathbf{J} = -\partial\rho/\partial t$, which is nonzero for a time-varying charge density. An example occurs in a capacitor circuit where time-varying charge densities $\partial\rho/\partial t$ exists on the plates. Second, there is an issue regarding the propagation of electromagnetic waves. In free space, where $\mathbf{J} = 0$, Ampère's law implies that $\nabla\times\mathbf{B} = 0$. **James Clerk Maxwell** conceived of displacement current $\partial\mathbf{D}/\partial t$ as a polarization current associated with the polarization of the individual molecules of the dielectric material, which he used to model the magnetic field as described later. He added this displacement current to Ampere's circuital law and established $\nabla\times\mathbf{B} = \partial\mathbf{D}/\partial t$ (*according to Maxwell's equation in a source-free region*). To treat these situations, the contribution of displacement current must be added to the source current term **J** in Ampere's law.

Applications of Ampere's Law

There are several applications of this law in common use. One can achieve a high inductance value concentrating magnetic field inside a long solenoid and toroidal core as shown in Fig. 2.13.

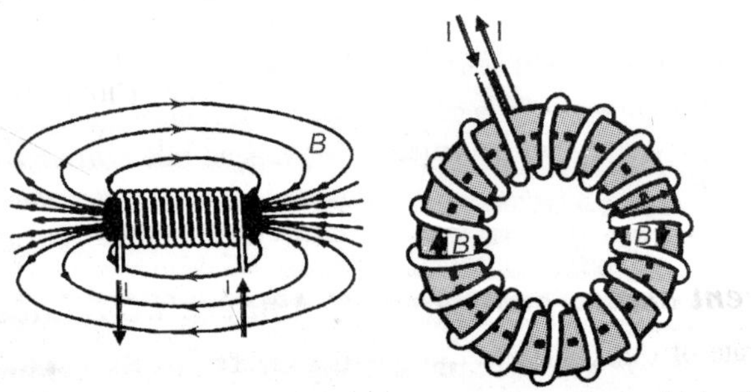

Fig. 2.13 Application of Ampere's law

Biot–Savart Law 2.3.9

In electromagnetism, the **Biot–Savart law** is described by an equation that relates the magnetic flux density **B** generated by an electric current I:

$$\mathbf{B} = \oint \frac{\mu_0}{4\pi} \frac{I\mathbf{dl} \times \hat{\mathbf{r}}}{|r|^2} \tag{2.28}$$

Here, I = Steady current, dl = Elemental length vector of the wire whose direction is the direction of current, $\hat{\mathbf{r}}$ = Unit vector in the direction pointing from the wire element towards the point at which the field is being computed, r = Distance of the field point from the current element Idl. Integration is done around the closed path of the current I.

The vector field **B** depends on the magnitude, direction, length, proximity of the electric current, and permeability of the medium. The application of the Biot–Savart law is to compute the magnetic field generated by a *steady current I* as shown in Fig. 2.14.

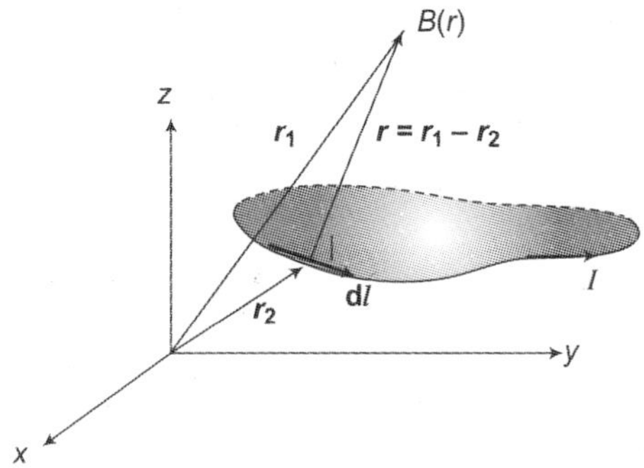

Fig. 2.14 Biot–Savart law

The law is valid in the magnetostatic field, and results in a **B**-field consistent with both Ampère's circuital law and Gauss's law for magnetism.

Faraday's Law 2.3.10

Michael Faraday established a law with the help of a classical experiment (1831) that *voltages or E-fields are produced in a circuit whenever time varying magnetic fields are associated with it.* This law is also called **Faraday's law of induction**. Faraday observed that a time-varying voltage $V_0(t)$ is produced at the open circuit of a secondary loop (Fig. 2.15) when (i) the transient current in an adjacent primary loop is time varying due to switching at $t = 0$, or (ii) there is a relative motion between the loops while steady current flows in the primary. In either of the two cases, a time-varying magnetic field is produced by the primary. This effect is mathematically described by the integrals:

$$\oint_l \mathbf{E} \cdot \mathbf{d}l = -\frac{d}{dt}\int_S \mathbf{B} \cdot \mathbf{dS} = -\int_S \frac{\partial \mathbf{B}}{\partial t} \cdot \mathbf{dS} \tag{2.29a}$$

Here S represents an open surface enclosed by a closed path l in a right handed sense. The line integral in left-hand side can be replaced by surface integral using the Stokes's theorem:

$$\oint_l \mathbf{E} \cdot \mathbf{d}l = \int_S \nabla \times \mathbf{E} \cdot \mathbf{dS} = -\int_S \frac{\partial \mathbf{B}}{\partial t} \cdot \mathbf{dS} \tag{2.29b}$$

In the limit l tending to 0, S reduces to a single point, so that the differential form of Faraday's law can be expressed by

$$\nabla \times \mathbf{E} = -\frac{\partial \mathbf{B}}{\partial t} \tag{2.29c}$$

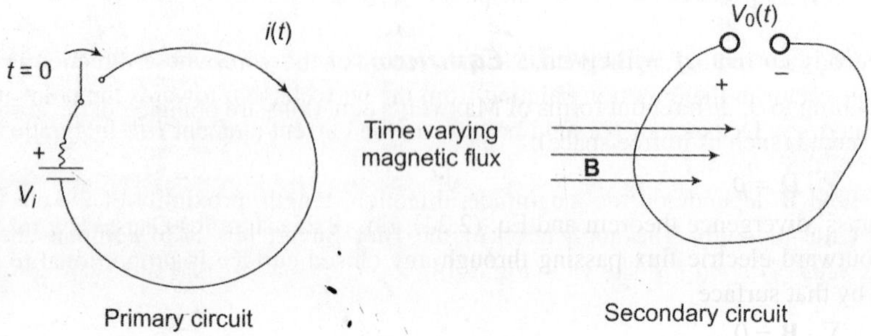

Fig. 2.15 Faraday's law of induction

Lenz's Law

Lenz's law predicts the direction of the induced current in a loop associated with the time varying magnetic flux and states that *the sign of the voltage polarities in the secondary is such that it tends to produce currents whose magnetic flux densities* **B** *opposes the time varying magnetic flux linkage.*

Maxwell's Equations 2.3.11

In the ***theory of the electromagnetic field***, the properties of electric and magnetic fields are described in terms of Maxwell's equations. Thus, the electromagnetic fields in wave propagation in a medium are the solution of Maxwell's equations as described below.

a. Integral Forms of Maxwell's Equations

In the absence of magnetic or polarizable media (such as in free space), **integral forms** of Maxwell's equations are given by

(I) $$\oint_S \mathbf{E}\cdot d\mathbf{S} = \frac{1}{\varepsilon_0}\int_V \rho dV \tag{2.30}$$

This signifies Gauss' law for static electricity that net electric flux passing through any closed surface is proportional to the total charge enclosed by that surface.

(II) $$\oint_S \mathbf{B}\cdot d\mathbf{S} = 0 \tag{2.31}$$

This signifies Gauss' law for static magnetism that net magnetic flux passing through any closed surface is zero since no isolated magnetic pole or charge exists.

(III) $$\oint_l \mathbf{E}\cdot d\mathbf{l} = -\frac{d}{dt}\int_S \mathbf{B}\cdot d\mathbf{S} \tag{2.32}$$

This signifies Faraday's law of induction for time varying fields that there is a net voltage or emf $V = \oint_l \mathbf{E}\cdot d\mathbf{l}$ developed around a closed path whenever a time varying magnetic flux passes through the path. The polarity of the voltage is, in accordance with Lenz's law, such that it tends to produce currents in the loop whose magnetic fields oppose the time varying flux linkage.

(IV) $$\oint \mathbf{H} \cdot \mathbf{d}l = \int_S \left(\mathbf{J} + \frac{\partial \mathbf{D}}{\partial t} \right) \cdot \mathbf{dS} \tag{2.33}$$

This signifies **Ampere's law** for ac current that the line integral of magnetic field **H** around any closed path l equals the current I enclosed by the path in a right-handed sense.

b. Differential Forms of Maxwell's Equations

In the limit l tending to 0, differential forms of Maxwell's equations are obtained in the absence of magnetic or polarizable media (such as in free space):

(I) $$\nabla \cdot \mathbf{D} = \rho \tag{2.34}$$

Using Gauss' divergence theorem and Eq. (2.24), Eq. (2.25) signifies Gauss' law for static electricity that net outward electric flux passing through any closed surface is proportional to the total charge enclosed by that surface

(II) $$\nabla \cdot \mathbf{B} = 0 \tag{2.35}$$

Using Gauss' divergence theorem and Eq. (2.25), Eq. (2.29) signifies Gauss' law for static magnetism that net magnetic flux passing through any closed surface is zero since no isolated magnetic pole or charge exists.

(III) $$\nabla \times \mathbf{E} = -\frac{\partial \mathbf{B}}{\partial t} \tag{2.36}$$

Using Stoke's theorem and Eq. (2.30) this signifies Faraday's law of induction for time varying fields.

(IV) $$\nabla \times \mathbf{H} = \mathbf{J} + \frac{\partial \mathbf{D}}{\partial t} \tag{2.37}$$

Using Stoke's theorem and Eq. (2.27) this signifies Ampere's law for time varying current,

The electrical properties (σ, ε, μ) of a general medium having losses offer constitutive relations.

$$\mathbf{D} = (\varepsilon' - j\varepsilon'')\,\mathbf{E} = \text{Electric displacement vector coul/m}^2 \tag{2.38a}$$

$$\mathbf{B} = \mu\,\mathbf{H} = (\mu' - j\mu'')\,\mathbf{H} = \text{Magnetic induction vector web/m}^2 \tag{2.38b}$$

$$\mathbf{J} = \sigma\mathbf{E} + \mathbf{J}_i = \text{Electric current density A/m}^3 \text{ or A/m}^2 \text{ or A/m} \tag{2.38c}$$

$$\varepsilon = \varepsilon_r\,\varepsilon_0 = \text{Electric permittivity (farad/m)} \tag{2.38d}$$

$$\mu = \mu_r\,\mu_0 = \text{Magnetic permeability (henry/m)} \tag{2.38e}$$

Here, $\mathbf{J}_i$ is the externally impressed current density producing the electromagnetic fields. σ, ε'' and μ'' account for the joule heating loss, dielectric loss and magnetic loss in the medium, respectively. The loss in a dielectric medium is accounted by a parameter, called the loss tangent of the medium and is given by

$$\tan\delta = \frac{\omega\varepsilon'' + \sigma}{\omega\varepsilon'} \tag{2.39a}$$

The δ is an angle in a complex plane between the resistive (lossy) component of an electromagnetic field and its reactive (lossless) component.

In a perfect dielectric $\sigma = 0$, and

$$\tan\delta = \frac{\varepsilon''}{\varepsilon'} \tag{2.39b}$$

In a lossy magnetic medium, a magnetic loss tangent is also defined by

$$\tan\delta_m = \frac{\mu''}{\mu'} \tag{2.39c}$$

c. Maxwell's Equations for Time-harmonic Fields

The fields in wave propagation are function of both space coordinates and time, and signals are composition of sum of infinite number of sine and cosine harmonic functions. Taking a particular component at frequency ω, time harmonic solution can be written as $e^{j\omega t}$.

For time-harmonic fields ($e^{j\omega t}$), the above equations reduce to ($\mathbf{E} = \mathbf{E}_0 e^{j\omega t}$; $\mathbf{H} = \mathbf{H}_0 e^{j\omega t}$)

$$\nabla \times \mathbf{E} = -j\omega\mu\, \mathbf{H} \tag{2.40a}$$

$$\nabla \times \mathbf{H} = -(\sigma + j\omega\varepsilon)\, \mathbf{E} \tag{2.40b}$$

$$\nabla \cdot \mathbf{D} = \rho \tag{2.40c}$$

$$\nabla \cdot \mathbf{B} = 0 \tag{2.40d}$$

$$\nabla \cdot \mathbf{J} = -j\omega\rho \tag{2.40e}$$

Skin Effect 2.3.12

Skin effect is the tendency of an ac or RF current to distribute itself within a conductor such that the current density near the surface of the conductor is greater than that at its core. That is, the electric current is maximum at the "skin" of the conductor, and decreases to a value equal to $1/e$ (about 0.37) times the surface value at an average depth called the **skin depth**. The skin effect causes the effective resistance of the conductor to increase with the frequency of the current because more part of the conductor does not conduct current. Skin effect is due to eddy currents set up by the ac current as shown in Fig. 2.16

At power frequency in copper, skin depth is about 8.5 mm. At high frequencies skin depth is much smaller. Skin effect may be minimized by using hollow pipe-shaped conductors.

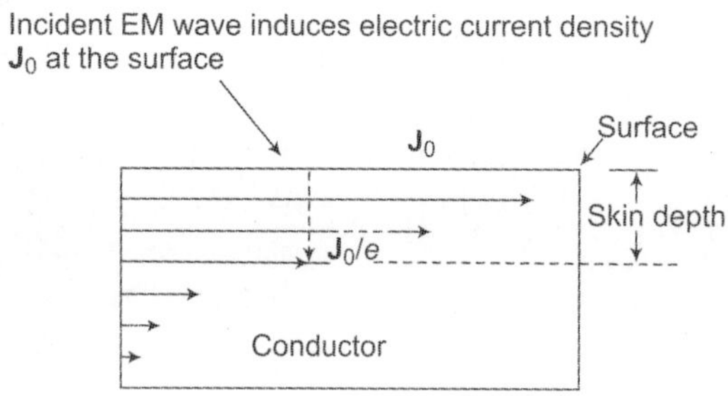

Fig. 2.16 Skin effect

When an electromagnetic wave interacts with a conductive material, mobile charges within the material are made to oscillate back and forth with the same frequency as the impinging fields. The movement of these charges, usually electrons, constitutes an alternating electric current density $\mathbf{J} = \hat{\mathbf{n}} \times \mathbf{H}$. The magnitude of $\mathbf{J}$ decreases exponentially with depth. The skin effect has practical consequences in the design of radio-frequency and microwave circuits and antennas.

The skin depth δ is expressed by

$$\delta = \frac{1}{\sqrt{\pi f \mu \sigma}} \tag{2.40f}$$

where σ = Conductivity of conductor, f = Frequency of current

Skin-depth values for some common good conductors at a frequency of 1 GHz (microwave region) are shown in Table 2.2. At microwave frequencies, most of the current in a good conductor flows in an extremely thin region near the surface (permeability of conductor = μ_0). The variation of skin depths of some conductors with frequency is computed using MATLAB and shown in Fig. 2.17. The extremely short skin depth at microwave frequencies shows that any bad conductor with surface coating using a good conductor like silver or gold becomes a good guiding conductor at high frequencies (VHF and microwaves).

Table 2.2

Conductor	Aluminum $\sigma = 3.817 \times 10^7$ ℧	Copper $\sigma = 5.8 \times 10^7$ ℧	Gold $\sigma = 4.1 \times 10^7$ ℧	Iron $\sigma = 1.03 \times 10^7$ ℧
Skin depth (μm) at 1 GHz	2.6	2.1	2.48	4.96

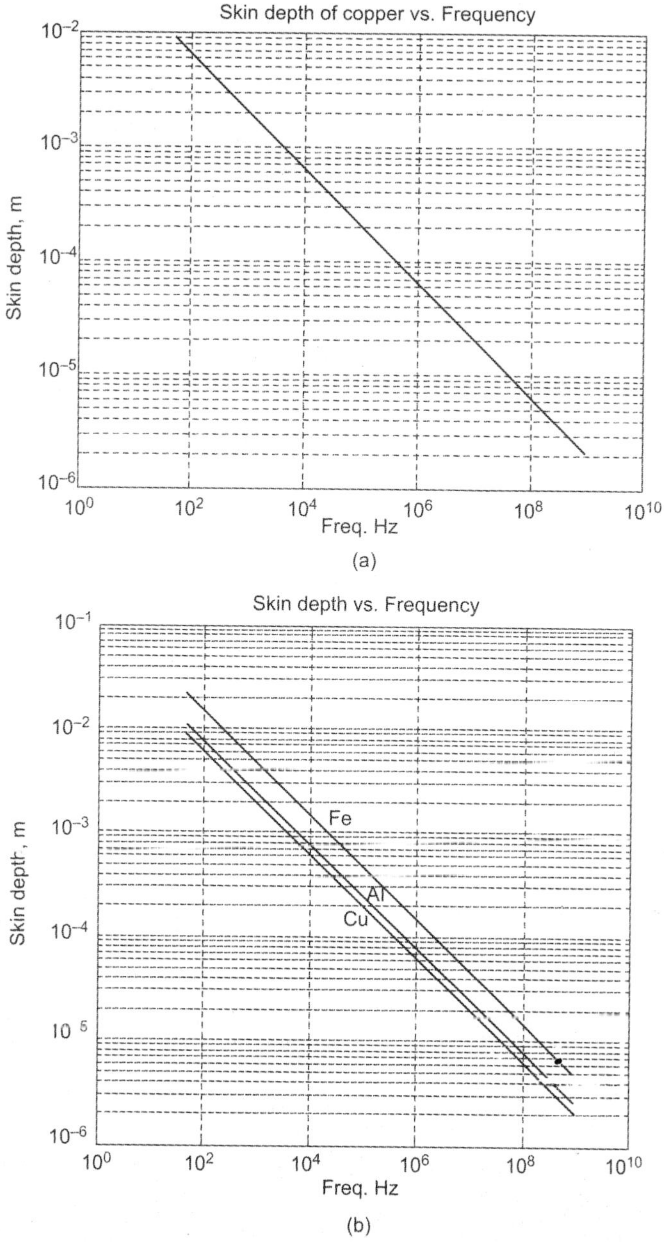

Fig. 2.17 Skin depths of different conductors vs. frequency

```
%MATLAB program for skin depth
clear all;
miu = 4.*pi.*10.^-7;
f = 50:1000:10^9;
cond = 5.8*10.^7;%Cu
skin = 1./(sqrt(pi.*miu.*f.*cond));
loglog(f,skin,'b');
grid on;
 hold on;
cond = 3.82*10.^7;%Al
skin = 1./(sqrt(pi.*miu.*f.*cond));
loglog(f,skin,'r');
grid on;
 hold on;
cond = 1.03*10.^7;%Fe
skin = 1./(sqrt(pi.*miu.*f.*cond));
loglog(f,skin,'g');
grid on;
 hold on;
 xlabel('frEq., Hz');
 ylabel('Skin depth, m');
 title('Skin depth of copper vs.Frequency');
```

WAVE EQUATION 2.4

Since both **E** and **H** are unknown, a solution for **E** and **H** for electromagnetic signal cannot be found either from a single equation or from two equations from Maxwell's equations since one is dependent on other. Hence, a single equation is constructed from Maxwell's equations involving only one parameter, **E** or **H** as follows. This equation will be called the **wave equation**. Derivation of wave equations for electromagnetic waves in a simple, source-free ($\sigma = 0$, $\rho = 0$, $\mathbf{J}_i = 0$), lossless medium is done by taking curl of first two of Maxwell's equations:

$$\nabla \times \nabla \times \mathbf{E} = \nabla (\nabla \cdot \mathbf{E}) - \nabla^2 \mathbf{E}$$

$$= -\mu \frac{\partial(\nabla \times \mathbf{H})}{\partial t} = -\mu\varepsilon \frac{\partial^2 \mathbf{E}}{\partial t^2} \tag{2.41a}$$

$$\nabla \times \nabla \times \mathbf{H} = \nabla(\nabla \times \mathbf{E}) - \nabla_2 \mathbf{H}$$

$$= \varepsilon \frac{\partial(\nabla \cdot \mathbf{E})}{\partial t} = -\mu\varepsilon \frac{\partial^2 \mathbf{H}}{\partial t^2} \tag{2.41b}$$

Since in a source-free region, $\nabla \cdot \mathbf{E} = 0$ and $\nabla \cdot \mathbf{H} = 0$, the above reduce to

$$\nabla^2 \mathbf{E} - \mu\varepsilon \frac{\partial^2 \mathbf{E}}{\partial t^2} = 0 \tag{2.42a}$$

$$\nabla^2 \mathbf{H} - \mu\varepsilon \frac{\partial^2 \mathbf{H}}{\partial t^2} = 0 \tag{2.42b}$$

These equations are called wave equations in a simple, source-free, lossless medium for **E**-fields, and **H**-fields, respectively. The solutions for **E** and **H** can be determined knowing given boundary conditions.

In presence of the exciting current $\mathbf{J}_i$, if the medium is charge free ($\rho = 0$) but has nonzero conductivity. It can be shown that the wave equations are

$$\nabla^2 \mathbf{E} + \gamma^2\mathbf{E} = j\omega\mu\mathbf{J}_i \tag{2.43a}$$

$$\nabla^2 \mathbf{H} + \gamma^2\mathbf{H} = -\nabla x\,\mathbf{J}_i \tag{2.43b}$$

Here, γ is called the propagation constant of the medium, and is given by

$$\gamma = \sqrt{-j\omega\mu(\sigma + j\omega\varepsilon)} = \sqrt{-j\omega\mu\sigma + \omega^2\mu\varepsilon}$$
$$= jk\sqrt{1 - \frac{j\sigma}{\omega\varepsilon}} = \alpha + j\beta \tag{2.44}$$

where α is called the attenuation constant due to σ, β is called the phase constant of the wave, and $k = \omega\sqrt{\mu\varepsilon} = \beta$, propagation constant in nonconducting lossless ($\sigma = 0$) unbounded medium.

Lossy Dielectric Medium 2.4.1

In a low-loss dielectric medium, $\sigma << \omega\varepsilon$. Hence,

$$\beta = \omega\sqrt{\mu\varepsilon} = k$$

$$\eta = \sqrt{\frac{j\omega\mu}{\sigma + j\omega\varepsilon}}$$

$$= \sqrt{\frac{\mu}{\varepsilon}}\left(1 - \frac{j\sigma}{\omega\varepsilon}\right)^{-1/2} \approx \sqrt{\frac{\mu}{\varepsilon}}\left(1 + \frac{j\sigma}{2\omega\varepsilon}\right) \tag{2.45a}$$

$$\tan\delta = \frac{\sigma}{\omega\varepsilon} \tag{2.45b}$$

$$\gamma = jk\sqrt{1 - \frac{j\sigma}{\omega\varepsilon}} \approx jk\left(1 - \frac{j\sigma}{2\omega\varepsilon}\right) = \frac{k\sigma}{2\omega\varepsilon} + jk = \alpha + j\beta \tag{2.45c}$$

Therefore, the attenuation and phase constants are given by

$$\alpha = \frac{\sigma}{2}\sqrt{\frac{\mu}{\varepsilon}} \tag{2.45d}$$

$$\beta = \omega\sqrt{\mu\varepsilon} = k \tag{2.45e}$$

Lossless Dielectric Medium 2.4.2

In a lossless or perfect dielectric medium, $\sigma = 0$, $\varepsilon'' = 0$, and $\varepsilon = \varepsilon'$. Therefore,

$$\eta = \sqrt{\frac{\mu}{\varepsilon}} \tag{2.46a}$$

$$\tan\delta = 0 \tag{2.46b}$$

$$\gamma = jk = j\beta \tag{2.46c}$$

$$\alpha = 0 \tag{2.46d}$$

$$\beta = \omega\sqrt{\mu\varepsilon} = k \tag{2.46e}$$

Free Space 2.4.3

Free space is characterized by $\sigma = 0$, $\varepsilon'' = 0$, and $\varepsilon = \varepsilon' = \varepsilon_0$, $\mu = \mu_0$, and other parameters will change accordingly:

$$\eta_0 = \sqrt{\frac{\mu_0}{\varepsilon_0}} = 120\pi \approx 377\text{ohm} \tag{2.47a}$$

$$\tan\delta = 0 \tag{2.47b}$$

$$\gamma = jk_0 = j\beta_0 \tag{2.47c}$$

$$\alpha = 0 \tag{2.47d}$$

$$\beta_0 = \omega\sqrt{\mu_0\varepsilon_0} = k_0 = \frac{2\pi}{\lambda_0} \tag{2.47e}$$

Conducting Medium 2.4.4

In a good conductor, conduction current is much greater than the displacement current even at microwave frequencies, so that $\sigma >> \omega\varepsilon$. Therefore,

$$\eta = \sqrt{\frac{j\omega\varepsilon}{\sigma}} = (1+j)R_S \tag{2.48a}$$

$$R_S = \text{Conductor surface resistance } (rf) = \sqrt{\frac{\omega\mu}{2\sigma}} \tag{2.48b}$$

$$\tan\delta = \infty \tag{2.48c}$$

$$\gamma = \sqrt{j\sigma\omega\mu} = (1+j)\sqrt{\pi f\mu\sigma} = \alpha + j\beta \tag{2.48d}$$

Therefore,

$$\alpha = \beta = \sqrt{\pi fm\sigma} \tag{2.48e}$$

Microwave signals can penetrate a conductor, falling to $1/e$ of its surface value at a depth called *skin depth* which is given by

$$\delta_s = \frac{1}{\sqrt{\pi f\mu\sigma}} = \frac{1}{\alpha} \tag{2.48f}$$

Example 2.1 A uniform plane wave at 1 GHz is propagating in a nonmagnetic unbounded dielectric medium of dielectric constant 2.5, and 0.01 S/m conductivity. Calculate the attenuation and phase constants.

Solution Given

$$f = 1\times10^9\text{ Hz} \Rightarrow \lambda_0 = c/f = 0.300\text{ m}$$

$$\varepsilon = 2.5\varepsilon_0$$

$$\mu = \mu_0$$

$$\sigma = 0.01\text{ S/m}$$

Hence,

$$\alpha = \frac{\sigma}{2}\sqrt{\frac{\mu}{\varepsilon}} = \frac{0.01}{2}\sqrt{\frac{4\pi\times10^{-7}}{2.5\times10^{-9}\times\frac{1}{36\pi}}} = 1.19\text{ Np/m}$$

$$\beta = \omega\sqrt{\mu\varepsilon} = k = 2\times3.14\times10^9\sqrt{4\times3.14\times10^{-7}\times\frac{10^{-9}}{36\pi}} = 33.16\text{ rad/m}$$

ELECTROMAGNETIC ENERGY AND POWER FLOW 2.5

When a potential source supplies time-varying currents in a conductor, it emits electromagnetic energy. The energy supplied by the source is partly stored in the electric and magnetic fields inside a closed volume in the immediate vicinity of the conductor and partly propagated away as an electromagnetic wave in associated medium such as free space, coaxial line, waveguides, etc.

Electric Stored Energy 2.5.1

It is shown in Eqs. (2.9a-d) that there is potential energy V at a point r associated with a point electric charge q placed at the origin. This energy is equal to the work done W per unit charge by an external source in moving a test charge from infinity to that point. So that

$$V = -\int_{\infty}^{r} \mathbf{E} \cdot d\mathbf{l} = \frac{W}{q} \tag{2.49a}$$

Electric energy stored per unit charge is therefore $W = qV$.

Discrete Charge Distribution

To determine the ***electric energy*** associated with an assembly of charges, let us first consider n points $P_1, P_2, P_3, \ldots P_n$ in a charge-free region V bounded by surface S as shown in Fig. 2.18. In a charge-free region, initially there is no electric field in V. Therefore, if a point charge q_1 is brought from infinity to the point P_1 (r_1), no work $(W_1 = 0)$ is done. Now in presence of q_1 in V, another charge q_2 is brought from infinity to the point P_2 (r_2), work is done $W_2 = q_2V_{21}$ in the presence of electric field due to the first charge q_1, where V_{21} is the potential at P_2 due to the charge q_1. Now when the third charge q_3 is moved from infinity to the point P_3, the work done is $W_3 = q_3(V_{32} + V_{31})$. Likewise, work done in bringing n^{th} charge to P_n is $W_n = q_n(V_{n(n-1)} + V_{n-1(n-2)} + \cdots V_{n1})$. Therefore, the total work done is given by

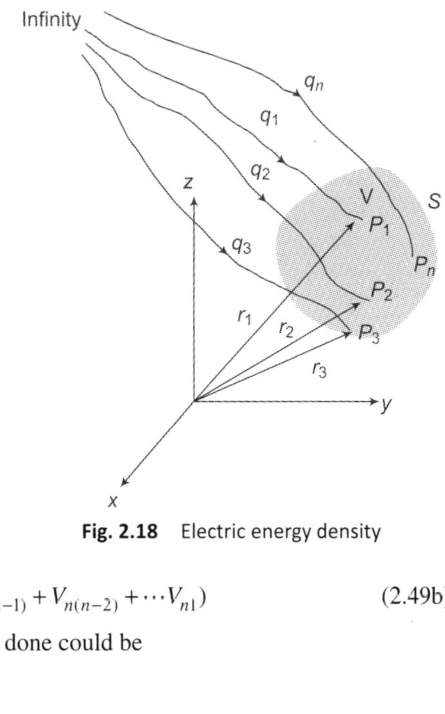

Fig. 2.18 Electric energy density

$$W_E = W_1 + W_2 + \cdots W_n$$

$$= 0 + q_2V_{21} + q_3(V_{32} + V_{31}) + \cdots q_n(V_{n(n-1)} + V_{n(n-2)} + \cdots V_{n1}) \tag{2.49b}$$

If the charges were moved in reverse order, the total work done could be

$$W_E = W_n + W_{n-1} + \cdots W_1$$

$$W_E = 0 + q_{n-1}V_{(n-1)n} + q_{n-2}(V_{(n-2)(n-1)} + V_{(n-2)(n+1)}) + \ldots q_1(V_{1(n-1)} + V_{1(n-2)} + \ldots V_{1n}) \tag{2.49c}$$

For mathematical simplification, the sum of (2.49b) and (2.49c) yields

$$2W_E = q_1V_1 + q_2V_2 + q_3V_3 + \ldots q_nV_n$$

Here, italic $V_{i\text{'s}}$ represent the total potentials at points $P_{i\text{'s}}$, $i = 1, 2, \ldots, n$

Therefore, the total work done is

$$W_E = \frac{1}{2}(q_1V_1 + q_2V_2 + q_3V_3 + \ldots q_nV_n) = \frac{1}{2}\sum_{i=1}^{i=n} q_iV_i \text{ Joule} \tag{2.49d}$$

Continuous Charge Distribution

If the region V contains continuous distribution of charge of volume density ρ, the electrical energy associated with it is expressed from (2.49d) in terms of integral over the volume region

$$W_E = \frac{1}{2}\int \rho V dv = \frac{1}{2}\int (\nabla \cdot \mathbf{D}) V dv; \text{ where } \nabla \cdot \mathbf{D} = \rho \tag{2.50a}$$

Using the vector identity

$$\nabla \cdot (\mathbf{D}V) = \mathbf{D} \cdot \nabla V + (\nabla \cdot \mathbf{D})V$$

Or

$$(\nabla \cdot \mathbf{D})V = \nabla \cdot (\mathbf{D}V) - \mathbf{D} \cdot \nabla V \tag{2.50b}$$

in (2.50a),

$$W_E = \frac{1}{2}\left[\int_v (\nabla \cdot \mathbf{D}V)dv - \int_v (\mathbf{D} \cdot \nabla V)dv\right] \tag{2.50c}$$

Using the divergence theorem to the first term of the right-hand side, $\mathbf{E} = -\Delta V$ and $\mathbf{D} = \varepsilon \mathbf{E}$, (2.50c) reduces to

$$W_E = \frac{1}{2}\varepsilon\left[\int_S V(\mathbf{E} \cdot d\mathbf{S}) + \int_v |\mathbf{E}|^2 dv\right] \tag{2.50d}$$

If it is assumed that V is a sphere, centred on the origin, and the limit in which the radius r of this sphere goes to infinity. It is known that the electric field at large distances from a bounded charge distribution varies as $1/r^2$ and the potential as $1/r$. The surface area of the sphere increases as T^2. Hence, the integrand in the first term of the right-hand side of (2.50d) varies as $1/r$. Hence, in the limit as $r \to \infty$, the surface integral in (2.50d) becomes zero. Thus, (2.50d) reduces to

$$W_E = \frac{1}{2}\varepsilon \int_V E^2 dv \tag{2.50e}$$

where the integral is over all space. Equation (2.50e) tells that the potential energy of a continuous charge distribution is stored in the electric field. The **energy density** in **E**-field is given by

$$W_E = \frac{1}{2}\varepsilon E^2 \tag{2.50f}$$

For time-varying fields, E is the rms value of the electric field.

Magnetic Stored Energy 2.5.2

When a time-varying (ac or RF) electric current is flowing in an inductor, there is energy stored in the magnetic field. The instantaneous power supplied for the current flow in the inductor is

$$P = P = IV = IL\frac{dI}{dt} \tag{2.51a}$$

The energy input to build to steady current I is given by the integral

$$W_m = \int_0^t P dt = L\int_0^I i di = \frac{1}{2}LI^2 \tag{2.51b}$$

An expression for the magnetic stored energy density can be obtained in terms of magnetic field **H** considering a long solenoid on a magnetic material of permeability μ with length l and a radius r consisting of N turns of wire.

Here,

$$L = \frac{\mu N^2 A}{l}, I = \frac{Bl}{\mu N} = \frac{Hl}{N} \tag{2.51c}$$

The **energy density** (energy/volume) stored in the magnetic field is

$$w_m = \frac{\frac{1}{2}LI^2}{Al} = \frac{\frac{1}{2}\frac{\mu N^2 A}{l}\left(\frac{Hl}{N}\right)^2}{Al} = \frac{1}{2}\mu H^2 \tag{2.51d}$$

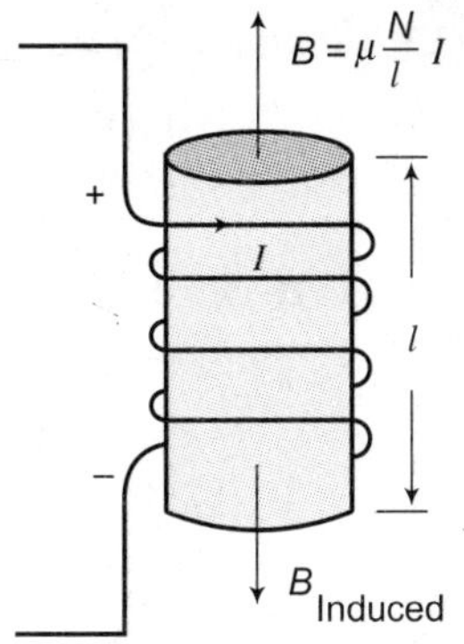

Fig. 2.19 Magnetic energy in solenoid

Power Flow and Poynting Vector 2.5.3

Electric Stored Energy

For electromagnetic wave propagation, the instantaneous power transmitted through a unit area (instantaneous **power density**) is called the **Poynting vector**, ***P***

$$\boldsymbol{P} = \boldsymbol{E} \times \mathbf{H} \tag{2.51e}$$

Time average power flow through a unit area is given by

$$P_{av} = \frac{1}{T}\int_0^T P dt \tag{2.51f}$$

For time-harmonic fields (2.51f) becomes

$$\mathbf{P}_{av} = \frac{1}{2}(\mathbf{E} \times \mathbf{H}) \tag{2.51g}$$

POYNTING THEOREM 2.6

Simple Media 2.6.1

The ***Poynting theorem*** states that in a medium having nonzero real quantities σ, μ, ε (no dielectric and magnetic losses), the total power flowing out of a given volume V is equal to the time rate of decrease of energy stored within V minus the conducting heat loss, and the mathematical expression of the theorem can be derived from the Maxwell's equations:

$$\nabla x \mathbf{E} = -\frac{\partial \mathbf{B}}{\partial t} = -\mu \frac{\partial \mathbf{H}}{\partial t} \tag{2.52a}$$

$$\nabla x \mathbf{H} = \mathbf{J} + \frac{\partial \mathbf{D}}{\partial t} = \sigma \mathbf{E} + \varepsilon \frac{\partial \mathbf{E}}{\partial t} \tag{2.52b}$$

From (2.52b),

$$\mathbf{E} \cdot (\nabla x \mathbf{H}) = \sigma E^2 + \varepsilon \mathbf{E} \cdot \frac{\partial \mathbf{E}}{\partial t} \tag{2.52c}$$

Using the vector identity

$$\nabla\cdot(\mathbf{H}\times\mathbf{E})=\mathbf{E}\cdot(\nabla\times\mathbf{H})-\mathbf{H}\cdot(\nabla\times\mathbf{E})$$

Equation (2.52c) becomes

$$\mathbf{H}\cdot(\nabla\times\mathbf{E})+\nabla\cdot(\mathbf{H}\times\mathbf{E})=\sigma E^2+\varepsilon\mathbf{E}\cdot\frac{\partial\mathbf{E}}{\partial t} \tag{2.52d}$$

From (2.52a),

$$\mathbf{H}\cdot(\nabla\times\mathbf{E})=-\mu\mathbf{H}\cdot\frac{\partial\mathbf{H}}{\partial t}=-\frac{1}{2}\mu\frac{\partial(\mathbf{H}\cdot\mathbf{H})}{\partial t} \tag{2.52e}$$

From (2.52d) and (2.52e),

$$-\frac{1}{2}\mu\frac{\partial H^2}{\partial t}-\nabla\cdot(\mathbf{E}\times\mathbf{H})=\sigma E^2+\frac{1}{2}\varepsilon\frac{\partial E^2}{\partial t}$$

or

$$\nabla\cdot(\mathbf{E}\times\mathbf{H})=-\frac{1}{2}\varepsilon\frac{\partial E^2}{\partial t}-\frac{1}{2}\mu\frac{\partial H^2}{\partial t}-\sigma E^2 \tag{2.52f}$$

Integrating over the region volume V and applying the divergence theorem,

$$\int_V\nabla\cdot(\mathbf{E}\times\mathbf{H})dv=\oint_S(\mathbf{E}\times\mathbf{H})\cdot d\mathbf{S}=-\frac{\partial}{\partial t}\int_V\left[\frac{1}{2}\varepsilon E^2+\frac{1}{2}\mu H^2\right]dv-\int_V\sigma E^2dv \tag{2.52g}$$

Power flow through S = Time rate of decrease in energy stored in V + Power dissipated in V.

Equation (2.52g) expresses the ***Poyinting theorem***. From the conservation of energy principle, ***the surface integral on the LHS giving the total power leaving the region V is equal to the sum of the time rate of decrease in energy stored in the E and H fields in V and the total power dissipated due to non zero conductivity of the medium.*** From the LHS of (2.52g), it is seen that the integration of the poynting vector over a closed surface represents the total power flowing out of that surface.

Media with Dielectric and Magnetic Losses 2.6.2

In a medium having σ, $\mu=\mu'-j\mu''$, $\varepsilon=\varepsilon'-j\varepsilon''$ (nonzero dielectric and magnetic losses), **Poynting theorem** states that the total power impressed by a current source within a region of volume V is equal to the sum of the time-average power dissipated as heat inside the region, the complex power transmitted from the region and $2\omega j$ times the difference between time-average magnetic and electric energies stored within the region. Using Gauss's theorem total power flow through a closed surface S enclosing a volume V can be written as

$$\oint_S\mathbf{E}\times\mathbf{H}^*\cdot d\mathbf{S}=\oint_V\nabla\cdot(\mathbf{E}\times\mathbf{H}^*)dV \tag{2.53a}$$

Using vector identity and Maxwell's equations with complex conjugate field $\mathbf{H}^*$, we can derive

$$\nabla\cdot(\mathbf{E}\times\mathbf{H}^*)=\mathbf{H}^*\cdot\nabla\times\mathbf{E}-\mathbf{E}\cdot\nabla\times\mathbf{H}^*=j\omega\mathbf{H}\cdot\mathbf{H}^*-j\omega\mathbf{D}^*\cdot\mathbf{E}-\mathbf{E}\cdot\mathbf{J}^* \tag{2.53b}$$

Substituting (2.53b) to (2.53a),

$$\oint_S\mathbf{E}\times\mathbf{H}^*\cdot d\mathbf{S}=-j\omega\int_V(\mathbf{B}\cdot\mathbf{H}^*+\mathbf{E}\cdot\mathbf{D}^*)dV-\int_V\mathbf{E}\cdot\mathbf{J}^*dV \tag{2.53c}$$

Separating real and imaginary parts,

$$Re\cdot\oint_S \mathbf{E}\times\mathbf{H}^*\cdot d\mathbf{S}+\omega\oint_V(\mu''H^2+\varepsilon''E^2)\,dV+\oint_V \sigma E^2 dV = -\int_V \mathbf{E}\cdot\mathbf{J}_i dV \tag{2.53d}$$

and

$$\text{Im}\cdot\oint_S \mathbf{E}\times\mathbf{H}^*\cdot d\mathbf{S} = -\omega(W_m - W_e) \tag{2.53e}$$

Here, the externally impressed current density $\mathbf{J}_i$ is assumed real. Equation (2.53d) shows that the real power P_i impressed by the current source within the volume V represented by the right-hand side term is the sum of real electromagnetic power flow P_r through the closed surface S from volume V (represented by the first term in the left-hand side), the power loss due to electric and magnetic polarization damping forces (represented by the second term in the left-hand side), and joule healing loss P_L (represented by the third term in the left-hand side). Equation (2.53e) states that the imaginary part of the complex power flow from V is equal to the ω times the total reactive energy stored in the magnetic and electric fields in V. In general, when the losses are small ($\varepsilon'' << \varepsilon'$ and $\mu'' << \mu'$).

$$P_i = P_r + j\,\omega(W_m - W_e) + P_L \tag{2.53f}$$

EM WAVE SOLUTIONS 2.7

Wavefront 2.7.1

The electromagnetic wave solutions are the **E** and **H** fields in a medium and are derived from Maxwell's equations or the wave equations in the three basic coordinate systems—rectangular (x, y, z), cylindrical (ρ, ϕ, z) and spherical (r, θ, ϕ), depending on the source geometries and the distance from the source (Fig. 2.20).

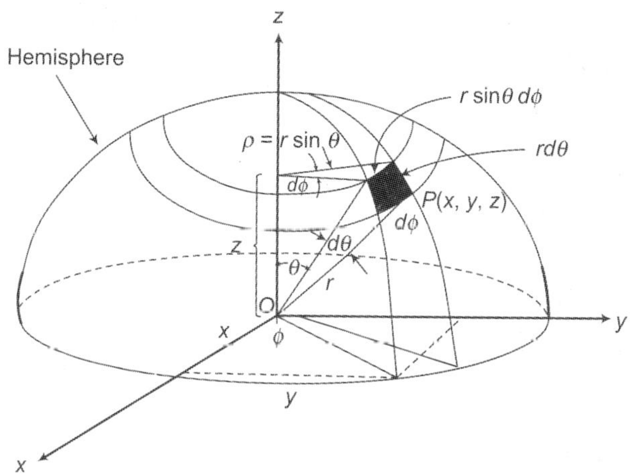

Fig. 2.20 Rectangular, cylindrical and spherical coordinate systems

When an electric current source emits electromagnetic energy, and it propagates in the medium, the field components are complex having both amplitude and phase which are also functions of space coordinates and time. The phase patterns in space coordinates determine the wave type as plane, or cylindrical, or spherical. The phase distribution in the medium exhibits locus of constant phase surfaces, called the **wavefront** as shown in Fig. 2.21. If the wavefront is plane, the wave is called the plane wave and the solutions are obtained in the rectangular coordinate system. If the wavefront is cylindrical, the wave is called a cylindrical wave and the solutions are obtained in the cylindrical coordinate system. If the wavefront is spherical, the wave is called a spherical wave and the solutions are obtained in the spherical coordinate system. It is natural that an ideal point source or a very small source generates waves equally in all directions and the wavefront is spherical. A linear current source generates cylindrical wavefronts. Both the wavefronts at a very large (ideally infinite) distance appear plane over a practical size of observation surface as explained in Fig. 2.21 (d). In this paragraph, solutions of the wave equations are described in these three coordinate systems.

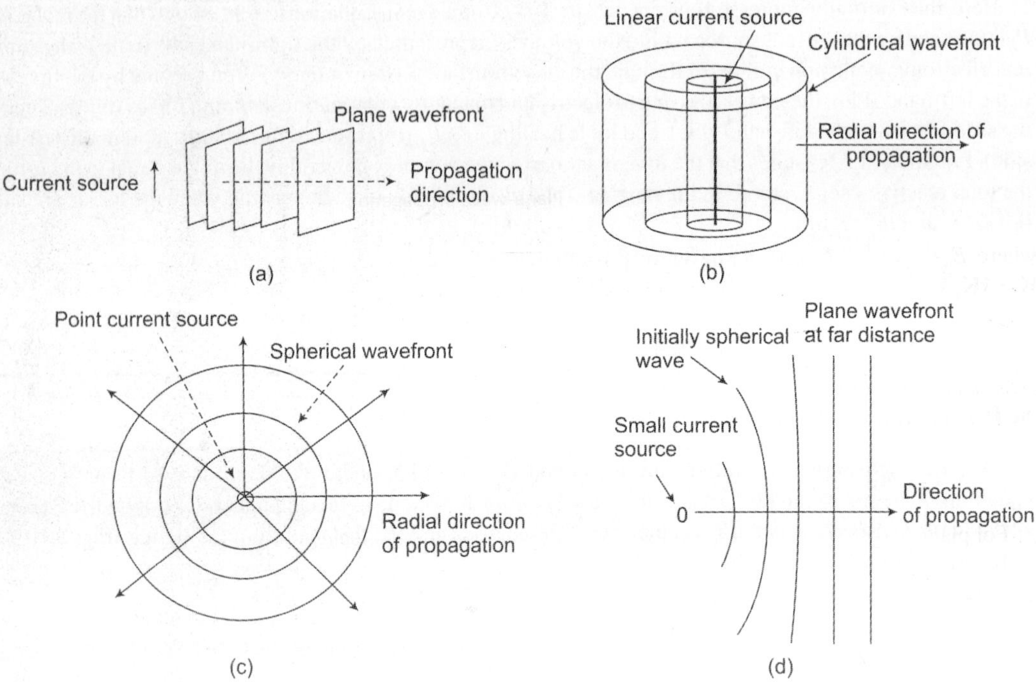

Fig. 2.21 Different natures of wavefronts

PLANE WAVES IN UNBOUNDED MEDIUM 2.8

When microwaves propagate with the phase remaining constant over a set of planes, it is called a plane wave. In addition, if the magnitude also remains constant, the wave is called a uniform plane wave having the following properties.

(a) The electric and magnetic fields are mutually perpendicular to each other and also perpendicular to the direction of propagation, i.e. the wave is transverse electromagnetic (TEM).

(b) **E** and **H** fields are always in time phase.

(c) The magnitudes of the two fields are always constant.

(d) The stored energies are equally divided between the **E** and **H** fields.

(e) The power transmitted by the two fields is in the direction of propagation.

The mathematical solution of a plane wave in an unbounded (ideally boundary less) source-free (source is at far away distance from the observation region) medium is obtained from the wave equation in rectangular coordinates (x, y, z) and can be written as

$$\mathbf{E}(r) = \mathbf{E}_0 e^{-\mathbf{K}\cdot r} \tag{2.54}$$

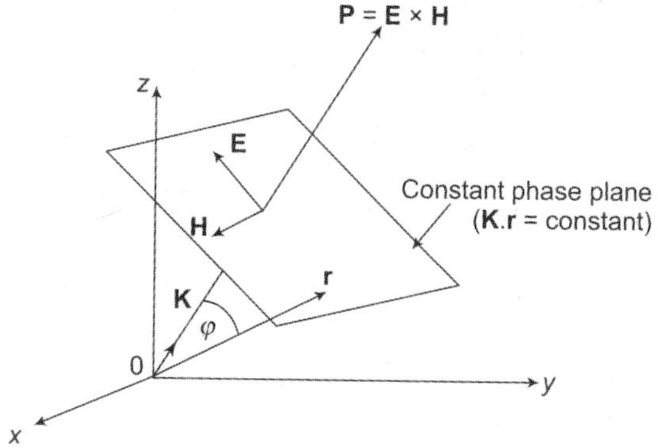

Fig. 2.22 Plane-wave propagation

where E_o is the amplitude of the wave propagating in the direction specified by the propagation vector $\mathbf{K} = \hat{\mathbf{x}}\mathbf{K}_x + \hat{\mathbf{y}}\mathbf{K}_y + \hat{\mathbf{z}}\mathbf{K}_z$, and the magnitude of **K** is

$$K = \sqrt{\omega^2 \mu\varepsilon - j\omega\mu\sigma} = \gamma \tag{2.55}$$

Here, **r** is the position vector of the point of observation. **E** remains constant on a wavefront. The solution for the **H**-field is obtained from Maxwell's equation in terms of **E**:

$$\mathbf{H} = \frac{1}{\omega\mu}(\mathbf{K} \times \mathbf{E}) \tag{2.56}$$

For plane waves, the wave impedance is the same as the characteristic impedance of the medium ($\mathbf{Z}_w = \eta$).

If waves propagate in a lossless medium ($\sigma = 0$, $\alpha = 0$) along the + z-axis, with component $\mathbf{E} = \hat{\mathbf{x}} E_x(z, t)$, the wave equation for this field simplifies to

$$\frac{\partial^2 E_x}{\partial z^2} - \mu\varepsilon \frac{\partial^2 E_x}{\partial t^2} = 0 \tag{2.57}$$

When the current source is sinusoidal, the general solution to this wave equation is

$$E_x(z, t) = C_1 \cos(\omega t - \beta z) + C_2 \cos(\omega t + \beta z) \tag{2.58a}$$

$$H_y(z, t) = \frac{1}{\eta}\{C_1 \cos(\omega t - \beta z) - C_2 \cos(\omega t + \beta z)\} \tag{2.58b}$$

Here, β is the propagation constant and $-\beta z$ and $+\beta z$ represent propagation in $+z$ and $-z$ directions, respectively, and the wave impedance

$$\eta = \sqrt{\frac{\mu}{\varepsilon}} = \frac{E_x}{H_y} \text{ for propagation along } z\text{-axis} \tag{2.58c}$$

The constant C's are determined by the source and the other boundary conditions. The physical meaning of Eqs. (2.58) is that the first term represents an incident wave travelling in the $+z$ direction and the second term represents a reflected wave in the $-z$ direction. The negative sign ($-C_2$) in the second term for magnetic field appears due to change in H_y direction of reflection wave.

Wave Velocities 2.8.1

In a lossless medium, $\sigma = 0$, $\alpha = 0$, and $\mathbf{K} = \gamma = jB$. If waves propagate along the $+z$-axis, the uniform plane-wave solution can be written as

$$E_x = E_0 \sin(\omega t - \beta z) \tag{2.59a}$$

$$H_y = (E_0/\eta) \sin(\omega t - \beta z) \tag{2.59b}$$

The argument of the sine functions of the above equations is the phase of the fields,

$$\phi = \omega t - \beta z \tag{2.59c}$$

The frequency of the wave is given by

$$f = \frac{\omega}{2\pi} = \frac{1}{T}$$

where T is the time period. The phase constant is defined by

$$\beta = \frac{\partial \phi}{\partial z} \tag{2.59e}$$

for a fixed value of time.

The wavelength λ is the distance, alone the z-direction, in which the phase changes by 2π radians for a fixed value of time: $\lambda = 2\pi/\beta$.

The speed with which a constant value of instantaneous phase travels is called the ***phase velocity*** u_p. For a lossless medium, it is equal to and denoted by the same symbol as the velocity of propagation, determined solely by the medium

$$u_p = \frac{dz}{dt} = \frac{\omega}{\beta} = \frac{1}{\sqrt{\mu\varepsilon}}$$

$$= \lambda f \tag{2.60}$$

The velocity at which the energy is propagated is called the *group velocity*, and is expressed by

$$u_g = \frac{d\omega}{d\beta} \tag{2.61}$$

From (2.58) and (2.60), we can write

$$\omega t \pm \beta z = \omega\left(t \pm \frac{z}{u_p}\right); \ \beta = \frac{\omega}{u_p},$$

In free space, $u_p = c = \dfrac{1}{\sqrt{\mu_0 \varepsilon_0}}$, the velocity of light $= 3 \times 10^8$ m/s

Example 2.2 A z-polarized plane wave is propagating in the xy plane in a direction making an angle ϕ with the x-axis. Find the expressions of **E** and **H**.

Solution If the direction of propagation is **r**, the E-field for z-polarized wave is expressed by

$$\mathbf{E} = \hat{\mathbf{z}} E_o e^{-j\mathbf{K}.\mathbf{r}} = \hat{\mathbf{z}} E_o e^{-jk(x\cos\varphi + y\sin\varphi)}$$

The magnetic field can be expressed by

$$\mathbf{H} = \frac{1}{\omega\mu}(\mathbf{K}\, x\, \mathbf{E}) = \frac{k}{\omega\mu}(\hat{\mathbf{x}}\cos\varphi + \hat{\mathbf{y}}\sin\varphi) \times \hat{\mathbf{z}} E_0 e^{-jk(x\cos\varphi + y\sin\varphi)}$$

$$= \frac{E_0}{\eta}(\hat{\mathbf{x}}\sin\varphi - \hat{\mathbf{y}}\sin\varphi) e^{-jk(x\cos\varphi + y\sin\varphi)}$$

Example 2.3 The electric field of a uniform plane wave is given by $\mathbf{E} = \hat{\mathbf{x}}\, 20 \sin(3\pi\, 10^8\, t - \pi z) + \hat{\mathbf{y}}\, 20 \cos(3\pi\, 10^8\, t - \pi z)$ V/m.

Find (a) phase velocity, (b) the corresponding magnetic field **H**, and (c) the polarization of the wave.

Solution

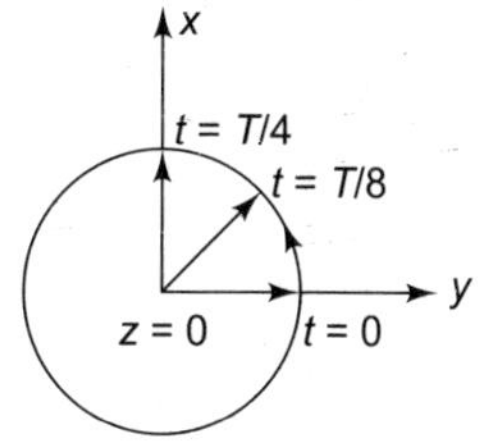

Fig. 2.23 +z is inward normal to the plane of paper

(a) $\omega = 3\pi 10^8$ rad/s, $\beta = \pi$ rad/m, $u_p = \omega/\beta = 3 \times 10^8$ m/s

(b) Given $\mathbf{E} = \hat{\mathbf{x}}E_x + \hat{\mathbf{y}}E_y$

$$\mathbf{H} = \frac{\hat{\mathbf{z}} \times \mathbf{E}}{\eta} = \frac{\hat{\mathbf{z}} \times (\hat{\mathbf{x}}E_x + \hat{\mathbf{y}}E_y)}{\eta}$$

$$= \hat{\mathbf{y}}\frac{20}{377}\sin(3\pi 10^8 t - \pi z) - \hat{\mathbf{x}}\frac{20}{377}\cos(3\pi 10^8 t - \pi z);\ \text{A/m}$$

$$= 0.0531[-\hat{\mathbf{x}}\cos(3\pi 10^8 t - \pi z) + \hat{\mathbf{y}}\sin(3\pi 10^8 t - \pi z)];\ \text{A/m}$$

(c) The E-field has two orthogonal components with equal amplitude and is out of phase by 90°. Thus, the wave is circularly polarized.

Now, at $z = 0, t = 0, \mathbf{E} = \hat{\mathbf{y}}20$ or $E_y = 20$ V/m

At $z = 0, t = T/2 = \dfrac{2\pi}{4\omega}$ or $\omega t = \dfrac{\pi}{2}$

$$\mathbf{E} = \hat{\mathbf{x}}\, 20 \sin\frac{\pi}{2} + \hat{\mathbf{y}}\, 20 \cos\frac{\pi}{2} = \hat{\mathbf{x}}20$$

Or, $E_x = 20$ V/m

At $z = 0, t = T/8 = \dfrac{2\pi}{8\omega}$ or $\omega t = \dfrac{\pi}{4}$

$$\mathbf{E} = \hat{\mathbf{x}}\, 20 \sin \pi/4 + \hat{\mathbf{y}}\, 20 \cos \pi/4 = (x + \hat{\mathbf{y}})\, 20/\sqrt{2}$$

Or, $|\mathbf{E}| = 20$ V/m

Thus, as time progress, the tip of **E** rotates from y-axis to the x-axis in the anti-clockwise direction when looked in the $+z$ direction. Therefore, the wave is left circularly polarized.

Example 2.4 The electric field of a plane wave is represented by $\boldsymbol{E} = 10\, y \cos(10^9 t + 30z)$ V/m. Determine the (a) magnetic field **H**, (b) phase velocity, and (c) dielectric constant of the medium, where $\mu = \mu_0$

Solution

E-field amplitude $E_0 = 10$V/m, $\mathbf{E} = \hat{\mathbf{y}}E_y$

$$\omega = 10^9 \text{ rad/sec}$$

$$\beta = 30 \text{ rad/m}$$

(a)
$$\nabla \times \mathbf{H} = \frac{\partial \mathbf{D}}{\partial t} = \varepsilon\frac{\partial \mathbf{E}}{\partial t} = -\hat{y}\varepsilon 10 \sin(10^9 t + 30z) \times 10^9$$

Therefore, $\nabla \times \mathbf{H}$ has only y component.

Therefore,

$$\hat{\mathbf{y}}\left[\frac{\partial H_z}{\partial \hat{\mathbf{x}}} - \frac{\partial H_x}{\partial \hat{\mathbf{z}}}\right] = -\hat{\mathbf{y}}10^{10}\varepsilon\sin(10^9 t + 30z)$$

For plane wave with E_y propagating in the z-direction, $H_z = 0$. Therefore,

$$-\hat{\mathbf{y}}\frac{\partial H_x}{\partial \hat{\mathbf{z}}} = -\hat{\mathbf{y}}10^{10}\ \varepsilon\sin(10^9 t + 30z)$$

or,
$$\frac{\partial H_x}{\partial \hat{z}} = \varepsilon 10^{10}\sin(10^9 t + 30z)$$

Therefore, $H_x = \int \varepsilon 10^{10}\sin(10^9 t + 30z)\, dz$

$$= -\varepsilon 10^9 \frac{\cos(10^9 t + 30z)}{3}\ \text{A/m}$$

or,
$$\mathbf{H} = -\hat{\mathbf{x}}\varepsilon\frac{10^9}{3}\cos(10^9 t + 30z)\ \text{A/m}$$

(b)
$$\mathbf{u}_p = \frac{\omega}{\beta} = \frac{10^9}{30} = \frac{10^8}{3}\ \text{m/s}$$

(c)
$$\mathbf{u}_p = \frac{1}{\sqrt{\mu\varepsilon}}$$

or,
$$\mu\varepsilon = \frac{1}{u_p^2} = \left(\frac{3}{10^8}\right)^2 = 9\times10^{-16}$$

Since
$$\mu = \mu_0, \varepsilon = \frac{9\times10^{-16}}{\mu_0} = \frac{9\times10^{-16}}{4\pi\times10^{-7}}$$

Dielectric constant of the medium is

$$\varepsilon_r = \frac{\varepsilon}{\varepsilon_0} = \frac{9\times10^{-16}\times36\pi}{4\pi\times10^{-7}\times10^{-9}} = 81$$

Plane Waves in a Lossy Dielectric 2.8.2

In a lossy dielectric, $\varepsilon = \varepsilon' - j\varepsilon''$ and $\sigma << \omega\varepsilon$.

The intrinsic impedance of the dielectric is given by

$$\eta = \sqrt{\frac{j\omega\mu}{\sigma + j\omega\varepsilon}} = \sqrt{\frac{\mu}{\varepsilon}}\left(1 - j\frac{\sigma}{\omega\varepsilon}\right)^{-1/2}$$

$$\cong \sqrt{\frac{\mu}{\varepsilon}}\left[1 + \frac{j\sigma}{2\omega\varepsilon}\right] \tag{2.62a}$$

where $\frac{\sigma}{\omega\varepsilon}$ is the *loss tangent* of the medium. For low-loss case $\left(\frac{\sigma}{\omega\varepsilon} \ll 1\right)$, the propagation constant can be expressed as

$$K = jk\left(1 - \frac{j\sigma}{\omega\varepsilon\tau}\right)^{1/2}$$

$$= jk\left(1 - \frac{j\sigma}{2\omega\varepsilon}\right) \tag{2.62b}$$

$$= \alpha + j\beta$$

where $k = \omega(\mu\,\varepsilon)^{1/2}$ is the propagation constant in dielectric $(\mu, \varepsilon) = \beta$

Therefore, the attenuation and phase constants are obtained from real and imaginary parts of (2.62b), respectively, and are given by

$$\alpha = \frac{\sigma}{2}\sqrt{\frac{\mu}{\varepsilon}} \tag{2.62c}$$

$$\beta = \omega(\mu\,\varepsilon)^{1/2} = k \tag{2.62d}$$

Plane Waves in a Lossless Dielectric 2.8.3

In the case of lossless dielectric medium, $\varepsilon = \varepsilon'$, $\varepsilon'' = 0$, $\sigma = 0$. Therefore, the intrinsic impedance and the propagation constant in lossless dielectric are given

$$\eta = \sqrt{\frac{\mu}{\sigma}} \tag{2.63a}$$

$$\beta = \omega\sqrt{\mu\varepsilon} \tag{2.63b}$$

For the free space (ε_0, μ_0), $\eta = \sqrt{\frac{\mu_0}{\varepsilon_0}} = 120\,\pi$ ohms ≈ 377 ohm (2.63c)

Plane Waves in a Good Conductor 2.8.4

In a good conductor, the conduction current is normally much greater than the displacement current even at microwave frequencies (below the optical region). It is useful to consider the approximation $\sigma >> \omega\varepsilon$. The propagation constant can be written from (Eq. 2.55) as

$$K = (j\sigma\omega\mu)^{1/2} = (1 + j)\sqrt{\pi f \mu\sigma} = \alpha + j\beta \tag{2.64a}$$

Hence, $\alpha = \beta = \sqrt{\pi f \sigma\mu}$ (2.64b)

The fields inside the conductor are then (Fig. 2.22) from (2.54)

$$\mathbf{E} = \mathbf{E}_0 e^{-\alpha\mathbf{K}\cdot\mathbf{r}}\, e^{-j\beta\mathbf{K}\cdot\mathbf{r}} \tag{2.64c}$$

$$\mathbf{H} = \frac{1+j}{\sqrt{2}}\sqrt{\frac{\sigma}{2\pi f\mu}}(\mathbf{K}\times\mathbf{E}) \tag{2.64d}$$

Both fields decrease with penetration inside the conductor falling to $1/e$ of their surface values in a depth equal to the skin depth:

$$\delta = \sqrt{\frac{1}{\pi f \mu \sigma}} = \frac{1}{\alpha} = \frac{1}{\beta} \tag{2.64e}$$

$\delta \to 0$ as $\sigma \to \infty$ and is smaller for higher frequencies for nonzero finite conductivity.

The intrinsic impedance of a good conductor is given by

$$\eta = \sqrt{\frac{j\omega\varepsilon}{\sigma}} = (1+j)\sqrt{\frac{\omega\mu}{2\sigma}} = (1+j)R_s \tag{2.64f}$$

where
$$R_s = \sqrt{\frac{\omega\mu}{2\sigma}} \tag{2.64g}$$

The time average of the electrical energy density and the magnetic energy density inside the conductor are

$$w_e = \frac{1}{2}\varepsilon E^2 \tag{2.64h}$$

$$w_e = \frac{1}{2}\frac{\sigma E^2}{\omega} \tag{2.64i}$$

The ratio of these energies is $\omega\varepsilon/\sigma$, which is a very small number. Thus, within the conductor, the electrical energy density is negligible in comparison to the magnetic energy, and good conductors are essentially not penetrated by electric fields but by magnetic fields, especially at lower frequencies. The phase velocity inside the conductor is

$$u_p = \frac{\omega}{\beta} = \omega\delta \tag{2.64j}$$

Example 2.5 A microwave signal at 2.45 GHz is transmitted through a medium having $\sigma = 2.17$ S/m, $\varepsilon = 47\ \varepsilon_0$ and $\mu = \mu_0$. Find K, λ, u_p, R_s and η.

Solution Given $\sigma = 2.17$ S/m, $\varepsilon_r = 47$, $\mu = \mu_0$, $f = 2.45$ GHz.

$$K = jk(1 - j\sigma/\varepsilon\omega)^{1/2}$$

$$k = \omega(\mu\varepsilon)^{1/2} = 2\pi f\,(\mu_0\,\varepsilon_0\,\varepsilon_r)^{1/2}$$

$$= 2 \times 3.14 \times 2.45 \times 10^9 \times (4\pi \times 10^{-7} \times 47 \times 10^{-9}/36\pi)^{1/2} = 351.6 \text{ rad/s}$$

$$\frac{\sigma}{\omega\varepsilon} = \frac{2.17 \times 36\pi}{2\pi \times 2.45 \times 10^9 \times 47 \times 10^{-9}} = 0.33921$$

$$K = j351.6\sqrt{(1 - j0.33921)} = 58.82 + j356.49 = \alpha + j\beta$$

Therefore,
$$\alpha = 58.82 \text{ Np/m}$$

$$\beta = 356.5 \text{ rad/m}$$

$$\lambda = \frac{2\pi}{\beta} = 0.0176\text{m}$$

$$u_p = \frac{\omega}{\beta} = \frac{2\pi f}{\beta} = \frac{2\pi \times 2.45 \times 10^9}{356.5} = 4.316 \times 10^7 \text{ m/s}$$

$$R_s = \sqrt{\frac{\omega\mu_0}{2\sigma}} = \sqrt{\frac{2\pi\times 2.45\times 10^9\times 4\pi\times 10^{-7}}{2\times 2.17}} = 66.75 \text{ ohm}$$

$$\eta = \sqrt{\frac{j\omega\mu_0}{\sigma + j\omega\varepsilon}} \cong \sqrt{\frac{\mu_0}{\varepsilon_r\varepsilon_0}}\left(1 - j\frac{\sigma}{\omega\varepsilon}\right)^{-1/2} = \frac{\eta_0}{\sqrt{\varepsilon_r}}\left(1 - \frac{j\sigma}{\omega\varepsilon}\right)^{-1/2}$$

$$= \frac{377}{(47)^{1/2}}\frac{1}{[1 - j0.33921]^{1/2}} = 52.80 + j8.71 \text{ ohm}$$

Example 2.6 The electric field of a uniform plane wave in a medium (σ = 4S/m, ε_r = 80, μ_0) propagating in +z direction is given by

$$\mathbf{E} = \hat{\mathbf{x}}\, 10 \cos (5 \times 10^4\, \pi)t, \text{ V/m}$$

at z = 0. Find the instantaneous power flow per unit area along the z-direction.

Solution

$$\omega = 5 \times 10^4\, \pi \text{ rad/s}$$

$$f = \frac{\omega}{2\pi} = 2.5\times 10^4 \text{Hz} = 25 \text{ kHz}$$

$$K = j\omega\sqrt{(\mu_0\varepsilon)}\sqrt{(1 - j\sigma/\varepsilon\omega)}$$

$$= j5\times 10^4\pi\sqrt{4\pi\times 10^{-7}\times 80\times\frac{10^{-9}}{36\pi}}\sqrt{1 - \frac{j\times 4\times 36\pi}{5\times 10^4\times 80\times 10^{-9}}}$$

$$= 0.63 + j\, 0.63$$

$$= \alpha + j\beta$$

Therefore, $\alpha = \beta = 0.63$

and
$$\eta = \sqrt{\frac{j\omega\mu_0}{\sigma + j\omega\varepsilon}} = \sqrt{\frac{j5\times 10^4\pi\times 4\pi\times 10^{-7}}{4 + j5\times 10^4\pi\times 80\times\frac{10^{-9}}{36\pi}}}$$

$$= 0.1571 + j0.1571 \text{ ohm} = 0.22 \angle 45°$$

Therefore, $\mathbf{E}(z, t) = \hat{\mathbf{x}}\, 10 \cos (5 \times 10^4\, \pi t - 0.63z)e^{-0.63z}$ V/m

$$\mathbf{H}(z, t) = \hat{\mathbf{y}}\frac{10}{4.24\angle 45°}\cos(5\times 10^4\pi t - 0.63z)\, e^{-0.63z} \text{ A/m}$$

$$= \hat{\mathbf{y}}\, 45 \cos [(5 \times 10^4\pi t - 0.63z - \pi/4)]e^{-0.63z} \text{ A/m}$$

The instantaneous power flow per unit area is the Poynting vector $\mathbf{P} = \mathbf{E} \times \mathbf{H}$

$$= \hat{\mathbf{z}} + 42.4 \cos (5 \times 10^4\, \pi t - 0.63z) \cos (5 \times 10^4\, \pi t - \pi/4 - 0.63z)e^{-1.26z} \text{ W/m}^2.$$

POLARIZATION OF WAVES 2.9

The polarization of electromagnetic waves is named according to the locus that the tip of **E** describes with time while propagation in a medium. The polarization of the wave is called linear or circular or elliptical when the tip of the **E** vector describes with time a straight line, a circle or an ellipse on the xy-plane, respectively, while looking towards the direction of propagation (+z), as shown in Fig. 2.24.

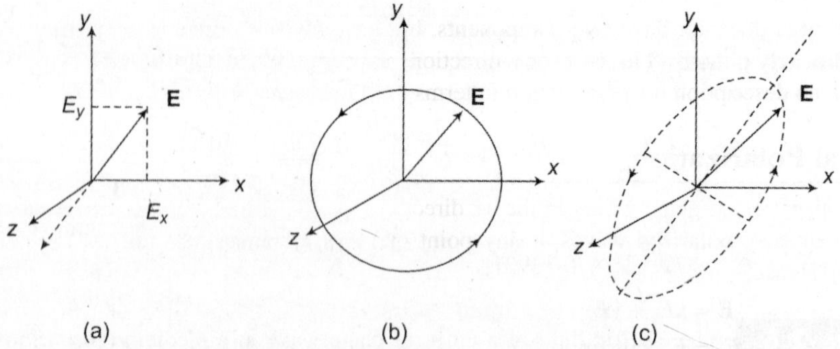

Fig. 2.24 Polarization of electromagnetic waves: (a) Linear polarization, (b) Circular polarization (c) Elliptical polarization

Linear Polarization 2.9.1

For the propagation of plane waves in the $+z$ direction, the electric field $\mathbf{E} = \mathbf{E}(z, t)$, in general has two orthogonal components at any point on the z = constant plane (Fig. 2.25) and can be represented by

$$E_x = E_{xo} \cos \omega t \tag{2.65a}$$

$$E_y = E_{yo} \cos(\omega t + \theta) \tag{2.65b}$$

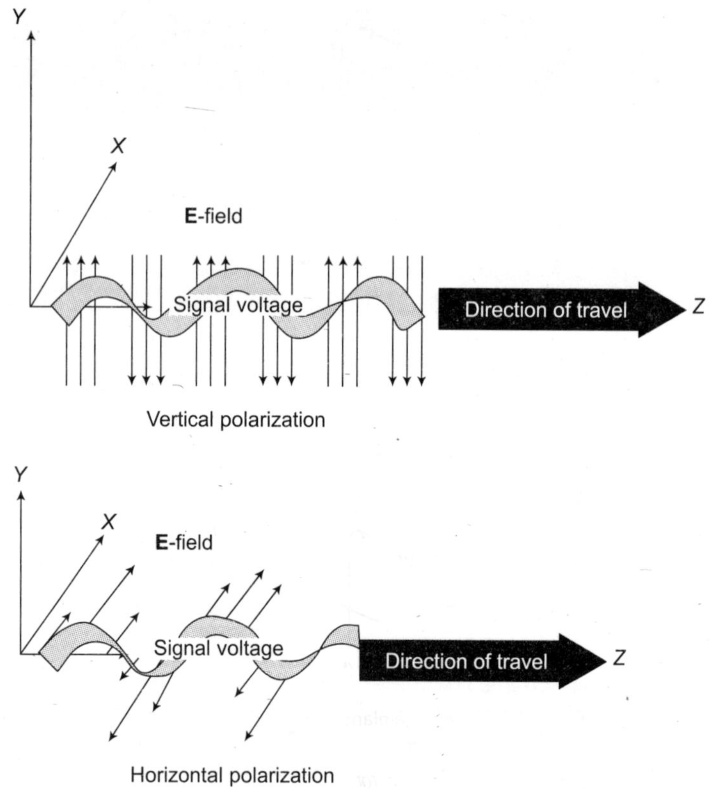

Fig. 2.25 Plane wave propagation in the $+z$ direction

If the **E** field does not have two components, but has only one, either *x*- or *y*-component, the wave is said to be **linearly polarized** in the *x*- or *y*-direction, respectively. Since for plane waves, **E** and **H** fields are orthogonal, no description for polarization in terms of **H** field is necessary.

Elliptical Polarization 2.9.2

For the propagation of plane waves in the +*z* direction, the **E** field, in general, is the superposition of two orthogonal linearly polarized waves at any point on the *z* = constant plane and can be represented by Eqs. (2.65). Hence,

$$\mathbf{E} = \hat{\mathbf{x}}E_x + \hat{\mathbf{y}}E_y \tag{2.65c}$$

$$E_x = E_{xo} \cos \omega t \tag{2.65d}$$

$$E_y = E_{yo} \cos (\omega t + \theta) \tag{2.65e}$$

where $-\pi \leq \theta \leq \pi$. As the time passes vector **E** describes an ellipse as shown in Fig. 2.26.

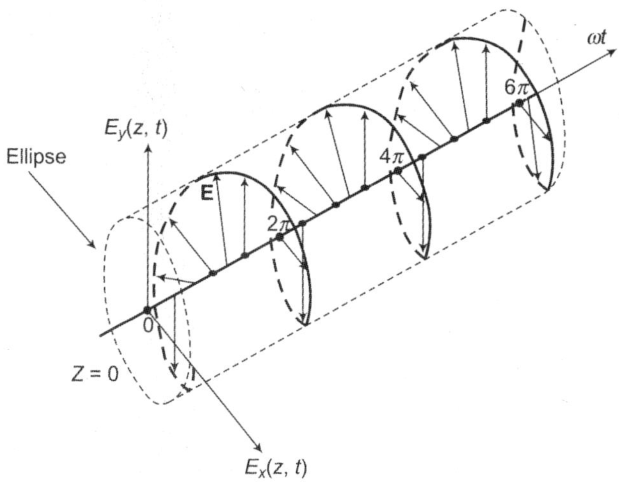

Fig. 2.26 Polarization of a plane electromagnetic wave at *z* = 0 as a function of time

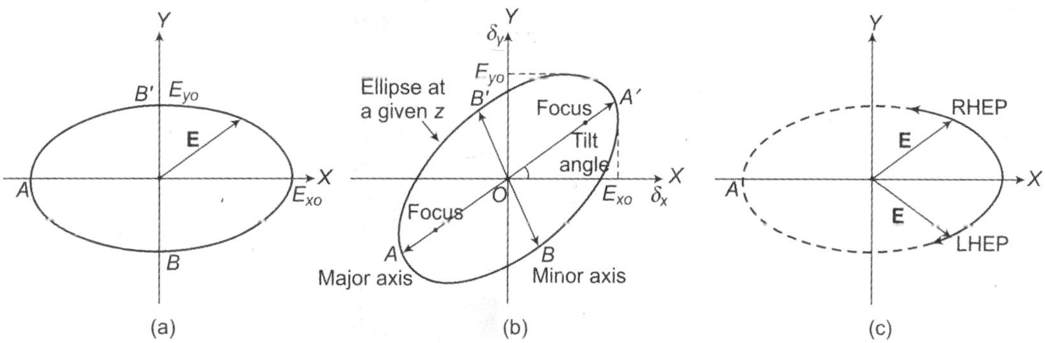

Fig. 2.27 Polarization ellipse of a plane electromagnetic wave at *z* = 0 on *xy* plane

(i) If $\theta = -\pi/2$, $\mathbf{E} = \mathbf{x}\, E_{xo} \cos \omega t + \mathbf{y}E_{yo} \sin \omega t$ (2.65f)

Now from the above,

$$\cos \omega t = E_x/E_{xo} \tag{2.65g}$$

$$\sin \omega t = E_y / E_{yo} \tag{2.65h}$$

Squaring from both sides,

$$\frac{E_x^2}{E_{xo}^2} + \frac{E_y^2}{E_{yo}^2} = 1 \tag{2.65i}$$

If $E_{xo} \neq E_{yo}$, the last equation represents an ellipse with semi-major axis E_{xo} and semi-minor axis E_{yo} as shown in Fig. 2.28.

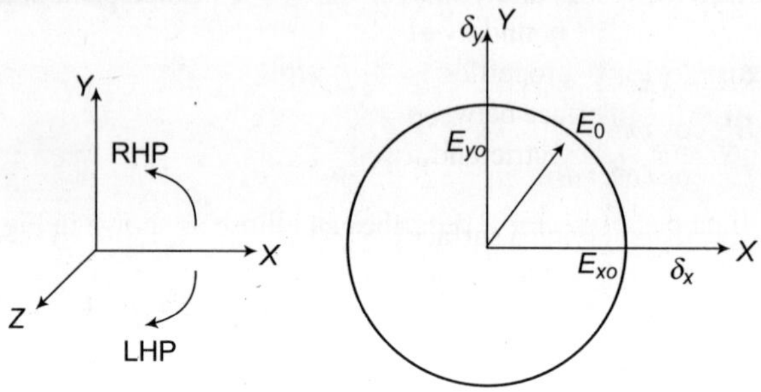

Fig. 2.28 Left and right-hand circular polarizations

(ii) If $\theta = +\pi/2$, $E = x\,E_{xo} \cos \omega t - y\,E_{yo} \sin \omega t$ (2.65j)

$$\cos \omega t = E_x / E_{xo} \tag{2.65k}$$

$$\sin \omega t = -E_y / E_{yo} \tag{2.65l}$$

Squaring from both sides,

$$\frac{E_x^2}{E_{xo}^2} + \frac{E_y^2}{E_{yo}^2} = 1 \tag{2.65m}$$

Again, if $E_{xo} \neq E_{yo}$, the last equation represents an ellipse with semi-major axis E_{xo} and semi-minor axis E_{yo}.

Therefore, it can be shown that for E_y lagging or leading E_x in time phase, **E** constantly changes both its magnitude and direction describing an **ellipse**. The rotation rate of **E** around the origin is ω radian/second.

If $\theta \neq \pm\,\pi/2$, the ellipse will be inclined to the *x* and *y*-axes. The tilt angle of major axis w.r.t. the *x*-axis is dependent on E_{xo}, E_{yo}, and θ, and can be shown that:

$$\text{Tilt} = \frac{1}{2}\tan^{-1}\left[\frac{2E_{xo}E_{yo}}{E_{xo}{}^2 - E_{yo}{}^2}\cos\theta\right] \tag{2.65n}$$

If $-\pi/2 \leq \theta < 0$, polarization is **Right-Handed Elliptical Polarization (RHEP)** and when $\pi/2 \geq \theta > 0$, it is **Left-Handed Elliptical Polarization (LHEP)**.

Circular Polarization 2.9.3

When $\theta = \pm\,\pi/2$ and $E_{ox} = E_{oy} = E_0$, say, then

$$E_x^2 + E_y^2 = E_0^2 \tag{2.65o}$$

This is an equation of a circle of radius E_0 and centre at (0, 0). If E_y lags E_x by $\pi/2$, the result is RHCP. If E_y leads E_x by $\pi/2$, the result is LHCP.

The polarization of electromagnetic waves will mostly depend on the source of generating the waves and how the source is excited.

BOUNDARY CONDITIONS 2.10

The solution to Maxwell's equations in a region of volume V is not unique unless the behaviour of the field on the boundary S of V is known, or at the boundary of material bodies with different electrical properties is known. These boundaries may be the interface between two different dielectrics or between a dielectric and a conducting media.

Consider a boundary surface S, containing a surface charge density ρ_s separating the regions 1 and 2 characterized by (σ_1, ε_1, μ_1) and (σ_2, ε_2, μ_2), respectively, as shown in Fig. 2.29.

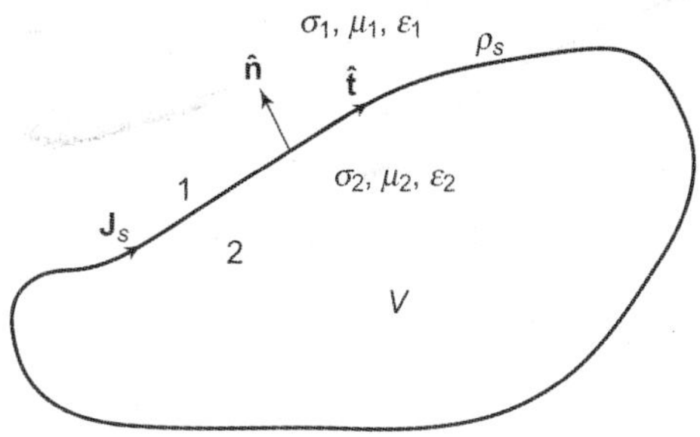

Fig. 2.29 Boundary conditions

The following four basic boundary conditions must be satisfied by the fields for a specific solution.

1. The tangential components of electric field intensity are continuous across the boundary:

$$E_{1t} = E_{2t} \tag{2.66a}$$

2. The normal components of magnetic flux density are continuous across the boundary:

$$B_{1n} = B_{2n} \tag{2.66b}$$

3. The normal components of electric flux density are discontinuous at the boundary by the surface charge density ρ_s:

$$D_{1n} - D_{2n} = \rho_s \tag{2.66c}$$

4. The tangential components of magnetic field are discontinuous at the boundary by the surface current density J_s:

$$H_{1t} - H_{2t} = J_s \tag{2.66d}$$

If one of the media is a perfect conductor ($\sigma = \infty$, $\varepsilon_r = 1$, $\mu_r = 1$) and the other one is a perfect dielectric ($\sigma = 0$, $\varepsilon_r > 1$, $\mu_r = 1$) then

$$\left.\begin{aligned} \hat{n} \times \mathbf{E} &= \mathbf{E}_t = 0 \\ \hat{n} \cdot \mathbf{B} &= \mathbf{B}_n = 0 \\ \hat{n} \cdot \mathbf{D} &= \mathbf{D}_n = \rho_s \\ \hat{n} \times \mathbf{H} &= \mathbf{H}_t = J_s \end{aligned}\right\} \tag{2.66e}$$

where $\hat{n}$ represents unit normal vector at the interface.

Proof of Boundary Conditions 2.10.1

Electric Boundary Conditions

❑ ***Interface between Two Dielectric Media*** The above boundary conditions can be proved using the interface between two dielectric media 1 and 2 as shown in Fig. 2.30. The electric field existing in the two regions 1 and 2 are represented by $\mathbf{E}_1$ and $\mathbf{E}_2$, respectively.

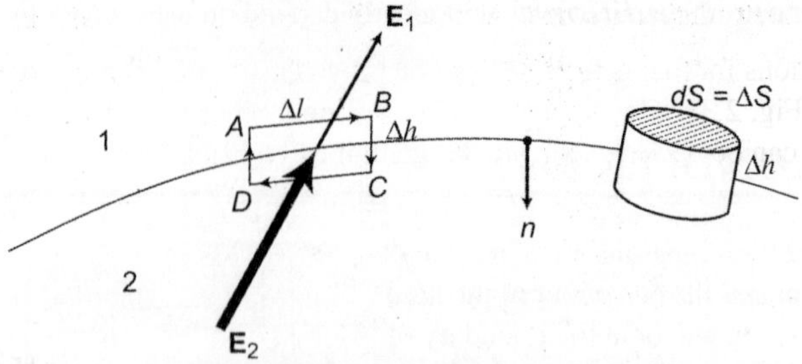

Fig. 2.30 Dielectric boundary

These fields can be decomposed into tangential (suffix t) and normal (suffix n) components to the interface as

$$\mathbf{E}_1 = \mathbf{E}_{1t} + \mathbf{E}_{1n} \tag{2.67a}$$

$$\mathbf{E}_2 = \mathbf{E}_{2t} + \mathbf{E}_{2n} \tag{2.67b}$$

According to the Maxwell's equation around the closed path *ABCDA*,

$$\oint \mathbf{E}\cdot d\boldsymbol{l} = 0 = E_{1t}\,\Delta l - E_{1n}\frac{\Delta h}{2} - E_{2n}\frac{\Delta h}{2} - E_{2t}\,\Delta l + E_{2n}\frac{\Delta h}{2} + E_{1n}\frac{\Delta h}{2} \tag{2.67c}$$

At the interface $\Delta h \to 0$, so that from (2.67c)

$$\boldsymbol{E}_{1t} = \boldsymbol{E}_{2t} \tag{2.67d}$$

$$\boldsymbol{D}_{1t}/\varepsilon_1 = \boldsymbol{D}_{2t}/\varepsilon_2 \quad \text{or,} \quad \varepsilon_2\boldsymbol{D}_{1t}/\varepsilon_2\boldsymbol{D}_{2t} \tag{2.67e}$$

which proved that tangential components of the *E*-field are **continuous** but the tangential components of electric flu*x* densities are **discontinuous** across the boundary.

Considering a small symmetrical cylindrical volume of height Δh and cross section dS enclosing surface electric charge density ρ_s at the interface (Fig. 2.30),

$$\oint \mathbf{D}.d\mathbf{S} = \Delta Q = \rho\Delta S = D_{1n}\Delta S - D_{2n}\Delta S\text{, in the limit } \Delta h \to 0$$

Or

$$\rho = D_{1n} - D_{2n} \tag{2.67f}$$

This proved that the normal components of electric flu*x* densities are discontinuous by the electric surface charge density at the interface.

If interface is charge free, $\rho = 0$, and

$$D_{1n} = D_{2n} \tag{2.67g}$$

or,

$$\varepsilon_1 E_{1n} = \varepsilon_2 E_{2n} \tag{2.67h}$$

Equations (2.67g) and (2.67h) proved that in charge free interface normal components of electric flu*x* densities are continuous but the normal components of electric fields are discontinuous at the interface.

❑ ***Interface between a Dielectric Medium and a Good Conductor*** A good conductor, such as copper, silver, gold, etc., is nearly perfect conductor ($\sigma \to \infty$) inside which $E = 0$. If the medium 2 is considered a perfect conductor then from (2.67c)–(2.67h) for $\Delta h \to 0$,

$$E_{1t} = 0, D_{1t} = 0 \tag{2.67i}$$

$$D_{1n} = \rho_s,\ E_{1n} = \rho_s/\varepsilon_1 \tag{2.67j}$$

Equations (2.67i) and (2.67j) are boundary conditions of the interface between a dielectric medium and a good conductor.

Magnetic Boundary Conditions

The boundary conditions for magnetic fields can be proved using the interface between two magnetic media 1 and 2 as shown in Fig. 2.31. The magnetic field existing in the two regions 1 and 2 are represented by $\mathbf{H}_1$ and $\mathbf{H}_2$. These fields can be decomposed into tangential and normal components to the interface as

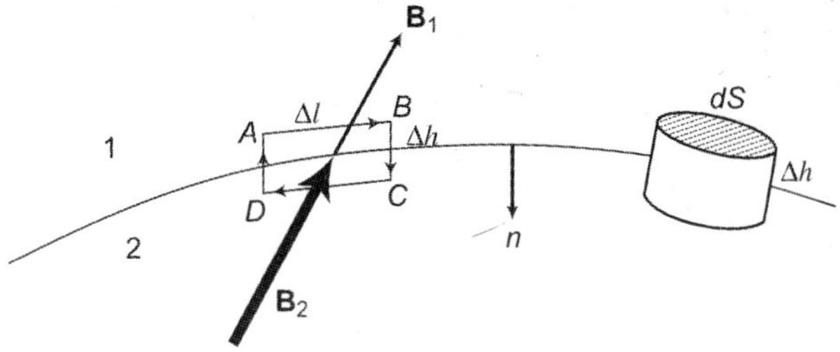

Fig. 2.31 Magnetic boundary

$$\mathbf{H}_1 = \mathbf{H}_{1t} + \mathbf{H}_{1n} \tag{2.68a}$$

$$\mathbf{H}_2 = \mathbf{H}_{2t} + \mathbf{H}_{2n} \tag{2.68b}$$

Applying Ampere's law

$$\oint \mathbf{H} \cdot dl = I$$

to the rectangular closed path $ABCDA$ while surface electric current density is assumed normal to the path, we obtained

$$\mathbf{J} \cdot \Delta l = \mathbf{H}_{1t} \cdot \Delta l + \mathbf{H}_{1n} \cdot \frac{\Delta h}{2} + \mathbf{H}_{2n} \cdot \frac{\Delta h}{2} - \mathbf{H}_{2t} \cdot \Delta l - \mathbf{H}_{2n} \frac{\Delta h}{2} - \mathbf{H}_{1n} \cdot \frac{\Delta h}{2} \tag{2.68c}$$

In the limit $\Delta h \to 0$, (2.61c) reduces to

$$\mathbf{J} = \mathbf{H}_{1t} - \mathbf{H}_{2t} = (\mathbf{H}_1 - \mathbf{H}_2) \times \hat{\mathbf{n}} \tag{2.68d}$$

Equation (2.68d) established that tangential components of magnetic fields are discontinuous by the electric surface current density on the boundary. $\hat{n}$ is a unit vector normal to the boundary and is directed from medium 1 to medium 2. If the boundary is free of current, $J = 0$ and tangential components of magnetic fields become continuous

$$H_{1t} = H_{2t} \tag{2.68e}$$

Considering a small symmetrical cylindrical volume of height Δh and cross section dS and applying Gauss's law for magnetic fields

$$\oint \mathbf{B} \cdot d\mathbf{S} = 0$$

in the limit, $\Delta h \to 0$,

$$B_{1n} = B_{2n} \tag{2.68f}$$

or

$$\mu_1 H_{1n} = \mu_2 H_{2n} \tag{2.68g}$$

Equations (2.68f) and (2.68g) established that normal components of magnetic fields are discontinuous but the normal components of magnetic flux densities are continuous on the boundary.

Example 2.17 Let a region 1 ($y < 0$) be a free space and a region 2 ($y > 0$) be a dielectric with no surface charge on the plane interface (xy plane at $y = 0$). In the region 1,

$$\mathbf{D}_1 = \hat{\mathbf{x}}10x^2 + \hat{\mathbf{y}}(6+y) + \hat{\mathbf{z}}10z$$

Find $\mathbf{D}_2$ and $\mathbf{E}_2$ in the region 2 at the boundary ($y = 0$)

Solution Since the normal component of **D** is continuous at $y = 0$, $\mathbf{D}_{2y} = \mathbf{D}_{1y} = 6$, at $y = 0$

Therefore, $E_{2y} = 6/\varepsilon_2$

Now, $\mathbf{E}_1 = \mathbf{D}_1/\varepsilon_1 = 1/\varepsilon_1 \cdot [\hat{\mathbf{x}}\,10x^2 + \hat{\mathbf{y}}(6+y) + \hat{\mathbf{z}}10z]$

Since tangential components of **E** are continuous at $y = 0$

$$E_{2x} = E_{1x} = \frac{10x^2}{\varepsilon_1} \text{ and } E_{2z} = E_{1z} = \frac{10z}{\varepsilon_1}$$

Therefore, at the boundary ($y = 0$)

$$\mathbf{E}_2 = \hat{\mathbf{x}}E_{2x} + \hat{\mathbf{y}}E_{2y} + \hat{\mathbf{z}}E_{2z}$$

$$= \hat{\mathbf{x}}\frac{10x^2}{\varepsilon_1} + \hat{\mathbf{y}}\frac{6}{\varepsilon_2} + \hat{\mathbf{z}}\frac{10z}{\varepsilon_1}$$

and

$$\mathbf{D}_2 = \varepsilon_2\mathbf{E}_2 = \hat{\mathbf{x}}\frac{\varepsilon_2}{\varepsilon_1}10x^2 + \hat{\mathbf{y}}6 + \hat{\mathbf{z}}\frac{\varepsilon_2}{\varepsilon_1}10z$$

REFLECTION AND TRANSMISSION OF A PLANE-WAVE INCIDENCE AT THE INTERFACE OF TWO DIELECTRIC MEDIA 2.11

When a plane wave incidents on an interface between two different media the wave undergoes reflection and transmission at the interface. The characteristics of the reflection and transmission will depend on the type of interface, polarization of the wave and the angle of incidence with respect to the interface as described below.

Normal Incidence 2.11.1

Let an x polarized uniform plane wave propagating in the $+z$ direction is incident normally at the interface at $z = 0$ between two dielectric media $1(z < 0;\ \sigma_1,\ \mu_1,\ \varepsilon_1)$ and $2(z > 0;\ \sigma_2,\ \mu_2,\ \varepsilon_2)$ as shown in Fig. 2.32. The incident (in $+z$ direction), reflected (in $-z$ direction) and transmitted (in $+z$ direction) waves are represented with subscripts i, r, and t, respectively, as

$$\mathbf{E}_i(z) = \hat{\mathbf{x}}E_{io}e^{-\gamma_1 z} \quad \mathbf{H}_i(z) = \hat{\mathbf{y}}\frac{1}{\eta_1}E_{io}e^{-\gamma_1 z} \tag{2.69a}$$

$$\mathbf{E}_r(z) = \hat{\mathbf{x}}E_{ro}e^{+\gamma_1 z} \quad \mathbf{H}_r(z) = -\hat{\mathbf{y}}\frac{1}{\eta_1}E_{ro}e^{+\gamma_1 z} \tag{2.69b}$$

$$\mathbf{E}_t(z) = \hat{\mathbf{x}}E_{to}e^{-\gamma_2 z} \quad \mathbf{H}_t(z) = \hat{\mathbf{y}}\frac{1}{\eta_2}E_{to}e^{-\gamma_1 z} \tag{2.69c}$$

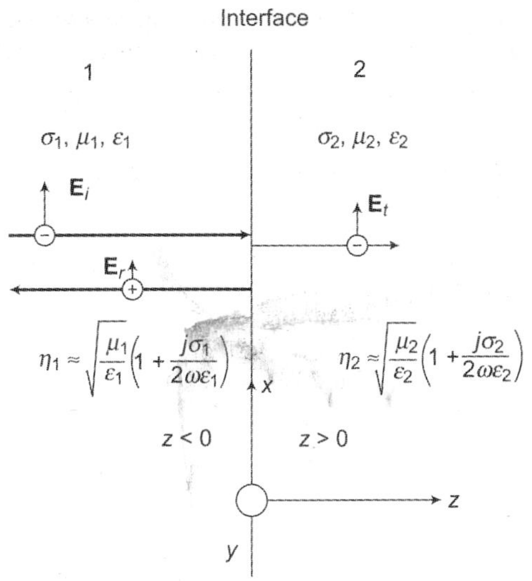

Fig. 2.32 Reflection and transmission of plane waves at an interface

where the –ve sign in (2.69b) indicates the direction of power flow according to Poynting vector.

The total fields in medium 1 and 2 are

$$\mathbf{E}_{1\text{total}} = \mathbf{E}_i + \mathbf{E}_r \qquad \mathbf{H}_{1\text{total}} = \mathbf{H}_i + \mathbf{H}_r \tag{2.69d}$$

$$\mathbf{E}_{2\text{total}} = \mathbf{E}_t \qquad \mathbf{H}_{2\text{total}} = \mathbf{H}_t \tag{2.69e}$$

For normal incident of plane waves, **E** and **H** fields are tangential to the interface and thus, continuous at the boundary:

$$\mathbf{E}_{1\text{total}}\big|_{z=0} = \mathbf{E}_{2\text{total}}\big|_{z=0} \qquad \mathbf{H}_{1\text{total}}\big|_{z=0} = \mathbf{H}_{2\text{total}}\big|_{z=0} \tag{2.69f}$$

or

$$\mathbf{E}_{io} + \mathbf{E}_{ro} = \mathbf{E}_{to} \qquad \frac{1}{\eta_1}(\mathbf{E}_{io} - \mathbf{E}_{ro}) \mp \frac{1}{\eta_2}\mathbf{E}_{to} \tag{2.69g}$$

From (2.69g), we obtained

$$\mathbf{E}_{ro}/\mathbf{E}_{io} = \frac{\eta_2 - \eta_1}{\eta_2 + \eta_1} = \Gamma = \text{reflection coefficient at the interface.} \tag{2.69h}$$

and
$$\mathbf{E}_{to}/\mathbf{E}_{io} = \frac{2\eta_2}{\eta_2 + \eta_1} = T = \text{transmission coefficient at the interface.} \tag{2.69i}$$

From (2.69h) and (2.69i), the following relations yield

$$1 + \Gamma = T \tag{2.69j}$$

$$= -1 \le \Gamma \le 1 \tag{2.69k}$$

for 0 (perfect conductor) $\le \eta_2 \le \infty$ (perfect dielectric). Equations (2.69j) and (2.69k) represent the extreme cases of wave incidence on a perfect conductor and a perfect dielectric. The total wave in the medium 1 is the sum of incident and oppositely directed reflected waves and produces **standing wave** as shown in Fig. 2.33 for different values of ωt. The **E**-field tangential to the conducting wall is zero and this null field repeats at interval of half-wavelength $z = 0, -\frac{\lambda}{2}, -\lambda$, ... Electric field shows maximum at $z = -\frac{\lambda}{4}, -\frac{3\lambda}{4}$, ...

The magnetic field possesses opposite values at these points. The waveform a, a'; b, b'; and c, c' represent the instantaneous forms at $t = 0, \frac{T}{8}, \frac{T}{4}$, respectively.

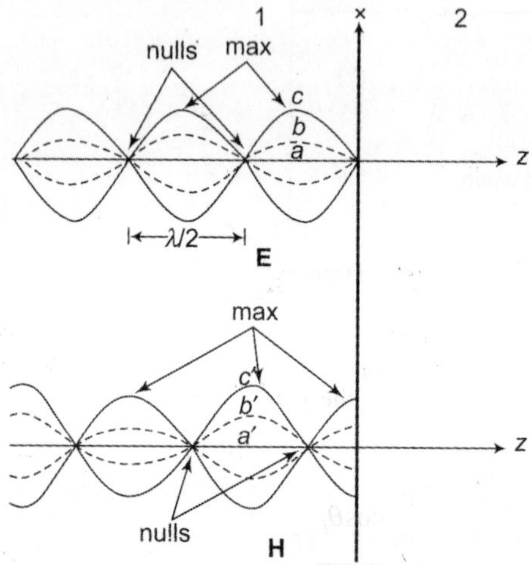

Fig. 2.33 Standing waves on reflection (a,a': $\omega t = 0$; b, b': $\omega t = \frac{\pi}{4}$; c, c' : $\omega t = \frac{\pi}{2}$)

There are three special cases:
(1) When both the two media match (same) $\eta_1 = \eta_1$, $\Gamma = 0$, $T = 1$. (2) When two media are lossless, $\sigma_1 = \sigma_2 = 0$, η_1 and η_2, and Γ, T all are real. (3) Depending on the both the media, Γ is +ve or –ve: $\eta_2 \geq \eta_1$, $\Gamma \geq 0$ and $\eta_2 \leq \eta_1$, $\Gamma \leq 0$.

Oblique Incidence 2.11.2

In general, when a uniform plane wave incident on a plane interface at the xy-plane at $z = 0$ lies between two lossless dielectric media 1 and 2, the wave is partially reflected back to the incident medium 1 and partially transmitted to the second medium as shown in Fig. 2.34. Let us assume that θ_i, θ_r and θ_t are the angles of incidence, reflection and transmission, respectively. Any arbitrary polarization of the field can be decomposed into two components, one in the plane of incidence (called parallel polarization) and another perpendicular to the plane of incidence (called perpendicular polarization). In both the cases, the reflection and refraction follow Snell's law:

$$\theta_i = \theta_r \tag{2.70a}$$

and

$$\frac{\sin\theta_t}{\sin\theta_i} = \frac{n_1}{n_2} = \sqrt{\frac{\mu_1\varepsilon_1}{\mu_2\varepsilon_2}} = \sqrt{\frac{\varepsilon_1}{\varepsilon_2}}, \text{ for nonmagnetic media } (\mu_1 = \mu_2 = \mu_0) \tag{2.70b}$$

where n_1, n_2 are the indices of refraction of the media 1 and 2 respectively. $\hat{\mathbf{k}}_i$, $\hat{\mathbf{k}}_r$, and $\hat{\mathbf{k}}_t$ are the unit vectors in the direction of propagation of incident, reflected, and transmitted waves, respectively.

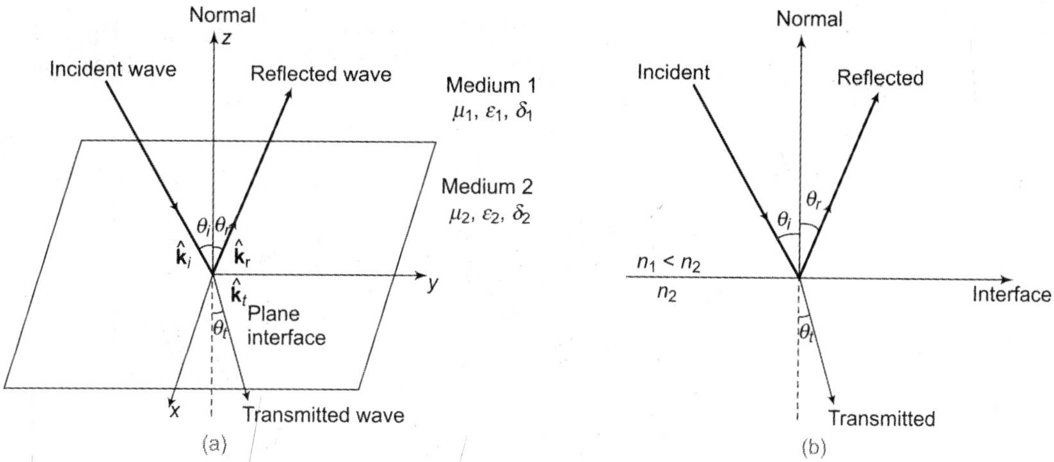

Fig. 2.34 Reflection and transmission at a plane interface

Parallel Polarization

For the parallel polarization of an incident wave, the reflection and transmission coefficients are given by Fresnel equations;

$$\Gamma_{\parallel} = \frac{\left(\frac{\varepsilon_2}{\varepsilon_1} - \sin^2\theta_i\right)^{1/2} - \frac{\varepsilon_1}{\varepsilon_2}\cos\theta_i}{\left(\frac{\varepsilon_2}{\varepsilon_1} - \sin^2\theta_i\right)^{1/2} + \frac{\varepsilon_1}{\varepsilon_2}\cos\theta_i} \tag{2.70c}$$

$$T_{\parallel} = \frac{2\sqrt{\frac{\varepsilon_1}{\varepsilon_2}\cos\theta_i}}{\left(\frac{\varepsilon_2}{\varepsilon_1} - \sin^2\theta_i\right)^{1/2} + \frac{\varepsilon_1}{\varepsilon_2}\cos\theta_i} \tag{2.70d}$$

Here,

$$1 - \Gamma_{\parallel} = \sqrt{\frac{\varepsilon_2}{\varepsilon_1}}\, T_{\parallel} \tag{2.70e}$$

The z-directed wave impedances in media 1 and 2 are given, respectively, by

$$Z_1 = \left.\frac{E_x^{(1)}}{H_y^{(1)}}\right|_{z=0} = \eta_1\cos\theta_i, Z_2 = \eta_2\cos\theta_t \tag{2.70f}$$

where η_1 and η_2 are the intrinsic impedance of media 1 and 2, respectively.

Perpendicular Polarization

For the perpendicular polarization of incident waves,

$$\Gamma_\perp = \frac{\cos\theta_i - \left(\dfrac{\varepsilon_2}{\varepsilon_1} - \sin^2\theta_i\right)^{1/2}}{\left(\dfrac{\varepsilon_2}{\varepsilon_1} - \sin^2\theta_i\right)^{1/2} + \cos\theta_i} \tag{2.71a}$$

$$T_\perp = \frac{2\cos\theta_i}{\left(\dfrac{\varepsilon_2}{\varepsilon_1} - \sin^2\theta_i\right)^{1/2} + \cos\theta_i} \tag{2.71b}$$

The z-directed wave-impedances in media 1 and 2 are given by

$$Z_1 = \frac{\eta_1}{\cos\theta_i},\ Z_2 = \frac{\eta_2}{\cos\theta_t} \tag{2.71c}$$

Γ and T are related in the following way:

$$1 + \Gamma_\perp = T_\perp \tag{2.71d}$$

Detail derivation of (2.70a)–(2.70e), and (2.71a)–(2.71d) are not included in this book. The reader may consult the books on Electromagnetic Theory for this purpose.

Total Transmission and Brewster Angle 2.11.3

If the angle of incidence is varied, there will exist an angle $\theta_i = \theta_b$ say, called the Brewster angle, when $\Gamma_{\parallel} = 0$ and total transmission of parallel polarized waves occurs at the interface. Under this condition from (2.70c)

$$\theta_i = \theta_b = \sin^{-1}\sqrt{\frac{\varepsilon_2}{\varepsilon_1 + \varepsilon_2}}\ for\ \mu_1 = \mu_2 = \mu_0 \tag{2.71e}$$

$\Gamma_\perp$ never becomes zero so long as the two media are different ($\eta_1 \neq \eta_2$). Hence, the Brewster angle does not exist for perpendicular polarization. Any arbitrarily polarized wave incident at θ_b, will be reflected with **E** polarized perpendicular to the plane of incidence and the other component of **E**, parallel to the plane of incidence, is totally transmitted for nonmagnetic media. The nature of variation of $\Gamma_{||}$ and $\Gamma_\perp$ with the incident angle θ_i is shown in Fig. 2.35(a).

This equation is known as **Brewster's law**. For a glass medium ($n_2 \approx 1.5$) in air ($n_1 \approx 1$), so that θ_B is around 56° for a glass medium in air or vacuum. For an air-water interface ($n_2 \approx 1.33$) and θ_B is approximately 53°. Note that this statement is only true when the refractive indices of both materials are real numbers, as is the case for materials like air and glass. For losses materials that absorb wave, like metals and semiconductors, n is complex, and $\Gamma_{\parallel}$ does not generally go to zero.

Total Internal Reflection and Critical Angle 2.11.4

When moving from a denser medium into a less dense one (i.e. $n_1 > n_2$), there will be an incidence angle known as the ***critical angle***, for which the wave is totally reflected and $\Gamma_\perp = \Gamma_{\parallel} = 1$. This phenomenon is known as **total internal reflection**. The critical angle is approximately 41° for glass in air. It is seen from (2.70) and (2.71) that this cannot occur for real values of θ_i and θ_t. However, if sin $\theta_i > 1$ for both polarizations, θ_i imaginary, and $\sin\theta_i > \sqrt{\frac{\mu_2\varepsilon_2}{\mu_1\varepsilon_1}}$. Therefore, there exists a critical angle θ_c given by

$$\sin\theta_c > \sqrt{\frac{\mu_2\varepsilon_2}{\mu_1\varepsilon_1}} \tag{2.72}$$

such that a wave incident on a plane interface at an angle equal to or greater than this angle θ_c will be totally reflected. This is shown in Fig. 2.35 (b). For glass medium in air, $\theta_c \approx 41°$.

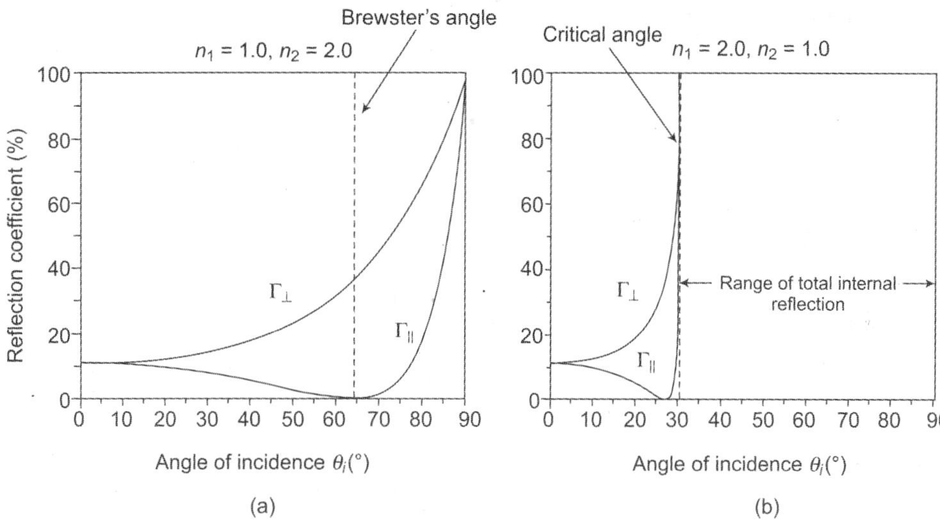

Fig. 2.35 Variation reflection coefficient with incident angle

When the wave is at near-normal incidence to the interface ($\theta_i \approx \theta_t \approx 0$), the reflection and transmission coefficient are given by

$$\Gamma = \Gamma_\perp = \Gamma_\parallel = \frac{n_1 - n_2}{n_1 + n_2} \tag{2.73}$$

$$T = T_\perp = T_\parallel = 1 - \Gamma = \frac{4n_1 n_2}{(n_1 + n_2)^2} \tag{2.74}$$

REFLECTION FROM AND TRANSMISSION THROUGH A PLANE INTERFACE BETWEEN A DIELECTRIC AND CONDUCTING MEDIA 2.12

The behaviour of a microwave signal incident at a plane interface between a conducting medium and a dielectric medium can be analyzed assuming TEM wave propagation for the incident wave in a lossless dielectric medium 1, where the medium 2 is a conductor. In the incident (dielectric) medium and the transmitted medium (conductor), the fields satisfy the wave equation. The propagation constants of the incident and reflected waves are $k = \omega\sqrt{\mu\varepsilon}$ and that of the transmitted waves in the conducting medium is

$$K = \omega\sqrt{\mu\varepsilon}\sqrt{(1 - j\sigma/\varepsilon\omega)} = \gamma_c$$

$$\cong (1 + j)\sqrt{\mu f \sigma \pi} \text{ for } \sigma >> \varepsilon\omega \tag{2.75}$$

$$= \alpha_c + j\beta_c$$

Here, the attenuation constant α_c and phase constants β_c are equal and are given by

$$\alpha_c = \beta_c = \sqrt{\mu f \sigma \pi} \tag{2.76}$$

Representing the unit vectors of propagation constants in the direction of incidence, reflected, and transmitted waves as $\hat{\mathbf{k}}_i$, $\hat{\mathbf{k}}_r$, and $\hat{\mathbf{k}}_t$, respectively.

The incident fields are represented by

$$\mathbf{E}_i = \mathbf{E}_{0i}\, e^{-j\beta_c \hat{\mathbf{k}}_i \cdot \mathbf{r}} \tag{2.77}$$

$$\mathbf{H}_i = \frac{1}{\omega\mu}(\beta \hat{\mathbf{k}}_i \times \mathbf{E}_i) \tag{2.78}$$

The reflected fields are

$$\mathbf{E}_r = \mathbf{E}_{0i} e^{-j\beta \hat{\mathbf{k}}_r \cdot \mathbf{r}} \tag{2.79}$$

$$\mathbf{H}_r = \frac{1}{\omega\mu}(\beta \hat{\mathbf{k}}_r \times \mathbf{E}_r) \tag{2.80}$$

The transmitted fields inside the conductor are, therefore,

$$\mathbf{E}_t = \mathbf{E}_{0t} e^{-\gamma j \mathbf{k} \cdot \mathbf{r}} = E_{0t} e^{-j\alpha_c \hat{\mathbf{k}}_t \cdot \mathbf{r}} \cdot e^{-j\beta_c \hat{\mathbf{k}}_i \cdot \mathbf{r}} \tag{2.81}$$

$$\mathbf{H}_t = \frac{\gamma_c \hat{\mathbf{k}}_t \times \mathbf{E}_t}{\omega\mu} = (\alpha_c + j\beta_c)\frac{\hat{\mathbf{k}}_t x \mathbf{E}_t}{\omega\mu} = \frac{1+j}{\sqrt{2}}\sqrt{\frac{\sigma}{\omega\varepsilon}}(\hat{\mathbf{k}}_t x\, \mathbf{E}_t) \tag{2.82}$$

Two features of Eqs (2.78) and (2.82) are immediately evident and are given as follows.

1. The conductivity gives rise to exponential damping of the wave in metal.
2. The **E** and **H** fields are no longer in phase.

Both **E** and **H** fields decrease with penetration, falling to $1/e$ of their surface values in a distance equal to skin depth $\delta = \sqrt{\dfrac{1}{\pi f \mu \sigma}}$. The attenuation coefficient in the conductor in surface normal direction is $\sqrt{\pi f \mu \sigma}$ cos θ_t. As seen from Eqs 2.78 and 2.82, the ratio of electric and magnetic energy densities inside the conductor is $\omega\varepsilon/\sigma << 1$ at frequencies below the optical region. Therefore, electric field is highly attenuated inside the conductor whereas the magnetic field is considerably transmitted through.

The laws of reflection and Snell's law are given by

$$\theta_i = \theta_r, \frac{\sin\theta_i}{\sin\theta_t} = \sqrt{\frac{\sigma}{2\omega\varepsilon}} \tag{2.83}$$

We can write

$$\cos\theta_t = \frac{\gamma_c}{\beta_c}(\beta_c{}^2 - k_i^2 \sin^2\theta_i)^{1/2} \approx \gamma_c \tag{2.84}$$

when $\sigma >> \omega\varepsilon$ or $\beta >> k_i$, It shows that to a very high degree of approximation, the propagation constant normal to the interface in the conductor is nearly $K = \gamma_c$ and is independent θ_i since cos $\theta_i \approx 1$, for all $\theta_i < 90°$.

The wave impedance in the conductor in the direction normal to the interface is

$$\eta_c = \frac{E_t}{H_t}\cos\theta_t \approx (1+j)\sqrt{\frac{\omega\mu}{2\sigma}} \tag{2.85}$$

For a nonmagnetic conductor ($\mu = \mu_0$), the ratio of the reflected energy to the incident energy for normal incidence is

$$|\Gamma|^2 \approx 1 - 2\sqrt{\frac{2\omega\varepsilon}{\sigma}} \tag{2.86}$$

Therefore, reflection loss is given by

$$R(dB) = 10\log_{10}\sqrt{\frac{\sigma}{8\omega\varepsilon}} \tag{2.87}$$

For a non-magnetic conductor ($\mu = \mu_0$). The ration of the reflected energy to the incident energy for normal incidence is the ratio of the transmitted energy at the interface to the incident energy for normal incidence is

$$|\Gamma|^2 \approx \frac{4\omega\varepsilon}{\sigma} \tag{2.88}$$

Therefore, the absorption loss for transmission up to a depth d is

$$A\ (dB) = 8.686\ \alpha d - 8.686\ d\ (\pi f \mu \sigma)^{1/2} \tag{2.89}$$

RECIPROCITY THEOREM 2.13

In classical electromagnetism, **reciprocity** refers to the interchange of time-harmonic electric current densities (sources) and the resulting electromagnetic fields satisfying Maxwell's equations for time-invariant linear media under certain constraints. The most common and general theorem is **Lorentz reciprocity** named after **Hendrik Lorentz** in 1896. It states that the "relationship between an ac current and the resulting electric field is unchanged if the source and the field points are interchanged." The reciprocity theorems are used in many electromagnetic applications, such as analyzing electrical networks and antenna systems. For example, antennas work reciprocally as transmitters or receivers and that an antenna's transmitting and receiving radiation patterns are identical.

Lorentz Reciprocity

Let us consider that a current density $\mathbf{J}_1$ that produces an electric field $\mathbf{E}_1$ and a magnetic field $\mathbf{H}_1$, where all three are periodic functions of time with angular frequency ω, and in particular they have time-dependence $\exp(-j\omega t)$. Suppose that a second current $\mathbf{J}_2$ at the same frequency ω produces fields $\mathbf{E}_2$ and $\mathbf{H}_2$. The Lorentz reciprocity theorem then states, under certain simple conditions on the materials of the medium described below, that for an arbitrary surface S enclosing a volume V:

$$\int_V \left[\mathbf{J}_1 \cdot \mathbf{E}_2 - \mathbf{E}_1 \cdot \mathbf{J}_2) \, dV = \int_S (\mathbf{E}_1 \times \mathbf{H}_2 - \mathbf{E}_2 \times \mathbf{H}_1\right] \cdot \mathbf{dS} \tag{2.90}$$

By the divergence theorem, in differential form,

$$\mathbf{J}_1 \cdot \mathbf{E}_2 - \mathbf{E}_1 \cdot \mathbf{J}_2 = \nabla \cdot [\mathbf{E}_1 \times \mathbf{H}_2 - \mathbf{E}_2 \times \mathbf{H}_1] \tag{2.91}$$

A special case of the Lorentz reciprocity theorem applies when the volume V entirely contains both of the sources (or alternatively, if V intersects neither of the sources) the volume integral is zero, so that

$$\int_S (\mathbf{E}_1 \times \mathbf{H}_2) \cdot \mathbf{dS} = \int_S (\mathbf{E}_2 \times \mathbf{H}_1) \cdot \mathbf{dS} \tag{2.92}$$

DISPERSION 2.14

For a signal (such as a pulse) comprising a band of frequencies, different frequency components propagate with different velocities according to the expression of phase velocity (2.68). This causes distortion of the signal. This phenomenon is called ***dispersion***.

If the medium is lossy, the wave propagation can be represented by a damped sinusoidal wave as a function of distance at a given time t, as explain in Fig. 2.36.

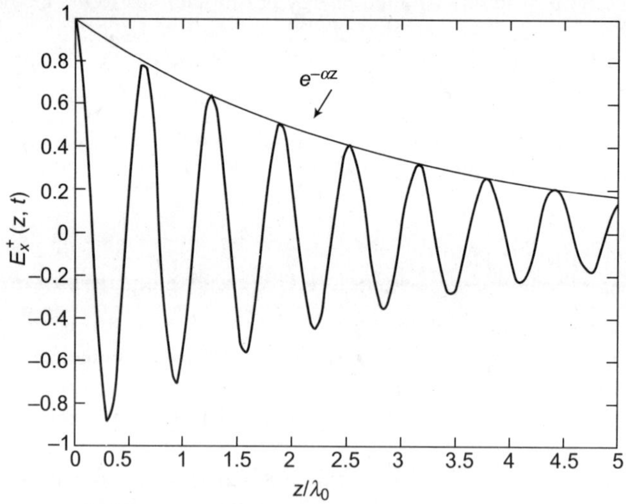

Fig. 2.36 Plane wave propagation in loss medium

DIFFRACTION AND SCATTERING 2.15

Diffraction is the phenomenon of bending of EM waves when they pass near the edge of an obstacle or through small openings. When an EM wave travels, each point on a wave front presents itself as an isotropic source. If it incidences on an obstacle some wave fronts pass beside or above the obstruction and radiate in the area beyond shadow region as shown in Fig. 2.37. Each point of an advancing wavefront acts as a secondary spherical waves. The propagating wave as a whole is the sum of all the secondary waves. Diffraction is more pronounced when dimensions of obstruction are small compared to wavelength. It is important as the potential interference between terrestrial/stations sharing the same frequency.

Electromagnetic **scattering** is the physical effect of an electromagnetic wave, in which the direction or polarization of the EM waves is changed when the waves encounter propagation medium discontinuities smaller than the wavelength, viz. foliage, troposphere air particles, etc. Instead of proceeding in a straight line, the energy is distributed in random or disordered way as shown in Fig. 2.38. Scattering and diffraction result in propagation of radio waves inside shadow region.

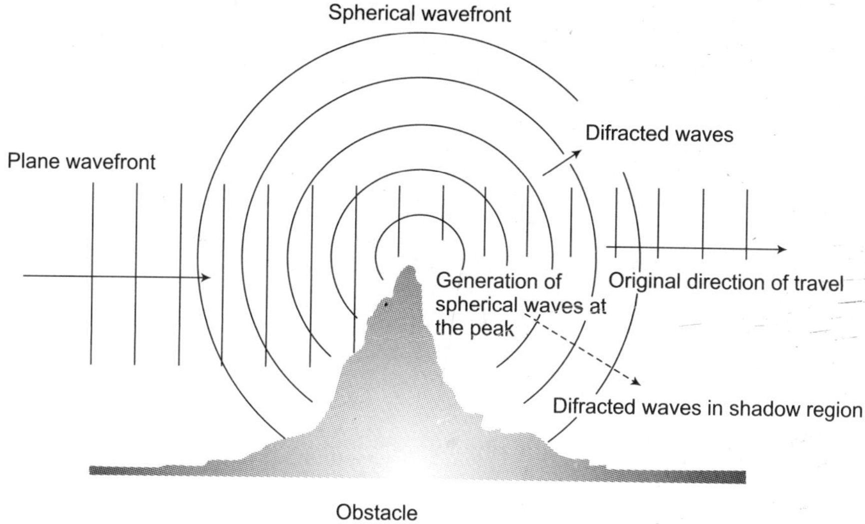

Fig. 2.37 Diffraction of EM waves

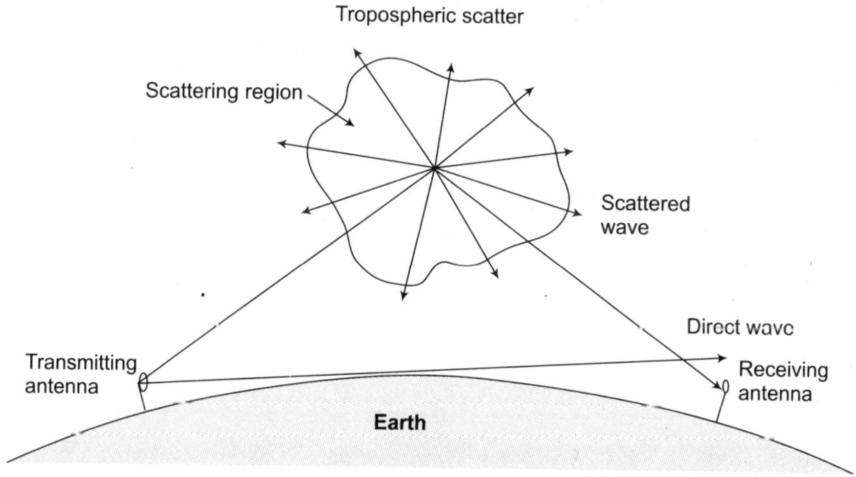

Fig. 2.38 Scattering of EM waves

CYLINDRICAL WAVE SOLUTIONS 2.16

There are microwave lines and devices which are circular cylindrical structures such as circular cylindrical waveguides. Analysis of these structures can be done by solving wave equation in cylindrical coordinate system. The solution will be the cylindrical wave function.

Let us consider the wave equation which is satisfied by the field solution in a circular cylindrical structure with axis as the z-axis, radius in the ρ-direction and azimuthal symmetry in the φ-direction:

$$\nabla^2\mathbf{E} + k^2\mathbf{E} = 0 \tag{2.93}$$

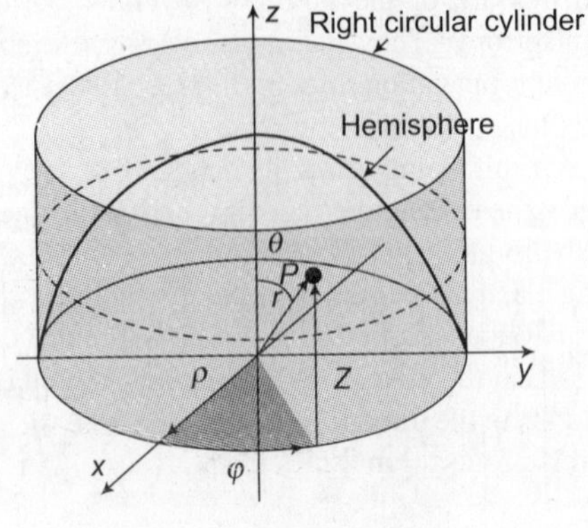

Fig. 2.39 Cylindrical waveform

For the propagation in the z-direction, scalar Helmholtz equation in cylindrical coordinates is

$$\frac{\partial^2 \psi}{\partial \rho^2} + \frac{1}{\rho}\frac{\partial \psi}{\partial \rho} + \frac{1}{\rho^2}\frac{\partial^2 \psi}{\partial \varphi^2} + \frac{\partial^2 \psi}{\partial z^2} + k^2 \psi = 0 \tag{2.94}$$

Here, ψ is a scalar function related to **E** or **H** field.

SPHERICAL WAVE SOLUTIONS 2.17

Spherical waves are generated from a small *RF* current element and the fields satisfy the wave equation in a spherical coordinate system:

$$\nabla^2 \mathbf{E} + k^2 \mathbf{E} = 0 \tag{2.95}$$

$$\frac{1}{r^2}\frac{\partial}{\partial r}\left(r^2 \frac{\partial \psi}{\partial r}\right) + \frac{1}{r^2 \sin\theta}\frac{\partial}{\partial \theta}\left(\sin\theta \frac{\partial \psi}{\partial \theta}\right) + \frac{1}{r^2 \sin^2\theta}\frac{\partial^2 \psi}{\partial \varphi^2} + k^2 \psi = 0 \tag{2.96}$$

PROPAGATION OF MICROWAVES IN FERRITE 2.18

The propagation characteristics of microwaves in a ferrite magnetic material are utilized to develop many microwave nonreciprocal devices. Ferrites are complex solids represented by $M^{+2}O.Fe_2O_3$, where M^{+2} is the ion of a divalent metal (cobalt, nickel, zinc, magnesium, cadmium, iron, manganese, chromium, copper, etc., or a mixture of these). The specific resistance of ferrites is very high (of the order of 10^7–10^8 ohm m), which is 10^{14} times as high as those of metals, relative permittivity is of the order of 10–15, and loss tangent tan δ = 10^{-4} (low loss at microwave frequencies). The permeability of the ferrite is an asymmetric tensor and can be represented by

$$\tilde{\mu} = \mu_0 \begin{bmatrix} \mu & -jK & 0 \\ jK & \mu & 0 \\ 0 & 0 & 1 \end{bmatrix} \tag{2.97}$$

for magnetically saturated ferrite, magnetized in z-direction in the presence of an RF field propagating in this direction. Here, μ and K are functions of frequency, field and other parameters.

The relative permeability of ferrites are of the order of several thousands. Thus, ferrites are good dielectrics, but exhibit magnetic anisotropy. They have nonreciprocal electrical properties, i.e. (1) the transmission coefficient for microwave propagation through ferrites is not the same for different directions, and (2) nonreciprocal rotation of the plane of polarization.

The magnetic properties of ferrites result mainly from the magnetic dipole moment **m** associated with the electron spin. In absence of any source, the net angular momentum of a full shell is zero. Only electron spins in the incomplete shell are effective to produce a net dipole moment per atom. Thus, a ferrite can be regarded as a collection of *N* effective spinning electrons per unit volume giving rise to total magnetic dipole moment $\mathbf{M} = N\mathbf{m}$. In the absence of an external magnetic field, the dipole moments are oriented at random and the combined effect is negligibly small so that the material does not show any magnetization.

In a steady external magnetic field $\mathbf{H}_0$ in the direction other than that of the **m** of an electron, **m** precesses gyromagnetically around $\mathbf{H}_0$ due to the torque $\boldsymbol{\tau} = \mathbf{m} \times \mu\,\mathbf{H}_0$ and tends to align the electron spin with $\mathbf{H}_0$ The frequency of precessions is given by

$$f_0(\text{MHz}) \cong 2.8 H_0(\text{oersted}) \tag{2.98}$$

where 2.8 MHz/oersted is called the gyromagnetic ratio γ.

The direction of precession is determined by $\mathbf{H}_0$ and is clockwise looking along $\mathbf{H}_0$, as shown in Fig. 2.40. The angle of precession ψ $(= \underline{|\mathbf{H}_0\mathbf{m}})$ decreases due to friction and the ferrite is magnetized with magnetization momentum **M** when the electron spins are aligned with $\mathbf{H}_0$.

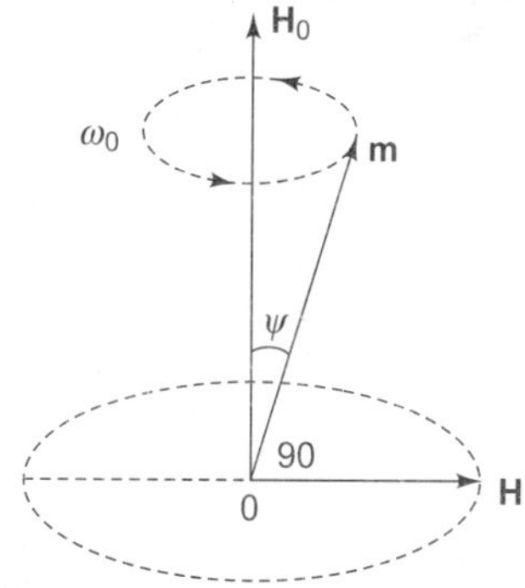

Fig. 2.40 Precession of spinning electron in a steady magnetic field

The behaviour of magnetization and demagnetization of ferrite shows hysteresis property of ferrite as shown in Fig. 2.41.

Ferrite shows a high remanance value $\mathbf{B}_r$ indicating that the sample remains magnetized even when the applied external dc magnetic field $\mathbf{H}_0$ is reduced to zero from its saturation magnetization $\mathbf{H}_s$.

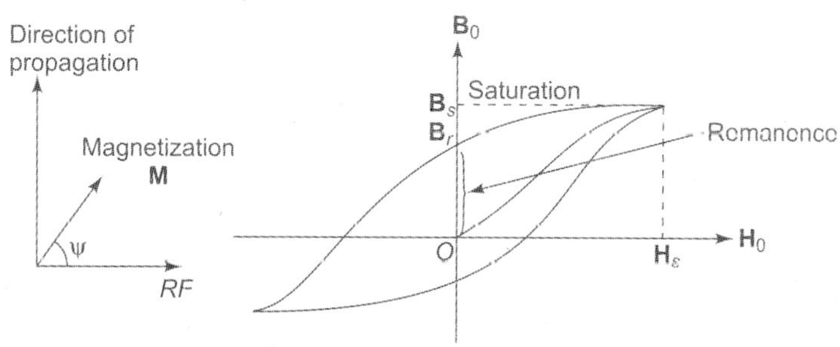

Fig. 2.41 Hysteresis in ferrite

TEM Wave in Ferrite 2.18.1

The behaviour of TEM wave propagation in a ferrite medium is very typical and is used for development of many microwave components. When the ferrite sample is magnetized to saturation in its plane in a direction parallel to the *RF* magnetic field $\psi = 0°$ (Fig. 2.41) then for small RF signal strengths there is a minimum interaction between the ferrite and the *RF* magnetic field. Consequently the effective *RF* relative permeability is approximately unity, $\mu_r \approx 1$. When the ferrite sample is magnetized to saturation in its plane in a direction perpendicular to the *RF* magnetic field $\psi = 90°$, maximum interaction takes place between the ferrite medium and the *RF* magnetic field. Under this situation, the effective *RF* relative permeability is given by

$$\mu_r = \frac{\mu^2 - K^2}{\mu} \tag{2.99}$$

where μ and K are the components of the permeability tensor, $K = \frac{\gamma_m 4\pi M_s}{\omega}$.

In the frequency range of interest over which $\gamma 4\pi M_s/\omega < 1$, and with small applied magnetic field, the effective *RF* relative permeability becomes

$$\mu_r \approx 1 - \left(\frac{\gamma_m 4\pi M_s}{\omega}\right)^2 \tag{2.100}$$

where $4\pi M_s$ = Ferrite magnetization. At frequency $\omega = \gamma_m 4\pi M_s$, the magnetic loss becomes significant and above equation modifies.

The propagation constant of the TEM wave in ferrite is, therefore, given by

$$\beta = \omega\sqrt{\mu_r \mu_0 \varepsilon_r \varepsilon_0} \tag{2.101}$$

where $\mu_r = 1$ when the ferrite is magnetized in the non-interacting state and μ_r is given by Eq. (2.99), when the ferrite is magnetized in the most interacting state. Therefore, the propagation constant changes its state with the direction of magnetization. This property is used in the development of ferrite reciprocal phase-shiflers and described in Chapter 5.

Gyromagnetic Resonance 2.18.2

In addition to external steady magnetic field, if an RF or microwave circularly polarized magnetic field **H** at frequency $f \approx f_0$ is applied perpendicular to $\mathbf{H}_0$, the angle of precession ψ of **M** and also the amplitude of induced procession will tend to increase at $f \approx f_0$ when the direction of rotation of **H** and **M** coincide. Due to magnetic frictional loss, the amplitude and the angle of **M** attain a steady state, describing a conical surface around $\mathbf{H}_0$. The energy continuously supplied by the RF field is dissipated as heat in the ferrite. When the interaction between the RF or microwave field and the electrons is reduced, the ferrites show lower losses. When **H** rotates opposite to **M**, the ferrite dissipates no time-average power and exhibits low loss.

Therefore microwave propagation in ferrites shows a gyromagnetic resonance with a peak of loss for clockwise polarization of **H** which coincides with that of **M** and a flat low loss for opposite polarization as shown in Fig. 2.42 where the horizontal scale represents $\mathbf{H}_0$ or frequency f_0 since both are linearly related.

The clockwise and anticlockwise polarizations of propagating waves, produce total ac fields

$$B_t^+ = \mu_0 M^+ + B^+ = \mu_+ H^+ \tag{2.102}$$

$$B_t^- = \mu_0 M^- + B^- = \mu_- H^- \tag{2.103}$$

respectively, and thus introduce two complex permeabilities denoted by plus and minus subscripts, as given by

$$\mu_+ = \mu'_+ - j\mu''_+ \tag{2.104}$$

$$\mu_- = \mu'_- - j\mu''_- \tag{2.105}$$

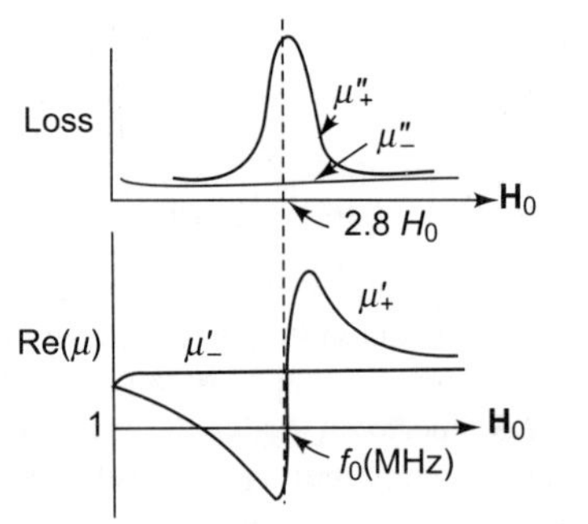

Fig. 2.42 Gyromagnetic resonance in ferrite

From Fig. 2.42, it is seen that the real and imaginary parts of the relative permeability μ_-: are independent of the externally applied steady magnetic field. But μ_+ shows a resonant behaviour at a value of $\mathbf{H}_0$ for a given frequency. Since a linearly polarized plane wave can be considered to be composed of two circularly polarized waves, the above analysis explains the nonreciprocal behaviour of ferrite materials to microwave propagation. This property is utilized in making nonreciprocal ferrite components.

Faraday Rotation in Ferrites 1.18.3

A plane circularly polarized wave propagating in $\mathbf{H}_0$ direction will have two different propagation constants given by

$$\beta^+ = \omega\sqrt{\varepsilon\mu_+} = \frac{2\pi}{\lambda^+} \text{ Clockwise}$$

$$\beta^- = \omega\sqrt{\varepsilon\mu_-} = \frac{2\pi}{\lambda^-} \text{ Anticlockwise}$$

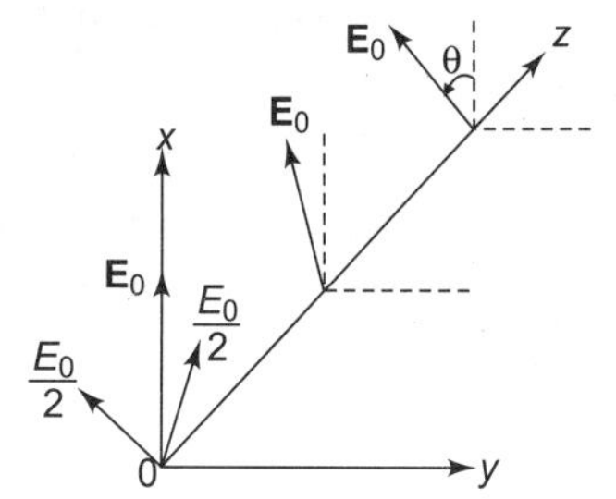

Fig. 2.43 Faraday rotation in ferrite

Therefore, for a linearly polarized wave propagating along $\mathbf{H}_0$, the plane of polarization rotates. This phenomenon is a nonreciprocal one. The rotation of the electric field of a linearly polarized wave passing through a magnetized ferrite is known as Faraday rotation as explained below with Fig. 2.43.

Let a linearly polarized TEM wave propagates in the ferrite along $+z$ axis with $\mathbf{E} = \hat{\mathbf{x}}Ex$ at $z = 0$. The linearly polarized wave may be decomposed into the sum of clockwise and anti-clockwise circularly polarized waves. The component waves propagate with different phase constants β^+ and β^-, respectively, and the two corresponding electric field vectors rotate at different rates. Over a distance λ, the resultant linearly polarized wave will undergo a phase delay.

$$(\beta^+ + \beta^-)\lambda = 4\pi \quad (2.106)$$

Therefore, $\lambda = 4\pi/(\beta^+ + \beta^-) = 2\lambda^- \lambda^+/(\lambda^+ + \lambda^-)$ (2.107)

and corresponding rotation angle.

$$(\beta^+ - \beta^-)\frac{\lambda}{2} = \left(\frac{\beta^+ - \beta^-}{\beta^+ + \beta^-}\right)2\pi \quad (2.108)$$

The rotation per unit distance is $\theta = (\beta^+ - \beta^-)/2$. This property is utilized in designing ferrite isolator and circulators which will be described in Chapter 5.

REVIEW QUESTIONS

2.1 The magnitude of an **E**-field for a plane wave travelling in free space is 200 volts per metre. Find the average power flow per unit area in the direction of propagation. What is the power flow through a circular surface of radius 1 m that is perpendicular to the direction of propagation?

2.2 The velocity of an electromagnetic wave travelling in a dielectric material is 2.5×10^8 m/s. What is the relative permittivity of the dielectric and the intrinsic impedance for the medium?

2.3 For a plane wave in free space, the magnitude of the **H**-field is 0.05 A/m. Find the average power flow per unit area. Determine the maximum power flow through an area of 4 cm^2.

2.4 A microwave oven is operated at 2.5 GHz with a field of 100 V/m magnitude. A cup of water has a volume of 100 cm^3 and relative permittivity $\varepsilon_r = 9$. Calculate the power being supplied to the water. $\sigma_{water} = 2 \times 10^{-4}$ S/m.

2.5 The half-space $z > 0$ is filled with a material with relative permeability 10 and dielectric constant 2.5. When an x-polarized plane wave is incident normally on this material from air, calculate the reflection and transmission coefficients.

2.6 A uniform plane wave of frequency 100 kHz is incident on the plane interface between free space ($z < 0$) and sea water ($z > 0$, $\sigma = 4$ S/m, $\varepsilon_r = 80$ and $m = m_0$). Determine (a) the amplitude of the incident wave electric field for which the amplitude of the transmitted wave electric field at $z = 1$ m is 1 mV/m; and (b) the depth at which the transmitted wave electric field is 1/e times the incident wave field.

2.7 Calculate the time average electric energy in a plane wave $E = 10 \cos 5 \times 10^{-4}\ \omega t$ V/m contained in a volume, one wavelength long in the direction of propagation with a cross-sectional area of 1 m^2. Find the Poynting vector.

2.8 A plane wave $\mathbf{E} = \hat{\mathbf{x}}\, \mathbf{E}_0 \exp\,[-\hat{j}(k_y y + k_z z)]$ is obliquely incident at a conducting plane xz at $y = 0$ at an angle θ_i with respect to the normal. Show that the H associated with the incident and reflected plane waves is $\mathbf{H} = +2\mathbf{E}_0\sqrt{\dfrac{\varepsilon_0}{\mu_0}}$ $(\cos\theta_i \cos k_y\, y\hat{z} - j \sin\theta_i \sin\theta_i \sin k_y \sin k_y\, y\, \hat{y}) \exp(-jk_z z)$.

2.9 A plane wave propagates at a frequency of 10 GHz in teflon ($\sigma = 10^{-5}$ S/m) and carries a total power of 10 watts. Calculate the total power lost in a 10 m^2 cross-section and 1 m path length in the material.

2.10 Calculate the phase velocity and attenuation constant for glass of dielectric constant 5 and conductivity of 10^{-12} mhos/m.

2.11 A surface *S* in the *xy*-plane is bounded by the paraboloid: $z = 4 - x^2 - y^2$. *Find the total flux through S* due the electric field: $\boldsymbol{E}(x, y, z) = \boldsymbol{x}\, 5z + \boldsymbol{y}\, x + \boldsymbol{z}\, y^2$.

2.12 Find the outward flux of $\boldsymbol{E} = x^2 x + y^3\, y + z^4\, z$ from a spherical surface S: $x^2 + y^2 + z^2 = a^2$.

2.13 In a medium the electric field *is given by* $\boldsymbol{E} = \boldsymbol{x}\, xy^2 z^3$. Calculate

(a) the electric charge density.

(b) outward flux from a region bounded by: $0 \le x \le 1$, $0 \le y \le 2$ and $0 \le z \le 3$.

2.14 Given the electric field $E_r = kr^2$, calculate the charge density ρ and the total charge and stored electric energy in a region bounded by: $0 \le r \le 1$, $0 \le \theta \le \pi$ and $0 \le \varphi \le 2\pi$.

2.15 A plane wave travelling in free space has magnitude of an *E*-field equal to 200 volts per meter. Find the average power flow per unit area in the direction of propagation. What is the power flow through a circular surface of radius 1 m that is perpendicular to the direction of propagation?

2.16 The velocity of an electromagnetic wave travelling in a dielectric material is 2.5×10^8 m/s. What is the relative permittivity of the dielectric and the intrinsic impedance for the medium?

2.17 A plane wave in free space, has the magnitude of the **H**-field 0.01 A/m. Calculate the average power flow per unit area. Determine the maximum power flow through an area of 4 cm^2.

2.18 A microwave oven is operated at 2.5 GHz with a field of magnitude 100 V/m. A cup of water has a volume of 100 cm^3 and relative permittivity $\varepsilon_r''' = 9$. Calculate the power being supplied to the water.

2.19 The half-space $z > 0$ is filled with a material with relative permittivity 10 and dielectric constant 2.5. when an *x*-polarized plane wave is incident normally on this material from air. Calculate the reflection and transmission coefficient.

2.20 A uniform plane wave of frequency 100 kHz is incident on the plane interface between free space ($z > 0$) and sea water ($z > 0$, $\sigma = 4$ S/m, $\varepsilon_r = 80$ and $\mu = \mu_0$). Determine (a) the amplitude of the incident wave electric field for which the amplitude of the transmitted wave electric field at $z = 1$ m is 1 Mv/m: and (b) the depth at which the transmitted wave electric field is 1/*e* times the incident wave field.

2.21 Calculate the total electric energy at an instant in a plane wave contained in a volume, one wavelength long in the direction of propagation with a cross-sectional area of 1 m^2. Find the Poynting vector.

2.22 A polar wave $E = \hat{x}E_0 \exp[-\hat{j}(k_y y + k_z z)]$ is obliquely incident at a conducting plane xz at $y = 0$ at an angle θ_I with respect to the normal. Show that the *H* associated with the incident and reflected plane waves is

$$\mathrm{H} = -2E_0\sqrt{\frac{\varepsilon_0}{\mu_0}}(\cos\theta_i \cos k_y y\bar{z} + j\sin\theta_i \sin k_y \sin k_y y\bar{y})\exp(-jk_z z)$$

2.33 A plane wave propagates at a frequency of 10 GHz in Teflon ($\sigma = 0$) and carries a total power of 10 watts. Calculate the total power lost in a 10 m^2 path length and section in the material.

2.24 Calculate the phase velocity and attenuation constant for glass of dielectric constant 5 and conductivity of 10^{-12} mhos/m.

2.25 Starting from Maxwell's equations, derive the wave equation in an infinite, homogeneous, isotopic, ideal dielectric medium. Find the solution of the wave equation for uniform plane waves. Show that the ratio of the electric and magnetic field strengths, is a constant of the medium.

2.26 Explain how the concept of displacement current was introduced by Maxwell to account for the production of magnetic fields in free space.

2.27 Using Faraday's and Ampere's laws derive Maxwell's equation in integral form. Obtain the corresponding' differential or point relations by applying Stoke's theorem.

2.28 Derive an expression for the skin-depth of penetration of an electromagnetic wave in a good conducting medium.

2.29 State and explain the Faraday's law in integral form.

2.30 State and explain the Ampere's circuital law in integral form.

2.31 State and prove the conditions on the tangential and normal components of D and E at the boundary between two dielectrics

2.32 A sinusoidally varying uniform plane electromagnetic wave is launched in a homogeneous good conducting medium. Derive expressions for the attenuation constant, the phase constant and intrinsic impedance. What is skin effect?

2.33 What is meant by the polarization of a wave? Find the condition of generating linearly polarized and circularly polarized waves.

2.34 (a) Derive Maxwell's equation from the first principle. (b) Prove the following: $J_s = n \times H$, $M = E \times n$

2.35 (a) Write the differential form of Maxwell's equations. Are all four Maxwell's equations independent? Explain. (b) Explain the significance of displacement current.

2.36 Explain what is meant by "Polarization" of electromagnetic wave propagation. Distinguish between "circular" and "elliptical" polarizations.

2.37 (a) Distinguish between conduction and displacement currents. (b) Compare the magnitudes of the conduction and displacement current densities in a good conductor for which $\sigma = 10^7$ mhos/ meter and $\varepsilon_r = 1$,when electric field intensity is expressed by sin 100 πt.

2.38 (a) Explain the term 'skin depth' and calculate its value for copper at 1 MHz, Assume the conductivity of 5.8×10^7 mhos metre.

2.39 Starting from Maxwell's equation, derive the wave equation for an electromagnetic wave in an unbound homogeneous, isotopic, source-free lossless medium.

2.40 Derive and explain the Maxwell's equation in point form and integral form.

2.41 Explain what is meant by the term displacement current. Deduce equation of continuity of current:

$$\text{div}\left(j + \frac{\partial D}{\partial t}\right) = 0$$

2.42 (a) Explain the Poynting theorem. For a time-harmonic electromagnetic field, write expressions of Instantaneous Poynting vector and time average Poyntic vector. (b) Prove that linear polarized plane wave can be resolved into a right-handed and left-handed circularly polarized waves of equal waves of equal amplitude.

2.43 (a) Explain the equation of continuity for time varying field. (b) Find the relation between **E** and **H** in a uniform plane wave. Hence, find the value of intrinsic impedance of free space. (c) An electromagnetic wave at 1 GHz is propagated through a material having $\mu_r = 3$ and $\varepsilon_r = 9$. Determine the (a) velocity of propagation, (b) intrinsic impedance of free space and of material, (c) wavelength in free space and in material.

2.44 Starting from Maxwell's field equations in their differential form, prove that the Poynting vector gives the power flow per unit area of cross section, at a point in the medium of propagation.

2.45 (a) Write down Maxwell's field equations in their integral form and explain their physical significance. (b) From Maxwell's equations, deduce the boundary conditions at the interface between two ideal dielectric media.

2.46 (a) State and prove Poynting's theorem for the electromagnetic field.
(b) Define Poynting vector and explain its significance.

REFERENCES

1. Aulock, V H; Wilhelm and Clifford, E F: *Linear Ferrite Devices for Microwave Applications*, Academic Press, New York, 1968.
2. Cheng, D K: *Field and Wave Electromagnetics,* Addison-Wesley Longman Inc., 2nd Ed.1999.
3. Collin, R E: *Foundation for Microwave Engineering*, McGraw-Hill, Inc., International Editions, 1992.
4. Clarricoats, P J B: *Microwave Ferrites*, Chapman and Hall Ltd., London, 1961.
5. Fox, A G; Miller and Weiss, M T: *Behaviour and Application of Ferrites in the Microwave Region*, Bell System, Tech. J, Vol. 34, 1955.
6. Harrigton, R F: *Time Harmonic Electromagnetic Fields*, McGraw-Hill Book Co., 1961.
7. Hayt, W H, Jr: *Engineering Electromagnetics*, 5th Ed., McGraw-Hill Book Company, New York, 1989.
8. Johnk, C T A: *Engineering Electromagnetic Fields and Waves*, 2nd Ed., John Wiley and Sons, Inc., New York, 1988.
9. Kenneth R D:,*Engineering Electromagnetics*, Prentice-Hall Inc., 1998.
10. Kraus, J D: *Electromagnetics*, 3rd Ed., McGraw-Hill Book Company, New York, 1984.
11. Marcuvitz, N: *Waveguide Handbook* (MIT Radiation Laboratory Series, Vol. 10), McGraw-Hill, New York, 1951.
12. Ramo, S; Whinnery, J R and Van Duzer, T: *Fields and Waves in Communication Electronics*, 2nd Ed., John Wiley, New York, 1984.
13. Sadiku M N O: *Elements of Electromagnetics*, 3rd Ed., Oxford University Press, 2004.
14. Shen, L C and Kong, J A: *Applied Electromagnetism*, Brooks-Cole, California, 1983.
15. Frey, J and Bhasin, K B: *Microwave Integrated Circuits*, Artech House, Inc.,1985.
16. Stratton, J A: *Electromagnetic Theory*, McGraw-Hill Book Company, New York, 1941.
17. Wait, J R: *Electromagnetic Wave Theory*, Harper & Row Publishers, Inc., New York, 1985.

3 RF AND MICROWAVE TRANSMISSION LINES

INTRODUCTION 3.1

When an RF or microwave signal is transmitted through conventional transmission lines consisting of two open wires of length and conductor separation larger than the wavelength, energy is lost due to electromagnetic radiation from the lines. This radiation loss is reduced or eliminated by making transmission of RF or microwave signals through some guiding structures. The guiding structures may consist of two concentric coaxial conductors called coaxial lines or two parallel plates such as strip and microstrip lines, or it may be a single hollow conductor called a waveguide. Many other forms are also possible which will be described in this chapter. The two wire-lines can be analyzed in terms of voltage, current and impedance using the distributed circuit theory. Waveguides can be analyzed from the solution of Maxwell's field equations.

The microwave circuits and devices constitute a section or sections of microwave transmission lines. Microwave signals are propagated through these lines as electromagnetic waves and scattered from the associated junction of these lines to travel in well-defined directions or ports. This chapter attempts to bridge the gap between electromagnetic theory and microwave circuits by highlighting some salient concepts for different configurations of these lines or guides. The structures considered here are (1) multiconductor lines, viz. coaxial lines, strip lines, microstrip lines, slot lines and coplanar lines, (2) single conductor lines such as rectangular waveguides, circular waveguides, ridge waveguides, and (3) open-boundary structures, viz. dielectric rods. In the first category of lines, the mode of transmission is a TEM or quasi-TEM wave where E and H fields have only single components, both perpendicular to each other and also perpendicular to the direction of propagation. In the second category of lines, the modes are either TE wave (components of E fields are perpendicular to the direction of propagation) or TM wave (components of H fields are perpendicular to the direction of propagation). The third category of lines support, in general, a combination of TE and TM waves called the hybrid HE modes, except possible axisymmetric modes which are either purely TE or TM waves. Since the general characteristics of signal propagation in all these lines are the same, the basic transmission-line theory concerning voltage standing waves, reflection and impedance is applicable to problems in all these guiding structures when excited by a single mode.

For microwave applications with wavelengths from 1 metre (300 MHz) to 1 millimetre (300 GHz), low-frequency circuit analysis techniques cannot be used. Here, distributed circuit or transmission-line analysis should be used where the voltage and current along a transmission line vary in magnitude and phase as a function of position.

In this section, the important *RF* and microwave engineering concepts and nomenclatures are summarized with particular attention given to *RF* and microwave transmission lines.

THEORY OF TRANSMISSION LINES 3.2

A transmission line is an electrical signal propagating medium between a transmitting end (source end) and a receiving end (load end). Consider a two-conductor uniform transmission line of length l is excited at $z = 0$ with a time harmonic voltage source of voltage $v_g(t) = V_g(z)e^{j\omega t}$ having source impedance Z_g and terminated at $z = l$ by a load impedance Z_L as shown in Fig. 3.1. The instantaneous voltage and current on the line at any arbitrary location z can be represented by

$$v(z, t) = V(z)e^{j\omega t} \tag{3.1a}$$

$$i(z, t) = I(z)e^{j\omega t} \tag{3.1b}$$

respectively. At any point z, the total voltage and current on the line are given by the superposition of incident or forward travelling signals and the reflected or reverse travelling signals:

$$V(z) = V_i e^{-\gamma z} + V_r e^{\gamma z} \tag{3.2a}$$

$$I(z) = I_i e^{-\gamma z} + I_r e^{\gamma z} \tag{3.2b}$$

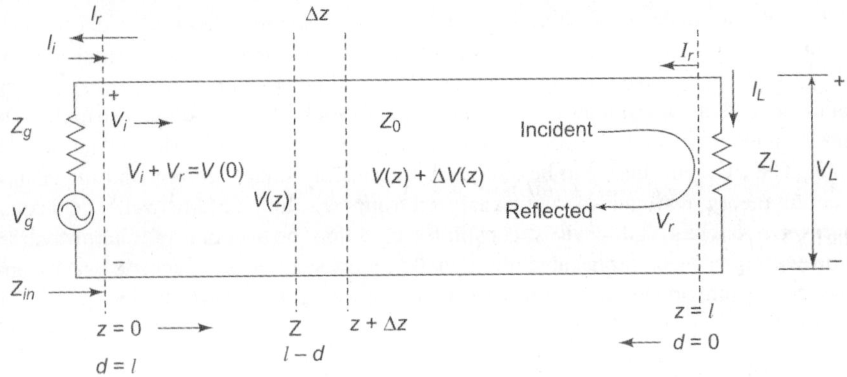

Fig. 3.1 General schematic of transmission line

Here, the time harmonic factor $e^{j\omega t}$ gets cancelled from both sides of (3.2a) and (3.2b) and not shown. γ is called the propagation constant on the line. (V_i, I_i) and (V_r, I_r) are the wave amplitudes of waves travelling along $+z$ and those travelling along $-z$, respectively.

Since at RF and microwave frequencies (small λ), the line parameters such as voltage, current and impedance vary from point to point along the whole length of the line in the direction of propagation, the transmission line is a distributed network. However, an infinitesimal small section $\Delta z << \lambda$ can be considered as a lumped network and the analysis of transmission lines are carried out for the infinitesimal element with lumped parameters R, L, G, and C per unit length in the following section.

Transmission Line Equations 3.2.1

A small length of the line can be represented by an equivalent L-section network with constant parameters R, L, G, and C per unit length as shown in Fig. 3.2, where R is the resistance per unit length which takes into account the ohmic loss in the line conductor, L is the inductance per unit length which takes into account the magnetic energy storage around the conductor, G is the conductance per unit length which takes into account the dielectric loss between the line conductors, and C is the capacitance per unit length which appears due to two conductors at different potentials and represents the electric energy storage. These parameters are independent of frequency.

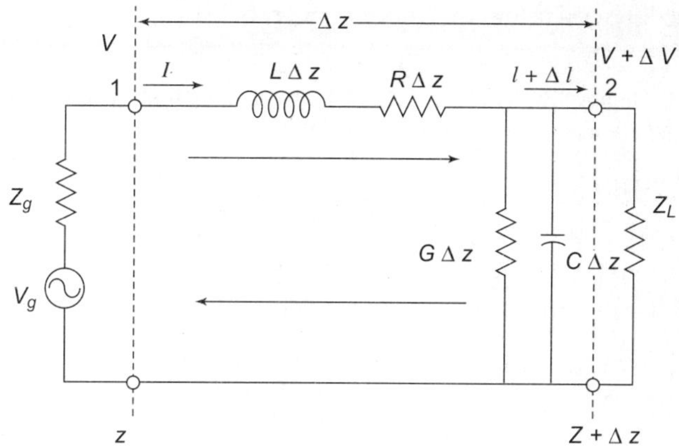

Fig. 3.2 Equivalent circuit of a small section of a uniform transmission line

Since, in general, the line voltage and current per unit length decrease as z increases, the differential values can be expressed by

$$\Delta V = -Z\Delta z.I$$

Or in the limit $\Delta z \to 0$,

$$\frac{dV}{dz} = -ZI = -(R + j\omega L)I \tag{3.3a}$$

Similarly,

$$\frac{dI}{dz} = -YV = -(G + j\omega C)V \tag{3.3b}$$

where

$$Z = R + j\omega L \quad \text{ohm/m} \tag{3.4a}$$

$$Y = G + j\omega C \quad \text{ohm/m} \tag{3.4b}$$

are called the **series impedance** and **shunt admittance** per unit length of the line. The time factor $e^{j\omega t}$ got cancelled from both sides of (3.3a) and (3.3b) so that only phasor values (functions of z) are considered. Eliminating I from (3.3a) and V from (3.3b), the voltage and current wave equations are generated as, respectively,

$$\frac{d^2V}{dz^2} = ZYV = (R + j\omega L)(G + j\omega C) = \gamma^2 V \tag{3.5}$$

$$\frac{d^2I}{dz^2} = ZYI = (R + j\omega L)(G + j\omega C) = \gamma^2 I \tag{3.6}$$

Here, $\gamma = \sqrt{ZY} = [(R + j\omega L)(G + j\omega C)]^{1/2} = \alpha + j\beta$ (unit less quantity per meter) (3.7)

and is called the **propagation constant**, α is called the **attenuation constant** (attenuation of signal per unit length of the line), and β is the **phase constant** (change of phase of the signal per unit length of the line).

Differential equations (3.5) and (3.6) are the general forms of transmission-line equations, the solutions of which are given in (3.2a) and (3.2b), respectively, which represent the total voltage and current at any point z on the line. The first term in the right-hand side represents the **incident wave** towards the load and the second term represents the **reflected wave** from the load. The values of (V_i, I_i) and (V_r, I_r) are to be determined.

Some special cases are described below:

a. Low Loss Line ($R << \omega L$, $G << \omega C$)

When the losses in the line are low, as in the case of high-frequency lines, $R << \omega L$, $G << \omega C$. Under this condition, Eq. (3.7) can be approximated as

❑ *Propagation Constant*

$$\gamma = \alpha + j\beta \cong j\omega\sqrt{LC}\left[1 + \frac{1}{j2\omega}\left(\frac{R}{L} + \frac{G}{C}\right)\right] \quad \text{m}^{-1} \tag{3.8a}$$

Separating real and imaginary parts from (3.8a), we get

❑ *Attenuation Constant*

$$\alpha \cong \frac{1}{2}\left(R\sqrt{\frac{C}{L}} + G\sqrt{\frac{L}{C}}\right) \quad \text{neper/m} \tag{3.8b}$$

❑ *Phase Constant*

$$\beta = \frac{\partial \phi}{\partial z} \cong \omega\sqrt{LC} \quad \text{radian/m} \tag{3.8c}$$

The velocity of the signal is called the ***phase velocity***, given by

❑ *Phase Velocity*

$$u_p = \frac{\partial z}{\partial t} = \frac{\omega}{\beta} \cong \frac{1}{\sqrt{LC}} = \lambda f \quad \text{m/s} \tag{3.8d}$$

and is independent of frequency.

The energy of the signal can be considered to be propagated with the group velocity given by

❑ *Group Velocity*

$$u_g = \frac{\partial \omega}{\partial \beta} \cong \frac{c^2}{u_p} \quad \text{m/s; } c = \text{velocity of light in the medium (m/s)} \tag{3.8e}$$

b. Lossless Line ($R = 0$, $G = 0$)

For lines with very good conductors ($\sigma \to \infty$) and associated with lossless dielectric medium ($\varepsilon'' = 0$), $R = 0 = G$. From (3.8b) and (3.8c),

$$\alpha = 0 \tag{3.9a}$$

$$\gamma = \alpha + j\beta = j\beta \cong j\omega\sqrt{LC} \quad \text{m}^{-1} \tag{3.9b}$$

$$u_p = \frac{1}{\sqrt{LC}} = \lambda f \quad \text{m/s} \tag{3.9c}$$

c. Distortion in Transmission Lines

From (3.7) without any assumption in losses, squaring both sides, one obtains

$$\alpha^2 - \beta^2 + j2\alpha\beta = (RG - \omega^2 LC) + j\omega(LG + RC) \tag{3.10a}$$

Equating real and imaginary parts from (3.10a),

$$\alpha^2 - \beta^2 = RG - \omega^2 LC \tag{3.10b}$$

$$2\alpha\beta = \omega(LG + RC) \tag{3.10c}$$

Squaring (3.10c) and substituting into the result for α^2 from (3.10b) yield

$$\beta^4 + (RG - \omega^2 LC)\,\beta^2 - \frac{\omega^2}{4}(LG + RC)^2 = 0 \tag{3.10d}$$

Obtaining the solution for β from (3.10c) and substituting in (3.10a),

we get
$$\beta = \left[\frac{\omega^2 LC - RG + \sqrt{(RG - \omega^2 LC)^2 + \omega^2 (LG + RC)^2}}{2}\right]^{1/2} \tag{3.11a}$$

$$\alpha = \left[\frac{RG - \omega^2 LC + \sqrt{(RG - \omega^2 LC)^2 + \omega^2 (LG + RC)^2}}{2}\right]^{1/2} \tag{3.11b}$$

From last two equations it is seen that attenuation is a function of frequency. Thus a signal consisting of a band of frequencies will attenuate differently at different frequencies. Therefore, the line will introduce what is called the ***frequency distortion***. Moreover, the phase constant is not a linear function of frequency. Therefore, the phase velocity $u_p = \omega/\beta$ of different frequency components of the signal will have different velocities and signal will suffer "**Dispersion**" which is a kind of distortion called "**Delay distortion**". The net result will be the distortion of the signal at the receiver end.

d. Distortionless Lines

A line is said to be distortion free when α and $u_p = (\omega/\beta)$ are both independent of frequency and thus β is a direct function of frequency.

❑ *Lossless Lines* From (3.11a) and (3.11b), it can be seen that if $R = G = 0$, then

$$\alpha = \left[\frac{-\omega^2 LC + \sqrt{(\omega^2 LC)^2}}{2}\right]^{1/2} = 0 \tag{3.12a}$$

$$\beta = \left[\frac{\omega^2 LC + \sqrt{(\omega^2 LC)^2}}{2}\right]^{1/2} = \left[\frac{\omega^2 LC + \sqrt{(\omega^2 LC)^2}}{2}\right]^{1/2}$$

$$= \omega\sqrt{LC} \tag{3.12b}$$

$$u_p = \frac{\omega}{\beta} = \frac{1}{\sqrt{LC}} = \text{constant} \tag{3.12c}$$

and β is a linear function of frequency. Therefore, *lossless lines are distortion-free*.

❑ *Lossy Lines* In case $R \neq 0 \neq G$, then if the inner square root term in (3.11a) become $(RG + \omega^2 LC)^2$

or,
$$\sqrt{(RG - \omega^2 LC)^2 + \omega^2 (LG + RC)^2} = \sqrt{(RG + \omega^2 LC)^2} \tag{3.12d}$$

Squaring both sides and equating the coefficients of corresponding powers of ω on both sides of the resulting equation, one gets

$$(LG)^2 + (RC)^2 - 2RG\,LC = 0 \tag{3.12e}$$

or,
$$(LG - RC)^2 = 0$$

or
$$\frac{R}{L} = \frac{G}{C} \tag{3.13}$$

This is the condition of no distortion in the line.

Substituting this last relation into (3.11a) and (3.11b), yield

$$\alpha = \sqrt{RG} = R\sqrt{\frac{C}{L}}, \text{ independent of frequency} \tag{3.14a}$$

$$\left.\begin{aligned} \beta &= \omega\sqrt{LC} = \text{linear function of frequency} \\ u_p &= \frac{1}{\sqrt{LC}} = \text{constant} \\ u_g &= \frac{1}{\sqrt{LC}} = \text{constant} \end{aligned}\right\} \tag{3.14b}$$

Therefore, condition (3.13) establishes that when this relation is satisfied, the attenuation, phase and group velocities are independent of frequency and the line will be distortion free.

It may be noted that line constants R, L, G, C are not really constant in frequency. They changes with frequency due to skin effect at high RF frequencies. However, in normal cases, (3.13)is the condition of no distortion.

Line Impedances 3.2.2

There are two impedances defined for transmission lines, the **characteristic impedance** and **input impedance**. These impedances are defined and formulated in the following sections.

a. Lossy Line

If the transmission line has a very long length $l \to \infty$, the incident voltage component becomes $V_i e^{-\gamma z} \to 0$ at the load end and thus the reflection of the signal does not arise. When the reflected wave in the line is zero, the ratio at any point z, $V(z)/I(z)$ is called the **characteristic impedance** of the line and is defined by

$$Z_0 = \frac{V(z)}{I(z)} = \frac{V_i e^{-\gamma z}}{I_i e^{-\gamma z}} = \frac{V_i}{I_i} = \sqrt{\frac{Z}{Y}} = \sqrt{\frac{R + j\omega L}{G + j\omega C}} = R_0 + jX_0 \text{ ohm} \tag{3.15a}$$

where R_0 and X_0 are the real and imaginary parts of Z_0. When $R << \omega L$ and $G << \omega C$ for low-loss lines and also for microwave frequencies,

$$Z_0 = \sqrt{\frac{L}{C}} \quad \text{and is real} \tag{3.15b}$$

The reciprocal of Z_0 is called the **characteristic admittance $Y_0 = 1/Z_0$.**

From Fig. 3.1, it is seen that the impedance of the line at any point z looking towards the load end can generally be expressed by

$$Z_{in}(z) = \frac{V(z)}{I(z)} = \frac{V_i e^{-\gamma z} + V_r e^{\gamma z}}{I_i e^{-\gamma z} + I_r e^{\gamma z}} = Z_0 \frac{V_i e^{-\gamma z} + V_r e^{\gamma z}}{V_i e^{-\gamma z} - V_r e^{\gamma z}} \tag{3.16}$$

Here, $I(z) = V(z)/Z_0$ is eliminated by using (3.15a) and –ve sign in the denominator arises due to reflected current direction opposite to that of the incident signal. The circuit in Fig. 3.1 can be now replaced by the equivalent circuit at Fig. 3.3.

The unknown values V_i, V_r can be determined from the terminal conditions at $z = 0$ and $z = l$: $V(0)$ and $I(0)$. Substituting these in (3.2a) and (3.2b) yields

$$V_i = \frac{1}{2}[V(0) + Z_0 I(0)] \tag{3.17a}$$

$$V_r = \frac{1}{2}[V(0) - Z_0 I(0)] \tag{3.17b}$$

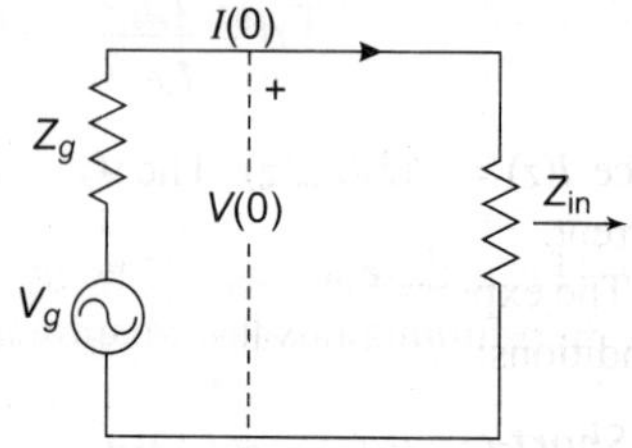

Fig. 3.3 Equivalent circuit of transmission line

From the equivalent circuit,

$$V(0) = \frac{V_g Z_{in}}{Z_{in} + Z_g} \quad \text{and} \quad I(0) = \frac{V_g}{Z_{in} + Z_g} \tag{3.18a}$$

Similarly, at load end, $I(l) = I_L$, and $V(l) = V_L = I_L Z_L$. Substituting in (3.2a) and (3.2b),

$$V_i = \frac{1}{2}(V_L + Z_0 I_L)e^{\gamma l} \text{ and } V_r = \frac{1}{2}(V_L - Z_0 I_L)e^{-\gamma l} \tag{3.18b}$$

Here, $V_L = I_L Z_L$. From (3.16–3.18), substituting the values of V_i and V_r and using the definition of sine and cosine hyperbolic functions in terms of exponential function $\left(\sinh x = \frac{e^{-x}+e^{x}}{2}, \cosh x = \frac{e^{-x}-e^{x}}{2}\right)$, the general expression of input impedance of a length l of line at the generator end is given by

$$Z_{in}(0) = Z_0 \frac{Z_0 \sinh \gamma l + Z_L \cosh \gamma l}{Z_L \sinh \gamma l + Z_0 \cosh \gamma l} = Z_0 \frac{Z_L + Z_0 \tanh \gamma l}{Z_0 + Z_L \tanh \gamma l} \tag{3.19a}$$

b. Lossless Lines

For a lossless line, $\alpha = 0$, $\gamma = j\beta$, $\tanh \gamma l = \tanh (j\beta l) = j \tan \beta l$, so that (3.19a) reduces to

$$Z_{in}(0) = Z_0 \frac{Z_L + jZ_0 \tanh \beta l}{Z_0 + jZ_L \tanh \beta l}; \beta = \frac{2\pi}{\lambda} \tag{3.19b}$$

The quantity βl is called the **electrical length**, in radian, of the line.

Reflection Coefficients 3.2.3

Since the transmission-line theory is described here in terms of propagation of voltage and currents along the line, reflections of these signals occur at a point where the impedance is not equal to the characteristic impedance Z_0.

a. Voltage Reflection Coefficient

Voltage reflection coefficient Γ_L of the load Z_L at the load end is defined by the ratio of reflected and incident voltages at the load:

$$\Gamma_L = \frac{V_r e^{\gamma l}}{V_i e^{-\gamma l}} \tag{3.20a}$$

Substituting the values of V_i and V_r from (3.18b) into the above equation, yields

$$\Gamma_L = \frac{V_L - Z_0 I_L}{V_L + Z_0 I_L} = \frac{Z_L I_L - Z_0 I_L}{Z_L I_L + Z_0 I_L} = \frac{Z_L - Z_0}{Z_L + Z_0} \tag{3.20b}$$

b. Current Reflection Coefficient

Current reflection coefficient Γ_{Li} of the load is defined by the ratio of reflected and incident currents through the load:

$$\Gamma_{Li} = \frac{I_r e^{\gamma l}}{I_i e^{-\gamma l}} = \frac{-V_r e^{\gamma l}}{V_i e^{-\gamma l}} = -\Gamma_L \tag{3.20c}$$

since $I(z) = V(z)/Z_{in}(z)$. The –ve sign indicates that return current is opposite in the direction of forward current.

The expressions of input impedance and load reflection coefficients are simplified for special terminating conditions:

❑ *Short-circuit Termination* $Z_L = 0$

$$Z_{inSC}(0) = Z_0 \tanh \gamma l \quad \text{for Lossy line} \tag{3.21a}$$

$$Z_{\text{inSC}}(0) = jZ_0 \tan \beta l \qquad \text{for Lossless line} \tag{3.21b}$$

Therefore, $\Gamma_{Lsc} = -1; \quad \Gamma_{Lisc} = +1$ (3.21c)

❑ ***Open-circuit Termination*** $Z_L = \infty$

$$Z_{\text{inOC}}(0) = Z_0 \coth \gamma l \qquad \text{for Lossy line} \tag{3.22a}$$

$$Z_{\text{inOC}}(0) = -jZ_0 \cot \beta l \qquad \text{for Lossless line} \tag{3.22b}$$

Therefore, $\Gamma_{LiOC} = -1; \quad \Gamma_{LOC} = +1$ (3.22c)

❑ ***Matched Termination*** $Z_L = Z_0$

$$Z_{in}(0) = Z_0 \qquad \text{for Lossy and lossless lines} \tag{3.23a}$$

Therefore, $\Gamma_L = 0 = \Gamma_{Li}$ (3.23b)

❑ ***Range of*** Γ_L From Eq. (3.20b), Eq. the following range of voltage reflection coefficient it is seen:

For $0 \le Z_L \le Z_0, \quad -1 \le \Gamma_L \le 0$ (3.24a)

$Z_0 \le Z_L \le \infty, \qquad 0 \le \Gamma_L \le +1$ (3.24b)

$-1 \le \Gamma_L \le 0$ (3.24c)

Therefore, For $0 \le Z_L \le \infty, -1 \le \Gamma_L \le +1$ (Total range).

❑ ***Reactive Load*** For reactive load termination in lines with Z_0 = real = R_0, $Z_L = \pm jX_L$.

Then
$$\Gamma_L = \frac{Z_L - R_0}{Z_L + R_0} = \frac{\pm jX_L - R_0}{\pm jX_L + R_0} = |\Gamma_L| e^{j\theta} \tag{3.25a}$$

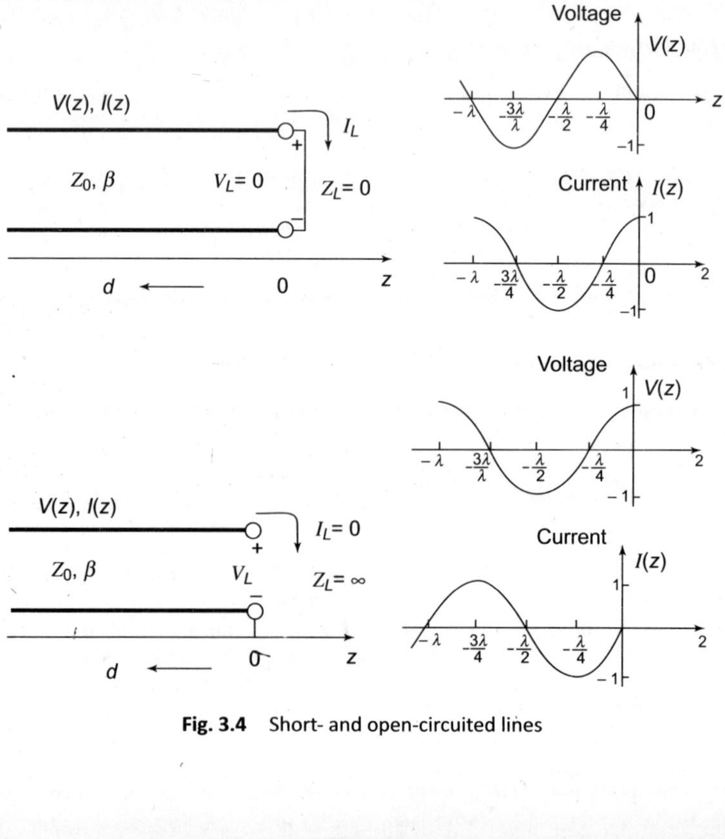

Fig. 3.4 Short- and open-circuited lines

Here, $$|\Gamma_L| = 1; \theta = \pm\tan^{-1}\left(\frac{X_L}{R_0}\right) \tag{3.25b}$$

Figure 3.5 shows the variation of input impedance of lossless line with the line length.

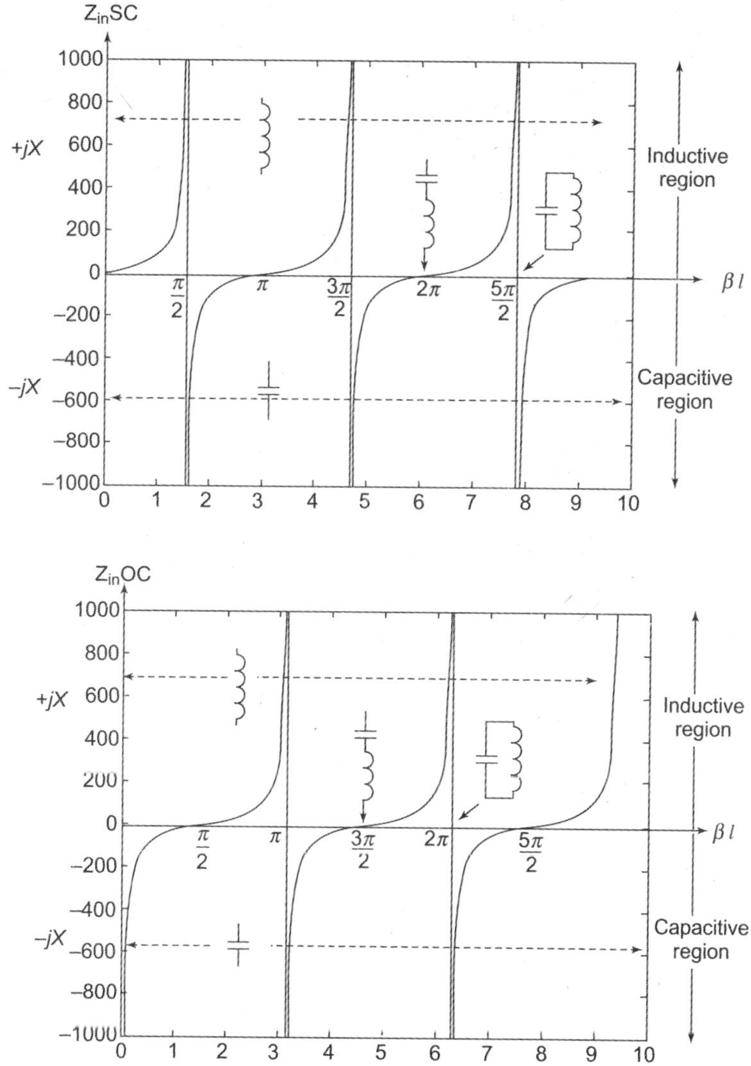

Fig. 3.5 Variation of input impedance of lossless line with the line length: (a) Z_L short (b) Z_L open

In most applications, the generator end of the transmission lines is well controlled by connecting the source having internal impedance equal to the characteristic impedance of the line ($Z_g = Z_0$). Under this condition, the source transfers maximum energy to the line due to impedance match between the source and the line. On the other hand, the load end cannot be so controlled due to different users connecting different devices not having impedance equal to Z_0. Therefore, the load end could be mismatched in impedances and some part of the signal is reflected back towards the source. Therefore, in transmission-line theory, the load end is the major analytical domain.

It is, therefore, convenient to measure the distance d in line from the load end such that for a length l of line, we can substitute $z = l - d$ as shown in Fig. 3.1. Accordingly, voltage reflection coefficient at any point $z = l - d$ is expressed by

$$\Gamma(z) = \frac{V_r e^{\gamma z}}{V_i e^{-\gamma z}} = \frac{V_r e^{\gamma(l-d)}}{V_i e^{-\gamma(l-d)}} = \frac{V_r e^{\gamma l} e^{-2\gamma d}}{V_i e^{-\gamma l}}$$

$$= \Gamma_L e^{-2\gamma d} = \Gamma_L e^{-2\alpha d} e^{-j2\beta d}$$

$$= |\Gamma_L| e^{j\theta_L} e^{-2\alpha d} e^{-j2\beta d} \tag{3.26}$$

where $$\Gamma_L = \frac{V_r e^{\gamma l}}{V_i e^{-\gamma l}} = |\Gamma_L| \angle \theta_L \tag{3.27}$$

Thus, for a line with losses, as we move from load end towards the generator (d increases), the complex reflection coefficient changes in an inward spiral way as shown in Fig. 3.6. For a lossless line ($\alpha = 0$), the complex reflection coefficient changes only in phase and describes a circle as shown in Fig. 3.7.

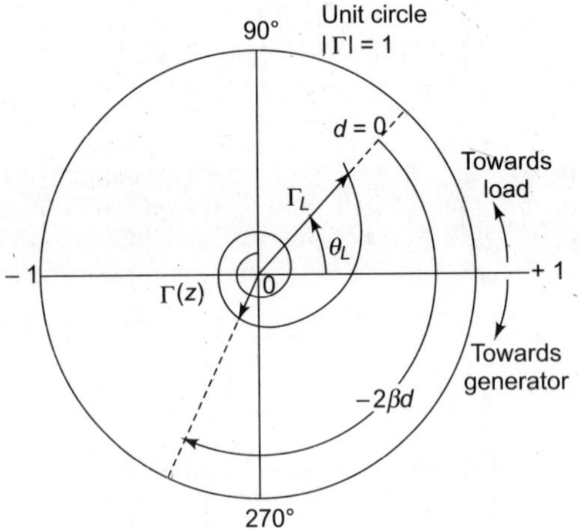

Fig. 3.6 Locus of reflection coefficient for a uniform line with losses

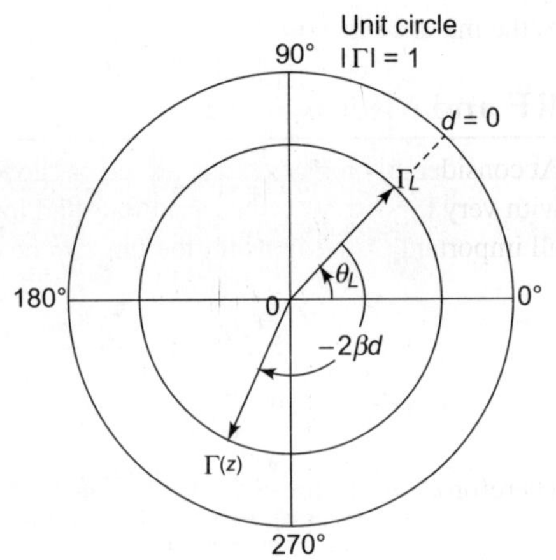

Fig. 3.7 Locus of reflection coefficient for a uniform lossless line

It must be noted that the reflected wave from the mismatched load will again be reflected from the source end if the source impedance is not matched to the line ($Z_g \neq Z_0$). Corresponding reflection coefficient,

$$\Gamma_g = \frac{Z_g - Z_0}{Z_g + Z_0} \tag{3.28}$$

From Eqs (3.19a), and (3.26),

$$Z_{in} = Z_0 \frac{1 + \Gamma_L e^{-2\gamma d}}{1 - \Gamma_L e^{-2\gamma d}} = Z_0 \frac{1 + \Gamma(z)}{1 - \Gamma(z)} \tag{3.29}$$

From Eq (3.2), the transmitted voltage and current at the load end are

$$V_L = V_i e^{-\gamma l} + V_r e^{\gamma l} \tag{3.30}$$

$$I_L = \frac{V_L}{Z_L} = \frac{V_i e^{-\gamma l}}{Z_0} - \frac{V_r e^{\gamma l}}{Z_0} \tag{3.31}$$

Solving Eqs (3.30) and (3.31), the voltage reflection and transmission coefficients are obtained as

$$\Gamma_L = \frac{V_r e^{\gamma l}}{V_i e^{-\gamma l}} = \frac{Z_L - Z_0}{Z_L + Z_0}; \tag{3.32a}$$

$$T = \frac{V_L}{V_i e^{-\gamma l}} = \frac{2Z_L}{Z_L + Z_0} = 1 + \Gamma_L; \tag{3.32b}$$

The power P_L transmitted to a mismatched load Z_L is given by

$$P_L = \text{Incident power} - \text{Reflected power}$$

or,

$$\frac{\text{Power transmitted to the load}}{\text{Incident power}} = 1 - \frac{\text{reflected power}}{\text{incident power}}$$

or,

$$\frac{|V_L|^2 / Z_L}{\left|V_i e^{-\gamma l}\right|^2 / Z_0} = \frac{Z_0}{Z_L}|T|^2 = 1 - |\Gamma_L|^2 \tag{3.33}$$

For matched termination, $Z_L = Z_0$, $\Gamma_L = 0$, so that $T = 1$ which indicates that total incident power is delivered to the matched load.

RF and Microwave Lines 3.2.4

At considerably high frequencies, i.e. in the RF and microwave region where the guiding structures are made with very high-conductivity materials and low-loss dielectric materials, $\omega L >> R$, and $\omega C >> G$. Therefore, all important expressions for the line can be considered for $\alpha = 0$ and then

$$Z_0 = \sqrt{\frac{L}{C}} = \text{real } R_0 \tag{3.34a}$$

$$\gamma = j\beta = j\omega\sqrt{LC}, \alpha = 0, \beta = \omega\sqrt{LC}, u_p = \frac{1}{\sqrt{LC}} \tag{3.34b}$$

Therefore, high-frequency lines are called low-loss or ideally lossless lines.

STANDING WAVES 3.3

When a line is terminated with a mismatched load, the incident and reflected signals interfere with each other at different phases to produce a resultant wave called the ***standing wave*** along the line. In this at regular intervals of half wavelength resultant wave produces maxima where the incident and reflected waves meet at same phase. Similarly it results in minima at regular intervals of half wave length where they meet at opposite phase.

The voltage and current at any point $z = l \quad d$ on the line $(0 < z < l)$ are given by

$$V(z) = V_i\,(e^{-\gamma z} + \Gamma_L\, e^{\gamma z}) = V_i\, e^{-\gamma z}(1 + \Gamma_L\, e^{-2\gamma z}) \tag{3.35a}$$

$$I\,(z) = \frac{V_i}{Z_0}(e^{-\gamma z} - \Gamma_L e^{\gamma z}) = \frac{V_i}{Z_0} e^{-\gamma z}(1 - \Gamma_L e^{2\gamma z}) \tag{3.35b}$$

where $\gamma = \alpha + j\beta$. For a complex load $Z_L = R_L + jX_L$, load reflection coefficient is also complex $\Gamma_L = |\Gamma_L| e^{j\theta_L}$. Therefore,

$$V(z) = V_i\, e^{-\alpha z}\, e^{-j\beta z}\left[1 + |\Gamma_L| e^{j\theta_L} e^{2(\alpha + j\beta)z}\right] = V_i e^{-\alpha z} e^{-j\beta z}\left[1 + |\Gamma_L| e^{2\alpha z} e^{j\varphi}\right] \tag{3.35c}$$

Similarly for current

$$I(z) = \frac{V_i}{Z_0} e^{-\alpha z} e^{-j\beta z}\left[1 - |\Gamma_L| e^{2\alpha z} e^{j\varphi}\right]; \tag{3.35d}$$

Here

$$\phi = \phi_L + 2\beta z \tag{3.35c}$$

The magnitude of the voltage is

$$|V(z)| = V_i e^{-\alpha z}\left|1 + |\Gamma_L| e^{2\alpha z}(\cos\phi + j\sin\phi)\right|$$

$$= \left|V_i e^{-\alpha z}\left[(1 + |\Gamma_L| e^{2\alpha z}\cos\varphi) + j|\Gamma_L| e^{2\alpha z}\sin\varphi)\right]\right|;$$

$$= V_i e^{-\alpha z}\sqrt{\left[(1 + |\Gamma_L| e^{2\alpha z}\cos\varphi)^2 + (|\Gamma_L| e^{2\alpha z}\sin\varphi)^2\right]};$$

or

$$|V_z| = V_i e^{-\alpha z}\sqrt{\left[1 + |\Gamma_L|^2 e^{4\alpha z} + 2|\Gamma_L| e^{2\alpha z}\cos\varphi\right]} \tag{3.36a}$$

Similarly for the current waves

$$|I_z| = \frac{V_i}{Z_0} e^{-\alpha z}\sqrt{\left[1 + |\Gamma_L|^2 e^{4\alpha z}(\cos^2\phi - \sin^2\phi) - 2|\Gamma_L| e^{2\alpha z}\cos\phi\right]}; \tag{3.36b}$$

(a) Standing Wave Maxima and Minima

From (3.36a) and (3.36b) it is clear that the line possesses a series of maxima and minima from the load end. These can be expressed as

$$V(z)_{\max} = V_i e^{-\alpha z}(1 + |\Gamma_L| e^{2\alpha z}); \qquad \text{for } \cos\phi = 1 \tag{3.36b}$$

$\phi = 0, 2\pi, 4\pi, ... = (n-1)2\pi; \quad n = 1, 2, 3, \ldots$

$$V(z)_{\min} = V_i^{-\alpha z}(1 - |\Gamma_L| e^{2\alpha z}); \qquad \text{for } \cos\phi = -1 \tag{3.36c}$$

$\phi = \pi, 3\pi, 5\pi, ... = (2n-1)\pi; \quad n = 1, 2, 3, \ldots$

For line with no attenuation ($\alpha = 0$)

$$V(z)_{\max} = V_i (1 + |\Gamma_L|); \qquad \text{for } \cos\phi = 1 \tag{3.36d}$$

$$V(z)_{\min} = V_i (1 - |\Gamma_L|); \qquad \text{for } \cos\phi = -1 \tag{3.36e}$$

Similarly for current

$$I(z)_{\max} = \frac{V_i}{Z_0}(1 + |\Gamma_L|); \qquad \text{for } \cos\phi = 1 \tag{3.36f}$$

$$I(z)_{\min} = \frac{V_i}{Z_0}(1 - |\Gamma_L|); \qquad \text{for } \cos\phi = -1 \tag{3.36g}$$

From above, the condition of voltage maxima but current minima is

$$\phi = \theta_L + 2\beta(l - d) = 0, 2\pi, 4\pi, ... = (n-1)2\pi; \tag{3.37a}$$

and the condition of voltage minima but current maxima is

$$\phi = \theta_L + 2\beta(l - d) = \pi, 3\pi, 5\pi, ... = (2n-1)\pi; \tag{3.37b}$$

where n (an integer) = 1, 2, 3, 4, 5, ...

(b) Positions of Voltage and Current Maxima and Minima

For a given length l of the line, terminated with a load of complex reflection coefficient $\Gamma_L = |\Gamma_L| e^{j\theta_L}$, the positions of ***voltage standing wave maxima d_{Vmax} and minima d_{Vmin}***, from the load end are given by the relations

$$d_{V\max} = l - \frac{(n-1)2\pi - \theta_L}{2\beta} \tag{3.38a}$$

$$d_{V\min} = l - \frac{(2n-1)\pi - \theta_L}{2\beta} \tag{3.38b}$$

Similarly, the positions of ***current standing wave maxima d_{Imax} and minima d_{Imin}***, are given by

$$d_{I\max} = l - \frac{(2n-1)\pi - \theta_L}{2\beta} \tag{3.38c}$$

$$d_{I\max} = l - \frac{(n-1)2\pi - \theta_L}{2\beta} \tag{3.38d}$$

(c) Distance between Successive Maxima and Minima

From (3.38a)-(3.38d) following results are obviously obtained for any load impedance.

(i) Distance between two successive voltage maxima, or two successive voltage minima = $\lambda/2$
(ii) Distance between two successive current maxima, or two successive current minima = $\lambda/2$
(iii) Distance between any two successive maximum and minimum voltage or current = $\lambda/4$

In a line with losses, Γ decreases exponentially by a factor of $e^{-2\alpha d}$ and a standing-wave pattern appears as shown in Fig. 3.8. Successive maxima and minima are spaced a distance $z = \pi/\beta = \lambda/2$.

Since the current reflection coefficient $\Gamma_{Li} = -\Gamma_L$, i.e., out-of-phase of voltage reflection coefficient, the standing-wave current maxima correspond to voltage minima and vice-versa. Therefore, Voltage Standing Wave (VSW) and Current Standing Wave (CSW) are 90° out of phase along the line. Other than short-circuit or open-circuit load, the amplitude of the incident wave and that of the reflected wave are not equal. Therefore, the minima of either wave is not zero. The standing-wave patterns of different loads on the line near the load end are shown in Fig. 3.9.

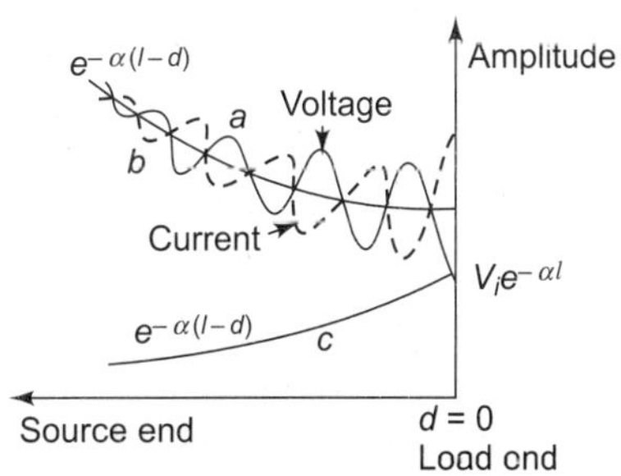

Fig. 3.8 (a) Standing wave pattern in a lossy line (b) Envelope of incident-wave amplitude (c) Envelope of reflected-wave amplitude

The magnitude of the standing waves is measured in terms of **Standing Wave Ratio** (SWR) defined by

$$\text{SWR} = \frac{\text{Max. voltage or current}}{\text{Min. voltage or current}} = S$$

$$S = \frac{1 + |\Gamma_L| e^{+2\alpha(b-d)}}{1 - |\Gamma_L| e^{+2\alpha(c-d)}}; \quad \alpha \neq 0 \text{ (lossy line)} \tag{3.39}$$

or,

$$S = \frac{1 + |\Gamma_L|}{1 - |\Gamma_L|}; \quad \alpha = 0 \text{ (lossless line)} \tag{3.40}$$

It can be written that, $|\Gamma_L| = \dfrac{S-1}{S+1}$ for lossless lines (3.41a)

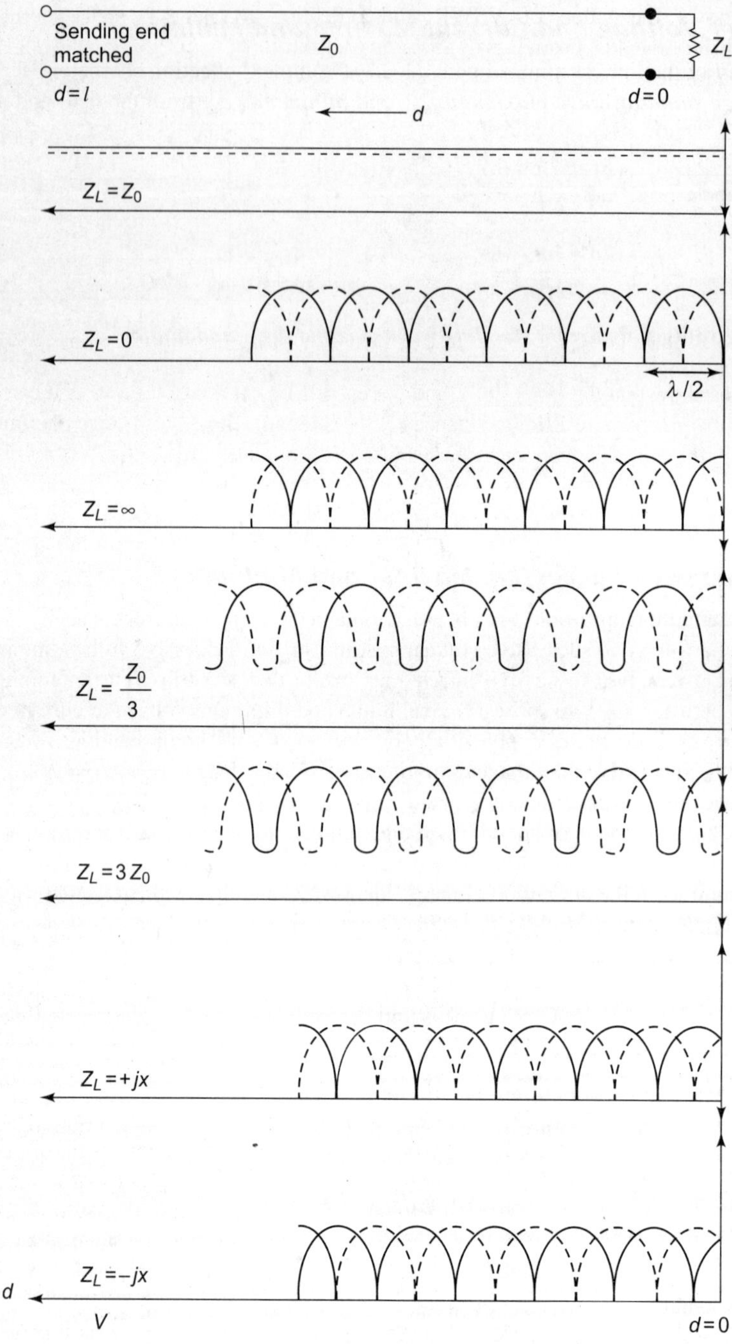

Fig. 3.9 Voltage standing-wave pattern for different loads in a lossless line (for resistive loads, minima are not very sharp)

Table 3.1 shows the value of VSWR and reflection coefficients for different loads. From Eqs (3.25a) and (3.40) for resistive load in a lossless line,

$$S = \frac{1+\dfrac{R_L \sim R_0}{R_L + R_0}}{1-\dfrac{R_L \sim R_0}{R_L + R_0}} = \frac{R_L + R_0 + R_L - R_0}{R_L + R_0 - R_L + R_0} = \frac{R_L}{R_0} \geq 1, \text{ for } R_L \geq R_0 \tag{3.41b}$$

$$= \frac{R_0}{R_L} \geq 1, \text{ for } R_L \leq R_0 \tag{3.41c}$$

Table 3.1 Reflection coefficient and VSWR for different lengths

Load	Voltage reflection coefficient	S
0	–1	∞
∞	1	∞
Z_0	0	1
$+jX$	$\exp[j\,2\tan^{-1}(X/Z_0)]$	∞
$-jX$	$\exp[-j\,2\tan^{-1}(X/Z_0)]$	∞

These relations show that the ranges of Γ and S are: $-1 \leq \Gamma_L < +1$ and $1 \leq S < \infty$. It can be shown that for a lossless line with real characteristic impedance Z_0, the line impedances at voltage maximum and minimum points are purely real, and also have maximum and minimum values, respectively. At voltage maximum or current minimum,

$$Z_{in} = Z_{max} = \frac{V_{max}}{I_{min}} = \frac{V_i(1+|\Gamma_L|)}{I_i(1-|\Gamma_L|)} = Z_0 S \tag{3.42a}$$

and at voltage minimum or current maximum

$$Z_{in} = Z_{min} = \frac{V_{min}}{I_{max}} = \frac{V_i(1-|\Gamma_L|)}{I_i(1+|\Gamma_L|)} = Z_0/S \tag{3.42b}$$

Equations (3.42a) and (3.42b) are very important in microwave measurements.

Example 3.1 A telephone line has $R = 6$ ohms/km, $L = 2.2$ mH/km, $C = 0.005$ mF/km, and $G = 0.05$ mmho/km. Determine Z_0, α, β at 1 kHz. If the line length is 100 km, determine the attenuation, phase shift and the phase velocity of the signal.

Solution

$$\omega = 2\pi\,1000 = 6280 \text{ rad/s}$$

$$Z_0 = \sqrt{\frac{R+j\omega L}{G+j\omega C}} = \sqrt{\frac{6+j6280\times 2.2\times 10^{-3}}{0.05\times 10^{-3}+j6280\times 0.005\times 10^{-3}}}$$

$$= 21.45 - j4.44 \text{ ohms}$$

$$\gamma = \sqrt{((R+j\omega L)(G+j\omega C))}$$

$$= \sqrt{(6+j6280\times 2.2\times 10^{-3})(0.05\times 10^{-3}+j6280\times 0.005\times 10^{-3})}$$

$$= 0.1404 + j\,6732 = \alpha + j\beta \text{ per km}$$

Therefore, $\alpha = 0.1404$ Np/km

$\beta = 0.6732$ rad/km

For a 100 km length,

Attenuation = $0.1404 \times 100 \times 8.686$ dB = 122 dB

Phase shift = $\beta l = 0.6732 \times 100$ rad/s = 67.32 rad/s

$$\text{Phase velocity} = \frac{\omega}{\beta} = \frac{6280}{0.6732} = 9.33\times 10^3 \text{ km/s}$$

Example 3.2 In Example 3.1, if a signal of 1 V supplies power to the line terminated in its characteristic impedance, find the power delivered at the receiving end.

Solution

Sending-end voltage $V_i = 1$ V, $l = 100$ km, Re $(Z_0) = 21.45$ ohm^{-1}

Load impedance $Z_L = Z_0$, $Z_{in} = Z_0$, $\gamma = 0.1404 + j\,0.6732$

Sending-end current $\quad I_i = \dfrac{V_i}{Z_{in}} = \dfrac{1.0}{21.45 - j4.44} = 0.0445 + j0.0093$ amp

Receiving-end current $\quad I_L = I_i\, e^{-\gamma l} = (0.0445 + \text{j}0.0093)\, e^{-(0.1404 + j0.6732) \times 100}$

$= -1.515 \times 10^{-8} + j3.3 \times 10^{-8}$ amp

Therefore, $\quad |I_L| = 3.632 \times 10^{-8}$ amp.

Power delivered at load $= |I_L^2| \times \text{Real}\,(Z_L) = |I_L^2| \times \text{Real}\,(Z_0) = (3.632 \times 10^{-8})^2 \times 21.45$

$= 2.83 \times 10^{-14}$ W

Example 3.3 A lossless transmission line with a characteristic impedance of 500 ohms is excited by a signal of voltage $10\angle 0°$ volts at 1.2 MHz. If the line is terminated by Z_L at a 1km distance, calculate (a) input impedance of the line for $Z_L = \infty$ and 0, and (b) the voltage at the midpoint of the line for $Z_L = Z_0$.

Solution

$$f = 1.2 \text{ MHz},\ \lambda = c/f = \frac{300}{1.2} = 250 \text{ m} \qquad V_i = 10 \text{ V}$$

$$\beta l = \frac{2\pi l}{\lambda} = \frac{2\pi \times 1000}{250} = 8\pi \text{ rad}$$

(a) For $Z_L = \infty$, $Z_{in} = -jZ_0 \cot \beta l = -j\,500 \cot (8\pi) = -j\,\infty$ ohm

(The line is essentially an *OC* because $l = 4\lambda$)

For $Z_L = 0$,

$$Z_{in} = jZ_0 \tan \beta l = j\,500 \tan (8\pi) = 0 \text{ ohm}$$

(b) For $Z_L = Z_0$,

$$V(z) = V_i\, e^{-\gamma z} \quad \alpha = 0$$

or $\quad V(l/2) = V_i\, e^{-j\beta l/2} = 10 e^{-j4\pi}$ V

Example 3.4 A 50-ohm lossless line connects a matched signal of 100 kHz to a load of 100 ohms. The load power is 100 mW. Calculate the

(a) Voltage reflection coefficient of the load
(b) VSWR of the load
(c) Position of the first V_{min} and V_{max}
(d) Impedance at V_{min} and V_{max} and values of V_{max} and V_{min}

Solution

$$Z_0 = 50 \text{ ohms},\ f = 100 \text{ kHz},\ Z_L = 100 \text{ ohms}$$

$$P_L = 100 \text{ mW},\ \lambda = \frac{c}{f} = \frac{3 \times 10^8}{100 \times 10^3} = 3000 \text{ m}$$

(a) $$\Gamma_L = \frac{Z_L - Z_0}{Z_L + Z_0} = \frac{100-50}{100+50} = \frac{1}{3}$$

(b) $$\text{VSWR} = \frac{1+\Gamma_L}{1-\Gamma_L} = \frac{1+1/3}{1-1/3} = 2$$

(c) For real $Z_L > Z_0$, first V_{max} is located at the load and first V_{min} is located at a distance $\lambda/4 = 750$ m from the load end.

(d) Load power $P_L = \dfrac{V_{max}^2}{Z_L} = 100 \text{ mW}$

or, $$V_{max} = \sqrt{(100 \times 100 \times 10^{-3})} = \sqrt{10} \text{ volts}$$

$$V_{min} = V_{max}/S = \sqrt{10}/2 \text{ volts}$$

At V_{max}, $Z_{in} = SZ_0 = 2 \times 50 = 100$ ohm

At V_{min}, $Z_{in} = Z_0/S = 50/2 = 25$ ohm

Mismatch Losses in Transmission Lines 3.3.1

Due to impedance mismatch between the input and output terminations of a lossy transmission line, five losses are often defined in microwave circuits as given below. They are also inter-related by expressions involving reflection coefficient, transmission coefficient, VSWR and attenuation constant.

The attenuation loss is a measure of the power loss due to signal absorption in the device and is defined at the output port by

Attenuation loss (dB) (in the line or device)
$$= 10 \frac{\text{Input energy} - \text{Reflected energy at the input}}{\text{Transmitted energy to the load}}$$

$$= 10\log\frac{|V_i|^2 - |V_r|^2}{(|V_i|^2 - |V_r|^2)e^{-2\alpha l}} = 8.686\alpha l \tag{3.43}$$

The reflection loss is a measure of power loss during transmission due to the reflection of the signal as a result of impedance mismatch and is defined at a plane by

Reflection loss (dB) (at a plane)
$$= 10\log\frac{\text{Input energy}}{\text{Input energy} - \text{Reflected energy}}$$

$$= 10\log\frac{1}{1-|\Gamma|^2} = 10\log\frac{(S+1)^2}{4S} \tag{3.44}$$

The transmission loss is a measure of loss of power due to transmission through the line or device and is defined at the output port by

Transmission loss (dB) (due to the line)
$$= 10\log\frac{\text{Input energy}}{\text{Transmitted energy}}$$

$$= 10\log\frac{|V_i|^2}{(|V_i|^2 - |V_r|^2)\,e^{-2\alpha l}} = 10\log\frac{e^{2\alpha l}}{1-|\Gamma|^2}$$

$$= 8.686\,\alpha l + 10\log\frac{1}{1-|\Gamma|^2}$$

$$= \text{Attenuation loss (dB)} + \text{Reflection loss (dB)} \tag{3.45}$$

The return loss is a measure of the power reflected by a line or network or device and is defined at the input port by

$$\textbf{Return loss (dB)} = 10\log\frac{\text{Input energy to the device}}{\text{Reflected energy at the input of the device}}$$

$$= -20\log|\Gamma| \tag{3.46}$$

The insertion loss is a measure of the loss of energy in transmission through a line or device compared to direct delivery of energy without the presence of the line or device. Let P_1 be the power received by the load when connected directly to the source without the line or device, and P_2 be the power received by the load when the line or the device is inserted between the source and the load, while the input power is held constant. Then

$$\textbf{Insertion loss (dB)} = 10\log\frac{P_1}{P_2} \text{ in the circuit or device} \tag{3.47}$$

The insertion loss is contributed by

(a) Mismatch loss at the input
(b) Attenuation loss through the device, and
(c) Mismatch loss at the output

Example 3.5 A lossless transmission line with a characteristic impedance of 300 ohms is fed by a generator of voltage $\angle 0°$ volts and an impedance of 100 ohms. The line is 100 m long and is terminated by a resistive load of 200 ohms. Calculate the reflection loss, the transmission loss and the return loss.

Solution Given, $Z_g = 100$ ohms, $Z_0 = 300$ ohms, $l = 100$ m, $Z_L = 200$ ohms.

$$\text{Load reflection coefficient } \Gamma_L = \frac{Z_L - Z_0}{Z_L + Z_0} = \frac{200-300}{200+300} = -1/5$$

$$\text{Reflection loss} = 10\log\frac{1}{1-|\Gamma|^2} = 10\log\frac{1}{1-(1/25)}$$

$$= 10\log(25/24) = 0.1773 \text{ dB}$$

Transmission loss = Attenuation loss (dB) + Reflection loss (dB); Attenuation loss = 0 for lossless line

$$= 10\log 25/24 = 0.1773 \text{ dB};$$

$$\text{Return loss} = -20\log|\Gamma| = -20\log(1/5) = 13.98 \text{ dB}$$

Smith Chart 3.2.2

In order to avoid the tedious mathematical solution methods used above to solve complicated transmission line problems, Phillip H Smith devised a simple graphic tool, named after him as the **Smith chart**, where the normalized impedance, or admittance or reflection coefficient is plotted to read the magnitude in the radial direction and phase in the angular direction directly. The chart is constructed from the equations for the input and load impedances normalized with respect to the characteristic impedance of the line in the following manner:

$$\text{Normalized input impedance} = \frac{Z_{in}}{Z_0} = \frac{1+\Gamma_L e^{-2\gamma d}}{1-\Gamma_L e^{-2\gamma d}} \tag{3.48}$$

$$\text{Normalized load impedance} = \frac{Z_L}{Z_0} = \frac{1+\Gamma_L}{1-\Gamma_L} = r + jx \tag{3.49}$$

where r and x are the real and imaginary parts of the normalized load impedance. Separating the complex load reflection coefficient into real and imaginary parts, Γ_r, and Γ_i, respectively, we can write

$$\Gamma_L = |\Gamma_L| e^{j\theta_L} = \frac{Z_L/Z_0 - 1}{Z_L/Z_0 + 1} = \Gamma_r + j\Gamma_i \tag{3.50}$$

Reflection coefficient at $z = l - d$ can be expressed as

$$\Gamma_d = |\Gamma_L| e^{-2\alpha d} e^{j(\theta_L - 2\beta d)} \tag{3.51}$$

From Eqs (3.49) and (3.50)

$$(\Gamma_r - 1)^2 + (\Gamma_i - 1/x)^2 = (1/x)^2 \tag{3.52}$$

$$\left(\Gamma_r - \frac{r}{1+r}\right)^2 + \Gamma_i^2 = \left(\frac{r}{1+r}\right)^2 \tag{3.53}$$

Equations (3.52) and (3.53) represent two families or circles in the complex reflection coefficient plane ($|\Gamma_L|$, θ_L) as shown in Fig. 3.10 which is called the *Smith chart*. The characteristics of the Smith chart are described below in terms of the normalized impedance $Z_L/Z_0 = r + jx$, the normalized admittance $Y_L/Y_0 = g + jb$ and the normalized length l/λ.

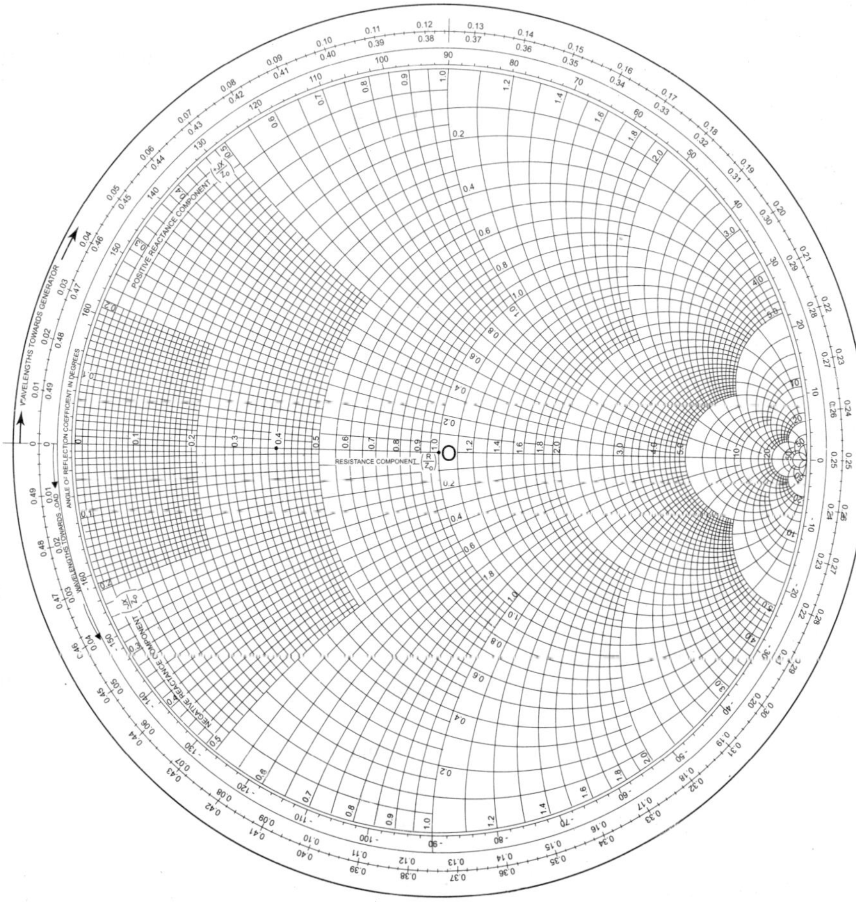

Fig. 3.10 Constant r and x circles in the reflection coefficient plane—Smith chart

1. Equation (3.52) represents the constant reactive (x) circles with radius $1/x$ and centres at 1, $1/x$, $-\infty \leq x \leq \infty$.
2. Equation (3.53) represents the constant resistance (r) circles with radius $\left[\frac{1}{1+r}, 0\right]$ and centres at on the Γ_r axis, $0 \leq r \leq \infty$.
3. Since the line impedance and the reflection coefficient change as we move towards the generator or towards the load, the distance along the line in terms of wavelength is given along the circumference of the chart, ($|\Gamma| = 1$), clockwise towards the generator or anticlockwise towards the load, respectively. One complete rotation covers a distance $\lambda/2$ along the line; the impedance and the reflection coefficient repeat themselves at these intervals.
4. The upper half of the chart represents inductive reactance jx and lower-half, capacitive reactance $-jx$.
5. Since admittance is the reciprocal of the impedance, the Smith chart can also be used for normalized admittance where the resistance scale reads the conductance and the inductive reactance scale reads capacitive reactance and vice-versa.
6. At a point of maximum voltage, line impedance $(Z_L/Z_0)_{max} = S$ and at minimum voltage, $(Z_L/Z_0)_{min} = 1/S$.
7. The centre O of the chart ($S = 1$) represents the matched impedance, the extreme right of the horizontal radius represents an open circuit ($S = \infty$, $Z_L/Z_0 = \infty$, $\Gamma = 1$) and the extreme left represents short circuit ($S = \infty$, $Z_L/Z_0 = 0$, $\Gamma = -1$).
8. The distances along the line are normalized with respect to the wavelength and are measured toward the generator and also toward the load along the periphery or unit circle ($|\Gamma| = 1$).
9. The circle passing through O ($S = 1$) is called the unit VSWR circle.

Applications of the Smith Chart 3.3.3

The Smith chart is conveniently used to determine the input and load impedances in a lossless transmission line from the values of the voltage standing wave ratio and the position of the voltage minimum on the line. The Smith chart also simplifies the procedure of finding stub positions and its length for impedance matching. These are described as follows.

To Determine Input Impedance of the Lossless Transmission Line of Length l and Terminated by a Load Z_L

The graphical solution involves the following steps using the Smith chart as shown in Fig. 3.11.

1. Locate the normalized load impedance on the Smith chart at, say A.
2. Join the centre point O (1, 0) and A. Extend OA up to the periphery to cut it at B.
3. A rotation towards the generator by l/λ from the load point A gives a point C on the outer circle which corresponds to the input point of the line.
4. In a movement along the line and the corresponding rotation on the Smith chart, the reflection coefficient and VSWR both remains constant in the lossless line.
5. A circular arc is drawn with O as the centre that passes through A and intersects the line OC at D.
6. The point of intersection D gives the required normalized input impedance of the line.

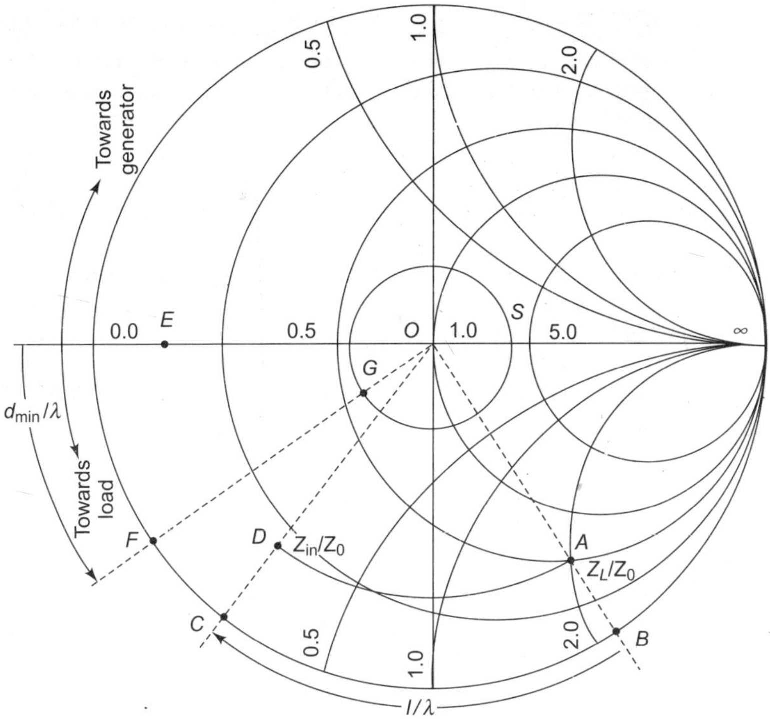

Fig. 3.11 Determination of the impedance of a lossless line using Smith chart

Example 3.6 A transmission line of 100 m length and a characteristic impedance of 100 ohms is terminated by a load $Z_L = 100 - j200$ ohms. Using the Smith chart, determine the line impedance and admittance at 25 m from the load end at a frequency of 10 MHz.

Solution

Given $f = 10$ MHz, $Z_0 = 100$ ohms, $Z_L = 100 - j200$ ohms,

$l = 100$ m, $d = 25$m.

$\lambda = c/f = 300/10 = 30$

$d/\lambda = 25/30 = 0.8333 = 0.5 + 0.333$

Normalized load impedance $Z_L/Z_0 = (100 - j200)/100 = 1 - j\,2$

The normalized load impedance is located at A on the Smith chart (Fig. 3.12). With origin O as the centre, a circle is drawn with radius OA. From the load point A, a normalized $d/\lambda = 0.833$ (= 0.5 + 0.25 + 0.083) is moved along the circumference towards the generator to reach the point B.

The normalized impedance at B is the $0.46 + j1.35$. Therefore, line impedance at a 25 m distance from the load is

$$Z_{in} = (0.46 + j1.35) \times 100 = 46 + j135 \text{ ohm}.$$

Draw a line from point B through O to cut the circle at point C. This point C gives normalized admittance $0.27 - j0.73$ of the line at a distance of 25 m from the load end. Therefore, admittance at a 25 m distance from the load is

$$Y_{in} = (0.27 - j\,0.73)/100 = 0.0027 - j\,0.0073 \text{ mho}$$

To Find the Load Impedance from the VSWR and Voltage Minimum Position

In a laboratory where the VSWR, S and position of the first voltage minimum d_{min} can be measured, the Smith chart in Fig. 3.12 gives a very simplified method of solution for load impedance in the following manner:

1. Draw a VSWR circle S with centre at O $(1 + j0)$.
2. When the line is shorted, the first voltage minimum occurs at the place of the load. This minimum shifts towards the generator by an amount d_{min}/λ when the line is loaded. The impedance at the voltage minimum is pure resistance of magnitude $1/S$. This corresponds to a point E on the left half of the real axis.
3. Move above distance d_{min}/λ from the minimum point E along the periphery towards the load and locate this position on the periphery as F.
4. Join O and F and find the intersection point G between the line OF and the VSWR circle.
5. Point G represents the normalized load impedance.

There are many other uses of the Smith chart in transmission-line theory and microwave engineering, viz. design of stubs for impedance matching, design of microwave filters, to determine the effects of discontinuities in a line, and so on. These will be discussed at appropriate places.

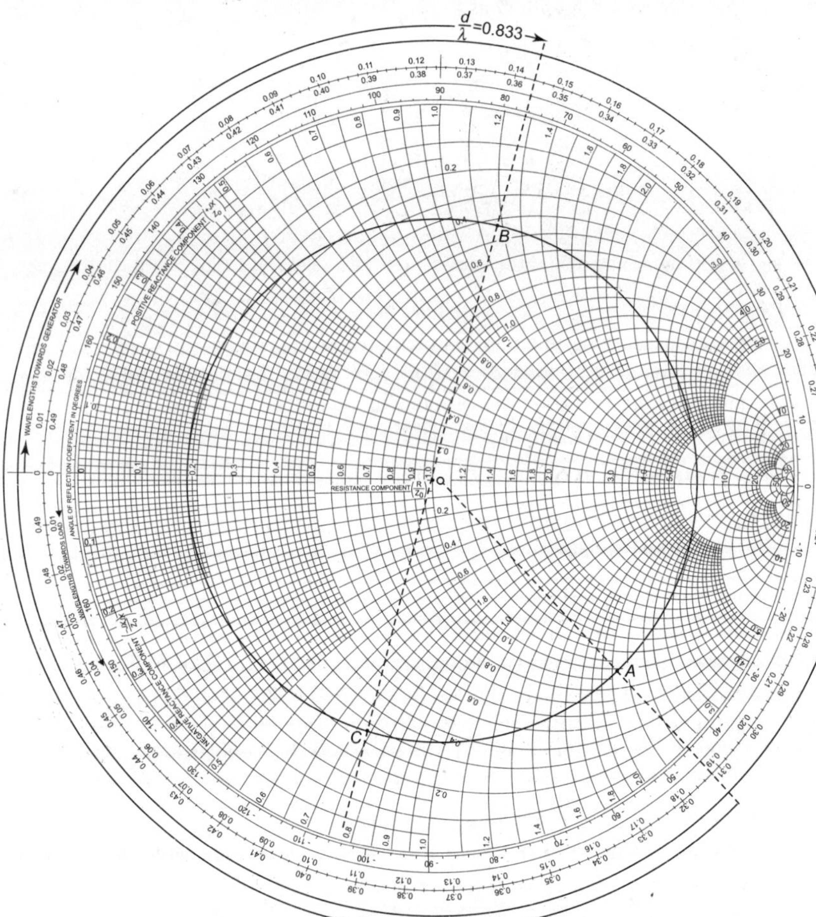

Fig. 3.12 Calculation of line impedance using the Smith chart

Example 3.7 A lossless line with a characteristic impedance of 50 ohms is terminated by an impedance Z_L. The voltage standing-wave maximum and minimum are found 2.5 V and 1 V, respectively and the distance between successive minima is 5 cm. The line is first terminated by a short and then the unknown load, so that a shift in the voltage minimum of 1.25 cm is observed towards the generator. Determine the load impedance using a Smith chart.

Solution

Given $\quad Z_0 = 50$ ohms, $V_{max} = 2.5$ V, $V_{min} = 1$ V

VSWR $\quad S = V_{max}/V_{min} = 2.5$

Distance between successive minima is $\lambda/2 = 5$ cm

Therefore, $\quad \lambda = 10$ cm

Under short-circuit condition, the first voltage minimum occurs at load point. Hence the shift in minimum of 1.25 cm = distance of the first voltage minimum d_{min} from the load.

$$d_{min}/\lambda = 1.25/10 = 0.125$$

Draw a constant S-circle for $S = 2.5$ with O as the centre and mark a V_{min} point A on the circle along the real axis on Smith chart (Fig. 3.13).

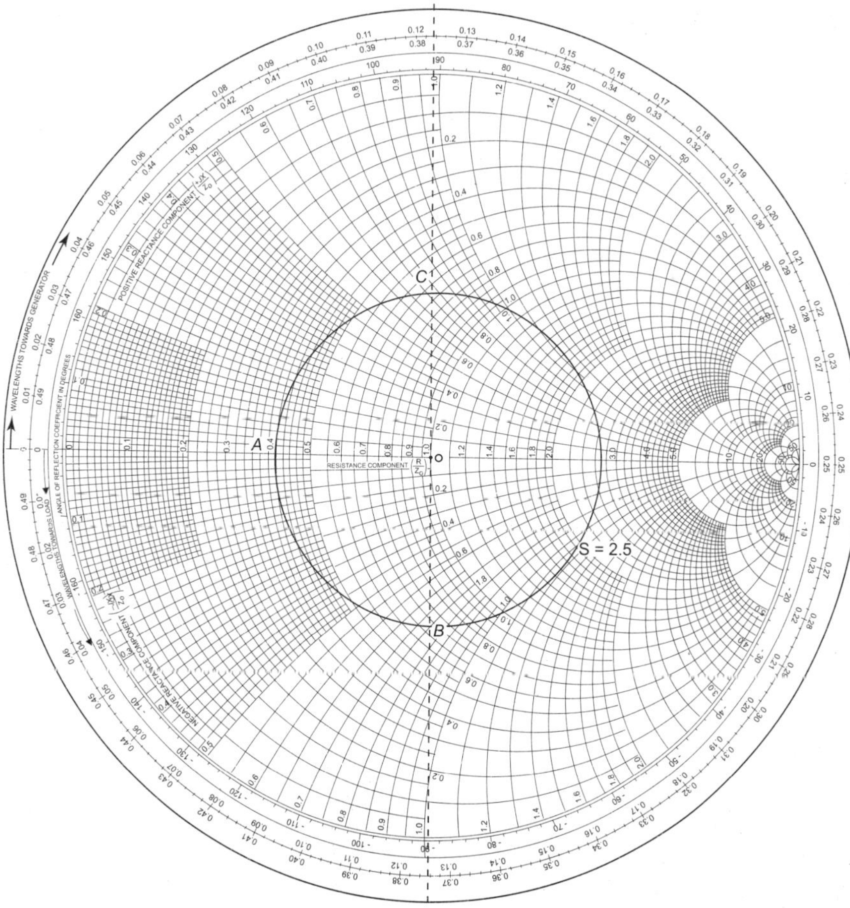

Fig. 3.13 Calculation of load impedance using the Smith chart

Travel $d_{min}/\lambda = 0.125$ from point A towards the load to reach point B, where normalized impedance is $0.69 - j\,0.72$. The diametrically opposite point C on the circle is the normalized load admittance point ($0.725 + j\,0.75$).

Therefore, load impedance at B is $Z_L = (0.69 - j\,0.72) \times 50 = 34.5 - j\,36$ ohms

Load admittance at C is $Y_L = 1/Z_L\,(0.725 + j\,0.75)/50 = 0.0145 + j\,0.015$ mho

3.4 GENERAL ANALYSIS OF TIME HARMONIC WAVES ON TRANSMISSION LINES

Let us consider a uniform transmission line of length l excited at $z = 0$ with a time harmonic voltage source of voltage $v_g(t) = V_g(z)e^{j\omega t}$ having source impedance Z_g and terminated at $z = l$ by a load impedance Z_L as shown in Fig. 3.14. The instantaneous voltage and current on the line at any arbitrary location z can be represented, respectively, by

$$V(z) = V_i e^{-\gamma z} + V_r e^{\gamma z} \quad (3.54a)$$

$$I(z) = I_i e^{-\gamma z} + I_r e^{\gamma z}$$

$$= \frac{1}{Z_0}(V_i e^{-\gamma z} - V_r e^{\gamma z}) \quad (3.54b)$$

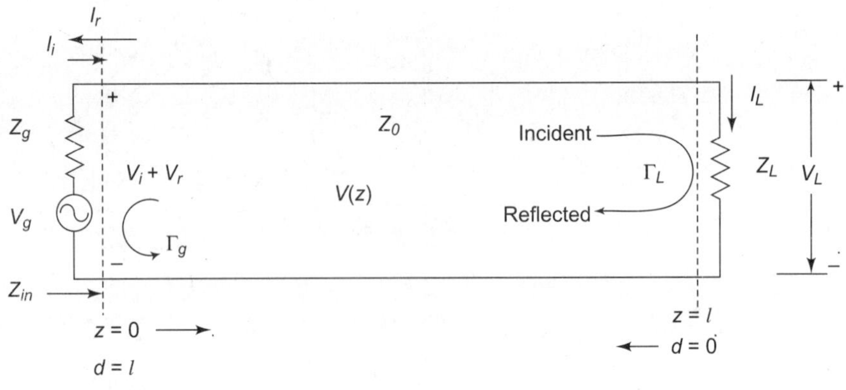

Fig. 3.14 General form of transmission line

The input voltage and current at $z = 0$ on the line are related by

$$V(0) = V_g - Z_g I(0) \quad (3.55)$$

At the load end,

$$\frac{V(l)}{I(l)} = Z_L \quad (3.56)$$

Substituting (3.54a) and (3.54b) in (3.55) and (3.56) yields

$$V(0) = V_g - \frac{Z_g}{Z_0}(V_i - V_r) \quad (3.57)$$

and

$$Z_0 \frac{V_i e^{-\gamma l} + V_r e^{\gamma l}}{V_i e^{-\gamma l} - V_r e^{\gamma l}} = Z_L \quad (3.58)$$

From (3.57) and (3.58),

$$V_r = \frac{V_g - \left(1 + \frac{Z_g}{Z_0}\right)V_i}{1 - \frac{Z_g}{Z_0}} \tag{3.59}$$

$$V_r = V_i \frac{\left(\frac{Z_L}{Z_0} - 1\right)}{\left(\frac{Z_L}{Z_0} + 1\right)} e^{-2\gamma l} \tag{3.60}$$

Substituting (3.60) in (3.59), and multiplying both sides of the resulting equation by $(1 - Z_g/Z_0)(1 + Z_g/Z_0)^{-1}$ and rearranging, we get

$$V_i = \frac{Z_0 V_g / (Z_0 + Z_g)}{1 - \Gamma_g \Gamma_L e^{-2\gamma l}} \tag{3.61a}$$

$$V_r = \frac{\Gamma_L Z_0 V_g / (Z_0 + Z_g)}{1 - \Gamma_g \Gamma_L e^{-2\gamma l}} e^{-2\gamma l} \tag{3.61b}$$

Here, $\Gamma_L = \dfrac{Z_L - Z_0}{Z_L + Z_0}$, the load-end voltage reflection coefficient (3.62a)

$\Gamma_g = \dfrac{Z_g - Z_0}{Z_g + Z_0}$, the generator-end voltage reflection coefficient (3.62b)

From (3.54a),(3.61a) and (3.61b),

$$V(z) = \frac{Z_0 V_g/(Z_0 + Z_g)}{1 - \Gamma_g \Gamma_L e^{-2\gamma l}}[e^{-\gamma z} + \Gamma_L e^{\gamma(z-2l)}] \tag{3.63a}$$

$$I(z) = \frac{V_g/(Z_0 + Z_g)}{1 - \Gamma_g \Gamma_L e^{-2\gamma l}}[e^{-\gamma z} - \Gamma_L e^{\gamma(z-2l)}] \tag{3.63b}$$

Substituting $\Gamma_g \Gamma_L e^{-2\gamma l} = a$, for $|a| < 1$, one obtains

$$\frac{1}{1 - \Gamma_g \Gamma_L e^{-2\gamma l}} = \frac{1}{1 - a} = 1 + a + a^2 + \cdots; \tag{3.64}$$

Substituting (3.64) in (3.63a), yields

$$V(z) = \frac{Z_0 V_g}{Z_0 + Z_g}[e^{-\gamma z} + \Gamma_L e^{-\gamma(2l-z)} + \Gamma_L \Gamma_g e^{-\gamma(2l+z)} + \Gamma_L^2 \Gamma_g e^{-\gamma(4l-z)} + \Gamma_L^2 \Gamma_g^2 e^{-\gamma(4l+z)} + \cdots]; \tag{3.65}$$

The right-hand side of (3.65), the first term represents the direct incident wave, the second term represents the wave reflected from the load, the third term represents the wave reflected first from the load and then from the generator, and so on. Thus, there is multiple reflections from the load and the generator due to impedance mismatch at both ends when $Z_g \neq Z_0 \neq Z_L$. Equation (3.65) is the general expression on voltage wave on the uniform transmission line at any point z along the length when excited with a time harmonic source.

Some special cases may be derived from (3.65) as follows:

❑ *Case I* Matched generator and matched load, $Z_g = Z_0 = Z_L$.
Here, $\Gamma_g = 0 = \Gamma_L$, so that from (3.65),

$$V(z) = \frac{V_g}{2}[e^{-\gamma z}]; \quad I(z) = \frac{V_g}{2Z_0}[e^{-\gamma z}]; \tag{3.66a}$$

$$V(0) = \frac{V_g}{2}; \tag{3.66b}$$

$$V(l) = \frac{V_g}{2}[e^{-\gamma l}]; \tag{3.66c}$$

$$Z_{in} = \frac{V(z)}{I(z)} = Z_o; \tag{3.66d}$$

❑ *Case II* Matched generator and mismatched load, $Z_g = Z_0 \neq Z_L$.
Here, $\Gamma_g = 0$, $\Gamma_L \neq 0$, so that from (3.65),

$$V(z) = \frac{V_g}{2}[e^{-\gamma z} + \Gamma_L e^{\gamma(z-2l)}]; \tag{3.67a}$$

$$I(z) = \frac{V_g}{2Z_0}[e^{-\gamma z} - \Gamma_L e^{\gamma(z-2l)}]; \tag{3.67b}$$

$$V(0) = \frac{V_g}{2}[1 + \Gamma_L e^{-2\gamma l}]; \tag{3.67c}$$

$$V(l) = \frac{V_g}{2}[e^{-\gamma l} + \Gamma_L e^{-\gamma l}] = \frac{V_g}{2}(1 + \Gamma_L)e^{-\gamma l} \tag{3.67d}$$

$$Z_{in}(z) = \frac{V(z)}{I(z)} = Z_o \frac{e^{-\gamma z} + \Gamma_L e^{\gamma(z-2l)}}{e^{-\gamma z} - \Gamma_L e^{\gamma(z-2l)}}; \tag{3.67e}$$

❑ *Case III* Matched generator and open-circuit load, $Z_g = Z_0$, $Z_L = \infty$.
Here $\Gamma_g = 0$, $\Gamma_L = +1$ in all (3.67a-e)

❑ *Case IV* Matched generator and short-circuit load, $Z_g = Z_0$, $Z_L = 0$.
Here, $\Gamma_g = 0$, $\Gamma_L = -1$ in all (3.67a-e)

TRANSIENT ANALYSIS OF TRANSMISSION LINES 3.5

When a pulse generator or battery connected to a transmission line is switched on, it takes some time for the current and voltage on the line to reach steady values due to presence of reactive elements L and C in which magnetic and electric energies, respectively, are stored. The storing of energies takes some time causing the delay in attaining steady values. This transitional period between the switching and the steady value is called the **transient**. For simplicity, let us consider a lossless line of length l and characteristic impedance Z_0 as shown in Fig. 3.15.

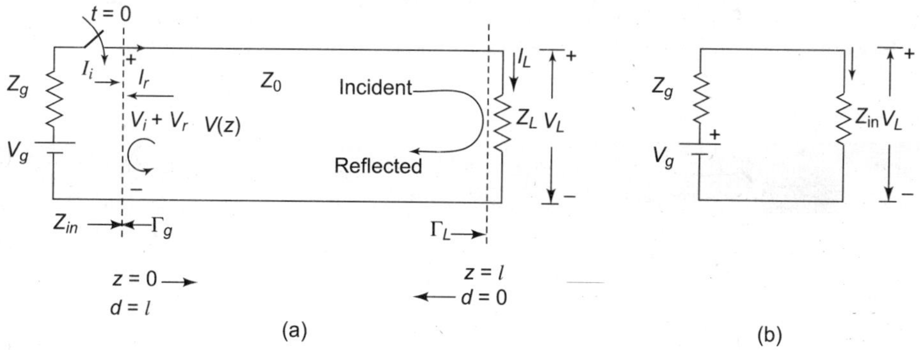

Fig. 3.15 Transient in transmission line

Step Response of Transmission Lines 3.5.1

Suppose that the line is driven by a dc generator of voltage v_g with internal impedance Z_g at $z = 0$ and terminated with a purely resistive load Z_L. When the switch is closed at $t = 0$, the starting current sees only Z_g and Z_0 as in Fig. 3.15(b), since the signal wave did not arrive at $z = l$ at that time when effect of load could arise. So, the starting current at $z = 0$, $t = 0^+$ is given by

$$I(0, 0^+) = I_0 = V_g/(Z_g + Z_0) \tag{3.68a}$$

and the initial voltage is

$$V(0, 0^+) = V_0 = I_0Z_0 = Z_0\, V_g\, /\, (Z_g + Z_0) \tag{3.68b}$$

After the switch is closed, wave $I^+ = I_0$ and $V^+ = V_0$ propagate towards the load at the phase velocity

$$u_p = \frac{1}{\sqrt{LC}} \tag{3.69}$$

Transit time is the time taken by the waves traveling in the positive z-direction to reach the load and interact with it. It is given by

$$t_1 = l/u_{\text{p}} \tag{3.70}$$

The voltage (or current) at the load end is the sum of the incident and reflected waves. Therefore,

$$V\,(l, t_1) = V_i + V_r = V_0 + \Gamma_L V_0 = (1+ \Gamma_L)V_0 \tag{3.71a}$$

and $$I(l, \text{t}_1) = I_i + I_r = I_0 - \Gamma_L I_0 = (1- \Gamma_L)I_0 \tag{3.71b}$$

where, load reflection coefficient $\Gamma_L = (Z_L - Z_0\,)/(Z_L + Z_0)$ (3.71c)

The reflected waves $V_r = \Gamma_L V_0$ and $I_r = -\Gamma_L I_0$ travel back toward the generator in addition to the waves V_0 and I_0.

At time $t_2 = 2t_1$, the reflected waves have reached the generator end so that

$$V(0, 2t_1) = V_i + V_r = \Gamma_g\Gamma_L V_0 + (1+ \Gamma_L)V_0 = (1+ \Gamma_L+ \Gamma_g\Gamma_L)V_0 \tag{3.72a}$$

and $$I(0, 2t_1) = I_i + I_r = -\Gamma_g(-\Gamma_L\, I_0) + (1 - \Gamma_L)I_0 = (1 - \Gamma_L+ \Gamma_g\Gamma_L)I_0 \tag{3.72b}$$

where, generator reflection coefficient $\Gamma_g = (Z_g - Z_0\,)/(\,Z_g + Z_0)$ (3.72c)

Again, the reflected waves $V_i = \Gamma_g\Gamma_L V_0$ and $I_i = \Gamma_g\Gamma_L\, I_0$ propagate towards the load and the process continues until the energy of the source is actually absorbed by the resistors Z_g and Z_L. Thus, the waves bounce back and forth as shown in Fig. 3.16. Therefore, if voltage vs time is plotted, the steady state will be attended after several bouncing as shown in Fig. 3.17.

Thus, the steady-state values attend after infinite number of reflections from both the ends for a lossless line of finite length. Table 3.2 shows the values of the voltages at different time intervals.

Table 3.2 Transient voltages on the line

Line voltage	Γ_g, Γ_L *at* $z = 0$	Γ_g, Γ_L *at* $z = 1$
V_1	V_0	0
V_2	$(1 + \Gamma_L + \Gamma_g\Gamma_L)\ V_0$ at $2t_1$	$(1 + \Gamma_L)\ V_0$ at t_1
V_3	$(1 + \Gamma_L + \Gamma_g\Gamma_L + \Gamma_g\Gamma_L^2 + \Gamma_g^2\Gamma_L^2)\ V_0$ at $4t_1$	$(1 + \Gamma_L + \Gamma_g\Gamma_L + \Gamma_g\Gamma_L^2)\ V_0$ at $3t_1$
V_4	$(1 + \Gamma_L + \Gamma_g\Gamma_L + \Gamma_g\Gamma_L^2 + \Gamma_g^2\Gamma_L^2 + \Gamma_g^2\Gamma_L^3 + \Gamma_g^3\Gamma_L^3)$ V_0 at $6t_1$	$(1 + \Gamma_L + \Gamma_g\Gamma_L + \Gamma_g\Gamma_L^2 + \Gamma_g^2\Gamma_L^2 + \Gamma_g^2\Gamma_L^3)\ V_0$ at $5t_1$
...	...	...
V_n	$(1 + \Gamma_L + \Gamma_g\Gamma_L + \Gamma_g\Gamma_L^2 + \Gamma_g^2\Gamma_L^2 + \Gamma_g^2\Gamma_L^3 + \cdots +$ $\Gamma_g^n\Gamma_L^n)\ V_0$ at $2nt_1$	$(1 + \Gamma_L + \Gamma_g\Gamma_L + \Gamma_g\Gamma_L^2 + \Gamma_g^2\Gamma_L^2 + \Gamma_g^2\Gamma_L^3 + \cdots +$ $\Gamma_g^{n-1}\Gamma_L^n)\ V_0$ at $2(n-1)\ t_1$

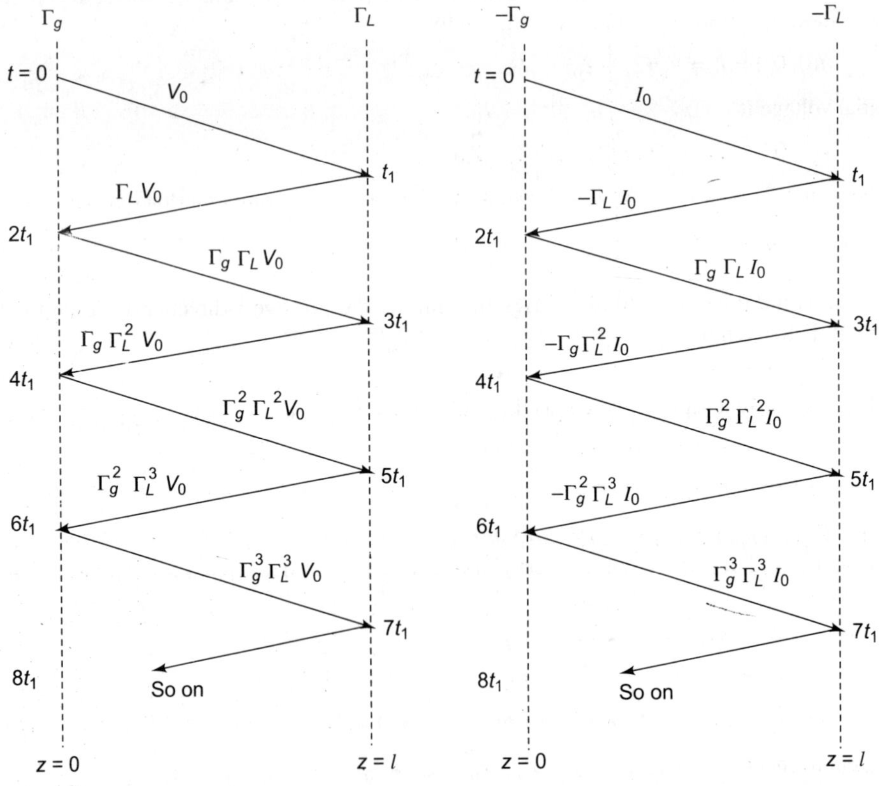

Fig. 3.16 Voltage and current wave bounce diagrams

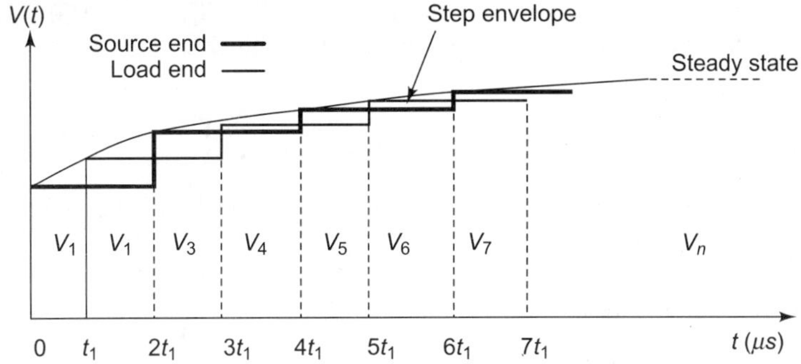

Fig. 3.17 Transient voltage vs time on transmission line for Γ_g, $\Gamma_L > 0$

The voltage and current waveforms change for different terminating impedances relative to Z_0 of the line. These are shown using the following examples.

Example 3.8 For the following transmission line of 100 m length, calculate and sketch

(a) the voltage at the load and generator ends for $0 < t < 6$ μs

(b) the current at the load and generator ends for $0 < t < 6$ μs.

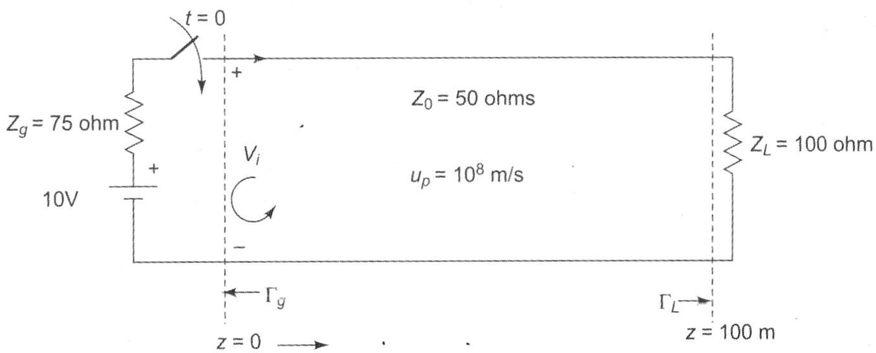

Fig. 3.18 Transmission line

Solution

(a) The voltage reflection coefficients at the generator and load ends.

$$\Gamma_g = (Z_g - Z_0)/ (Z_g + Z_0) = (75 - 50)/ (75 + 50) = 1/5$$

$$\Gamma_L = (Z_L - Z_0)/ (Z_L + Z_0) = (100 - 50) / (100 + 50) = 1/3$$

Transit time $t_1 = l/u_p = 100/10^8 = 1$ μs

Initial voltage at the generator end is

$$V_0 = \left(\frac{Z_0}{Z_0 + Z_g}\right)V_g = (50/125) \times 10 = 4 \text{ V is sent out to the load.}$$

The leading edge of the pulse arrives at the load at $t_1 = 1$ μs. Then $4 \times (1/3) = 1.33$ V is reflected back and reaches the generator at 2 μs and the process continues.

When t tends to infinity then,

$$V_\infty = Z_L V_g/(Z_L + Z_g) = (100/175) \times 10 = 5.71\text{V}$$

The current reflection coefficients at the generator and load ends are $-\Gamma_g = -1/5$ and $-\Gamma_L = -1/3$

Initial current $I_0 = V_0/Z_0 = 4/50 = 80$ mA.

When t tends to infinity then,

$$I_\infty = V_g/(Z_g + Z_L) = 10/175 = 57.15 \text{ mA}$$

Therefore, the equivalent circuits for the line and the bounce diagram are as shown in Fig. 3.19 to 3.21.

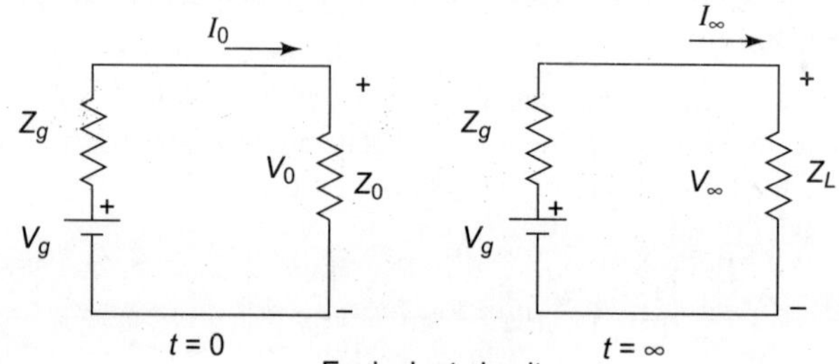

Equivalent circuit

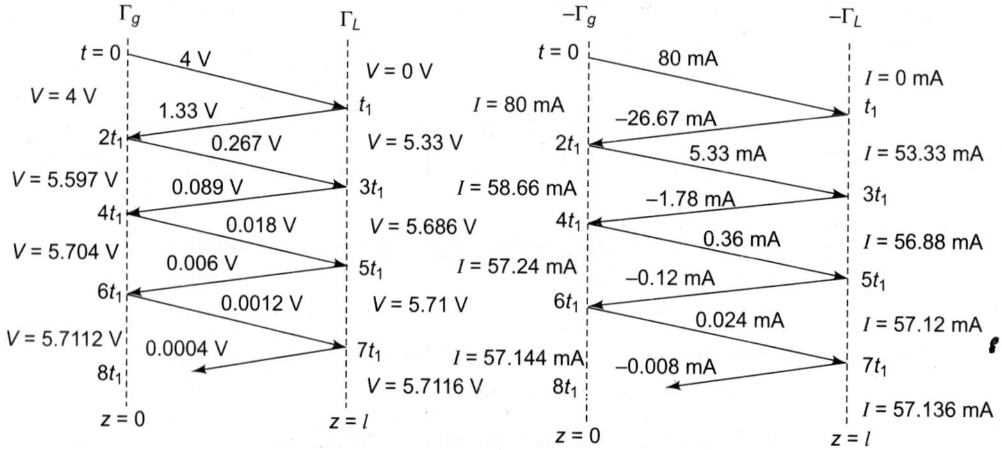

(a) Voltage (b) Current

Fig. 3.19 Voltage and current wave bounce diagram

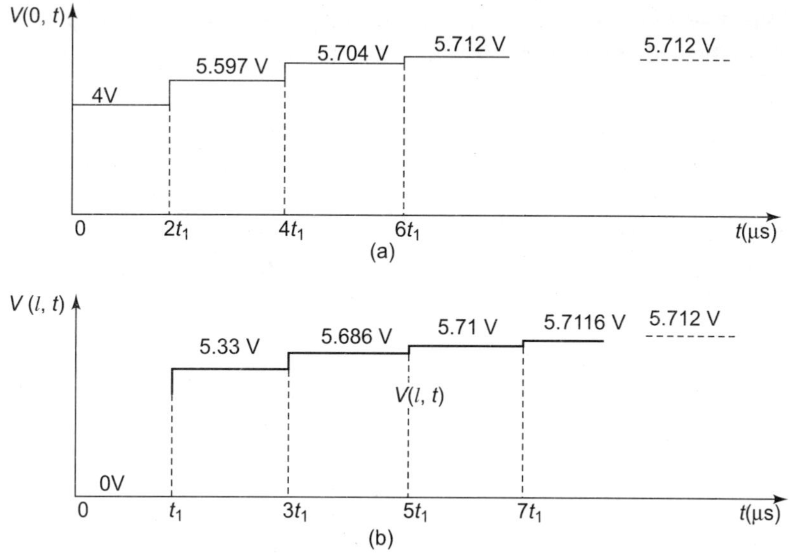

Fig. 3.20 Voltage (a) at generator end, and (b) at load end

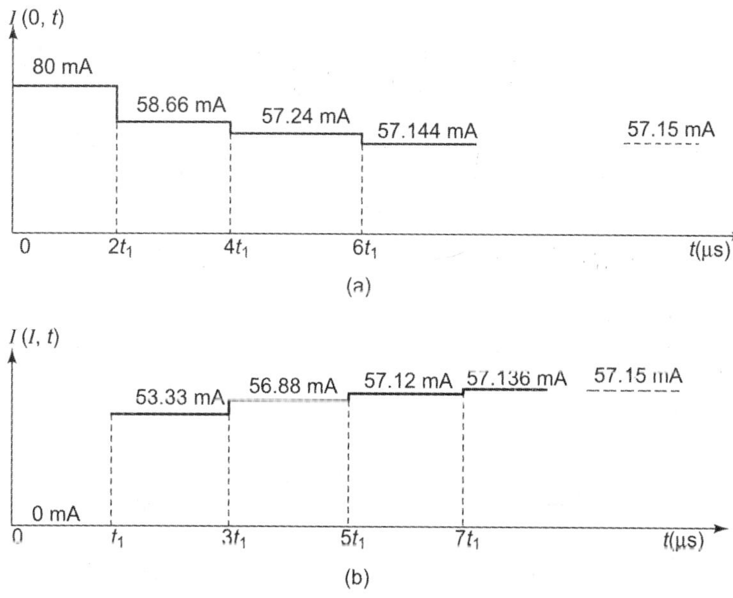

Fig. 3.21 Current (a) at generator end, and (b) at load end

As a result, the step response of a transmission line approaches a constant value along the entire line as t tends to ∞.

Rise Time Since every trace on a high-frequency or high-speed Printed Circuit Board (PCB) is a microwave transmission-line, transmission line behaviour is important when the rise time of the digital signal is less than or comparable to the propagation delay between the source and the load.

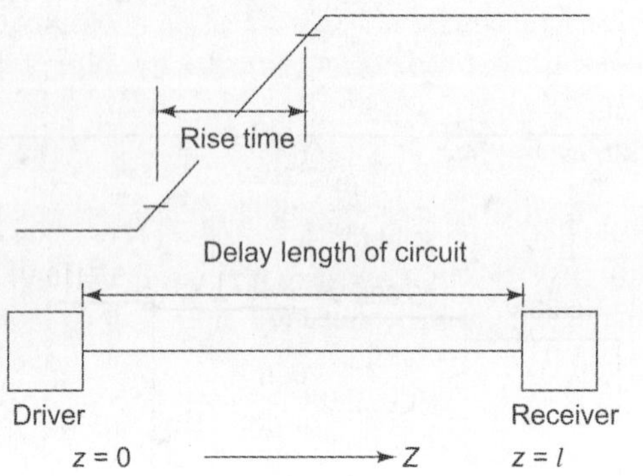

Fig. 3.22 Rise time and circuit length

The following equations show the relationships between the dielectric constant, the propagation velocity, the propagation delay, and the time delay:

$$u = \frac{c}{\sqrt{\varepsilon_r}}$$

$$t_p = \frac{l}{u} = \frac{\sqrt{\varepsilon_r}}{c}$$

$$t_d = \frac{l\sqrt{\varepsilon_r}}{c} = \sqrt{LC} \quad (3.73)$$

where L = Total series inductance for the length of the line
C = Total shunt capacitance for the length of the line
u = Propagation velocity, in metres/second
c = Speed of light in a vacuum (3×10^8 m/s)
ε_r = Dielectric constant
t_p = Propagation delay, in seconds per metre
t_d = time delay for a signal to propagate down a transmission line of length l, m

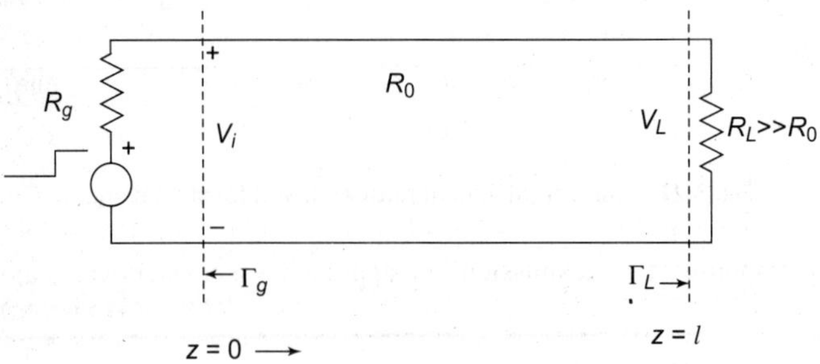

Fig. 3.23 Check pulse

Figure 3.23 shows a clock pulse given to a high-impedance logic gate through PCB trace of length l. Since the load impedance is very high, output load can be modeled as open circuit. If the switching time of the clock pulse is t', the voltage waveform at the source and load ends can be shown as in Fig. 3.23 for three cases: $R_g > R_0$, $R_g = R_0$, $R_g < R_0$, respectively. Three cases produce different voltage waves. For step input, if $R_g >> R_o$, the envelope of input and output step-voltage wave increases exponentially to steady-state value showing high rise time. Under this condition, a transmission line approximately behaves like a shunt capacitance Cl. Here, C is the line capacitance per unit length, so that the (10% to 90%) rise time of output voltage of RC circuit can be expressed by

$$t_r \approx 2.2\tau = 2.2\, R_g Cl \tag{3.74}$$

where $\tau = RC$ time constant.

Thus, when $R_g >> R_o$, the rise time of output voltage wave is proportional to the line length. The rise can be reduced by making $R_g \rightarrow R_o$. If $R_g = R_o$, steady state is reached with zero rise time. When $R_g < R_o$, the output voltage wave shows ringing with overshoot due to $\Gamma_g < 0$, $\Gamma_L > 0$. Such a case is undesirable.

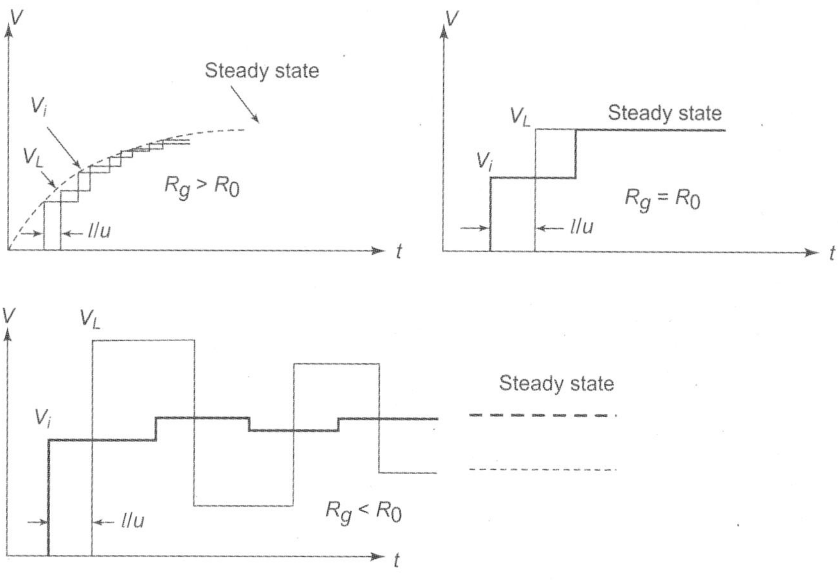

Fig. 3.24 Output voltage waves

COAXIAL LINE 3.6

At RF and higher range of frequencies, simple two-wire telephone-type transmission lines radiate electromagnetic waves by the RF currents flowing through the conductors. Therefore, at load end signal strength becomes low due to this radiation loss. One possibility of eliminating radiation is to guide the signal through closed conductors. This type of transmission lines are coaxial lines, strip lines, microstrip lines, co-planar strip lines, waveguides, and dielectric fibres (optical waveguides).

A coaxial line consists of an inner circular conductor of very high conductivity (ideally infinite) and an outer circular conductor as shown in Fig. 3.25.

The space between the two conductors is filled with a low-loss uniform homogeneous solid semirigid dielectric material. When the line is excited by an RF signal source as shown in Fig. 3.25(a), the two conductors are at two different potentials. Both the fields, $\boldsymbol{E}$ and $\boldsymbol{H}$, are transverse to the direction of propagation (z).

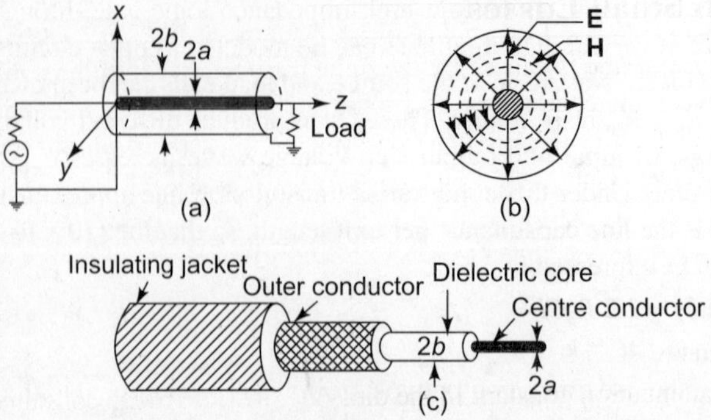

Fig. 3.25 Coaxial lines and TEM wave field

Thus, the mode of propagation in a symmetric coaxial line is the TEM wave. The self configurations of the mode are shown in Fig. 3.25 (b). Since a TEM wave does not have a cut-off frequency, a coaxial line is a broadband device and the characteristic impedance Z_0 of the line is equal to the wave impedance. For propagation in the $+z$ direction, the fields inside the line cross section are expressed by

$$\mathbf{E} = \mathbf{E}_t = \mathbf{E}_0(\rho,\phi)e^{-jkz} = \frac{V_0}{\ln(b/a)}\frac{\hat{\boldsymbol{\rho}}}{\rho}e^{-jkz} \tag{3.75a}$$

$$\mathbf{H} = \mathbf{H}_t = \pm\frac{\hat{\mathbf{z}}\times\mathbf{E}_0(\rho,\phi)}{\eta}e^{-jkz} = \frac{V_0}{\eta\ln(b/a)}\frac{\hat{\boldsymbol{\phi}}}{\rho}e^{-jkz} \tag{3.75b}$$

where $k = \omega\sqrt{\mu\varepsilon}$. Here, a and b are the outer and inner radii of the inner and outer conductors, respectively. Due to the skin effect, the signal currents flow through the outer surface of inner conductor and return through the inner surface of the outer conductor. The current density is given by

$$\mathbf{J}_s = \hat{\mathbf{n}}\times\mathbf{H} = \hat{\boldsymbol{\rho}}\times\mathbf{H}_t = \hat{\mathbf{z}}\frac{V_0e^{-jkz}}{\eta a\ln(b/a)} \tag{3.76}$$

The power flow through the coaxial line is expressed by

$$P = \frac{1}{2}\mathrm{Re}\int_a^b\int_0^{2\pi}\mathbf{E}\times\mathbf{H}^*\cdot\mathrm{dS} = \frac{\pi V_0^2}{\eta\ln(b/a)} \tag{3.77a}$$

where $\mathrm{dS} = \rho d\rho d\phi$. For TEM mode $\mathbf{E} = \eta\mathbf{H}$, $P = \frac{\eta}{2}\mathrm{Re}\int_a^b\int_0^{2\pi}|\mathbf{H}|^2\,\mathrm{dS}$; (3.77b)

The characteristic impedance of the symmetric coaxial line for a nonmagnetic dielectric is given by

$$Z_0 = \frac{1}{2\pi}\sqrt{\frac{\mu}{\varepsilon}}\ln\left(\frac{b}{a}\right) = \frac{60}{\sqrt{\varepsilon_r}}\ln\frac{b}{a}\text{ ohm} \tag{3.78}$$

where $\varepsilon_r = \varepsilon\varepsilon_0$, the dielectric constant of the dielectric medium between the conductors. The wavelength inside this TEM line is

$$\lambda_g = \frac{\lambda_0}{\sqrt{\varepsilon_r}} \tag{3.79}$$

Coaxial Line with Small Losses 3.6.1

The power loss in the coaxial line is caused by the finite conductivity σ of the conductors and the dielectric loss in the medium between the conductors. In a lossy dielectric, $\varepsilon = \varepsilon' - j\varepsilon''$ and $\sigma_d/\omega\varepsilon \ll 1$, where σ_d is the conductivity of the dielectric. Under such conditions, the propagation constant is complex, given by, $K = \alpha + j\beta \approx jk\,(1 - j\sigma_d/2\omega\varepsilon)$. Thus,

$$\alpha_d = \frac{\sigma_d}{2}\sqrt{\frac{\mu}{\varepsilon}} \tag{3.80a}$$

$$\beta = \omega\sqrt{\mu\varepsilon} = k \tag{3.80b}$$

where α_d and β are the attenuation constant in the dielectric and the phase constant, respectively.

Due to the finite conductivity σ of the conductors, the conductors exhibit a non-zero surface impedance given by

$$\eta_c = (1+j) = \sqrt{\frac{\omega\mu}{2\sigma}} = \frac{1+j}{\sigma\delta} \tag{3.81}$$

and a surface density $\mathbf{J}_s = \hat{\mathbf{n}} \times \mathbf{H}$. Here $\delta = \dfrac{1}{\sqrt{\pi f \mu\sigma}}$ is the skin depth.

Therefore, the electric field must have a non-zero tangential component given by $\eta_c\, J_s$ at the surface. Since the surface current is axial, an axial component of electric field must be present, resulting in a non-TEM wave mode of propagation. This axial component of **E** field gives rise to component of the Poynting vector directed into the conductor, resulting in power loss in the conductor, given per unit length, by

$$P_l = \frac{1}{2} Re\,\eta_c \oint_{s_1+s_2} (\mathbf{J}_s \cdot \mathbf{J}_s^*)dl = \frac{1}{2} R_s \oint_{s_1+s_2} |\mathbf{H}|^2\, dl \tag{3.82}$$

where integration is taken around the periphery $s_1 + s_2$ of the two conductors. Due to this power loss, the power propagated along the line decreases according to a factor $(e^{-\alpha_c z})^2 = e^{-2\alpha_c z}$. The rate of decrease of power with distance is given by

$$-\frac{\partial P}{\partial z} = 2\alpha_c P = \text{Power loss per unit length}$$
$$= P_l$$

Therefore, from (3.77a) and (3.82) the attenuation constant due to conductor loss is given by

$$\alpha_c = \frac{P_l}{2P} = \frac{R_s \displaystyle\int_{s_1+s_2} |\mathbf{H}|^2\, dl}{\sqrt{\dfrac{\mu}{\varepsilon}} \displaystyle\int_s |\mathbf{H}|^2\, d\dot{s}}$$

The total attenuation constant of the line becomes $\alpha = \alpha_c + \alpha_d$.

The distributed line parameters of symmetric coaxial transmission lines can be computed from magnetostatic and electrostatic field considerations. These parameters are given by the following relations:

$$R = \frac{R_s}{2\pi}\left(\frac{1}{a} + \frac{1}{b}\right) \tag{3.83a}$$

$$L = \sqrt{\mu_0 \varepsilon' Z_0} \approx \frac{\mu_0}{2\pi} \ln \frac{b}{a} \tag{3.83b}$$

$$G = \frac{\omega \varepsilon'' C}{\varepsilon'} \tag{3.83c}$$

$$C = \frac{\sqrt{\mu_0 \varepsilon'}}{Z_0} \cong \frac{2\pi\varepsilon'}{\ln \frac{b}{a}} \tag{3.83d}$$

For low loss lines, $$\alpha_d = \frac{GZ_0}{2}, \alpha_c = \frac{R}{2Z_0} \tag{3.84a}$$

$$\alpha = \frac{1}{2}\left(R\sqrt{\frac{C}{L}} + G\sqrt{\frac{L}{C}} \right) \tag{3.84b}$$

and $$Z_0 \cong \sqrt{\frac{L}{C}} \tag{3.84c}$$

For power transmission, coaxial lines are used up to about 3 GHz and for small signal transmission, up to about 18 GHz. For cables with $G \to 0$, the attenuation is minimum for $b/a = 3.6$. For an air dielectric, $\varepsilon_r = 1$ and $b/a = 3.6$ gives $Z_0 = 76.86$ ohms. With a solid polyethylene dielectric, $\varepsilon_r = 2.3$ and $Z_0 = 50.67$ ohms for minimum attenuation. Since the attenuation coefficient does not vary rapidly with b/a, 50 ohm lines are very commonly used for minimum attenuation.

Example 3.9 An air-filled coaxial transmission line has outer and inner conductor radii equal to 6 cm and 3 cm, respectively. Calculate the values of (a) inductance per unit length, (b) capacitance per unit length, and (c) characteristic impedance of the line.

Solution Given, $b = 6$ cm, $a = 3$ cm

(a) $L = (\mu_0/2\pi) \ln (b/a) = (4\pi \times 10^{-7}/2\pi) \ln (6/3) = 1.386 \times 10^{-7}$ h/m

(b) $C = 2\pi\varepsilon_0/\ln (b/a) = 2\pi \times 10^{-9}/[36\pi \ln (6/3)] = 0.08 \times 10^{-9}$ f/m

(c) $$Z_0 = \sqrt{\frac{L}{C}} = \sqrt{\left(\frac{\mu_0}{\varepsilon_0}\right)\left[\frac{\ln\left(\frac{b}{a}\right)}{2\pi}\right]^2} = 120\pi \times \frac{\ln\left(\frac{b}{a}\right)}{2\pi} = 41.59 \text{ ohm}$$

Higher-Order Modes 3.6.2

At sufficiently high frequencies, higher-order modes are generated in a coaxial line. The lower cut-off frequencies of the higher-order modes in a coaxial line are very high. Thus, higher-order mode interference is not encountered at the normal operating (lower) frequencies in a conventional coaxial lines. The first two higher-order modes in coaxial lines are TE_{11} and TM_{01}. The cut-off wavelength of these modes are

$$\lambda_C (TE_{11}) \cong \pi (a + b) \qquad \lambda_C (TM_{01}) \cong 2 (b - a) \tag{3.85}$$

Therefore, the average circumference of the inner and outer conductors in the propagating cross section of the coaxial line should be less than the operating wavelength in order to prevent higher-order mode interference. The field configuration of these higher order modes are shown in Fig. 3.26.

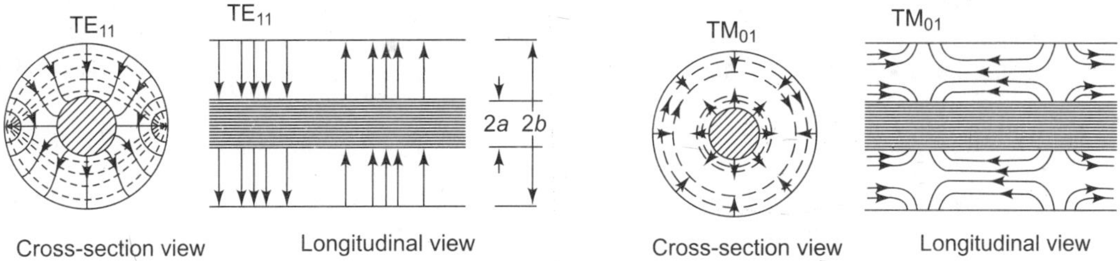

Fig. 3.26 Higher-order modes in coaxial lines—E-field lines ... H-field lines

Unbalance Characteristics of Coaxial Lines 3.6.3

A coaxial line is said to be an unbalanced line because the two conductors are not at equal and opposite potentials with respect to the ground. In that case, the outer conductor is commonly grounded and the propagation path on this side ideally has the ground potential. Whereas, the other one has different potential with respect to ground. Any interconnected equipment between the lines introduces unequal impedances in both lines and would present unequal impedances to ground. Therefore, when a coaxial cable is used to feed balanced circuits, such as dipole antennas or ordinary parallel-wire lines, impedance mismatch occurs.

EQUIVALENT CIRCUIT PARAMETERS OF A TRANSMISSION LINE 3.7

The distributed electrical circuit parameters *R*, *L*, *G* and *C* of a microwave trans-mission line which account for the joule heating loss, magnetic energy stored, the loss in the medium and the electric energy stored, respectively, can be expressed by

$$R = \frac{2P_L}{II^*} = \frac{\int_V \sigma E^2 dV}{I^2} \tag{3.86a}$$

$$L = \frac{4W_m}{II^*} = \frac{\mu' \int_V H^2 dV}{I^2} \tag{3.86b}$$

$$G = \sigma C/\varepsilon'$$

$$C = \frac{II^*}{4\omega^2 W_e} = \frac{I^2}{\omega^2 \varepsilon' \int_V E^2 dV} \tag{3.86c}$$

where σ is the electrical conductivity of the medium inside the line. The characteristic impedance of the line is given by

$$Z_0 = \sqrt{\frac{R + j\omega L}{G + j\omega C}} \approx \sqrt{\frac{L}{C}} \tag{3.87}$$

Since at microwave frequencies ($\omega L >> R$, $\omega C >> G$) reactive parts predominate the loss parameters R and G.

Example 3.10 The primary constants for a coaxial cable at 1 GHz are $L = 250$ nh/m, $C = 95$ pf/m. $R = 0.06$ ohm/m, and $G = 0$. Determine (a) the attenuation coefficient α, (b) the phase constant β, (c) the phase velocity u_p, (d) relative permittivity ε_r, and (e) power loss for a length of 10 m, when the input power is 500 watts.

Solution

$$l = 10 \text{ m}$$

$$\omega L = 2\pi \times 1 \times 10^9 \times 250 \times 10^{-9} = 1570.8 \text{ ohm/m}$$

$$\omega C = 2\pi \times 1 \times 10^9 \times 95 \times 10^{-12} = 596.6 \times 10^{-3} \text{ mho/m}$$

Therefore, $\omega L \gg R,\ \omega C \gg G = 0$

Hence, $Z_0 \simeq (L/C)^{1/2} = [(250 \times 10^{-9})/(95 \times 10^{-12})]^{1/2} = 51.3 \text{ ohm}$

(a) $\alpha = R/(2Z_0) = 0.06/(2 \times 51.3) = 5.85 \times 10^{-4} \text{ Np/m}$

$= 5.08 \times 10^{-3} \text{ dB/m}$

(b) $\beta = \omega(\mu\varepsilon)^{1/2} = \omega(LC)^{1/2}$

$= 2\pi \times 1 \times 10^9 \times (250 \times 95 \times 10^{-9} \times 10^{-12})^{1/2} = 30.0 \text{ rad/m}$

(c) $$u_p = \frac{1}{(LC)^{1/2}} = \frac{1}{(250 \times 95 \times 10^{-9} \times 10^{-12})^{1/2}} = 2.05 \times 10^8 \text{ m/s}$$

(d) $c = \sqrt{\varepsilon_r} u_p$

or, $\varepsilon_r = (c/u_p)^2 = [(3 \times 10^8)/(2.05 \times 10^8)]^2 = 2.14$

(e) $P_{\text{loss}} = P_{in} \times 2\alpha \times l = 500 \times 2 \times 5.85 \times 10^{-4} \times 10 = 5.85 \text{ watts}$

PLANAR TRANSMISSION LINES 3.8

Miniaturization of microwave circuits has taken place through the development of planar transmission lines which are flat, having low profile and light weight. This geometry allows control of the characteristic impedance of the line by defining the line dimensions in a single plane and is, therefore, suitable for microwave integrated circuits. The complete transmission line circuit can be fabricated in one step by thin-film technology and photolithography techniques. Among the several configurations, there are four basic forms which are widely used in microwave integrated circuits. These are (1) strip lines, (2) microstrip lines, (3) slot lines, and (4) coplanar strip lines as shown in Fig. 3.27.

Strip Lines 3.8.1

Strip lines consist of a central thin conducting strip of width $W \gg$ its thickness t placed inside a low-loss dielectric substrate of thickness b between two wide ground plates as shown in Fig. 3.27(a). The propagation characteristics in such a line is nearly TEM mode where most of the electric field lines are perpendicular to the centre and its parallel ground conductors and are concentrated over the width of the centre conductor. There are also fringing field lines at the edges of the centre strip as shown in Fig. 3.27(a). The fringing fields are extended up to a certain distance from the edges of the centre strip, beyond which the fields are practically zero. When $b < \lambda/2$, the field cannot propagate in the transverse direction and decreases exponentially. The energy is confined inside the line cross-section provided the width a of the ground plane is at least five times greater than the spacing b between the plates. Therefore, no vertical side walls are

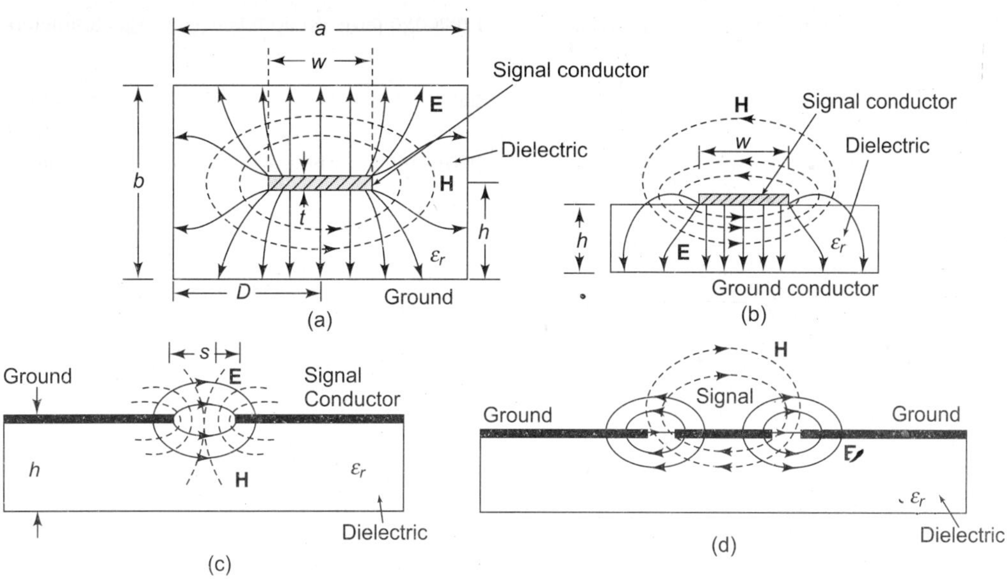

Fig. 3.27 Planar transmission lines: (a) Strip line (b) Microstrip line (c) Slot line (d) Coplanar strip line

necessary at the two transverse ends. The commonly used dielectrics were Teflon, polyolefine, polystyrene, etc. These lines are used over the frequency range from 100 MHz to 30 GHz. For a symmetric homogeneous strip line having centre strip thickness $t \to 0$, the characteristic impedance and the field distributions can be analysed accurately by the conformal transformation technique. The characteristic impedance of the symmetric stripline having zero strip thickness can be expressed as follows:

(1) For $w/b \leq 0.5$,

$$Z_0 = \frac{30}{\sqrt{\varepsilon_r}} \ln\left[\frac{2(1+\sqrt{k})}{(1-\sqrt{k})}\right] \text{ohm} \tag{3.88a}$$

(2) For $w/b > 0.5$,

$$Z_0 = \frac{30\pi^2}{\sqrt{\varepsilon_r}\,\ln\left(\frac{2(1+\sqrt{k'})}{(1-\sqrt{k'})}\right)} \text{ ohm} \tag{3.88b}$$

where $$k = \text{sech}\left(\frac{\pi w}{2b}\right) \tag{3.88c}$$

$$k' = \sqrt{1-k^2} = \tanh\left(\frac{\pi w}{2b}\right) \tag{3.88d}$$

For a finite nonzero thickness of the centre strip of a symmetrical strip line, an exact analysis using conformal transformation techniques is possible but this results in a complicated formula. An accurate value of the impedance can also be obtained by applying numerical techniques, such as the method of moments.

For a wide centre strip having nonzero thickness of a symmetric stripline ($w/b >> 0.35$), the characteristic impedance is given by

$$Z_0 = \frac{94.15}{\sqrt{\varepsilon_r}}\left[\frac{wK}{b} + \frac{C_f}{8.854\varepsilon_r}\right]^{-1} \text{ ohm} \tag{3.88e}$$

where
$$K = \frac{1}{1 - t/b} \tag{3.88f}$$

C_f = Fringing capacitance in pF/m, which arises due to fringing electric field at the edges of the strip

$$= \frac{8.854\ \varepsilon_r}{\pi}\ [2K \ln (K + 1) - (K - 1) \ln (K^2 - 1)]\text{PF/m} \tag{3.88g}$$

The characteristic impedance Z_0 for a partially shielded strip line vs w/b with t/b as a parameter is plotted in Fig. 3.28. It is seen that the value of Z_0 decreases with the increase of w/b and also increases with t/b. In practice, MICs use thickness t of the order of 1.5 to 3 mils and error introduced for the zero thickness assumption is negligible.

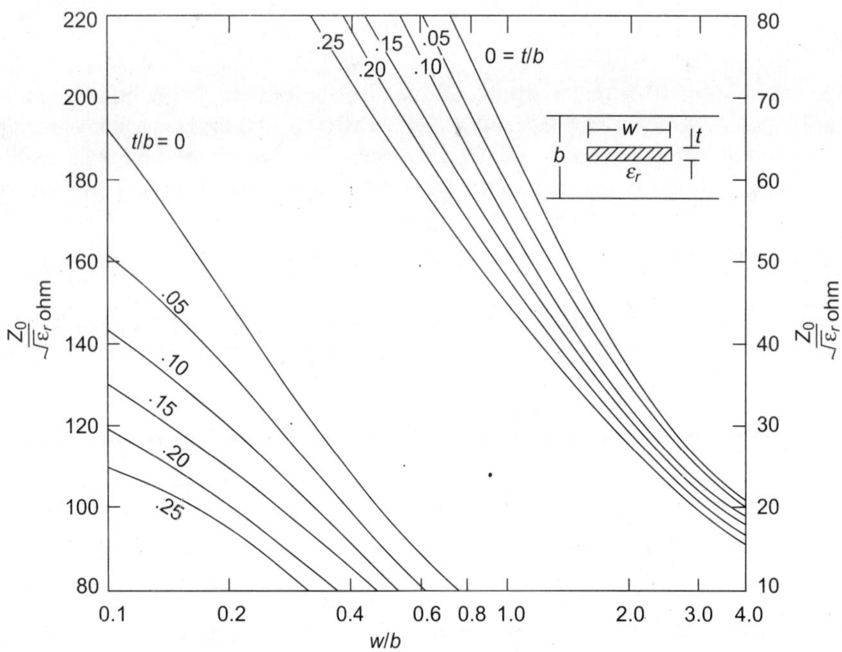

Fig. 3.28 Z_0 vs w/b

For large w/b, the strip line can be completely shielded to avoid radiation. A shielded strip line consists of a very thin strip centre conductor ($t/w \ll 1$) embedded in a low-loss or lossless dielectric medium enclosed in a rectangular metallic box as shown in Fig. 3.27(a). For a completely closed strip line with large w, there is more uniform TEM field region between the centre conductor and the horizontal plates. However, due to proximity between the edges of the centre plate and the sidewalls, fringing fields increase at the edges of the strip and some field lines terminate on the sidewalls. Here, fringe capacitance increases and the characteristic impedance decreases considerably. Since w is large, it may be necessary that the thickness of the strip is not negligible for requirement of mechanical strength to support the central conductor when there is air-dielectric. The variations of Z_0 with different dimensional changes are shown in Fig. 3.29 for a wide strip enclosed in a shielded box and arbitrarily placed in the cross section.

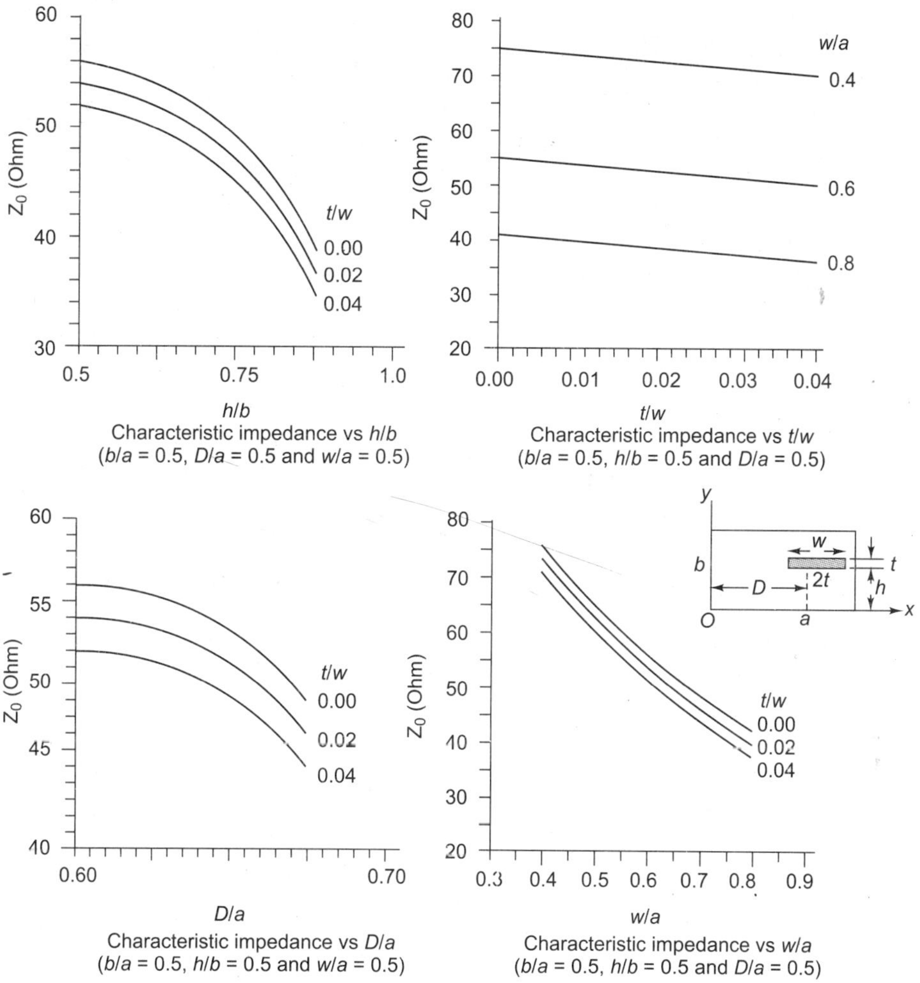

Fig. 3.29 Completely shielded strip line impedances

Figures 3.30 and 3.31 show electric field distributions in the cross section of a typical shielded strip line where the fields are nearly uniform within the range of the centre conductor. Since the propagation mode in a strip line is nearly TEM mode, the guide wavelength in the line is $\lambda_0/\sqrt{\varepsilon_r}$.

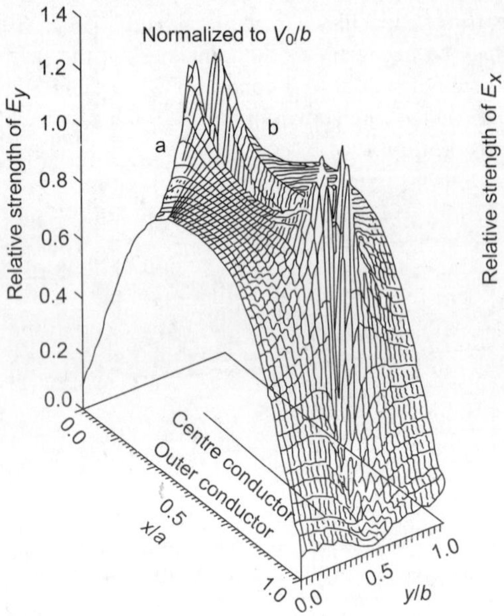

Fig. 3.30 Field distribution in the cross section of a shielded strip line ($h/b = 0.5$, $W/a = 0.6$)

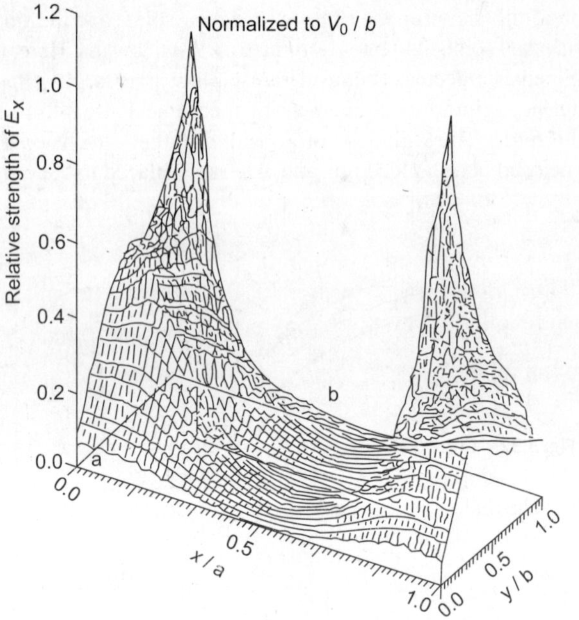

Fig. 3.31 Field distribution in the cross section of a shielded strip line ($h/b = 0.5$, $W/a = 0.6$)

Higher-order Modes in Strip Lines

The operating bandwidth of a strip line is considerably large for TEM mode of propagation. The upper frequency limit is set by the presence of the nearest higher-order TE_{10} and TM_{11} modes. The approximate expressions of cut-off wavelength for these modes are

$$\lambda_c(\text{TM}_{11}) = 2b\sqrt{\varepsilon_r} \tag{3.89a}$$

$$\lambda_c(\text{TE}_{10}) = \left(2w + \frac{\pi b}{2}\right)\sqrt{\varepsilon_r} \tag{3.89b}$$

For a considerably wide strip, the cut-off conditions can be estimated by analyzing the cross section using conformal transformation or numerical methods, viz., finite element method. The cut-off wavelengths of higher-order modes for typical shielded strip lines are shown in Fig. 3.32.

Losses in Strip Lines

For a low-loss dielectric substrate, the attenuation factor in the strip line arises from conductor losses and can be expressed by

$$\alpha_c = \frac{R_s}{Z_0 b}\,\frac{\dfrac{\pi w}{b} + \ln\left(\dfrac{4b}{\pi t}\right)}{\ln 2 + \dfrac{\pi w}{2b}} \text{ Np/m} \tag{3.90a}$$

$$w \geq 2b,\ t \leq b/10$$

$$\text{where, } R_s = \sqrt{\frac{\pi f \mu}{\sigma}} \tag{3.90b}$$

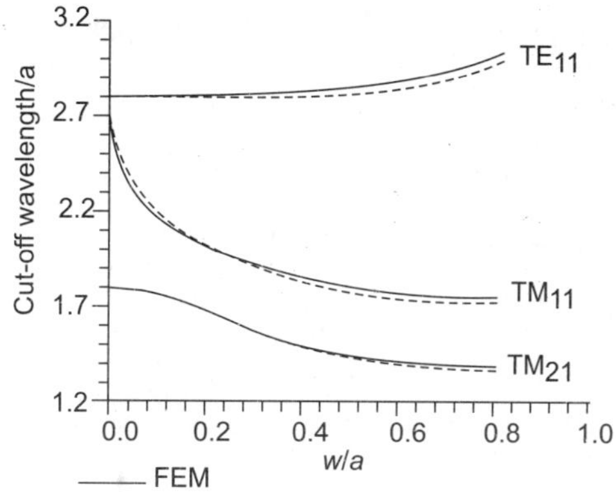

Fig. 3.32 Cut-off wavelength vs centre conductor width for the higher order TE and TM modes (b/a = 1.0, ε_r = 1)

The dielectric loss coefficient can be expressed as

$$\alpha_d = \frac{27.3 \tan\delta}{\lambda_g} \text{dB/m} \tag{3.90c}$$

Excitation of Strip Lines

Strip lines are excited by a coaxial line which interfaces the strip lines by means of a special launcher or connector. This connector consists of a thin, flat, small-conductor which forms the centre conductor of the co-axial line and two rectangular outer conductors joined together to the ground planes as shown in Fig. 3.33. Strip lines are suitable for the design of passive circuits but inconvenient for mounting active components. Hence, the concept of microstrip line come as described below.

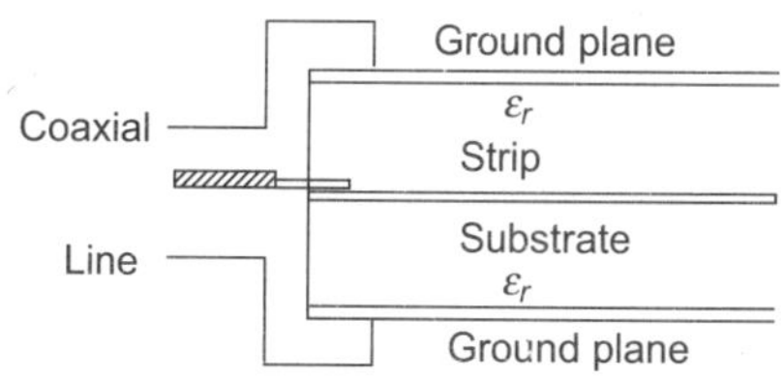

Fig. 3.33 Strip line launcher connector

Microstrip Lines 3.8.2

A microstrip line consists of a single ground plane and a thin strip conductor on a low-loss dielectric substrate above the ground plate. Since the size of the microwave solid-state devices is very small (of the order 0.008–0.08 mm^3), the technique of signal input to these devices and extracting output power from them uses microstrip lines on the surface on which they can be easily mounted. Figure 3.27(b) shows a typical cross section of a microstrip line. Due to absence of a top ground plate and the dielectric substrate above the strip, the electric field lines remain partially in the air and partially in the lower dielectric substrate. This makes the mode of propagation not pure TEM but what is called quasi-TEM. Due to open structure and any presence of discontinuity, the microstrip line radiates electromagnetic energy. The radiation loss is proportional to the square of the frequency. The use of thin and high dielectric materials reduces the radiation loss of the open structure where the fields are mostly confined inside the dielectric.

Effective Dielectric Constant

Since the propagation field lines in a microstrip lie partially in air and partially inside the homogenous dielectric substrate, the propagation delay time for a quasi-TEM mode is related to an effective dielectric constant ε_{eff}, given by

$$\varepsilon_{\text{eff}} = \frac{\varepsilon_r + 1}{2} + \frac{\varepsilon_r - 1}{2}\left[\left(1 + \frac{12h}{w}\right)^{-1/2} + 0.04\left(1 - \frac{w}{h}\right)^2\right]; w/h \le 1 \tag{3.91a}$$

$$\varepsilon_{\text{eff}} = \frac{\varepsilon_r + 1}{2} + \frac{\varepsilon_r - 1}{2}\left(1 + \frac{12h}{w}\right)^{-1/2} ; w/h \gg 1 \tag{3.91b}$$

where ε_r is the relative dielectric constant of the substrate material.

Characteristic Impedance and Guide Wavelength

The characteristic impedance of microstrip lines can be expressed by

$$Z_0 = \frac{60}{\sqrt{\varepsilon_{\text{eff}}}} \ln\left[\frac{8h}{w} + \frac{w}{4h}\right] \text{ ohm, } w/h \le 1 \tag{3.92a}$$

$$Z_0 = \frac{376.7}{\sqrt{\varepsilon_{\text{eff}}}\left[\frac{w}{h} + 1.4 + 0.667 \ln\left(\frac{w}{h} + 1.444\right)\right]} \text{ ohm; for } w/h > 1 \tag{3.92b}$$

$$Z_0 = \frac{376.7}{\sqrt{\varepsilon_{\text{eff}}}} \frac{h}{w} \text{ ohm; for } w/h \gg 1 \tag{3.92c}$$

The guide wavelength for the propagation of quasi-TEM mode is given by

$$\lambda_g = \frac{\lambda_0}{\sqrt{\varepsilon_{\text{eff}}}} \tag{3.92d}$$

Synthesis of Microstrip Lines

A microstrip line can be designed for given impedance Z_0 and substrate dielectric constant from the following formulae[1]:

$$\frac{W}{h} = \begin{cases} \dfrac{8e^A}{e^{2A} - 2} & \text{for } W/h < 2 \\ \dfrac{2}{\pi}\left[B - 1 - \ln(2B - 1) + \dfrac{\varepsilon_r - 1}{2\varepsilon_r}\left\{\ln(B - 1) + 0.39 - \dfrac{0.61}{\varepsilon_r}\right\}\right] & \text{for } W/h > 2 \end{cases} \tag{3.93a}$$

where

$$A = \frac{Z_0}{60}\sqrt{\frac{\varepsilon_r + 1}{2}} + \frac{\varepsilon_r - 1}{\varepsilon_r - 1}\left(0.23 + \frac{0.11}{\varepsilon_r}\right) \tag{3.93b}$$

$$B = \frac{377\pi}{2Z_0\sqrt{\varepsilon_r}} \tag{3.93c}$$

The variations of effective dielectric constant and the characteristic impedance vs w/h are shown in Fig. 3.34 and 3.35. A simple MATLAB programme for computation of $\varepsilon_{\text{reff}}$ and Z_0 is given in Table 3.3. It is seen that the value of Z_0 decreases with w/h and also with increase in ε_r.

1 Pozar D.M., " Microwave Engineering", second edition, John Wiley and Sons, Inc. 1999.

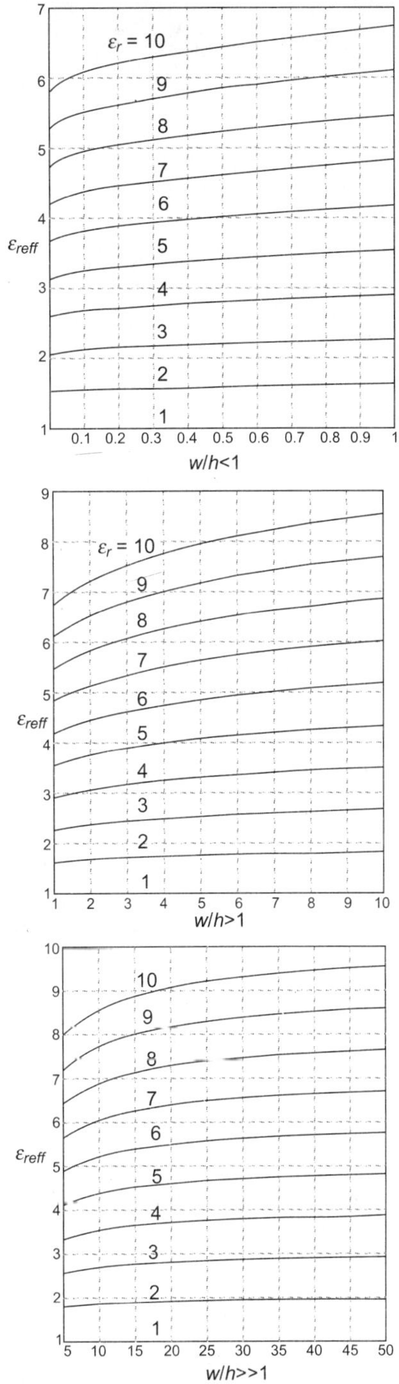

Fig. 3.34 Variations of ε_{reff} vs. w/h

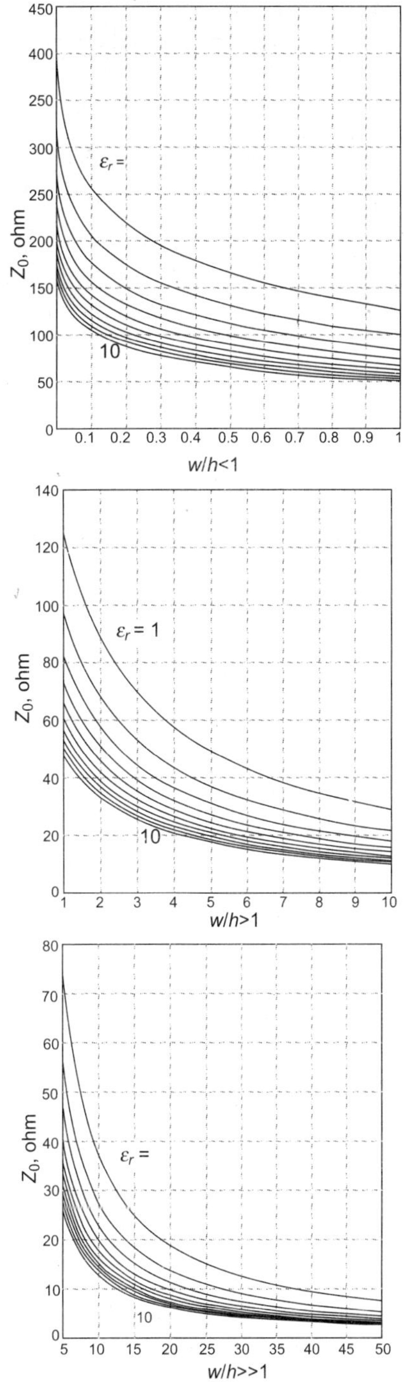

Fig. 3.35 Variations of Z_0 vs. w/h

Table 3.3 %Zo vs W/h for Microstrip lines W/h > 1

```
%**********************************************
for EPR = 1:1:10;
wn = 1:0.01:10;%wn = W/h
E2 = ((EPR+1)./2)+((EPR-1)./2).*(1+(12./wn)).^(-.5);% for W/h > 1
Z = (376.7./(sqrt(E2)))./(wn+1.4+0.667.*log(1.444+wn));% for W/h > 1
 subplot(1,2,1);
 plot(wn,E2,'r');%for W/h > 1
 hold on;
 xlim([1,10]);
 xlabel('W/h > 1');
 ylabel('epreff for W/h > 1');
 grid on;
 title('epreff vs W/h > 1 of microstrip line for different epr');


     subplot(1,2,2);
plot(wn,Z,'g');% for W/h > 1
hold on;
xlim([1,10]);
xlabel('W/h > 1');
ylabel('Zo for W/h > 1');
grid on;
title('Zo vs W/h > 1 of microstrip line for different epr');
end;
```

- The **power handling** capacity of a microstrip is limited by heating caused because of ohmic and dielectric losses and by dielectric breakdown.

  An increase in temperature due to conductor and dielectric losses limits the **average power** of the microstrip line, while the breakdown between the strip conductor and ground plane limits the *peak power*.

A metallic enclosure is normally required for most microstrip circuit applications, such as microstrip filters. The presence of conducting top and side walls will affect both, the characteristic impedance Z_c and the effective dielectric constant ε_{eff}.

- In practice, a rule of thumb may be applied in the microstrip filter design to reduce the effect of metallic enclosure: the height up to the cover should be more than eight times the substrate thickness, and the distance to walls more than five times the substrate thickness.

Losses in Microstrip Lines

For the nonmagnetic substrate material, two types of losses exist in microstrip lines which provide attenuation of signal:

1. dielectric loss in the substrate, and

2. ohmic loss in the strip conductor and the ground plane due to finite nonzero conductivity.

The total attenuation constant α can be expressed as $\alpha = \alpha_d + \alpha_c$ where α_d, α_c are the dielectric and ohmic attenuation constants, given by

$$\alpha_d = \frac{\sigma_d}{2}\sqrt{\frac{\mu}{\varepsilon}} = 27.3\frac{(\varepsilon_{\text{eff}}-1)}{(\varepsilon_r-1)}\frac{\varepsilon_r}{\varepsilon_{\text{eff}}}\cdot\frac{\tan\delta}{\lambda_g}\text{dB/m} \tag{3.94}$$

Here, σ_d is the conductivity of the dielectric substrate and $\tan\delta = \dfrac{\sigma_d}{\omega\varepsilon''}$ is the dielectric loss tangent.

For a low-loss dielectric substrate, the major attenuation factor at microwave frequencies arises due to finite conductivity of the strip conductor. This contributes to the ohmic losses due to the current on the strip. The current distribution in the transverse plane is fairly uniform with minimum value at the central axis and shoots up to a maximum at the edges of the strip.

For simplicity, assuming uniform current distribution in the region $-W/2 < x < W/2$, the attenuation constant due to ohmic loss of a wide line $(W/h > 1)$ is

$$\alpha_c \approx \frac{8.686}{Z_0 W}\sqrt{\frac{\pi f \mu}{\sigma}}\ \text{dB/m} \tag{3.95}$$

From Wheeler's work, the values of α_c for a wide range of W/h are plotted in Fig. 3.35. Microstrip lines also have radiation loss due to their thin open structure and any current discontinuities in the strip conductor.

The commonly used substrate materials for strip and microstrip lines are polytetrafluoroethylene (PTFE)/Teflon, RT Duroid 5880, alumina, and sapphire, etc., whose electrical characteristics are given in Table 3.4. Alumina is most widely used for frequencies up to 20 GHz. At higher frequencies, sapphire is used.

Table 3.4 Properties of substrate materials

Material	ε_r	*tan δ*
PTFE/glass	2.2	0.0002–0.0005
RT/Duroid 5880	2.26	0.001
Alumina	9.6 – 10	0.0002 0.0005
Sapphire	9.4	0.0001
GaAs	11–13	0.0016
Si	12	0.015

Excitation of Microstrip Line

The radiation loss of microstrip lines is eliminated by enclosing the microstrip within a metallic box having the first resonance frequency much above the signal frequency. Microstrip lines are excited by a coaxial line through a connector launcher as shown in Fig. 3.36. Microstrip lines are suitable for the design of passive circuits and series mounting of active components across a gap in the strip.

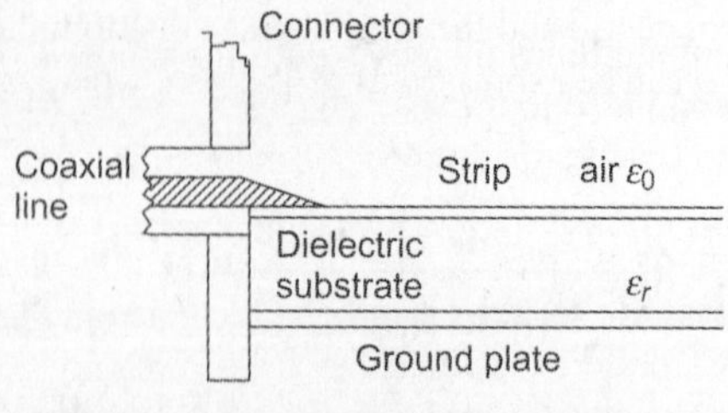

Fig. 3.36 Microstrip line connector

Slot Lines 3.8.3

A truly planar configuration used as transmission line in MICs includes slot lines as shown in Fig. 3.37(a), where the conductors are in one plane on a dielectric substrate. The slot line was first described by S. B. Cohn, and subsequently by others. Unlike strip lines and microstrip lines, active and passive components can easily be shunt mounted across the lines from the top. The propagating fields are concentrated in the dielectric regions at the gap between the two adjacent conductors. In contrast to the microstrip case, the slot line is a full planar transmission line since there is no bottom ground plane. The ground plane and the signal line plane are coplanar. The slot line has a zero cutoff frequency but does not support a **TEM** mode. Although the electric field components are transverse, the magnetic field has both longitudinal and transverse components as shown in Fig. 3.37(b) and 3.37(c). Thus, the fundamental propagating mode is called **quasi-TEM** having zero cut-off frequency. The magnetic field is found circularly polarized in the longitudinal plane at periodical locations where ferrite could be utilized to design ferrimagnetic devices such as ferrite isolators.

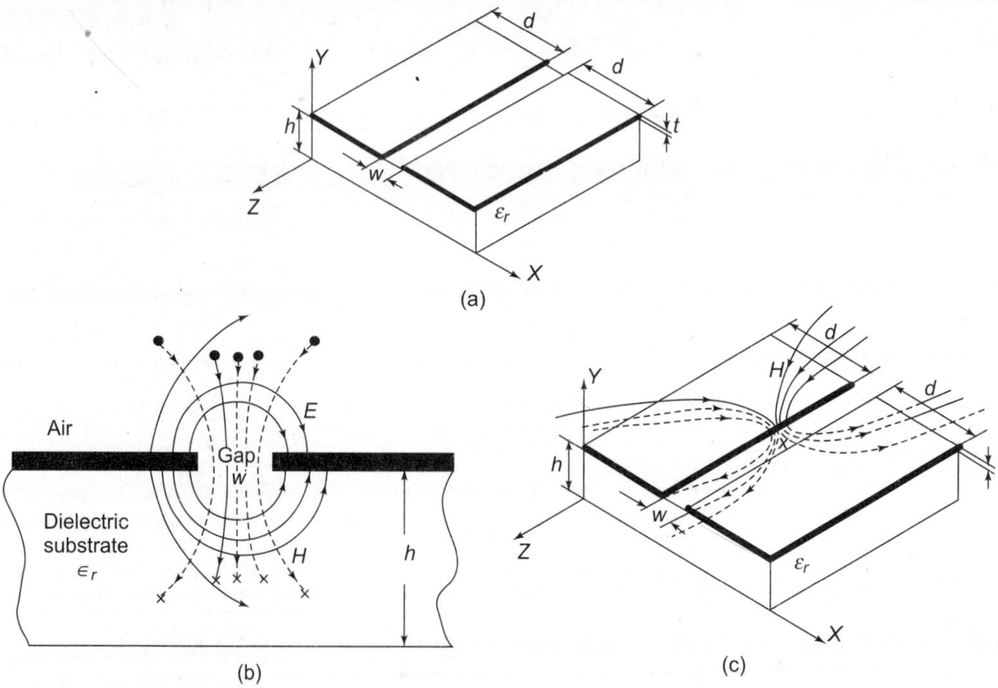

Fig. 3.37 Planar Transmission lines (a) Slot line, (b) **E** and **H** field lines in slot line cross-section (c) isometric view of slot line fields

The characteristic impedance Z_0 of slot lines increases with the increase of width of the slot. The Z_0 also varies rapidly with frequency as compared to microstrip lines.

Since there is no bottom ground plane, the number of the electric and magnetic field lines in the air is higher than the number of the same lines in air for the microstrip case. The effective dielectric constant of slot line is lower than in the microstrip. Consequently, the maximum realizable characteristic impedance values are higher than in the microstrip case. To reduce the electromagnetic radiation loss in the air, substrates with high dielectric constants (>10) are preferred so that the field is mainly concentrated inside the dielectric substrate. Instantaneous electric and magnetic field lines for the fundamental mode in the slot line are shown in Fig. 3.38 in a cross section.

Since the ground and the signal planes are placed side by side, magnetic field lines cannot be closed around conductors. They are distorted and moving into the plane of the figure at points marked **cross** [Fig. 3.37(b)] from the centre of the slot, and coming out from the plane of the figure at points marked **dot**. A 3D representation of the magnetic field lines for the fundamental mode is shown in Fig. 3.37(c).

SB Cohn reported that the intensity of H_z and H_y vs. y in the centre of the slot shows maximum and since H_z and H_y are never equal in magnitude,. the resulting magnetic field is never perfectly circular polarized, but elliptically polarized.

According to Cohn, the effective relative permittivity of the slot line is

$$\varepsilon_{\text{reff}} = (\varepsilon_r + 1)/2 \tag{3.96}$$

Closed-form approximate expressions for slot wavelength and characteristic impedance for a slot line are presented by Ramesh Garg and K.C. Gupta having an accuracy of about 2 percent for substrate permittivity ranging between 9.7 and 20. These are obtained by curve fitting the numerical results, based on Cohn's analysis and are given as follows.

In the ranges

$$9.7 \le \varepsilon_r \le 20 \qquad 0.02 \le w/h \le 1.0 \qquad 0.01 \le h/\lambda \le (h/\lambda)_0 \qquad (h/\lambda)_0 = 0.25/\sqrt{\varepsilon_r - 1}$$

where $(h/\lambda)_0$ is equal to the cutoff for the TE_{10} surface-wave mode on the slotline.

1. For 0.02 <, $w/h < 0.2$:

$$\lambda_g/\lambda = 0.923 - 0.448 \log \varepsilon_r + 0.2\, w/h - (0.29\, w/h + 0.047) \log (h/\lambda \times 10^2) \tag{3.97a}$$

$$\begin{aligned} Z_0 &= 72.62 - 35.19 \log \varepsilon_r + 50 \frac{(w/h - 0.02)(w/h - 0.1)}{w/h} \\ &+ \log (w/h \times 10^2)\,[44.28 - 19.58 \log \varepsilon_r] \\ &- [0.32 \log \varepsilon_r - 0.11 + \mathrm{w/h}\,(1.07 \log \varepsilon_r + 1.44)]\,(11.4 - 6.07 \log \varepsilon_r - \mathrm{h}/\lambda \times 10^2)^2 \end{aligned} \tag{3.97b}$$

2. For $0.2 \le w/h \le 1.0$

$$\begin{aligned} \lambda_g/\lambda &= 0.987 - 0.483 \log \varepsilon_r + w/h\,(0.111 - 0.0022\, \varepsilon_r) \\ &- (0.121 + 0.094 w/h - 0.0032\, \varepsilon_r) \log (h/\lambda \times 10^2); \end{aligned} \tag{3.97c}$$

$$\begin{aligned} Z_0 &= 113.19 - 53.55 \log \varepsilon_r + 1.25 w/h\,(114.59 - 51.88 \log \varepsilon_r) + 20(w/h - 0.2)(1 - w/h) \\ &- [0.15 + 0.23 \log \varepsilon_r + w/h\,(-0.79 + 2.07 \log \varepsilon_r)] \\ &[(10.25 - 5 \log \varepsilon_r + w/h\,(2.1 - 1.42 \log \varepsilon_r) - h/\lambda \times 10^2)^2] \end{aligned} \tag{3.97d}$$

Elio A Mariani, et. al., analyzed the other slot line parameters such as relative slot wavelength ratio λ_g/λ_0 and the characteristic impedance Z_0 and obtained results as a function of slot width w, substrate thickness h, dielectric constant $\in_r$ and frequency f. Based on this work Z_0 and λ_g/λ_0 vs h/λ plotted for different values of w/h as shown in Fig. 3.38. It is observed that the λ_g/λ_0 for different w/h ratios converge in the neighborhood of which corresponds to the thickness of the substrate as a quarter-wavelength in the dielectric. For larger h/λ ratios, surface-wave propagation could become a problem. The impedance variation with frequency is relatively small and presented similarly for the dielectric constant 11.

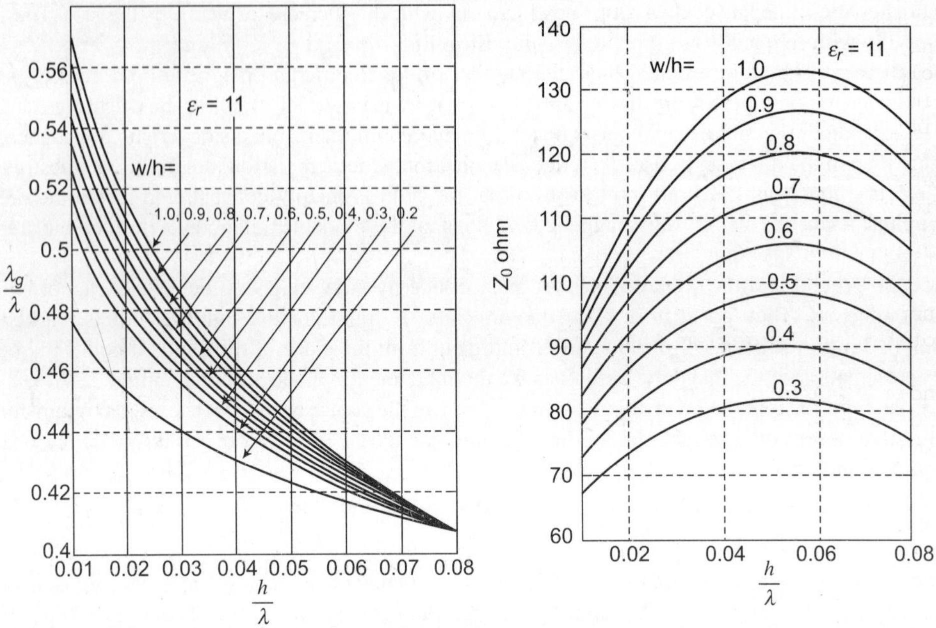

Fig 3.38 Variation of normalized slot wavelength (λ_g/λ) and Z_0 as function of normalized thickness of the substrate (h/λ) for different slot width (w/h) for a substrate of dielectric constant 11

```
%*****************************************************
% %SLOT line Zo and lamg vs h/lam for different W/h
% %*****************************************************
 clear all;
epr = 11;
hbylam0 = 0.25/sqrt(epr-1);
for wbyh = 0.1:0.1:1.0;
 hbylam = 0.01:0.001:0.08;
 if (wbyh < = 0.2)

lamgbylam = 0.923-0.448*log10(epr)+0.2*wbyh-(0.29*wbyh+0.047).*...
 log10(hbylam.*100);

z0 = 72.62-35.19*log10(epr)+50-(((wbyh-0.02).*(wbyh-0.1))./wbyh)+...
 (log10(wbyh.*100)).*(44.28-19.58*log10(epr))-(0.32*log10(epr)-...
 0.11+wbyh.*(1.07*log10(epr)+1.44)).*(11.4-6.07*log10(epr)-...
 hbylam.*100).^2;
 else
lamgbylam = 0.987-0.483*log10(epr)+wbyh.*(0.111-0.0022*epr)-
(0.121+0.094*...
 wbyh-0.0032*epr).*log10(hbylam.*100);
z0 = 113.19-53.55*log10(epr)+1.25*wbyh.*(114.59-51.88*log10(epr))+20.*...
 (wbyh-0.2).*(1-wbyh)-(0.15+0.23*log10(epr)+wbyh.*(-0.79+2.07*...
 log10(epr))).*((10.25-5.*log10(epr)+wbyh.*(2.1-1.42*log10(epr))-...
 hbylam.*100).^2);
subplot(1,2,1);
```

```
plot(hbylam,z0,'r');% for different W/h
 end;
 hold on;
xlim([0.01,0.08]);
ylim([60,140]);
xlabel('h/lam');
ylabel('Zo');
grid on;

title('Zo vs h/lam of slot line for different w/h');

 subplot(1,2,2);
plot(hbylam,lamgbylam,'b');%for diff hbylam

 end;
hold on;
xlim([0.01,0.08]);
ylim([0.4,0.58]);
xlabel('h/lam');
ylabel('lamg/lam');
title(' lamg/lam vs h/lam of slot line for different w/h');
grid on;
```

Excitation of Slot Line

Slot line can be excited with co-axial line probe on the plane of the slot line or with a microstrip line from the opposite side of the substrate as explained in Fig. 3.39 (a)–(c). The outer conductor of the coaxial cable should be connected to one plane as ground of the slot line. The centre conductor of the coaxial cable is connected to the other plane as signal line. In microstrip excitation, on one side there is the microstrip, while on the other side, there is the slot line. The microstrip is open terminated at a distance approximately a quarter of wavelength long after the crossing. The end of slot line is short terminated at a distance of corresponding quarter wavelength from the common overlap region indicating total phase shift of 360° in E-field yielding maximum E-field excitation at the common overlap region of excitation.

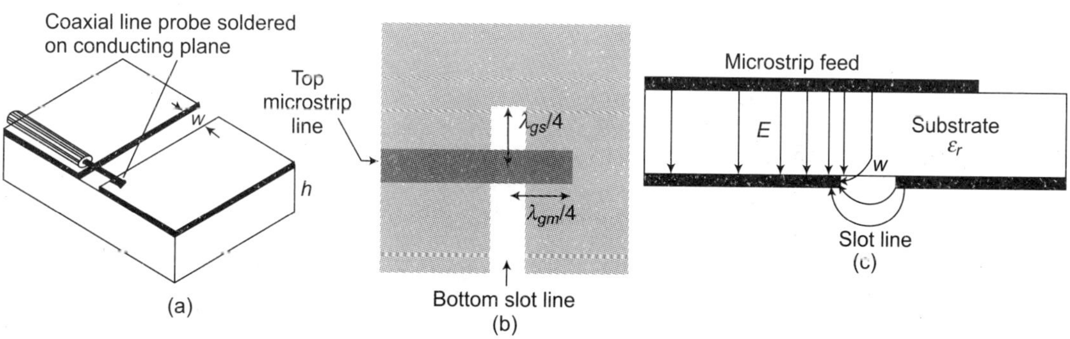

Fig. 3.39 Excitation of slot line

Coplanar Waveguide 3.8.4

A coplanar waveguide is a parallel three-conductor line consisting of a thin centre strip (w) and two coplanar thin ground strips (G) on a solid dielectric substrate is shown in Fig. 3.40. Coplanar lines are very convenient for both series and shunt mounting of components across the gap between the centre strip and the ground conductors. Like slot lines, the propagating modes in coplanar lines are TE modes and ferrite components can be designed due to existence of circular polarization of the magnetic field at intervals of half wavelength locations.

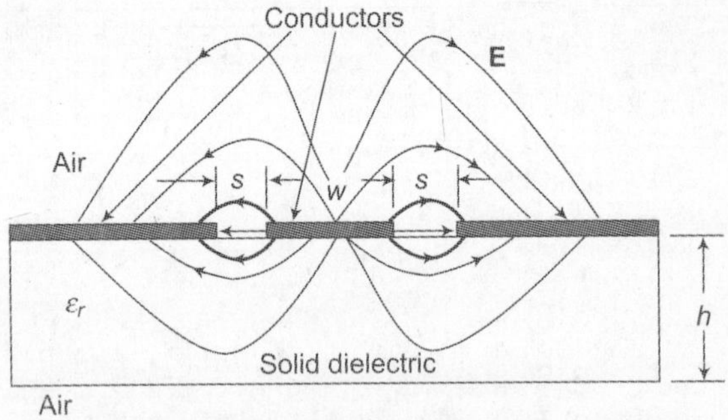

Fig. 3.40 Coplanar waveguide (CPW)

A quasi-static analysis of CPW can be done by the use of conformal mappings K.C. Gupta et. al. and R.E. Collin. With that assumption, the capacitance per unit length is the sum of the top (air filled) and bottom (dielectric filled) partial capacitances,

$$C_a = 2\cdot\varepsilon_0\frac{K(k_1)}{K'(k_1)};\quad C_d = 2\cdot\varepsilon_0\cdot\varepsilon_r\cdot\frac{K(k_1)}{K'(k_1)} \tag{3.98}$$

respectively.

Here, $K(k)$ and $K'(k)$ represent the complete elliptic integral of the first kind and its complement, and $k_1 = \frac{w}{w+2s}$. The ratio K/K' can be computed through die following formulae:

$$\frac{K(k)}{K'(k)} = \frac{\pi}{\ln\left(2\frac{1+\sqrt{k'}}{1-\sqrt{k'}}\right)} \quad \text{for} \quad 0\le k\le\frac{1}{\sqrt{2}} \tag{3.99}$$

$$\frac{K(k)}{K'(k)} = \frac{\ln\left(2\frac{1+\sqrt{k}}{1-\sqrt{k}}\right)}{\pi} \quad \text{for} \quad \frac{1}{\sqrt{2}}\le k\le 1 \tag{3.100}$$

with k' being the complementary modulus: $k' = \sqrt{1-k^2}$. The total line capacitance is, thus, the sum of C_d and C_a. The effective permittivity is, therefore,

$$\varepsilon_{\text{reff}} \cong \frac{\varepsilon_r+1}{2} \tag{3.101}$$

and the impedance

$$Z_0 = \frac{30\pi}{\sqrt{\varepsilon_{\text{reff}}}}\cdot\frac{K'(k_1)}{K(k_1)} \tag{3.102}$$

The impedance of the CPW vs w/s is shown in Fig. 3.41.

In real cases, the substrate has a finite thickness h and the effective permittivity is altered to

$$\varepsilon_{\text{reff}} = 1+\frac{\varepsilon_r-1}{2}\cdot\frac{K(k_2)}{K'(k_2)}\cdot\frac{K'(k_1)}{K(k_1)} \tag{3.103a}$$

$$\text{where } k_1 = \frac{w}{w+2s} \text{ and } k_2 = \frac{\sinh\left(\frac{\pi w}{4h}\right)}{\sinh\left(\frac{\pi\cdot(w+2s)}{4h}\right)} \tag{3.103b}$$

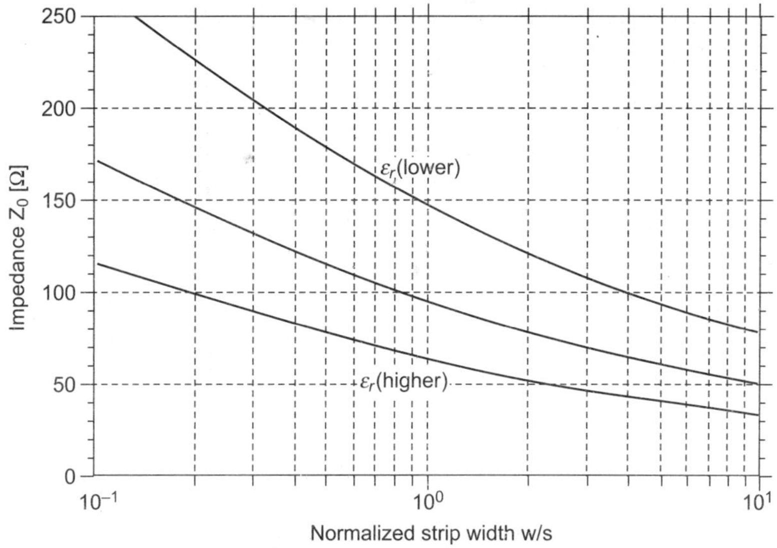

Fig. 3.41 Impedance of CPW vs w/s

To avoid microstrip line quasi TEM modes, it is recommended that $h \gg w + 2s$ and that the signal-line-side ground planes extend away from the track on each side much more than $w + s$ such that there is no field beyond this end and the bottom surface of the substrate. Under this condition, there could be a ground plane on the bottom surface of the substrate which does not affect the field distribution of TE modes.

Coplanar Waveguide with an Infinite Ground Plane at the Bottom

The characteristic impedance and effective dielectric constant of coplanar waveguide with an infinite ground plane at the bottom of the substrate is given by Brian C Wadell [10] and is expressed by

$$Z_0 = \frac{60\pi}{\sqrt{\varepsilon_{\text{reff}}}}\left[\frac{K(k)}{K(k')} + \frac{K(k_1)}{K(k_1')}\right]^{-1} \tag{3.104a}$$

$$\varepsilon_{\text{reff}} = \left\{1 + \varepsilon_r\left[\frac{K(k')}{K(k)} + \frac{K(k_1)}{K(k_1')}\right]\right\}\left\{1 + \frac{K(k')}{K(k)}\frac{K(k_1)}{K(k_1')}\right\}^{-1} \tag{3.104b}$$

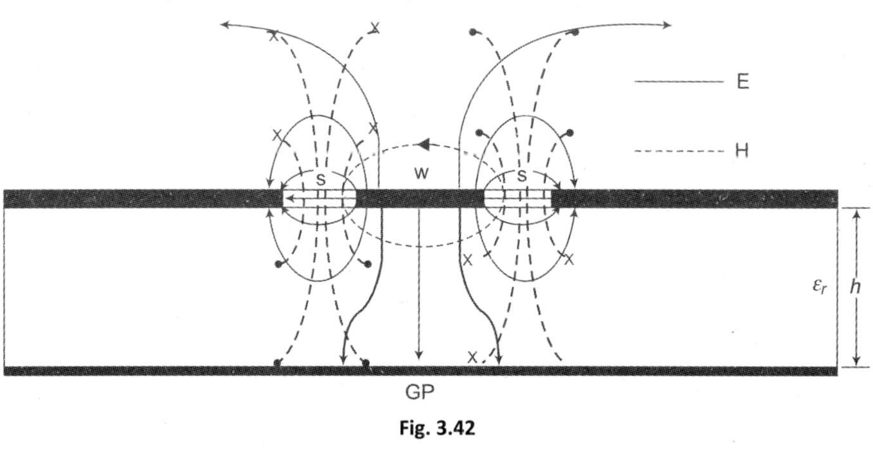

Fig. 3.42

Here,

ε_r = Dielectric constant of the substrate
s = Width of gap between signal and ground traces
w = Width of signal line
h = Thickness of dielectric

$$k = w/(w + 2s) \tag{3.105a}$$

$$k' = \sqrt{1-k^2} \tag{3.105b}$$

$$k_1 = \frac{\tanh\left(\dfrac{\pi w}{4h}\right)}{\tanh\left(\dfrac{\pi(w+2s)}{4h}\right)} \tag{3.105c}$$

$$k_1' = \sqrt{1-k_1^2} \tag{3.105d}$$

$K(k)$ = Elliptical integrals of the first kind.

Advantages and Disadvantages of Planar Transmission Lines 3.8.5

The basic advantages of planar transmission lines are (i) small size and weight, (ii) can be flash mounted on a metallic body, (iii) increased reliability, (iv) low cost, (v) easy access for component mounting, (vi) Z_0 can be controlled by defining the dimensions in a single plane, and (vii) passive circuit design is possible easily by changing the dimensions of the line in one plane only.

There are certain disadvantages in these lines, one of which is the low power-handling capability due to small size. There are also radiation losses from the open structures like microstrip, slot and coplanar lines. This loss is reduced by confining the field lines more in the dielectric substrate of a high dielectric constant, viz.

alumina ($\varepsilon_r = 9 - 10$),

sapphire ($\varepsilon_r = 9.3 - 11.7$), and

beryllia ($\varepsilon_r = 6.6$).

Strip lines also use polystyrene ($\varepsilon_r = 2.54$) as substrate. The circuit design using these configurations should be very accurate since matching screws and short-circuit plungers cannot be used in planar transmission line circuits. The design of some active devices is also limited because of low Q (of the order 100) obtainable with ordinary microstrip configurations.

Example 3.11 A microstrip line uses a substrate with $\varepsilon_r = 9.7$ (alumina) and $d = 0.5$ mm. The strip width is $w = 0.5$mm. Find $\varepsilon_{\text{reff}}$, Z_0 and the microstrip wavelength at $f = 2$ GHz.

Solution

Polar:

$$\varepsilon_{\text{reff}} = \frac{\varepsilon_r+1}{2} + \frac{\varepsilon_r-1}{2} \cdot \frac{1}{\sqrt{1+12\dfrac{d}{w}}}$$

$$= \frac{9.7+1}{2} + \frac{9.7-1}{2} \cdot \frac{1}{\sqrt{1+12}} = 6.556$$

$$Z_0 = \frac{60}{\sqrt{\varepsilon_{\text{reff}}}} \ln\left(\frac{8d}{w} + \frac{w}{4d}\right)$$

$$= \frac{60}{\sqrt{6.556}} \ln\left(8 + \frac{1}{4}\right) = 49.449\ \Omega$$

$$\lambda = \frac{2\pi}{\beta} = \frac{2\pi}{k_0\sqrt{\varepsilon_{\text{reff}}}} = \frac{\lambda_0}{\sqrt{\varepsilon_{\text{reff}}}}$$

$$\lambda_0 = \frac{c}{f} = \frac{3\times10^8}{2\times10^9} = 15 \text{ cm}$$

$$\lambda = \frac{\lambda_0}{\sqrt{\varepsilon_{\text{reff}}}} = \frac{15}{\sqrt{6.556}} = 5.858 \text{ cm}$$

Example 3.12 Consider the microstip line in the above example. Assume tan $\delta = 2 \times 10^{-4}$ and t = 0.2 mm. If f = 4 GHz, calculate the total attenuation (assume copper).

Solution

$$\varepsilon_{\text{reff}} = 6.556,\ Z_0 = 49.449\ \Omega,\ \lambda_0 = 1.5 \text{ cm}$$

$$\alpha_d = \frac{\pi}{\lambda_0}\frac{\varepsilon_r}{\sqrt{\varepsilon_{\text{reff}}}}\left(\frac{\varepsilon_{\text{reff}}-1}{\varepsilon_r-1}\right)\tan\delta = \frac{\pi}{7.5\sqrt{6.556}}\left(\frac{6.556}{9.7-1}\right)2\times10^{-4} = 2.02\times10^{-4} \text{ (Np/cm)}$$

$$\alpha_c = \frac{R_s}{Z_0 w}$$

$$R_s = \sqrt{\frac{\omega\mu}{2\sigma}} = \sqrt{\frac{\pi f \mu_0}{\sigma}} = \sqrt{\frac{\pi\times4\times10^9\times4\pi\times10^{-7}}{5.7\times10^7}} = 16.64\times10^{-3}\Omega$$

$$\alpha_c = \frac{16.64\times10^{-3}}{49.449\times0.5\times10^{-3}} = 0.673 \text{ Np/m} = 67.3\times10^{-4}\text{(Np/cm)}$$

$$\alpha = \alpha_c + \alpha_d = 69.32\times10^{-4} \text{ (Np/cm)}$$

METAL WAVEGUIDES 3.9

Introduction 3.9.1

Since the losses in two conductor lines increase with frequency, and the power handling capacity of MIC lines described above are low, high power microwave signals are transmitted through hollow metallic tubes, called waveguides, of uniform cross section as shown in Fig. 3.43. Signal is propagated as time harmonic electromagnetic waves ($e^{j\omega t}$) having multiple field components. Since the propagation medium inside the waveguides is normally air and the conductivity of the metal tube is very large for a very good conductor, such as Cu, (ideally assumed ∞), the signal propagates with minimum loss. Waveguides are formed with different cross-sections – rectangular, circular, and elliptical and correspondingly they are called rectangular, circular, elliptical waveguides, respectively.

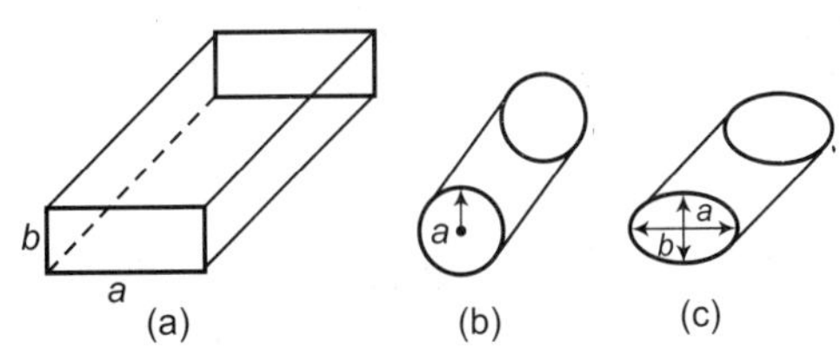

Fig. 3.43 Waveguides: (a) Rectangular (b) Circular (c) Elliptical

All waveguides have a number of common properties. Most important property is that they support an infinite number of field configurations, called **modes**. These are described below.

Rectangular Waveguides 3.9.2

A rectangular waveguide is a simplest waveguiding structure which is widely used and is shown in Fig. 3.44. The guide is fed at one end through a co-axial to waveguide adaptor where the inner conductor is projected inside the guide as a probe. The signal-current discontinuity in the probe end results in radiation of fields inside the guide to propagate along the length. The fields inside the guide can be obtained from the solutions to the wave equation which satisfies the boundary conditions inside the guide.

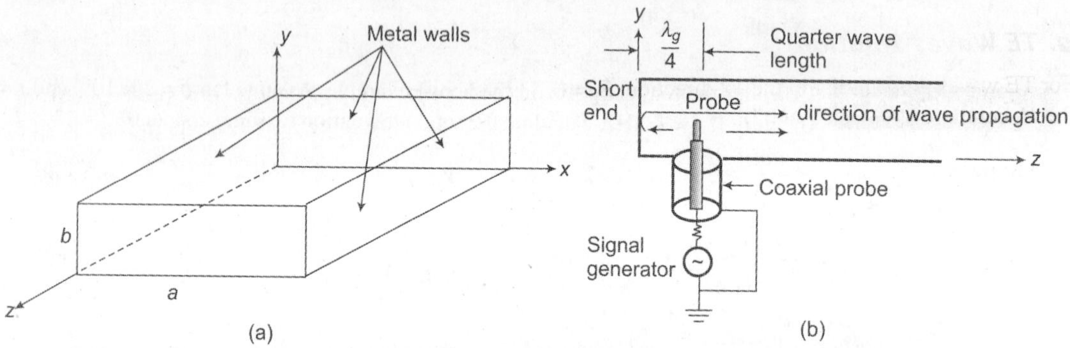

Fig. 3.44 Rectangular waveguide and its excitation by coaxial probe

To simplify the analysis, a rectangular waveguide of infinite length and field with a material (μ, ε, $\delta = 0$) is considered to find the fields in the regions away from the exciting probe so that the source-free wave equation can be used in the solution region inside the guide.

The time harmonic ($e^{j\omega t}$) fields inside the guide satisfy the Maxwell's equations

$$\nabla x\mathbf{E} = -\frac{\partial \mathbf{B}}{\partial t} = -j\omega\mu\mathbf{H} \text{ and } \nabla x\mathbf{H} = -\frac{\partial \mathbf{D}}{\partial t} = -j\omega\varepsilon\mathbf{E} \tag{3.106}$$

Since the wave is propagating in the z-direction, z-dependent solution for the fields is a travelling wave represented by the solution $e^{-\gamma z}$, where $\gamma = \alpha + j\beta$ is the propagation constant. These two curl equations can be written in terms of the x, y, and z components of the **E** and **H** fields. For simplification mathematical manipulations are used to express the transverse components of electric and magnetic fields in terms of the longitudinal components as follows:

$$E_x = -\frac{\gamma}{k_c^2}\frac{\partial E_z}{\partial x} - \frac{j\omega\mu}{k_c^2}\frac{\partial H_z}{\partial y} \tag{3.107a}$$

$$E_y = -\frac{\gamma}{k_c^2}\frac{\partial E_z}{\partial y} + \frac{j\omega\mu}{k_c^2}\frac{\partial H_z}{\partial x} \tag{3.107b}$$

$$H_x = \frac{j\omega\varepsilon}{k_c^2}\frac{\partial E_z}{\partial y} - \frac{\gamma}{k_c^2}\frac{\partial H_z}{\partial x} \tag{3.107c}$$

$$H_y = -\frac{j\omega\varepsilon}{k_c^2}\frac{\partial E_z}{\partial x} - \frac{\gamma}{k_c^2}\frac{\partial H_z}{\partial y} \tag{3.107d}$$

where

$$k_c^2 = \gamma^2 + \omega^2\mu\varepsilon \tag{3.107e}$$

Equations (3.107a–e) suggest that two basic sets of solutions exist, each defining a set of modes. Thus, there are two modes of propagation along z inside a hollow metal waveguide:

❑ *TE Modes* Transverse electric (TE) waves-in which the **E** field is wholly transverse ($E_z = 0$), and

❑ *TM Modes* Transverse magnetic (TM) waves in which the **H** field is wholly transverse ($H_z = 0$)

❑ *TEM Modes* Another mode, transverse electromagnetic (TEM) mode ($E_z = 0$, $H_z = 0$, and $k_c^2 = \gamma^2 + \omega^2\mu\varepsilon = 0$), which is possible in coaxial lines, is not possible in metal waveguides since to support this mode it requires an axial conductor for axial current flow as per Ampere's law or an axial displacement current (axial E field). Both are not present in metal wave guides.

a. TE Waves Solution

For TE waves propagating in the $+z$-direction, $E_z = 0$. Therefore, the field components may be derived from the H_z component of Eqs. (3.107):

$$E_x = (-)\frac{j\omega\mu}{k_c^2}\cdot\frac{\partial H_z}{\partial y} \tag{3.108a}$$

$$E_y = (+)\frac{j\omega\mu}{k_c^2}\frac{\partial H_z}{\partial x} \tag{3.108b}$$

$$H_x = (-)\frac{\gamma}{k_c^2}\frac{\partial H_z}{\partial x} \tag{3.108c}$$

$$H_y = (-)\frac{\gamma}{k_c^2}\frac{\partial H_z}{\partial y} \tag{3.108d}$$

At every point in the waveguide, the longitudinal component of magnetic field H_z satisfies the scalar wave equation

$$\nabla^2 H_z + k^2 H_z = 0 \tag{3.109}$$

where $k = \omega\sqrt{\mu\varepsilon}$ is called the wave number of the propagating medium (μ, ε) inside the waveguide and

$$\nabla^2 = \frac{\partial^2}{\partial x^2} + \frac{\partial^2}{\partial y^2} + \frac{\partial^2}{\partial z^2} \tag{3.110}$$

is the Laplacian operator. Since the z-dependent solution is of the form $e^{-\gamma z}$,

$$\frac{\partial^2}{dz^2}e^{-\gamma z} = \gamma^2 e^{-\gamma z} \tag{3.111}$$

Substituting (3.111) into (3.109), we get

$$\nabla_t^2 H_z + k_c^2 H_z = 0 \tag{3.112}$$

where $\nabla_t^2 = \frac{\partial^2}{\partial x^2} + \frac{\partial^2}{\partial y^2}$, is the transverse Laplacian operator.

The solutions of (3.112) must satisfy the boundary conditions imposed by the metal walls only for some discrete values of k_c, called the **eigenvalues**. For an infinitely high conducting metal (ideal assumption), each eigenvalue is real and independent of frequency. Solving for (3.107e),

$$\gamma = \alpha + j\beta = \sqrt{k_c^2 - \omega^2\mu\varepsilon} = jk\sqrt{1-(f_c/f)^2} \text{ per m} \tag{3.113a}$$

where $k = \omega\sqrt{\mu\varepsilon}$ and $f_c = \dfrac{k_c}{2\pi\sqrt{\mu\varepsilon}}$ Hz (3.113b)

From (3.113), three special cases are seen:

❑ *Propagation* When $f > f_c$, the propagation constant γ becomes imaginary ($= j\beta$) so that the wave will propagates along the length (z) and is represented by

$$e^{-\gamma z} = e^{-j*\text{realfactor}*z} = e^{-j\beta z}$$

and the wave propagation is called **propagating mode.**

❑ *Cut-off* When $f = f_c$, $\gamma = 0$ and no propagation takes place. The parameter f_c is called the ***cut-off frequency*** given by (3.113b).

❑ *Evanescent* When $f < f_c$, γ becomes real ($= \alpha$), which means that the fields decay exponentially as $e^{-\alpha z}$ with increasing values of z. The f_c represents the lowest frequency below which no propagation takes place in the guide and the fields generated at the source point or exciting point get highly attenuated within a distance of a wavelength. The mode of this type is called **nonpropagating modes** or **evanescent modes** generated near the excitation and disc down as thin even $\lambda_g/4$ distance.

❑ *Solution to the Wave Equation* Since the conducting walls of the waveguide are parallel to the coordinate axis, the solution of the wave equation (3.112) can be obtained by separation of variable method in which the solution can be written as the product of functions, each depending on a single variable, satisfying the boundary conditions on the walls:

$$H_z(x, y, z) = X(x)Y(y)Z(z) = X(x)Y(y)e^{-\gamma z} \tag{3.114}$$

since the z-dependent solution, $Z(z) = e^{-\gamma z}$, represents wave propagation along the length of the waveguide in the direction z. Substituting (3.114) into (3.112) and dividing both sides of the resultant equation by $X(x)$ $Y(y)$ e^{-yz}, we obtain

$$\frac{1}{X}\frac{d^2X}{dx^2} + \frac{1}{y}\frac{d^2y}{dy^2} + k_c^2 = 0 \tag{3.115}$$

Since the RHS of (3.115) is zero, the term $\dfrac{1}{X}\dfrac{d^2X}{dx^2}$ only is a function of x, and none of the other terms are functions of x. Therefore, the first term of (3.115) is a constant. In a similar way, the second term of (3.115) only is a function of y and is another constant. Choosing these constants as $-k_x^2$, and $-k_y^2$, respectively, (3.115) yields

$$-k_x^2 - k_y^2 + k_c^2 = 0$$

or

$$k_c^2 = k_x^2 + k_y^2 \tag{3.116a}$$

and

$$\frac{d^2X}{dx^2} + k_x^2X = 0 \tag{3.116b}$$

$$\frac{d^2Y}{dy^2} + k_y^2Y = 0 \tag{3.116c}$$

Equations (3.116b) and (3.116c) both represent homogeneous, second-order differential equations in a single variable. Their general solutions are

$$X(x) = A\cos k_x x + B\sin k_x x \tag{3.117a}$$

$$Y(y) = C\cos k_y y + D\sin k_y y \tag{3.117b}$$

where A, B, C, and D are unknown constants to be determined from the boundary conditions.

Thus, the general solution of (3.112) becomes

$$H_z(x,y,z) = [A\cos k_x x + B\sin k_x x][C\cos k_y y + D\sin k_y y]e^{-\gamma z} \tag{3.118}$$

The unknown constants to be determined from the boundary conditions that **tangential components of electric field** and the **normal component of magnetic field** on the surface of the waveguide walls (assuming perfect conductor) are zero when the walls are free of current source (**J** = 0). In this case for TE mode, solutions for all field components have to be derived from H_z as per (3.118) and from (3.108a)–(3.108d). For TE mode, the tangential components of electric field on the waveguide walls:

$$y = 0, b; \qquad E_x = 0 = -\frac{j\omega\mu}{k_c^2}\frac{\partial H_z}{\partial y}; \text{ or } \frac{\partial H_z}{\partial y} = 0 \text{ at } y = 0, b \tag{3.119a}$$

and

$$x = 0, a; \qquad E_y = 0 = \frac{j\omega\mu}{k_c^2}\frac{\partial H_z}{\partial x}; \text{ or } \frac{\partial H_z}{\partial x} = 0 \text{ at } x = 0, a \tag{3.119b}$$

Using (3.118), (3.119a) and (3.119b), we get

$$\left.\frac{\partial H_z}{\partial y}\right|_{y=0} = [A\cos k_x x + B\sin k_x x]Dk_y e^{-\gamma z} = 0; \text{ This gives } D = 0, \text{ since the factor} \tag{3.120a}$$

within [ ] cannot be zero at all x. Similarly,

$$\left.\frac{\partial H_z}{\partial x}\right|_{x=0} = [-B][C\cos k_y y + D\sin k_y y]e^{-\gamma z} = 0; \text{ This gives } B = 0 \tag{3.120b}$$

Therefore, the solution reduces to

$$\begin{aligned} H_z(x,y,z) &= AC\cos k_x x\cos k_y y e^{-\gamma z} \\ &= H_0\cos k_x x\cos k_y y e^{-\gamma z} \end{aligned} \tag{3.120c}$$

where, $H_0 = AC$ – constant field amplitude.

Now,

$$\left.\frac{\partial H_z}{\partial y}\right|_{y=b} = (-)H_0 k_y \cos k_x x\sin k_y b e^{-\gamma z} = 0 \tag{3.120d}$$

$$\left.\frac{\partial H_z}{\partial x}\right|_{x=a} = (-)H_0 k_x \sin k_x a\cos k_y y e^{-\gamma z} = 0 \tag{3.120e}$$

Equations (3.120d) and (3.120e) are satisfied when

$$k_y = \frac{n\pi}{b},\ k_x = \frac{m\pi}{a}, \text{ and } k_c = \sqrt{\left(\frac{m\pi}{a}\right)^2 + \left(\frac{n\pi}{b}\right)^2} \tag{3.121}$$

where $m = 0, 1, 2, 3, \ldots$; and $n = 0, 1, 2, 3, \ldots$; but m and n both together are not zero. Here, m and n denote the number of half-cycle variations in the x and y directions in the cross sectional plane of the guide. To each set of integer's m and n, a solution exists, and these modes are designated as the TE_{mn} modes. Therefore, the solution (3.118) becomes

$$H_z(x,y,z) = H_0\cos(m\pi x/a)\cos(n\pi y/b)e^{-\gamma z} \tag{3.122a}$$

Substituting (3.122a) in (3.108a-e), all field components of TE_{mn} mode are obtained as

$$H_y = -\frac{\gamma}{k_c^2}\frac{\partial H_z}{\partial y} = \frac{\gamma}{k_c^2}\left(\frac{n\pi}{b}\right)H_0\cos(m\pi x/a)\sin(n\pi y/b)e^{-\gamma z} \tag{3.122b}$$

$$H_x = -\frac{\gamma}{k_c^2}\frac{\partial H_z}{\partial x} = -\frac{\gamma}{k_c^2}\left(\frac{m\pi}{a}\right)H_0\sin(m\pi x/a)\cos(n\pi y/b)e^{-\gamma z} \tag{3.122c}$$

$$E_x = -\frac{j\omega\mu}{k_c^2}\frac{\partial H_z}{\partial y} = \frac{j\omega\mu}{k_c^2}\left(\frac{n\pi}{b}\right)H_0\cos(m\pi x/a)\sin(n\pi y/b)e^{-\gamma z} \tag{3.122d}$$

$$E_y = \frac{j\omega\mu}{k_c^2}\frac{\partial H_z}{\partial x} = -\frac{j\omega\mu}{k_c^2}\left(\frac{m\pi}{a}\right)H_0\sin(m\pi x/a)\cos(n\pi y/b)e^{-\gamma z} \tag{3.122e}$$

$$E_z(x, y, z) = 0 \tag{3.122f}$$

From (3.113a),(3.113b), and (3.121), some important parameters of the TE_{mn} wave are defined as cut-off wave number

$$k_c = \frac{2\pi}{\lambda_c} = \frac{2\pi f_c}{u} \tag{3.123a}$$

where $u = \frac{1}{\sqrt{\mu\varepsilon}}$, the phase velocity in an unbounded dielectric medium (μ, ε). For a lossless waveguide, $\alpha = 0$, so that, $\gamma = j\beta$, and:

Cut-off frequency

$$f_c = \frac{1}{2\sqrt{\mu\varepsilon}}\sqrt{\left(\frac{m}{a}\right)^2 + \left(\frac{n}{b}\right)^2} = \frac{u}{2}\sqrt{\left(\frac{m}{a}\right)^2 + \left(\frac{n}{b}\right)^2} \tag{3.123b}$$

Cut-off wavelength

$$\lambda_c = \frac{2}{\sqrt{(m/a)^2 - (n/b)^2}} \tag{3.123c}$$

The propagation constant

$$\gamma = \alpha + j\beta = \sqrt{k_c^2 - k^2} = \sqrt{\left(\frac{m\pi}{a}\right)^2 + \left(\frac{n\pi}{b}\right) - \omega^2\mu\varepsilon}$$

$$= j\omega\sqrt{\mu\varepsilon}\sqrt{1-\left(\frac{f_c}{f}\right)^2}\ ;\ \beta = \omega\sqrt{\mu\varepsilon}\sqrt{1-\left(\frac{f_c}{f}\right)^2}\ ;\ \text{when } \alpha = 0 \text{ (lossless guide)} \tag{3.123d}$$

The phase velocity of the waves

$$u_p = \frac{\omega}{\beta} = \frac{u}{\sqrt{1-(f_c/f)^2}} \tag{3.123e}$$

The group velocity

The velocity of energy propagation given by

$$u_g = \left(\frac{\partial\beta}{\partial\omega}\right)^{-1} = u^2/u_p = u\sqrt{1-(f_c/f)^2} = \frac{1}{\sqrt{\mu\varepsilon}}\sqrt{1-(f_c/f)^2} \tag{3.123f}$$

The guide wavelength

The propagating wavelength inside the guide is given by

$$\lambda_g = \frac{\lambda}{\sqrt{1-(\lambda/\lambda_c)^2}} \tag{3.123g}$$

where $\lambda = \mu/f$, the wavelength in an unbounded dielectric. $\lambda = \lambda_0$ in air or space.

The characteristic wave impedance in the guide can be derived as, for TE_{mn} mode

$$Z_w = \frac{E_x}{H_y} = -\frac{E_y}{H_x} = \frac{\omega\mu}{\beta} = \frac{\eta k}{\beta} = \frac{\eta}{\sqrt{1-(f_c/f)^2}} = 377\frac{\lambda_g}{\lambda_0}\text{ohm}\ ;\ \gamma = j\beta,\ \alpha = 0 \tag{3.123h}$$

where $\eta = \sqrt{\mu/\varepsilon}$ is the intrinsic impedance in an unbounded dielectric (μ, ε). $\eta = 120\,\pi$ in free space.

❑ *Dispersion in Waveguide* From (3.123e) and (3.123f), we see that since β is not a linear function of frequency for waveguide propagations, u_p and u_g are different for any given dielectric filling. The variations of phase and group velocities vs. frequency for a typical waveguide mode are shown in Fig. 3.45. Therefore, a real-life complex signal having finite bandwidth consists of different frequency components. These components will propagate at different velocities and reach the load end at different instants of time. This exhibits differential propagation delay and the original signal at load end cannot be reconstructed, i.e., distortion of the original signal will occur. This phenomenon is called the ***waveguide dispersion***. The differential propagation delay per unit length of the guide is given by

$$\Delta t = \left.\frac{1}{u_g}\right|_{max} - \left.\frac{1}{u_g}\right|_{min} = \frac{1}{u}\left[\frac{1}{\sqrt{1-\left(\frac{f_c}{f_{min}}\right)^2}} - \frac{1}{\sqrt{1-\left(\frac{f_c}{f_{max}}\right)^2}}\right];\ \text{s/m} \tag{3.123i}$$

Here, suffixes *max* and *min* are for the *maximum* and the *minimum* frequencies in the band.

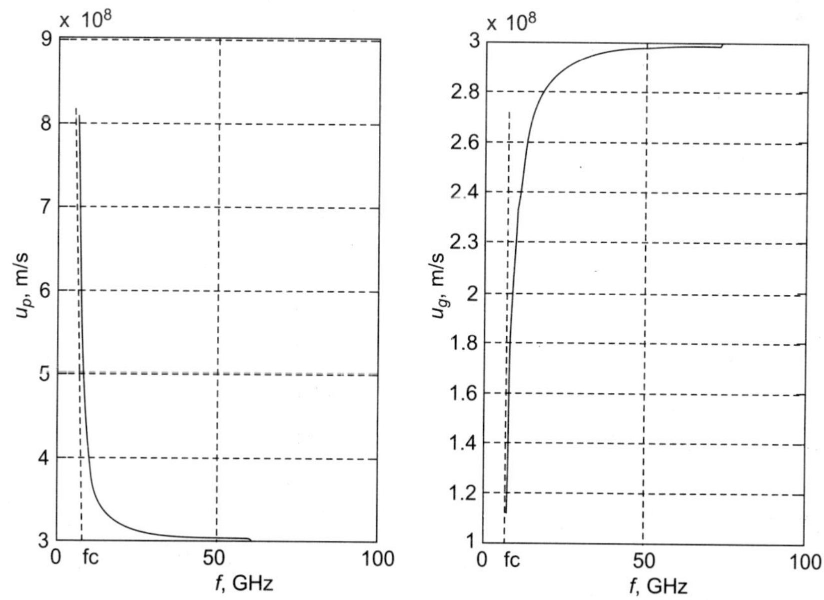

Fig. 3.45 Variation of phase and group velocities with frequency in rectangular waveguide

It is seen that as $f \to \infty$, both phase and group velocities approach to the velocity u of the wave of TEM mode in the unbounded dielectric material which fills the waveguide. This indicates that waveguide modes appear more and more like TEM modes at high enough frequencies. However, near the lowest cut-off, both u_p and u_g behave oppositely—u_p increases to infinity whereas, u_g decreases to zero.

Similarly, the guide wavelength vs frequency variation also shows that at very high frequencies, it is converging to the wavelength in the unbounded dielectric medium, which fills the guide (Fig. 3.46)

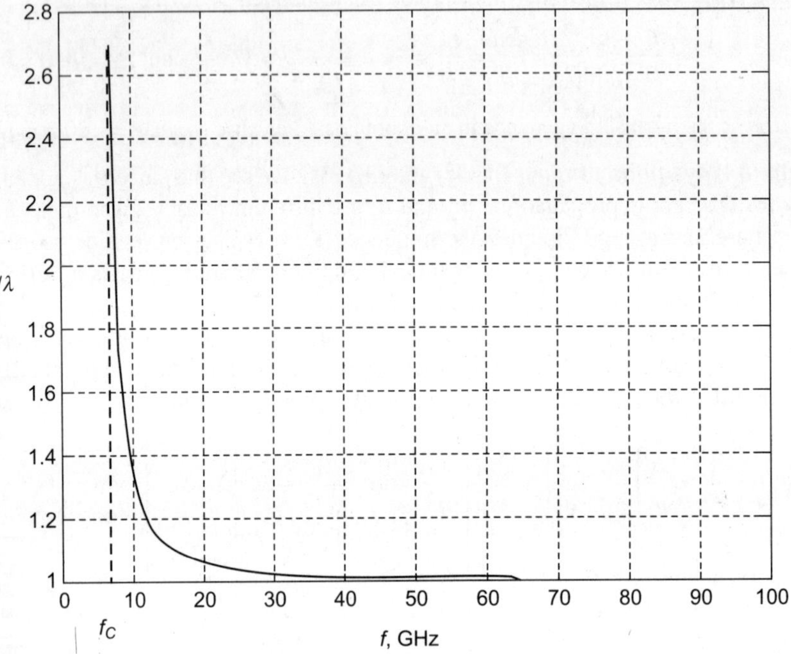

Fig. 3.46 Variation of normalized guide wavelength with frequency in rectangular waveguide

❑ ***Surface Electric Current on the Conducting Walls*** While wave propagation through the cross section of the guide, the magnetic fields induce electric currents on the inner surfaces of the guide walls. The surface current density in given by

$$\mathbf{J}_s = \hat{\mathbf{n}} x \mathbf{H}_t \tag{3.123j}$$

where $\hat{\mathbf{n}}$ is the unit normal vector outward to the inner surface of the guide and $\mathbf{H}_t$ is the tangential magnetic field on the walls.

TM Waves Solution

For TM waves propagating in the $+z$-direction, $H_z = 0$. Therefore, the field components can be derived from the E_z component of Eqs. (3.107)

$$E_x = -\frac{\gamma}{k_c^2}\frac{\partial E_z}{\partial x} \tag{3.124a}$$

$$E_y = -\frac{\gamma}{k_c^2}\frac{\partial E_z}{\partial y} \tag{3.124b}$$

$$H_x = \frac{j\omega\varepsilon}{k_c^2}\frac{\partial E_z}{\partial y} \tag{3.124c}$$

$$H_y = -\frac{j\omega\varepsilon}{k_c^2}\frac{\partial E_z}{\partial x} \tag{3.124d}$$

where $$k_c^2 = \gamma^2 + \omega^2\mu\varepsilon \tag{3.124e}$$

At every point in the waveguide, the longitudinal electric field E_z satisfies the scalar wave equation

$$\nabla^2 E_z + k^2 E_z = 0 \tag{3.125}$$

where $k = \omega\sqrt{\mu\varepsilon}$ is called the wave number of the propagating medium (μ, ε). The solutions of (3.125) must satisfy the boundary conditions imposed by the metal walls only for some discrete values of **eigenvalues** k_c. For infinitely high conducting metal (ideal assumption) each eigenvalue is real and independent of frequency. Equation (3.125) can be solved by the method of separation of variables in the similar way of TE mode case and applying the following boundary conditions that tangential components of electric field are zero on the waveguide walls:

$$E_z = 0 \text{ at } x = 0, a; \text{ and } y = 0, \text{b} \tag{3.126a}$$

$$E_x = 0 \text{ at } y = 0, b \tag{3.126b}$$

$$E_y = 0 \text{ at } x = 0, a \tag{3.126c}$$

The solutions are obtained as

$$E_z = \varepsilon_0 \sin\left(\frac{m\pi x}{a}\right)\sin\left(\frac{n\pi y}{b}\right) \tag{3.127a}$$

$$E_y = \frac{\gamma}{k_c^2}\cdot\frac{\partial E_z}{\partial y} = \frac{\gamma}{k_c^2}\frac{n\pi}{b}E_0 \sin\left(\frac{m\pi x}{a}\right)\cos\left(\frac{n\pi y}{b}\right) \tag{3.127b}$$

$$E_x = \frac{\gamma}{k_c^2}\cdot\frac{\partial E_z}{\partial x} = \frac{\gamma}{k_c^2}\frac{m\pi}{a}E_0 \cos\left(\frac{m\pi x}{a}\right)\sin\left(\frac{n\pi y}{b}\right) \tag{3.127c}$$

$$H_z = 0 \tag{3.127d}$$

$$H_y = -j\frac{\omega\varepsilon}{k_c^2}\frac{\partial E_z}{\partial x} = -j\frac{\omega\varepsilon}{k_c^2}\frac{m\pi}{a}E_0 \cos\left(\frac{m\pi x}{a}\right)\sin\left(\frac{n\pi y}{b}\right) \tag{3.127e}$$

$$H_x = j\frac{\omega\varepsilon}{k_c^2}\frac{\partial E_z}{\partial y} = j\frac{\omega\varepsilon}{k_c^2}\frac{n\pi}{b}E_0 \sin\left(\frac{m\pi x}{a}\right)\cos\left(\frac{n\pi y}{b}\right) \tag{3.127f}$$

where m = 1, 2, 3, ... and n = 1, 2, 3 .... Here, for TM_{mn} modes m and n are not equal to zero, since either $m = 0$ or $n = 0$ leads to zero-field intensities inside the excited waveguide. Here, m and n denote the number of half-cycle variations in the x and y directions in the cross-sectional plane of the guide. To each set of integers m and n, a solution exists, and these modes are designated as the TM_{mn} modes. For TM_{mn} modes, the equations for the cut-off frequency f_c, propagation constant b, the phase and group velocities u_p and u_g, and the guide wavelength λ_g are the same as those for TE_{mn} modes given in (3.123a)–(3.123g).

The characteristic wave impedance for TM_{mn} mode is different from that of TE_{mn} mode and is given by

$$Z_w = \frac{E_x}{H_y} = \frac{-E_y}{H_x} = \frac{\beta}{\omega\varepsilon} = \frac{\eta\beta}{k} = \eta\sqrt{1-\left(\frac{f_c}{f}\right)^2} = \frac{377\lambda_0}{\lambda_g} \text{ ohms} \tag{3.128}$$

From (3.123h) and (3.128), it is seen that the characteristic wave impedances of TE_{mn} and TM_{mn} modes are different and they exhibit a relation

$$Z_{w(\text{TE})}\cdot Z_{w(\text{TM})} = \eta^2 \tag{3.129}$$

When the mode does not propagate, Z_w becomes imaginary and the mode is called *evanescent mode*, characterized by no net energy flow through the guide—but energy is stored in the guide. The variation of wave impedances of TE and TM modes are shown in Fig. 3.47. It is seen that as $f \to \infty$, both TE and TM mode impedances approach to the intrinsic impedance of the dielectric medium filling the waveguide i.e., impedance of travelling wave of TEM mode in the unbounded dielectric material which fills the waveguide. This indicates again that waveguide modes appear more and more like TEM modes at high enough frequencies. However, near the lowest cut-off, both behave oppositely—Zw_{TE} increases to infinity, whereas, Zw_{TM} decreases to zero when frequency decreases.

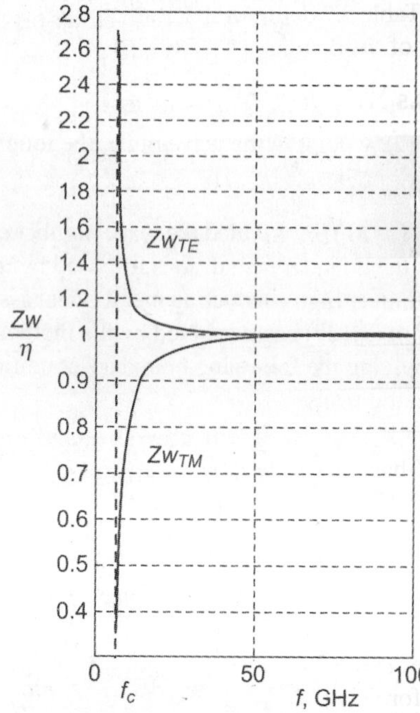

Fig. 3.47 The variation of wave impedances of TE and TM modes vs frequency

Dominant Mode

The cut-off frequency expressed by (3.123b) shows that the physical size of the waveguide will determine the propagation of the modes of specific orders determined by m and n. The minimum cut-off frequency is obtained for a guide having dimension $a > b$, for $m = 1$, $n = 0$. Since for TM_{mn} modes, $n \neq 0 \neq m$, the lowest order mode possible is TE_{10}, and is called the ***dominant mode*** in a rectangular waveguide for $a > b$. The field components of dominant mode are therefore, given by ($m = 1, n = 0$)

$$E_y = -\frac{j\omega\mu}{k_c^2}\left(\frac{\pi}{a}\right)H_0 \sin(\pi x/a)e^{-\gamma z}$$

$$H_x = \frac{\gamma}{k_c^2}\left(\frac{\pi}{a}\right)H_0 \sin(\pi x/a)e^{-\gamma z}$$

$$H_z(x, y, z) = H_0 \cos(\pi x/a)e^{-\gamma z} \tag{3.130}$$

From (3.121), $k_c = \pi/a$ for dominant TE_{10} mode. Substituting this value in (3.130) and normalizing above field components by letting

$$E_0 = -j\frac{\omega\mu}{k_c^2}\left(\frac{\pi}{a}\right)H_0 = -j\frac{\omega\mu a}{\pi}H_0,$$

the components of dominant TE_{10} mode becomes

$$E_y = E_0 \sin\left(\frac{\pi x}{a}\right)e^{-j\beta z} \tag{3.131a}$$

$$H_x = j\frac{\gamma}{\omega\mu}E_0 \sin\left(\frac{\pi x}{a}\right)e^{-\gamma z} \tag{3.131b}$$

$$H_z = j\frac{\pi}{\omega\mu a}E_0 \cos\left(\frac{\pi x}{a}\right)e^{-\gamma z} \tag{3.131c}$$

When $a < b$, the lowest cut-off frequency is for $m = 0$ and $n = 1$. Consequently, the dominant mode is TE_{01}. Table 3.5 describes the ratio of cut-off frequency of some higher-order modes normalized with respect to that of the dominant TE_{10} mode.

Table 3.5 $(f_c)_{mn}/(f_c)_{TE10}$ for $a > b$

Dimension	$(f_c)_{mn}/(f_c)_{TE10}$								
a/b	TE_{10}	TM_{01}	TE_{11} TM_{11}	TE_{20}	TE_{02}	TE_{21} TM_{21}	TE_{12} TM_{12}	TE_{22} TM_{22}	TE_{30}
1	1	1	1.414	2	2	2.236	2.236	2.828	3
1.5	1	1.5	1.803	2	3	2.500	3.162	3.606	3
2	1	2	2.236	2	4	2.828	4.123	4.472	3
3	1	3	3.162	2	6	3.606	6.083	6.325	3

In X-band, a standard rectangular waveguide has dimensions $a = 0.9''$, and $b = 0.4''$. Therefore, $a/b = 2.25$. From (3.123b) the general expression of cut off frequency for air-filled guide is

$$f_c = \frac{1}{2\sqrt{\mu\varepsilon_0}}\sqrt{\left(\frac{m}{a}\right)^2 + \left(\frac{n}{b}\right)^2} = \frac{c}{2}\sqrt{\left(\frac{m}{a}\right)^2 + \left(\frac{n}{b}\right)^2} \tag{3.131d}$$

Here, c = velocity of light in space. Table 3.6 shows the cut-off frequencies for different successive modes for

$$a/b = 2.286\,/1.016 \text{ cm} = 2.25$$

Table 3.6 Cut off frequencies of rectangular waveguide with $a = 2.25b$

m	n	f_c(GHz)	Mode	m	n	f_c(GHz)	Mode
1	0	6.5617	TE	0	2	29.5276	TE
2	0	13.1234	TE	1	2	30.2478	TE/TM
0	1	14.7638	TE	2	2	32.3125	TE/TM
1	1	16.1563	TE/TM	3	2	35.4877	TE/TM
3	0	19.6850	TE	2	3	46.1946	TE/TM
2	1	19.7533	TE/TM	0	3	44.2913	TE
3	1	24.6063	TE/TM	3	3	48.4688	TE/TM
				1	3	44.7748	TE/TM

Degenerate Modes

There exist some higher-order modes having the same cut-off frequencies. These are called ***degenerate modes***. It is seen that in a rectangular waveguide ($a \neq b$), the possible TE_{mn} and TM_{mn} modes having both $m \neq 0$ and $n \neq 0$, are always **degenerate**. In a square guide ($a = b$), all the TE_{pq}, TE_{qp}, TM_{pq} and TM_{qp} modes are together degenerate. For example, let: TE_{12}, TE_{21}, TM_{12}, TM_{21} be four degenerate modes having same cut-off frequency. Waveguide dimensions are, therefore, selected such that higher-order modes are not supported in the operating frequency band and, thus, only the desired mode can be propagating through the guide.

In most of the applications, only dominant-mode TE_{10} is used since it is easy to excite and receive through simple signal-coupling method as described below.

Dominant Range

In a waveguide, the **dominant range** is the range of frequencies for which only the dominant mode can propagate. It was seen earlier that for X-band waveguide, the dominant mode cut-off frequency is given by

$$f_c = \frac{c}{2}\sqrt{\left(\frac{1}{a}\right)^2} = \frac{c}{2a} = \frac{30}{2a} = \frac{30}{2\times 0.9\times 2.54} = 6.5617\text{ GHz}$$

For the next higher order mode, TE_{20}, $f_{cTE_{20}} = 13.12\text{ GHz} \approx 2\,f_{cTE_{10}}$.

Therefore, the dominant range is $f_{cTE_{10}} < f < 2\,f_{cTE_{10}}$. Since the propagation parameters of any mode are rapid functions of frequency near cut-off, some safe margin may be given, viz., $1.25f_{cTE_{10}} < f < 0.95f_{cTE_{20}}$.

It may be noted that

- If $a/b > \sqrt{3}$, second mode is TE_{20} and $f_{cTE_{20}} = 2 * \text{dominant } f_{cTE_{10}}$.
  Dominant range is $f_{cTE_{10}} < f < 2\,f_{cTE_{10}}$.
- If $a/b = \sqrt{3}$, second mode is TE_{01} and third modes are $TE_{20}/TE_{11}/TM_{11}$
  Dominant range is $f_{cTE_{10}} < f < f_{cTE_{01}}$.
- If $a/b < \sqrt{3}$, second mode is TE_{01} and third modes are TE_{11}/TM_{11}, and fourth mode is TE_{20}.
  Dominant range is $f_{cTE_{10}} < f < f_{cTE_{01}}$.

Example 3.13 Find the expressions for the wall currents of the TE_{10} mode in a rectangular waveguide ($a \times b$) for wave propagation along the z-axis.

Solution For dominant TE_{10} mode, nonzero field components are

$$E_y = E_0 \sin\left(\frac{\pi x}{a}\right)e^{-j\beta z}$$

$$H_x = j\frac{\gamma}{\omega\mu}E_0 \sin\left(\frac{\pi x}{a}\right)e^{-\gamma z}$$

$$H_z = j\frac{\pi}{\omega\mu a}E_0 \cos\left(\frac{\pi x}{a}\right)e^{-\gamma z}$$

The wall currents are contributed by tangential components of **H** field H_x and H_z.

(a) For the narrow wall at $x = 0$, $\hat{n} = \hat{x}$, so that

$$J_s\big|_{x=0} = \hat{x}\times(\hat{x}H_x + \hat{z}H_z) = \hat{y}H_z\big|_{x=0} = \hat{y}\; j\frac{\pi}{\omega\mu a}E_0 e^{-\gamma z}$$

Since the fields are symmetric,

$$J_s\big|_{x=a} = J_s\big|_{x=0}$$

(b) For the broad wall at $y = 0$, $\hat{n} = \hat{y}$, so that

$$J_s\big|_{y=0} = \hat{y}\times(\hat{x}H_x + \hat{z}H_z) = (-\hat{z}H_x + \hat{x}H_z)\big|_{y=0}$$

$$= -\hat{z}\frac{j\gamma}{\omega\mu}E_0 \sin\left(\frac{\pi x}{a}\right)e^{-\gamma z} + \hat{x}\frac{j\pi}{\omega\mu a}E_0 \cos\left(\frac{\pi x}{a}\right)e^{-\gamma z}$$

Since the fields are symmetric,

$$J_s\big|_{y=b} = J_s\big|_{y=0}$$

The field configurations of different TE_{mn} modes in a rectangular waveguide along with the surface current densities are shown in Fig. 3.48.

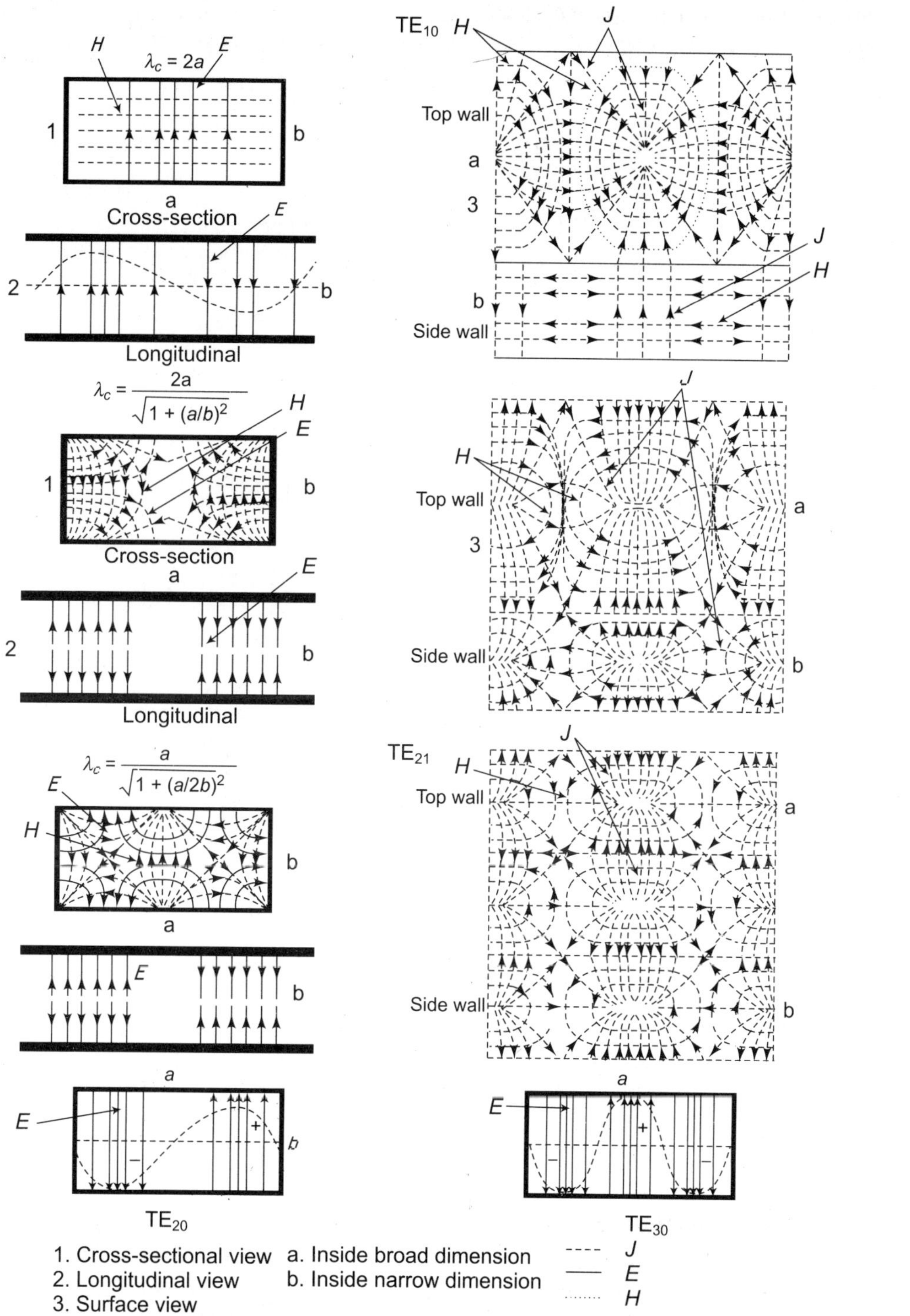

Fig. 3.48 Field configurations of higher order TE modes

The dominant TE_{10} mode current distribution and the field configuration are shown in Fig. 3.48. The field and surface current configurations of different TM_{mn} modes are shown in Fig. 3.49.

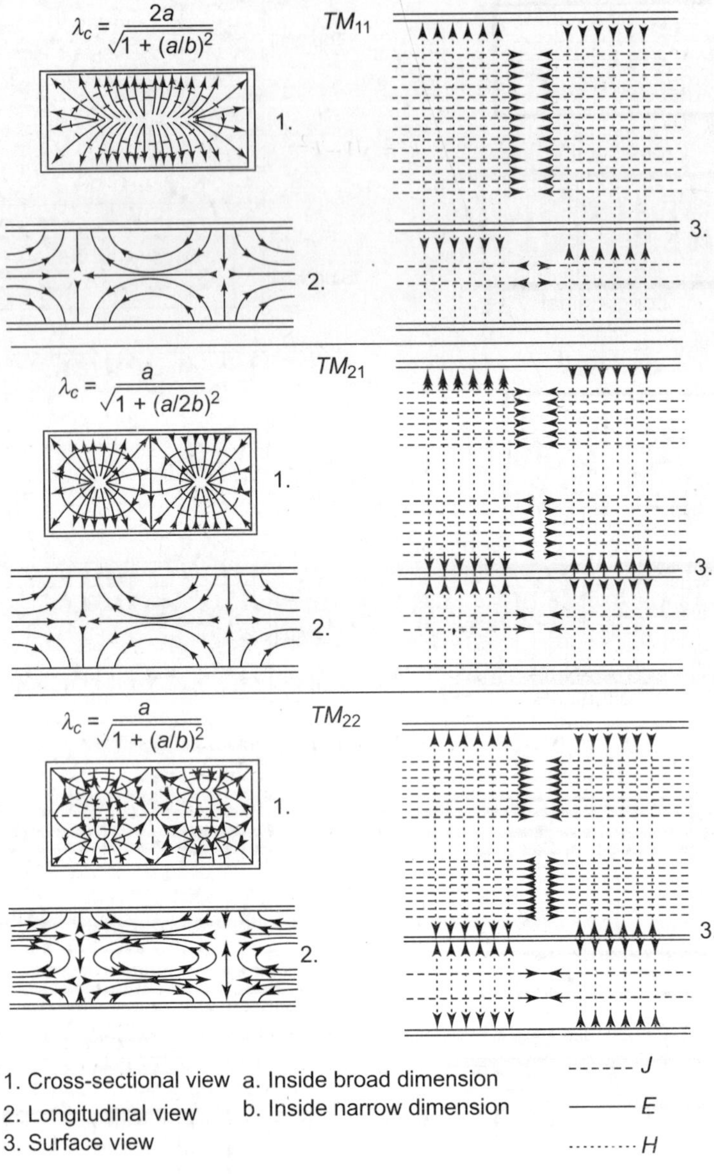

Fig. 3.49 Field configuration of TM Modes

Methods of Excitation of Modes

The waveguide modes are usually excited from a signal source through a coaxial cable. The outer conductor of the cable makes a 360º connection with the body of the waveguide and the centre conductor is projected inside the guide as a small probe as shown in Fig. 3.50. All TE_{mo} modes with an odd value of m have a maximum field at the centre of the cross section in transverse direction. These modes can be excited by a single probe at the centre as shown in Fig. 3.50(a). All TE_{mo} modes with an even m have a zero field at the centre and field polarities reverse alternately in the transverse direction as shown in Fig. 3.50(d). These last modes can be excited by two probes as shown in Fig. 3.50(d) having out-of-phase currents. In above two cases, end of the guide is short circuited at a distance $\lambda_g/4$ from the exciting probe so that the waves in the forward direction are enhanced by the waves which travel from the probe to the short end and reflected towards the forward direction with a phase change of 360° for reverse ($\lambda_g/4$) and forward reflected path ($\lambda_g/4$) + 180° phase change at reflection on the short $\left(\equiv \frac{\pi}{2} + \frac{\pi}{2} + \pi = 2\pi\right)$. Similarly, depending on the field configuration of the higher-order modes, excitation methods change. Excitation of other higher-order modes, TM_{11} and TM_{21}, are also shown in Fig. 3.50 (b) and (c).

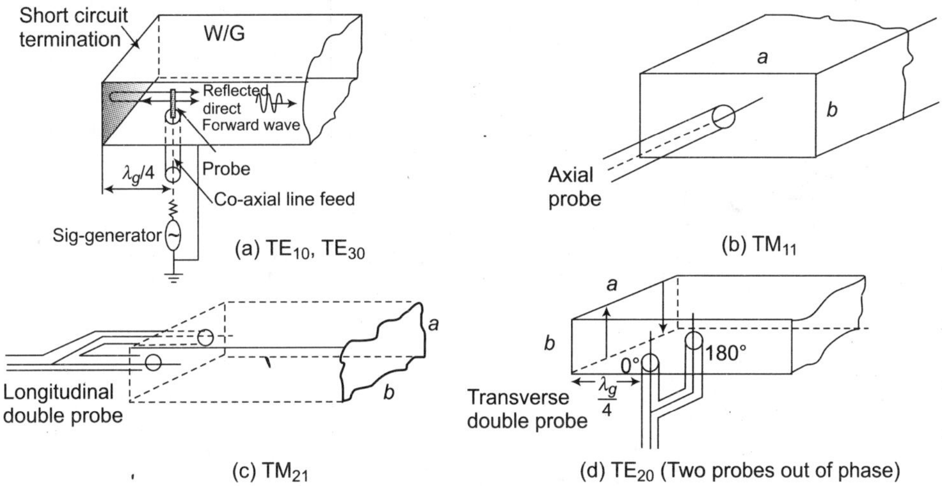

Fig. 3.50 Methods of excitation of modes in a rectangular waveguide

Example 3.14 A rectangular waveguide with $a = 5$ cm, $b = 2$ cm cross section is used to propagate TM_{11} mode at 10 GHz. Determine the cut-off wavelength, guide wavelength and the characteristic impedance.

Solution For TM_{11} mode, $m = 1, n = 1$

Given, $b = 2$ cm, $a = 5$ cm, $f = 10$ GHz

Therefore, $\lambda = c/f = 30/10 = 3$ cm

$$\lambda_c = \frac{2}{\sqrt{[(m/a)^2 + (n/b)^2]}} = \frac{2ab}{\sqrt{a^2 + b^2}} = \frac{2 \times 5 \times 2}{\sqrt{5^2 + 2^2}} = 3.714 \text{ cm}$$

$$\lambda_g = \frac{\lambda}{\sqrt{1 - (\lambda/\lambda_c)^2}} = \frac{3}{\sqrt{1 - (3/3.714)^2}} = 5.089 \text{ cm}$$

$$Z_0 = 120\pi\sqrt{[1 - (\lambda/\lambda_c)^2]} = 120\pi\sqrt{[1 - (3/3.714)^2]} = 222.24 \text{ ohm}$$

Example 3.15 Find the dominant frequency range of a rectangular waveguide filled with air, having inner dimensions $a = 1.9$ cm and $b = 0.9$ cm. What is the usable range?

Solution

The dominant mode TE_{10}, $f_c = \frac{30}{2a} = \frac{30}{2*1.9} = 7.8947$ GHz

Since $a/b = 1.9/.9 = 2.11 > \sqrt{3}$, the dominant range is

$$7.8947 < f < 2*7.8947 \text{ GHz}$$

or $$7.8947 < f < 15.7894 \text{ GHz}$$

The usable frequency range is

$$1.25*7.8947 < f < 0.95*15.7894 \text{ GHz}$$

or $$9.8684 < f < 15 \text{ GHz}$$

Power Flow in Rectangular Waveguide

The power transmitted through a waveguide from the input to output ends or ports in longitudinal direction (z) is calculated by assuming that the waveguide is infinitely long or a waveguide of finite length is terminated with matched loads at both the ends so that the propagation of signal takes place only from the input to output ends without any reflection from the both ends. The **time average power flow** can be expressed by

$$P_{mn} = \frac{1}{2}\text{Re}\int_0^a\int_0^b \mathbf{E}\times\mathbf{H}^*\cdot\mathbf{dS} \tag{3.132a}$$

where **E** and $\mathbf{H}^*$ are the transverse components of fields. Since the wave impedance for both TE_{mn} and TM_{mn} modes is given by

$$Z_w = \frac{E}{H_y} = -\frac{E_y}{H_x}$$

$$E_x = Z_w H_y \text{ and } E_y = -Z_w H_x \tag{3.132b}$$

Eliminating electric field components from (3.132a) using (3.132b), the time average power flow can be expressed by

$$P_{mn} = \frac{Z_w}{2}\int_0^a\int_0^b \left(|H_x|^2 + |H_y|^2\right) dx\, dy \tag{3.132c}$$

It is assumed that the waveguide is filled with lossless dielectric.

Substituting appropriate expressions for field components for TE_{mn} or TM_{mn} modes and the corresponding characteristic wave impedance into (3.132c), the time average power flow through the cross section of the waveguide can be evaluated.

The component of the fields appeared in the integrand of (3.132c) are expressed in terms of orthogonal functions (eigen function) given in (3.122) and (3.127). Therefore, power orthogonalities also apply and each possible mode of propagation in a lossless guide carries energy independently of all other modes that may be present. Even for small losses, power orthogonality applies with negligible error except for degenerate modes where coupling between the modes occur due to having same propagation constant.

Dominant Mode Power Flow (TE_{10})

For dominant-mode TE_{10}, $H_y = 0$. Therefore, the only transverse component of **H** is $\mathbf{H}_x$ given by (3.131b)

$$H_x = \left(\frac{j\gamma}{\omega\mu}\right) E_0 \sin\left(\frac{\pi x}{a}\right) e^{-\gamma z} \tag{3.133}$$

Therefore, the time average power flow through the guide excited in dominant TE_{10} mode is given by

$$P_{10} = \frac{Z_{w(TE_{10})}}{2}\int_0^a\int_0^b |H_x|^2 dxdy$$

$$= \frac{Z_{w(TE_{10})}}{2}\left|\frac{\gamma E_0}{\omega\mu}e^{-\gamma z}\right|^2 \int_0^a\int_0^b \sin^2\left(\frac{\pi x}{a}\right)dxdy$$

$$= \frac{Z_{w(TE_{10})}}{2}\left|\frac{\gamma E_0}{\omega\mu}e^{-\gamma z}\right|^2 b\int_0^a \sin^2\left(\frac{\pi x}{a}\right)dxdy$$

$$= \frac{Z_{w(TE_{10})}}{2}\left|\frac{\gamma E_0}{\omega\mu}e^{-\gamma z}\right|^2 b\frac{a}{2}$$

$$= \frac{Z_{w(TE_{10})}}{4}\left|\frac{\gamma E_0}{\omega\mu}e^{-\gamma z}\right|^2 ab$$

Substituting $Z_{w(TE_{10})} = \dfrac{\omega\mu}{\gamma}$ from (3.133h)

$$P_{10} = \frac{\omega\mu}{4\gamma}\left|\frac{\gamma E_0}{\omega\mu}e^{-\gamma z}\right|^2 ab\,; \tag{3.134}$$

For a lossless guide (filled with lossless dielectric and having perfect conducting walls), $\gamma = \alpha + j\beta = j\beta$, $\beta = (2\pi/\lambda_g)$. Hence, using the expression (3.123d) for β, the time average power flow in a lossless guide is

$$P_{10} = \frac{\omega\mu}{4\beta}\left(\frac{\beta E_0}{\omega\mu}\right)^2 ab = \frac{\beta}{4\omega\mu}abE_0^2$$

$$= \frac{1}{4\eta}\sqrt{1-\left(\frac{f_c}{f}\right)^2}\,E_0^2 ab \tag{3.135}$$

where $\eta = \sqrt{\mu/\varepsilon}$ = intrinsic impedance of unbounded medium of dielectric which fills the waveguide.

Attenuation in a Rectangular Waveguide

Attenuation of signal transmission in a metallic waveguide occurs due to the following two types of power losses:

1. **Ohmic loss** in the highly conductive guide walls due to finite conductivity ($\sigma \neq \infty$) or nonzero surface resistance $R_s = \sqrt{\dfrac{\omega\mu}{2\sigma}}$.
2. **Dielectric loss** in the material which fills the waveguide.

❑ *Ohmic Loss* Since the magnitudes of both **E** and **H** fields decrease according to $e^{-\alpha_c z}$, the power decreases proportionately with $|\mathbf{E}\times\mathbf{H}|$, i.e. according to $P(0)\,e^{-2\alpha_c z}$, where $P(0)$ is the input power at the sending end $z = 0$ and α_c is the attenuation constant due to the conductor loss. The rate of decrease of propagating power P_{mn} with the travelling distance z for a given mode, TE_{mn} or TM_{mn}, is given by

$$(-)\frac{dP_{mn}}{dz} = 2\alpha_c P_{mn} = P_L \tag{3.136a}$$

where P_L is the power loss per unit length of the guide and P_{mn} is the power transmitted through the guide at any point z. Therefore, the attenuation constant α_c is given by

$$\alpha_c = \frac{P_L}{2\,P_{mn}} \tag{3.136b}$$

The power loss per unit length of the guide is

$$P_L = \frac{1}{2} R_s \int_{\text{guide walls}} \mathbf{J}_s \cdot \mathbf{J}_s^* dl$$

$$= \frac{1}{2} R_s \oint_{\text{walls}} |\mathbf{J}_s|^2 dl \tag{3.136c}$$

Here, integration is taken over all the conducting walls of the waveguide. From (3.132c),

$$P_{mn} = \frac{Z_w}{2} \int_0^a \int_0^b \left(|H_x|^2 + |H_y|^2 \right) dx\, dy$$

Substituting P_{mn} from above and P_L from (3.136c) to (3.136b), we get the conductor attenuation constant for TE_{mn} or TM_{mn} modes

$$\alpha_c = \frac{\dfrac{R_s}{2} \oint_{\text{walls}} |\mathbf{J}_s|^2\, dl}{2\dfrac{Z_w}{2} \int_0^a \int_0^b \left(|H_x|^2 + |H_y|^2 \right) dx\, dy} \tag{3.136d}$$

For dominant TE_{10} mode, from (3.131), nonzero field components are

$$E_y = E_0 \sin\left(\frac{\pi x}{a}\right) e^{-j\beta z} \tag{3.137a}$$

$$H_x = \frac{j\gamma}{\omega\mu} E_0 \sin\left(\frac{\pi x}{a}\right) e^{-\gamma z} \tag{3.137b}$$

$$H_z = \frac{j\pi}{\omega\mu a} E_0 \cos\left(\frac{\pi x}{a}\right) e^{-\gamma z} \tag{3.137c}$$

Since $H_y = 0$, from (3.136d), we can write

$$\alpha_c = \frac{\dfrac{R_s}{2} \oint_{\text{walls}} |\mathbf{J}_s|^2\, dl}{Z_w \int_0^a \int_0^b |H_x|^2\, dx\, dy} \tag{3.137d}$$

❑ ***Computation of Denominator of (3.137d)*** From (3.135), the denominator of (3.137d) is computed as

$$Z_{w(TE_{10})} \int_0^a \int_0^b |H_x|^2 dxdy = \frac{\gamma E_0^2\, e^{-2\gamma z}}{2 \quad \omega\mu} ab \tag{3.138}$$

❑ ***Computation of Numerator of (3.137d)*** Assuming that the currents on the lossy walls are the same as the currents on the lossless walls, the surface electric current density on the guide wall is expressed by $\mathbf{J}_s = \hat{\mathbf{n}} x \mathbf{H}$, where $\hat{\mathbf{n}}$ is the unit outward normal vector at the inner side of the guide walls. The wall currents are contributed by tangential components of **H** field H_x and H_z.

1. For the narrow wall at $x = 0$, $\hat{\mathbf{n}} = \hat{\mathbf{x}}$, so that

$$\mathbf{J}_s\big|_{x=0} = \hat{\mathbf{x}} \times (\hat{\mathbf{x}}H_x + \hat{\mathbf{z}}H_z) = \hat{\mathbf{y}}H_z\big|_{x=0}$$

$$= \hat{\mathbf{y}}\frac{j\pi}{\omega\mu a}E_0 e^{-\gamma z} \qquad (3.139a)$$

2. For the narrow wall at $x = a$,
since the fields are symmetric, $\mathbf{J}_s\big|_{x=a} = \mathbf{J}_s\big|_{x=0}$ (3.139b)

3. For the broad wall at $y = 0$, $\hat{\mathbf{n}} = \hat{\mathbf{y}}$, so that

$$\mathbf{J}_s\big|_{y=0} = \hat{\mathbf{y}} \times (\hat{\mathbf{x}}H_x + \hat{\mathbf{z}}H_z) = (-\hat{\mathbf{z}}H_x + \hat{\mathbf{x}}H_z)\big|_{y=0}$$

$$= -\hat{\mathbf{z}}\left(\frac{j\gamma}{\omega\mu}\right)E_0 \sin\left(\frac{\pi x}{a}\right)e^{-\gamma z} + \hat{\mathbf{x}}\frac{j\pi}{\omega\mu a}E_0 \cos\left(\frac{\pi x}{a}\right)e^{-\gamma z} \qquad (3.139c)$$

4. For the broad wall at $y = b$, since the fields are symmetric,

$$\mathbf{J}_s\big|_{y=b} = \mathbf{J}_s\big|_{y=0} \qquad (3.139d)$$

Therefore, the numerator of (3.137b) representing the power loss P_L is given by

$$P_L = P_L\big|_{x=0} + P_L\big|_{x=a} + P_L\big|_{y=0} + P_L\big|_{y=b}$$

$$= 2(P_L\big|_{x=0} + P_L\big|_{y=0}) \qquad (3.139e)$$

since for symmetries, the same amount of power is dissipated in the walls, $x = 0$ and $x = a$ or $y = 0$ and $y = b$.

For the $x = 0$ wall

$$P_L\big|_{x=0} = \frac{R_s}{2}\left[\int_0^b \left|\frac{\pi}{\omega\mu a}E_0 e^{-\gamma z}\right|^2 dy\right]_{x=0} = \frac{R_s}{2}\left(\frac{\pi}{\omega\mu a}\right)^2 E_0^2\, b\, e^{-2\gamma z} \qquad (3.139f)$$

$$P_L\big|_{y=0} = \frac{R_s}{2}\int_0^a \left|-\hat{\mathbf{z}}\left(\frac{j\gamma}{\omega\mu}\right)E_0 \sin\left(\frac{\pi x}{a}\right)e^{-\gamma z} + \hat{\mathbf{x}}\frac{j\pi}{\omega\mu a}E_0\cos\left(\frac{\pi x}{a}\right)e^{-\gamma z}\right|^2 dx$$

$$= \frac{R_s}{2}\frac{E_0^2}{\omega^2\mu^2}e^{-2\gamma z}\left[\int_0^a \gamma^2 \sin^2\left(\frac{\pi x}{a}\right)dx + \left(\frac{\pi}{a}\right)^2\int_0^a \cos^2\left(\frac{\pi x}{a}\right)dx\right]$$

$$= \frac{R_s}{2}\frac{E_0^2}{\omega^2\mu^2}e^{-2\gamma z}\left[\frac{a\gamma^2}{2} + \left(\frac{\pi}{a}\right)^2\frac{a}{2}\right]$$

$$= \frac{R_s}{2}\frac{E_0^2}{\omega^2\mu^2}\frac{a}{2}e^{-2\gamma z}\left[\gamma^2 + \left(\frac{\pi}{a}\right)^2\right] \qquad (3.139g)$$

From (3.139e)–(3.139g), total power loss is

$$P_L = 2(P_L\big|_{x=0} + P_L\big|_{y=0})$$

$$= 2\left[\frac{R_s}{2}\left(\frac{\pi E_0}{\omega\mu a}\right)^2 b + \frac{R_s}{2}\left(\frac{E_0}{\omega\mu}\right)^2 \frac{a}{2}\left\{\gamma^2 + \left(\frac{\pi}{a}\right)^2\right\}\right] e^{-2\gamma z}$$

$$= \frac{R_s}{2}\left(\frac{\pi E_0}{\omega\mu a}\right)^2 \left[2b + \left(\frac{a}{\pi}\right)^2 a\gamma^2 + a\right] e^{-2\gamma z} \tag{3.139h}$$

From (3.137d), (138), and (3.139h),

$$\alpha_c = \frac{\frac{R_s}{2}\left(\frac{\pi E_0}{\omega\mu a}\right)^2 \left[2b + \left(\frac{a}{\pi}\right)^2 a\gamma^2 + a\right] e^{-2\pi}}{\frac{\gamma E_0^2}{2}\frac{e^{-2\gamma z}}{\omega\mu} ab}$$

or
$$\alpha_c = \frac{R_s\left(\frac{\pi}{a}\right)^2 \left[2b + \left(\frac{a}{\pi}\right)^2 a\gamma^2 + a\right]}{\omega\mu\gamma \cdot ab} \tag{3.139i}$$

Substituting from (3123d),

$$\gamma = j\omega\sqrt{\mu\varepsilon}\sqrt{1 - \left(\frac{f_c}{f}\right)^2}$$

into (3.139i), yields

$$\alpha_c = R_s \frac{[1 + (2b/a)(f_c/f)^2]}{\eta_d b\sqrt{1 - (f_c/f)^2}}; \quad \eta_d = \sqrt{\frac{\mu_d}{\varepsilon_d}} \tag{3.140a}$$

Here, η_d is the intrinsic impedance in the dielectric medium inside the waveguide. Since waveguide wall surface resistance $R_s = \sqrt{\frac{\omega\mu}{2\sigma}}$, (3.140a) reduces to

$$\alpha_c = \sqrt{\frac{\omega\mu}{2\sigma}} R_s \frac{[1 + (2b/a)(f_c/f)^2]}{\eta_d b\sqrt{1 - (f_c/f)^2}}$$

or
$$\alpha_c = \frac{1}{\eta_d b}\left[\frac{\pi\mu f}{\sigma\left[1 - (f_c/f)^2\right]}\right]^{1/2} [1 + (2b/a)(f_c/f)^2]$$

$$= \frac{\sqrt{\varepsilon_d}}{b}\left[\frac{\pi f}{\sigma\left[1 - (f_c/f)^2\right]}\right]^{1/2} [1 + (2b/a)(f_c/f)^2] \text{ Np/m} \tag{3.140b}$$

Expressions (3.140a) and (3.140b) represent attenuation constant in rectangular waveguide propagation for dominant-mode TE_{10}.

❑ *Dielectric Loss* Let us assume that the guide is completely filled with a low-loss dielectric material having conductivity σ_d. The material's polarization does not respond instantaneously to an applied microwave field. The response can be represented by a phase difference for which the permittivity is often treated as a complex function of the frequency of the applied field. In fact, when a medium is lossy, both the propagation constant γ and the intrinsic impedance η are complex, because of (1) **conduction loss**, where σ_d is nonzero and finite, and (2) because of **polarization loss**, where the permittivity is complex.

From the Maxwell's equation, for time harmonic fields,

$$\nabla \times \mathbf{H} = J + \frac{\partial \mathbf{D}}{\partial t} = \sigma_d \mathbf{E} + \varepsilon_d \frac{\partial \mathbf{E}}{\partial t} = \sigma_d \mathbf{E} + j\omega\varepsilon_d \mathbf{E}$$

$$= j\omega\left(\varepsilon_d + \frac{\sigma_d}{j\omega}\right)\mathbf{E} = j\omega\left[\varepsilon_d - j\frac{\sigma_d}{\omega}\right]\mathbf{E} = j\omega\varepsilon_c \mathbf{E} \tag{3.141a}$$

Here, $$\varepsilon_c = \varepsilon_d - j\frac{\sigma_d}{\omega} = \varepsilon' - j\varepsilon'' \tag{3.141b}$$

$$\varepsilon' = \varepsilon_d$$

$$\varepsilon'' = \frac{\sigma_d}{\omega} \tag{3.141c}$$

so that $$\sigma_d = \omega\varepsilon''$$

The physical mechanisms responsible for these losses are different, but their effect on the behaviour of **E** and **H** fields are same. For this reason, it is customary to combine all the loss parameters of a material into a single parameter $\varepsilon_d = \varepsilon' - j\varepsilon''$, that yields the correct propagation constant γ_d and the intrinsic impedance η_d as given below:

$$\gamma_d = \sqrt{j\omega\mu(\sigma_d + j\omega\varepsilon_d)}\ \text{m}^{-1} \tag{3.141d}$$

$$\eta_d = \frac{j\omega\mu}{\gamma_d} = \sqrt{\frac{j\omega\mu}{\sigma_d + j\omega\varepsilon_d}}\ \text{ohms} \tag{3.141e}$$

where σ_d and ε_d are the actual conductivity and permittivity of the dielectric material, respectively. For the nonmagnetic dielectric, $\mu = \mu_0$. The real part ε' of the permittivity is related to the stored energy within the medium, and the imaginary part ε'' is related to the dissipation (or loss) of energy within the medium. Here, δ and $\tan\delta$ are called the dielectric loss angle and the loss tangent of the medium, respectively:

$$\tan\delta = \frac{\varepsilon''}{\varepsilon'} = \frac{\sigma_d}{\omega\varepsilon'} \tag{3.141f}$$

The dielectric constant of the material is defined by $\varepsilon_r = \varepsilon'/\varepsilon_0$.

For a low-loss dielectric having conductivity $\sigma_d << \omega\varepsilon'$ at microwave frequencies, boundary condition $\mathbf{E}_{\tan} = 0$ on the walls is still valid and all the field expressions derived for the TE and TM modes for the lossless case still apply. Under this condition, $\tan\delta \approx \frac{\sigma_d}{\omega\varepsilon'} << 1$; $\varepsilon_d \approx \varepsilon'$..

Substituting (3.141f) into (3.141e), the intrinsic impedance of the low loss dielectric medium is given by

$$\eta_d = \sqrt{\frac{j\omega\mu_0}{\sigma_d + j\omega\varepsilon'}} = \sqrt{\frac{\mu_0}{\varepsilon'}}\left(1 - j\frac{\sigma_d}{\omega\varepsilon'}\right)^{-1/2}$$

$$\cong \sqrt{\frac{\mu_0}{\varepsilon'}}\left[1 + \frac{j\sigma_d}{2\omega\varepsilon'}\right] = \sqrt{\frac{\mu_0}{\varepsilon'}}\left[1 + \frac{j}{2}\tan\delta\right]; \text{ for } \frac{\sigma_d}{\omega\varepsilon'} << 1$$

or $$\eta_d = \frac{\eta_0}{\sqrt{\varepsilon_r}}\left(1 + j\frac{\tan\delta}{2}\right); \varepsilon_r = \varepsilon_0\varepsilon' \tag{3.141g}$$

The dielectric attenuation constant above cut-off frequency can be determined from the real part of the propagation constant:

$$\alpha_d = \text{Re}\,\gamma_d = \text{Re}\left[\sqrt{\left(\frac{m\pi}{a}\right)^2 + \left(\frac{n\pi}{b}\right)^2 - \omega^2\mu(\varepsilon' - j\varepsilon'')}\right]$$

or $$\alpha_d = \text{Re}\left[\sqrt{k_c^2 - \omega^2\mu(\varepsilon' - j\varepsilon'')}\right];\ k_c^2 = \left(\frac{m\pi}{a}\right)^2 + \left(\frac{n\pi}{b}\right)^2 \tag{3.141h}$$

Substituting $k_c = \frac{2\pi}{\lambda_c} = \frac{2\pi f_c}{u}$, where u = velocity of the wave in the dielectric medium, and using binomial expansion of (3.141h), the low-loss dielectric attenuation constant above cut-off is given by

$$\alpha_d = \pi\varepsilon''\eta_d \frac{f}{\sqrt{1-\left(\frac{f_c}{f}\right)^2}};\ \text{when } \frac{\varepsilon''}{\varepsilon'} << 1 - \left(\frac{f_c}{f}\right)^2 \text{ and } f > f_c$$

or $$\alpha_d = \frac{\pi\sigma_d\eta_0}{\omega\sqrt{\varepsilon_r}}\left(1 + j\frac{\tan\delta}{2}\right)\frac{f}{\sqrt{1-\left(\frac{f_c}{f}\right)^2}};\ \sigma_d = \omega\varepsilon'\tan\delta$$

or $$\alpha_d = 120\pi^2\varepsilon_0\sqrt{\varepsilon_r}\tan\delta\left(1 + j\frac{\tan\delta}{2}\right)\frac{f}{\sqrt{1-\left(\frac{f_c}{f}\right)^2}}\ \text{Np/m} \tag{3.141i}$$

Here, f_c is the cut-off frequency of the mode when no loss is present.

In the frequency region $f >> f_c$, the dielectric attenuation constant in the rectangular waveguide approaches that of α_0 for the unbounded dielectric medium.

From (3.140b) and (3.141i), the total attenuation constant is

$$\alpha = \alpha_c + \alpha_d \text{ Np/m} = 8.686\,\alpha \text{ dB/m} \tag{3.142}$$

The variation of attenuation constants α_c and α_d with frequency above cut-off for dominant TE_{10} mode are shown in Fig. 3.51 for a rectangular waveguide of dimensions a = 2.286 cm and b = 1.016 cm made of Cu walls and filled with RT/Duroid 5880 (ε_r = 2.26) and tan δ = 0.001. It is seen that the ohmic attenuation near cutoff is very high and decreasing with frequency to a lowest value and slowly increasing again. The dielectric attenuation constant is small above cut-off and increases at much higher rate as the frequency increases above cut-off. The results are obtained from the MATLAB programme given below.

```
%*****************************************
%Attenuation constant vs f in rect. W/g in dominant mode
%*****************************************
b = 1.016/100;
a = 2.25*b;
miu = 4.*pi.*10.^(-7);
eps = (10.^(-9))./(36.*pi);
sigma = 5.8*10.^7;
fc = (1./sqrt(miu.*eps))./(2.*a);
f = fc:0.1*10^9:30.*10^9;

alpc = (sqrt(eps)./b).*sqrt(pi.*f./(sigma.*(1-(fc./f).^2))).*(1+(2.*b./a)
 *(fc./f).^2);
alphac = 8.686*alpc;

plot(f,alphac);
grid;
%*****************************************
%Attenuation constant vs f in rect. W/g filled with lossy dielectric RT Duroidin
dominant mode
```

```
%*********************************************
b = 1.016/100;
a = 2.25*b;
i = sqrt(-1);
miu = 4.*pi.*10.^(-7);
epr = 2.2;
eps0 = ((10.^(-9))./(36.*pi));

eps = eps0.*epr;
tandel = 0.0006;
fc = (1./sqrt(miu.*eps))./(2.*a)
f = fc:0.1*10^9:30.*10^9;
sigmad = 2.*pi.*f.*eps.*tandel;
alpd = abs((120.*pi.^2*eps0.*sqrt(epr).*tandel.*(1+i.*0.5*tandel).*f./...
 (sqrt(1-(fc./f).^2))));
alphad = 8.686*alpd;

plot(f,alphad);
grid;
```

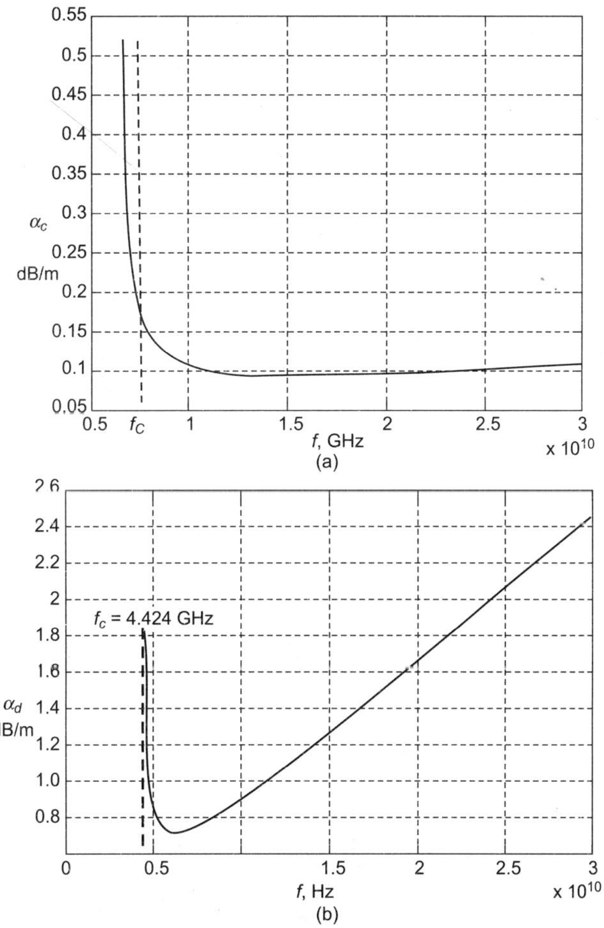

Fig. 3.51 Rectangular waveguide attenuation constants vs frequency (a) α_c (b) α_d

Example 3.16 A rectangular air-filled copper waveguide with a 0.9 inch × 0.4 inch cross section and 12 inch length is operated at 9.2 GHz with a dominant mode. Find (a) cut-off frequency, (b) guide wavelength, (c) phase velocity, (d) characteristic impedance, and (e) the loss.

Solution

Given $a = 0.9$ inch $= 2.28$ cm $\quad b = 0.4$ inch $= 1.01$ cm

$f = 9.2$ GHz $\quad \lambda = 30/9.2 = 3.26$ cm

$l = 12$ inch $= 30.48$ cm

Dominant mode is TE_{10}, $m = 1, n = 0$

(a) $$f_c = c/\lambda_0 = c/2a = 30/(2 \times 2.28) \text{ GHz} = 6.579 \text{ GHz}$$

(b) $$\lambda_g = \frac{\lambda}{\sqrt{[1-(f_c/f)^2]}} = \frac{3.26}{\sqrt{[1-(6.579/9.2)^2]}} = 4.664 \text{ cm}$$

(c) $$u_p = \frac{1}{\sqrt{\mu_0 \varepsilon_0 [1-(f_c/f)^2]}} = \frac{3\times 10^8}{\sqrt{[1-(6.579/9.2)^2]}} = 4.29 \times 10^8 \text{ m/sec}$$

(d) $$Z_0 = \frac{377}{\sqrt{[1-(f_c/f)^2]}} = 539.3 \text{ ohm}$$

(e) $$\alpha_c = \frac{R_s[1+(2b/a)(f_c/f)^2]}{b\sqrt{\mu_0/\varepsilon_0}\sqrt{\left[1-(f_c/f)^2\right]}} \text{ Np/m}$$

where $$R_s = \sqrt{\pi f \mu/\sigma} = \sqrt{\frac{[3.14\times 9.2\times 10^9 \times 4\pi \times 10^{-7}]}{5.8\times 10^7}} = 2.5 \times 10^{-2} \text{ ohm}$$

Therefore,

$$\alpha_c = \frac{2.5\times 10^{-2}}{1.01\times 377\times\sqrt{[1-(6.579/9.2)^2]}}[1+(2\times 1.01/2.286)(6.579/9.2)^2]$$

$$= 1.364 \times 10^{-4} \text{ Np/cm}$$

Therefore, total attenuation $= 1.364 \times 10^{-4} \times 30.48 = 4.1566 \times 10^{-3}$ Np

$= 8.686 \times 4.1566 \times 10^{-3}$ dB $= 0.036$ dB

Circular Waveguides 3.9.3

A circular waveguide is a cylindrical hollow metallic pipe with uniform circular cross section of a finite radius *a* as shown in Fig. 3.52. The general properties of the modes, TE_{mn} and TM_{mn}, excited in the circular waveguides are similar to those for the rectangular guide except that the coordinate system will be changed to a cylindrical one due to geometry involved. A unique property of the TM_{mn} modes in circular waveguides is rapid decrease in attenuation with increase in frequency. For this reason, the applications of a circular waveguide is important in long-length low-loss communication links.

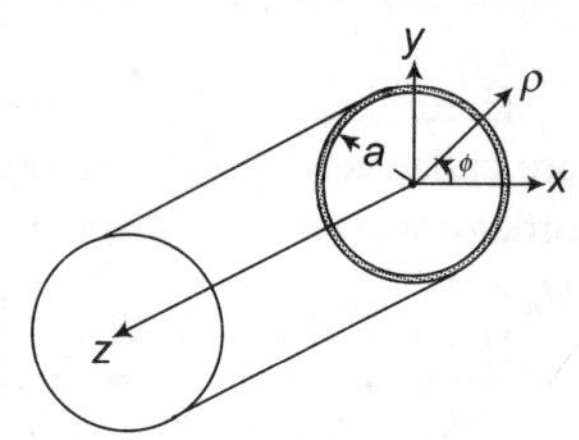

Fig. 3.52 Uniform circular waveguide

Ideally, an infinite length of a lossless uniform circular waveguide, filled with lossless homogenous dielectric material (ε, μ), can be assumed to be excited by a current source at one end located at infinity so that the solutions at an axial region are obtained from the following source-free homogeneous wave equation in the cylindrical coordinates (ρ, ϕ, z), since the geometry has a cylindrical symmetry:

$$\nabla_t^2\psi + k_c^2\psi = 0$$

or $$\frac{\partial^2\psi}{\partial\rho^2} + \frac{\partial\psi}{\rho\partial\rho} + \frac{1}{\rho^2}\frac{\partial^2\psi}{\partial\varphi^2} + k_c^2\psi = 0 \tag{3.143}$$

Here, $\psi = H_z$ for TE_{mn} modes

$= E_z$ for TM_{mn} modes

The z-dependent solution is $e^{-\gamma z}$ for propagation along the z-axis. The propagation constant in the guide is given by $\gamma = \pm\sqrt{k_c^2 - \omega^2\mu\varepsilon}$. The differential equation (3.143) can be solved by applying the method of separation of variables using initial solution for the propagation of wave in the z-direction as

$$\psi(\rho, \phi, z) = R(\rho)\,\Phi(\phi)\,e^{-\gamma z} \tag{3.144a}$$

Substituting $\psi(\rho, \phi)$ from (3.144a) in (3.143) and dividing by the product $R(r)\,\Phi(\phi)$, we obtain

$$\frac{\rho^2}{R(\rho)}\frac{d^2R(\rho)}{d\rho^2} + \frac{\rho}{R(\rho)}\frac{dR(\rho)}{d\rho} + k_c^2\rho^2 = -\frac{1}{\Phi(\phi)}\frac{d^2\Phi(\phi)}{d\phi^2} \tag{3.144b}$$

Since ρ and ϕ are independent variables, (3.144b) holds for all values of ρ and ϕ, both sides must be equal to the same constant, say, n^2. Therefore, we can separate (3.144b) into two ordinary differential equations:

$$\frac{d^2\Phi(\phi)}{d\phi^2} + n^2\Phi(\phi) = 0 \tag{3.144c}$$

and $$\frac{d^2R(\rho)}{d\rho^2} + \frac{1}{\rho}\frac{dR(\rho)}{d\rho} + \left(k_c^2 - \frac{n^2}{\rho^2}\right)R(\rho) = 0 \tag{3.144d}$$

Equation (3.144c) yields a simple harmonic solution

$$\Phi(\phi) = A_n\cos n\phi + B_n\sin n\phi \tag{3.144e}$$

where A_n and B_n are constants to be determined from the boundary conditions. Since all field components are periodic with respect to ϕ with period 2π, n is an integer. Choosing the reference $\phi = 0$ on the x-axis, nonzero solution exists when $B_n = 0$. Therefore,

$$\Phi(\phi) = A_n\cos n\phi \tag{3.144f}$$

Equation (3.144d) is known as **Bessel's differential equation**. The general solution of (3.144d) is

$$R(\rho) = C_nJ_n(k_c\rho) + D_nN_n(k_c\rho) \tag{3.144g}$$

Here, C_n and D_n are arbitrary constants to be determined from the boundary conditions. $J_n(k_c\rho)$ and $N_n(k_c\rho)$ are called the **Bessel function of the first kind** and **Bessel function of the second kind,** respectively, of order n with argument $k_c\rho$. Since the solution at $\rho = 0$ should be finite, whereas $N_n(k_c\rho) = \infty$ at $\rho = 0$, we take $D_n = 0$ for all n.

Hence,

$$R(\rho) = C_nJ_n(k_c\rho) \tag{3.144h}$$

Therefore, from (3.144f) and (3.144h), the general solution becomes

$$\psi(\rho, \phi, z) = R(\rho)\,\Phi(\phi)\,e^{-\gamma z} = A_nC_n\cos n\phi\,J_n(k_c\rho)e^{-\gamma z}$$

or $$\psi(\rho, \phi, z) = \psi_{0n}\,J_n(k_c\rho)\cos n\phi e^{-\gamma z} \tag{3.145}$$

where $\psi_{0n} = A_n C_n$, a new constant. $J_n(k_c\rho)$ represents the standing wave in the radial direction for $\rho \le a$ with finite field values at $\rho = 0$, and cos $n\phi$ represents the ϕ symmetric harmonic solution.

Similar to the rectangular waveguide case, the time harmonic ($e^{j\omega t}$) fields inside the guide satisfy the Maxwell's equations

$$\nabla \times \mathbf{E} = -\frac{\partial \mathbf{B}}{\partial t} = -j\omega\mu\mathbf{H} \text{ and } \nabla \times \mathbf{H} = -\frac{\partial \mathbf{D}}{\partial t} = -j\omega\varepsilon\mathbf{E} \tag{3.146a}$$

These two curl equations can be written in terms of the ρ, ϕ, z components of the **E** and **H** fields. For simplification, mathematical manipulations are used to express the transverse components of electric and magnetic fields in terms of the longitudinal components as follows.

Time harmonic wave propagating along $+z$ has solution in t and z variations as $e^{j\omega t}, e^{-\gamma z}$. Therefore, from ***Maxwell's curl equations***, in source free region (**J** = σ**E** = 0)

1. $$\nabla\times\mathbf{H} = (\sigma + j\omega\varepsilon)\mathbf{E} = j\omega\varepsilon\mathbf{E} = j\omega\varepsilon(\hat{\boldsymbol{\rho}}E_\rho + \hat{\boldsymbol{\phi}}E_\phi + \hat{\mathbf{z}}E_z) \tag{3.146b}$$

2. $$\nabla\times\mathbf{E} = -j\omega\mu\mathbf{H} = -j\omega\mu(\hat{\boldsymbol{\rho}}H_\rho + \hat{\boldsymbol{\phi}}H_\phi + \hat{\mathbf{z}}H_z) \tag{3.146c}$$

For the curl in cylindrical coordinates (ρ, ϕ, z),

$$\nabla \times \mathbf{A} = \left(\frac{1}{\rho}\frac{\partial A_z}{\partial \phi} - \frac{\partial A_\phi}{\partial z}\right)\hat{\boldsymbol{\rho}} + \left(\frac{\partial A_\rho}{\partial z} - \frac{\partial A_z}{\partial \rho}\right)\hat{\boldsymbol{\phi}} + \frac{1}{\rho}\left[\frac{\partial}{\partial \rho}(\rho A_\phi) - \frac{\partial A_\rho}{\partial \phi}\right]\hat{\mathbf{z}} \tag{3.146d}$$

Assuming $\sigma = 0$, and equating the component terms of (3.146b) and (3.146d),

(i) $$j\omega\varepsilon E_\rho = \frac{1}{\rho}\frac{\partial Hz}{\partial \phi} - \frac{\partial H_\phi}{\partial z} = \frac{1}{\rho}\frac{\partial H_z}{\partial \phi} + \gamma H_\phi \tag{3.146e}$$

(ii) $$j\omega\varepsilon E_\phi = -\gamma H_\rho - \frac{\partial H_z}{\partial \rho} \tag{3.146f}$$

(iii) $$j\omega\varepsilon E_z = \frac{1}{\rho}\frac{\partial}{\partial \rho}(\rho H_\phi) - \frac{1}{\rho}\frac{\partial H\rho}{\partial \phi} \tag{3.146g}$$

Similarly, equating the component terms of (3.146c) and (3.146d),

(iv) $$-j\omega\mu H_\rho = \frac{1}{\rho}\frac{\partial Ez}{\partial \phi} + \gamma E_\phi \tag{3.146h}$$

(v) $$-j\omega\mu H_\phi = -\gamma E_\rho - \frac{\partial E_z}{\partial \rho} \tag{3.146i}$$

(vi) $$-j\omega\mu H_z = \frac{1}{\rho}\frac{\partial}{\partial \rho}(\rho E_\phi) - \frac{1}{\rho}\frac{\partial E_\rho}{\partial \phi} \tag{3.146j}$$

Equations (3.146e)–(3.146j) are the **Maxwell's equations in scalar form** in cylindrical coordinates. Using the above equations, all cylindrical field components can be expressed in terms of E_z and H_z.

1. From (3.146e) and (v),

$$E_\rho = -\frac{j\omega\mu}{k_c^2}\frac{1}{\rho}\frac{\partial H_z}{\partial \phi} - \frac{\gamma}{k_c^2}\frac{\partial E_z}{\partial \rho} \tag{3.146k}$$

2. From (ii) and (iv),

$$H_\rho = -\frac{\gamma}{k_c^2}\frac{\partial H_z}{\partial \rho} + \frac{j\omega\varepsilon}{k_c^2}\frac{1}{\rho}\frac{\partial E_z}{\partial \phi} \tag{3.146m}$$

3. From (1) and (i),

$$H_\phi = -\frac{\gamma}{k_c^2}\frac{1}{\rho}\frac{\partial H_z}{\partial \phi} - \frac{j\omega\varepsilon}{k_c^2}\frac{\partial E_z}{\partial \rho} \tag{3.146n}$$

4. From (1) and (ii),

$$E_\phi = \frac{j\omega\mu}{k_c^2}\frac{\partial H_z}{\partial \rho} - \frac{\gamma}{k_c^2}\frac{1}{\rho}\frac{\partial E_z}{\partial \phi} \tag{3.146p}$$

Here, $k_c^2 = k^2 + \gamma^2$

Equations (3.146e-h) suggest that two basic sets of solutions exist, each defining a set of modes. Thus, there are two modes of propagation along z inside a hollow metal waveguide: TM and TE.

a. TM Modes in Circular Waveguides

Let a uniform circular waveguide of inner radius a, and infinitely long be made of perfect conductor and completely filled with lossless homogenous dielectric material (μ, ε). For TM waves propagating in axial (z) direction, $\psi = E_z$, and $H_z = 0$. We can write from (3.145),

$$\psi(\rho, \phi, z) = E_z = E_{0n}\, J_n(k_c\rho) \cos n\phi e^{-\gamma z} \tag{3.147a}$$

where n = 0,1,2,3,... From (3.146b), the other field components are determined from the following equations:

$$E_\rho = \frac{-\gamma}{k_c^2}\frac{\partial E_z}{\partial \rho} = -\frac{\gamma}{k_c}E_{0n}J_n'(k_c\rho)\cos n\phi e^{-\gamma z} \tag{3.147b}$$

$$E_\phi = \frac{-y}{k_c^2}\frac{\partial E_z}{\rho\partial \phi} = \frac{\gamma}{k_c^2}\frac{n}{\rho}E_{0n}J_n(k_c\rho)\sin n\phi e^{-\gamma z} \tag{3.147c}$$

$$H_\rho = \frac{j\omega\varepsilon}{k_c^2}\frac{\partial E_z}{\rho\partial \phi} = -\frac{j\omega\varepsilon}{k_c^2}\frac{n}{\rho}E_{0n}J_n(k_c\rho)\sin n\phi e^{-\gamma z} \tag{3.147d}$$

$$H_\phi = \frac{-j\omega\varepsilon}{k_c^2}\frac{\partial E_z}{\partial \rho} = -\frac{j\omega\varepsilon}{k_c}E_{0n}J_n'(k_c\rho)\cos n\phi e^{-\gamma z};\ J' = \frac{\partial \mathbf{J}}{\partial \rho} \tag{3.147e}$$

$$H_z = 0 \tag{3.147f}$$

In lossless guides, ($\alpha = 0$), $\gamma = j\beta$, and

$$k_c = \sqrt{(\gamma^2 + k^2)} = \sqrt{(-\beta^2 + \omega^2\mu\varepsilon)} \tag{3.147g}$$

Here, the coefficient depends on the field strength of excitation. The admissible values of k_c are called the **eigenvalues**, and they are determined from the boundary condition that $E_{\tan} = E_z = 0$ at $\rho = a$, that is from (3.147a).

$$J_n(k_c a) = J_n(x_{np}) = 0 \qquad \text{for TM}_{\text{np}} \text{ modes} \tag{3.148a}$$

Here, $x_{np} = k_c a$ represents the roots of (3.148a), where $p = 1, 2, 3, ...,$ denotes the p_{th} zero of (3.148a). There are infinite number of roots possible for (3.148a). It is given in standard mathematical books, $J_0(0) = 1$, $J_n(0) = 0$ for $n = 1,2,3,...$ Table 3.7 shows lower-order roots $x_{np}$of $J_n(x_{np}) = 0$. These roots determine the eigenvalues of the circular waveguide TM$_{np}$ modes, $n = 1, 2, 3, ...; \phi = 1, 2, 3, ...$

Table 3.7 Roots of $J_n(X) = 0$

p \ n	0	1	2	3	4	5
1	2.405	3.832	5.136	6.380	7.588	8.771
2	5.520	7.016	8.417	9.761	11.065	12.339
3	8.654	10.173	11.620	13.015	14.372	

The cut-off frequency is obtained from these roots:

$$k_c a = \frac{2\pi}{\lambda_c} a = \frac{2\pi f_c}{u} a = x_{np}$$

Or
$$f_c = \frac{x_{np} u}{2\pi a} = \frac{x_{np}}{2\pi a\sqrt{\mu\varepsilon}} \tag{3.148b}$$

From (3.148b) and Table 3.7, the TM$_{np}$ mode sequence from lowest to higher orders are

TM$_{01}$: $$f_c = \frac{2.405}{2\pi a\sqrt{\mu\varepsilon}}$$

TM$_{11}$: $$f_c = \frac{3.832}{2\pi a\sqrt{\mu\varepsilon}}$$

TM$_{21}$: $$f_c = \frac{5.136}{2\pi a\sqrt{\mu\varepsilon}}$$

TM$_{02}$: $$f_c = \frac{5.520}{2\pi a\sqrt{\mu\varepsilon}} \tag{3.148c}$$

and so on. The other parameters are determined in a similar way as done in the case of rectangular waveguide:

$$\beta = k\sqrt{1-(f_c/f)^2} \text{ rad/m} \tag{3.148d}$$

$$\lambda_g = \frac{\lambda}{\sqrt{1-(\lambda/\lambda_c)^2}} \text{ m} \tag{3.148e}$$

$$u_p = \frac{\omega}{\beta} = \frac{u}{\sqrt{1-(f_c/f)^2}} = \frac{\lambda_g}{\lambda} u > u \text{ ; m/s} \tag{3.148f}$$

$$u_g = \frac{1}{d\beta/d\omega} = u\sqrt{1-(f_c/f)^2} = \frac{\lambda}{\lambda_g} u < u \text{ ; m/s} \tag{3.148g}$$

$$Z_{TM} = \eta\sqrt{1-(f_c/f)^2} < \eta \text{ ohm} \tag{3.148h}$$

b. TE Modes in Circular Waveguides

For TE waves propagating in axial (z) direction, $\psi = H_z$, and $E_z = 0$. We can write from (3.145),

$$\psi(\rho, \phi, z) = H_z = H_{0n} J_n(k_c\rho)\cos n\phi e^{-\gamma z}; \tag{3.149a}$$

where n = 0, 1, 2, 3, …. From (3.146b), the other field components are determined from the following equations:

$$H_\rho = -\frac{\gamma}{k_c^2}\frac{\partial H_z}{\partial \rho} = -\frac{\gamma}{k_c}H_{0n}J_n'(k_c\rho)\cos n\phi e^{-\gamma z} \tag{3.149b}$$

$$H_\phi = -\frac{\gamma}{k_c^2}\frac{1}{\rho}\frac{\partial H_z}{\partial \varphi} = \frac{\gamma}{k_c^2}\frac{n}{\rho}H_{0n}J_n(k_c\rho)\sin n\phi e^{-\gamma z} \tag{3.149c}$$

$$E_\rho = -\frac{j\omega\mu}{k_c^2}\frac{1}{\rho}\frac{\partial H_z}{\partial \phi} = \frac{j\omega\mu}{k_c^2}\frac{n}{\rho}H_{0n}J_n(k_c\rho)\sin n\phi e^{-\gamma z} \tag{3.149d}$$

$$E_\phi = \frac{j\omega\mu}{k_c^2}\frac{\partial H_z}{\partial \rho} = \frac{j\omega\mu}{k_c}H_{0n}J_n'(k_c\rho)\cos n\phi e^{-\gamma z} \tag{3.149e}$$

$$E_z = 0 \tag{3.149f}$$

In lossless guides, ($\alpha = 0$) $\gamma = j\beta$, and

$$k_c = \sqrt{(\gamma^2 + k^2)} = \sqrt{(-\beta^2 + \omega^2\mu\varepsilon)} \tag{3.149g}$$

Here, the coefficient H_{0n} depends on the field strength of excitation. The admissible values of k_c are called the **eigenvalues**, and they are determined from the boundary condition that $|\mathbf{E}_{\tan}| = E\phi = 0$ at $\rho = a$, that is from (3.149e),

$$J'_n(k_c a) = J'_n(x'_{np}) = 0 \text{ for TE}_{np} \text{ modes} \tag{3.150a}$$

Here, $x'_{np} = k_c a$ represents the roots of (3.148a), where p = 1, 2, 3, ..., denotes the p_{th} zero of (3.150a). There is infinite number of roots possible for (3.150a). Table 3.8 shows lower-order roots x'_{np} of $J'_n(x'_{np})$ = 0. These roots determine the eigenvalues of the circular waveguide TE modes, n = 0, 1, 2, 3, ..., p = 1, 2, 3, ...

Table 3.8 Roots of $J'_n(X) = 0$

p \ n	0	1	2	3	4	5
1	3.832	1.841	3.054	4.201	5.317	6.416
2	7.016	5.331	6.706	8.015	9.282	10.520
3	10.174	8.536	9.970	11.346	12.682	13.987

The cut-off frequency is obtained from these roots:

$$k_c a = \frac{2\pi}{\lambda_c}a = \frac{2\pi f_c}{u}a = x'_{np}$$

or $$f_c = \frac{x'_{np}u}{2\pi a} = \frac{x'_{np}}{2\pi a\sqrt{\mu\varepsilon}} \tag{3.150b}$$

From (3.150b) and Table 3.8, the TE_{np} mode sequence from lowest to higher orders are

TE_{11}: $$f_c = \frac{1.841}{2\pi a\sqrt{\mu\varepsilon}}$$

TE_{21}: $$f_c = \frac{3.054}{2\pi a\sqrt{\mu\varepsilon}}$$

TE_{01}: $$f_c = \frac{3.832}{2\pi a\sqrt{\mu\varepsilon}}$$

TE_{31}: $$f_c = \frac{4.201}{2\pi a\sqrt{\mu\varepsilon}} \tag{3.150c}$$

and so on. The other parameters are determined in a similar way as done in the case of rectangular waveguide:

$$\beta = k\sqrt{1-(f_c/f)^2} \text{ rad/m} \tag{3.150d}$$

$$\lambda_g = \frac{\lambda}{\sqrt{1-(\lambda/\lambda_c)^2}} \text{ m} \tag{3.150e}$$

$$u_p = \frac{\omega}{\beta} = \frac{u}{\sqrt{1-(f_c/f)^2}} = \frac{\lambda_g}{\lambda}u > u \text{ m/S} \tag{3.150f}$$

$$u_g = \frac{1}{d\beta/d\omega} = u\sqrt{1-(f_c/f)^2} = \frac{\lambda}{\lambda_g}u < u \text{ m/S} \tag{3.150g}$$

$$Z_{TE} = \frac{\eta}{\sqrt{1-(f_c/f)^2}} > \eta \text{ ohm} \tag{3.150h}$$

Here, $k = \frac{2\pi}{\lambda} = \omega\sqrt{\mu\varepsilon}, \lambda = \frac{1}{f\sqrt{\mu\varepsilon}}$, and $\eta = \sqrt{\frac{\mu}{\varepsilon}}$ (3.150i)

The nature of variation of these parameters with frequency is similar to that for rectangular waveguide.

c. Dominant Mode

From (3.148c) and (3.150c), it can be seen that in a circular waveguide of internal radius a, the dominant mode is TE_{11} due to its lowest cut-off frequency for $m = n = 1$.

d. Degenerate Modes

It is seen from Tables 3.7 and 3.8, $x_{0p}' = x_{1p}$ and hence, all the TE_{0p} and TM_{1p} modes are degenerate in a uniform circular waveguide, $p = 1, 2, 3, ...$ The mode sequence from lowest to higher order modes is TE_{11}, TM_{01}, TE_{21}, TE_{01} and TM_{11} (degenerate modes), TE_{31}, TM_{21}, TE_{41}, TE_{12}, TM_{02}, and so on.

e. Mode Field Configurations

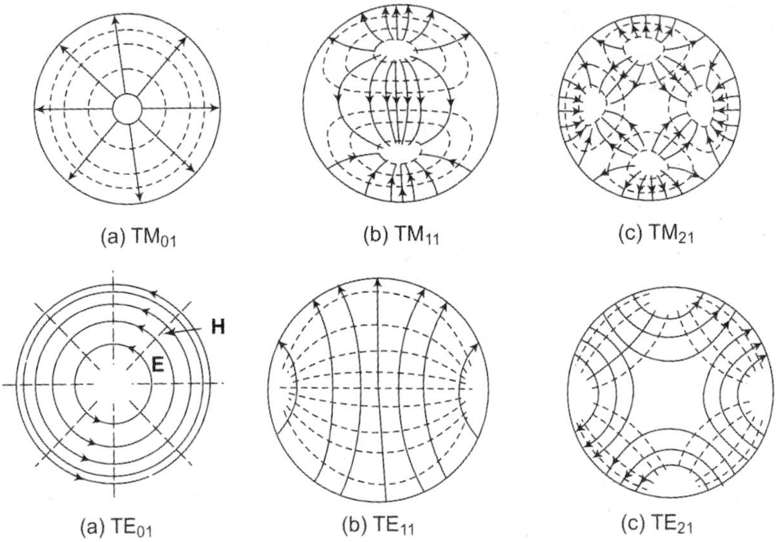

Fig. 3.53 Field configurations of different TE and TM modes

Excitation of Modes in Circular Waveguides

Since the electric field is generated with a polarisation parallel to the electric current and that magnetic field perpendicular to a loop of electric current, various TE and TM modes can be generated in a circular waveguide by coaxial line probes or loops as shown in Fig. 3.54. In Fig. 3.54(a) coaxial line probe excites the TE_{10} dominant mode in a rectangular waveguide which is converted to TE_{11} dominant mode in the circular waveguide through the transition length in between them. In Fig. 3.54(b) longitudinal coaxial line probe directly excites the symmetric TM_{01} mode. TE_{01} mode is excited by means of two diametrically oppositely placed longitudinal narrow slots parallel to the broad wall of the connected rectangular waveguides as shown in Fig. 3.54(c).

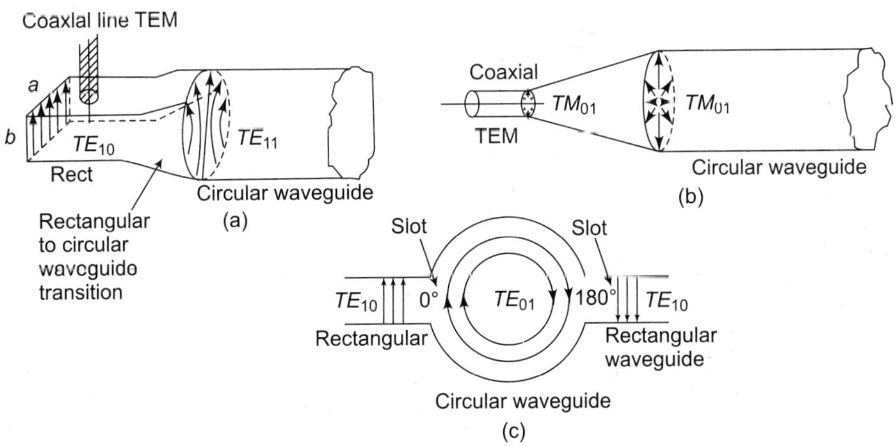

Fig. 3.54 Methods of excitation of TE and TM modes in circular waveguides: (a) TE_1 mode excitation (b) TM_{01} mode excitation (c) TE_{01} mode excitation

Attenuation in a Circular Waveguide

The attenuation in a circular waveguide for TE and TM modes can be found in the same manner as in rectangular waveguides. For an air-filled guide the attenuation is due to finite conductivity of the waveguide walls and can be expressed by

$$\alpha = \frac{\text{power loss per unit length}}{2 \times \text{average power transmitted}}$$

Average power transmitted is given by

$$P_{np} = \frac{1}{2Z_w} \int_0^{2\pi} \int_0^{a} \left(|E_\varphi|^2 + |E_\rho|^2 \right) \rho d\rho \, d\phi \tag{3.151a}$$

or

$$P_{np} = \frac{Z_w}{2} \int_0^{2\pi} \int_0^{a} \left(|H_\phi|^2 + |H_\rho|^2 \right) \rho d\rho \, d\phi \tag{3.151b}$$

From (3.150h) and (3.151a) for TE_{np} modes,

$$P_{np} = \frac{\sqrt{1-(f_c/f)^2}}{2\eta} \int_0^{2\pi} \int_0^{a} \left(|E_\rho|^2 + |E_\varphi|^2 \right) \rho d\rho \, d\phi \tag{3.152}$$

From (3.1148h) and (3.151a) for TM_{np} modes,

$$P_{np} = \frac{1}{2\eta\sqrt{1-(f_c/f)^2}} \int_0^{2\pi} \int_0^{a} \left(|E_\rho|^2 + |E_\phi|^2 \right) \rho d\rho \, d\phi \tag{3.153}$$

Power loss P_L per unit length of the guide is given by

$$P_L = \frac{R_s}{2} \oint_{\text{guide walls}} \mathbf{J}_s \cdot \mathbf{J}_s^* \, dl \tag{3.154}$$

The final expressions for the attenuation constant are obtained as

$$\alpha_{TE} = \frac{R_s}{aZ_0} (1 - f_c^2/f^2)^{-1/2} \left(\frac{f_c^2}{f^2} + \frac{n^2}{x_{np}^2 - n^2} \right) \tag{3.155}$$

$$\alpha_{TM} = \frac{R_s}{aZ_0} (1 - f_c^2/f^2)^{-1/2} \tag{3.156}$$

For TE_{op} modes attenuation falls off as $f^{-3/2}$ according to

$$\alpha = \frac{R_s}{aZ_0} \frac{f_c^2}{f\,(f^2 - f_c^2)^{-1/2}} \tag{3.157}$$

as shown in Fig. 3.55. Such rapid decrease in attenuation with increasing frequency makes it possible to use TE_{01} mode for very long low-loss waveguide communication links. But there are practical difficulties while operating the guide at a frequency well above the dominant mode TE_{11}. At these frequencies, any small discontinuities in the guide converts the power in the TE_{01} mode to other modes with different propagation phase constants. At a sufficient distance, away from the discontinuities; these additional modes are again converted back into a TE_{01} mode leading to signal distortion.

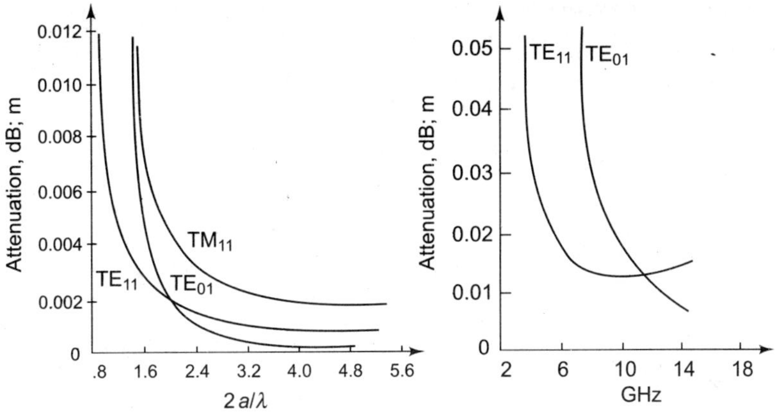

Fig. 3.55 Attenuation characteristics in a circular waveguide

Example 3.17 An air-filled circular waveguide having an inner radius of 1 cm is excited in dominant mode at 10 GHz. Find (a) the cut-off frequency of dominant mode, (b) guide wavelength, (c) wave impedance. Find the bandwidth for operation in dominant mode only.

Solution Given $a = 1$ cm, $f = 10$ GHz. The dominant mode is TE_{11}, $m = 1$, $n = 1$. $\lambda = c/f = 30/10 = 3$ cm.

(a)
$$f_c = cx'_{11}/2\pi a$$
$$= \frac{3 \times 10^{10} \times 1.841}{2 \times 3.14 \times 1} = 8.795 \text{ GHz}$$

(b)
$$\lambda_g = \frac{\lambda}{\sqrt{[1-(f_c/f)^2]}} = \frac{3}{\sqrt{[1-(8.795/10)^2]}} = 6.303 \text{ cm}$$

(c)
$$Z_w = 377 \times \lambda_g/\lambda$$
$$= 377 \times 6.303/3$$
$$= 792 \text{ ohm}$$

$$\text{Bandwidth} = f_c \text{ of next higher order-mode } TM_{01} - f_c \text{ of } TE_{11}$$
$$= c \times x_{01}/2\pi a - 8.795$$
$$= \frac{3 \times 10^{10} \times 2.405}{2 \times 3.14 \times 1} - (8.795 \times 10^9)$$
$$= 11.49 - 8.795 = 2.695 \text{ GHz}$$

Ridge Waveguides 3.9.4

A ridge waveguide is formed with a rectangular ridge projecting inward from one or both of the wide walls in a rectangular waveguide as shown in Fig. 3.56. The ridge has the effect of increasing the capacitance between the wide walls by reducing the dimension at the maximum **E**-field region parallel to **E**. This capacitance lowers the cut-off frequency of the dominant mode TE_{10} and increases the operating frequency range. Of course, the ridge lowers the effective impedance of the guide. The ridge does not however, disturb the next higher-order mode, TE_{20}.

The disadvantages of a ridge guides over a normal rectangular waveguide are (1) the reduction in the power it can handle, and (2) higher attenuation. An approximate expression of cut-off frequency for the ridge guide is

$$f_c = \frac{1}{2\pi}\sqrt{\frac{4g}{\mu\,\varepsilon bd(a-d)}}$$

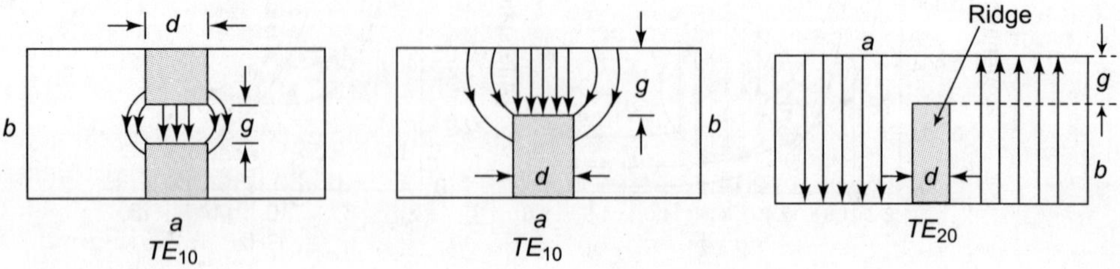

Fig. 3.56 Ridge waveguides of **E**-field

Surface Waveguides 3.9.5

A surface waveguide is an open boundary structure, such as a dielectric rod, dielectric-coated conducting wire, dielectric sheet on a metal plane, a corrugated conducting plane or cylinder as shown in Fig. 3.57.

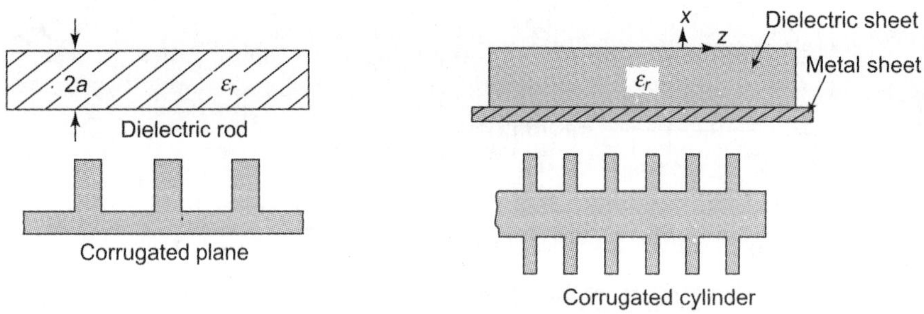

Fig. 3.57 Surface waveguides

These structures can guide an electromagnetic wave in longitudinal direction with a propagation function $e^{-j\beta z}$, but the field strength decays from the surface exponentially in transverse direction. Such waveguides are characterised by

1. Hybrid mode HE_{mn} of propagation (TE_{mn} + TM_{mn}), and axisymmetric pure TM_{on} or TE_{on} modes
2. Zero frequency cut-off for non-axisymmetric dominant hybrid mode HE_{11}
3. A finite number of discrete modes at a given frequency, together with a continuous eigen value spectrum

Dielectric Rod Waveguides

A uniform cylindrical low-loss high dielectric constant dielectric rod is commonly used for dielectric surface waveguide. Wave propagation takes place by total internal reflection of an obliquely incident wave at the interface between the dielectric and the air medium outside. The field intensity drops off exponentially with an increase in distance from the outer surface of the rod. Thus, the propagation energy is confined inside and at the surface of the rod with decrease in radiation loss as the ratio diameter/wavelength and the dielectric constant increase. The axisymmetric modes are pure TM or TE modes having a nonzero cut-off frequency whereas all modes with angular dependence are a combination of a TM and a TE mode, called the hybrid *HE* or *EH* modes, depending on whether the TE or TM mode predominates, respectively. Hybrid modes consist of all components of **E** and **H** fields. The HE_{11} mode is the dominant mode having zero cut-off frequency as shown in Fig. 3.58. The propagation constant $\beta > k_0$ inside the dielectric so that the phase velocity is less than that of light in free space.

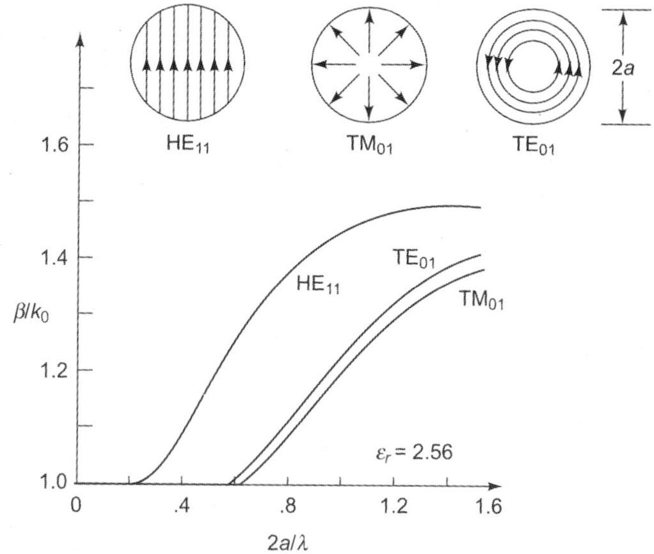

Fig. 3.58 Propagation characteristic

The cut-off wavelength for the first higher order-mode in a dielectric rod of radius a is given by

$$\lambda_c \cong 2.6a\sqrt{\varepsilon_r - 1} \tag{3.158}$$

Thus for higher frequency (mm waves), dielectric waveguides must be very thin. At optical frequencies, flexible dielectric waveguide or optical fibre is manufactured in the form of a thin filament of dielectric which is coated successively by a number of other dielectrics of lower dielectric constants to transmit optical signals (light).

❑ *Excitation of a Dielectric Waveguide* The easiest way to excite the dominant mode in a dielectric rod is to use a circular waveguide with flared flanges as shown in Fig. 3.59. The rod is tapered inside the guide for impedance matching.

For launching of waves from a circular waveguide to the dielectric rod, there is a discontinuity at the junction between the waveguide and the rod. This causes reflection of waves. A flared horn makes this launching smooth with gradual transition from the waveguide to the dielectric rod with minimum reflection. The launching efficiency is given by

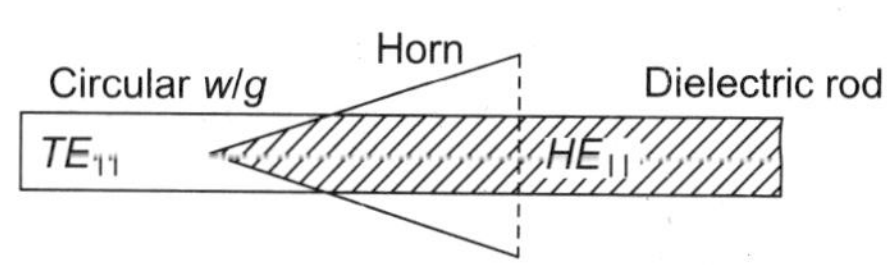

Fig. 3.59 Excitation of hybrid modes in a dielectric guide

$$\eta_L = \frac{\text{Surface Wave Power}}{\text{Input Power} - \text{Reflected Power}} = \frac{|S_{12}|^2}{1-|S_{11}|^2} \tag{3.159}$$

η_L can be determined by measuring *S*-parameters using *Deschamp's method.* This utilises measured values of the complex input reflection coefficient for eight different lengths of the uniform dielectric rod changing in steps of 1/8th guided wavelength with the free end of the rod short-circuited by a shorting plate of dimensions much larger than the free space wavelength.

Power-Handling Capability of Microwave Transmission Lines 3.9.6

The maximum power that can be transmitted through a microwave transmission line is limited by the *RF* electric breakdown in the region where the electric field intensity is at maximum E_{bd} in the guide or line. At atmospheric pressures, electric breakdown occurs in the form of a spark discharge with a loud sound. At low pressures in the upper atmospheres, a glow discharge occurs. Large reflections of the propagating signal occur at the discharge location. The reflected wave produces instability of the signal source at the sending end and can even damage source. At the breakdown point the conductor of the line or guide becomes oxidized or even burnt out. The level of breakdown voltage or field depends on the nature of the medium (gas), gas pressure, frequency and degree of initial gas ionisation.

Rectangular Waveguide

The electric and magnetic fields of dominant TE_{10} mode in a rectangular waveguide section are given by the following components.

$$E_y = -\frac{j\omega\mu}{k_c^2}\left(\frac{\pi}{a}\right)H_0 \sin(\pi x/a)e^{-\gamma z}$$

$$H_x = \frac{\gamma}{k_c^2}\left(\frac{\pi}{a}\right)H_0 \sin(\pi x/a)e^{-\gamma z}$$

$$H_z(x, y, z) = H_0 \cos(\pi x/a)e^{-\gamma z}$$

❑ ***Wall Currents*** The electric current distributions on the waveguide walls are obtained from the relation $\mathbf{J} = \hat{\mathbf{n}} \times \mathbf{H}$ and are shown in Fig. 3.60 for the dominant TE_{10} mode.

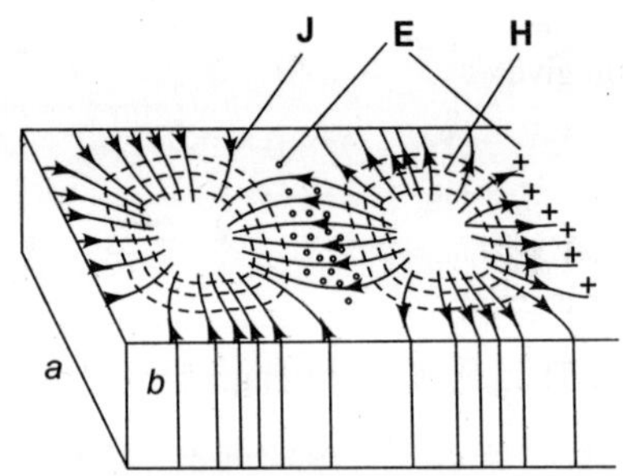

Fig. 3.60 Rectangular waveguide wall currents

❑ ***Electric Breakdown*** The time average power flow through the rectangular waveguide for dominant TE_{10} mode is given by

$$P_{10} = \frac{1}{2}\text{Re}\int \mathbf{E}\times\mathbf{H}^*.\mathbf{dS}$$

$$= \frac{E_0^2\, ab}{4Z_w}\left(\frac{\beta}{k_c}\right)^2$$

$$= \frac{E_0^2\, ab}{4}\frac{[1-(f_c/f)^2]^{1/2}}{\eta} \tag{3.160}$$

where $\eta = \sqrt{\dfrac{\mu_0}{\varepsilon_0}}$, the intrinsic impedance of free space and Z_w is the characteristic wave impedance for dominant mode. It is known experimentally that in dry air at normal atmospheric pressure, the microwave electric breakdown takes place when the electric field level at the middle of the broad wall is of the order of 30 kV/cm. Substituting this value of field in Eq. 3.160, the corresponding breakdown power is obtained for the dominant TE_{10} mode as

$$(P_{bd})_{10} \approx 597\, ab[1 - (\lambda_0/2a)^2]^{1/2} \text{ kW} \tag{3.161}$$

where a, b and λ_0 are in cm.

For a rectangular air-filled waveguide (2.3 cm × 1.0 cm) the breakdown power in dominant mode at 9.375 GHz is approximately 987 kW. The power handing capacity of the guide is given by

$$P_{\max} = 27\left(\frac{E_{bd}}{f_{\max}}\right)\sqrt{1-\left(\frac{f_c}{f}\right)^2} \text{ watts}$$

$f_{\max}$ = maximum operable frequency

Circular Waveguide

For the dominant TE_{11} mode and the minimum attenuation TE_{01} mode in circular air-filled waveguide, the breakdown powers for a maximum field of 30 kV/cm, are respectively given by

$$(P_{bd})_{11} \approx 1790\, a^2\, [1 - (f_{c11}/f)^2]^{1/2} \text{ kW} \tag{3.162}$$

$$(P_{bd})_{01} \approx 1805\, a^2\, [1 - (f_{c01}/f)^2]^{1/2} \text{ kW} \tag{3.163}$$

where a is the radius of the circular waveguide in cm, and f_{c11} and f_{c01} are the cut-off frequencies of TE_{11} and TE_{01} modes, respectively.

Coaxial Line

In coaxial lines, the maximum field strength (breakdown) is determined by the cable dielectric breakdown and in given by

$$E_{bd} = \frac{V_{\text{peak}}/2a}{\ln(b/a)} \tag{3.164}$$

where V_{peak} is the peak voltage at breakdown. For harmonic time variation in a matched line, the breakdown power is given by

$$P_{bd} = \left[\frac{V_{\text{peak}}}{\sqrt{2}}\right]^2 / Z_0 = 4a^2 E_{bd}^{\ 2} \ln(b/a) \tag{3.165}$$

For E_{bd} = 30 kV/cm, the breakdown power for air-filled coaxial line for TEM mode is given by

$$P_{bd} = 3600\, a^2 \ln(b/a), \text{ kW} \tag{3.166}$$

where all dimension are expressed in cm.

Since the cross-sectional dimensions of coaxial lines are restricted to satisfy the conditions $(b + a) < \lambda/\pi$ to exclude higher-order modes, and for a 50 ohm line, $b/a \approx 2.3$, for the single dominant mode propagation, $a = 0.3\ \lambda/\pi$, $b = 0.7\ \lambda/\pi$ and the breakdown power becomes

$$(P_{bd})_{\text{TEM}} = 398 \text{ kW} < (P_{bd})_{\text{TE10}} \text{ at } 9.375 \text{ GHz.} \tag{3.167}$$

Therefore, the breakdown power of coaxial lines is always lower than that of dominant waveguides operating at same frequency due to reduced separation distance between the conductors compared to that between waveguide walls. Due to a similar reason, microstrip lines have further lower breakdown voltage. The smaller the dimensions of a transmission line, the lower the breakdown power.

The breakdown is reduced at points of discontinuities in the line where intensities of the electric field is higher. In many applications, transmission line ends are fitted with a connector of various dimensions. Due to inherent constructional discontinuities in the connector, the breakdown power in the line is limited not by the guide or line but by the connectors.

The breakdown field is highly dependent on the gas pressure inside the guide as shown schematically in Fig. 3.61 as a function of frequency and pressure.

When the high-power microwave equipment is designed to operate at high altitude (low pressure), the waveguide is usually pressurised to increase the breakdown voltage.

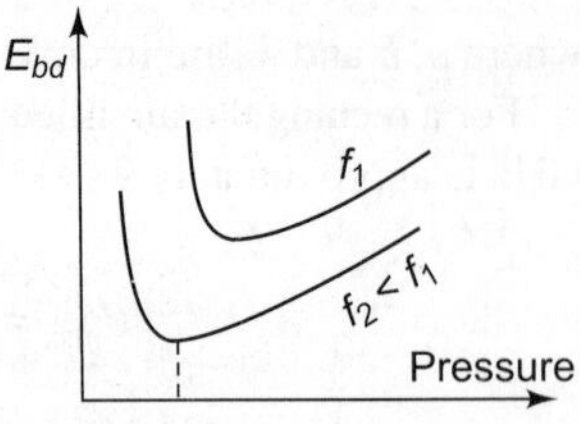

Fig. 3.61 E_{bd} vs gas pressure

Example 3.18 A rectangular waveguide of 2.3 cm × 1 cm dimensions is excited in dominant mode at 9.375 GHz. Calculate the breakdown power.

Solution Dominant mode is TE_{10}

$a = 2.3$ cm, $b = 1.0$ cm

$f = 9.375$ GHz, $\lambda_0 = 3.2$ cm

Assuming breakdown field = 30 kV/cm.

$$P_{bd} = 597\, ab \sqrt{1-(\lambda/2a)^2} \text{ kW}$$

$$= 597 \times 2.3 \times 1.0 \times \sqrt{1-\left(\frac{3.2}{2\times 2.3}\right)^2} = 986.4 \text{ kW}$$

Example 3.19 An air-filled coaxial line is operating at $\lambda = 3.2$ cm in TEM mode. Assuming the ratio of outer and inner radii of the conductors is $b/a = 3$ and $a = \lambda/4\pi$, calculate the breakdown power.

Solution

$$P_{bd} = 3600\, a^2 \ln \frac{b}{a} \text{ kW} = 256.5 \text{ kW}$$

Microstrip Line

Since thin dielectric substrates are normally used in microstrip lines, the power handling capability of these lines is lower compared to those of coaxial lines and waveguides. The average power handling is limited by (1) heat generated due to ohmic loss and loss in the dielectric substrate and, (2) by dielectric breakdown due to close proximity between ground and the trace conductors at different potentials. The increase in temperature due to conductor and dielectric losses limits the average power of the microstrip line while the breakdown between the strip conductor and ground plane limits the peak power. Calculations for average power-handling capabilities of microstrip lines are presented by Bahl and Gupta [xx]. The maximum average power handling capability for a given line is expressed by

$$P_{av} = (T_{max} -- T_{amb})/\Delta T$$

where ΔT denotes the rise in temperature per watt, T_{max} is the maximum operating temperature of the substrate material, and T_{amb} is the ambient temperature.

The maximum operating temperature of microstrip circuits is the temperature up to which the electrical and physical characteristics remain unchanged.

For polystyrene, maximum operating temperature is 100° C whereas, for most of the other dielectrics, it is much more than 100° C, depending on the change of substrate properties with temperature, the change of physical dimensions with temperature, and the type of connectors. Taking 100° C as a reasonable maximum temperature and 25° C as the ambient value, and line impedance Z_0 = 50 ohm, average power handling capacity may be calculated. For typical substrate thickness of 0.6 mm (25 mil), variation of average power handling with frequency is approximately as shown in Fig. 3.62 for alumina and GaAs substrates. It is seen that the value decreases with increase in frequency.

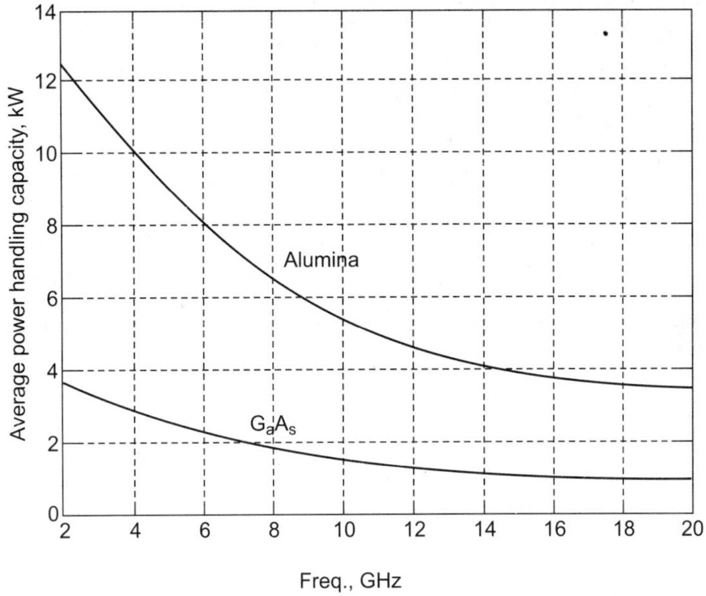

Fig. 3.62 Variation of average power-handling capacity vs. frequency

Example 3.20 Express Maxwell's equations (1) and (2) in scalar form in cylindrical coordinate system for time harmonic field in a medium $\sigma = 0$.

(1) $\nabla \times \mathbf{H} = \mathbf{J} + \dfrac{\partial \mathbf{D}}{\partial t}$ (2) $\nabla \times \mathbf{E} = -\dfrac{\partial \mathbf{B}}{\partial t}$

Solution Time harmonic wave propagating along $+z$ has solution in t and z variations as $e^{j\omega t}, e^{-\gamma z}$. Therefore,

(a) $$\nabla \times \mathbf{H} = (\sigma + j\omega\varepsilon)\mathbf{E} = j\omega\varepsilon\mathbf{E} = j\omega\varepsilon(\hat{\boldsymbol{\rho}} E_\rho + \hat{\boldsymbol{\phi}} E_\phi + \hat{\mathbf{z}} E_z) \quad (1)$$

(b) $$\nabla \times \mathbf{E} = -j\omega\mu\mathbf{H} = -j\omega\mu(\hat{\boldsymbol{\rho}} H_\rho + \hat{\boldsymbol{\phi}} H_\phi + \hat{\mathbf{z}} H_z) \quad (2)$$

For the curl in cylindrical coordinates (ρ, ϕ, z),

$$\nabla \times \mathbf{A} = \left(\frac{1}{\rho}\frac{\partial A_z}{\partial \phi} - \frac{\partial A_\phi}{\partial z}\right)\hat{\boldsymbol{\rho}} + \left(\frac{\partial A_\rho}{\partial z} - \frac{\partial A_z}{\partial \rho}\right)\hat{\boldsymbol{\phi}} + \frac{1}{\rho}\left[\frac{\partial}{\partial \rho}(\rho A_\phi) - \frac{\partial A_\rho}{\partial \phi}\right]\hat{\mathbf{z}}$$

Assuming $\sigma = 0$, and equating the component terms of Eqs. (1) and (3):

(i) $$j\omega\varepsilon E_\rho = \frac{1}{\rho}\frac{\partial Hz}{\partial \phi} - \frac{\partial H_\phi}{\partial z} = \frac{1}{\rho}\frac{\partial H_z}{\partial \phi} + \gamma H_\phi \quad (3)$$

(ii) $$j\omega\varepsilon E_\phi = -\gamma H_\rho - \frac{\partial H_z}{\partial \rho}$$

(iii) $$j\omega\varepsilon E_z = \frac{1}{\rho}\frac{\partial}{\partial \rho}(\rho H_\phi) - \frac{1}{\rho}\frac{\partial H\rho}{\partial \phi}$$

Equating the component terms of Eqs. (2) and (3):

(iv) $$-j\omega\mu H_\rho = \frac{1}{\rho}\frac{\partial E_z}{\partial \phi} + \gamma E_\phi$$

(v) $$-j\omega\mu H_\phi = -\gamma E_\rho - \frac{\partial E_z}{\partial \rho}$$

(vi) $$-j\omega\mu H_z = \frac{1}{\rho}\frac{\partial}{\partial \rho}(\rho E_\phi) - \frac{1}{\rho}\frac{\partial E_\rho}{\partial \phi}$$

Equations from serial number (i) to (vi) are the **Maxwell's equations in scalar forms.**

Example 3.21 Using the equations of Example 3.20, find all cylindrical field components in terms of E_z and H_z.

Solution From (i) and (v),

$$E_\rho = -\frac{j\omega\mu}{k_c^2}\frac{1}{\rho}\frac{\partial H_z}{\partial \phi} - \frac{\gamma}{k_c^2}\frac{\partial E_z}{\partial \rho}$$

From (ii) and (iv),

$$H_\rho = -\frac{\gamma}{k_c^2}\frac{\partial H_z}{\partial \rho} + \frac{j\omega\varepsilon}{k_c^2}\frac{1}{\rho}\frac{\partial E_z}{\partial \phi}$$

From (1) and (i),

$$H_\phi = -\frac{\gamma}{k_c^2}\frac{1}{\rho}\frac{\partial H_z}{\partial \phi} - \frac{j\omega\varepsilon}{k_c^2}\frac{\partial E_z}{\partial \rho}$$

From (1) and (ii),

$$E_\phi = \frac{j\omega\mu}{k_c^2}\frac{\partial H_z}{\partial \rho} - \frac{\gamma}{k_c^2}\frac{1}{\rho}\frac{\partial E_z}{\partial \phi}$$

Here, $$k_c^2 = k^2 + \gamma^2$$

Example 3.22 Calculate Z_{TE} from the field components in cylindrical coordinates.

Solution For TE mode $E_z \equiv 0$. Wave impedance is defined as $Z_{TE} \equiv \frac{|\mathbf{E}_t|}{|\mathbf{H}_t|}$, where suffix t represents total transverse components of fields. Example 3.21 gives:

$$E_\rho = -\frac{j\omega\mu}{k_c^2}\frac{1}{\rho}\frac{\partial H_z}{\partial \phi}$$

$$H_\rho = -\frac{\gamma}{k_c^2}\frac{\partial H_z}{\partial \rho}$$

$$H_\phi = -\frac{\gamma}{k_c^2}\frac{1}{\rho}\frac{\partial H_z}{\partial \phi}$$

$$E_\phi = \frac{j\omega\mu}{k_c^2}\frac{\partial H_z}{\partial \rho}$$

Now $$|\mathbf{H}_t| = \sqrt{H_\rho^2 + H_\phi^2} = \sqrt{\left(\frac{\gamma}{k_c^2}\frac{\partial H_z}{\partial \rho}\right)^2 + \left(\frac{\gamma}{k_c^2}\frac{1}{\rho}\frac{\partial H_z}{\partial \phi}\right)^2}$$

$$= \frac{\gamma}{k_c^2}\sqrt{\left(\frac{k_c^2}{j\omega\mu}\right)^2 E_\phi^2 + \left(\frac{k_c^2}{j\omega\mu}\right)^2 E_\rho^2} = \left(\frac{\gamma}{\omega\mu}\right)\sqrt{E_\rho^2 + E_\phi^2} = \frac{\gamma}{\omega\mu}|\mathbf{E}_t|$$

Therefore,

Wave impedance $$Z_{TE} = \frac{|\mathbf{E}_t|}{\left|\frac{\gamma}{\omega\mu}\mathbf{E}_t\right|} = \frac{\omega\mu}{\gamma}$$

Example 3.23 Calculate Z_{TM} from the field of components in rectangular coordinates x,y,z.

Solution For TM mode, $H_z \equiv 0$. Wave impedance is defined as $Z_{TM} \equiv \frac{|\mathbf{E}_t|}{|\mathbf{H}_t|}$, where suffix t represents total transverse components of fields.

(2a) $$-j\omega\varepsilon E_x = jkH_y + \frac{\partial H_x}{\partial y}$$

(2b) $$-j\omega\varepsilon E_y = -jkH_x - \frac{\partial H_z}{\partial x}$$

Solution

$$|E_x|^2 + |E_y|^2 = \left(\frac{k_{TM}}{\omega\varepsilon}\right)^2 \left(|H_y|^2 + |H_x|^2\right) \text{ or } |\mathbf{E}_t|^2 = \frac{k_{TM}}{\omega\varepsilon}|\mathbf{H}_t|$$

$$\eta_{TE} \equiv \frac{|\mathbf{E}_t|}{|\mathbf{H}_t|} = \frac{k_{TM}}{\omega\varepsilon}$$

REVIEW QUESTIONS

3.1 Distinguish between the characteristic impedance and the input impedance of a line. Under what conditions will they be the same? In a lossless line how can the input impedance be made purely inductive, capacitive, infinite, and zero?

3.2 Sketch the standing wave patterns for voltage and current along a line when the termination is (a) open circuit, (b) short circuit, (c) purely inductive load, (d) purely capacitive load, (e) matched load Z_0, (f) resistive impedance $> Z_0$ and $< Z_0$.

3.3 A lossless line with a characteristic impedance of 50 ohms is terminated by a load Z_L of VSWR 1.2 and fed by a matched generator of 100 mW power at the input. Calculate the minimum and maximum values of voltages along the line.

3.4 A lossless transmission line with the characteristic impedance of 100 ohm produces a voltage minimum at a distance of 20 cm from the load end when short-circuited. When the short is replaced by a load, the same voltage minimum is shifted by 9 cm towards the load. If the VSWR of the load is 3.0, using the Smith chart, find the value of the load.

3.5 A transmission line of 10 m length with a characteristic impedance of 50 ohm, $C = 65$ pF/m, and $\alpha = 0.06$ dB/m is fed by a matched generator and terminated by a matched load. If ac source voltage is $10/\angle 0°$ at 100 MHz, find the voltage at the load.

3.6 A lossless transmission line with a characteristic impedance of 50 ohm is terminated by a load of $40 + j$ 30 ohm. If the magnitude of the incident wave is 10 V, find the maximum and minimum values of voltage on the line and VSWR at the load.

3.7 Calculate the incident power for the transmission line of Example 2.6 and also the net power delivered to the load.

3.8 A lossless transmission line has $C = 95$ pF/m, $Z_0 = 100$ ohms. If the operating frequency is 3 GHz, calculate

(a) phase constant (b) phase velocity (c) line wavelength

3.9 An air-filled rectangular copper waveguide of 0.9 inch × 0.4 inch cross section is operated at 10 GHz. Calculate the attenuation constant in dB/m. Find the frequency and attenuation at which the attenuation is minimum for TE_{10} mode.

3.10 A circular waveguide is filled with a lossless dielectric of $\varepsilon_r = 9$. If the cut-off frequency is 6 GHz, calculate the diameter and find the upper frequency limit over which only the dominant mode will propagate.

3.11 An air-filled X-band waveguide (0.9 inch × 0.4 inch) is fed from a signal source at 8 GHz. Calculate the wave impedance. Explain about the characteristic of single propagation through the guide.

3.12 Design a rectangular waveguide with filling by a dielectric of $\varepsilon_r = 4$ so that the cut-off frequency for the dominant mode is 14 GHz and the cut-off frequency for the TM_{11} mode is 30 GHz.

3.13 A rectangular waveguide (0.9 inch × 0.4 inch) made of brass ($\sigma = 1.1 \times 10^7$ mho/m) is excited at 9 GHz in dominant mode. If the guide is filled with Teflon ($\varepsilon_r = 2.2$. tan $\delta = 0.0005$), calculate α_c and α_d.

3.14 Calculate the cut-off frequencies for TE_{11} TM_{01} and TE_{01} modes in a circular waveguide of radius 1 cm filled with a dielectric of $\varepsilon_r = 2.5$.

3.15 A microstrip line with $W = 2$ mm, $h = 0.4$ mm has a quartz substrate of $\varepsilon_r = 3.8$ and tan $\delta = 0.0001$. Find the (a) effective permittivity (b) Z_0 (c) α_c and α_d at 9 GHz.

3.16 Design a 50 ohm strip line with PTFE substrate material of $\varepsilon_r = 9.6$ and $h = 1/8$". Calculate the cut-off wavelengths for higher order TM_{11} and TE_{10} modes.

REFERENCES

1. Collin, R. E., *Foundation for Microwave Engineering*, McGraw-Hill, Inc., International Editions, 1992.
2. Collin, R. E., *Field Theory of Guided Waves*, McGraw-Hill Book Co., 1960.
3. Lebedev, I., *Microwave Engineering*, MIR Publishers, Moscow, 1973.
4. Edwards, T., *Foundations for Microstrip Circuit Design*, Sec. Edition, John Wiley and Sons, 1992.
5. Ramo, J. R., Whinnery, and T. V. Duzer, *Firleds and Waves in Communication Electronics*, 2nd Ed. John Wiley and Sons, 1984.
6. Liboff, R. L. and G. C. Dalmon, *Transmission lines. Waveguides, and Smith Charts*, McMillan Publishing Co., New York, 1985.
7. Bharathi Bhat and S. K. Koul, *Stripline-like Transmission Lines for Microwave Integrated Circuits*, Wiley Eastern Ltd., 1989.
8. Marcuvitz, N., *Waveguide Handbook*, Peter Peregrinus Ltd. (IEE), 1986.
9. Gupta, K. C., G. Ramesh, and I. J. Bahl, *Microstrip lines and slotlines*. Artech House, Inc., Dedham, Massachusetts, 1979.
10. Sander, K. F. and G. A. Reed, *Transmission and Propagation of Electromagnetic Waves*, Cambridge University Press, Cambridge, 1978.
11. Johnson, W. C., *Transmission Line & Networks*, McGraw Hill, New York, 1950.
12. Wheeler, H. A., *Reflection charts Relating to impedance Matching*. IEEE Trans. on MTT, no. 9 Sept.,z 1984.

13. Bahl, I.J., and Gupta, KC: Average Power-handling Capability of Microstrip Lines, *IEEE Microwaves, Optics and Acoustics,* January 1979, Vol. 3, No.1.

14. Bahl, Average Power-handling capability of multiplayer Microstrip Lines," *Int. J. RF Microw. CAE*, vol. 11, pp. 385–395, 2001.

15. Wen-Yan Yin, Dong, XT Junfa Mao, and Le-Wei Li: *Average Power Handling Capability of Finite-Ground Thin-Film Microstrip Lines Over Ultra-Wide Frequency Ranges*, IEEE Microwave and Wireless Components Letters, Vol. 15, no. 10, October 2005

16. *Gupta, KC, Garg, R Bahl, I.J. and Bhartia P:* Microstrip Lines and Slotlines, *2nd Ed.Artech House,* Inc., *1996.*

17. *Collin, RE:* Foundations for Microwave Engineering, *2nd Ed. New York: Mc Graw-Hill, 1992*

18. *Ryde, John D.* Networks, lines and Fields Asia Pub. House, 1955.

4 MICROWAVE INTEGRATED CIRCUIT DESIGN AND MANUFACTURING

INTRODUCTION 4.1

Microwave Integrated Circuits (MICs) are miniature microwave circuits. They are fabricated such that all circuit components are interconnected to each other in a planar transmission line, such as strip lines, microstrip lines, slot lines or coplanar lines, and the active components are built into the substrate. The propagating modes, excitation method, various losses and the characteristic impedance, are described in Chapter 3. Different types of MICs and their design technology are described. Design and fabrication of lumped elements and active devices in these structures are also described.

TYPES OF MICs AND THEIR TECHNOLOGY 4.2

MICs are divided into two categories, viz. hybrid and monolithic. In **hybrid technology**, the active devices are manufactured in the form of chips which are bonded to a substrate that carries the rest of the circuit and is commonly known as Hybrid Microwave Integrated Circuit (HMIC). On the other hand, **monolithic** is a "single solid integrated" structure and is commonly known as Monolithic Microwave Integrated Circuit (MMIC). In this technology, various active devices and their interconnection lines or patterns are fabricated on to a single crystal substrate of semiconductor material (mostly GaAs, InP) by IC manufacturing techniques.

HYBRID MIC (HMIC) TECHNOLOGY 4.3

In the **hybrid technology**, semiconductor devices are make in package or chip form and bonded into the passive devices. Microwave active devices are in large variety, such as transistors, mixer diodes, varactors, PIN diodes, Gunn diodes, IMPATT diodes, etc. It is not so easy to make common specifications of a standard process to manufacture several kinds of devices simultaneously in an optimum way. The interconnections between the active devices have relatively large dimensions, which requires a large area of the substrate. This is one disadvantage of HMIC. The **monolithic technology is** relatively simple but expensive. The **hybrid circuits** again can be divided into thick-film and thin-film circuits [4,7].

Substrates Used

RT duroid glass epoxy is a soft material used below 10 GHz. At higher frequencies, hard substrate materials are used—alumina, sapphire, GaAs—in which line definition is very good for use at high ends of microwave frequencies.

Thick-film Manufacture 4.3.1

Thick-film hybrid technology is well established for the low-cost manufacture of MIC at low microwave frequencies (< 10 GHz). The frequency limitation arises due to the increasing losses associated with the thick-film conductors and the tolerancing difficulties with the increase in frequency. Satisfactory circuits such as parallel-coupled and open-circuit stub filters, couplers, attenuators, power dividers, are realised using this technology.

Any of the following two basic methods are adopted for the manufacture of 'thick-film' MICs:

(a) Thick-film patterns are deposited/printed as conductive, resistive or insulating layers and fired on to the ceramic substrate—usually alumina or sometimes quartz.

(b) A printed circuit technique is used to etch the desired pattern in the copper cladding of usually polyolefin substrate (plastic).

Each method is much simpler and has fewer requirement of both equipment and environment than thin-film technology. The method (b) is a very common PCB fabrication method. The method (a) is mostly used for MIC manufacture and elaborated in the following paragraph.

Deposition Method in Thick-film Technology

A paste of metal, usually gold, is kept in a jar. A fine-mesh screen with areas open for the circuit pattern is placed rigidly on the top of the substrate. A few millimetres of the gold paste is placed on the screen and some of the paste is squeezed through these open areas and on to the surface of the substrate. Settling, drying, and firing sequences complete the 'thick' deposit. The actual film thickness is usually of the order of 10 mm. The details of the fabrication steps are as follows:

(a) A metal (Au) paste in a jar is kept in a refrigerator for substantial shelf life.

(b) The artwork for the circuit definition is prepared and photographic processing used to obtain a positive transparency.

(c) Fine stainless steel or polyester mesh is tightly stretched over a rigid frame fitting into the screen printer. This screen in also coated with a suitable photoresist layer.

(d) The positive transparency is held in intimate contact with the coated surface of the screen.

(e) Exposure of standard UV light and wash-bake processes make a screen with apertures for the required circuit. All other areas are made opaque with a durable photoresist layer.

(f) The screen is then placed in a printing jig. The substrate is placed below the aperture region of the screen firmly by vacuum suction.

(g) A few millimetres of conductor paste is placed on the screen.

(h) After optimizing the parameters, the aperture region of the screen is wet, and a wet deposite of paste is transferred on to the substrate.

(i) The wet-circuit paste is left horizontally to settle in a clean room for about 15 minutes and then dried at nearly 100°C temperature for about 20 minutes using an infrared drying machine.

(j) The deposit is fired at 900–1000°C to form a metallic substance.

(k) Finally, after firing the circuit, either laser trimming or an etch-back process is conducted to achieve the precise definition of the circuit.

Disadvantages

In thick film technology accuracy is poor, size is not very small, parasitic effect is high at higher frequencies, and batch production is inconvenient.

Thin-film Manufacture 4.3.2

In order to meet the stringent demand on the accuracy and repeatability of the circuit technology and performance, a controlled thin-film manufacturing process in a clean room is used for the majority of MIC fabrication. Very high-conductivity metals such as Au, Cu or Al are used for traces. Cu is an optimum choice for its high conductivity and low cost. But to protect Cu from oxidation, a final Au film is evaporated over the Cu as an environmental protective layer. The thickness of the metal film is kept at least to four skin depth at the lowest frequency of operation.

There are four main steps involved:

(a) Substrates are cleaned to a very high specification.
(b) Evaporation or sputtering—a very thin layer (< 0.1 mm) of chromium on to the surface of the substrate.
(c) Evaporation or sputtering—a very thin layer (< 0.5 mm) Cu or Au of similar thickness on to this layer of (b).
(d) Electroplating the bulk conductor Cu of approximately 5 mm thickness on to the layer (c).

The circuits are defined using photolithographic techniques. The very thin layers are produced by magnetron sputtering where the combined *E* and *H* fields produce ionization of the Cu charge. The ions are attracted towards the substrate used as anode.

Accomplishment of Circuit Definition

The circuit definition is accomplished by either of the following methods as described in Fig. 4.1(a) and 4.1(b).

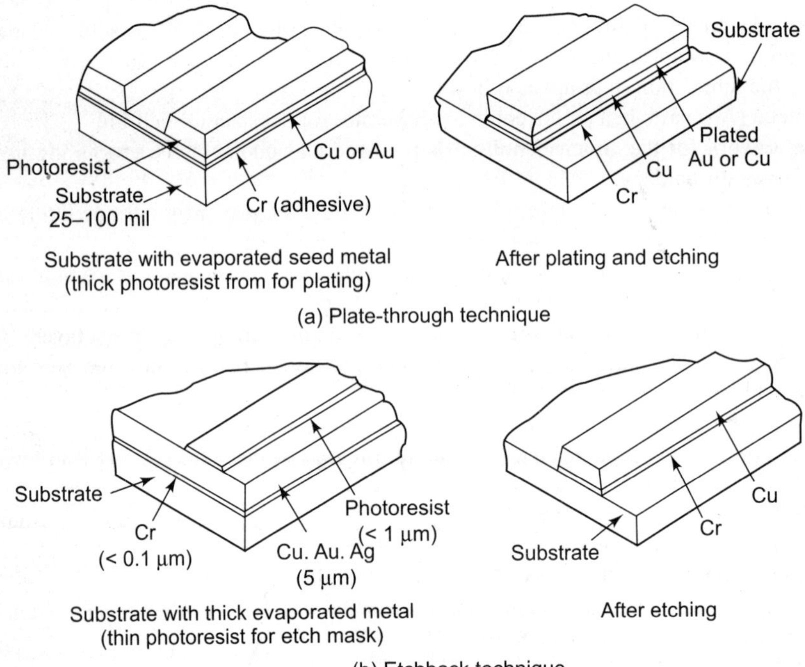

Fig. 4.1 Circuit traces accomplishment [Source: Ref. 4.7]

❑ ***Plate-through Technique*** After sputter coating of Cu or Au on a very clean surface of the substrate, a photoresist layer of similar thickness to the final metal film is spun on to the surfaces. UV light exposure through a precision contact photomask as per required circuit pattern is followed by subsequent processing to yield a pattern of slots in the photoresist layer. The circuit pattern is formed by plating Cu through these slots and then the photoresist layer is washed and any excess sputter coating is etched off to obtain the metal trace on the substrate.

❑ *Etch-back Technique* The very clean surface of the substrate is coated with thin films of metal by sputtering and a photoresist layer is spun on to the surfaces. The photomask exposure is done such that the larger areas of the surface where metal is not required are left unprotected by the photoresist layer. The circuit pattern is practically defined by a fine pattern of the photoresist layer across the very thin sputtered layers. The unwanted areas of metal which were originally sputtered on are then removed. Electroplating is done to yield the desired final thickness.

The advantages of 'etch-back technique' over 'plate-through technique' are

(a) A thinner (~0.5 mm) photoresist layer may be used.
(b) Plating is more uniformly distributed.
(c) Edges of the microstrip are nearly square.

Care must be taken such that the substrate surface must be extremely smooth like an optical surface finish and free from dust or grit to avoid perforation in the layer. **Sapphire substrates** are a very good choice.

❑ *Working Environment and Equipment* The following laboratory environment and equipment are recommended for the precise and repeatable circuit production.

(a) Class 10,000 clean room
(b) Coordinatograph for 25 x size of artwork
(c) Precision 25 x reduction camera
(d) Magnetron sputtering in high vacuum
(e) Vacuum spinner (1000–7000 per/min) and timer
(f) High intensity collimated UV light source
(g) Ultra-pure water (18 M-ohm resistivity)

Resistive Film 4.3.3

For terminations, bias networks, attenuators and many types of MICs, resistive films are also manufactured in thin and thick-film forms. These resistive films should have a stable low temperature coefficient of resistivity. Nichrome and tantalum are commonly used resistive materials. A typical resistance value of 50–5000 ohms is required in MIC and can be realized by the deposition of films of these materials on the substrate and then laser trimming the edges to achieve good resistance tolerance in thin-film technology.

In thick-film technology, the resistive materials are deposited and processed in a similar way as given for the conductive films except that the firing profile is more critical for the control of the final resistivity.

MONOLITHIC TECHNOLOGY 4.4

Monolithic Microwave Integrated Circuits (MMICs) are conceptually true integrated circuits in which all active and passive circuit elements or components and interconnections are formed and processed simultaneously within a single semi-insulating semiconductor substrate by some deposition scheme such as epitaxy, ion implantation, sputtering, evaporation, diffusion, or a combination of these processes and others. Many drawbacks of hybrid technology are eliminated and the operation is very well extended to millimetre wave region, where dimensions and sizes of the substrate become much smaller. Two types of semiconductor substrates are normally used, viz. Si and GaAs. Si is generally used in the low frequency ICs. However, GaAs substrates are used for MMICS, which have the election mobility much higher than that in Si. The high mobility in GaAs, coupled with shot length of gate, results in external, small transit time which makes it possible to use at microwave and higher frequencies. High resistivities (> 2000 ohms-cm) maintain the sufficiently low transmission losses and to provide enough isolation between components on the chip. GaAs, though more expensive, has the advantage that it possesses higher resistivity than that of Si by a factor of 10^4 and some of the active devices such as TEDs, GaAs transistors, can be grown right into the substrates.

For achieving accurate device geometries and circuit patterns, substantially higher resolution electron-beam photolithography is used, particularly for the manufacture of submicron gates for GaAs MES FET to operate at the higher microwave and mm-wave frequencies. This is because with normal photolithography, diffraction effects limit the resolution of planar lines to dimensions in the order of a light wavelength. A combined processing technique involving photolithography and electron-beam lithography is advantageous in the fabrication of commercial MMICs.

It is important that the MMIC substrates should have capability to retain their intrinsic properties during the annealing cycle in the production process at a temperature range of 800–900°C. Si has a drawback since its intrinsic insulating properties degrade at temperatures > 800°C. Chrome-compensated semi-insulating GaAs retain resistivity at 800–900°C during the fabrication process.

However, Si is possible to process using a combination of ion implantation and laser annealing which avoids increasing the temperature of the substrate. Table 4.1 presents a comparison of hybrid MIC vs. MMIC technologies.

Table 4.1 Hybrid MIC vs. MMIC Technology

Technology	*mm-wave capabilities*	*Application volume*
Thick-film hybrid MIC	No (only up to 10 GHz)	Low
Thin-film hybrid MIC	Yes (up to 30 GHz)	Medium
MMIC	Yes up to 100 GHz	High

The monolithic technology has several advantages over HMIC and is given below:

(a) Improved reliability and reproducibility,
(b) Small size and lightweight in producing complex circuits,
(c) Broad-band performance; due to minimum parasitics
(d) Circuit design and multifunction performance on a single chip.
The undesired parasitic effects are minimized in MMIC due to absence of wire bonding and embedment of active components within a printed circuit. Thus, higher frequency performance of MMIC becomes better than hybrid MIC

Some of the monolithic approach disadvantages and problem areas are

(a) Circuit tuning is difficult
(b) Troubleshooting is difficult
(c) Cross-talk interference suppression may be a problem for dense packaging
(d) Integration of high power sources (IMPATT) is difficult for heat dissipation
(e) Difficult processing
(f) Very high foundry cost

However, careful design using Computer-Aided Design (CAD) and technical judgement of the expert minimizes these problems. This is beyond the scope of this book at present. The following paragraphs describe the details of design and manufacturing of MMIC.

Design and Fabrication of MMIC 4.4.1

MMIC design and fabrication involve a choice of materials and fabrication processes. The material selection of MMIC design and manufacturing is an important criteria. Once the circuit design is made and simulated in the integrated circuit form, the fabrication process starts using the best selected materials.

A. Material Selection

There are four categories to select materials for MMICs.

(a) Substrate materials on which electronic devices are built—Si at lower frequencies, and GaAs InP at higher frequencies.

(a) Conductor materials to form both the conductor pattern and the bottom ground plane—aluminium, copper, gold and silver for very good conductivity.

(b) Dielectric materials for blockers, capacitors—SiO, Si_3N_4, Al_2O_3, Ta_2O_5.

(c) Resistive materials for bias networks, terminations, and attenuators—Cr, NiCr, Ta, and Ti.

❑ ***Substrate Materials*** The substrate materials ideally should have the following characteristics and the selection of the materials will depend on the thermal dissipation, the circuit function, and the type of the circuits.

(a) High dielectric constant,
(b) Low power dissipation or loss tangent
(c) Temperature and frequency-invariant dielectric constant
(d) High purity and surface smoothness
(e) High resistivity and dielectric strength
(f) High thermal conductivity
(g) Uniform thickness

The properties of some common substrates used for MMIC fabrication are given in Table 4.2.

Table 4.2 Properties of substrate materials

Material	ε_r	*Loss tangent at 10 GHz*	*Thermal Conductivity (W/cm °C)*	*Usage*
Alumina	9.8	2×10^{-4}	0.3	Microstrip
Ferrite/Garnet	12–16	2×10^{-4}	0.03	Microstrip, coplanar
Sapphire	9.3–11.7	10^{-4}	0.4	Microstrip, lumped element

❑ ***Conductor Materials*** The conductor materials for MMIC should have the following properties. The selection of the conductors is determined by the compatibility with other materials to be used in the circuit and the process of fabrication.

(a) High conductivity
(b) Low temperature coefficient of resistance
(c) Good adhesion to the substrate
(d) Easy to deposit or electroplate
(e) Good etchability
(f) Good solderability

In order to reduce the losses, the conductor thickness should be kept more than 3 skin depth (~10 μm) which can be achieved by evaporation or plating and photoetching. Table 4.3 shows the properties of some commonly used conductors.

It is seen that good electrical conductors have poor substrate adhesion. Hence, for good adhesion, poorer conductivity is compromised. Aluminium is a good choice for its relatively good conductivity and also good adhesion. However, high conductivity materials can be used in combination with a very thin film of good adhesive but poorer conductivity material layers between the substrate and the good conductors. Such combinations are Cr-Au, Cr-Cu and Ta-Au. An adhesion layer has a typical surface resistivity of 500–1000 ohms/square metre without loss.

Table 4.3 Properties of commonly used conductors, both HMIC and MMIC

Material	*Skin depth at 1 GHz (mm)*	*Surface resistivity ohms/sq*	*Adhesion to substrate*	*Method of deposition*
Ag	1.4	2.5×10^7	Poor	Evaporation, screening
Cu	1.5	2.6×10^7	Very poor	Evaporation, plating
Au	1.7	3.0×10^7	Very poor	Evaporation, plating
Al	1.9	3.3×10^7	Very good	Evaporation
Mo	2.7	4.7×10^7	Good	Electron-beam evaporation, sputtering
Cr	2.7	4.7×10^7	Very Good	Evaporation
Ti	4.0	7.2×10^7	Very good	Electron-beam sputtering

❑ ***Dielectric Materials*** The dielectric materials used in MMIC should have the following properties:

(a) Low RF dielectric loss
(b) High voltage-withstanding capability
(c) Good reproducibility
(d) Ability to undergo process

The properties of some dielectric materials are given in Table 4.4. SiO, SiO_2 and Ta_2O_5 are the most commonly used.

Table 4.4 Properties of dielectric materials

Material	*Method of deposition*	e_r	*Dielectric strength (kV/cm)*
SiO	Evaporation	6–8	400
SiO_2	Deposition, sputtering	4	10,000
Al_2O_3	Anodization, evaporation	7–10	4000
Ta_2O_5	Anodization, sputtering	22–25	6000

Thin-film SiO is not very stable and is used in noncritical applications, viz. by-pass capacitors. SiO_2 capacitors can have $Q > 100$ and $C \sim 0.02–0.05$ pF/sq. mil.

❑ ***Resistive Materials*** Resistive materials are used in MMIC for bias networks, termination and attenuators. The properties of such materials should be

(a) Good stability
(b) Low temperature coefficient of resistance
(c) Good heat-dissipation capability
(d) High surface resistivity 10–1000 ohm/square

The properties of some resistive materials are given in Table 4.5.

Table 4.5 Properties of resistive materials

Material	*Deposition method*	*Resistivity ohm/ square*	*Stability*
Cr	Evaporation	10–1000	Poor
NiCr	Evaporation	40–400	Good
Ta	Sputtering	5–100	Excellent
Cr-SiO	Evaporation	600	Fair
Ti	Evaporation	5–2000	Fair

It is seen that for better stability, NiCr and Ta are the most commonly used resistive materials.

B. MMIC Fabrication

In MMIC, active devices are grown in a semiconducting substrate material, usually GaAs and passive elements of the circuit are grown in the same substrate simultaneously. Different techniques are used in the fabrication of MMICs such as epitaxial growth, diffusion and ion implantation, photolithography, metal deposition and etching. These are briefly described below:

❑ *Diffusion and Ion Implantation* These two processes are used in controlling the doping amounts in active device fabrications. In the diffusion process, impurities are diffused into the pure material at elevated temperatures to produce *n* or *p*-type layers. In the ion-implantation process, a substrate is doped with high-energy ion impurities and the doping concentration is more precisely controlled at lower temperature with better reproducibility and the implanted layer is usually anealed at a specific temperature for uniformity in doping.

❑ *Epitaxial Growth* In the epitaxial process, single-crystal semiconductor layers are grown on a single crystal semiconductor substrate by any of the three types of **Epitaxy-Vapour-Phase Epitaxy (EVPE)**, **Molecular-Beam Epitaxy (MBE)** and **Liquid-Phase Epitaxy (LPE)**. This process provides means to optimize the device and circuit performances by controlling the doping profiles.

(a) For Si and GaAs devices, vapour-phase epitaxy and MBE are used.

(b) Molecular-beam-epitaxy process uses the reaction of one or more thermal beams of atoms or molecules with a crystalline surface under ultra-high vaccum conditions. This process has very precise control in both chemical composition and doping profiles.

(c) In liquid-phase epitaxy, layers are grown on crystalline substrates by direct precipitation from the liquid phase. This process is very useful for **GaAs devices**. Because of its slow growth rate, control of epitaxial layer thickness (≥ 0.2 mm) and of doping and composition in multilayered structures is precise.

❑ *Photolithography* Lithography is a process of converting patterns of geometric shapes on a mask to a thin layer of UV radiation-sensitive material (photoresist) to cover the surface of a semiconductor wafer by means of electron-beam lithography, or ion beam or laser beam.

❑ *Photo-etching* Most commonly, the photolithography technique is used to coat the substrate with a uniform film of a photosensitive emulsion (photoresist) and is shown in Fig. 4.2. A mask for the designed opening patterns is placed over the photoresist. UV light exposes the photoresist through the mask to form a polymerized photoresist.

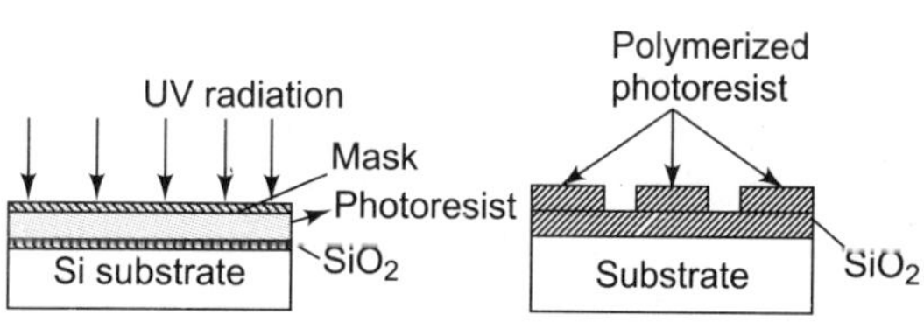

Fig. 4.2 Photo-etching process

The mask is then removed and unpolymerized portions of the photoresist are removed by dissolving it using trichloro-ethylene. The portion of SiO_2 which is not covered by the photoresist is removed by HCl acid.

◆ **Deposition** The semiconductor impurities are deposited on the substrate by any of the following three methods:

(a) **Vacuum Evaporation**—Impurity material is kept in a metallic container which is heated by passing a high electric current. A mask is put over the substrate. The masked substrate and the heated metallic container are placed in a vacuum glass tube. The impurity is evaporated and deposited on the substrate to form a layer of polycrystal.

(b) **Electron-Beam Evaporation**—The impurity in the metallic container is vaporized by seaming the substrate in the metallic container by a narrow beam of electrons.

(c) **Sputtering**–A diode system is formed in a vacuum chamber with a crucible containing the impurity as a cathode and the substrate as an anode under dc operation. A small trace of argon gas is injected into the vacuum so that a glow discharge of argon occurs under a high dc voltage between the anode and cathode. The accelerated positive ions moving towards the cathode displace the atoms from the impurity. The impurity atoms attains high energy to reach the substrate and are deposited on it.

A Typical Fabrication Procedure Involves Four Basic Steps for MMIC.

An SiO_2 layer is deposited on the semiconductor substrate over which a photoresist layer is formed. Ultraviolet light is made to pass through a photographic mask to the photoresist. Selected oxide regions are removed by means of chemical etching with HCl acid. An organic solvent is used to dissolve the photoresist in the oxide leaving the desired opening undissolved.

Fabrication of Active Devices

There are a large number of circuits used in MMIC. The most important transistors are MESFET and HEMT. The fabrication processes follow some basic methods which are described below using examples.

❑ *Fabrication of MOS Devices* Metal-Oxide-Semiconductor (MOS) technology is important for Very Large-Scale Integrated Circuits (VLSIC). The most important MOS devices are MOSFETs and their fabrication processes in MMIC is described below. MOSFETs are four-terminal devices grown on an n-channel or p-channel Si-substrate into which two highly doped *n*+ or *p*+ sections are diffused, respectively. These highly doped sections are the source and drain separated by 0.5 mm. A thin layer of insulating SiO_2 is grown over the surface of the structure.

MOSFET fabrication technology involves the following steps as explained in Fig. 4.3.

(a) **SiO_2 layer formation**—An SiO_2 layer is made to form on the surface of a *p*-type Si-substrate by exposing the top surface to dry oxygen.

(b) ***n*+ layer diffusion**—Two n^+ layers are diffused into a *p*-region through two openings by using the photoresist and UV ray technique.

(c) **Photo-etching**—The centre portion of SiO_2 is removed by masking and photo-etching method.

(d) **Re-oxidation**—The total top surface of the final material is exposed to dry oxygen and as a result, an SiO_2 layer is formed to completely cover the top surface.

(e) **Deposition**—Phosphorus glass is deposited to cover the surface of SiO_2.

(f) **Photo-etching**—By UV ray photo-etching, two windows are again opened on the SiO_2 above the n^+ diffused regions.

(g) **Metallization**—The entire top surface of the device is metalized using Al.

(h) **Masking and Etching**—By masking and photo-etching, Al metal contacts are retained to the diffused gate, drain and source regions. The unwanted metal regions are etched away.

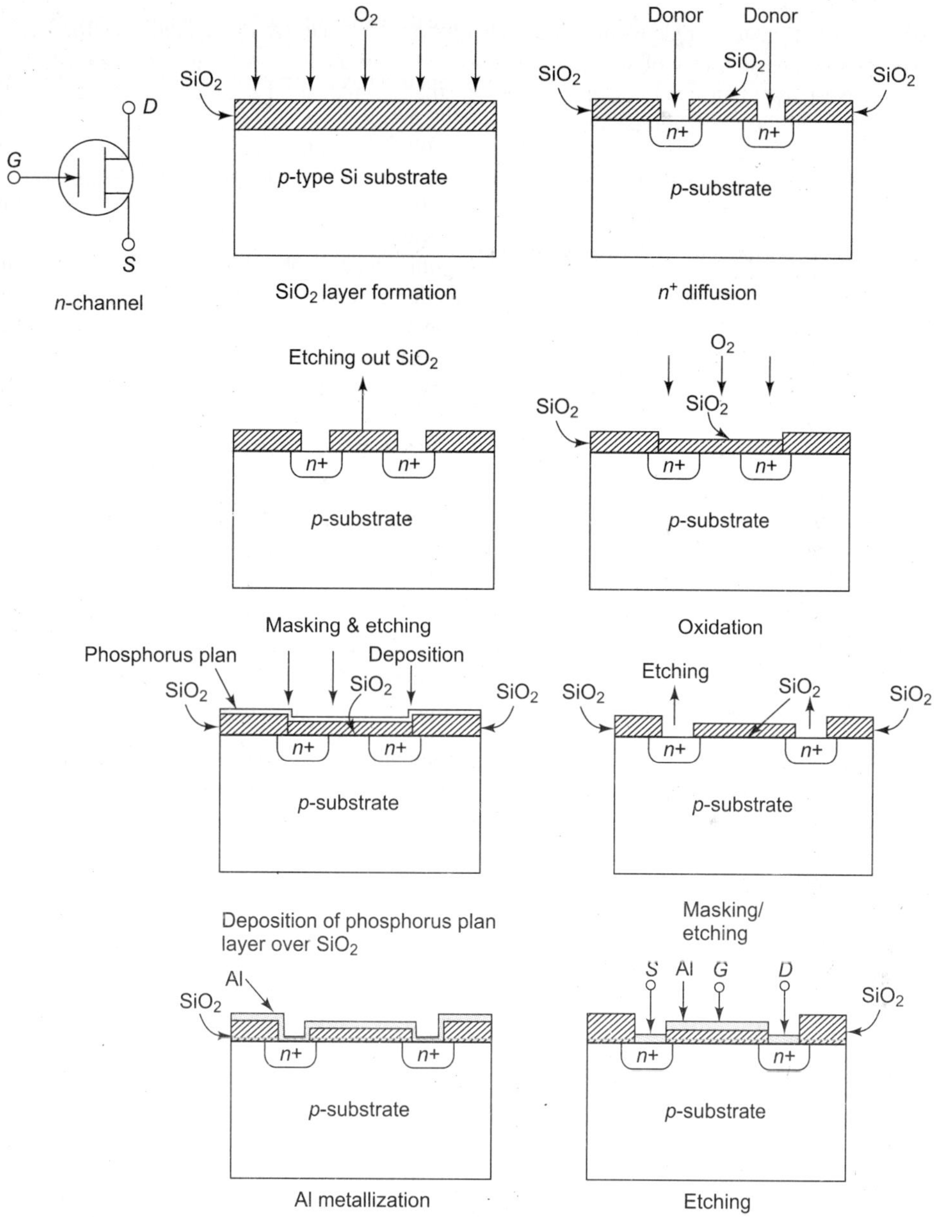

Fig. 4.3 MOSFET fabrication

MIC LUMPED AND DISTRIBUTED ELEMENTS 4.5

At microwave frequencies, conventional lumped elements *R*, *L*, *C* cannot be used since they produce parasitic effects due to lead inductance, loss resistance and capacitance. They also act as radiating elements causing loss of signal. The RF and microwave equivalent circuit of such elements are shown in Fig. 4.4 along with their typical impedance response vs. frequency.

Thus, Fig. 4.4 shows that a single element behaves resistive or inductive or capacitive as the frequency changes. Therefore, special forms of lumped elements are fabricated for MIC using sections of planar transmission lines of less than a wavelength. A significant size reduction is achieved by shortening the guide wavelength λ_g of these distributed circuits through high dielectric constant of the substrate materials $\left(\lambda_g = \dfrac{\lambda}{\sqrt{\varepsilon_r}}\right)$, but subsequently other problems may occur, such as excitation of higher-order surface waves, too narrow line width to fabricate for 50-ohm impedance, and temperature variation of ε_r.

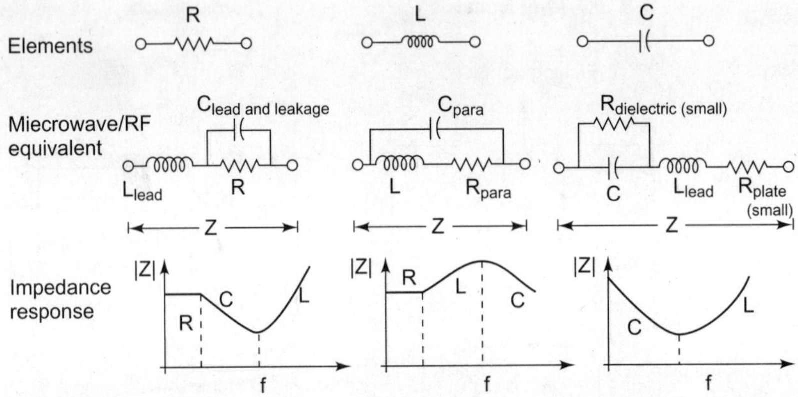

Fig. 4.4 Conventional lumped elements, their higher frequency equivalent and impedance response

Lumped elements are used in a variety of circuits. Lumped resistive terminations are needed for couplers, lumped large capacitors are used essentially for bias bypass applications, and planar inductors are useful for RF isolation and for matching purposes, especially at the lower end at the microwave frequency where stub conductors are very large. By keeping the size of these components much smaller than a wavelength through the use of photolithography techniques, the elements remain lumped through the *X*-band. Beyond the *X*-band, distributed elements are preferred. However, it is difficult to realize a truly lumped element because of lead inductance and parasitic effects. We shall review the design principles of planar lumped elements that are used at microwave frequencies. The lumped elements have been fabricated on polished alumina, quartz, sapphire and GaAs substrates.

Planar Resistors from Lossy Lines 4.5.1

It is known that for a lossy line of length l whose characteristic impedance is Z_1, the input impedance of the line for a terminating load Z_L can be written as

$$Z_{in} = Z_1 \frac{Z_L + Z_1 \tan \gamma l}{Z_1 + Z_L \tan \gamma l} \tag{4.1a}$$

For small length $l \ll \lambda_g/4$, $\tan \gamma l \approx \gamma l$. Therefore, (4.1a) reduces to

$$Z_{in} = Z_1 \frac{Z_L + Z_1 \gamma l}{Z_1 + Z_L \gamma l} \tag{4.1b}$$

(i) If this line is short circuited, $Z_L = 0$, so that

$$Z_{in} = Z_1 \gamma l = l\sqrt{\frac{R + j\omega L}{G + j\omega C}}\sqrt{(R + j\omega L)(G + j\omega C)}$$

$$= l(R + j\omega L) = Rl + j\omega Ll \tag{4.2}$$

Thus, a short length of short circuited line acts as

- shunt resistance $Z_{in} = Rl$ when loss is high ($R >> L$)
- shunt inductance $Z_{in} = j\omega Ll$ when loss is low ($R << L$)

(ii) When the lossy line is terminated with a matched load, $Z_L = Z_1$,

$Z_{in} = Z_1 = \sqrt{\dfrac{R + j\omega L}{G + j\omega C}} \simeq \sqrt{\dfrac{R}{G}}$, a real element independent of $l \ll \dfrac{\lambda_g}{4}$. Under this condition, the short length of the line will act as a series resistance. Figure 4.5 shows some configurations of series and shunt resistances and their techniques of realization.

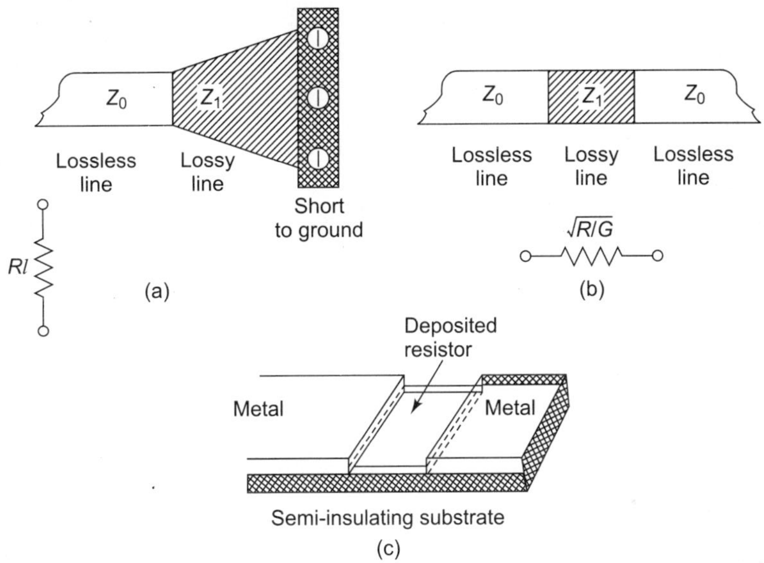

Fig. 4.5 Planar lumped resistors: (a) shunt resistor (b) series resistor (c) resistor-formation technique

For the design of planar resistors, some factors are considered: (1) the material resistivity; (2) temperature coefficient of the resistive materials; (3) the thermal resistance of the material; and (4) the frequency bandwidth.

The planar resistor can be realized by a thin resistive film deposited on a semi-insulating substrate. The resistive materials are Al, Cu, Au, Ta, etc., having resistivity in the range 30 to 1000 ohms per square. They have the following categories:

(a) Semiconductor film resistor which is fabricated by forming an isolated land of conducting epitaxial film on a semi-insulating substrate.

(b) Deposited metal film resistor which is formed by evaporating a metal layer over the semi-insulating substrate and forming the desired pattern by photolithography.

The parasitic capacitance and distributed inductance of the film produce frequency dependency of planar resistors at high frequencies. To minimize thermal resistance, the area of the film should be kept large. Since the width w of the resistive film should not differ from the width of the feeder line to minimize discontinuity effect, the length l is kept large to minimize thermal resistance, but limited to a value specified by the sheet resistivity ρ_s of the film, the line width w, desired resistance value R, and film thickness t and is expressed by

$$l = \frac{wRt}{\rho_s} << \lambda_g/4 \tag{4.3}$$

The ac resistance of the element is given by

$$R_{ac} = Rl = \frac{KR_s l}{2(w+t)} \tag{4.4}$$

Here, $R_s = \sqrt{\frac{\omega\mu}{2\sigma}}$ is called the surface resistance per sq.m of the film material.

K = Correction factor to consider current distribution on the film and depends on w/t.
= 1.3 for $w/t = 1$
= 1.5 for $w/t = 10$
= 2.0 for $w/t = 100$

The dc resistance does not have much importance at microwave frequencies.

Planar Inductors from Lossless Lines 4.5.2

For a short length of lossless high impedance line of characteristic impedance Z_1, and terminated by a low impedance line $Z_0 << Z_1$, input impedance

$$Z_{in} = Z_1 \frac{Z_0 + jZ_1 \tan\beta l}{Z_1 + jZ_0 \tan\beta l}; \quad \gamma = j\beta$$

$$\approx Z_1 \frac{Z_0 + jZ_1\beta l}{Z_1 + jZ_0\beta l}; \text{ where } \tan\beta l \approx \beta l \text{ for } l << \lambda_g/4$$

$$= Z_1 \frac{Z_0/Z_1 + j\beta l}{1 + j\,Z_0} \approx Z_1 \cdot j\beta l = jX, X - Z_1\beta l$$

Since $Z_0 << Z_1$,

Hence $Z_{in} = jX$, inductive reactance (4.5)

Therefore, when a short length of high impedance lossless line is terminated by a low-impedance line, the high-impedance line behaves like an inductor to its input feeder line. Figure 4.6 illustrates several forms of planar lumped inductors for MIC which are realized in single-layer metallization technique. In all forms, the total line length is a small fraction of a wavelength to satisfy lumped condition. Typical inductance values for monolithic circuits are in the range from 0.5 to 10 nH. Higher values are difficult to achieve due to appearance of intersegment fringing capacitance for meander line and spiral forms and parasitic capacitance to ground. Figures 4.6 (a, b, c, f) are used as series elements and Fig. 4.6 (d and e) as shunt elements.

A single metallic strip on a dielectric substrate without a ground plane can be used as fabrication of lumped inductors. The inductance of a linear strip (Fig. 4.6 (a) and (b) of length l (mil), thickness t (mil) and width w (mil) is given by

$$L(\text{nH/mil}) = 5.08 \times 10^{-3} l \left[\ln\left\{\frac{l}{(w+t)}\right\} + 1.193 + 0.022\left(\frac{w+t}{l}\right)\right]; 1 \text{ mil} = 10^{-3} \text{ inch} \tag{4.6}$$

An approximate expression for circular spiral inductors with n turns is given by

$$L = \frac{a^2 n^2}{8a + 11c} \tag{4.7}$$

where $c = \dfrac{d_o - d_i}{2}$

$$a = \frac{d_o + d_i}{4}$$

d_i = Inner diameter of the spiral in mil
d_o = Outer diameter of the spiral in mil

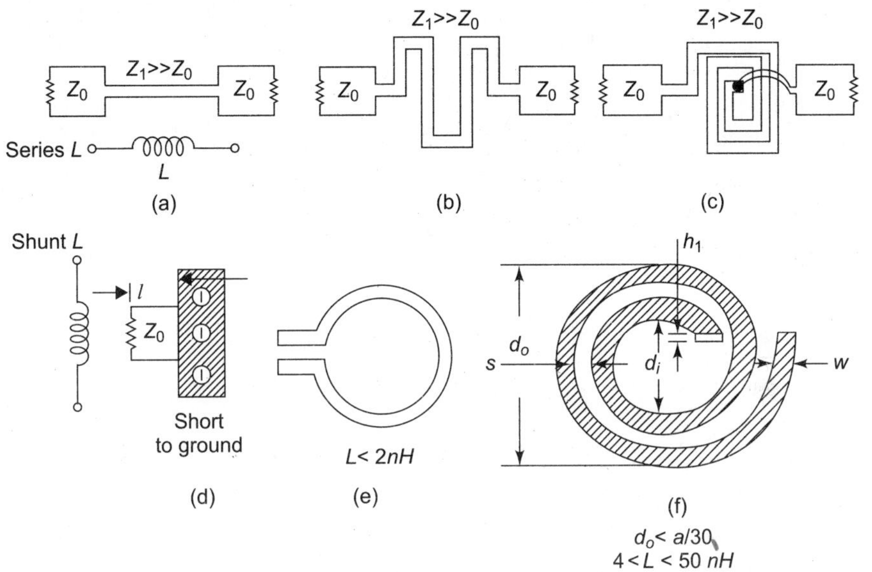

Fig. 4.6 Planar lumped inductor configurations: (a) High-impedance line section (b) Meander line (c) Square spiral (d) Short-circuited line (e) Ring (f) Circular spiral

Circular spirals have been used in preference to square ones as tuning elements for achieving higher Q since Q is proportional to the volume-to-surface ratio and circular geometry has less area than square for the same volume. Considering RF current flow within skin depth on the top and bottom surfaces of the metal film, the Q of a spiral inductor is given by

$$Q = \frac{\omega L}{R_{ac}} \tag{4.8}$$

The ac resistance R_{ac}, ignoring the side walls due to thickness of the strip and losses in the substrate, is given by

$$R_{ac} = \frac{an\pi R_s}{w} \tag{4.9}$$

Thus, from Eqs (4.8) and (4.9),

$$Q = \frac{2Lw}{na}\sqrt{\frac{\sigma f}{\mu \pi}} \tag{4.10}$$

Therefore, Q increases with the frequency f and width w. The maximum w, however, is limited by the self-resonance of the inductor with the parasitic inter-turn capacitance. Moreover, increasing w would increase the coil size, violating the lumped condition. Typical inductors can be made with $Q > 50$ at 2 GHz with diameter < 0.05 inch. The spiral strip thickness is chosen several skin depths (about 6 microns for Cu or 8 microns for Al at 2 GHz). Typical inductance values for monolithic circuits fall in the range of 0.5 nH to 10 nH. The inductor design consideration is, therefore,

(a) spiral width w should be large
(b) overall diameter d_0 should be small, $d_i/d_0 = 1.5$ so that Q (circular) > Q (square) by 10%, $L_{circular} < L_{square}$ by 20%.
(c) spiral separations between the turns should be very small

Circular ring inductor

$$L = 5.08 \times 10^{-3} l \left[\ln\left(\frac{t}{w+t}\right) - 1.76 \right] \tag{4.11a}$$

Square spiral inductor

$$L = 8.5\sqrt{A}\ n^{5/3}\ \text{nH/mil} \tag{4.11b}$$

Here, A = area of the spiral, sq.cm
n = number of turns

Planar Capacitors from Lossless Line 4.5.3

When a short length of lossless low impedance Z_1 line is terminated by a higher impedance Z_0, the input impedance of the low impedance line is given by

$$Z_{in} = Z_1 \frac{Z_0 + jZ_1 \tan\beta l}{Z_1 + jZ_0 \tan\beta l} \approx \frac{Z_1}{j\beta l} = -\frac{j}{\beta l/Z_1} = -jX$$

Here, the capacitive reactance is $X = \dfrac{1}{\beta l/Z_1} = \dfrac{Z_1}{\beta l}$ (4.12)

Therefore, the low impedance line Z_1 will behave as a capacitor to its input feeder line. There are a variety of planar capacitors that can be realised in MIC—those achieved with a single metallization scheme, and those achieved using a two-level metallization technology on a dielectric film as shown in Fig. 4.7.

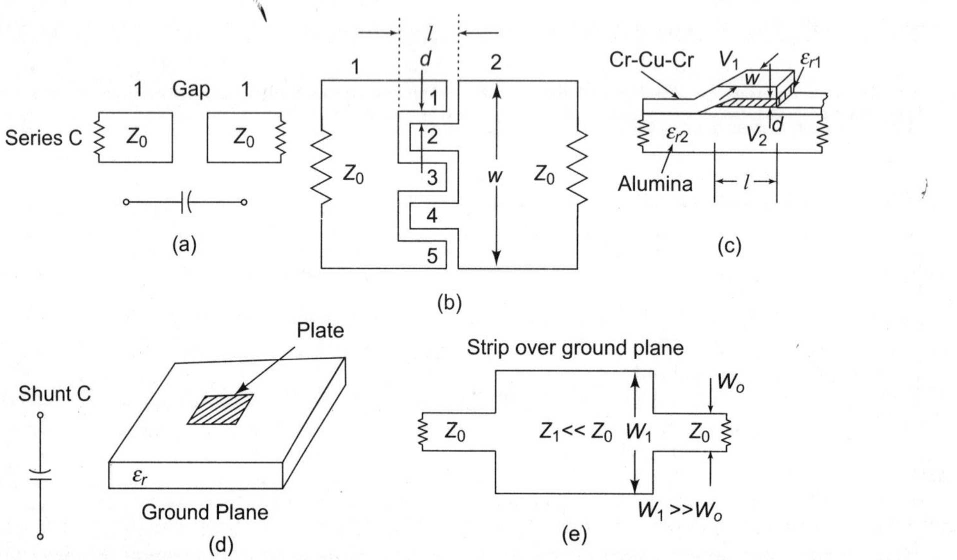

Fig. 4.7 Planar lumped capacitors: (a) End coupled (0.001–0.05 pF) (b) Interdigital (0.01–0.5 pF), (c) Dielectric overlay (0.05–100 pF) over ground plane (d) Top plate over ground plane (e) Low impedance MS line section

The first two [Fig. 4.7(a) and (b)] use no dielectric film and depend on electrostatic coupling via the substrate and are suitable for applications where low values of capacitance (< 1 pF) are required, such as high-impedance matching circuits. The third one [(Fig. 4.7(c)] is overlay structure which uses a dielectric film and is suitable for low-impedance (power) circuits, by-pass and dc blocking circuits with higher values (0.05–100 pF).

The sources of loss in planar capacitors are conductor loss in the metallization, and dielectric loss of the film. For the first two categories, there are high charge and current concentration near the edges which limit the Q factors. A typical Q factor is of the order of 50 at the X-band can be achieved. In the remaining three categories, the current distribution is more uniform over the metal plates and higher Q values can be achieved in the range 50–100 at the X-band. The losses in dielectric film restrict the upper limit of Q. The technology used for film deposition should be compatible with the technology used for the active devices, such as FET in MMIC. The dielectric films, SiO_2 and Si_3N_4, easily satisfy this criteria.

The performance of the dielectric film is expressed in terms of figure of merits which are defined by any of the following:

- Capacitance-breakdown voltage product

$$CV_b = \varepsilon_r \, \varepsilon_0 \, E_b = 10^4 \text{ pF.V/mm}^2 \tag{4.13}$$

- Capacitance-dielectric Q-factor product

$$CQ_d = \frac{C}{\tan \delta_d} \tag{4.14}$$

where C is the capacitance per unit area, V_b is the breakdown voltage, E_b is the corresponding breakdown field (1–2 MV/cm), ε_r is the dielectric constant (4–20) and tan δ_d is the dielectric loss tangent (0.1–0.001). The highest figure of merit is desirable for these lumped elements.

The metal thickness to be used in the lumped circuits depend on the frequency greatly. For the elements in the form of sections of microstrip lines and sandwich-type capacitors, where RF currents lie on the dielectric side of the conductors, a film thickness of 2–3 skin depth is needed. For sections of slot line, coplanar line, spiral inductors, interdigitated capacitors, where RF current flows on both sides of the conductors and for inductors where RF current distribution is non-uniform, thicker metal of the order of 10 skin depths is needed.

Capacitor Design Considerations 4.5.4

Interdigital Capacitor

As shown in Fig. 4.7(b), interdigital capacitors do not require any special steps in fabrication other than MIC technology. The capacitance value depends on the w/l ratio, gap distance between two conductors, and the number of fingers N. A metal–SiO_2–metal structure is fabricated on the top surface of the alumina substrate. The capacitor value increases with decrease of gap distance d between two successive conductors. The capacitor value changes with frequency depending on the w/l ratio. Hence, w/l needs to be optimised such that the variation of C with frequency is minimum. For example, for $w/l = 1.66$, $N = 24$, C = constant in frequency up to 3 GHz.

Sandwich Capacitor

Sandwich type capacitors require special steps to fabricate as follows with reference to Fig. 4.7(c).

(a) Layers of Cr–Cu–Cr are deposited on an alumina substrate in single-vacuum evaporation. Cr layers are very thin and are used for the adhesion of metal to the dielectric substrate. Alternate combination is Ti–Au–Ti.

(b) Photoresist etches the bottom layer and a metal layer is etched over the substrate.

(c) An SiO_2 film is deposited by growing with voltage breakdown of 10 MV/m field. SiO_2 thickness is nearly 0.5–1.0 mm. SiO_2 is cleaned from the unwanted portion of the substrate by selective photo-etching. The SiO_2 is then densified by heating (400°C) to increase the Q from about 60 to 600. Q = 5000 can be achieved by densifying at 600°C.

The capacitance of dielectric overlay metal-oxide–metal-capacitor of Fig. 4.7(c) can be expressed by

$$C = \varepsilon_0 \varepsilon_{r_1} \frac{lw}{d}, \text{F}$$

The capacitance of an interdigital capacitor can be expressed by

$$C = \frac{\varepsilon_r + 1}{w} l[(N - 3)\, 0.089 + 0.10] \text{ pF/cm}$$

where w = finger-base width (cm), l = finger length (cm), N = number of fingers.

CHOICES IN FILM TECHNOLOGY 4.6

Thick vs Thin Films for MICS 4.6.1

In addition to some difficulties in obtaining well-defined traces with both thick and thin films, there are loss considerations of losses in the choice of dielectric substrate and their surface finish. For aluminas, the conductor loss increases with surface roughness and the dielectric loss (tan δ) also increases due to the increased impurities. The use of thick or thin films is also connected with the choice of the substrate. Thick-film materials will not stick to highly polished material because the chemical bond is weak and in hazardous environments, such as hot moist atmospheres, the traces will peel from the substrate. On the other hand, bounding of the thick-film with rough alumina is excellent, and thick films are suitable. At high frequencies, there is no choice other than thin films on a highly polished surface of substrates.

Monolithic vs Hybrid 4.6.2

To provide resistive isolation for good circuit performance, monolithic techniques usually involve a semi-insulating, semi-conducting substrate (> 1000 ohm cm). Active devices are fabricated locally by growing them epitaxially in the insulating material. Si cannot maintain resistivity during fabrication process and thus, is not a very successful substrate. GaAs does have the required characteristics and is suitable. Monolithic technique is better for higher frequency circuits (>18 GHz) where hybrid parasitics limit performance. Since hybrid techniques use high-quality substrates on which active devices and chip capacitors are mounted and bonded in place, servicing of the circuits becomes easier.

Example 4.1 Find the planar resistance for the following dimensions: resistive film length, width and thickness of 10 mm, 5 mm, 0.1 mm respectively. The sheet resistivity of the film is 2.4×10^{-8} ohm-m.

Solution

$$R = \frac{l\rho_s}{wt} = \frac{10 \times 10^{-3} \times 2.4 \times 10^{-8}}{5 \times 10^{-3} \times 0.1 \times 10^{-6}} = 0.48 \text{ ohm}$$

Example 4.2 Calculate the inductance of a circular spiral inductor having number of turns = 10, spiral separation = 100 mil, inner diameter = 100 mil, outer diameter = 500 mil.

Solution $n = 10$, $s = 100$ mil, $d_i = 100$ mil, $d_0 = 500$ mil

$$L = \frac{a^2n^2}{8a+11C} = \frac{\left(\frac{600}{4}\right)^2 \times 10^2}{8\times\frac{600}{4}+11\times 200} = 661.7 \text{ nH/mil}$$

$$a = \frac{do+di}{4} = \frac{500+100}{4} = \frac{600}{4} \text{ mil}$$

$$c = \frac{do-di}{2} = \frac{500-100}{2} = 200 \text{ mil}$$

Example 4.3 Determine the capacitance of an interdigitated capacitor fabricated on a substrate of $\varepsilon_r = 13$. Other parameters are N = 10, substrate height $h = 0.1$ inch, finger length = 0.001 inch, finger base width = 0.02 inch.

Solution Given, $w = 0.02$ inch, $l = 0.001$ inch, $n = 10$, $h = 0.1$ inch, $\varepsilon_r = 13$

Capacitance $$C = \frac{\varepsilon_r+1}{w} l\,[(N-3)\times 0.089+0.1] \text{ pF/cm}$$

$$= \frac{13+1}{0.02} 0.001[(10-3)\times 0.089+0.1] = 0.5061 \text{ pF/cm}$$

REVIEW QUESTIONS

4.1 Distinguish between thick-film and thin-film technology. Describe the fabrication process of MMIC and hybrid MICs.

4.2 Describe why MMICs are superior than hybrid MICs. Discuss their relative advantages and disadvantages.

4.3 Explain how semiconductor and dielectric layers are grown while fabricating MMICs.

4.4 Explain using electrical equivalent circuits and their responses why normal lumped elements *R*, *L* and *C* cannot be used at high *RF* and microwave frequencies. How are they realized in microstrip form?

4.5 State the various design considerations for fabrication of lumped inductors and capacitors in MIC. What are the additional criteria for fabrication of sandwich-type capacitors?

4.6 Describe the different steps involved in MIC fabrication. What are the dielectric substances used as substrates? What are the different etching processes?

4.7 A 5-turn circular spiral inductor is implemented in microstrip form with inner and outer diameters of 5 mm and 15 mm respectively. Find the inductance value.

4.8 A 10 pF lumped capacitor is formed by overlay structure using alumina dielectric with $\varepsilon_r = 10$. If the dielectric breakdown field is 1 MV/cm, find the figure of merit of dielectric film.

4.9 Describe the basic fabrication process of MOSFETs.

REFERENCES

1. Daly, D A; Stanley, P; Knight, M Caulton, and R. Ekholdt: *Lumped Elements in Microwave Integrated Circuits*, IEEE Trans. on MTT, Dec. 1967, pp. 713–721.
2. Frey J, and K Bhasin: *Microwave Integrated Circuits*, Artech House, Inc., 1985.
3. Gupta, K C, and Singh, A: *Microwave Integrated Circuits*, 1974.
4. Caulton, Martin: *Film Technology in Microwave Integrated Circuits*, Proc. IEEE, V–59, Oct. 1971, pp. 1481–1489.
5. Pucel, R A: *Design Considerations for Monolithic Microwave Circuits*, IEEE Trans. 1981.
6. Liao, Samuel Y: *Microwave Devices and Circuits*, 3rd Edition, Pearson Education, 2003.
7. Edwards, Terry: *Foundations for Microstrip Circuit Design*, John Wiley and Sons, 2nd Edition 1992.

5 IMPEDANCE MATCHING TRANSFORMERS

INTRODUCTION 5.1

When a transmission line is not terminated by its characteristic impedance (Fig. 5.1), according to the ***maximum power transfer*** theorem, maximum power is not delivered to the load and power will be wasted due to reflection from the load. This reduces the efficiency of transmission. A microwave circuit involves many different types of transmission lines and devices, each characterized by its own impedance. Impedance transformation between two reference planes along a microwave transmission line is necessary for the matching of arbitrary load impedance at one plane to the line or the matching of two lines having different characteristic impedances. There are numerous methods for matching a mismatched load or line with the other connected circuit. These techniques are described below.

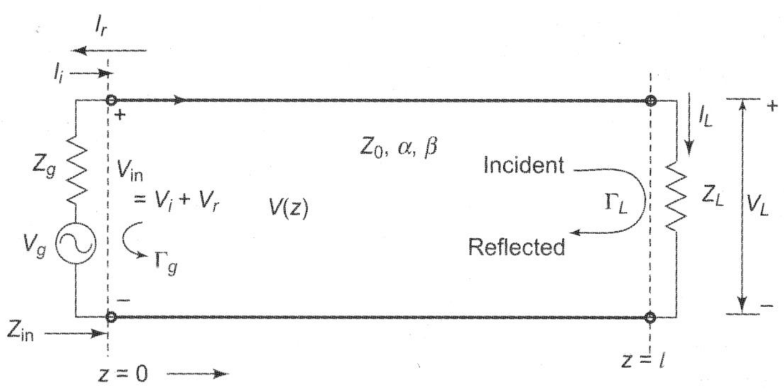

Fig. 5.1 Transmission line with mismatched termination

GENERAL CONDITION FOR IMPEDANCE MATCHING 5.2

Consider a lossless line ($\alpha = 0$) with real characteristic impedance Z_0 as shown in Fig. 5.1, where the generator-end and load-end voltages and currents are given by

$$V(0) = V_i \qquad V(l) = V_r \tag{5.1}$$

$$I(0) = I_i \qquad I(l) = I_r \tag{5.2}$$

For any terminating impedance, $Z_L = R_L = jX_L$, Eq. (3.19b) in Chapter 3 gives

$$Z_{in}\Big|_{z=0} = Z_{in}(0) = Z_0 \frac{Z_L + jZ_0 \tan \beta l}{Z_0 + jZ_L \tan \beta l} = \frac{V_i}{I_i} \tag{5.3}$$

We can write $V_i = V_r \cos \beta l + jI_r Z_0 \sin \beta l$ (5.4a)

$$I_i = I_r \cos \beta l + j(V_r/Z_0) \sin \beta l \quad (5.4b)$$

or

$$|V_i| = |V_r| \left[\left(\cos \beta l + \frac{Z_0 X_L}{Z_L^2} \sin \beta l \right)^2 + \left(\frac{Z_0 R_L}{Z_L^2} \sin \beta l \right)^2 \right]^{1/2} \quad (5.5)$$

$$|I_i| = |I_r| \left[\left(\cos \beta l + \frac{X_L}{Z_0} \sin \beta l \right)^2 + \left(\frac{R_L}{Z_0} \sin \beta l \right)^2 \right]^{1/2} \quad (5.6)$$

From the above equations, it is apparent that the line is matched to produce reflection-free uniform magnitudes of voltage and current throughout the line when

$$X_L = 0 \text{ and}$$

$$Z_L = R_L = Z_0 \quad (5.7)$$

Therefore, the reactive component of the mismatched terminating impedance should be eliminated or cancelled by introducing a suitable element in the line and at the same time, the real part of the mismatched impedance should be transformed to Z_0. There are basically two types of matching considered here: one is **narrowband matching** and the other one is **broadband matching**.

NARROWBAND MATCHING 5.3

Impedance matching over a narrow band of frequencies can be accomplished by inserting reactive elements between the two mismatched impedances. The method uses shunt reactive elements, such as reactive windows/irises, tuning screws, and series and shunt-stub matching. These are explained below.

Waveguide Windows/Irises/Diaphragms 5.3.1

When a flat thin metal obstacle partially fills the waveguide, excited in dominant mode, transversely as shown in Fig. 5.2, it offers inductive or capacitive reactances across the rectangular waveguide, depending on the extension of the obstacle across the narrow or the broad dimensions, respectively. Such obstacles are called irises or windows or diaphragms. When a TE_{10} dominant wave is incident on such a junction, higher-order evanescent TE and TM waves are excited near the junction to satisfy the boundary condition of zero total tangential electric field on the irises. These higher-order modes are of nonpropagating or evanescent type, i.e., they die down within a distance of quarter wavelength from the junction and store reactive energy at the junction.

The irises of Fig. 5.2 (a) and (b) affect mainly the magnetic field of the dominant TE_{10} mode and store predominantly magnetic energy, resulting in inductive characteristics. The irises of Fig. 5.2 (c) and (d) mainly affect the E field of the TE_{10} mode, concentrating it in the gap and storing predominantly electric energy from the evanescent modes. Thus, these irises attain capacitive characteristics. The normalized susceptances of these commonly used thin irises are given below:

Inductive Irise

❑ *Symmetrical Rectangular Case (Fig. 5.2a)*

$$B/Y_0 = -(\lambda_g/a) \cot^2 (\pi d/2a)/[1 + 1/6\,(\pi d/\lambda_g)^2]; \text{ for } d/a << 1 \quad (5.8)$$

$$= -(\lambda_g/a) \cot^2 [2\pi (a - d)/a]/[1 + (8/3)\, \pi^2 (a - d)^2/\lambda_g^2]; \text{ for } (a - d)/a << 1 \quad (5.9)$$

$$\approx -(\lambda_g/a) \cot^2 \left(\frac{\pi d}{2a} \right)$$

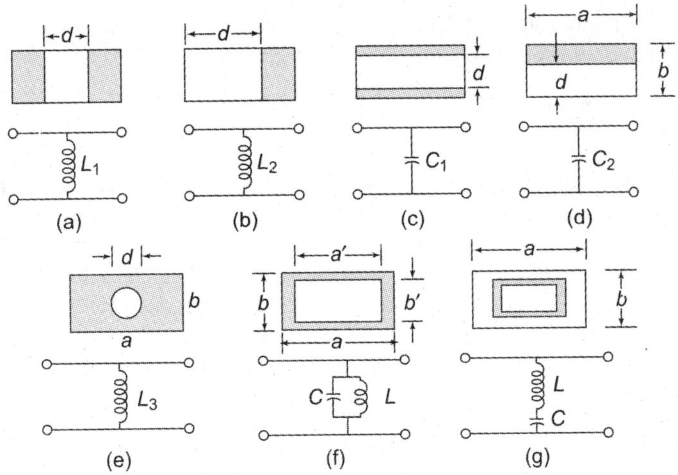

Fig. 5.2 Irises in rectangular waveguide: (a) Symmetrical inductive (b) Asymmetrical inductive (c) Symmetrical capacitive (d) Asymmetrical capacitive (e) Circular inductive (f) Parallel resonant iris (g) Series resonant iris (shaded area is metal and unshaded area is open)

❑ *Asymmetrical Rectangular Case (Fig. 5.2b)*

$$B/Y_0 = -[\lambda_g \pi^2 (a-d)^2/2a^3][1 + (\pi^2/\lambda_g^2)(a-d)^2 \ln(\pi/2)(a-d)/a]; \text{ for } (a-d)/a \ll 1 \quad (5.10)$$

Capacitive Irise

❑ *Symmetrical Case (Fig. 5.2c)*

$$B/Y_0 = (4b/\lambda_g)[\ln(2b/\pi d) + b^2/(2\lambda_g^2)] \approx \frac{4b}{\lambda_g}\ln\left[\operatorname{cosec}\left(\frac{\pi d}{2b}\right)\right]; \quad \text{for} \quad d/b \ll 1 \quad (5.11)$$

$$= \pi^2(b-d)^2/(2b\lambda_g); \quad \text{for} \quad (b-d)/b \ll 1 \quad (5.12)$$

❑ *Asymmetrical Case (Fig. 5.2d)*

$$B/Y_0 = (8b/\lambda_g)[\ln(2b/\pi d) + 2b^2/\lambda_g^2]; \quad \text{for} \quad d/b \ll 1 \quad (5.13)$$

$$= \pi^2(b-d)^2/(b\lambda_g); \quad \text{for} \quad (b-d)/b \ll 1 \quad (5.14)$$

Figure 5.3 shows the variations of normalized susceptance with iris dimension. Irises are employed as impedance-matching elements where the reactive impedance is needed to cancel the opposite reactance in the mismatched line.

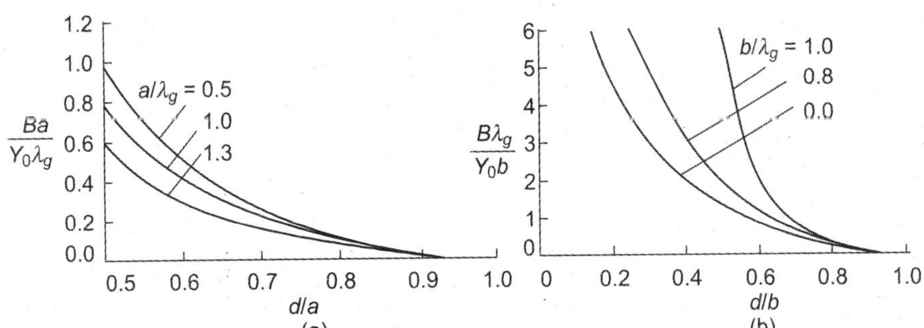

Fig. 5.3 Normalized susceptances of irises: (a) Symmetrical inductive (b) Capacitive

Inductive irises are also made with a small centred circular hole of diameter d as shown in Fig. 5.2(e). The normalized inductive susceptance of a small circular hole is given by

$$\frac{B}{Y_0} = \frac{3ab\lambda_g}{2\pi d^3}; \quad \text{for} \quad d << b \tag{5.15}$$

When the iris is extended both across the narrow and broad walls, it behaves as a parallel L-C circuit in shunt across the guide as shown in Fig. 5.2(f). By choosing the dimensions of a' and b', resonant condition (zero susceptance) can be obtained at selective frequencies in the waveguide band. Quality factor of the resonator decreases with increase in aperture size. Q of the order of 10 can be achieved for a small aperture size. The series resonant iris is a metal ring supported by a lossless dielectric as shown in Fig. 5.2(g).

Tuning Screw 5.3.2

A tuning screw is a metal rod or probe inserted into a rectangular waveguide through the centre of a broad wall perpendicularly. This probe provides a reactance across the guide which can be varied from capacitive to inductive by changing the depth of penetration as shown in Fig. 5.4. The theoretical and experimental analysis shows that a thin screw of diameter << $\lambda_g/4$ possesses susceptance of the following nature:

(a) Capacitive when $h < \lambda_g/4$,

(b) Infinite when $h \sim \lambda_g/4$, and

(c) Inductive when $h > \lambda_g/4$.

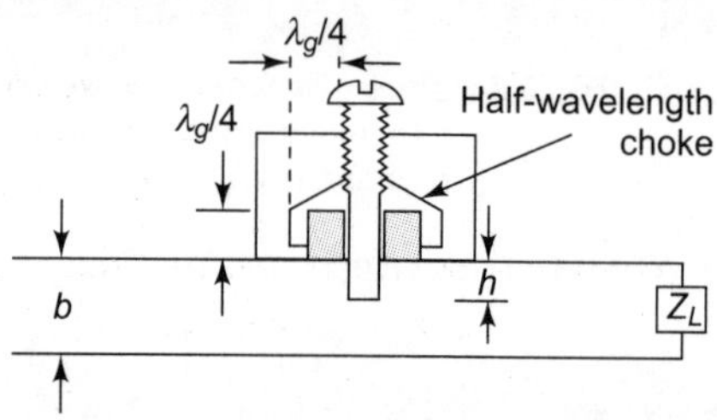

Fig. 5.4 Tuning screw in waveguide

A tuning screw is used as a tuning device for impedance matching with waveguide components with a complex load Z_L on account of its shunt reactance. To avoid power leakage through the screw gap, a half-wavelength choke is used at the screw insertion junction, where the leakage signal sees zero impedance due to $\lambda_g/2$ short-circuited line. The tuning screw can also be slided along the axis of the waveguide through a narrow longitudinal slot at the centre of the broad wall. This helps varying both the penetration and the position of the tuning screw along a longitudinal distance of the half-guide wavelength for better matching with ease. Such devices are called *slide-screw tuners*.

A single screw reactance provides very narrowband matching. A greater frequency range of impedance matching can be achieved with three fixed screws placed 3 $\lambda_g/8$ apart (Fig. 5.5) whose depth of penetration can be varied independently for matching.

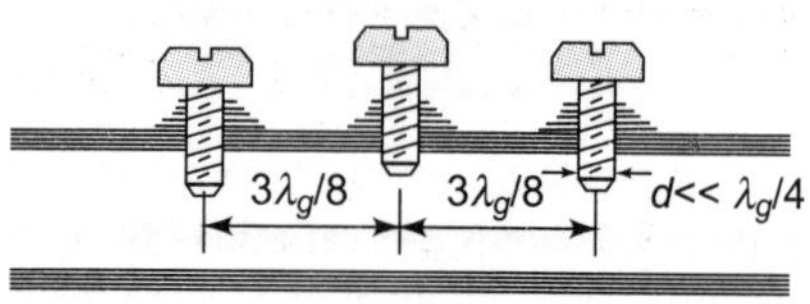

Fig. 5.5 Three-screw tuner

Inductive and Capacitive Posts 5.3.3

A thin cylindrical metal post, placed parallel to **E** of the TE_{10} mode, and extending completely across the narrow width of the waveguide at the centre of the broad wall, provides an inductive susceptance across the guide operated in the dominant TE_{10} mode. When the post is placed across the waveguide at right angles to **E** of the TE_{10} mode, it offers capacitive susceptance across the guide. The value of the susceptance increases with the diameter of the post. Thus, these posts can be used as matching elements. Figure 5.6 shows the post and its susceptance characteristics.

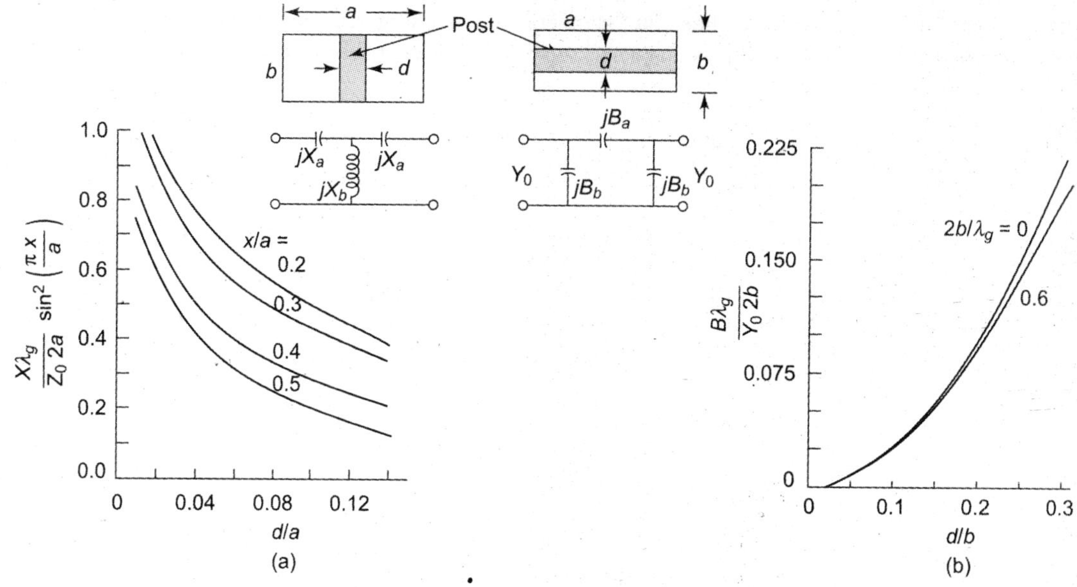

Fig. 5.6 Post and its susceptance characteristics

Stub Matching in Lossless Lines 5.3.4

It is seen that the impedance of an open-circuited or a short-circuited lossless line of length l is a function of cot βl or tan βl, respectively. From the tan and cot curves shown in Fig. 5.7, it is seen that such terminations results in a wide range of inductive and capacitive reactances, depending on the length of the line. Thus, any value of reactance ranging from $-\infty$ to ∞ can be obtained by proper choice of the length of OC or SC lines, called the stubs. Short-circuited sections are preferable to open-circuited sections, because a good short is easily achieved in all kinds of transmission lines.

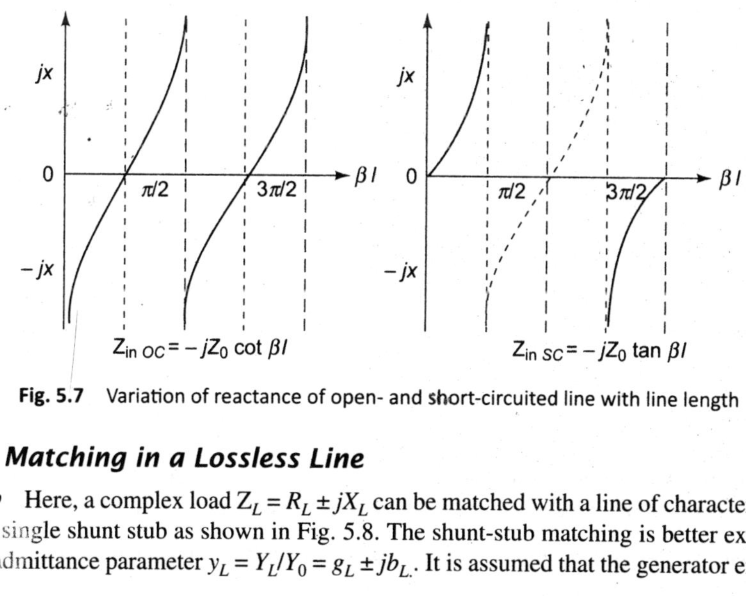

Fig. 5.7 Variation of reactance of open- and short-circuited line with line length

Single-stub Matching in a Lossless Line

❑ ***Shunt Stub*** Here, a complex load $Z_L = R_L \pm jX_L$ can be matched with a line of characteristic impedance $Z_0 = R_0$ using a single shunt stub as shown in Fig. 5.8. The shunt-stub matching is better explained in terms of normalized admittance parameter $y_L = Y_L/Y_0 = g_L \pm jb_L$. It is assumed that the generator end is matched to the characteristic impedance of the line. The following section describes the matching procedure for single-stub matching. In order to avoid complicated mathematics, graphical solutions using **Smith chart** are widely used in the matching design.

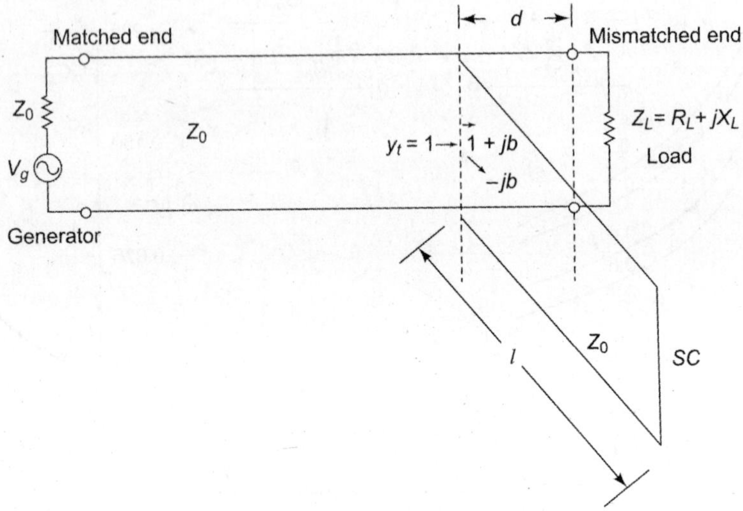

Fig. 5.8 Single shunt-stub matching

In general, when only load is connected, a voltage standing wave is formed in the line. If we move away from the complex load towards the generator along the line, there are voltage minima at which the reflection coefficient is a negative real quantity and the input admittance is pure conductance of value $Y_{in} = (1 - G)/(1 + G) = g = S$, where S is the load VSWR. Similarly, there are voltage maxima at which the admittance is pure conductance of value $Y_{in} = (1 + G)/(1 - G) = 1/S$. Therefore, in between these two points, there must be a point at a certain distance d(say) towards the generator from the first voltage minimum where the real part of the normalized admittance is unity, so that $y_{in} = 1 \pm jb$. The reactive component $\pm jb$ of this admittance can be cancelled by an equal and opposite susceptance $\mp jb$ of a short-circuited shunt stub of length l, located at that point, and having characteristic impedance equal to that of the line, i.e. Z_0.

The total normalized admittance of the terminated line and the stub looking towards the load is then

$$y_t = 1 \pm jb \mp j \cot \beta l = 1; \text{ so that } \cot \beta l = b \tag{5.16a}$$

for matching. Here,

$$d = \frac{\lambda}{4\pi} \cos^{-1}[(S - 1)/(S + 1)];$$

and

$$l = \frac{\lambda}{2\pi} \tan^{-1}\left[\sqrt{S}/(S-1)\right] \tag{5.16b}$$

The following methods can be used in the design of a single shunt stub with the help of the Smith chart of Fig. 5.9.

1. The load admittance point is located at a point A corresponding to *VSWR S*. Join the centre O with A.
2. The location of the stub is found by rotating OA clockwise towards the generator with O as centre, to intersect the unit resistance circle at B, where the real part of the normalized admittance is 1.0. The distance $\frac{d}{l}$ corresponding to the arc AB gives the stub position as per Fig. 5.9.

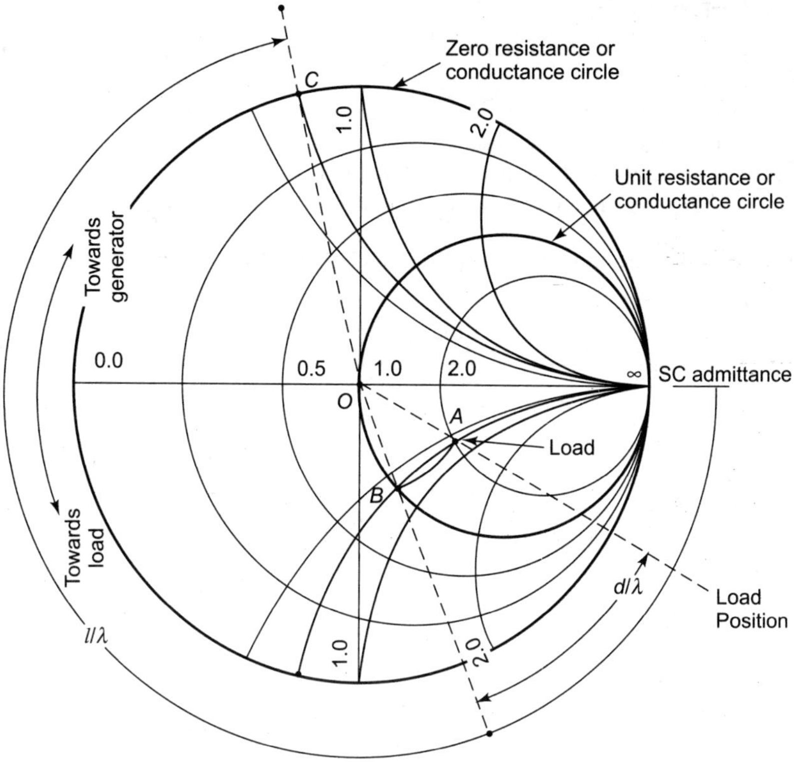

Fig. 5.9 Single shunt-stub matching using Smith chart

3. The normalized susceptance at *B* is read and an equal and opposite pure susceptance value of the stub is located at *C* on the outer circle ($g = 0$).
4. The stub length is found by rotating in a clockwise direction around the unit circle to *C*, starting from the right-hand intersection between the unit circle and the real axis which corresponds to the short-circuit admittance position on the stub.

❑ ***Series Stub*** At voltage minimum, input impedance of the line is $Z_{in} = 1/S$, and at the maximum, $Z_{in} = S$. Therefore, in between these two points, at a distance d from the load towards the generator, normalized input impedance $z_{in} = 1 \pm jx$. A short-circuited series stub of length l can cancel the reactive part to provide matching condition $z_{in} = \dfrac{Z_{in}}{Z_0} = 1$.

Here,
$$d = \frac{\lambda}{4\pi}\cos^{-1}[(1 - S)/(1 + S)];$$
$$l = \frac{\lambda}{2\pi}\tan^{-1}[(1 - S)/\sqrt{S}] \quad (5.17)$$
$$\pm jx = \mp j\tan\beta l$$

The series stub is convenient for waveguide and stripline circuits. The shunt stub is easy to construct for coaxial lines (also for two-wire lines).

Double-stub Matching

It is stated in the previous section that a single stub must be located at a fixed point near the load, where the real part of the line admittance is Y_0. In microwave transmission lines, such as coaxial lines, or waveguides, it is not practical to find such an exact position to place the stub. Moreover, this position changes for different loads and frequencies. Under such circumstances, matching can be better achieved by the use of double stubs, placed at fixed positions and having variable lengths as shown in Fig. 5.10. The two stubs are spaced at a distance of $s = 3\lambda_g/8$ where the first stub B is arbitrarily placed either at the load point or at a distance of $d \geq 0$ from the load end A. The second stub is placed at C, so that $BC = \dfrac{3\lambda_g}{8}$.

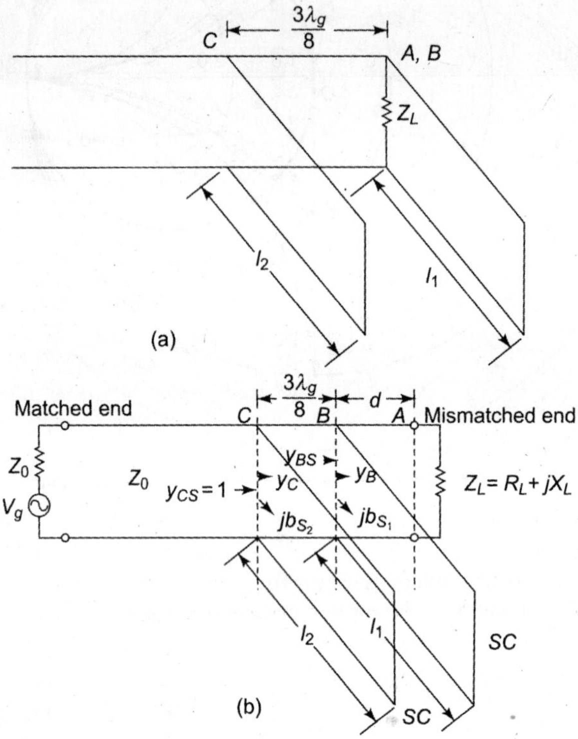

Fig. 5.10 Double-stub matching: (a) First stub at load end (b) First stub at a distance from load

Double-stub matching is conveniently designed and positioned by the use of the Smith chart and from the knowledge of the load Z_L and its VSWR. The illustration of double-stub matching is shown in Fig. 5.11 where it is desired that the input admittance at the second stub position C should be Y_0 corresponding to a point on the unit circle ($g = 1$) of the Smith chart having its centre at O'.

The following steps are involved.

1. The normalized load admittance y_L is located at A, diametrically opposite to load impedance point $z_L = Z_L/Z_0$.
2. From the load admittance point A, a rotation towards generator on the load VSWR circle by an arbitrary amount d/λ_g (= 0.06 say), gives the position B of the first stub where the normalized admittance towards load for the line alone may be expressed as

$$y_B = g_B + jb_B \tag{5.18a}$$

3. The unit conductance circle ($g = 1$) is rotated around the chart centre O towards the load by $3\lambda_g/8$, where the new centre of the circle becomes O'' on the line passing $+j$ and $-j$ susceptance points.
4. The point of intersection B' between g_B circle and the rotated unit conductance circle with centre O'' gives the total normalized admittance y_{BS} of the combination of the line at the right of the point B and the short-circuited first stub of length l_1:

$$y_{BS} = y_B + jb_{S1} = g_B + jb_B + jb_{S1} \tag{5.18b}$$

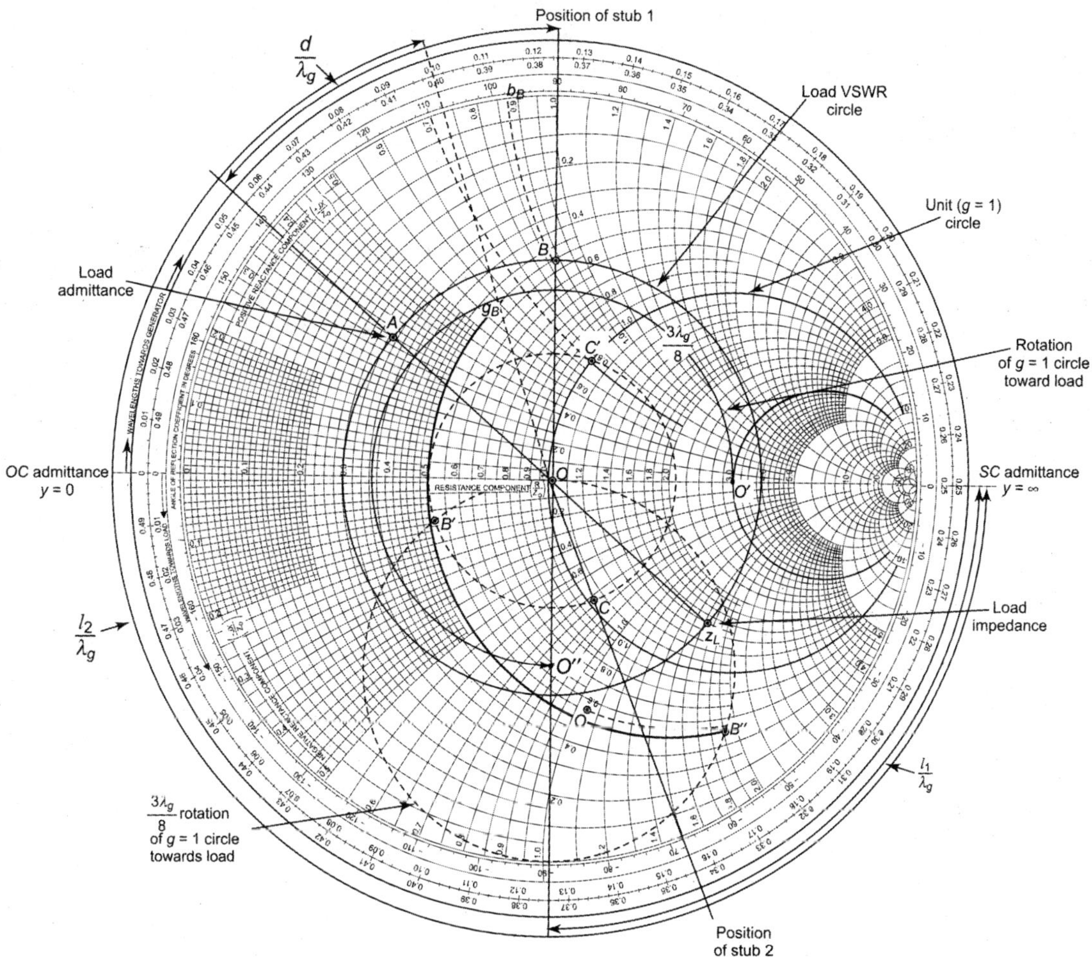

Fig. 5.11 Double-stub design using the Smith chart

Here, jb_{S1} is the susceptance of the first stub alone which can be determined by knowing y_{BS} and y_B from the Smith chart and length l_1/λ_g is determined from the short circuit admittance point of stub 1.

5. From the point B', rotates towards generator by $3\lambda_g/8$ to reach the point C on the original $g = 1$ circle. The total normalized admittance at C for the original line and first stub is then

$$y_C = 1 + jb_C \tag{5.18c}$$

6. For matching, the total admittance at C with the combination of (i) line, (ii) first stub, and (iii) second stub should be

$$y_{CS} = 1.0 \quad \text{where reactive point is cancelled.} \tag{5.18d}$$

This is achieved by adjusting the length l_2 of the second stub to provide an equal and opposite susceptance of $jb_{S2} = -jb_C$ corresponding to the point C' on the Smith chart.

7. The stub lengths are found by rotating towards the generator around the unit reflection coefficient outer most circle, starting from the right-hand intersection of the unit circle and the real axis, which corresponds to the origin of infinite susceptance ($y = \infty$) of the short termination of the second stub, up to the point of corresponding stub susceptance $-jb_C$ corresponding to the point C'.

Thus, finally the impedance matching is done from the values of d, l_1 and l_2 read from the Smith chart.

Limitation

The double-stub system cannot match all impedances. If the normalized load conductance exceeds $\frac{1}{\sin^2 \beta s}$, double stub cannot be used, since the matching condition will not be satisfied for those values of load conductance, where s is the separation distance between the stubs. Also alternatively, if $g_L > 2$, there will be no point of intersection between the rotated circle and $g = 1$ circle, and no solution will be possible.

Realization of Stubs in Microwave Circuits 5.3.5

Coaxial Line

In two-conductor coaxial lines, the stubs are realized by T-junctions at the stub positions as shown in Fig. 5.12, where moveable shorts are used to adjust the lengths of the stubs. The stub position on the line can be adjusted by a sliding arrangement, where a longitudinal narrow slot on the outer conductor of the line allows movement of the centre conductor of the stub. However, this arrangement is a complicated process and, therefore, the coaxial line matching requires two fixed double stubs of variable lengths l_1 and l_2. The double-shunt stubs are easy to construct for coaxial lines. A series stub cannot be realized easily in coaxial lines, because the circuit cannot be closed using a short in this configuration.

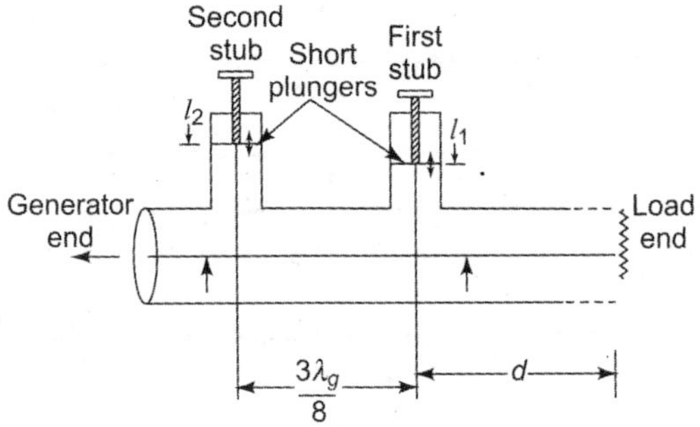

Fig. 5.12 Coaxial line stubs

Waveguide Stub

In a rectangular waveguide, the stub may be connected in the E or H plane by placing a waveguide short-circuit plunger, across the wide dimensions or across the narrow dimensions, respectively, as shown in

Fig. 5.13. For an *E*-plane stub, the electric field lines penetrate from the main waveguide into the stub, whereas, in the *H*-plane stub the magnetic field lines penetrate from the waveguide into the stub. Thus, the *E*-stub offers reactance in series with the equivalent main line and the *H*-stub presents a parallel susceptance across the main line. Since a single stub in a waveguide cannot be located very accurately at any place, two fixed stubs are preferred in waveguide circuits. The combined *E* and *H* stubs of Fig. 5.13(c), called the *E-H* tuner, matches a wide range of load impedances in which the positions of the short-circuit plungers in the *E* and *H*-arms are simultaneously varied for matching.

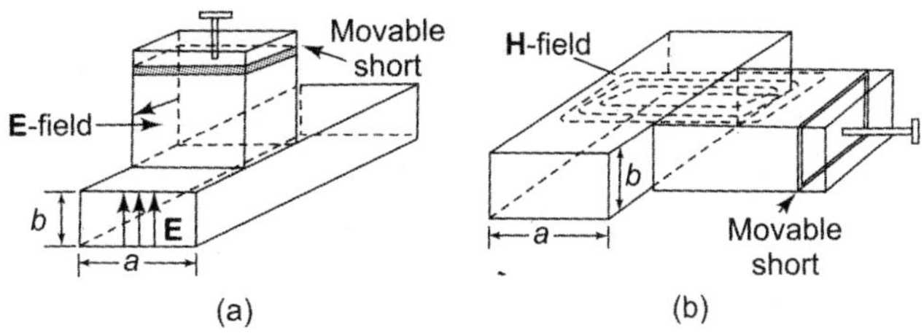

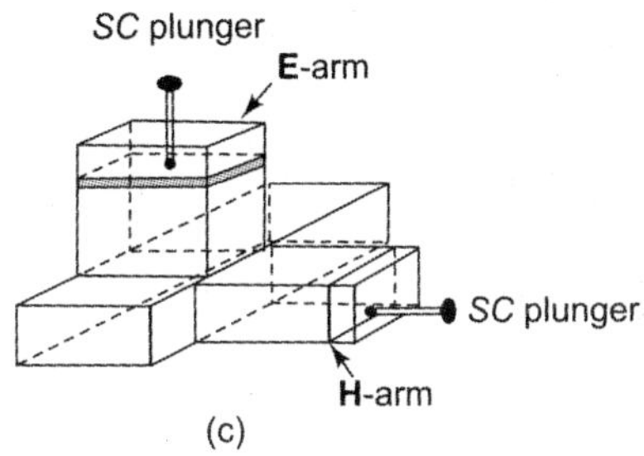

Fig. 5.13 Waveguide stubs: (a) E stub (b) H stub (c) E-H tuner

Example 5.1 A load impedance of 73–*j*80 ohms is required to be matched to a 50-ohm coaxial line having a lossless dielectric of dielectric constant 4, using a short-circuited shunt stub at 500 MHz. Determine the position and length of the stub.

Solution The guide wavelength in TEM mode coaxial line with dielectric $\varepsilon_r = 4$ is

$$\lambda_g = \frac{30}{0.5\sqrt{4}} = 30 \text{ cm}$$

Let the stub characteristic impedance also be 50 ohms.

Normalized load impedance $z_L = (73 - j80)/50 = 1.46 - j1.6$. The position and the length of the stub are found from the Smith chart of Fig. 5.14 as follows.

1. z_L is located on the Smith chart.
2. A load VSWR circle is drawn through z_L, with centre O at unity of real axis.
3. The normalized load admittance y_L is located at A, diametrically opposite to z_L on the circle.
4. A distance $\frac{d}{\lambda_g}$ is moved from A towards generator to reach a point B (point of intersection between $g = 1$ circle and the load VSWR circle) on the circle where $y = 1 + jb$. The stub distance $\frac{d}{\lambda_g} = 0.12$ and line susceptance at B towards load $jb = j\,1.38$ are read from the Smith chart.

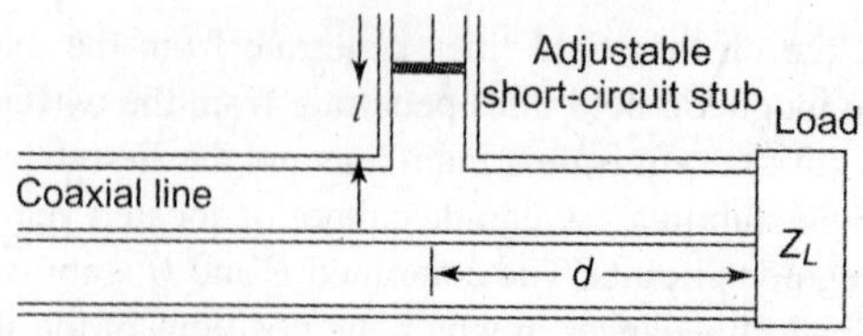

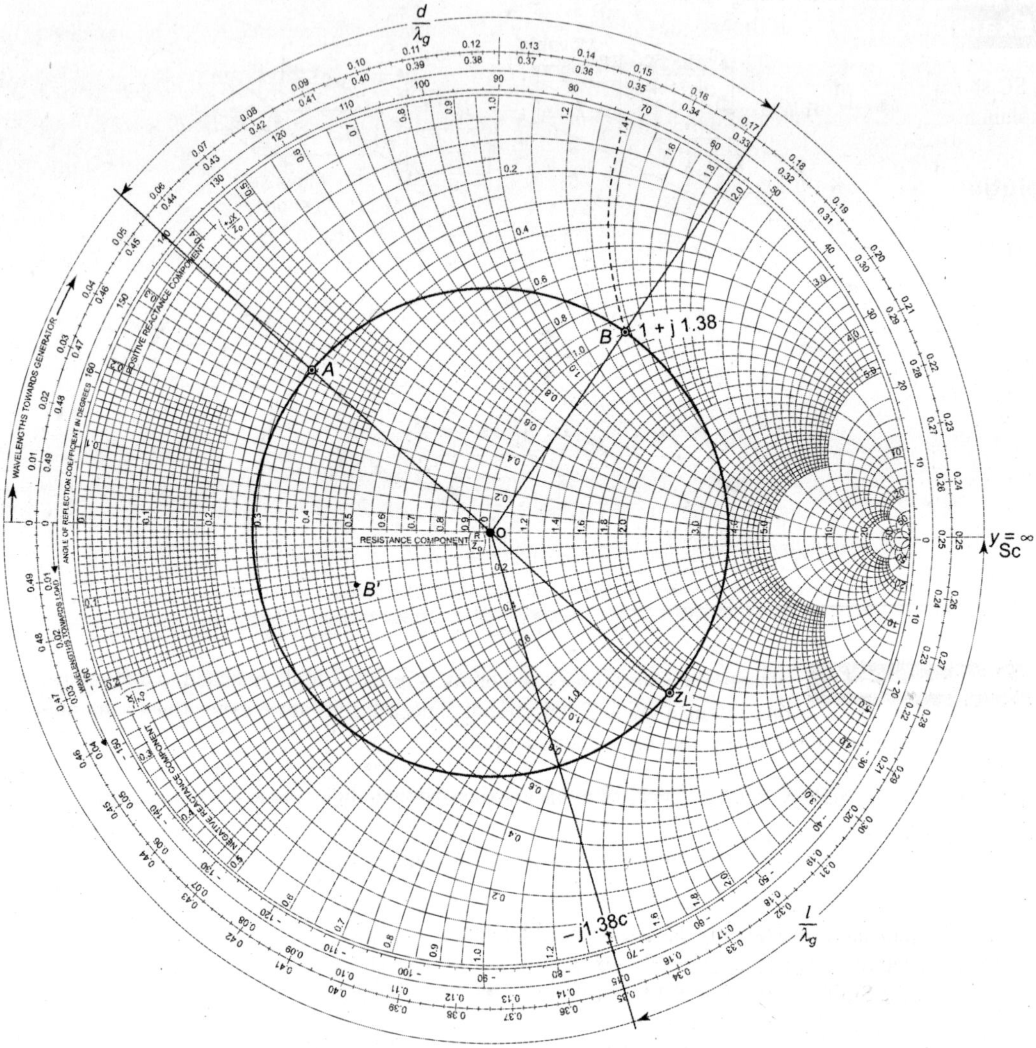

Fig. 5.14 Single coaxial stub design

5. To cancel the susceptance *j* 1.38 by the stub at *B* of length *l*, the stub should provide a susceptance of $-j$ 1.38 corresponding to *C*. To find the stub length *l*, a distance of $\frac{l}{\lambda_g} = 0.1$ is moved towards

the generator from the short-circuit susceptance point $y = \infty$ to reach the stub susceptance point C, where $y = -j\,1.38$.

Therefore, the required distance and length of the stub are

$$d = 0.12\,\lambda_g = 3.6 \text{ cm.}$$

$$l = 0.1\,\lambda_g = 3.0 \text{ cm.}$$

The coaxial line configuration of the stub is shown in Fig. 5.14.

Example 5.2 A 100-ohm coaxial line with air as dielectric is terminated by a load impedance of $75 + j40$ ohms and is excited at 1 GHz by a matched generator. Find the position of an SC shunt single-matching stub of 100-ohm impedance on the line, and determine the length of the stub. Explain how such a stub is realized in a rectangular waveguide circuit, excited in the dominant mode.

Solution The *SC* shunt normalized load impedance $z_L = \dfrac{75 + j40}{100} = 0.75 + j0.4$

The line wavelength $\lambda_g = c/f = 30$ cm.

The stub design is made using the Smith chart of Fig. 5.15 as follows.

1. z_L is located on the Smith chart and the corresponding VSWR circle is drawn through z_L.
2. The normalized load admittance y_L is located at a diametrically opposite point A on the VSWR circle to obtain the value $y_L = 1.0 - j\,0.55$.
3. Since the real part at A is 1.0, the stub may be located at the load point which may be difficult to place. Therefore, a distance $\dfrac{d}{\lambda_g}$ is moved towards the generator from y_L to reach a normalized admittance point of intersection B between the VSWR circle and the unit conductance circle where the normalized admittance is $y_B = 1 + j\,0.55$. This gives $\dfrac{d}{\lambda_g} = 0.292 \approx 0.3$.
4. For matching, the normalized stub susceptance is $-j\,0.55$ which is located on the outer circle.
5. To find the length of the short-circuited stub, a distance $\dfrac{l}{\lambda_g}$ is moved towards the generator from the short-circuit infinite admittance point to reach C so that $\dfrac{l}{\lambda_g} = 0.17$.

Therefore,

$$d = 0.292 \times 30 = 8.76 \text{ cm}$$

$$l = 0.17 \times 30 = 5.1 \text{ cm}$$

$$Z_{0\text{stub}} = 100 \text{ ohms.}$$

A shorted shunt stub can be realized in rectangular waveguide by joining it to the waveguide's narrow dimension so that magnetic loops can penetrate from the main waveguide into the stub as shown in Fig. 5.15. It is assumed that the tee junction formed between the waveguide and the stub do not create any field discontinuities, d and l should be determined in terms of guide wavelength in place of TEM line λ_g above.

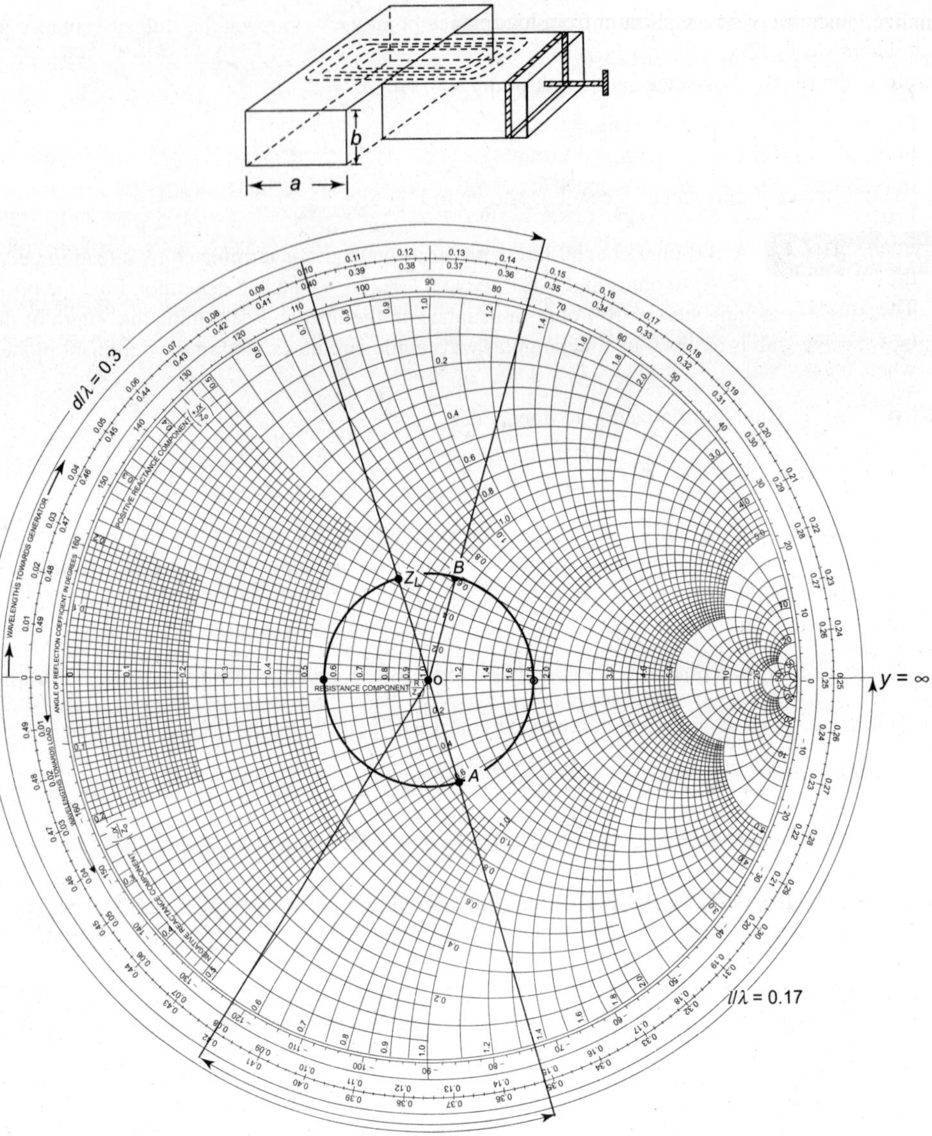

Fig. 5.15 Single waveguide stub design

Example 5.3 A lossless, 50-ohm air-filled coaxial line has V_{max} = 2.5 V and V_{min} = 1 V when terminated with an unknown load. The distance between the successive voltage minima is 5 cm and the first voltage minimum from the load end is 1.25 cm. Design a short-circuited single stub for impedance matching.

Solution Distance between two successive minima is

$$\lambda/2 = 5 \text{ cm} \qquad \therefore \quad \lambda = 10 \text{ cm}$$

$$\text{VSWR} = V_{max} / V_{min} = 2.5/1 = 2.5$$

Normalized distance of first minimum from load end is

$$d_{\min} / \lambda = 1.25/10 = 0.125$$

1. Using Smith chart of Fig. 5.16, a VSWR circle $S = 2.5$ is drawn.
2. Locate first minimum point A on the real axis where normalized impedance is $1/S = 1/2.5 = 0.4$ (real).
3. From A, a distance $d_{\min}/\lambda$ is moved towards load to reach the load point B where the normalized load impedance $z_L = 0.7 - j\,0.72$. An opposite point C is marked for load admittance $y_L = 0.7 + j\,0.72$.
4. From C, a distance d/λ is moved towards the generator to reach the point D where S and $g = 1$ circle intersects. At D, the normalized line admittance $= 1 + jb = 1 + j\,0.95$. The distance of the stub from the load is $d/\lambda = 0.035$.
5. The stub must provide a susceptance of $-j0.95$ for matching ($y_{in} = 1$). From the short-circuit admittance ($g = \infty$) at the point E, a distance l/λ is moved towards the generator to reach a point F on the outer circle where the normalized susceptance is $-j\,0.95$. $l/\lambda = 0.13$, the required stub length.
   Therefore,

$$d = 0.035 \times 10 = 0.35 \text{ cm, and } l = 0.13 \times 10 = 1.3 \text{ cm}$$

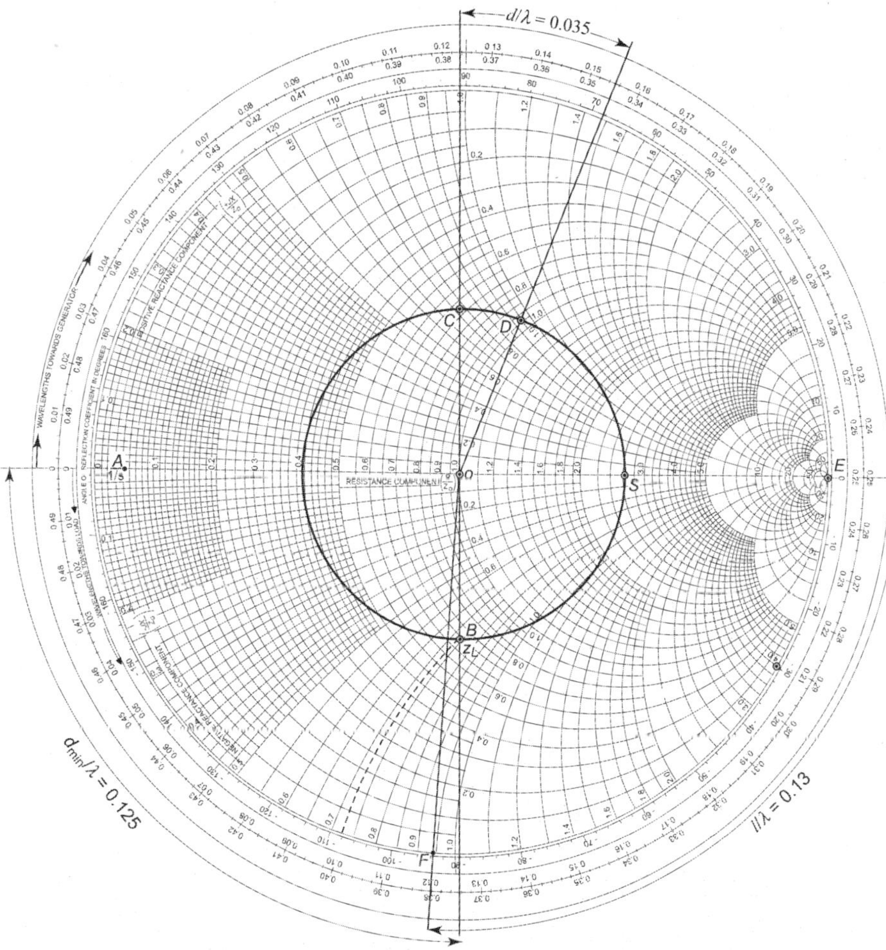

Fig. 5.16 Single stub design from load VSWR and voltage minimum position

Example 5.4 A 75-ohm, air-filled coaxial line is terminated with a complex load of 109.5 – j120 ohms. Design a double-stub matching system using short-circuited coaxial lines of 75-ohm characteristic impedance.

Solution The normalized load impedance

$$z_L = (109.5 - \text{j}120)/75 = 1.46 - \text{j}1.6;$$

The load reflection coefficient

$$|\Gamma| = \left|\frac{Z_L - Z_0}{Z_L + Z_0}\right| = \left|\frac{109.5 - j120 - 75}{109.5 - j120 + 75}\right| = \left|\frac{34.5 - j120}{184.5 - j120}\right| = 0.567$$

$$\text{VSWR} = \frac{1+|\Gamma|}{1-|\Gamma|} = \frac{1+0.567}{1-0.567} = 3.6$$

The double-stub system is designed using the Smith chart of Fig. 5.17 and the following steps.

1. Normalized load impedance z_L is located on the Smith chart. A circle is drawn through z_L with centre at 0(y = 1) and normalized load admittance y_L is located on the circle at A diametrically opposite to z_L, where $y_L = 0.31 + j\,0.35$.
2. At an arbitrary distance $d/\lambda = 0.067$ from y_L towards the generator along the circle of radius $S = OA$ gives the position B of the first stub where the normalized admittance is

   $$y_B = g_B + jb_B = 0.5 + \text{j}\,0.85$$

3. The unit conductance circle (g = 1) with centre O' is rotated towards the load by $3\lambda/8$ to reach the new centre O'' of the rotated circle on the line passing through the points $+j$ and $-j$.
4. The point of intersection B' between g_B circle and the rotated unit conductance circle of the centre O'' gives the total admittance of line shunted by first stub:

   $$y_{BS} = g_B + jb_B + jb_{S1} = 0.5 + j\,0.85 + jb_{S1}$$
   $$= 0.5 - j\,0.14, \text{ Therefore, } j\,0.85 + jb_{S1} = -j0.14$$

   or, the susceptance of the first stub is $jb_{S1} = -j\,0.14 - j\,0.85 = -j0.99 \simeq -j1.0$
5. The length l_1 of the first stub is found by rotating towards the generator along the periphery of the outer circle from the right-hand intersection of the unit circle and the real axis, which corresponds to the short-circuit admittance of the stub to its input susceptance value – j0.99 on the outer circle to obtain

   $$l_1/\lambda = 0.125$$

6. The point B' is rotated towards the generator with radius OB' by $3\lambda/8$ to C on the original g = 1 circle (centre O') to get the total normalized admittance at C due to the line and the first stub:

   $$y_C = 1 - jb_C = 1 - j\,0.70$$

7. Matching by second stub is obtained by its susceptance $jb_{S2} = +jb_C = +j\,0.70$ corresponding to the opposite point C' on the circle of radius OB' or OC.
8. The length of the second stub l_2 is found by rotation along the outer circle towards the generator from the right-hand intersection of the unit circle and the real axis to the point $+j\,0.70$:

   $$l_2/\lambda = 0.347$$

   Hence, the required stub lengths are

   $$l_1/\lambda = 0.125$$

$l_2/\lambda = 0.347$

for distance of the first stub from the load $d/\lambda = 0.067$.

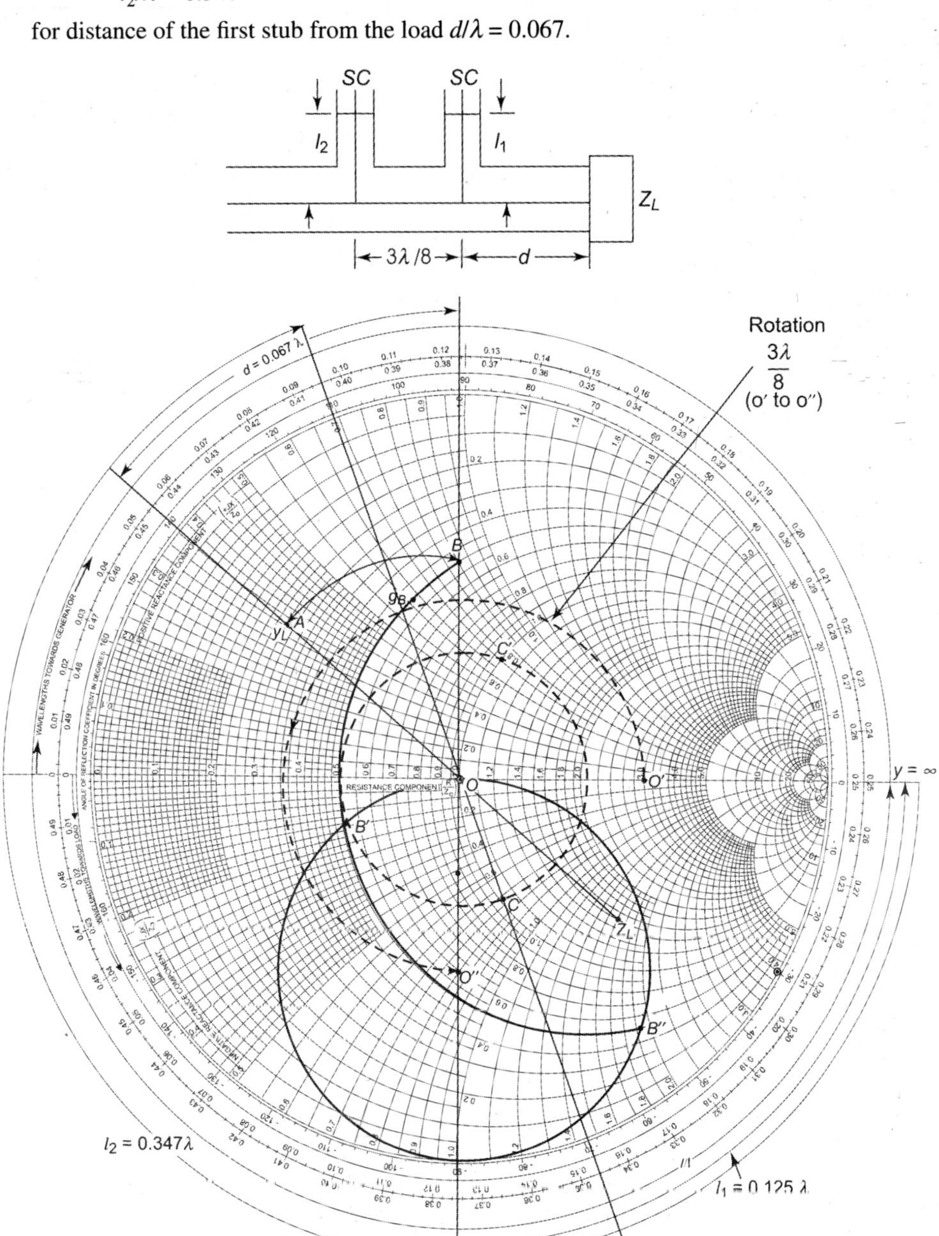

Fig. 5.17 Coaxial line double-stub design

Microstrip Line Impedance Matching 5.3.6

Lumped Element Matching

In microwave integrated circuits, realization of stubs is difficult because good short- or open-circuit configurations of the lines are not easily achieved. However, MIC stub matching is used by approximate realization of short and open.

When the size of RF/microwave circuits are < $\lambda/10$, lumped element resistors, capacitors and inductors are conveniently used for impedance matching up to 1–2 GHz. Furthermore, lumped elements of size < $\lambda/10$ are built in MMIC and hybrid MIC which can be used up to even 60 GHz.

These electrical elements are realized via several methods discussed in Chapter 4. Some are shown in Fig. 5.18.

Resistors

1. Planar resistor [Fig. 5.18(a)] is formed in a transmission line using lossy film of nichrome or tantalum nitride in thin-film technology.
2. Chip resistor [Fig. 5.18(b)] is made with a lossy film deposited on a ceramic chip.

Inductors

1. A loop of a transmission line [Fig. 5.18(c)] with radius < $\lambda/10$ is a realization of small inductance.
2. A spiral inductor [Fig. 5.18(d)] with air bridge provides L values up to about 10 nH. Total length of the spiral conductor must be a small fraction of the wavelength to function as a lumped element.

Capacitors

1. A single gap [Fig. 5.18(e)] in the strip can be used for $C < 0.5$ pF.
2. A short microstrip stub [Fig. 5.18(f)] can be used for $C < 0.1$ pF.
3. An interdigital gap capacitor [Fig. 5.18(g)] in a microstrip line provides $C < 0.5$ pF. The C values depend on the number of fingers and their length.
4. A chip capacitor [Fig. 5.18(h)] is used for $C < 25$ pF.

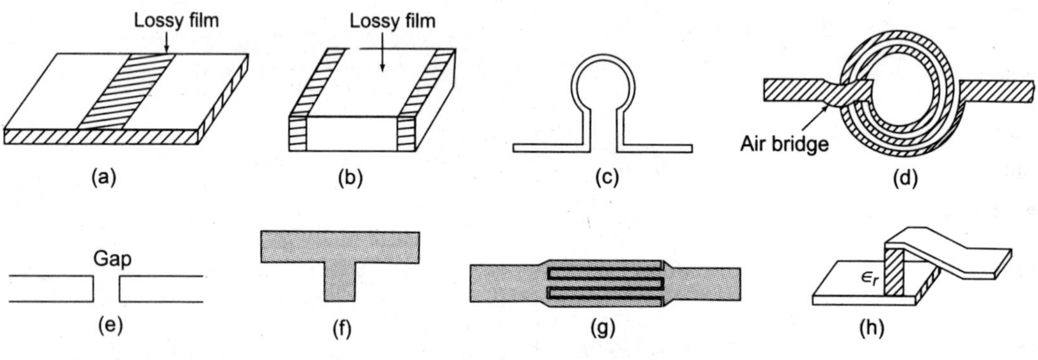

Fig. 5.18 MIC Lumped matching elements

L-Section Matching Networks

L-section networks consist of L and C. A section like this acts as an impedance transformer between resistances R_{in} and R_{out} that terminate the input and output respectively, of the section. Two cases are important: (1) the capacitor shunts R_{in} and, (2) the capacitor shunts R_{out} as shown in Fig. 5.19. These are analysed in the following sections.

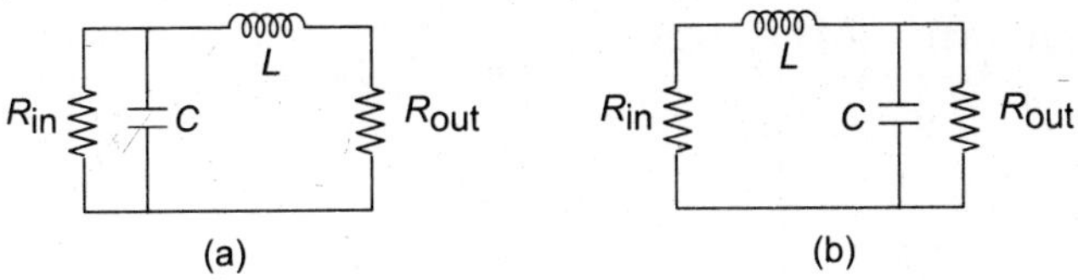

Fig. 5.19 L-section impedance matching (a) $R_{in} > R_{out}$ (b) $R_{out} > R_{in}$

Single L-Section Matching

It is important to recall the parameters for *rf* lines:

$$Z_o = \sqrt{\frac{L}{C}} \quad \text{and} \quad \omega_o = \frac{1}{\sqrt{LC}} \tag{5.19}$$

❑ *Case 1* The input impedance of the L-section when terminated in R_{out} with capacitor C shunts R_{in} can be expressed as

$$Z_{in} = \frac{(R_{out} + j\omega L)\dfrac{1}{j\omega C}}{R_{out} + j\omega L + \dfrac{1}{j\omega C}} = \frac{Z_o^2}{R_{out}} \frac{1 - j\dfrac{\omega_o}{\omega}\dfrac{R_{out}}{Z_o}}{1 - j\dfrac{Z_o}{R_{out}}\left(\dfrac{\omega_o}{\omega} - \dfrac{\omega}{\omega_o}\right)} \tag{5.20a}$$

If we equate

$$\frac{\omega_o}{\omega}\frac{R_{out}}{Z_o} = \frac{Z_o}{R_{out}}\left(\frac{\omega_o}{\omega} - \frac{\omega}{\omega_o}\right)$$

or

$$\frac{\omega}{\omega_o} = \sqrt{1 - \frac{R_{out}}{R_{in}}} \tag{5.20b}$$

In (5.20a), $$Z_{in} = \frac{Z_o^2}{R_{out}} = \text{Real } Z_{in} = R_{in} \rightarrow \text{resistive} \tag{5.20c}$$

It is seen from (5.20b) that the operating frequency $f = \dfrac{\omega}{2\pi}$ is real when $R_{out} \le R_{in}$.

Therefore, the shunt capacitor should be placed in the input side when $R_{out} \le R_{in}$ for physically realisable matching using LC network having $Z_0 = \sqrt{R_{in}R_{out}}$.

◆ **Example** For matching between $R_{out} = 50$ ohms and $R_{in} = 75$ ohms, we find $Z_0 = \sqrt{R_{in}R_{out}} = 25\sqrt{6}$ $= 61.24$ ohm, and $\omega = \omega_0\sqrt{1 - \dfrac{R_{out}}{R_{in}}} = \omega_0\sqrt{1 - \dfrac{50}{75}} = \omega_0\sqrt{\dfrac{1}{3}}$, or $\omega_0 = \omega\sqrt{3}$. If the design frequency $f = 1$ MHz, $\omega_0 = 2\pi * 10^6\sqrt{3}$ radian/s. Therefore, $L = \dfrac{Z_o}{\omega_o} = 5.63$ μH, and $C = \dfrac{1}{\omega_o Z_o} = 1500$ pF

❑ *Case 2* The input impedance of the L-section when terminated in R_{out} with capacitor shunts R_{out} can be expressed as

$$Z_{in} = j\omega L + \frac{R_{out}\dfrac{1}{j\omega C}}{R_{out} + \dfrac{1}{j\omega C}} = \frac{Z_o^2}{R_{out}} \frac{1 - j\dfrac{R_{out}}{Z_o}\left(\dfrac{\omega_o}{\omega} - \dfrac{\omega}{\omega_o}\right)}{1 - j\dfrac{\omega_o}{\omega}\dfrac{Z_o}{R_{out}}} \tag{5.21a}$$

If we equate in (5.21a)

$$\frac{\omega_o}{\omega}\frac{Z_o}{R_{\text{out}}} = \frac{R_{\text{out}}}{Z_o}\left(\frac{\omega_o}{\omega} - \frac{\omega}{\omega_o}\right)$$

or
$$\frac{\omega}{\omega_o} = \sqrt{1 - \frac{R_{\text{in}}}{R_{\text{out}}}} \quad (5.21b)$$

Then
$$Z_{\text{in}} = \frac{Z_o^2}{R_{\text{out}}} = R_{\text{in}} \quad (5.21c)$$

It is seen that the operating frequency $f = \dfrac{\omega}{2\pi}$ is real when $R_{\text{out}} \geq R_{\text{in}}$.

Therefore, the shunt capacitor should be placed in the output side when $R_{\text{out}} \geq R_{\text{in}}$ and LC network will match R_{in} with R_{out} when the characteristic impedance of matching portion is $Z_0 = \sqrt{R_{\text{in}}R_{\text{out}}}$.

◆ **Example** For matching between $R_{\text{out}} = 75$ ohms and $R_{\text{in}} = 50$ ohms, we find $Z_0 = \sqrt{R_{\text{in}}R_{\text{out}}} = 25\sqrt{6}$ = 61.24 ohms, and $\omega = \omega_0\sqrt{1-\dfrac{R_{\text{in}}}{R_{\text{out}}}} = \omega_0\sqrt{1-\dfrac{50}{75}} = \omega_0\sqrt{\dfrac{1}{3}}$, or $\omega_0 = \omega\sqrt{3}$. Again, if the design frequency f = 1 MHz, $\omega_0 = 2\pi * 10^6\sqrt{3}$ radian/s. Thus, the same values of L and C are obtained, but only C should be placed on the load side.

It is, therefore, clear that at the operating frequency ω, impedance matching using L-section imposes that the **capacitor shunts the higher of the two resistances** at the terminals.

Microstrip Stubs

A microstrip line can be terminated by using coaxial open or short-circuit SMA connectors at the end of the board. But in case of impedance matching at an interior location of the board of MIC, such terminations are not practical and hence, usage of short or open stubs for impedance matching in MIC requires special considerations as explained below.

❑ ***Microstrip Open Circuit*** A physical open circuit of a microstrip line does not represent a true electrical open circuit because of three phenomena:

1. Fringing fields at the physical end of the strip cause the line to behave as if there was a capacitance that was connected at the open end or equivalently, a line length that extended past the physical open end to an ideal open circuit as shown in Fig. 5.20. Because of this, the line behaves as if it was longer than its geometrical length. Experimentally, the matched condition is observed at a lower frequency than intended by design.

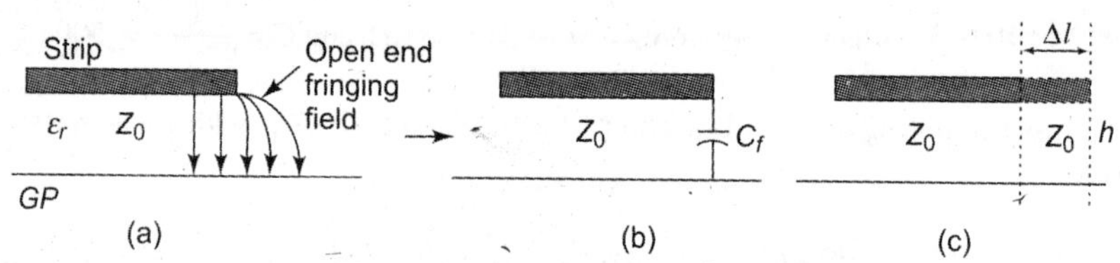

Fig. 5.20 End effect of microstrip open circuit (line width = w, substrate height = h) (a) Actual (b) Equivalent fringe capacitance C_f (c) Equivalent extended length Δl replacing C_f

2. Radiation from the open end-discontinuity produces an equivalent shunt conductance at the open end.
3. Launching of surface waves from the open end of the strip results in an equivalent shunt conductance at the open end (not shown in Fig. 5.20).

Expression for end-effect length extension can be given by

$$\Delta l = 0.412\, h \left(\frac{\varepsilon_{\text{eff}} + 0.3}{\varepsilon_{\text{eff}} - 0.258}\right)\left(\frac{w/h + 0.262}{w/h + 0.813}\right) \tag{5.22a}$$

This is further simplified from Cohn's result as

$$(\Delta l)_{\max} \simeq 0.441 h \tag{5.22b}$$

Microstrip Short Circuits

A microstrip short at the stub end can be realized by a short wire bonded to the strip and ground plane through a hole in the substrate. The wire will have negligible parasitic reactive effects at lower range of microwave frequencies < 3 GHz. At higher frequencies (> 3 GHz), a shorting pin produces parasitic reactance depending on the line width and the pin diameter. Much work has been carried out by researchers and an effective shorting method is adopted by bonding a short hollow metallized cylinder, called via, of diameter ≈ line width/2.3 bonded to the strip and the ground plane through the substrate as shown in Fig. 5.21. The MIC stub-matching rules for single and double stubs are same as described before. The schematic representation of microstrip short and open stubs is shown in Fig. 5.22.

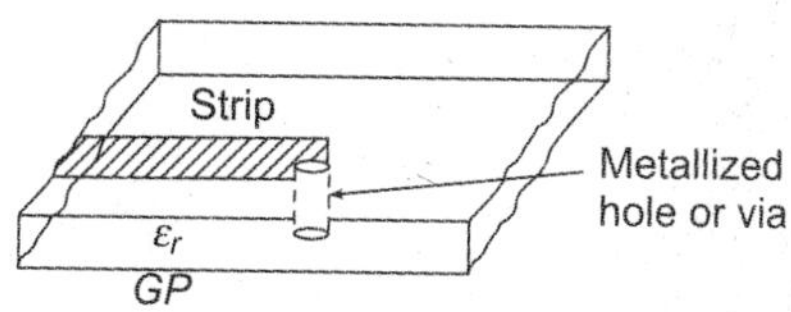

Fig. 5.21 Short-circuiting hollow metallized cylinder

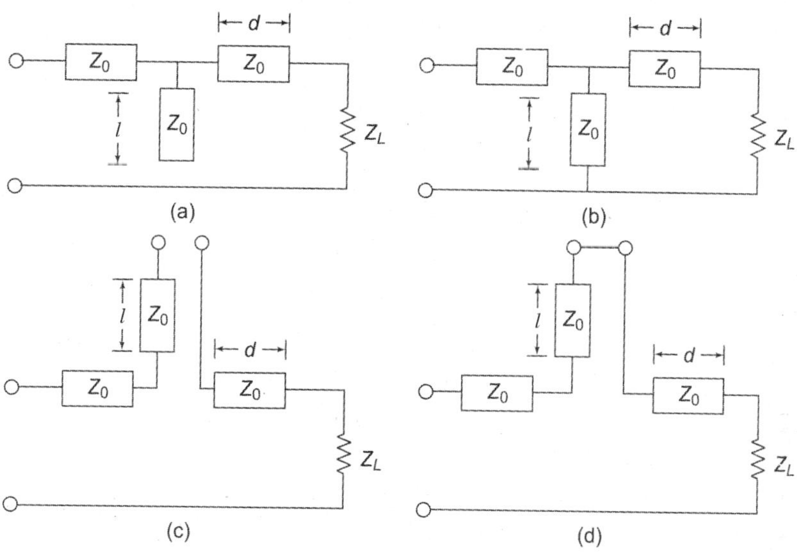

Fig. 5.22 Microstrip shunt and series stubs: (a) Shunt open stub (b) Shunt short stub (c) Series open stub (d) Series short stub

BROADBAND MATCHING 5.4

Impedance-matching techniques described above results in matching over a narrow band of frequencies since the dimensions of matching elements ideally satisfy the condition only at a given frequency. A small band of frequencies around the desired value may give VSWR within the tolerable range $1 \le S \le 2$. Matching over a broader range can be achieved using special methods as described here.

Quarterwave Impedance Transformers 5.4.1

Design

Transmission line equations (5.3), (5.4a) and (5.4b) show that a lossless line can be used as voltage and current transformers with the secondary values V_r or I_r at the load end and the primary values V_i or I_i at the sending end, by the proper selection of l and Z_0. If a line of electrical length $\beta l = \pi/2$ having characteristic impedance Z_2 is terminated by an impedance Z_3, the input impedance of the line is given by

$$Z_{in} = Z_2 \frac{Z_3 + jZ_2 \tan \beta l}{Z_2 + jZ_3 \tan \beta l} = Z_2^2/Z_3 \text{ when } \beta l = \pi/2 \quad \text{or} \quad l = \lambda/4 \tag{5.23a}$$

or,

$$Z_2 = \sqrt{Z_{in} Z_3} \tag{5.23b}$$

Thus, Z_3 is matched to the generator of impedance $Z_1 = Z_{in}$ by a $\frac{\lambda_g}{4}$ line of impedance

$$Z_2 = \sqrt{Z_1 Z_3} \tag{5.24}$$

Therefore, a generator of impedance Z_1 can be matched to a load $Z_3 \ne Z_1$ by inserting a length $l = \lambda_g/4$ of the transmission line between them, whose characteristic impedance is $[Z_1 \, Z_3]^{1/2}$. This matching section of line is called the **quarterwave matching transformer** as shown in Fig. 5.23 (a) in microstrip form.

When Z_1 and Z_3 are widely different, better matching can be achieved by using multisection quarterwave transformers, each having characteristic impedance equal to the square root of the product of those on both sides as shown in Fig. 5.23 (b) and length of quarter wavelength each.

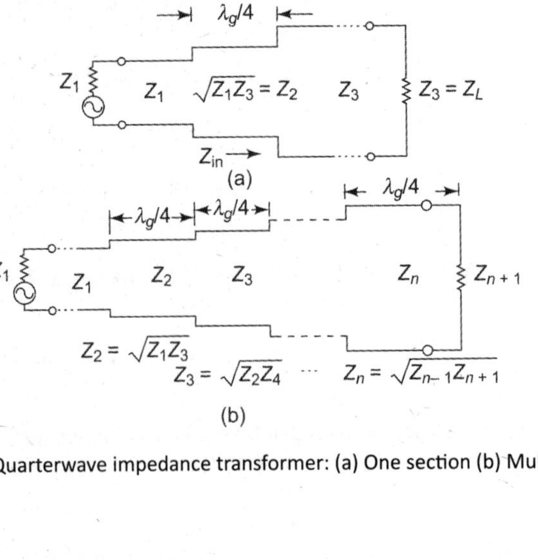

Fig. 5.23 Quarterwave impedance transformer: (a) One section (b) Multisection

Frequency Response of Single Section

A single-section quarterwave transformer provides a perfect matching at the single designed frequency f_0 (guide wavelength λ_{g0}) and at all the odd harmonics for which the length becomes equal to an odd multiple of $\lambda_{g0}/4$. The band-width of such a section can be determined by considering the electrical length $\beta l = \theta$ of the transformer at any other frequency $f \neq f_0$ and finding the input reflection coefficient Γ. Let a quarterwave transformer Z_2 match a load $Z_3 = Z_L$ to a line of impedance Z_1. Then the reflection coefficient at the input of the line Z_1 is given by (Fig. 5.23a)

$$\Gamma = \frac{Z_{in} - Z_1}{Z_{in} + Z_1} \tag{5.25}$$

where input impedance of the transformer Z_2 is

$$Z_{in} = Z_2 \frac{Z_L + jZ_2 \tan\theta}{Z_2 + jZ_L \tan\theta} \tag{5.26}$$

Substituting this Z_{in} and $Z_2 = \sqrt{Z_1 Z_L}$, Eq. (5.25) becomes

$$|\Gamma| = \frac{1}{\left[1 + \left(\frac{2\sqrt{Z_1 Z_L}}{Z_L - Z_1}\sec\theta\right)^2\right]^{1/2}} = \frac{\zeta\cos\theta}{\sqrt{1+\zeta^2\cos^2\theta}}; \; \zeta = \left|\frac{Z_L/Z_1 - 1}{2\sqrt{Z_L/Z_1}}\right| \tag{5.27}$$

when $\theta = 0, |\Gamma(0)| = \left|\frac{Z_L - Z_1}{Z_L + Z_1}\right| = \frac{1}{\sqrt{1+\zeta^2}}$ since $Z_{in} = Z_L$ (5.28)

and $\theta = \pi/2, |\Gamma| = 0,$ since $Z_{in} = Z_2^2/Z_L$ (5.29)

Thus, |Γ| varies periodically with θ as shown in Fig. 5.24 which in turn shows the variation of the magnitude of reflection coefficient with frequency. The bandwidth of the transformer is defined for the range of |Γ| having the maximum allowable limit $|\Gamma_m|$ around the midband frequency f_0, where $|\Gamma| = 0$ and $\theta = \pi/2$ or $f = f_0$. Therefore, for $|\Gamma| = |\Gamma_m|$,

$$\theta = \theta_m = \cos^{-1}\left|\frac{2|\Gamma_m|\sqrt{(Z_1 Z_L)}}{(Z_L - Z_1)\sqrt{1-|\Gamma_m|^2}}\right| \tag{5.30}$$

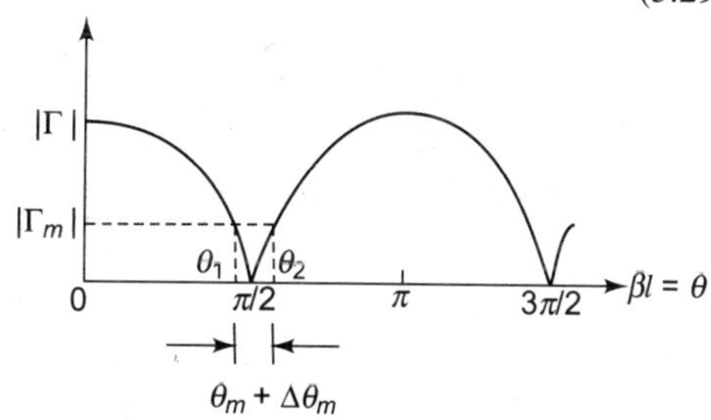

Fig. 5.24 Response of a single-section quarterwave transformer

For TEM line, $\theta = \beta l$, so that at θ_m, $f_m = 2f_0\theta_m/\pi$. The bandwidth of the single-section transformer is, therefore,

$$\Delta f = 2(f_0 - f_m) = 2f_0(1 - 2\theta_m/\pi) \tag{5.31}$$

or,

$$\frac{\Delta f}{f_0} = 2 - \frac{4}{\pi}\cos^{-1}\left|\frac{\Gamma_m \Gamma(0)}{\sqrt{\left[1-|\Gamma_m|^2\right]\left[1-|\Gamma(0)|^2\right]}}\right| \tag{5.32}$$

At the band edge, the electrical length, frequency and wavelength parameters are defined, respectively, by

$$\theta = \theta_1 = 0.5\pi f_1/f_0,\ \lambda_g = \lambda_{g1} \text{ and } \theta = \theta_2 = \pi - \theta_1 = 0.5\pi f_2/f_0,\ \lambda_g = \lambda_{g2}$$

The lengths of the transformer and the fractional bandwidth are then given by

$$l = \frac{\lambda_{g0}}{4} = \frac{\lambda_{g_1}\lambda_{g_2}}{2(\lambda_{g_1}+\lambda_{g_2})} \quad (5.33)$$

and

$$\frac{\Delta f}{f_0} = 2\frac{\lambda_{g_1}-\lambda_{g_2}}{\lambda_{g_1}+\lambda_{g_2}} = 2\frac{f_2-f_1}{f_2+f_1} \quad (5.34)$$

For a non-dispersive transmission line such as air-filled coaxial line, $\lambda_g = \lambda_0$, the free space wavelength. The above analysis shows that the bandwidth of a single section of a quarterwave transformer is small. This analysis assumes that the characteristic impedances Z_1 and Z_L are frequency independent. This is a good approximation for TEM wave lines. But for waveguides, these impedances vary with frequency and the bandwidth characteristic of a quarterwave transformer becomes complicated, and is not discussed here. The equivalent shunt susceptances which appear at the junctions of the transformer ends, due to sudden change of dimensions of the lines with different impedances, are neglected in this theory.

Realization of a Waveguide Quarterwave Impedance Transformer 5.4.2

Waveguide impedance transformers can be easily designed with steps either in the magnetic plane or in the electric plane, for which the mathematical expressions for impedance ratio is simpler. Figure 5.25 shows a typical waveguide transformer consisting of two waveguides of narrow wall dimensions b_1 and b_3 joined by a quarterwave-long, third waveguide of narrow wall dimension.

$$b_2 = \sqrt{(b_1/b_3)} \quad (5.35)$$

For an E-plane transformer of Fig. 5.21(b), the impedance matching condition is given by

$$a_2/\lambda_g = \sqrt{\left(\frac{a_1 a_3}{\lambda_{g_1}\lambda_{g_3}}\right)} \quad (5.36)$$

where λ_{g2}, λ_{g1}, and λ_{g3} are the guide wavelengths of the transformers and the two waveguide sections, respectively.

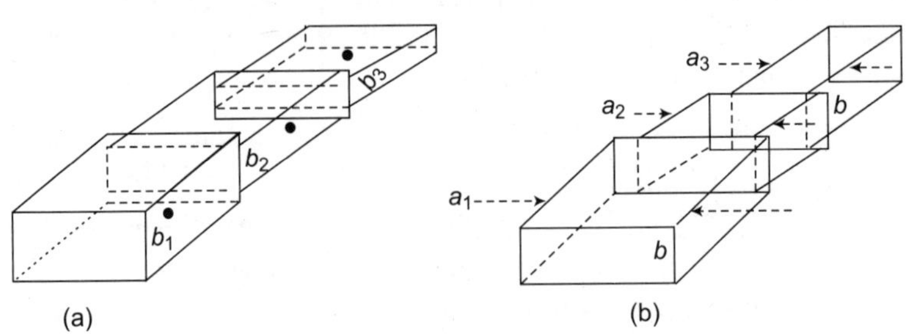

Fig. 5.25 Quarterwave guide transformers: (a) H-plane (b) E-plane

Effect of Multiple Reflections at Junctions of Transformers 5.4.3

Although the matching sections result in minimum reflection at the design frequency, each step or junction contributes to reflection. The overall reflection coefficient at the input can be evaluated by considering all the multiple reflections at the junction of the transformer. It can be shown that for small reflections, the total reflection coefficient can be obtained only from the first-order reflections.

Figure 5.26 shows two discontinuities of a single quarterwave transformer at which the reflection and transmission coefficients are Γ and T respectively, which are indicated with appropriate suffices. These are expressed by

$$\Gamma_1 = \frac{Z_2 - Z_1}{Z_2 + Z_1}, \Gamma_2 = -\Gamma_1,\ \Gamma_3 = \frac{Z_3 - Z_2}{Z_3 + Z_2} \tag{5.37}$$

and
$$T_{21} = 1 + \Gamma_1,\ T_2 = 1 + \Gamma_{12} = 1 - \Gamma_1 \tag{5.38}$$

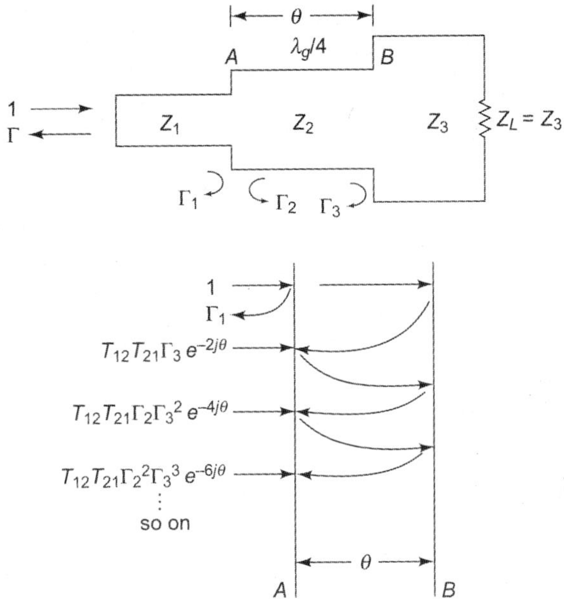

Fig. 5.26 Multiple reflection in a quarterwave transformer

Let us assume that a unit amplitude wave is incident at the input. Inspection of the figure shows that the net reflected wave amplitude due to infinite number of to and fro multiple reflections is

$$\Gamma = \Gamma_1 + T_{21}\, T_{12}\, \Gamma_3\, e^{-j2\theta} + T_{12}\, T_{21}\, \Gamma_3^2\, \Gamma_2\, e^{-j4\theta} + \ldots;$$

$$= \Gamma_1 + T_{12}\, T_{21}\, \Gamma_3\, e^{j2\theta} \sum_{n=0}^{\infty} \Gamma_2^n\ \Gamma_3^n\, e^{-j2n\theta};$$

$$= \Gamma_1 + \frac{T_{12} T_{21} \Gamma_3 e^{-j2\theta}}{1 - \Gamma_2 \Gamma_3 e^{-j2\theta}}; \tag{5.39}$$

Since $\sum_{n=0}^{\infty} x^n = \frac{1}{1-x}$

Eliminating T_{12} and T_{21} by $1 - \Gamma_1$ and $1 + \Gamma_1$, respectively,

$$\Gamma = \frac{\Gamma_1 + \Gamma_3\, e^{-j2\theta}}{1 + \Gamma_1 \cdot \Gamma_3\, e^{-j2\theta}} \tag{5.40}$$

Assuming small reflections, $|\Gamma_1|$ and $|\Gamma_3| \ll 1$, so that

$$\Gamma \approx \Gamma_1 + \Gamma_3 e^{-j2\theta} \tag{5.41}$$

Thus, for small reflections, the total reflection coefficient can be obtained only from the first-order reflection with a maximum error of 4% for $|\Gamma_1| = |\Gamma_3| \leq 0.2$.

Example 5.5 What are the required length and impedance of a $\lambda_g/4$ transformer that will match a 100-ohm load to a 50-ohm, air-filled line at 10 GHz? Consider both rectangular waveguide (2.286 cm × 1.016 cm) and coaxial line cases. What is the frequency band of operation for a coaxial line over which the reflection coefficient remains less than 0.1?

Solution Given, $Z_0 = 50$ ohms, $Z_L = 100$ ohms, $f_0 = 10$ GHz, and $\Gamma_m \leq 0.1$, $\lambda_0 = c/f_0 = 30/10 = 3$ cm.

(a) For *X*-band rectangular *w/g* in the TE_{10} mode,

$$\lambda_g = \frac{\lambda_0}{\sqrt{1-\left(\dfrac{\lambda_0}{2a}\right)^2}}$$

$$= \frac{3}{\sqrt{1-\left(\dfrac{3}{2\times 2.286}\right)^2}} \approx \frac{3}{1-0.43} \approx 4 \text{ cm}$$

Therefore, the length of the matching section is $l = \lambda_g/4 = 1$ cm.

(b) For the air-filled coaxial line, $\lambda_g = \lambda_0 = c/f_0 = 30/10 = 3$ cm. Therefore, the length of the matching section is

$$l = \lambda_g/4 = 3/4 = 0.75 \text{ cm.}$$

The characteristic impedance of the matching section

$$Z_2 = \sqrt{(Z_1 \cdot Z_L)} = \sqrt{(50\times 100)} = 50\sqrt{2} = 70.71 \text{ ohms}$$

The fractional bandwidth for TEM line is

$$\Delta f/f_0 = 2-(4/\pi)\cos^{-1}\left|\frac{2|\Gamma_m|\sqrt{Z_1 Z_L}}{(Z_L - Z_1)\sqrt{(1-\Gamma_m^2)}}\right|$$

$$= 2-\frac{4}{180}\cos^{-1}\left|\frac{2\times 0.1\sqrt{(100\times 50)}}{(100-50)\sqrt{[1-(0.1)^2]}}\right|$$

$$= 2-(4/180)\cos^{-1}\left|\frac{0.2\times 50\sqrt{2}}{50\times\sqrt{0.99}}\right| = 2-(4/180)\cos^{-1}(0.282/0.99) = 0.367$$

Therefore, $\Delta f = 0.367 \times 10 \text{ GHz} = 3.67 \text{ GHz}$

Example 5.6 An empty rectangular waveguide is matched to a dielectric filled waveguide (ε_r = 2.56) operating in TE_{10} mode at 10 GHz, by means of a quarterwave transformer. Find the length and dielectric constant of the matching section. The broader dimension of the waveguide is a = 2.5 cm.

Solution Let l and K_1 be the length and dielectric constant for the matching section guide as shown in Fig. 5.27. Given,

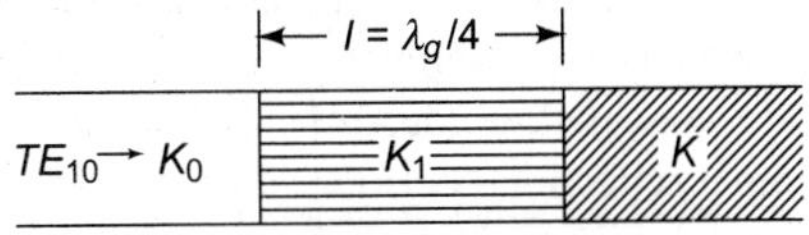

Fig. 5.27 Quarterwave waveguide matching section

$$a = 2.5 \text{ cm}$$

$$f = 10 \text{ GHz}$$

$$K = 2.56$$

$$K_0 = 1, \text{ empty guide dielectric constant}$$

$$\lambda_0 = 30/10 = 3 \text{ cm}$$

Let Z, Z_1, Z_0 are the wave impedances of the sections for the dominant TE_{10} mode. We will assume that due to introduction of the matching section, no higher-order modes are generated.

Wave impedances of the dielectric filled waveguide and air-filled waveguide, excited in TE_{10} mode, are

$$Z = \frac{377}{\sqrt{2.56 - \left(\frac{\lambda_0}{2a}\right)^2}} = \frac{377}{\sqrt{2.56 - \left(\frac{3}{2 \times 2.5}\right)^2}} = 254.17 \text{ ohms}$$

$$Z_0 = \frac{377}{\sqrt{1 - \left(\frac{\lambda_0}{2a}\right)^2}} = \frac{377}{\sqrt{1 - \left(\frac{3}{2 \times 2.5}\right)^2}} = 471.25 \text{ ohms}$$

Therefore, impedance of the matching section at the designed frequency of 10 GHz, is

$$Z_1 = \sqrt{[Z_0\, Z]} = \sqrt{[254.17 \times 471.25]} = 346.09 \text{ ohms}$$

Dielectric constant ε_{r_1} of the matching section guide is obtained from

$$Z_1 = \frac{377}{\sqrt{\varepsilon_{r_1} - \left(\frac{3}{2 \times 2.5}\right)^2}} \text{ ohms} = 346.09 \text{ ohms.}$$

or, $$\sqrt{\varepsilon_{r_1} - \left(\frac{3}{5}\right)^2} = 377/346.09$$

or, $$\varepsilon_{r_1} = 1.5466$$

Length of matching section at the designed frequency 10 GHz is 1/4 of guide wavelength in this section, i.e.,

$$l = \lambda_{g1}/4 = \frac{1}{4} \frac{\lambda_0}{\sqrt{K_1 - \left(\frac{\lambda_0}{2a}\right)^2}}$$

$$= \frac{3}{4} \frac{1}{\sqrt{1.5466 - \left(\frac{3}{2 \times 2.5}\right)^2}} = 0.6885 \text{ cm.}$$

Multisection Quarterwave Transformer 5.4.4

In order to increase the bandwidth of quarterwave impedance matching transformer, multisections are used with gradual change of step sizes. The input reflection coefficient can be obtained from partial reflection coefficients at various junctions per step of the transformer as shown in Fig. 5.28.

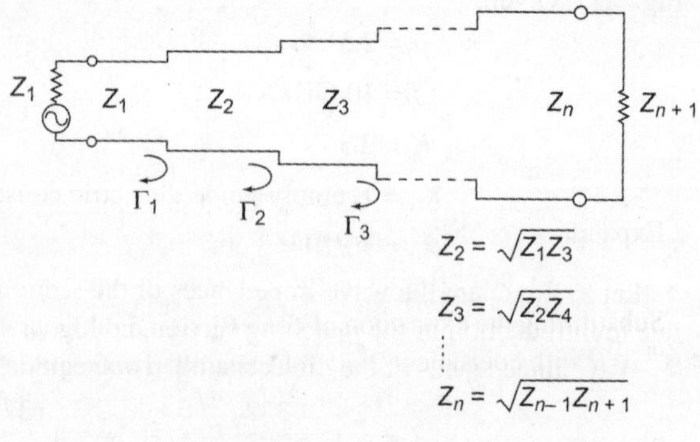

Fig. 5.28 Multisection quarterwave transformer

The problem here is to match the two impedances Z_1 and Z_{n+1} by $N = n - 1$ sections of the quarterwave transformers. There are now n steps which generate n reflection coefficients $\Gamma_1, \Gamma_2, ..., \Gamma_n$. Because of the discontinuities at the junctions, each component wave will undergo multiple reflections back and forth, between any two successive steps. For small reflections, multiple reflections within a section are neglected and only primary reflection at each discontinuity is considered. Therefore, for $n - 1$ sections, the total reflection coefficient Γ at any frequency f, can be written as

$$\Gamma = \Gamma_1 + \Gamma_2 e^{-j2\theta} + \Gamma_3 e^{-j4\theta} + ... + \Gamma_n e^{-j2(n-1)\theta} \tag{5.42}$$

where $\Gamma_i = (Z_{i+1} - Z_i)/(Z_{i+1} + Z_i)$, the reflection coefficient at the junction between $i + 1$ and ith sections for a small reflection.

Assuming that the transformer is symmetrical with respect to a transverse reference plane at the centre,

$$\Gamma_1 = \Gamma_n, \Gamma_2 = \Gamma_{n-1}, \Gamma_3 = \Gamma_{n-2}, \text{etc.} \tag{5.43}$$

Then,

$$\Gamma = 2 [\Gamma_1 \cos (n - 1) \theta + \Gamma_2 \cos (n - 3) \theta + \Gamma_3 \cos (n - 5) \theta + ...] \tag{5.44}$$

$$= 2 \sum_{k=1}^{m} \Gamma_k \cos [n - (2k - 1)] \theta; m = n/2 \text{ for even } n \tag{5.45}$$

$$= 2 \sum_{k=1}^{m} \Gamma_k \cos [n - (2k - 1)] \theta + \Gamma_{(n+1)/2}; m = (n - 1)/2 \text{ for odd } n \tag{5.46}$$

At $f = f_0$, $\theta = \pi/2$, for quarterwave sections,

so that $\Gamma = 0$ for n even (5.47)

$$= \Gamma_{(n+1)/2} - 2\Gamma_{(n-1)/2} + 2\Gamma_{(n-3)/2} + ... \pm 2\Gamma_1, \text{ for } n \text{ odd} \tag{5.48}$$

By a proper choice of the reflection coefficient Γ_k, $k = 1, 2, ..., n$, a variety of passband characteristics of a multisection quarterwave transformer can be obtained. The impedances Z_k's for various sections are calculated from Γ_k's. The most common passband characteristics are obtained from the *binomial* (maximally flat) and *Chebyshev* (equal ripple) distribution functions which are described below.

Maximally Flat Quarterwave Transformer 5.4.5

Maximally flat passband characteristic is obtained if Γ_k's are made proportional to the binomial coefficients c, that is,

$$\Gamma_k = \Gamma_1 c_{k-1}^{n-1} \tag{5.49}$$

The reflection coefficient Γ can now be expressed as

$$\Gamma = 2\Gamma_1 \sum_{k=1}^{m} c_{k-1}^{n-1} \cos[n-(2k-1)]\theta; \; m = \frac{n}{2} \text{ for even } n \tag{5.50}$$

$$= 2\Gamma_1 \sum_{k=1}^{m} c_{k-1}^{n-1} \cos[n-(2k-1)]\theta + c_{(n-1)/2}^{n-1}; \; m = \frac{n-1}{2} \text{ for odd } n \tag{5.51}$$

Expanding $\cos p\theta$ in terms of $\cos\theta$, where $p = n - (2k-1)$, we get

$$\cos p\theta = 2\cos(p-1)\theta \cos\theta - \cos(p-2)\theta \tag{5.52}$$

Substituting the expansion of $\cos p\theta$ into the expression of reflection coefficient, only one term containing $\cos^{(n-1)}\theta$ with coefficient 2^{n-2} for expansion of $\cos(n-1)\theta$ remains, to give

$$\Gamma = 2\Gamma_1 2^{n-2} \cos^{n-1}\theta = \Gamma_1 2^{n-1} \cos^{n-1}\theta \tag{5.53}$$

Since $|\Gamma|$ and its first $(n-2)$ derivatives with respect to frequencies (or θ) vanish at the centre frequency $f = f_0$, where $\theta = \pi/2$, the binomial distribution provides a maximally flat response as shown in Fig. 5.29.

In the design of the transformer, a maximum permissible value of $|\Gamma| = |\Gamma_m|$, may be prescribed over a specified band of frequencies from f_1 to f_2 with $f_0 = (f_1 + f_2)/2$ and a given number of sections n. The in-band reflection coefficient Γ is then computed from the equation

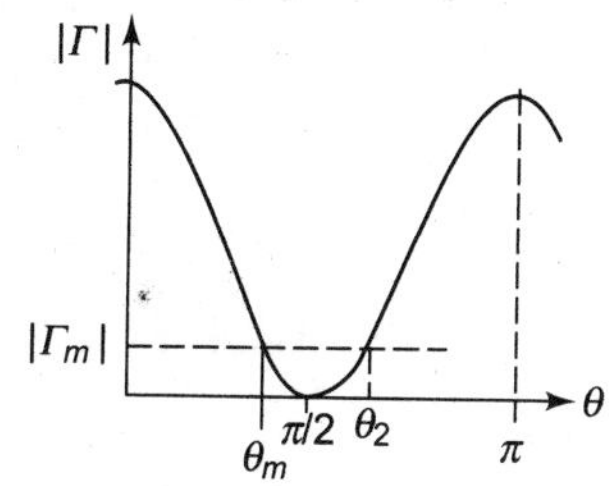

Fig. 5.29 Maximally flat response

$$|\Gamma| = \Gamma_1 2^{n-1} \cos^{n-1}\theta \le |\Gamma_m|$$

or,

$$|\Gamma| \le \frac{|\Gamma_m|}{2^{n-1}\cos^{n-1}\theta}; \; \theta = \theta_1 \text{ or } \theta_2 \text{ at } |\Gamma| = |\Gamma_m| \tag{5.54}$$

$$|\Gamma_m| = \Gamma_1 2^{n-1} \cos^{n-1}\theta_1 = \Gamma_1 2^{n-1}\cos^{n-1}\theta_2; \tag{5.55}$$

For symmetry of the response curve, $\cos\theta_1 = |\cos\theta_2|$, or $\theta_2 = \pi - \theta_1$. If r is the ratio of the wavelengths at the band edge frequencies f_1 and f_2, some of the design formulae are given by

$$r = \frac{\lambda_{g1}}{\lambda_{g2}} = \frac{\theta_2}{\theta_1} = \frac{\pi - \theta_1}{\theta_1} = \frac{\theta_2}{\pi - \theta_2} \tag{5.56}$$

$$\theta_1 = \frac{\pi}{1+r} = \frac{2\pi l}{\lambda_{g1}} = \frac{\pi\lambda_{g0}}{2\lambda_{g1}} \tag{5.57}$$

$$\theta_2 = \frac{\pi r}{1+r} = \frac{2\pi l}{\lambda_{g2}} = \frac{\pi\lambda_{g0}}{2\lambda_{g2}} \tag{5.58}$$

$$l = \frac{\lambda_{g0}}{4} = \frac{\theta_1\lambda_{g1}}{2\pi} = \frac{\theta_2\lambda_{g2}}{2\pi} \tag{5.59}$$

$$\lambda_{g0} = \frac{2\lambda_{g1}\lambda_{g2}}{\lambda_{g1} + \lambda_{g2}} \tag{5.60}$$

If n is not specified, it is selected by trial and error method to find $|\Gamma| \le |\Gamma_m|$ over the passband. The characteristic impedances of the sections are determined from the relation

$$\Gamma_k = \frac{Z_{k+1} - Z_k}{Z_{k+1} + Z_k} = c_{k-1}^{n-1}\Gamma_1 \tag{5.61}$$

where Z_{i+1} is the characteristic impedance of the ith section and Γ_i is the reflection coefficient at the input of the ith section as shown in Fig. 5.28. Further, if the reflections are small ($\Gamma_k \leq 0.35$ for an error of less then 4%), we can write

$$\ln\left(\frac{Z_{k+1}}{Z_k}\right) = 2\left[\frac{Z_{k+1}-Z_k}{Z_{k+1}+Z_k} + \frac{1}{3}\left(\frac{Z_{k+1}-Z_k}{Z_{k+1}+Z_k}\right)^3 + \ldots\right]$$

$$= 2\left[\Gamma_k + \frac{1}{3}\Gamma_k^3 + \ldots\right] \tag{5.62}$$

$$\approx 2\Gamma_k$$

Therefore, $$\Gamma_k = \frac{1}{2}\ln\left(\frac{Z_{k+1}}{Z_k}\right) = c_{k-1}^{n-1}\frac{1}{2}\ln\left(\frac{Z_2}{Z_1}\right) \tag{5.63}$$

or, $$\ln(Z_{k+1}/Z_k) = c_{k-1}^{n-1}\ln\left(\frac{Z_2}{Z_1}\right) \tag{5.64}$$

Thus the impedance can be computed from the corresponding reflection coefficients

$$\Gamma_1 = (1/2)\ln(Z_2/Z_1),\ Z_2 = Z_1\,[\text{anti}\ln(2\Gamma_1)]$$

$$\Gamma_2 = (1/2)\ln(Z_3/Z_2),\ Z_3 = Z_2\,[\text{anti}\ln(2\Gamma_2)]$$

$$\vdots$$

$$\Gamma_k = (1/2)\ln(Z_{k+1}/Z_k),\ Z_n = Z_{n-1}\,[\text{anti}\ln(2\Gamma_{n-1})] \tag{5.65a}$$

By adding,

$$\Gamma_1 + \Gamma_2 + \cdots + \Gamma_n = \frac{1}{2}\ln\left(\frac{Z_{n+1}}{Z_1}\right) \tag{5.65b}$$

or, $$2(\Gamma_1 + \Gamma_2 + \ldots + \Gamma_{n/2}) = \frac{1}{2}\ln\left(\frac{Z_{n+1}}{Z_1}\right) \text{ for even } n \tag{5.66}$$

and $$2(\Gamma_1 + \Gamma_2 + \ldots) + \Gamma_{(n+1)/2} = \frac{1}{2}\ln\left(\frac{Z_{n+1}}{Z_1}\right) \text{ for odd } n \tag{5.67}$$

Example 5.7 Design a two-section air-filled coaxial line binomial transformer to match the original line of characteristic impedance 50 ohm to a 100 ohm load given in Example 5.5. What bandwidth is obtained for $\Gamma_m = 0.1$ for matching at 10 GHz?

Solution Given number of sections, $N = 2$, number of steps $n = N + 1 = 3$.

$Z_1 = 50$ ohms, $Z_L = 100$ ohms, $\Gamma_m = 0.1$, $f_0 = 10$ GHz, $\lambda_0 = 3$ cm.

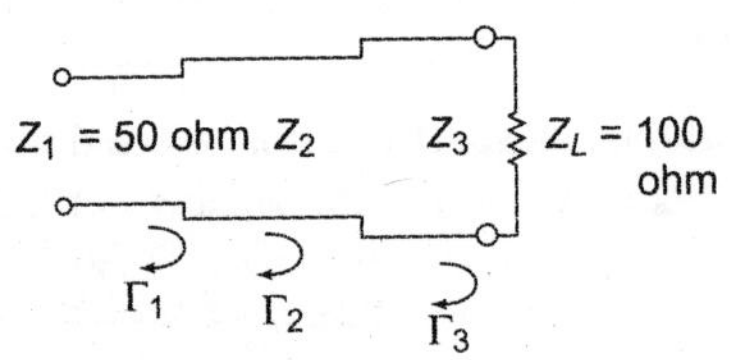

Fig. 5.30

Design: For odd $n = 3$, Eq. 5.65a gives

$$\Gamma_1 + \Gamma_2 + \Gamma_3 = (1/2)\ln(100/50)$$

$$= (1/2)\ln 2 \approx 0.6931/2$$

$$= 0.3465$$

For symmetrical binomial transformer $\Gamma_k = \Gamma_1 c_{k-1}^{n-1}$, so that

$$\Gamma_1 = \Gamma_3;$$

$$\Gamma_2 = c_1^2 \Gamma_1 = 2\Gamma_1$$

Therefore, $\Gamma_1 + 2\Gamma_1 + \Gamma_1 = 0.3465$

or, $\Gamma_1 = 0.3465/4 = 0.0866$

$$\Gamma_2 = 2\Gamma_1 = 2 \times 0.0866 = 0.1732$$

Therefore, $Z_2 = Z_1 e^{2\Gamma_1} = 50\, e^{0.1732} = 59.46$ ohms

$$Z_3 = Z_2 e^{2\Gamma_2} = 59.46\, e^{2 \times 0.1732} = 84.1 \text{ ohms}$$

For air-filled coaxial line, length of the sections is $\lambda_g/4 = \lambda_0/4 = 3/4 = 0.75$ cm

Bandwidth: $|\Gamma_m| = \Gamma_1 2^{n-1} \cos^{n-1}\theta_1 = \Gamma_1 2^{n-1}\cos^{n-1}\theta_2$

or, $0.1 = \Gamma_1 2^2 \cos^2\theta_1 = 0.0866 \times 4 \times \cos^2\theta_1$

or, $\cos^2\theta_1 = 0.1/(0.0866 \times 4) = 0.2887$

or, $\cos\theta_1 = 0.5373$

$$\theta_1 = 57.5°$$

$$\theta_2 = \pi - \theta_1 = 180 - 57.5 = 122.5°$$

Therefore, $\Delta\theta = \theta_2 - \theta_1 = 122.5 - 57.5 = 65°$;

$$r = \theta_2/\theta_1 = 122.5/57.5 = 2.13$$

For TEM line the bandwidth

$$\Delta f/f_0 = (2/\pi)(\theta_2 - \theta_1) = (2/180°) \times 65° = 130/180 = 0.7222$$

or, $\Delta f = 0.722 \times 10$ GHz $= 7.22$ GHz which is nearly double the bandwidth of Example 5.5. Therefore, multisection matching results in larger bandwidth.

Example 5.8 Design a five-section quarterwave, impedance transformer, to match the two lines of impedances of 100 ohm and 1000 ohm. Assume binomial distribution passband response of the reflection coefficients for the wavelength ratio of 2, corresponding to the band edges. Find the maximum VSWR at the input of the transformer.

Solution For five sections, number of steps $n = 6$ (even), wavelength ratio, $\lambda_{g1}/\lambda_{g2} = r = 2$, original input and output line impedances are,

$$Z_1 = 100 \text{ ohms}$$

$$Z_{n+1} = 1000 \text{ ohms}; \quad n + 1 = 6 + 1 = 7$$

Assuming small reflections at each step from Eq. 5.66 (even section)

$$2(\Gamma_1 + \Gamma_2 + \Gamma_3) = \frac{1}{2}\ln\frac{1000}{100} - 1.15$$

For binomial distribution,

$$\Gamma_k = c_{k-1}^{n-1}\Gamma_1 = c_{k-1}^5 \Gamma_1$$

$$\Gamma_1 = c_0^5 \Gamma_1 = \Gamma_1 = \Gamma_6$$

$$\Gamma_2 = c_1^5 \Gamma_1 = 5\Gamma_1 = \Gamma_5$$

$$\Gamma_3 = c_2^5 \Gamma_1 = 10\Gamma_1 = \Gamma_4$$

Therefore, $2(\Gamma_1 + 5\Gamma_1 + 10\Gamma_1) = 1.15$

or, $\Gamma_1 = \frac{1.15}{32} = 0.036 = \Gamma_6$

$\Gamma_2 = 5 \times 0.036 = 0.180 = \Gamma_5$

$\Gamma_3 = 10 \times 0.036 = 0.360 = \Gamma_4$

The characteristic impedances of the sections are

$Z_2 = 100$ [anti ln (2×0.036)] = 107.5 ohms

$Z_3 = 107.5$ [anti ln (2×0.180)] = 154 ohms

$Z_4 = 154$ [anti ln (2×0.360)] = 316 ohms

$Z_5 = 316$ [anti ln (2×0.360)] = 648 ohms

$Z_6 = 648$ [anti ln (2×0.180)] = 930 ohms

At the band edge,

$$\theta_1 = \frac{\pi}{1+r} = \frac{\pi}{1+2} = \frac{\pi}{3}$$

Therefore, $\cos\theta_1 = 0.5 = x$ (say)

$\therefore$ $|\Gamma_m| = \Gamma_1 2^{n-1} \cos^{n-1}\theta_1 = 0.036 \times 2^5 \times \cos^5\theta_1$

or, $\cos^5\theta_1 = (0.5)^5 = 0.03125$

Therefore, $|\Gamma_m| = 0.072 \times 16 \times 0.03125 = 0.036$

$$\text{Maximum VSWR} = \frac{1+|\Gamma_m|}{1-|\Gamma_m|} = \frac{1.036}{0.964} = 1.075$$

If a condition that maximum allowable $|\Gamma_m| \le 0.03$ or less is imposed, more matching sections are needed and can be found by trial and error method with increasing $n > 6$.

Example 5.9 Design a three-section binomial transformer to match a 100-ohm load to a 50 ohm line. The maximum VSWR that can be tolerated is 1.1. What bandwidth can be obtained ?

Solution Given $Z_1 = 50$ ohms, $Z_L = Z_5 = 100$ ohms, $N = 3$, $n = N + 1 = 4$

$VSWR_{max} = [1 + |\Gamma_m|]/[1 - |\Gamma_m|] = 1.1$

or, $|\Gamma_m| = (S-1)/(S+1)$

$= (1.1 - 1)/(1.1 + 1)$

$= 0.1/2.1 = 0.0476$

Sum of reflection coefficients at the four steps:

$\Gamma_1 + \Gamma_2 + \Gamma_3 + \Gamma_4 = (1/2)\ln(Z_5/Z_1) = (1/2)\ln(100/50)$

$= 0.3465$

For symmetrical binomial transformer: $\Gamma_1 = \Gamma_4$, $\Gamma_2 = \Gamma_3$, so that

$\Gamma_2 = c_1^3\,\Gamma_1 = 3\,\Gamma_1 = \Gamma_3$

Therefore, $\Gamma_1 + 3\,\Gamma_1 + 3\Gamma_1 + \Gamma_1 = 0.3465$

or $8\,\Gamma_1 = 0.3465$

or, $\Gamma_1 = 0.0434$

$\Gamma_2 = 3\,\Gamma_1 = 0.1302$

Therefore, $Z_2 = Z_1\, e^{2\Gamma_1} = 50 \times e^{0.0868} = 50 \times 1.09 = 54.5$ ohm

$Z_3 = Z_2\, e^{2\Gamma_2} = 54.5 \times e^{0.26} = 54.5 \times 1.297 = 70.5$ ohm

$Z_4 = 70.5\, e^{2\Gamma_3} = 91.2$ ohm

$l = \lambda_g/4$ at centre frequency f_0

Bandwidth

$$|\Gamma_m| = \Gamma_1\, 2^{n-1} \cos^{n-1}\theta_1$$

or, $0.0476 = 0.0434 \times 2^3 \times \cos^3\theta_1$

or, $\cos^3\theta_1 = 0.0476/(0.0434 \times 8) = 0.1378$

or, $\cos\theta_1 = (0.1378)^{1/3}$

$= 0.5165$

Therefore, $\theta_1 = 58.9°$

$\theta_2 = \pi - \theta_1 = 121.1°$

Therefore, $\Delta f/f_0 = (2/\pi)(\theta_2 - \theta_1)$ for TEM waves

$$= \frac{2}{180}(121.1 - 58.9) = 0.69 = 69\%$$

Example 5.10 Design a two-section rectangular waveguide binomial transformer to match an empty rectangular waveguide with a dielectric-filled guide in Example 5.6 excited in dominant mode.

Solution Number of steps, $n = 2 + 1 = 3$

$N = n - 1 = 2$

$a = 2.5$ cm

$f = 10$ GHz

$K_4 = 2.56$

$K_1 = 1$

$Z_4 = 254.17$ ohms, $Z_1 = 471.25$ ohms (from Examplc 5.6)

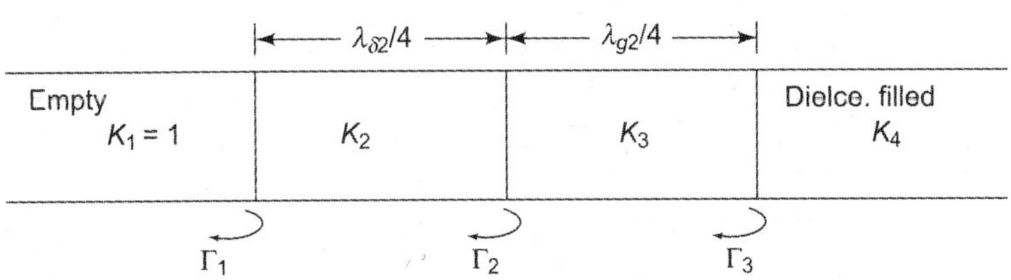

Fig. 5.31 Multisection waveguide transformer design

Assuming first-order reflection at the steps

$$\Gamma_1 + \Gamma_2 + \Gamma_3 = (1/2)\ln(Z_4/Z_1) = \frac{1}{2}\ln(254.17/471.25) = -0.3087$$

Assuming a symmetrical transformer

$$\Gamma_1 = \Gamma_3$$

$$\Gamma_2 = c_1^2\,\Gamma_1 = 2\Gamma_1$$

Therefore, $\Gamma_1 + 2\,\Gamma_1 + \Gamma_1 = 4\,\Gamma_1 = (1/2)\ \ln\ (Z_4/Z_1) = -\,0.3087$

or,
$$\Gamma_1 = -\,0.03087/4$$
$$= -\,0.07717 = \Gamma_3$$
$$\Gamma_2 = \Gamma_1 \times 2 = -\,0.1543$$

Now,
$$Z_2 = Z_1\ [\text{anti}\ \ln\ (2\Gamma_1)] = 403.725\ \text{ohms}$$
$$Z_3 = Z_2\ [\text{anti}\ \ln\ (2\Gamma_2)] = 295.79\ \text{ohms}$$

Therefore,
$$403.725 = \frac{377}{\sqrt{K_2 - \left(\dfrac{3}{2\times 2.5}\right)^2}}\ \text{ohms}$$

and
$$295.79 = \frac{377}{\sqrt{K_3 - \left(\dfrac{3}{2\times 2.5}\right)^2}}\ \text{ohms}$$

or,
$$K_2 = 1.232$$
$$K_3 = 1.984$$

The lengths of the sections are

$$l_2 = \frac{\lambda_{gK_2}}{4} = \frac{1}{4}\frac{\lambda_0}{\sqrt{[K_2 - (\lambda_0/2a)^2]}}$$

$$= \frac{3}{4\sqrt{[1.232 - (3/5)^2]}} = 0.804\ \text{cm}$$

$$l_3 = \frac{\lambda_{gK_3}}{4}$$

$$= \frac{1}{4}\frac{\lambda_0}{\sqrt{[K_3 - (\lambda_0/2a)^2]}}$$

$$= \frac{3}{4\sqrt{[1.984 - (3/5)^2]}} = 0.588\ \text{cm}$$

Equal Ripple Transformer 5.4.6

Equal ripple characteristics of a multisection transformer are obtained by choosing the reflection coefficient Γ proportional to a *Chebyshev polynomial* $T_{n-1}(x)$ of order $n-1$ for $N = n-1$ number of sections:

$$T_{n-1}(x) = \cos\ [(n-1)\ \cos^{-1}(x)];\ -1 \le x \le 1$$
$$= \cosh\ [(n-1)\ \cosh^{-1}\ (x);\ x < -1,\ x > 1 \qquad (5.68)$$

where x is the parameter for change of variable according to

$$x = \frac{\cos\theta}{\cos\theta_1} = \frac{\cos\theta}{|\cos\theta_2|} \tag{5.69}$$

and is a function of frequency. The polynomials have the following properties as shown in Fig. 5.32 and are as follows.

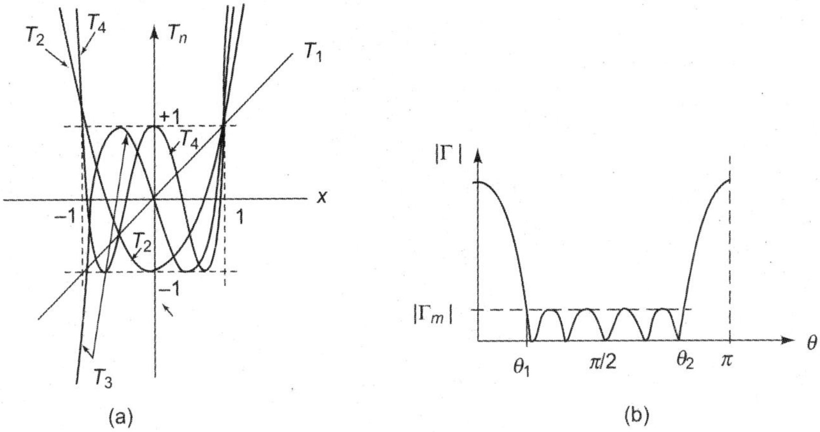

Fig. 5.32 Chebyshev response: (a) Chebyshev polynomial (b) reflection coefficient characteristics

1. All polynomials pass through the point (1, 1).
2. $-1 \le T(x) \le 1$ for $-1 \le x \le 1$
3. All roots occur within $-1 \le x \le 1$, and all maxima and minima have values of 1 and – 1, respectively.
4. $T_k(x) = 2x\,T_{k-1}(x) - T_{k-2}(x)$

An equal ripple characteristic results in a larger bandwidth compared to a binomial transformer for the same number of sections. The total reflection coefficient Γ behaves according to a Chebyshev response. This makes Γ to have equal ripple characteristic over the range θ_1 to $\pi - \theta_1 = \theta_2$ with a specified maximum ripple height $|\Gamma_m|$. Thus with reference to a transverse plane at the centre of the transformer,

$$\Gamma = |\Gamma_m|\, T_{n-1}(x) = 2\sum_{k=1}^{m} \Gamma_k \cos[n-(2k-1)]\theta;\ m = n/2 \text{ for even } n \tag{5.70}$$

$$= 2\sum_{k=1}^{m} \Gamma_k \cos[n-(2k-1)]\theta + \Gamma_{(n+1)/2};\ m = (n-1)/2 \text{ for odd } n \tag{5.71}$$

This ensures that $|\Gamma_m T_{n-1}(x)| \le |\Gamma_m|$, since $|T_{n-1}(x)| \le 1$. The Γ_k's are obtained by expanding the right-hand side of Eqs 5.70 and 5.71 in terms of powers of cos θ and substituting $x \cos \theta_1$ for cos θ and equating the coefficient of like powers of x with those on the left-hand side. Chebyshev transformer gives the largest bandwidth for a specified $|\Gamma_m|$ and number of sections and for a given bandwidth gives the lowest ripple $|\Gamma_m|$. Design will be clear in the following example.

Example 5.11 Design a two-section Chebyshev quarterwave impedance transformer to match a line with characteristic impedance of 100 ohms to a 200 ohm line. The maximum tolerable reflection coefficient is 0.05.

Solution

Here, n = no. of sections + 1 = 2 + 1 = 3; (odd) → Eq. 5.71 to be followed:

$$T_{n-1}(x) = T_2(x) = 2x^2 - 1$$

Therefore, $\Gamma = 2\,[\Gamma_1 \cos(3-1)\,\theta] + \Gamma_2 = 2\Gamma_1 \cos 2\theta + \Gamma_2$

$$= |\Gamma_m|\, T_2(x) = 0.05\, T_2(x)$$

or, $2\Gamma_1 \cos 2\theta + \Gamma_2 = 0.05\,(2x^2 - 1)$

or, $2\Gamma_1\,(2\cos^2\theta - 1) + \Gamma_2 = 0.1x^2 - 0.05$

Since, $x = \dfrac{\cos\theta}{\cos\theta_1}$, $\cos\theta = x\cos\theta_1$

$$2\Gamma_1\,(2\cos^2\theta_1\, x^2 - 1) + \Gamma_2 = 0.1\, x^2 - 0.05$$

Equating the coefficients of equal power

$$4\,\Gamma_1 \cos^2\theta_1 = 0.1,\ \Gamma_2 - 2\Gamma_1 = -\,0.05$$

Therefore, $\Gamma_1 = \dfrac{0.1}{4\cos^2\theta_1}$

$$\Gamma_2 = 2\Gamma_1 - 0.05 = \frac{0.1}{2\cos^2\theta_1} - 0.05$$

For small reflections,

$$2\Gamma_1 + \Gamma_2 = \frac{1}{2}\ln\frac{200}{100} = \frac{1}{2}\ln 2 = 0.3466$$

or $2 * \dfrac{0.1}{4\cos^2\theta_1} + \dfrac{0.1}{2\cos^2\theta_1} - 0.05 = 0.3466$

or, $\dfrac{0.1}{2\cos^2\theta_1} = 0.3466 + 0.05 = 0.3966$

or, $\cos^2\theta_1 = \dfrac{0.1}{0.3966} = 0.2521$

or, $\cos\theta_1 = 0.5021$

Therefore, $\theta_1 = 59.86°$

$$\Gamma_1 = \frac{0.1}{4 * 0.2521} = 0.09916 = \Gamma_3$$

$$\Gamma_2 = \frac{0.1}{2 * 0.2521} - 0.05 = 0.1483$$

$$Z_2 = 100\,[\text{anti ln}\,(2 * 0.09916)] = 121.94 \text{ ohms}$$

$$Z_3 = 121.94\,[\text{anti ln}\,(2 * 0.1483)] = 164.04 \text{ ohms}$$

Summary

Therefore, the sequence of impedances are

Z_1 = 100 ohm

Z_2 = 121.9 ohm

Z_3 = 164 ohm

$$Z_4 = Z_L = 200 \text{ ohm}$$

Maximum reflection coefficient at band edges is $|\Gamma_m| = 0.05$.

$$\text{VSWR} = \frac{1+|\Gamma_m|}{1-|\Gamma_m|} = 1.105$$

Wavelength ratio $r = (\pi - \theta_1)/\theta_1 \cong 2.$

Example 5.12 Design a two-section Chebyshev transformer to match a 100 ohm load to a 50 ohm line. The maximum VSWR allowed is 1.2. Find the corresponding bandwidth.

Solution Number of sections $N = n - 1 = 2$, number of steps $n = 3$, $Z_L = Z_4 = 100$ ohms, $Z_1 = 50$ ohms

$$|\Gamma_m| = \frac{1.2-1}{1.2+1} = \frac{0.2}{2.2} = 0.0909$$

Since n is odd, from Eq. 5.71 total reflection coefficient for symmetrical transformer (n odd)

$$\Gamma = |\Gamma_m|\, T_{n-1}(x) = 0.0909\, T_2(x) = 2\,[\Gamma_1 \cos 2\theta] + \Gamma_2$$

or, $0.0909[2x^2 - 1] = 2[\Gamma_1 (2\cos^2\theta - 1)] + \Gamma_2$; $\cos\theta = x\cos\theta_1$

$$= 2\Gamma_1 (2x^2\cos^2\theta_1 - 1)] + \Gamma_2$$

Equating the coefficient of x^2, x, etc.

$$2\times 0.091 = 4\,\Gamma_1 \cos^2\theta;\ \text{or, } \Gamma_1 = 2\times\frac{0.091}{4\cos^2\theta_1} = \frac{0.0455}{\cos^2\theta_1}$$

$$-0.091 = -2\Gamma_1 + \Gamma_2; \quad \text{or,} \quad \Gamma_2 = 2\Gamma_1 - 0.091$$

For symmetry $\Gamma_1 = \Gamma_3$

Therefore, $\Gamma_1 + \Gamma_2 + \Gamma_3 = 2\Gamma_1 + \Gamma_2 = \frac{1}{2}\ln\left(\frac{100}{50}\right) = 0.3466$

or, $$\frac{2\times 0.0455}{\cos^2\theta_1} + \frac{2\times 0.0455}{\cos^2\theta_1} - 0.091 = 0.3466$$

or, $\cos^2\theta_1 = 0.4159$

or, $\cos\theta_1 = 0.6449$, $\theta_1 = 49.84°$

Therefore, $$\Gamma_1 = \frac{(2\times 0.091)}{(4\times 0.4159)} = 0.1094$$

$$\Gamma_2 = 2\Gamma_1 - 0.091 = 0.2088 - 0.091 = 0.1122$$

Now, $$\Gamma_1 = \frac{(Z_2 - Z_1)}{(Z_2 + Z_1)},\quad \Gamma_2 = \frac{(Z_3 - Z_2)}{(Z_3 + Z_2)}$$

$$Z_2 = 50e^{2\Gamma_1} = 62.3 \text{ ohms}$$

$$Z_3 = 62.3\times e^{2\Gamma_2} = 78 \text{ ohms}$$

The fractional bandwidth

$$\Delta f/f_0 = \frac{2}{\pi}(\theta_2 - \theta_1)$$

$$= \frac{2}{\pi}(\pi - 2\theta_1) = \frac{2}{180}(180 - 2 \times 49.84) = 89.24\%$$

TAPERED TRANSMISSION LINES 5.5

The error introduced in neglecting the equivalent shunt reactive elements and higher-order reflections at the discrete steps of multisection quarterwave impedance transformers can be minimized by using a continuously tapered transmission line between the two mismatch impedances Z_s and Z_L as shown in Fig. 5.33.

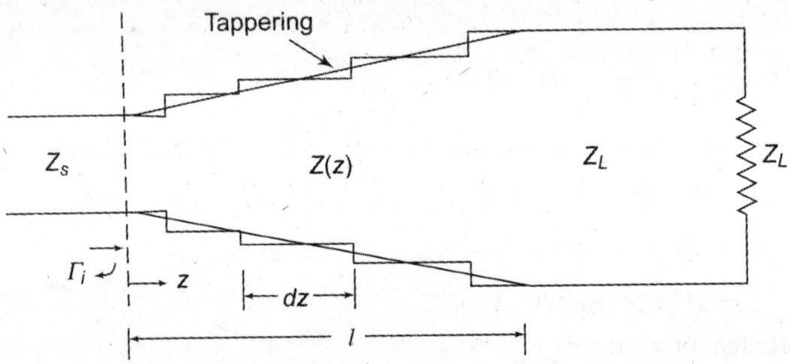

Fig. 5.33 Tapered transmission line matching section

Because of continuous tapering of the transverse dimensions of the line in the longitudinal direction z, the impedance of the line is a function of distance z. Assuming that the taper is gradual, it can be approximated by a large number of elementary sections dz with incremental impedance change dZ and a differential reflection coefficient

$$d\Gamma = \frac{(Z + dZ) - Z}{(Z + dZ) + Z} = \frac{dZ}{2Z + dZ} \cong \frac{dZ}{2Z} \approx \frac{1}{2}d(\ln Z) \tag{5.72}$$

Since $dZ \ll Z$,

The input reflection coefficient for a tapered lossless line of length l can then be given by

$$\Gamma_i = (1/2)\int_0^l d(\ln Z)\, e^{-j\beta z} \tag{5.73}$$

Γ_i can be determined if $Z(z)$ is known, or for a given response characteristic of Γ_i as a function of frequency, $Z(z)$ can be determined.

Exponential Taper Line Section 5.5.1

For the exponential taper, Z varies exponentially and hence ln Z varies linearly from Z_s to Z_L with z. Thus,

$$\ln Z = (z/l) \ln Z_L$$

or

$$Z = e^{(z/l)} \ln Z_L \tag{5.74}$$

Therefore, assuming that $\beta = 2\pi/\lambda$, is independent of z,

$$\Gamma_i = (1/2)\int_0^l d(\ln Z)e^{-j2\beta z} \tag{5.75}$$

$$= \frac{\ln Z_L}{2l}\int_0^l e^{-j2\beta z}\cdot dz$$

$$= \frac{\ln Z_L}{2l}\left[\left(\frac{e^{-j2\beta z}}{-j2\beta}\right)\right]_0^l$$

$$= \frac{\ln Z_L}{2\beta l}e^{-j\beta l}.\frac{e^{j\beta l}-e^{-j\beta l}}{2j}$$

$$= \frac{\ln Z_L}{2}e^{-j\beta l}.\frac{\sin\beta l}{\beta l}$$

or,
$$|\Gamma_i| = \frac{\ln Z_L}{2}\left|\frac{\sin\beta l}{\beta l}\right| \tag{5.76}$$

Since $\beta\left(=\frac{2\pi l}{c}f\right)$ is proportional to frequency, the variation of $|\Gamma_i|$ with frequency is a $\frac{\sin x}{x}$ pattern. It is seen that at specific frequencies for which $\beta l = n\pi\ (n = 1, 2, 3, ...)$, $\Gamma_i = 0$ and impedance matching becomes perfect for lossless lines. At any frequency when the length of the taper section $l > \lambda/2$, Γ_i is very small and decreases with increase in length l. This shows that a minimum length of $\lambda/2$ is preferred for good matching at lowest frequency of interest (Fig. 5.34).

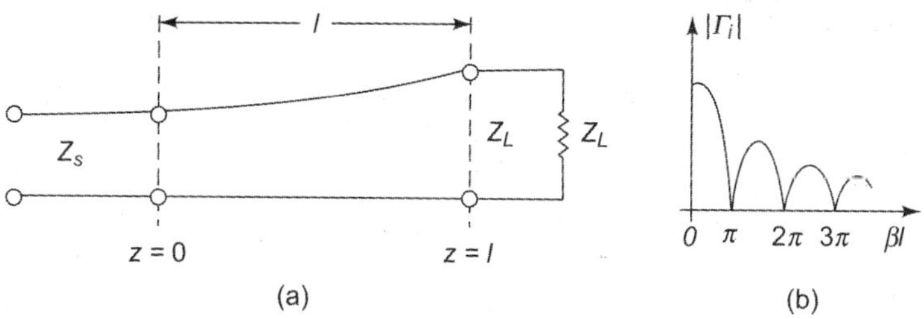

Fig. 5.34 (a) Exponential taper line (b) $|\Gamma_i|$ response along line $\left(\left|\frac{\sin x}{x}\right| \text{form}\right)$

Waveguide Taper Matching 5.5.2

When two waveguide sections of different shapes are required to be joined together, such as a rectangular waveguide at TE_{10} mode and a circular waveguide at TE_{11} mode, a gradual taper as shown in Fig. 5.35 is required over a distance of at least $2\lambda_g$ for impedance matching between them. This also reduces abrupt discontinuity at the junction. A taper is also used to connect two similar waveguide sections of different dimensions.

The characteristic impedance of a rectangular ($a \times b$) waveguide is defined by

$$Z_0 = \frac{\eta_0 \pi b \lambda_g}{2a\lambda_0} \tag{5.77}$$

The impedance match due to two different dimensions can be achieved by using a taper to adjust the values of *a* and *b* until the characteristic impedances of the two sections become equal.

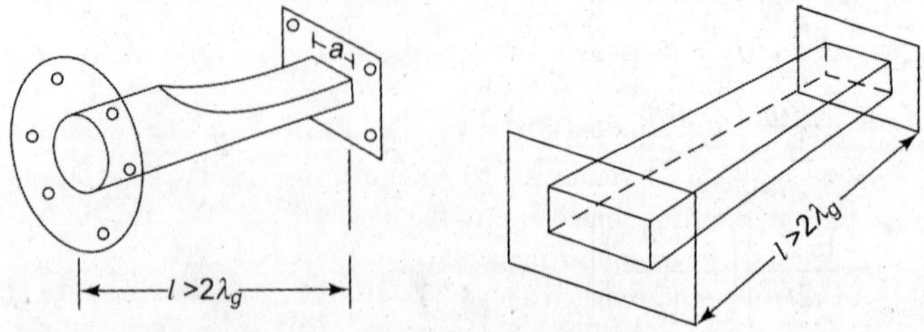

Fig. 5.35 Waveguide tapers

Matching a waveguide to free-space impedance can also be achieved by flaring the guide in the form of a horn until the factor $\pi b\lambda_g/(2a\lambda_0)$ becomes unity so that $Z_0 = \eta_0$. However, the final dimension will be reasonably large.

REVIEW QUESTIONS

5.1 A 50 ohm lossless line is terminated with a load of $30 - j40$ ohms. A short-circuited 50 ohm stub is connected parallel to this line at a distance *d* from the load. Find *d* and the length of the stub for matching of the load.

5.2 A lossless line is terminated with a load of VSWR = 3 and the first voltage minimum and maximum are located at distance 50 cm and 250 cm from the load, respectively. Design a single short-circuited stub for making VSWR = 1.

5.3 Design a single-stub tuner to match a load $Z_L = 110 - j70$ ohms to a lossless 50 ohm transmission line.

5.4 The 50 ohm lossless transmission line is terminated with a loading impedance of 75 ohms. Design a single-stub matching system using a Smith chart.

5.5 A complex load $Z_L = 100 + j100$ ohms is connected to a 50 ohm lossless line. Design a double-stub, 50 ohm system for matching. The first stub is placed 0.4λ away from load and the spacing between the stubs is $3\lambda/8$.

5.6 Explain what terminations cannot be matched with a lossless line by the double-stub system. Why?

5.7 A lossless 100 ohm transmission line is terminated with an admittance $0.003 - j0.003$ mho. Design a double-stub tuner with spacing $\lambda/8$ when the first stub is at the position of the load.

5.8 A 50 ohm lossless transmission line is required to be matched with the load impedance $75 - j60$ ohms, by a double-stub tuner with separation distance $3\lambda/8$ and the distance of the first stub from the load is 0.01λ. Determine the lengths of the matching stubs.

5.9 A load is required to be matched to a 100 ohm line by means of fixed double stub tuner separated by $\lambda/4$. At the position of the first stub, the line admittance is $0.008 - j0.014$ mho. Find the length of the stubs when the line is excited at 100 MHz. Explain if there is any limitation on the use a of double stub $\lambda/4$ apart.

5.10 A lossless 100 ohm coaxial line is terminated by a load of admittance $0.016 - j0.008$ mho. The line is matched by means of two fixed stubs separated by $3\lambda/8$, with the first stub placed at the plane of the load. Calculate the length of each stub by using the Smith chart.

5.11 Design a microstrip shunt open stub matching network to match a load impedance $20 + j\,0.5$ ohm to a 50 ohm microstrip line.

5.12 Design a two-section Chebyshev quarterwave transformer for overall impedance ratio 2.5 to have a VSWR < 1.03 over a 20 per cent bandwidth.

5.13 A three-section Chebyshev quarterwave transformer with $\Gamma_m = 0.1$ has a bandwidth 80 per cent and overall impedance ratio 20. Find the VSWR at each step and the maximum passband VSWR.

5.14 Design a three-section Chebyshev quarterwave transformer of 40 per cent bandwidth for an impedance ratio of 10, VSWR $\leq$ 1.01 in the equiripple region.

5.15 Design a six-step binomial quarterwave transformer between lines of impedances of 200 ohms and 1000 ohms.

5.16 Design a four-section maximally flat transformer for load ratio 3 and a normalized frequency band 2.

5.17 A quarterwave section is used to match a 100 ohm resistive load to a 50 ohm line at a given frequency. (a) Calculate the characteristic impedance of the matching section, and (b) the VSWR on the main line with matching transformer when the frequency is increased by 20%.

5.18 Design an *L*-section network to match a load 200–*j*100 ohm to a 100 ohm line at 500 MHz.

REFERENCES

1. Cohn, S B: *Optimum Design of Stepped Transmission Line Transformers*, IRE trans: Vol. MTT-3, pp. 16–21, April, 1955.
2. Cohn, S B: *Problems in Strip Transmission Lines*, IRE trans: PGMTT-3, No. 2, March 1955, pp. 119–126.
3. Collin, R E: *General Synthesis of Quarterwave Impedance Transformers*, IRE trans: PGMIT–7, pp. 233–237, April 1959, and PGMIT-8, pp. 243–244 March, 1960.
4. Collin, R E: *The Optimum Tapered Transmission Line Matching Section*, Proc: IRE, Vol.44, pp. 539–548, April 1956.
5. Collin, R E: *Theory and Design of Wideband Multisection Quarterwave Transformers*, Proc. IRE, Vol. 43, pp. 179–185, February 1955.
6. Hammerstead, E O, and Bekkadal F: *A Microstrip Handbook ELAB Report*, STF 44A74169 N 7034, Univ. of Trondheim–NTH, Norway, 1975.
7. Howe, H: *Stripline Circuit Design*, Artech House, 1974.
8. Johnson, R C: *Design of Linear Double Tapers in Rectangular waveguides*, IRE trans: Vol. MTT-7, pp. 374–378, July 1959.
9. Young, Leo: *Inhomogeneous Quarterwave*, IRE trans, PGMIT-8, pp. 478–482, September, 1960.
10. Young, Leo: *Practical Design of a Wideband Quarterwave Transformer in Waveguide*, The Microwave Journal, 6, No. 10, pp. 76–79, (October, 1963).
11. Young, Leo: *Stepped Impedance Transformers and Filter Prototypes*, IRE trans., PGMIT-10, pp. 339–359 Sept. 1962.
12. Levy, R: *Directional Couplers, Advances in Microwave*, Vol. 1, Academic Press, 1966, p. 186.
13. Matthaei, G L, Young and Jones, E M T: *Microwave Filters, Impedance Matching Networks and Coupling Structures,* McGraw-Hill Book Company, 1984.
14. Riblet, H J: *General Synthesis of Quarterwave Impedance Transformers*, IRE trans., Vol. MTT-5, pp. 36–43, January, 1957.
15. Young, L: *Tables for Cascaded Homogeneous Quarterwave Transformers*, IRE trans., Vol. MTT-7, April, 1959.

6 MICROWAVE NETWORK THEORY AND PASSIVE DEVICES

INTRODUCTION 6.1

A microwave network is formed when several microwave devices and components such as sources, attenuators, resonators, filters, amplifiers, etc., are coupled by transmission lines or waveguides for the desired transmission of a microwave signal. The point of interconnection of two or more devices is called a junction.

For a low-frequency network, a port is a pair of terminals whereas for a microwave network, a port is a reference plane transverse to the length of the microwave transmission line or waveguide. At low frequencies, the physical length of the network is much smaller than the wavelength of the signal transmitted. Therefore, the measurable input and output variables are voltage and current which can be related in terms of the impedance *Z*-parameters, or admittance *Y*-parameters, or hybrid *h*-parameters, or *ABCD* parameters. For a two-port network as shown schematically in Fig. 6.1, these relationships are given by

$$\begin{bmatrix} V_1 \\ V_2 \end{bmatrix} = \begin{bmatrix} Z_{11} & Z_{12} \\ Z_{21} & Z_{22} \end{bmatrix} \begin{bmatrix} I_1 \\ I_2 \end{bmatrix} \quad \text{Impedance } (Z) \text{ parameters} \tag{6.1}$$

$$\begin{bmatrix} I_1 \\ I_2 \end{bmatrix} = \begin{bmatrix} Y_{11} & Y_{12} \\ Y_{21} & Y_{22} \end{bmatrix} \begin{bmatrix} V_1 \\ V_2 \end{bmatrix} \quad \text{Admittance } (Y) \text{ parameters} \tag{6.2}$$

$$\begin{bmatrix} V_1 \\ I_2 \end{bmatrix} = \begin{bmatrix} h_{11} & h_{12} \\ h_{21} & h_{22} \end{bmatrix} \begin{bmatrix} I_1 \\ V_2 \end{bmatrix} \quad \text{Hybrid } (h) \text{ parameters} \tag{6.3}$$

$$\begin{bmatrix} V_1 \\ I_1 \end{bmatrix} = \begin{bmatrix} A & +B \\ C & +D \end{bmatrix} \begin{bmatrix} V_2 \\ I_2 \end{bmatrix} \quad \text{Transmission } (ABCD) \text{ parameters} \tag{6.4a}$$

In Eq. (6.4a), $+I_2$ for current at Port 2 is outward and $-I_2$ taken for current at Port 2 is inward. Here, Z_{ij}, Y_{ij}, and *A*, *B*, *C* and *D* are suitable constants that characterize the junction. *A*, *B*, *C* and *D* parameters are convenient to represent each junction when a number of circuits are connected together in cascade. Here the resultant matrix, which describes the complete cascade connection, can be obtained by multiplying the matrices describing each junction as follows:

$$\begin{bmatrix} A & B \\ C & D \end{bmatrix} = \begin{bmatrix} A_1 & B_1 \\ C_1 & D_1 \end{bmatrix} \begin{bmatrix} A_2 & B_2 \\ C_2 & D_2 \end{bmatrix} \cdots \begin{bmatrix} A_n & B_n \\ C_n & D_n \end{bmatrix} \tag{6.4b}$$

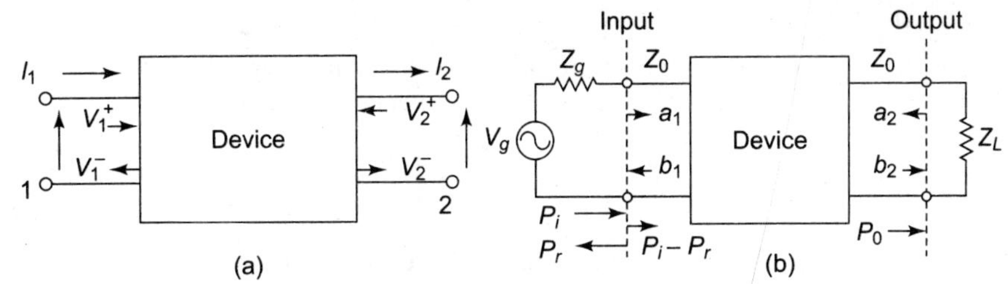

Fig. 6.1 A two-port network

These parameters can be measured under short- or open-circuit condition for use in the analysis of the circuit.

At microwave frequencies, the physical length of the component or line is comparable to or much larger than the wavelength. Furthermore, the voltage and current cannot be uniquely defined at a given point in a single conductor waveguide. Furthermore, measurement of *Z, Y, h* and *ABCD* parameters is difficult at microwave frequencies due to the following reasons:

1. Absence of unique definition of voltage and current as noted above and non-availability of terminal voltage and current-measuring equipment, even in the cases of TEM lines (coaxial, strip and microstrip lines) where such quantities can be uniquely defined.
2. Short-circuit, and especially open-circuit, are not easily achieved for a wide range of frequencies.
3. Presence of active devices makes the circuit unstable for short or open circuit.

Therefore, microwave circuits are analyzed using scattering or *S*-parameters which linearly relate the amplitudes of scattered (reflected and transmitted) waves with those of incident waves. However, many of the circuit-analysis techniques and circuit properties that are valid at low frequencies are also valid for microwave circuits. Thus, for circuit analysis, *S*-parameters can be related to the *Z* or *Y* or *ABCD* parameters. The properties of the parameters are described in the following sections.

PROPERTIES OF *Z* AND *Y* MATRICES FOR RECIPROCAL NETWORKS 6.2

In a reciprocal network, the impedance and the admittance matrices are symmetrical and the junction media are characterized by scalar electrical parameters *m* and *e*. For a multiport (*N* port) network, let the incident wave amplitude V_n^+ be so chosen that the total voltage $V_n = V_n^+ + V_n^-$ at all ports $n = 1, 2, ..., N$, except the *i*th port where the fields are E_i, H_i. Similarly, let $V_n = 0$ at all ports except jth one where the fields are E_j, H_j. Then from the Lorentz reciprocity theorem,

$$\int_S (\mathbf{E}_i \times \mathbf{H}_j - \mathbf{E}_j \times \mathbf{H}_i) \cdot dS = 0 \tag{6.5}$$

where *S* is the closed surface area of the conducting walls enclosing the junction and *N* ports in the absence of any source. Since the integral over the perfectly conducting walls vanishes, the only non-zero integrals are those taken over the reference planes of the corresponding ports, so that

$$\sum_{n=1}^{N} \int_{t_n} (\mathbf{E}_i \times \mathbf{H}_j - \mathbf{E}_j \times \mathbf{H}_i) \cdot dS = 0 \tag{6.6}$$

Since V_n except V_i and V_j are zero, $\mathbf{E}_{ti} = \mathbf{n} \times \mathbf{E}_i$ and $\mathbf{E}_{tj} = \boldsymbol{n} \times \mathbf{E}_j$ are zero on all reference planes at the corresponding ports except t_i and t_j, respectively. Therefore, Eq. (6.6) reduces to

$$\int_{t_i} (\mathbf{E}_i \times \mathbf{H}_j) \cdot dS = \int_{t_j} (\mathbf{E}_j \times \mathbf{H}_i) \cdot dS \tag{6.7}$$

or $$P_{ij} = P_{ji} \tag{6.8}$$

where P_{ij} represents the power at the reference plane i due to an input voltage at the plane j.

From the admittance matrix representation $[I] = [Y][V]$ and power relation $P = VI$, Eq. (6.8) reduces to

$$V_i V_j Y_{ij} = V_j V_i Y_{ji} \quad \text{or,} \quad Y_{ij} = Y_{ji} \tag{6.9}$$

and $$Z_{ij} = Z_{ji} \tag{6.10}$$

This proves that the impedance and admittance matrices are symmetrical for a reciprocal junction.

SCATTERING OR *S*-MATRIX REPRESENTATION OF MULTIPORT NETWORK 6.3

As discussed in Section 6.1, the incident and reflected wave amplitudes of microwaves at any port are used to characterize a microwave circuit. The amplitudes are normalized in such a way that the square of any of these variables gives the average power in that wave in the following manner:

Input power at the nth port, $$P_{in} = 1/2\, |a_n|^2 \tag{6.11}$$

Reflected power at the nth port $$P_{rn} = 1/2\, |b_n|^2 \tag{6.12}$$

where a_n and b_n represent the normalized incident wave peak amplitude and normalized reflected wave peak amplitude at the nth port. The concept of scattering(s) parameters comes from the fact that *rf* and microwave circuit may contain some discontinuity or discontinuities in the signal propagation path. At an discontinuity the wave scattered in different directions as evanescent waves containing infinite number of higher order modes. These modes attenuated very fast after a short distance from the point of discontinuity within about a quarter wavelength. Then only the executed mode comes out from the different ports. All these emerging waves are considered as reflected waves at the corresponding parts. The waves entering the parts are considered the input or incident wave.

In a two-port network, we can express the normalized waves in terms of normalized voltages:

$$a_1 = \frac{V_1^+}{\sqrt{Z_0}} = \frac{V_1 - V_1^-}{\sqrt{Z_0}}, \quad a_2 = \frac{V_2^+}{\sqrt{Z_0}} = \frac{V_2 - V_2^-}{\sqrt{Z_0}} \tag{6.13}$$

$$b_1 = \frac{V_1^-}{\sqrt{Z_0}} = \frac{V_1 - V_1^+}{\sqrt{Z_0}}, \quad b_2 = \frac{V_2^-}{\sqrt{Z_0}} = \frac{V_2 - V_2^+}{\sqrt{Z_0}} \tag{6.14}$$

where a's represent normalized incident wave amplitude and b's represent normalized reflected wave amplitude at the corresponding ports. Here, the total voltage wave is the sum of incident and emergent voltage waves V^+ and V^- respectively:

$$V_1 = V_1^+ + V_1^- \qquad I_1 = I_1^+ - I_1^- \tag{6.15}$$

$$V_2 = V_2^+ + V_2^- \qquad I_2 = I_2^+ - I_2^- \tag{6.16}$$

The numeric suffices represent the port number. The total or net power flow into any port is given by r.m.s. value

$$P = P_i - P_r = \frac{1}{2}(|a|^2 - |b|^2) \tag{6.17}$$

Factor $\frac{1}{2}$ comes from the relation $V_{\text{rms}} = \frac{V_{\text{peak}}}{\sqrt{2}}$ or $|a|$ and $|b|$ are magnitor of peak values.

Therefore, in this normalization process, the characteristic impedance is normalized to unity. For a two-port network (Fig. 6.1), the relation between incident and reflected waves are expressed in terms of scattering parameters S_{ij}'s:

$$b_1 = S_{11} a_1 + S_{12} a_2 \tag{6.18}$$

$$b_2 = S_{21} a_1 + S_{22} a_2 \tag{6.19}$$

The normalization process leads to a symmetrical scattering matrix for reciprocal structures. The physical significance of S-parameters can be described as follows:

$S_{11} = (b_1/a_1)_{a_2=0}$ = Reflection coefficient Γ_1 at Port 1 when Port 2 in terminated with a matched load ($a_2 = 0$)

$S_{22} = (b_2/a_2)_{a_1=0}$ = Reflection coefficient Γ_2 at Port 2 when Port 1 in terminated with a matched load ($a_1 = 0$)

$S_{12} = (b_1/a_2)_{a_1=0}$ = Attenuation of wave travelling from Port 2 to Port 1 when $a_1 = 0$

$S_{21} = (b_2/a_1)_{a_2=0}$ = Attenuation of wave travelling from Port 1 to Port 2 when $a_2 = 0$

In general, since the incident and reflected waves have both amplitude and phase, the S-parameters are complex numbers.

For multiport (N) networks or components, the S-parameter equations are expressed by

$$\begin{bmatrix} b_1 \\ b_2 \\ \vdots \\ b_N \end{bmatrix} = \begin{bmatrix} S_{11} & S_{12} & \cdots & S_{1N} \\ S_{21} & S_{22} & \cdots & S_{2N} \\ \vdots & & & \\ S_{N1} & S_{N2} & \cdots & S_{NN} \end{bmatrix} \begin{bmatrix} a_1 \\ a_2 \\ \vdots \\ a_N \end{bmatrix} \tag{6.20}$$

Figure 6.2 shows a linear multiport network having N ports. We can write for any arbitrary n-th port of this junction the following:

$$a_n = V_n^+/\sqrt{Z_{on}} = I_n^+\sqrt{Z_{on}} \tag{6.20a}$$

$$b_n = V_n^-/\sqrt{Z_{on}} = I_n^-\sqrt{Z_{on}} \tag{6.20b}$$

Net input power at port n

$$P_{ninp} = \frac{1}{2}(|a_n|^2 - |b_n|^2) \tag{6.20c}$$

Power dissipated in a termination at port n is

$$P_{ndissp} \doteq \frac{1}{2}(|b_n|^2 - |a_n|^2) = \frac{1}{2}|b_n|^2(1 - |\Gamma_n|^2) \tag{6.20d}$$

$$\Gamma_n = \frac{b_n}{a_n} \tag{6.20e}$$

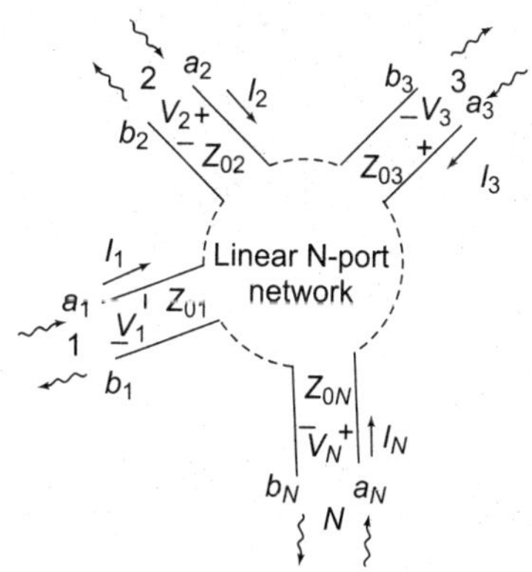

Fig. 6.2 A multiport (N) linear network

is the magnitude of the load reflection coefficient at port n. Above relations hold for all the ports with proper suffices. The total voltage and current at port n are

$$V_n = V_n^+ + V_n^- = \sqrt{Z_{on}}(a_n + b_n) \tag{6.20f}$$

$$I_n = I_n^+ - I_n^- = \frac{1}{\sqrt{Z_{on}}}(a_n - b_n) \tag{6.20g}$$

Solving the above equations for a_n and b_n in terms of the terminal voltage and current yields

$$a_n = \frac{1}{2}\left(\frac{V_n}{\sqrt{Z_{on}}} + I_n\sqrt{Z_{on}}\right) \tag{6.20h}$$

$$b_n = \frac{1}{2}\left(\frac{V_n}{\sqrt{Z_{on}}} - I_n\sqrt{Z_{on}}\right) \tag{6.20i}$$

In microwave devices or circuits, it is important to express several losses in terms of S-parameters when the ports are match terminated. In a two-port network, if power fed at Port 1 is P_i, power reflected at the same port is P_r and the output power at Port 2 is P_o then following losses are defined in terms of S-parameters:

$$\text{Insertion loss (dB)} = 10 \log \frac{P_i}{P_o} = 10 \log \frac{|a_1|^2}{|b_2|^2}$$

$$= 20 \log \frac{1}{|S_{21}|} = 20 \log \frac{1}{|S_{12}|} \tag{6.21}$$

$$\text{Transmission loss or attenuation (dB)} = 10 \log \frac{P_i - P_r}{P_o} = 10 \log \frac{1-|S_{11}|^2}{|S_{12}|^2} \tag{6.22}$$

$$\text{Reflection loss (dB)} = 10 \log \frac{P_i}{P_i - P_r} = 10 \log \frac{1}{1-|S_{11}|^2} \tag{6.23}$$

$$\text{Return loss (dB)} = 10 \log \frac{P_i}{P_r}$$

$$= 20 \log \frac{1}{|\Gamma|} = 20 \log \frac{1}{|S_{11}|} \tag{6.24}$$

Example 6.1 Find the S-matrix of a length l of a lossless transmission line terminated by matched impedance.

Solution For a length l of a transmission line, there is no discontinuity at the two ends, so that $S_{11} = S_{22} = 0$. The output signals arise due to input:

$$b_1 = a_2 e^{-j\beta l}$$

$$b_2 = a_1 e^{-j\beta l}$$

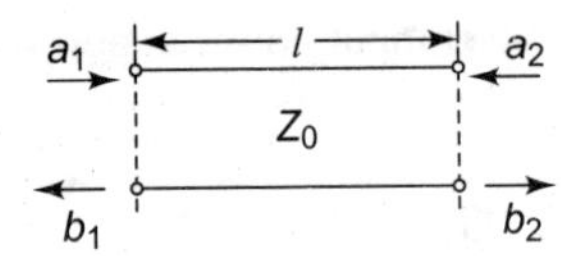

Fig. 6.3 Transmission lines of Example 6.1

or,

$$\begin{bmatrix} b_1 \\ b_2 \end{bmatrix} = \begin{bmatrix} 0 & e^{-j\beta l} \\ e^{-j\beta l} & 0 \end{bmatrix} \begin{bmatrix} a_1 \\ a_2 \end{bmatrix} = [S][a]$$

$$\therefore \quad [S] = \begin{bmatrix} 0 & e^{-j\beta l} \\ e^{-j\beta l} & 0 \end{bmatrix}$$

6.3.1 Properties of S-Parameters for a Junction of Ports having Common Characteristic Impedance

In general, the scattering parameters are complex quantities having the following properties for different characteristics of the microwave network or junction where characteristic impedance is same at all ports.

1. S-matrix is always a square matrix of order $N \times N$ for an N-port network and its elements are complex quantities (real and imaginary parts).

2. Phase-shift property of S-Matrix

At any given frequency and for a given positions of the reference planes, the elements of S-matrix, S_{ij}, have definite values.

(i) If the frequency is changed, these elements change values also.

(ii) At a given frequency, the complex S-parameters of a network are defined with respect to the positions of the reference plan (ports). If these positions change, S-parameters of the network also change. For an outward shift of reference plane (Fig. 6.4) by a distance l_j, the phase shift occurs is $\phi_j = \beta_j l_j$ and correspondingly new elements of the S-matrix become

$$S'_{ij} = S_{ij}e^{-j\phi_j}; i \neq j$$

$$S'_{ij} = S_{ij}e^{-j2\phi_j}; i = j$$

Power of exponent is doubles for $i = j$, since reflected wave travels $2l_j$ distance.

Let the original S-matrix equation for the two-port network with reference planes 1, 2 be

$$[b] = [S][a] \tag{6.25a}$$

Equation (6.25a) is modified to

$$[b] = [S'][a] \tag{6.25b}$$

where $[S]$ is changed to $[S']$ for outward shift in reference planes to $1'$, $2'$ but the incident and reflected wave voltage can be still denoted by $[a]$ and $[b]$, respectively:

$$S'_{11} = S_{11}\,e^{-j2\phi_1},\ S'_{22} = S_{22}\,e^{-j2\phi_2},\ S'_{12} = S_{12}\,e^{-j(\phi_1+\phi_2)},\ S'_{21} = S_{21}\,e^{-j(\phi_1+\phi_2)}.$$

These can be proved as follows. For the above shift in reference planes, the new wave variables becomes

$$a_1' = a_1\,e^{-j\phi_1},\quad a_2' = a_2\,e^{-j\phi_2}$$

$$b_1' = b_1\,e^{-j\phi_1},\quad b_2' = a_2\,e^{-j\phi_2} \tag{6.25c}$$

where $\phi = \beta_1 l_1$, $\phi_2 = \beta_2 l_2$ and –ve sign in the power of exponent arises due to outward shift of ports. For inward shift, this sign would be +ve. Hence, we can write

$$\begin{bmatrix} b_1' \\ b_2' \end{bmatrix} = \begin{bmatrix} b_1\,e^{-j\phi_1} \\ b_2\,e^{-j\phi_2} \end{bmatrix} = \begin{bmatrix} e^{-j\phi_1} & 0 \\ 0 & e^{-j\phi_2} \end{bmatrix}\begin{bmatrix} b_1 \\ b_2 \end{bmatrix} \tag{6.25d}$$

Similarly,

$$\begin{bmatrix} a_1' \\ a_2' \end{bmatrix} = \begin{bmatrix} e^{-j\phi_1} & 0 \\ 0 & e^{-j\phi_2} \end{bmatrix}\begin{bmatrix} a_1 \\ a_2 \end{bmatrix} \tag{6.25e}$$

From (6.25a), (6.25b), (6.25d) and (6.25e), the new S-matrix becomes

$$[S'] = \begin{bmatrix} e^{-j\phi_1} & 0 \\ 0 & e^{-j\phi_2} \end{bmatrix}[S]\begin{bmatrix} e^{-j\phi_1} & 0 \\ 0 & e^{-j\phi_2} \end{bmatrix} = \begin{bmatrix} S_{11}\,e^{-j2\phi_1} & S_{12}\,e^{-j(\phi_1+\phi_2)} \\ S_{21}\,e^{-j(\phi_1+\phi_2)} & S_{22}\,e^{-j2\phi_2} \end{bmatrix} \tag{6.26}$$

This property valid for any number of ports and is called the phase shift property.

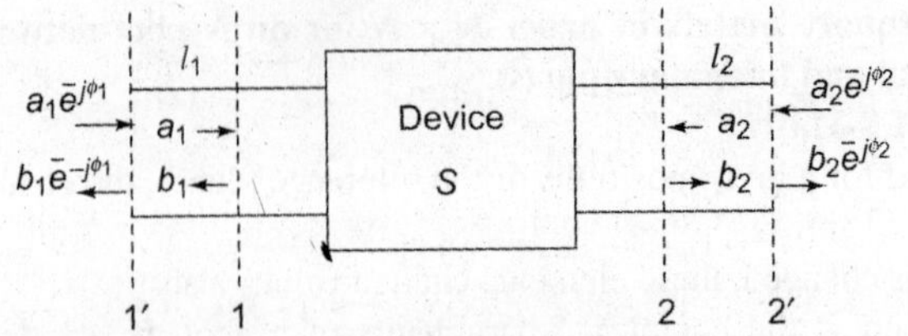

Fig. 6.4 Phase-shift property of S

3. Diagonal elements are zero for perfect matched network.

For an ideal N-port matched network with matched termination at all the ports, $S_{ii} = 0$ since there is no reflection from any port. Therefore, under perfect matched conditions, the diagonal elements of [S] are zero.

4. S-matrix is symmetric for a reciprocal network

A reciprocal device has the same transmission characteristics in either direction of a pair of ports and is characterized by a symmetric scattering matrix,

$$S_{ij} = S_{ji} \ (i \neq j) \tag{6.27}$$

which results in transpose of S-matrix:

$$[S]_t = [S] \tag{6.28}$$

This condition can be proved in the following manner.

❑ *Proof*

For a reciprocal network with the assumed normalization

$$V_n = V_n^+ + V_n^- \qquad I_n = I_n^+ - I_n^-$$

$$\sqrt{Z_{\text{on}}}\, I_n = a_n - b_n; \quad I_n = \frac{1}{\sqrt{Z_{\text{on}}}}(a_n - b_n)$$

$$\frac{V_n}{\sqrt{Z_{\text{on}}}} = a_n + b_n; \quad V_n = \sqrt{Z_{\text{on}}}\,(a_n + b_n) \tag{6.28a}$$

The z-matrix equation is given by

$$[V] = [z]\,[I] \tag{6.28b}$$

With the assumed normalization (6.28b) can be written using (6.28a) as

$$\sqrt{Z_0} = ([a]+[b]) = [z]\frac{1}{\sqrt{Z_0}}([a]-[b])$$

or $\quad Z_0[b] + [z][b] = [z][a] - z_0[a]$

or $\quad [b](z_0[u] + [z]) = [a]\,([z] - z_0[U]) \qquad (6.29c)$

where [U] is the unit matrix having unity diagonal elements and zero off-diagonal elements. (6.28c) can now becomes

$$[b] = ([z] - z_0[U])\,(z_0[U] + [z])^{-1}\,[a] \tag{6.29d}$$

The S-matrix equation for the network is

$$[b] = [S]\,[a] \tag{6.30}$$

Comparing Eqs (6.29) and (6.30), we have

$$[S] = ([Z] + [U])^{-1}\,([Z] - [U]) \tag{6.31}$$

For reciprocal network, the Z-matrix is symmetric. Hence, from the symmetric of Z-matrix

$$([Z] - [U])_t = [Z] - [U] \tag{6.32}$$

$$([Z] + [U])_t = [Z] + [U] \tag{6.33}$$

Therefore, from (6.31), (6.32), and (6.33)

$$[S]_t = ([Z] - [U])_t \; ([Z] + [U])_t^{-1} = ([z] - [U]) \; ([z] + [U])^{-1} = [S] \tag{6.34}$$

Thus, it is proved that $[S]_t = [S]$ for a symmetrical junction.

5. *S*-matrix has unitary property for a lossless network

For a lossless *N*-port network S-matrix satisfies the following *Unitary Property*:

(i) The sum of the products of each term of any one row or any column of the *S*-matrix multiplied by its complex conjugate (dot product) is unity

$$\sum_{n=1}^{N} S_{ni} \cdot S_{nj}^{*} = 1; \quad i = j$$

(ii) The sum of the products of each term of any one row or any column of the *S*-matrix multiplied by the complex conjugate of a different row or column, respectively, is zero (columns are ortho-normal)

$$\sum_{n=1}^{N} S_{ni} \cdot S_{nj}^{*} = 0; \quad i \neq j$$

❑ *Proof*

Let us consider a 4-port lossless Network ($N = 4$).

The *S*-matrix equation can be written as

$$[b] = [S][a]$$

or,

$$\begin{aligned} b_1 &= S_{11}a_1 + S_{12}a_2 + S_{13}a_3 + S_{14}a_4 \\ b_2 &= S_{21}a_2 + S_{22}a_2 + S_{23}a_3 + S_{24}a_4 \\ &\vdots \qquad \vdots \qquad \vdots \\ b_4 &= S_{41}a_1 + S_{42}a_2 + S_{43}a_3 + S_{44}a_4 \end{aligned} \tag{6.35}$$

For a loss-less network, the total power leaving N-ports must be equal to the total power input to these ports, so that

$$\sum_{n=1}^{N=4} |b_n|^2 = \sum_{n=1}^{N=4} |a_n|^2$$

or,

$$\sum_{n=1}^{N=4} |a_n|^2 = \sum_{n=1}^{N=4} \left| S_{n_1} a_1 + S_{n_2} a_2 + \cdots + S_{n_n} a_n \right|$$

or,

$$|a_1|^2 + |a_2|^2 + |a_3|^2 + |a_4|^2 = \sum_{n=1}^{N=4} \left| S_{n_1} a_1 + S_{n_2} a_2 + S_{n_3} a_3 + S_{n_4} a_4 \right|^2 \tag{6.36}$$

If only one port (*i*th say) is excited and all other ports are match terminated, $a_n = 0, n \neq i$.

In this case, $|a_i|^2 = \sum_{n=1}^{N} \left| S_{n_i} a_i \right|^2 = |a_i|^2 \sum_{n=1}^{N} \left| S_{n_i} \right|^2$

or,

$$\sum_{n=1}^{N} \left| S_{n_i} \right|^2 = 1 \tag{6.37}$$

If S_{n_i} is complex, $\left|S_{n_i}\right|^2 = S_{ni} \cdot S_{ni}^*$ (6.38)

From Eqs (6.37) and (6.38),

$$\boxed{\sum_{n=1}^{N} S_{ni} \cdot S_{ni}^* = 1} \tag{6.39a}$$

or, $S_{1i} \cdot S_{1i}^* + S_{2i} \cdot S_{2i}^* + \cdots + S_{ni} \cdot S_{ni}^* = 1;$ (6.39b)

For $N = 4$, $i = 1$

$$S_{11}\, S_{11}^* + S_{21} \cdot S_{21}^* + \cdots + S_{41} \cdot S_{41}^* = 1$$

For $N = 4$, $i = 2$

$$S_{12}\, S_{12}^* + S_{22} \cdot S_{22}^* + \cdots + S_{42} \cdot S_{42}^* = 1$$

For $N = 4$, $i = 4$

$$S_{14}\, S_{14}^* + S_{24} \cdot S_{24}^* + \cdots + S_{44} \cdot S_{44}^* = 1 \tag{6.39c}$$

Equations (6.39a)–(6.39c) prove the statement 3(i).

If two ports i and j are excited and all others are match terminated, $a_n = 0$ when $n \neq i$, $n \neq j$. Therefore, from Eq. (6.36), we can write

$$\left|a_i\right|^2 + \left|a_j\right|^2 = \sum_{n=1}^{N=4} \left|S_{n_i} a_i + S_{n_j} a_j\right|^2$$

$$= \left|S_{1i}\, a_i + S_{1j}\, a_j\right|^2 + \left|S_{2i}\, a_i + S_{2j}\, a_j\right|^2 + \left|S_{3i}\, a_i + S_{3j}\, a_j\right|^2 + \left|S_{4i}\, a_i + S_{4j}\, a_j\right|^2 \tag{6.40a}$$

Since for any complex quantity $|S|^2 = S \cdot S^*$, the above expression may be multiplied out to give the following results:

$$\begin{aligned}\left|a_i\right|^2 + \left|a_j\right|^2 = & \left\{\left|S_{11}\right|^2 + \left|S_{21}\right|^2 + \left|S_{31}\right|^2 + \left|S_{41}\right|^2\right\} \left|a_i\right|^2 \\ & + \left\{\left|S_{12}\right|^2 + \left|S_{22}\right|^2 + \left|S_{32}\right|^2 + \left|S_{42}\right|^2\right\} \left|a_j\right|^2 \\ & + \{S_{11}\, S_{12}^* + S_{21}\, S_{22}^* + S_{31}\, S_{32}^* + S_{41}\, S_{42}^*\}\, a_i\, a_j^* \\ & + \{S_{12}\, S_{11}^* + S_{22}\, S_{21}^* + S_{32}\, S_{31}^* + S_{42}\, S_{41}^*\}\, a_j\, a_i^*\end{aligned}$$

Using Eq. (6.39a), the above equation reduces to

$$\begin{aligned}\left|a_i\right|^2 + \left|a_j\right|^2 = & \left|a_i\right|^2 + \left|a_j\right|^2 + \{S_{11}S_{12}^* + S_{21}\, S_{22}^* + S_{31}S_{32}^* + S_{41}S_{42}^*\}\, a_i\, a_j^* \\ & + \{S_{12}S_{11}^* + S_{22}\, S_{21}^* + S_{32}S_{31}^* + S_{42}S_{41}^*\}\, a_j\, a_i^*\end{aligned} \tag{6.40b}$$

Equating LHS with RHS of Eq. (6.40b) yields

$$\{S_{11}S_{12}^* + \cdots + S_{41}S_{42}^*\}\, a_i\, a_j^* + \{S_{12}S_{11}^* + \cdots + S_{42}S_{41}^*\}\, a_j\, a_i^* = 0 \tag{6.40c}$$

Since a_i and a_j are independent signals, they may be chosen in any convenient manner, let $a_i = a_j$. Then from (6.40c),

$$\{S_{11}S_{12}^* + \cdots + S_{41}S_{42}^*\} + \{S_{12}S_{11}^* + \cdots + S_{42}S_{41}^*\} = 0 \tag{6.40d}$$

Let $a_i = ja_j$ where multiplier $j = \sqrt{-1}$, then

$$\{S_{11}S_{12}^* + \cdots + S_{41}S_{42}^*\} - \{S_{12}S_{11}^* + \cdots + S_{42}S_{41}^*\} = 0 \tag{6.40e}$$

Since $a_i^* = (ja_j)^* = -ja_j^*$

Combining Eqs (6.40d) and (6.40e) results in

$$\left.\begin{aligned} &S_{11}S_{12}^* + S_{21}S_{22}^* + S_{31}S_{32}^* + S_{41}S_{42}^* = 0 \\ \text{or,}\quad &S_{12}S_{11}^* + S_{22}S_{21}^* + S_{32}S_{31}^* + S_{42}S_{41}^* = 0 \end{aligned}\right\} \tag{6.40f}$$

In general, for an N-port network, above equations yield

$$\boxed{\sum_{n=1}^{N} S_{ni} \cdot S_{nj}^* = 0}\ ; \quad i \neq j \tag{6.40g}$$

Equation (6.40g) proves the statement 3(ii).

Equation (6.39a) and (6.40g) can be written in a single form:

$$\boxed{[S^*][S]_t = [U]} \quad \text{or} \quad \boxed{[S^*] = [S]_t^{-1}} \tag{6.41}$$

Therefore, (6.39a) and (6.40g) or (6.41) is the necessary condition for the network to be lossless and is called unitary property of S-matrix. [S] is called a unitary matrix.

S-parameters to a Two-port Network with Mismatched Load 6.3.2

A two-port network or junction is formed when there is a discontinuity between the input and output ports of a transmission line. Many configurations of such junctions practically exist, some of which are shown in Fig. 6.5.

During propagation of a microwave through the junction from one port to another, evanescent modes are excited at each discontinuity which contains reactive energy. Evanescent modes decay very fast away from the junction and become negligible after a distance of the order of one wavelength. The terminal reference planes 1 and 2 are chosen beyond this distance so that the equivalent voltage and currents at these positions are proportional to the total transverse electric and magnetic fields, respectively, for the propagating mode only. These circuits are analyzed using S-matrix formulation.

Consider a two-port network of Fig. 6.5(c) terminated by normalized load and generator impedances $\frac{Z_L}{Z_0}$ and $\frac{Z_g}{Z_0}$. Then the load reflection coefficient

$$\Gamma_2 = \frac{a_2}{b_2} = \frac{Z_L/Z_0 - 1}{Z_L/Z_0 + 1} \tag{6.42}$$

Now,

$$b_1 = S_{11}a_1 + S_{12}a_2 = S_{11}a_1 + S_{12}b_2\Gamma_2 \tag{6.43}$$

$$b_2 = S_{21}a_1 + S_{22}a_2 = S_{21}a_1 + S_{22}b_2\Gamma_2 \tag{6.44}$$

Solving for the input reflection coefficient,

$$\Gamma_1 = b_1/a_1 = S_{11} + \frac{S_{12}S_{21}\Gamma_2}{1 - S_{22}\Gamma_2} \tag{6.45}$$

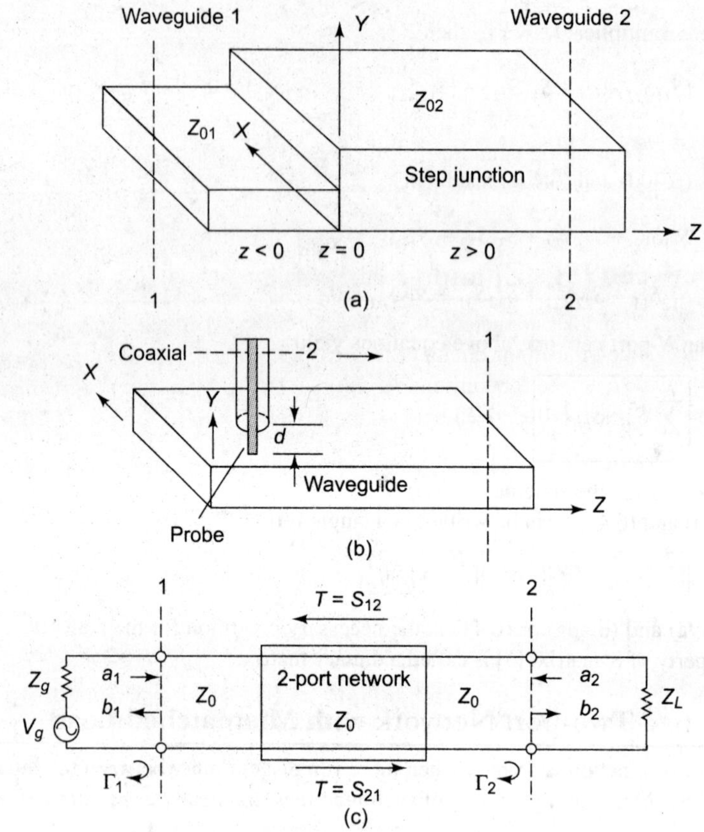

Fig. 6.5 Two-port junctions: (a) waveguide step junction, (b) coaxial to waveguide transition, and (c) a two-port network model

Therefore, for a mismatch load, input reflection coefficient $\Gamma_1 \neq S_{11}$. For a reciprocal network, $S_{12} = S_{21}$ so that

$$\Gamma_1 = S_{11} + \frac{S_{12}^2\,\Gamma_2}{1 - S_{22}\,\Gamma_2} \tag{6.46}$$

Further, if the junction is lossless, from Eqs (6.39) and (6.40),

$$S_{11}\,S_{11}^* + S_{12}\,S_{12}^* = 1 \tag{6.47}$$

$$S_{22}\,S_{22}^* + S_{12}\,S_{12}^* = 1 \tag{6.48}$$

$$S_{11}\,S_{12}^* + S_{12}\,S_{22}^* = 0 \tag{6.49}$$

Therefore, for a lossless, reciprocal two-port network, terminated by a mismatch load, Eqs 6.47) and (6.48) give

$$|S_{11}| = |S_{22}| \tag{6.50}$$

From Eqs (6.49) and (6.50),

$$|S_{12}| = \sqrt{(1 - |S_{11}|^2)} \tag{6.51}$$

and the input reflection coefficient

$$\Gamma_1 = S_{11} + \frac{S_{12}^2\,\Gamma_2}{1 - S_{22}\,\Gamma_2} \tag{6.52}$$

The last equation is the working equation for the computation of the S-parameters. By measuring Γ_1 for known values of Γ_2 (0, –1 and 1) a set of simultaneous equations are obtained which will give the S-parameters of a reciprocal junction.

Comparison between [S], [Z] and [Y] Matrices 6.3.3

From the analysis of S, Z and Y parameters, it is known that $[S]$, $[Z]$ and $[Y]$ represent unique intrinsic properties of the device, supplying all the circuit characteristics of the device at a given frequency. Therefore, terminal conditions may be changed for convenience without affecting the matrices themselves. The $[S]$ can be expressed in terms of $[Z]$ and $[Y]$ as given below.

Relation between [S] and [Z] Matrices

For a multiport network, $[V] = [Z]\,[I]$. The steady-state total voltage and the current at the kth port are

$$V_k = V_k^+ + V_k^- = Z_{k_1} I_1 + Z_{k_2} I_2 + \cdots + Z_{kn} I_n$$

$$= \sum_{j=1}^{n} Z_{kj}\, I_j \tag{6.53a}$$

and

$$I_k = I_k^+ - I_k^- = \frac{1}{Z_0}(V_k^+ - V_k^-) \tag{6.53b}$$

or,

$$Z_0 I_k = (V_k^+ - V_k^-)$$

or,

$$V_k^+ = V_k^- + Z_0 I_k = V_k - V_k^+ + Z_0 I_k$$

or,

$$V_k^- = \frac{1}{2}(V_k + Z_0 I_k) = \frac{1}{2}\left(\sum_{j=1}^{n} Z_{kj} I_j + Z_0 I_k\right); \tag{6.53c}$$

Similarly,

$$V_k^- = \frac{1}{2}\left(\sum_{j=1}^{n} Z_{kj} I_j - Z_0 I_k\right); \tag{6.53d}$$

Therefore,

$$[V^+] = \frac{1}{2}\begin{bmatrix} Z_{11} + Z_0 & Z_{12} & \cdots & Z_{1n} \\ Z_{21} & Z_{22} + Z_0 & \cdots & Z_{2n} \\ \vdots & & & \\ Z_{n_1} & Z_{n_2} & \cdots & Z_{nn} + Z_0 \end{bmatrix} [I]$$

$$= \frac{1}{2}([Z] + Z_0[U])\,[I] \tag{6.53e}$$

or,

$$[I] = 2\,([Z] + Z_0[U])^{-1}\,[V^+] \tag{6.53f}$$

Similarly,

$$[V^-] = \frac{1}{2}([Z] - Z_0[U])\,[I]$$

$$= ([Z] - Z_0[U])([Z] + Z_0[U])^{-1}[V^+] \quad (6.53g)$$

or, $$\left[\frac{V^-}{\sqrt{Z_0}}\right] = ([Z] - Z_0[U])([Z] + Z_0[U])^{-1}\left[\frac{V^+}{\sqrt{Z_0}}\right]$$

or, $$[b] = ([Z] - Z_0[U])([Z] + Z_0[U])^{-1}[a] = [S][a] \quad (6.53h)$$

Therefore,

$$[S] = ([Z] - Z_0[U])([Z] + Z_0[U])^{-1} = ([Z]/Z_0 - [U])([Z]/Z_0 + [U])^{-1} \quad (6.53i)$$

Relation between [S] and [Y] Matrices

Similar to the relation between [S] and [Z] matrices, we can write

$$V_k = V_k^+ + V_k^- \quad (6.54a)$$

and $$I_k = V_k^+ Y_0 - V_k^- Y_0 = \sum_{j=1}^{n} V_j Y_{kj} \quad (6.54b)$$

Therefore, $$V_k^+ = \frac{1}{2Y_0}(V_k Y_0 + I_k) = \frac{1}{2Y_0}\left[V_k Y_0 + \sum_{j=1}^{n}(V_j Y_{kj})\right]; \quad (6.54c)$$

and $$V_k^- = \frac{1}{2Y_0}(V_k Y_0 - I_k) = \frac{1}{2Y_0}\left[V_k Y_0 - \sum_{j=1}^{n}(V_j Y_{kj})\right]; \quad (6.54d)$$

Hence, $$[V^+] = \frac{1}{2Y_0}([Y] + Y_0[U])[V];$$

and $$[V^-] = \frac{-1}{2Y_0}([Y] - Y_0[U])[V]; \quad (6.54e)$$

$$= \frac{1}{2Y_0}\left(Y_0[U] - 2Y_0[Y]\right)\left([Y] + Y_0[U]\right)^{-1}[V^+] \quad (6.54f)$$

or, $$[b] = (Y_0[U] - [Y])([Y] + Y_0[U])^{-1}[a] = [S][a] \quad (6.54g)$$

Therefore, $$[S] = ([U] - [Y]/Y_0)([U] + [Y]/Y_0)^{-1} \quad (6.54h)$$

Similarly, $[\bar{Z}] = (U - S)^{-1}(U + S)$; $\bar{Z} = Z/Z_0$, a normalized parameter (6.54i)

Thus, the following properties are found common for [S], [Z] and [Y]:

1. Number of elements are equal.
2. For reciprocal devices, both [Z] and [S] satisfy reciprocity properties.

$$Z_{ij} = Z_{ji},\ S_{ij} = S_{ji}$$

3. If [Z] is symmetrical, [S] is also symmetrical.
4. The following are the advantages of [S] over [Z] or [Y].
   (a) In microwave techniques, the source remains ideally constant in power, regardless of circuit changes. Besides frequency measurements, the only other possible measurement parameters are VSWR, power and phase. These are essentially measurements of b/a, $|a|^2$ and $|b|^2$. Such a direct correspondence is not possible with [Z] or [Y] representations.

(b) The unitary property of [*S*] helps a quick check of the power balance of lossless structures. No such immediate check is possible with [*Z*] or [*Y*].

(c) [*S*] is defined for a given set of reference planes only. If the reference planes are changed, the S-coefficients vary only in phase. This is not the case in [*Z*] or [*Y*], because voltage and current are functions of complex impedance and, therefore, both magnitude and phase change in [*Z*] and [*Y*].

When the characteristic impedances of two ports are different, the *S*-matrix of the junction can be derived as given in Examples 6.2 and 6.3.

Example 6.2 A shunt impedance *Z* is connected across a transmission line with characteristic impedance Z_0. Find the *S*-matrix of the junction.

Solution For *S*-parameters, two ports are considered matched. Let the output line be match terminated. So that $a_2 = 0$.

Therefore,

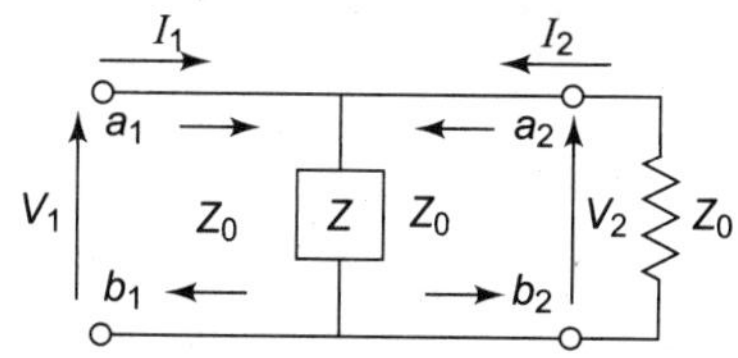

Fig. 6.6 Transmission lines for Example 6.2

$$b_1 = S_{11}a_1, \quad b_2 = S_{12}a_1$$

$$S_{11} = \frac{Y_0 - Y_{in}}{Y_0 + Y_{in}} = \frac{Y_0 - (Y_0 + Y)}{Y_0 + (Y_0 + Y)}$$

$$= \frac{-Y}{2Y_0 + Y} = \frac{-1}{1 + 2Z/Z_0} = S_{22} \text{ for symmetry}$$

Therefore, $2Z/Z_0 = -1/S_{11} - 1$

or $$Z = -\frac{Z_0}{2}(1/S_{11} + 1)$$

If $Z = j\omega L$, $$L = j\frac{Z_0}{2\omega}\left(1 + \frac{1}{S_n}\right)$$

Now for the pure shunt element, the transmitted wave amplitude (for $a_2 = 0$) can be expressed by $b_2 = a_1 + b_1 = a_1 + S_{11}\, a_1 = a_1\,(1 + S_{11})$.

Therefore,

$$S_{21} = \frac{b_2}{a_1} = 1 + S_{11} = S_{12} = \frac{2Y_0}{2Y_0 + Y} = \frac{2Z/Z_0}{1 + 2Z/Z_0}$$

$$\therefore \quad [S] = \frac{1}{2Y_0 + Y}\begin{bmatrix} -1 & 2Y_0 \\ 2Y_0 & -1 \end{bmatrix} = \frac{1}{1 + 2Z/Z_0}\begin{bmatrix} -1 & 2Z/Z_0 \\ 2Z/Z_0 & -1 \end{bmatrix}$$

Alternative

$$V_1 = Z I_1 + Z I_2; \quad [Z] = \begin{bmatrix} Z & Z \\ Z & Z \end{bmatrix}$$

$$V_2 = ZI_1 + ZI_2;$$

$$[S] = \left(\frac{1}{Z_0}[Z] - [U]\right)\left(\frac{1}{Z_0}[Z] + [U]\right)^{-1}$$

$$= [\bar{Z} + U - 2][\bar{Z} + U]^{-1} = [U] - 2[\bar{Z} + U]^{-1}$$

$$= \frac{1}{1+2\bar{Z}}\begin{bmatrix} -1 & 2\bar{Z} \\ 2\bar{Z} & -1 \end{bmatrix}; \quad S_{11} - S_{12} = -1$$

If $Y = jB$, S matrix is Unitary.

Example 6.3 Two transmission lines of characteristic impedance Z_1 and Z_2 are joined at plane pp'. Express S-parameters in terms of impedances when each line is matched terminated.

Solution The incident and scattered wave amplitude are related by $[b] = [S]\,[a]$.

(i) Since the output line is matched ($a_2 = 0$), the input impedance Z_{in} at the junction = Z_2 = load for line Z_1. Therefore,

$$S_{11} = \frac{Z_2 - Z_1}{Z_2 + Z_1} = \text{reflection coefficient on the input side.}$$

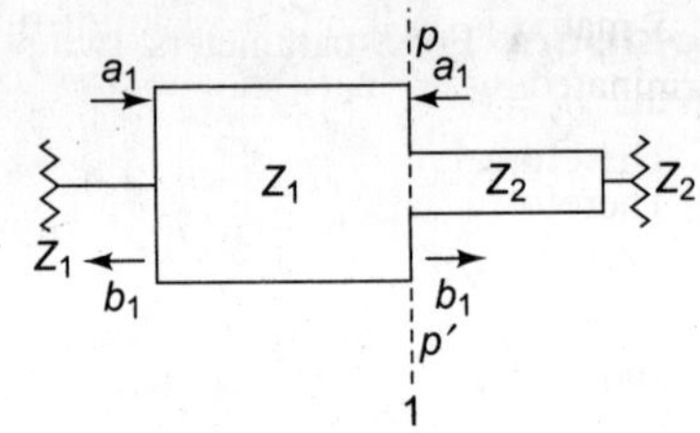

Fig. 6.7 Transmission lines for Example 6.3

(ii) Similarly, for symmetry, assuming input side is matched ($a_1 = 0$),

$$S_{22} = \frac{Z_1 - Z_2}{Z_1 + Z_2} = -S_{11}$$

(iii) In general,

$$b_1 = S_{11}\,a_1 + S_{12}\,a_2$$

$$b_2 = S_{21}\,a_1 + S_{22}\,a_2$$

With output line matched ($a_2 = 0$) for lossless line, Z_{in} at the junction = Z_2, a pure shunt element. Therefore, $b_1 = S_{11}\,a_1$. Due to different impedances of the two arms,

$$\sqrt{Z_2}\,b_2 = (a_1 + b_1)\sqrt{Z_1} = (a_1 + S_{11}\,a_1)\sqrt{Z_1} = a_1(1 + S_{11})\sqrt{Z_1}$$

or, $$S_{21} = b_2/a_1 = (1 + S_{11})\sqrt{Z_1/Z_2}$$

or, $$S_{21} = \left(1 + \frac{Z_2 - Z_1}{Z_2 + Z_1}\right)\sqrt{\frac{Z_1}{Z_2}} = \frac{2\sqrt{Z_2 Z_1}}{Z_2 + Z_1}$$

(iv) With the input line matched ($a_1 = 0$), $b_2 = a_2 S_{22}$

$$\sqrt{Z_1}\,b_1 = (a_2 + b_2)\sqrt{Z_2} = (a_2 + S_{22}\,a_2)\sqrt{Z_2} = [a_2\,(1 + S_{22})\sqrt{Z_2}\,]$$

or, $$S_{12} = b_1/a_2 = (1 + S_{22})\sqrt{Z_2/Z_1}$$

$$= \left(1 + \frac{Z_1 - Z_2}{Z_1 + Z_2}\right)\sqrt{\frac{Z_2}{Z_1}} = \frac{2\sqrt{Z_1 Z_2}}{Z_1 + Z_2}$$

Therefore, $$[S] = \begin{bmatrix} \dfrac{Z_2 - Z_1}{Z_2 + Z_1} & \dfrac{2\sqrt{Z_1 Z_2}}{Z_1 + Z_2} \\ \dfrac{2\sqrt{Z_1 Z_2}}{Z_1 + Z_2} & \dfrac{Z_1 - Z_2}{Z_1 + Z_2} \end{bmatrix}$$

Example 6.4 A series reactance $Z = jX$ is connected between two lines with different characteristic impedances Z_1 and Z_2. Find the S-matrix of the junction.

Solution The normalized voltages and power inputs at Port 1 and 2 are

$$a_1 = \frac{V_1^+}{\sqrt{Z_1}}, a_2 = \frac{V_2^+}{\sqrt{Z_2}}$$

$$P_1 = \frac{a_1^2}{2}, P_2 = \frac{a_2^2}{2}$$

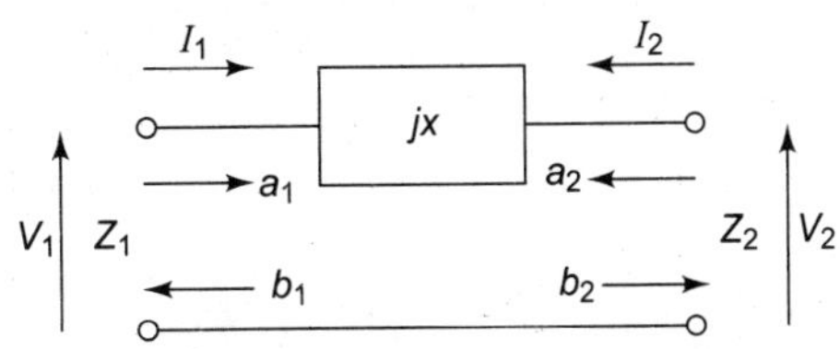

Fig. 6.8 Transmission lines for Example 6.4

S-matrix is determined assuming that each of the lines are terminated by its characteristic impedances.

Therefore, $$\frac{V_1^-}{V_1^+} = \frac{b_1}{a_1} = S_{11} = \frac{Z_{in} - Z_1}{Z_{in} + Z_1} = \frac{jX + Z_2 - Z_1}{jX + Z_2 + Z_1}$$

and $$\frac{V_2^-}{V_2^+} = \frac{b_2}{a_2} = S_{22} = \frac{jX + Z_1 - Z_2}{jX + Z_1 + Z_2}$$

Net input voltage for output line matched ($a_2 = 0$)

$$a_1 + b_1 = a_1 (1 + S_{11})$$

Input current $$I_1 = \frac{V_1^+}{Z_1} - \frac{V_1^-}{Z_1} = (1 - S_{11})\frac{V_1^+}{Z_1}$$

For the continuity of current in a lossless series element jX, with output port matched ($a_2 = 0$),

$$I_2 = -I_1 = -\frac{V_1^+}{Z_1}(1 - S_{11}) = -I_2^-$$

Also $$I_2^- = \frac{V_2^-}{Z_2}$$

$$\therefore \quad \frac{V_1^+}{Z_1}(1 - S_{11}) = \frac{V_2^-}{Z_2}$$

Thus, for symmetry

$$S_{21} = S_{12} = \frac{b_2}{a_2} = \frac{V_2^-}{\sqrt{Z_2}} \Big/ \frac{V_1^+}{\sqrt{Z_1}} = \frac{V_2^-}{V_1^+}\sqrt{\frac{Z_1}{Z_2}}$$

$$= \left(\frac{Z_2}{Z_1}\right)(1 - S_{11})\sqrt{\frac{Z_1}{Z_2}} = \sqrt{\frac{Z_2}{Z_1}}(1 - S_{11})$$

or, $$S_{21} = S_{12} = \sqrt{\frac{Z_2}{Z_1}} \frac{2\sqrt{Z_1 Z_2}}{Z_1 + Z_2 + jX}$$

If $Z_1 = Z_2 = Z_0$, $S_1 = S_{21} = (1 - S_{11}) = \dfrac{2Z_0}{2Z_0 + jX} = 1 - \dfrac{jX}{jX + 2Z_0}$

Example 6.5 A series reactance $Z = jX$ is inserted in a ∞ length of transmission line of characteristic impedance Z_0. Find S-parameters for the junction. Prove that $[S]$ is unitary.

Solution

Here, $Z = jX$

Let $Y = \dfrac{1}{jX}$

$\therefore$ $I_1 = YV_1 - YV_2$

$I_2 = -YV_1 + YV_2$

$$[Y] = \begin{bmatrix} Y & -Y \\ -Y & Y \end{bmatrix}$$

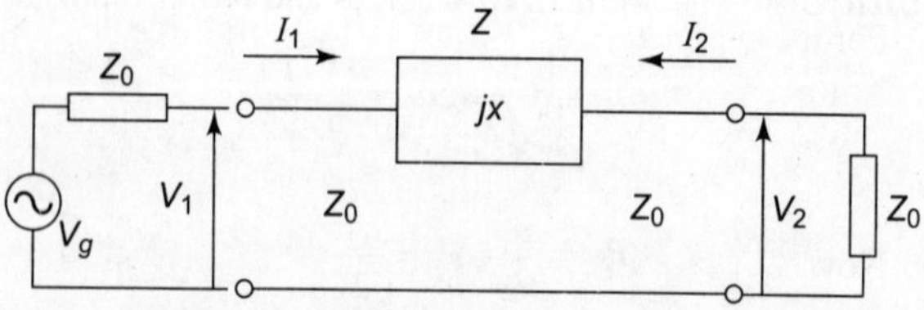

Fig. 6.9 Transmission lines for Example 6.5

Now $[S] = (U - Y/Y_0)(U + Y/Y_0)^{-1}$

$$= (U - \bar{Y})(U + \bar{Y})^{-1} = \frac{1}{1 + 2\bar{Y}}\begin{bmatrix} 1 & 2\bar{Y} \\ 2\bar{Y} & 1 \end{bmatrix}; S_{11} + S_{12} = 1$$

$$S_{11} = \Gamma_1\big|_{a_2=0} = \frac{(Z + Z_0) - Z_0}{(Z + Z_0) + Z_0} = \frac{Z}{Z + 2Z_0}$$

$$S_{22} = \Gamma_2\big|_{a_1=0} = \frac{Z}{Z + 2Z_0}$$

$$S_{12} = \frac{2V_1}{V_g} = \frac{2Z_0}{Z + 2Z_0} = 1 - \frac{Z}{Z + 2Z_0}$$

$$S_{21} = \frac{2V_2}{V_g} = \frac{2Z_0}{Z + 2Z_0} = 1 - \frac{Z}{Z + 2Z_0}$$

Hence,
$$[S] = \begin{bmatrix} \Gamma_1 & 1 - \Gamma_1 \\ 1 - \Gamma_1 : & \Gamma_1 \end{bmatrix}$$

If $Z = jX$, $|S_{11}|^2 + |S_{21}|^2 = 1$, $|S_{12}|^2 + |S_{22}|^2 = 1$, $S_{11}S_{12}^* + S_{21}S_{22}^* = 0$ and $S_{12}S_{11}^* + S_{22}S_{21}^* = 0$. The last equations satisfy unitary conditions. Therefore, S matrix of a series reactance (lossless) circuit is unitary.

Example 6.6 Determine the S-matrix of a 3 dB T-network attenuator shown in Fig. 6.10 terminated in a 50-ohm matched load with $Z_1 = 17.12$ ohms, $Z_2 = 141.78$ ohms.

Solution

$$\frac{Z_1}{2} = \frac{17.12}{2} = 8.56$$

$$S_{11} = \frac{V_1^-}{V_1^+}\bigg|_{V_2^+=0} = \frac{b_1}{a_1} = \frac{Z_{in1} - Z_0}{Z_{in1} + Z_0}$$

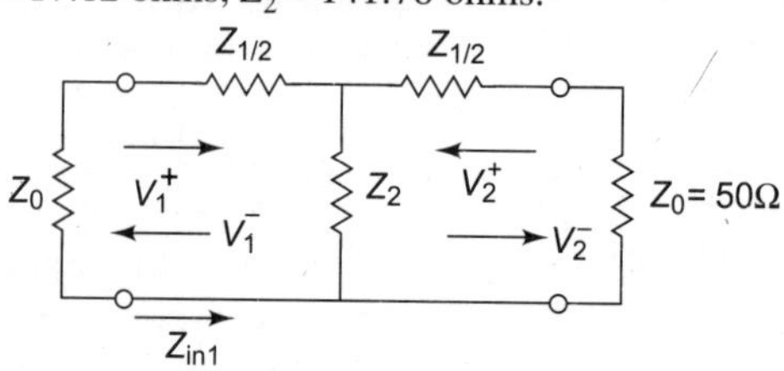

Fig. 6.10 Network attenuator for Example 6.6

$$Z_{in1} = 17.12/2 + 141.78 \times (17.12/2 + 50)/(141.78 + 17.12/2 + 50)$$
$$= 50 \text{ ohms}$$

$$\therefore \quad S_{11} = 0$$

For the symmetry, $S_{22} = S_{11} = 0$, $V_1^- = 0$, $V_2^+ = 0$

$$S_{21} = \left.\frac{V_2^-}{V_1^+}\right|_{V_2^+=0}$$

Now $\quad V_1 = V_1^+ + V_1^- = V_1^+$

$$V_2 = V_2^+ + V_2^- = V_2^-$$

Again, $\left(50 + \dfrac{17.12}{2}\right) \parallel 141.78 = \dfrac{141.78 \times 58.56}{141.78 + 58.56} = 41.44$

Therefore, $\quad V_2^- = V_2 = V_1\left(\dfrac{41.44}{41.44 + 8.56}\right)\left(\dfrac{50}{50 + 8.56}\right) = 0.707\ V_1$

$$\therefore \quad S_{21} = \frac{V_2^-}{V_1^+} = \frac{V_2^-}{V_1} = 0.707 \quad \text{or} \quad [S] = \begin{bmatrix} 0 & 0.707 \\ 0.707 & 0 \end{bmatrix}$$

Example 6.7 The S-parameters of a two-port network are given by

$$[S] = \begin{bmatrix} 0.2\angle 0^\circ & 0.6\angle 90^\circ \\ 0.6\angle 90^\circ & 0.1\angle 0^\circ \end{bmatrix}$$

(a) Prove that the network is reciprocal but not lossless. (b) Find the return loss at Port 1 when Port 2 is short-circuited.

Solution

(a) $$[S] = \begin{bmatrix} 0.2\angle 0^\circ & 0.6\angle 90^\circ \\ 0.6\angle 90^\circ & 0.1\angle 0^\circ \end{bmatrix}$$

Since $[S]$ is symmetric, the network is reciprocal. For lossless network, unitary property must be satisfied:

$$S_{1i} \cdot S_{1i}^* + S_{2i} \cdot S_{2i}^* = 1 = |S_{11}|^2 - |S_{21}|^2$$

$$S_{1k} \cdot S_{1i}^* + S_{2k} \cdot S_{2i}^* = 0$$

$$[S^*]\,[S]_t = [U]$$

Here,

$$S_{21} = S_{12} \text{ leads to } |S_{11}|^2 + |S_{12}|^2 = (0.2)^2 + (0.6)^2 = 0.04 + .36 = 0.40 < 1$$

Therefore, the network is not lossless.

(b) The input reflection coefficient for short-circuit termination at Port 2 ($\Gamma_2 = -1$)

$$\Gamma_1 = S_{11} + \frac{S_{12}^2 \Gamma_2}{1 - S_{22}\Gamma_2} = 0.2 + \frac{(0.6e^{j\pi/2})^2(-1)}{1 - 0.1 \times (-1)}$$

$$= 0.2 - \frac{(j0.6)^2}{1 + 0.1} = 0.2 + \frac{0.36}{1.1} = 0.5273$$

Therefore, return loss

$$= -20 \log |\Gamma_1| = -20 \log_{10} (0.5273)$$
$$= 5.56 \text{ dB}$$

Example 6.8 An ideal transformer of 1000 : 100 turns operates at 500 MHz and excited by voltage generator of impedance by 50 ohms. Find the *S*-matrix.

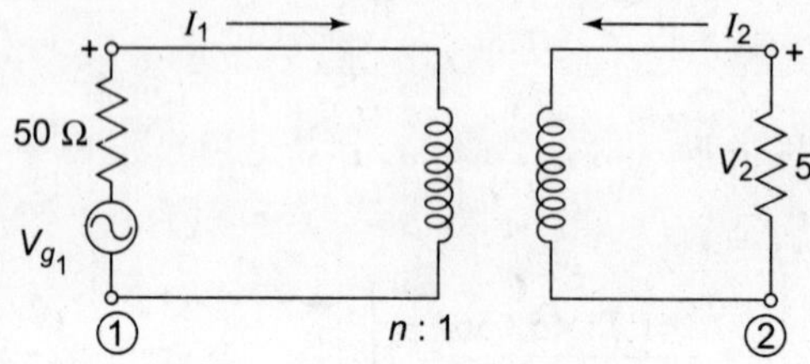

Fig. 6.11 Transformer for Example 6.8

Solution For an ideal transformer,

$$\frac{V_1}{V_2} = \frac{-I_2}{I_1} = n$$

$$\frac{V_1(-I_2)}{V_2(I_1)} = \frac{Z_1}{Z_2} = n^2$$

$$S_{11} = \Gamma\Big|_{a_2=0} = \frac{Z_0 n^2 - Z_0}{Z_0 n^2 + Z_0} = \frac{n^2-1}{n^2+1} = \frac{99}{101};$$

$$\frac{V_2}{V_{g_1}} = \frac{n^2}{n^2+1} = \frac{nV_2}{V_{g_1}}$$

$$\therefore \quad S_{21} = \frac{2V_2}{V_{g_1}}\Big|_{a_2=0} = \frac{2n}{n^2+1} = \frac{20}{101}$$

Similarly, $$S_{22} = \Gamma_2\Big|_{a_1=0} \frac{(Z_0/n^2) - Z_0}{(Z_0/n^2) + Z_0} = \frac{1-n^2}{1+n^2} = -\frac{99}{101}$$

$$\frac{V_2}{V_{g_2}} = \frac{Z_0/n^2}{(Z_0/n^2)+Z_0} = \frac{1}{1+n^2} = \frac{V_1/n}{V_{g_2}}$$

$$S_{12} = \frac{2V_1}{V_{g_2}}\Big|_{a_1=0} = \frac{2n}{n^2+1} = \frac{20}{101}$$

$$\therefore \quad [S] = \begin{bmatrix} \frac{99}{100} & \frac{20}{101} \\ \frac{20}{101} & -\frac{99}{101} \end{bmatrix}$$

Here we see that

1. $S_{11} \neq S_{22}$ since the network is not symmetrical.
2. $S_{12} = S_{21}$ (reciprocal)

3. Ideal transformer is lossless, since [S] is

$$\text{Unitary}\begin{cases} |S_{11}|^2+|S_{21}|^2=1 \\ |S_{12}|^2+|S_{22}|^2=1 \\ S_{11}\,S_{12}^*+S_{21}S_{22}^*=0 \\ S_{12}\,S_{11}^*+S_{22}S_{21}^*=0 \end{cases}$$

Example 6.9 A two-port network has the following S-matrix

$$[S]=\begin{bmatrix} 0.1\angle 0^\circ & 0.8\angle -45^\circ \\ 0.8\angle 45^\circ & 0.2\angle 0^\circ \end{bmatrix}$$

(a) Determine whether the network is reciprocal and/or lossless.
(b) If Port 2 in terminated with a matched load, calculate return loss at Port 1.
(c) If Port 2 is terminated with a short calculate return loss at Port 1.

Solution

(a) Since [S] is nonsymmetric, it is not a reciprocal network.

Here, $|S_{11}|^2+|S_{21}|^2=(0.1)^2+(0.8)^2=0.01+0.64=0.65\neq 1$

or, $\sum_{n=1}^{N=2}|S_{ni}|^2\neq 1$

Therefore, the network is not lossless.

(b) For match terminated port 2,

$$|S_{11}|=|\Gamma_1|=0.1$$

Therefore, return loss at port 1 is

$$RL_1=-20\log_{10}[\Gamma_1]=-20\log_{10}(0.1)$$
$$-20\text{ dB}$$

(c) For short circuited port 2, $b_2=-a_2$ (or $\Gamma_2=-1$)

The network equation is

$$[b]=[S]\,[a]$$

or, $$b_1=S_{11}a_1+S_{12}a_2=S_{11}a_1-S_{12}b_2 \qquad \text{(i)}$$
$$b_2=S_{21}a_1+S_{22}a_2=S_{21}a_1-S_{22}b_2 \qquad \text{(ii)}$$

From (ii),

$$b_2=\frac{S_{21}\,a_1}{1+S_{22}}\quad\text{or,}\quad\frac{b_2}{a_1}=\frac{S_{21}}{1+S_{22}}$$

Dividing (i) by a_1,

$$\frac{b_1}{a_1}=\Gamma_1=S_{11}-S_{12}\frac{b_2}{a_1}=S_{11}-S_{12}\cdot\frac{S_{21}}{1+S_{22}}$$

$$=0.1-\frac{(0.8\angle -45^\circ)(0.8\angle 45^\circ)}{1+0.2}$$

or $$\Gamma_1 = 0.1 - \frac{0.64}{1.2} = 0.1 - 0.53 = -0.43$$

Therefore, return loss at port 1 is

$$RL_1 = -20\log(|\Gamma_1|) = -20\log(0.43) = 7.33 \text{ dB}.$$

The above example shows that the reflection coefficient Γ_i at any port i is not equal to S_{ii}:

$$\Gamma_i \neq S_{ii}$$

Therefore, unless all ports are matched $S_{ii} \neq \Gamma_i$. Similarly, it can be shown that the transmission coefficient T_{ij} from port j to Port i is not equal to S_{ij} unless all other ports are matched. The scattering parameters of a linear network are exclusive properties of the network itself and are defined under the conditions that all ports are matched. If the terminations of the ports changed, S-parameters of the network do not change. Only the external characteristics of the ports, such as reflection coefficient at a given port and transmission coefficients between two ports are changed.

Example 6.10 (a) If a lossless two-port network is reciprocal, show that $|S_{21}|^2 = 1 - |S_{11}|^2$

(b) If the lossless two-port network is non-reciprocal, show that it is impossible to have unidirectional transmission, where $S_{12} = 0$ and $S_{21} \neq 0$.

Solution

(a) For reciprocal network,

$$[S] = \begin{bmatrix} S_{11} & S_{12} \\ S_{12} & S_{22} \end{bmatrix}; S_{12} = S_{21}$$

For lossless network

1st row: $|S_{11}|^2 + |S_{12}|^2 = 1$

$\therefore \quad |S_{12}|^2 = 1 - |S_{11}|^2$ or $|S_{21}|^2 = 1 - |S_{11}|^2$. Since $S_{21} = S_{12}$

(b) For lossless non-reciprocal network ($S_{12} \neq S_{21}$). If $S_{12} = 0 \neq S_{21}$

$$[S] = \begin{bmatrix} S_{11} & 0 \\ S_{21} & S_{22} \end{bmatrix}$$

First row: $|S_{11}|^2 = 1$; $\therefore |S_{11}| = 1$

First column: $|S_{11}|^2 + |S_{21}|^2 = 1$

$\therefore \quad |S_{21}|^2 = 1 - |S_{11}|^2 = 1 - 1 = 0$

$\therefore$ for lossless nonreciprocal network, $S_{12} = 0 = S_{21}$

Therefore, it is impossible to have unidirectional transmission.

Relations of Z, Y and $ABCD$ Parameters with S-Parameters 6.3.4

Relationship between ABCD Parameters and Z-Parameters

For the two-port network shown in Fig. 6.12, current–voltage equation in terms of $ABCD$ parameters is

$$\begin{bmatrix} V_1 \\ I_1 \end{bmatrix} = \begin{bmatrix} A & B \\ C & D \end{bmatrix} \begin{bmatrix} V_2 \\ I_2 \end{bmatrix} \tag{6.55a}$$

The V–I equation in term of Z-parameters is

$$\begin{bmatrix} V_1 \\ V_2 \end{bmatrix} = \begin{bmatrix} Z_{11} & Z_{12} \\ Z_{21} & Z_{22} \end{bmatrix} \begin{bmatrix} I_1 \\ -I_2 \end{bmatrix} \tag{6.55b}$$

Solving these two Eqs (6.55a) and (6.55b), we get

$$A = \left.\frac{V_1}{V_2}\right|_{I_2=0} = Z_{11}/Z_{21}$$

$$B = \left.\frac{V_1}{I_2}\right|_{V_2=0} = \frac{Z_{11}Z_{22} - Z_{12}Z_{21}}{Z_{21}}$$

$$C = \left.\frac{I_1}{V_2}\right|_{I_2=0} = \frac{1}{Z_{21}}$$

$$D = \left.\frac{I_1}{I_2}\right|_{V_2=0} = \frac{Z_{22}}{Z_{21}} \tag{6.55c}$$

For a reciprocal network, $Z_{12} = Z_{21}$ and $AD - BC = 1$. Equation (6.55c) show the relationship between *ABCD* parameters and *Z*-parameters.

Relationship between ABCD Parameters and Y-Parameters

For the two-port network of Fig. 6.12, the current–voltage equation in terms of *Y*-parameters:

$$I_1 = Y_{11}V_1 + Y_{12}V_2$$
$$I_2 = Y_{21}V_1 + Y_{22}V_2 \tag{6.56a}$$

Solving Eqs (6.55a) and (6.56b), we get the relationship between *ABCD* and *Y*-parameters:

$$A = -\frac{Y_{22}}{Y_{21}}$$

$$B = -\frac{1}{Y_{21}}$$

$$C = \frac{-(Y_{11}Y_{22} - Y_{12}Y_{21})}{Y_{21}}$$

$$D = \frac{-Y_{11}}{Y_{21}} \tag{6.56b}$$

Relationship between [S] and [ABCD]

Let the two-port network have current and voltage configurations as shown in Fig. 6.12. The steady state voltage and current are expressed by

$$V = V^+ + V^-; \qquad a = \frac{V^+}{\sqrt{Z_0}}; \tag{6.57a}$$

$$I = \frac{V^+}{Z_0} - \frac{V^-}{Z_0}; \quad b = \frac{V^-}{\sqrt{Z_0}}; \tag{6.57b}$$

Therefore, $V = \sqrt{Z_0}(a+b)$ (6.57c)

$$I = \frac{1}{\sqrt{Z_0}}(a-b) \tag{6.57d}$$

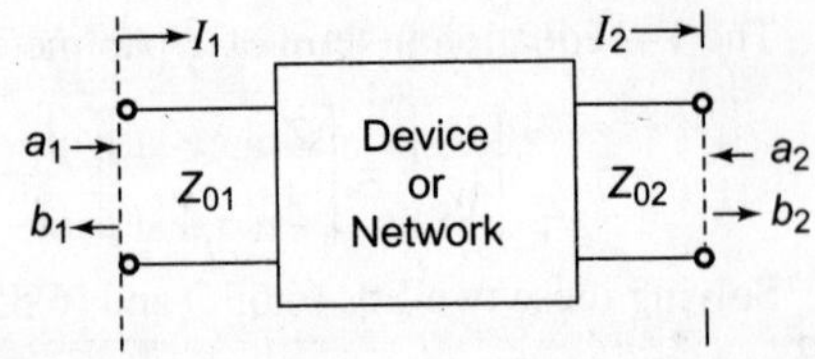

Fig. 6.12 Two-port network between two lines of unequal impedance Z_{01} and Z_{02}

Now for the two-port network shown

$$V_1 = AV_2 + BI_2$$

and $I_1 = CV_2 + DI_2$ (6.57e)

Considering direction of I_2 in Fig. 6.12,

$$\text{or, } \sqrt{Z_{01}}\,(a_1+b_1) = A\sqrt{Z_{02}}\,(a_2+b_2)+\frac{B}{\sqrt{Z_{02}}}(b_2-a_2) \tag{6.57f}$$

$$\text{and } \frac{1}{\sqrt{Z_{01}}}(a_1-b_1) = C\sqrt{Z_{02}}\,(a_2+b_2)+\frac{D}{\sqrt{Z_{02}}}(b_2-a_2) \tag{6.57g}$$

where Z_{01} and Z_{02} are the characteristic impedances of input and output lines, respectively.

Therefore,

$$a_1+b_1 = A\sqrt{\frac{Z_{02}}{Z_{01}}}(a_2+b_2)+\frac{B}{\sqrt{Z_{01}Z_{02}}}(b_2-a_2) \tag{6.58a}$$

$$a_1-b_1 = C\sqrt{Z_{01}Z_{02}}\,(a_2+b_2)+D\sqrt{\frac{Z_{01}}{Z_{02}}}(b_2-a_2) \tag{6.58b}$$

Using following normalizations,

$$A\sqrt{\frac{Z_{02}}{Z_{01}}} = \bar{A}$$

$$\frac{B}{\sqrt{Z_{01}Z_{02}}} = \bar{B},\ C\sqrt{Z_{01}Z_{02}} = \bar{C} \text{ and } D\sqrt{\frac{Z_{01}}{Z_{02}}} = \bar{D} \tag{6.58c}$$

we can express

$$a_1+b_1 = \bar{A}(a_2+b_2)+\bar{B}(b_2-a_2) \tag{6.58d}$$

$$a_1-b_1 = \bar{C}(a_2+b_2)+\bar{D}(b_2-a_2) \tag{6.58e}$$

The S-parameter equation is for the network given by

$$b_1 = S_{11}a_1 + S_{12}a_2 \tag{6.58f}$$

$$b_2 = S_{21}a_1 + S_{22}a_2 \tag{6.58g}$$

Solving for Eqs (6.58d – 6.58g),

$$S_{11} = \frac{\bar{A}+\bar{B}-\bar{C}-\bar{D}}{\bar{A}+\bar{B}+\bar{C}+\bar{D}} \tag{6.58h}$$

$$S_{22} = \frac{-\bar{A}+\bar{B}-\bar{C}+\bar{D}}{\bar{A}+\bar{B}+\bar{C}+\bar{D}} \tag{6.58i}$$

$$S_{12} = \frac{2(\bar{A}\bar{D}-\bar{B}\bar{C})}{\bar{A}+\bar{B}+\bar{C}+\bar{D}} \tag{6.58j}$$

$$S_{21} = \frac{2}{\bar{A}+\bar{B}+\bar{C}+\bar{D}} \tag{6.58k}$$

(i) If the network is reciprocal,

$$S_{12} = S_{21}, \text{ so that } \bar{A}\bar{D} - \bar{B}\bar{C} = 1$$

$$\text{and } S_{12} = S_{21} = \frac{2}{\bar{A}+\bar{B}+\bar{C}+\bar{D}} \tag{6.59a}$$

(ii) If the network is symmetrical,

$$S_{11} = S_{22} \quad \text{so that} \quad \bar{A} = \bar{D}$$

$$\text{and } S_{11} = S_{22} = \frac{\bar{B}-\bar{C}}{2\bar{A}+\bar{B}+\bar{C}} \tag{6.59b}$$

(iii) If the symmetrical and reciprocat is placed in a common transmission line of characteristic impedance $Z_0 = Z_{01} = Z_{02}$, then

$$\bar{A} = A,\ \bar{B} = B/Z_0,\ \bar{C} = CZ_0,\ \bar{D} = D$$

and

$$S_{11} = \frac{A + B/Z_0 - CZ_0 + D}{A + B/Z_0 + CZ_0 + D} = S_{22} \tag{6.59c}$$

$$S_{21} = \frac{2}{A + B/Z_0 + CZ_0 + D} = S_{12} \tag{6.59d}$$

Example 6.11 Find *ABCD* parameters of the shunt admittance *Y* in the transmission line shown in Fig. 6.13.

Solution

$$\begin{bmatrix} V_1 \\ I_1 \end{bmatrix} = \begin{bmatrix} A & B \\ C & D \end{bmatrix} \begin{bmatrix} V_2 \\ I_2 \end{bmatrix}$$

Here, $V_1 = V_2$; $\therefore$ $A = 1$, $B = 0$, for the common node 1

and $V_1 = V_2 = \dfrac{I_1 - I_2}{Y}$, or $I_1 = V_2 Y + I_2$

$\therefore$ $C = Y$, $D = 1$ from (6.55c)

Hence, $[ABCD] = \begin{bmatrix} 1 & 0 \\ Y & 1 \end{bmatrix}$

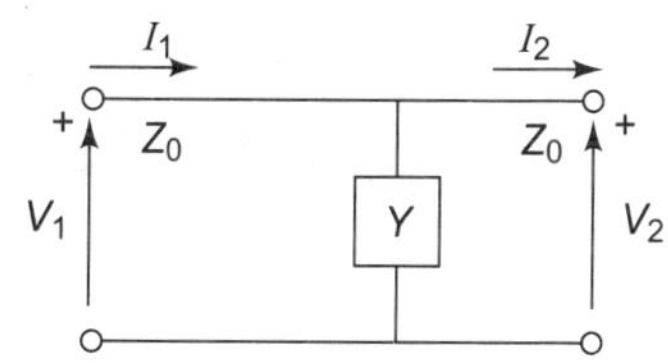

Fig. 6.13 Transmission lines for Example 6.11

Example 6.12 Find the *ABCD* parameters of the series reactance *jX* placed in the following transmission line.

Solution For the current continuity, $I_1 = I_2$.
Therefore, $C = 0$, $D = 1$ from (6.55c)
Also,

$$V_1 = I_2\,(jX) + V_2 \text{ leads to } A = 1, B = jX.$$

$$\therefore \quad [ABCD] = \begin{bmatrix} 1 & jX \\ 0 & 1 \end{bmatrix}$$

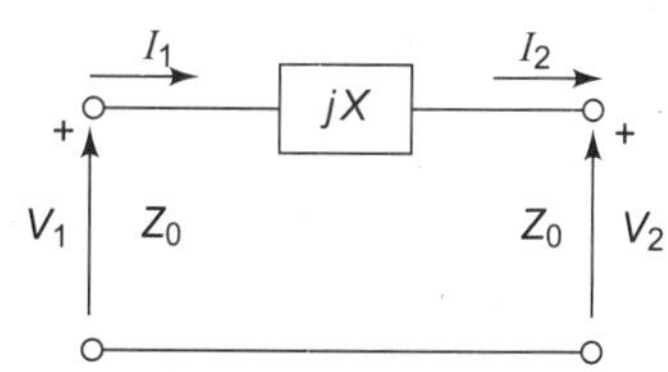

Fig. 6.14 Transmission line for Example 6.12

Example 6.13 In a lossless transmission line of characteristic impedance Z_0, a series reactance jX and two shunt susceptances jB are placed as shown in Fig. 6.15. Find the $ABCD$ parameters for the network.

Solution Here, the network is symmetric. The $ABCD$ matrix of the network is the result of cascaded $ABCD$ matrices for each element determined in Examples 6.11 and 6.12.

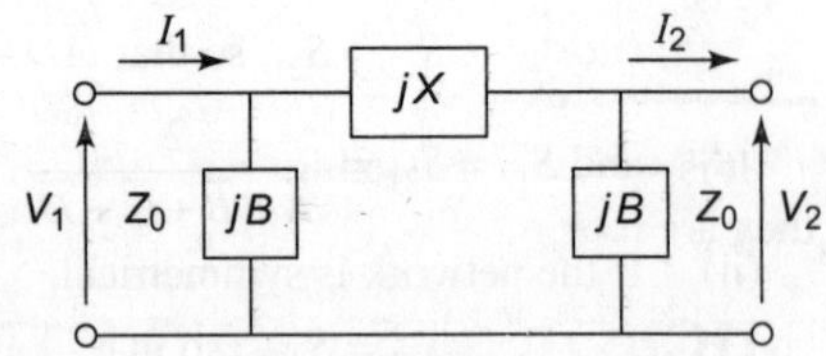

Fig. 6.15 Transmission lines for Example 6.13

$$\therefore \quad [ABCD] = \begin{bmatrix} 1 & 0 \\ jB & 1 \end{bmatrix} \begin{bmatrix} 1 & jX \\ 0 & 1 \end{bmatrix} \begin{bmatrix} 1 & 0 \\ jB & 1 \end{bmatrix}$$

$$= \begin{bmatrix} 1 - BX & jX \\ jB(2 - BX) & (1 - BX) \end{bmatrix}$$

Example 6.14 In a lossless transmission line, a symmetric T-junction is formed as shown in Fig. 6.16. Find the resultant $ABCD$ matrix.

Solution For the cascaded $ABCD$ matrices formed by individual elements,

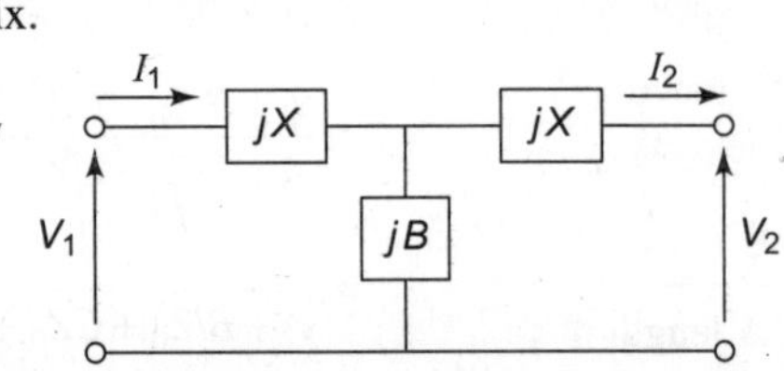

Fig. 6.16 Transmission lines for Example 6.14

$$\therefore \quad [ABCD] = \begin{bmatrix} 1 & jX \\ 0 & 1 \end{bmatrix} \begin{bmatrix} 1 & 0 \\ jB & 1 \end{bmatrix} \begin{bmatrix} 1 & jX \\ 0 & 1 \end{bmatrix}$$

$$= \begin{bmatrix} 1 - BX & jX(2 - BX) \\ jB & 1 - BX \end{bmatrix}$$

Example 6.15 Find the $ABCD$ matrix of a length of lossless microstrip transmission line terminated in its characteristic impedance Z_0.

Solution

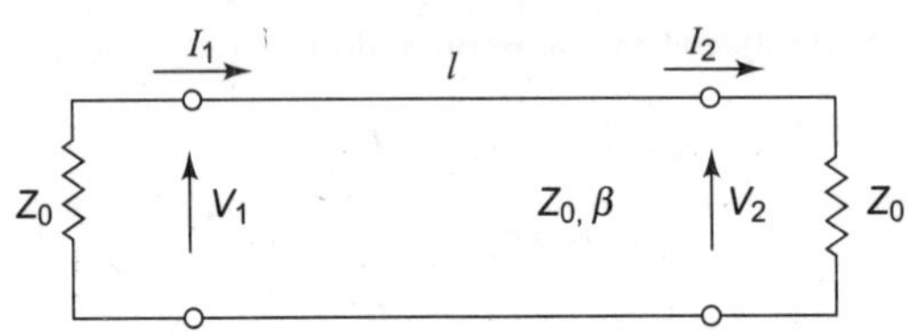

Fig. 6.17 Transmission lines for Example 6.15

Let length = l and $\gamma = j\beta$

With the chosen current and voltage in the attached figure, the transmission line equation from (3.18b) and (6.55a) can be expressed as

$$\therefore \quad V_1 = V_2 \cosh(\gamma l) + I_2 Z_0 \sinh(\gamma l) = AV_2 + BI_2$$

$$I_1 = \frac{V_2}{Z_0} \sinh(\gamma l) + I_2 \cosh(\gamma l) = CV_2 + DI_2$$

Now comparing

$$\begin{bmatrix} A & B \\ C & D \end{bmatrix} = \begin{bmatrix} \cos h\gamma l & \sin h\gamma l \\ \dfrac{\sin h\gamma l}{Z_0} & \cos h\gamma l \end{bmatrix}$$

For lossless line, $\gamma = \alpha + j\beta = j\beta$, $\alpha = 0$ so that

$$\begin{bmatrix} A & B \\ C & D \end{bmatrix} = \begin{bmatrix} \cos\beta l & jZ_0 \sin\beta l \\ \dfrac{j\sin\beta l}{Z_0} & \cos\beta l \end{bmatrix}$$

It is seen that the condition $AD–BC = 1$ holds for all the above five circuits (Example 6.11–6.15) because they are reciprocals.

MICROWAVE PASSIVE COMPONENTS 6.4

Microwave passive devices and components are designed using sections of co-axial lines, waveguides, strip lines and microstrip lines for use in both laboratory and in microwave communication and radar systems. These components can be considered as one-port or multiport networks characterized by the basic parameters like the VSWR, reflection coefficient, and various losses defined by Eqs (6.21)–(6.24), under output matched conditions. In this section, the basic operating principles of a number of most commonly used devices such as line sections, connectors, terminators, attenuators, phase shifters, directional couplers, power dividers, T-junctions, hybrids, etc., are described. In addition to coaxial and waveguide versions, important passive devices in MIC configurations are also included.

Coaxial Cables 6.4.1

A length of coaxial cable is used for interconnecting several microwave components. The theory of coaxial lines was described in Chapter 3. In this section some practical aspects of these lines are described. The outer conductor of the coaxial line is used to guide the signal through TEM mode and shields the external or internal signal leakage through it. The standard characteristic impedance of these cables are 50 ohms and 75 ohms. There are three basic types of coaxial cables with increasing order shielding, i.e., flexible, semi-rigid and rigid. Flexible coaxial cables use low-loss solid or foam polyethylene dielectrics. The outer single braid or double braid of the flexible cable is constructed for electromagnetic shielding by using knitted metal wire mesh as shown in Fig. 6.18. Rigid cables have air dielectric and conductors are supported by small dielectric spacers such that they do not produce significant discontinuities to the signal flow. Semi-rigid cables have solid dielectric and use a thin copper outer conductor so that it can be bent for convenient routing. Coaxial cables are used in the frequency range from dc to microwaves. Since the attenuation in these cables increases with frequency, the upper frequency of operation is limited.

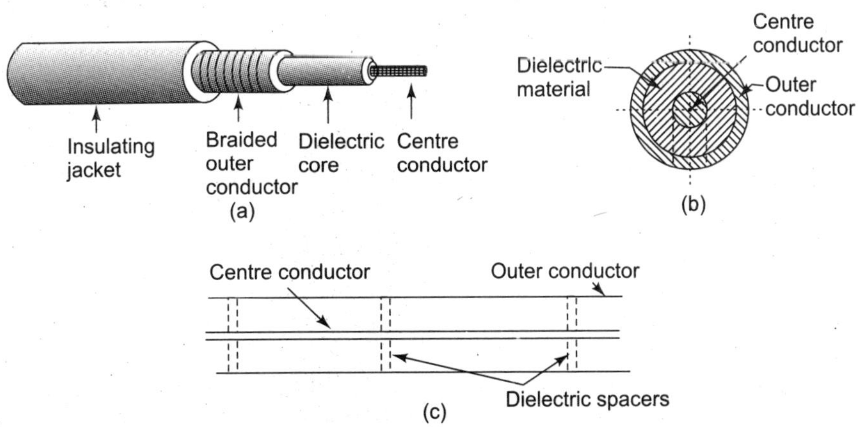

Fig. 6.18 Coaxial cable: (a) flexible (b) semi-rigid (c) rigid

The shielding effectiveness of the outer conductor is expressed in terms of transfer impedance Z_T of the cable, which is defined as

$$Z_T = \frac{\text{Longitudinal voltage } V_i \text{ induced per unit length on one side of the shield (outside)}}{\text{The leakage current } I_s \text{ flowing on the other side (inside) of the shield}} = \frac{V_i}{I_s} \quad (6.60)$$

If both ends of a short cable are terminated into matched loads, as shown in Fig. 6.19, the load voltage is

$$V_0 = \frac{V_i}{2} = \frac{I_s Z_T}{2} \quad (6.61)$$

Therefore,

$$Z_T = 2V_0/I_s \quad (6.62)$$

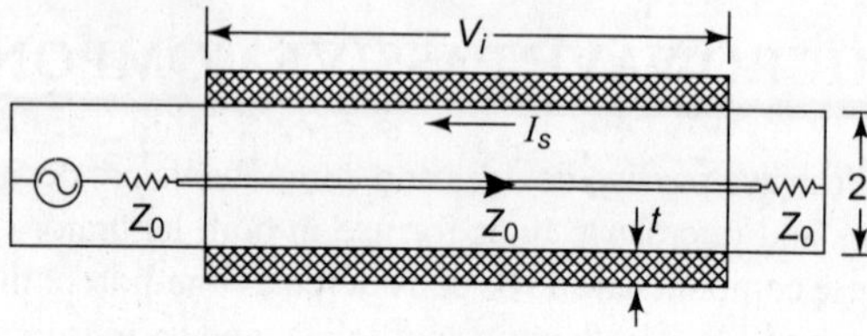

Fig. 6.19 Transfer impedance of a coaxial cable

For a thin-walled tubular shield, the transfer impedance is given by

$$Z_T = \frac{1}{2\pi b \sigma t} \frac{(1+j)t/\delta}{\sinh\left[(1+j)t/\delta\right]} \quad \text{for } t << b << \delta \quad (6.63)$$

where b is the inner radius of the shield, t is the shield wall thickness, σ is the conductivity of the shield, δ is the skin depth in the shield. For a better shielding, Z_T should be smaller.

Some standard coaxial cables are given in Table 6.1 with their *radio guide* (RG) and *universal* (U) numbers and specifications.

Table 6.1 Some standard coaxial cables

Number RG	Z_0 *ohms*	*Inner and outer conductor*
9/U	50	Silver-plated copper
11/U	75	Tinned copper
58A/U	50	Tinned copper
59/U	75	Copper
141A/U	50	Silver-plated copper
179B/U	75	Silver-plated copper
214/U	50	Silver-plated copper

Coaxial Connectors and Adapters 6.4.2

Coaxial cables are terminated or connected to other cables and components by means of shielded standard connectors. The outer shield makes a 360 degree, extremely low-impedance joint to maintain shielding integrity. These connectors are of various types depending on the frequency range and the cable diameter. Commonly used microwave connectors are type *N* (male/female), BNC (male/female), TNC (male/female), APC (sexless), etc. Adapters, having different connectors at the two ends, are also made for interconnection between two different ports in a microwave system. The basic schematic diagrams of these connectors and adapters are shown in Fig. 6.20. The type *N* (Navy) connector is a 50 and 75-ohm connector which was designed for military system applications during World War II. This is suitable for flexible or rigid cables in the frequency range of 1–18 GHz. The BNC (Bayonet Navy Connector) is suitable for 0.25 inch, 50-ohm or

75-ohm flexible cables used up to 1 GHz. The TNC (Threaded Navy Connector) is like BNC, except that, the outer conductor has a thread to make firm contact in the mating surface to minimise radiation leakage at higher frequencies. These connectors are used up to 12 GHz.

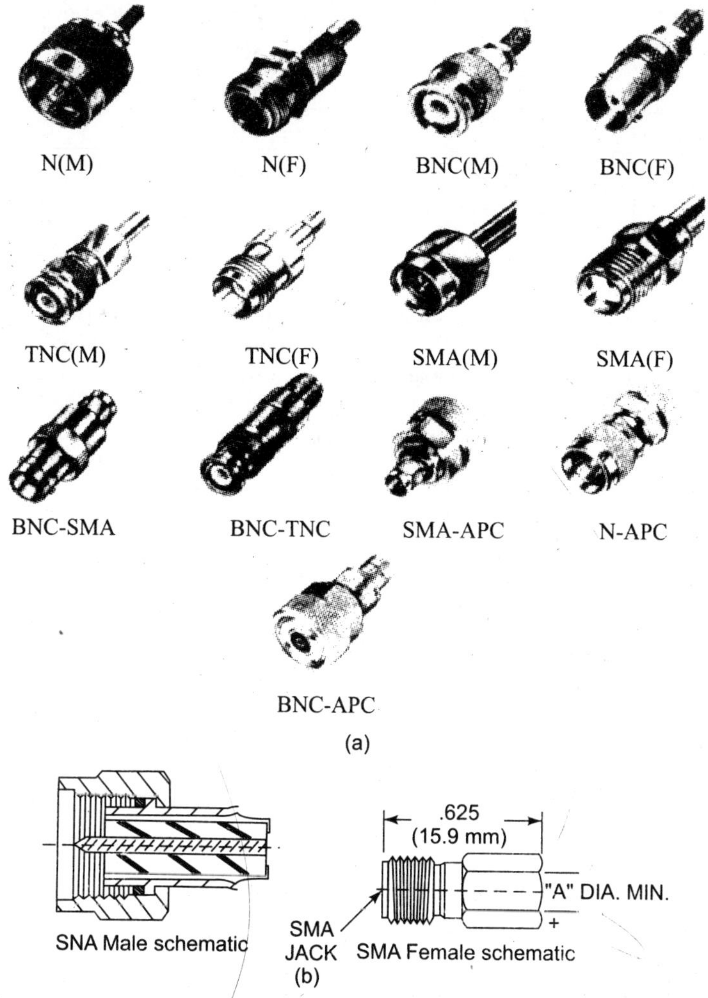

Fig. 6.20 Coaxial connectors and adapters

The SMA (Sub-Miniature A) connectors are used for thin flexible or semirigid cables. The higher frequency is limited to 24 GHz because of generation of higher-order modes beyond this limit. All the above connectors can be of male or female configurations except the APC-7(Amphenol Precision Connector-7 mm) which provides coupling without male or female configurations. The APC-7 is a very accurate 50-ohm, low VSWR connector which can operate up to 18 GHz.

Another APC-3.5 connector is a high-precision 50-ohm, low VSWR connector which can be either the male or female and can operate up to 34 GHz. It can mate with the oppositely sexed SM connector. Table 6.2 shows the type, dielectric in mating space and impedance of some of the above standard connectors.

Table 6.2 *Mating connectors*

Type	*Sex*	*Dielectric in mating space*	*Impedance (ohms)*
N	M/F	Air	50/75
BNC	M/F	Solid	50/75
TNC	M/F	Solid	50/75
SMA	M/F	Solid	50
APC-7	Sexless	Air	50
APC-3.5	Sexed	Air	50

Waveguide Sections 6.4.3

Waveguide sections are used for low-loss transmission, change of direction and polarization of the signal. Sections may be straight with rectangular or circular cross sections, rectangular twisted sections in either **E** or **H** planes, rectangular bends, circular flexible section, rectangular or circular tapered section, and transition from circular to rectangular cross sections. Some of them are as shown in Fig. 6.21. The standard waveguides are made of brass, bronze or aluminium. For higher frequency bands, the surface is finished with silver coatings to minimise ohmic loss. The *S*-matrix for ideal sections are

$$[S] = \begin{bmatrix} 0 & 1 \\ 1 & 0 \end{bmatrix} \tag{6.64}$$

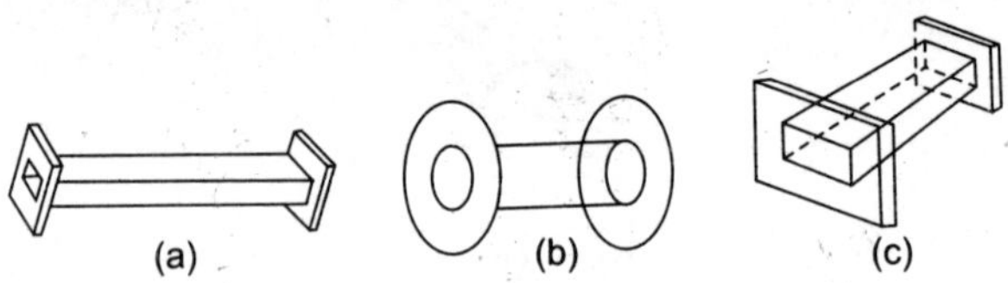

Fig. 6.21 Waveguide sections: (a) rectangular (b) circular (c) rectangular taper

The general theory of waveguides is given in Chapter 3. This section describes some practical aspects of rectangular and circular waveguides for the construction of waveguide components.

Rectangular Waveguide Sections

It has been shown that attenuation due to finite conductivity of waveguide walls decreases with increase of frequency at $f > f_c$ to a minimum and increases again for a given b/a ratio. The attenuation also increases when b/a reduces. The design requirements of a rectangular waveguide for the dominant mode are as follows:

1. Energy transmission through a single mode TE_{10} (dominant)
2. The breakdown power should be higher than the transmitted power
3. Low attenuation constant

For a waveguide of dimension $b/a < 0.5$, the successive ascending-order modes are TE_{10}, TE_{20}, TE_{01}, etc. Therefore, the dimensions of the cross section which ensure only the dominant mode propagation are determined as follows.

$$\begin{aligned} &\lambda_{max} < \lambda_c = 2a \text{ of } TE_{10} \text{ dominant mode} \\ &\lambda_{min} > \lambda_c = a \text{ of } TE_{20} \text{ mode} \\ &\lambda_{min} > \lambda_c = 2b \text{ of } TE_{01} \text{ mode} \end{aligned} \tag{6.65}$$

Hence, the dimensions for dominant mode rectangular waveguides are to be selected according to

$$0 < b < \lambda/2,\ \lambda/2 < a < \lambda \tag{6.66}$$

To provide adequate power handling capability, usually the dimensions are chosen as $b = 0.3\lambda$ to 0.4λ, $a = 0.7\lambda$ to 0.8λ. For a standard X-band guide $a = 2.286$ am, $b = 1.016$ cm.

Circular Waveguide Sections

Circular waveguide sections are used in some waveguide components where circularly polarized waves are involved. Both TE and TM modes can be excited depending on the requirements. The dominant mode in a circular waveguide is the TE_{11} mode. Out of the multitude of possible modes in circular waveguides, the modes of highest practical interest are TE_{11}, TE_{01} and TM_{01}. The field configurations of these modes are shown in Fig. 6.22.

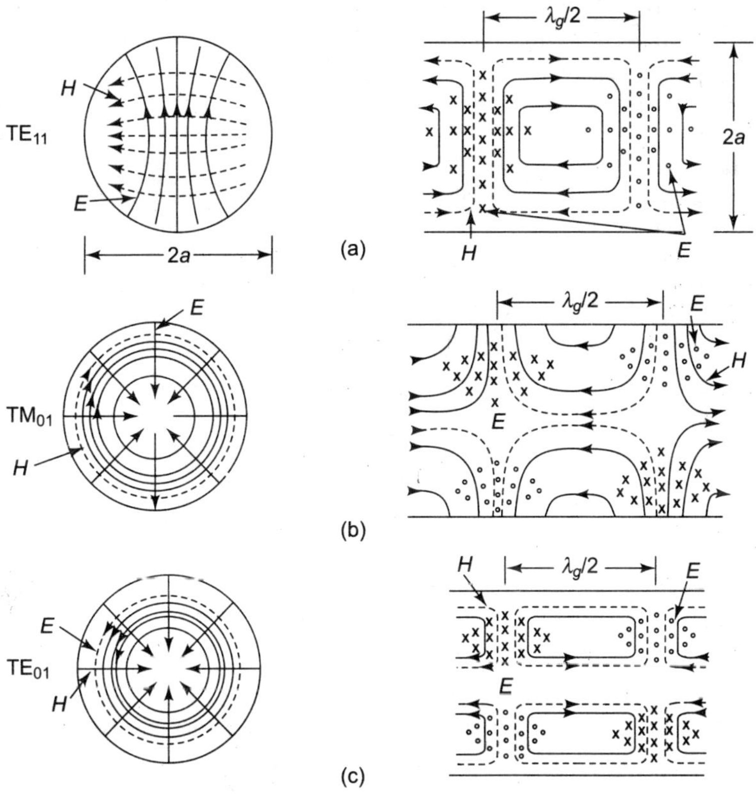

Fig. 6.22 Circular waveguide modes (a) TE_{11} (b) TM_{01} (c) TE_{01}

The electric currents on the waveguide walls for the TM_{01} mode are purely longitudinal, whereas for the TE_{01} mode it flows along closed circular paths and does not have longitudinal components.

The attenuation in the circular waveguide for TE_{01} mode is very low and decreases with frequency as $f^{-3/2}$. Therefore, although TE_{01} mode is not dominant, it is used for long telecommunication lines and also for resonant cavities displaying very high Q wavemeters. Figure 6.23 shows the distribution of the cut-off wavelengths of a circular waveguide for a few modes.

Selection of the guide radius is made for propagation of a single dominant TE_{11} mode such that

$$a \le \frac{1.841}{2\pi f_{c(\text{Hz})}\sqrt{\mu\varepsilon}} \qquad f \ge f_{c\text{TE}_{11}} = \frac{1.841}{2\pi a\sqrt{\mu\varepsilon}} \tag{6.67}$$

For other higher modes, the dimensions are selected based on the corresponding root of Bessel functions as described in Chapter 3 and by an appropriate excitation method.

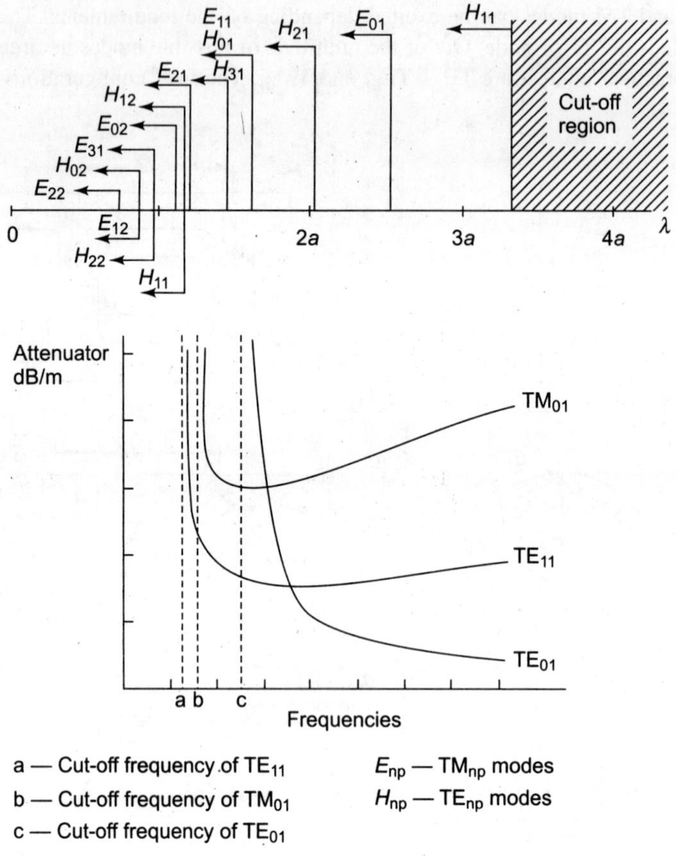

Fig. 6.23 Distribution of cut-off wavelength in a circular waveguide and attenuation characteristics

Waveguide Flanges 6.4.4

Waveguide sections are fitted with flanges as shown in Fig. 6.24, which are connected using nut–bolt arrangement with other sections or waveguide devices. In order to avoid leakage of the high-power signal between two flange joints, choke flanges are sometimes used which consist of a circular groove of depth *d* in the flange such that the depth *d* plus the distance to the waveguide joint on the wide wall is half the guide wavelength. This arrangement makes an ideally short-circuited half wavelength line providing zero resistance contact between the inner walls of the two waveguide sections at the joint. Inadequate contact at the joint result in power loss due to leakage, Joule heating, partial reflection of waves and even burn-out of the contact for high-power applications.

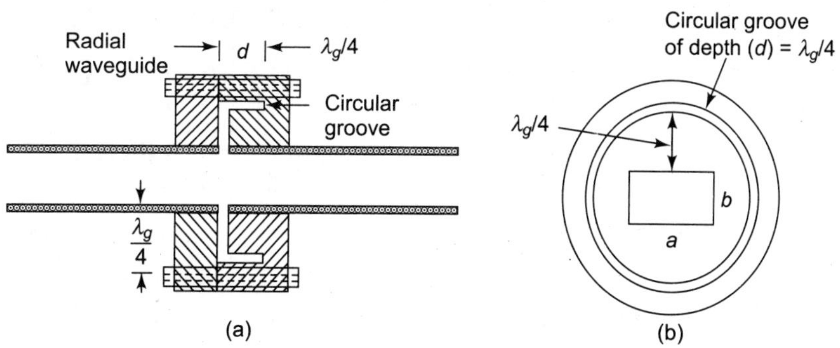

Fig. 6.24 Choke flange: (a) side view (b) cross section

At the joint, the first waveguide is equipped with a choke flange and the second one with a plane flange. The choke flange is frequency selective. At the edge of a 10–15% bandwidth, VSWR of the order of 1.02 to 1.05 can be achieved. Choke connections are used in microwave oscillator and amplifier tubes and coaxial rotary joints.

Rotary Joints 6.4.5

Rotary joints are used to change the direction of microwave propagation between two guides by rotating one with respect to the other. Both coaxial line and circular waveguide versions are used. In radar applications a fixed waveguide is connected to a horn antenna by a rotary joint for feeding a paraboloid reflector which rotates for tracking. The most important design considerations of rotary joints are low reflection, negligible power leakage through, and mechanical strength at the joint. Figure 6.25 shows coaxial line and waveguide rotary joints.

The sliding joint is designed by incorporating half wavelength short-circuited lines or chokes in such a manner that the actual point of sliding contact is at a current minimum with zero impedance.

In the waveguide rotary joint, input and output rectangular waveguides operated in the dominant TE_{10} mode are connected through a section of circular waveguide operated in ϕ-symmetric TM_{01} mode. The diameter of the circular waveguide is selected in such a way that modes higher than TM_{01} cannot propagate. The lowest dominant mode TE_{11} is suppressed by a ring filter. Moreover, the diameter of the circular waveguide and the end length h are chosen such that h is an odd multiple of the quarter guide wavelength for the TE_{11} mode, but an even multiple of the quarter guide wavelength for the TM_{01} mode. Then the series junction between the rectangular and circular guides produces infinite input impedance of the section h at the junction for the TE_{11} mode. This establishes weak TE_{11} mode in the circular guide. The section h effectively produces zero input impedance at the junction for the TM_{01} mode to be strongly excited in the circular guide. The distance between the input and output rectangular guides is selected such that resonance in the TE_{11} mode is avoided.

The junctions between the rectangular and the circular guides are matched by using a reactive tuner, such as inductive irises in the rectangular waveguide side. Such rotary joints produce a good VSWR < 1.1. h is odd multiple of $\lambda_g/4$ for TE_{11} mode and even multiple of $\lambda_g/4$ for TM_{01} mode.

Strip and Microstrip Line Sections 6.4.6

Strip and microstrip line sections are used for interconnection and impedance matching in microwave integrated circuits. Impedance and attenuation characteristics of such lines are described in Chapter 3. In this section several forms of discontinuity are described for circuit design. Some of these sections are described below.

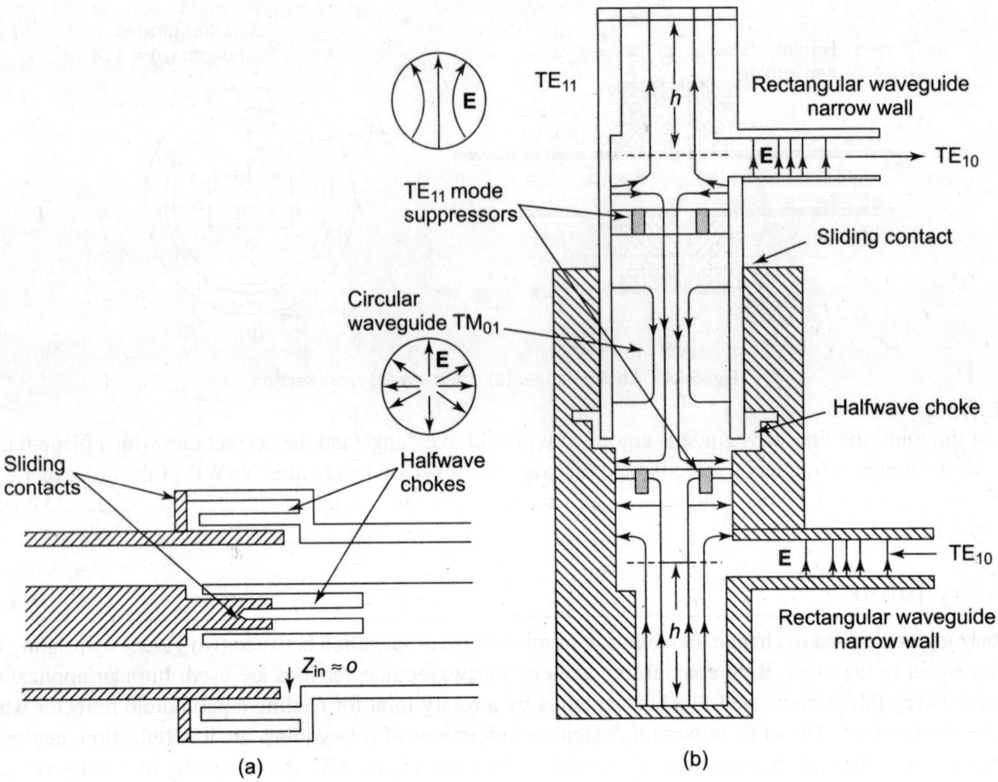

Fig. 6.25 Rotary joints: (a) coaxial (b) waveguide

Open-Ended Section

The following three phenomena are associated with a physical open end of a microstrip line:

(a) Fringing fields extended beyond the physical end of the strip

(b) Radiation from the open end due to discontinuity of current

(c) Launching of surface waves from the open end of the strip which disturbs the quasi-TEM mode of propagations

The phenomenon (a) can be accounted for by modelling the fringing fields by means of an equivalent shunt capacitance C_f. This capacitance is also equivalent to an extra length $\Delta l << \lambda_g$ of the same line with electrically open circuit end as shown in Fig. 6.26.

For a given structure the end effect length is given by

$$\Delta l = 0.412h\left(\frac{\varepsilon_{\text{eff}} + 0.3}{\varepsilon_{\text{eff}} - 0.258}\right)\left(\frac{w/h + 0.262}{w/h + 0.813}\right) \tag{6.68}$$

Short Circuit

In a microstrip line, a short circuit can be achieved by a short-circuiting post or called via between the strip and the ground plane as shown in Fig. 6.27, place through the substrate.

At frequencies below 3 GHz, this shorting post could be a short metallic wire bonded to the microstrip and the ground plane.

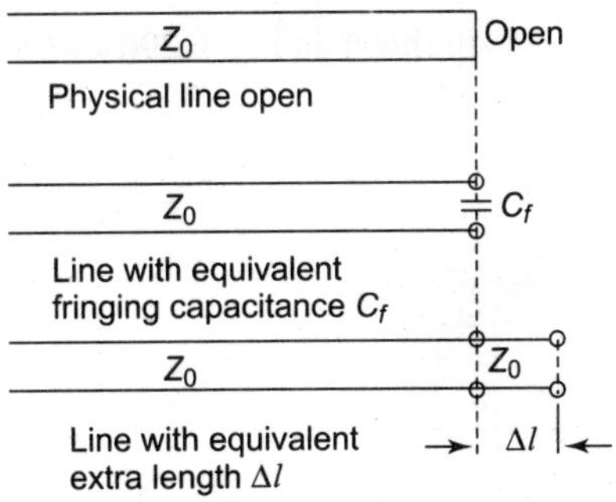

Fig. 6.26 Equivalent end effect length

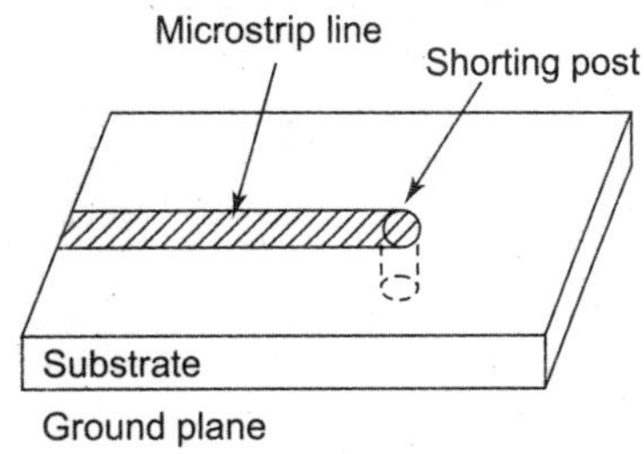

Fig. 6.27 Short circuit in a microstrip line

Right-angled Bend

In order to divert the microwave signal path, a right-angled microstrip bend is often used. The microstrip bend is a discontinuity where it is seen that current flow around the corner is critical. The skin effect pushes the current to the outside edges of the microstrip, and the current crowding towards the inside edge of the bend. A schematic diagram of such a bend with electric current distributions is shown in Fig. 6.28 and 6.29(a). The charge accumulation at the outer corner and interruption of current result in null current at the outer corner and the current maximum on the inner corner [Fig. 6.29(a)]. The result is an equivalent T-network with series inductance L_b and a shunt capacitance C_b.

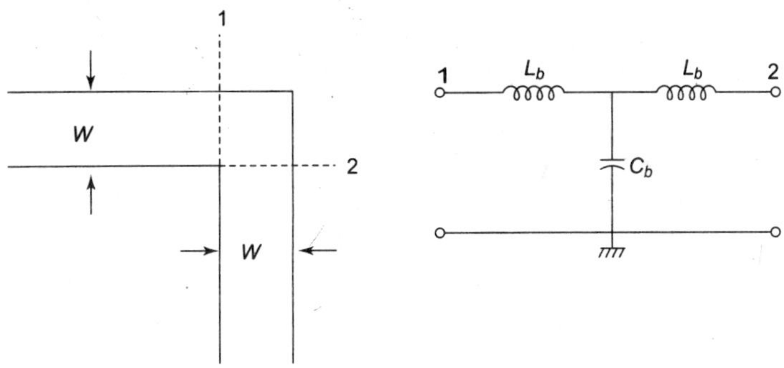

Fig. 6.28 Right angled bend and its electrical equivalent circuit

Different authors have given the expression of the bend capacitance and inductances based on their works. The expressions given by K C Gupta, et al. are

$$\frac{C_b}{h} = \frac{w}{h}[(9.5\,\varepsilon_r - 1.25)\,w/h + 5.2\,\varepsilon_r + 7.0];\ \text{pF/m} \tag{6.69a}$$

$$\frac{L_b}{h} = 4\sqrt{\frac{w}{h}} - 4.21;\ \text{nH/m} \tag{6.69b}$$

The expressions given by Konpang, et al. are modified and are given by

$$C_b/h = (10.35\varepsilon_r + 2.5)\left(\frac{w}{h}\right)^2 + (2.6\,\varepsilon_r + 5.64)\left(\frac{w}{h}\right)\ \text{pF/m} \tag{6.70a}$$

$$L_b/h = 0.7\left[1.0 - 1.35e^{-0.18\left(\frac{w}{h}\right)^{1.39}}\right]\ \text{nH/m} \tag{6.70b}$$

A typical variations of these equivalent parameters with w/h are shown in Fig. 6.29(b) where the results of both the expressions given here are agreed well.

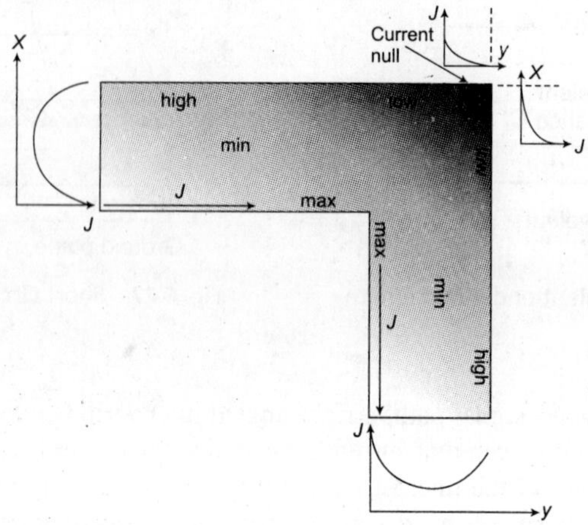

Fig. 6.29 (a) Current distributions near right-angle bend

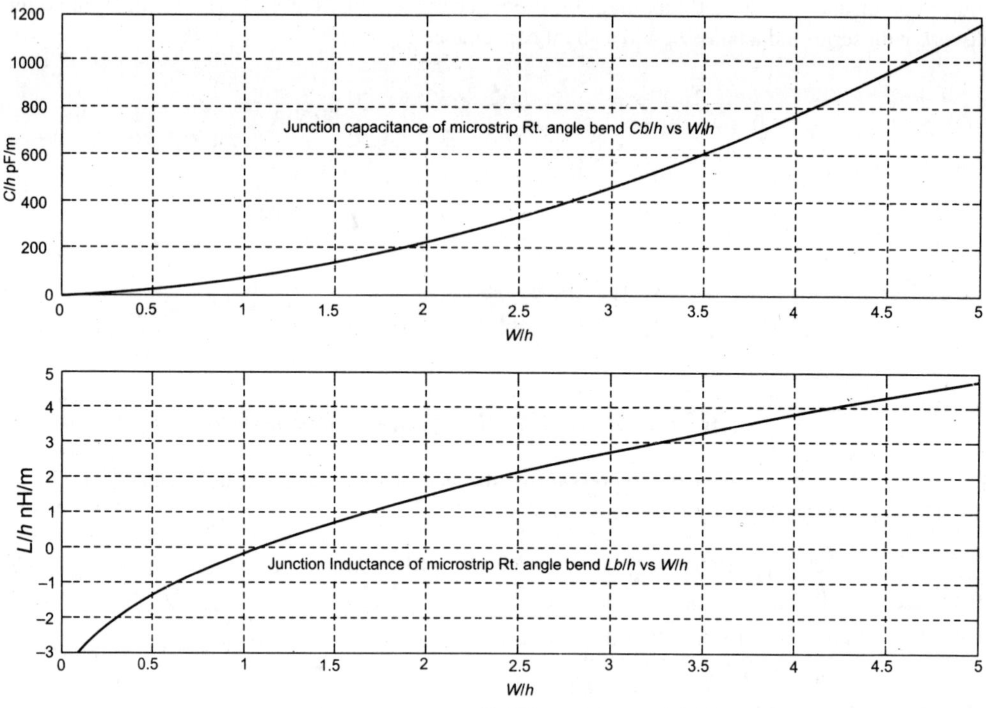

Fig. 6.29 (b) Variation of bend capacitance and inductance

The bend can be metered at the outer corner with a simple 45° cut to reduce the bend effect without affecting any major current distributions [Fig. 6.30(a)]. The capacitance has been reduced by 50% and inductance may be doubled than those in the unmetered case. A very good VSWR is obtained for a cut where the range of cut is $b = 0.6w - 0.8w$ for $\varepsilon_r = 4.4$, $h = 1.6$ mm. The results of VSWR vs frequency are shown in Fig. 6.30(b).

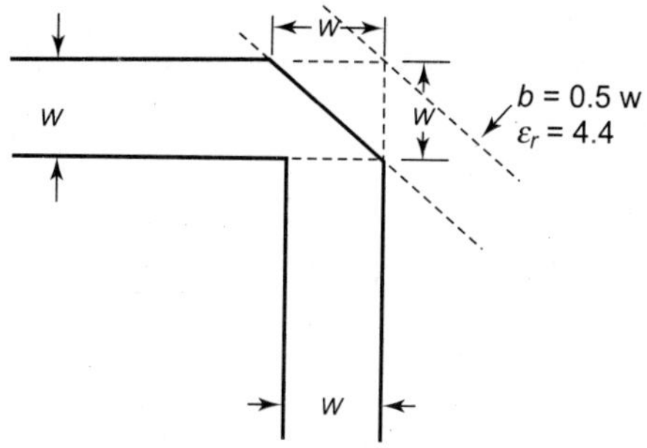

Fig. 6.30 (a) Metered bend with a simple 45° cut

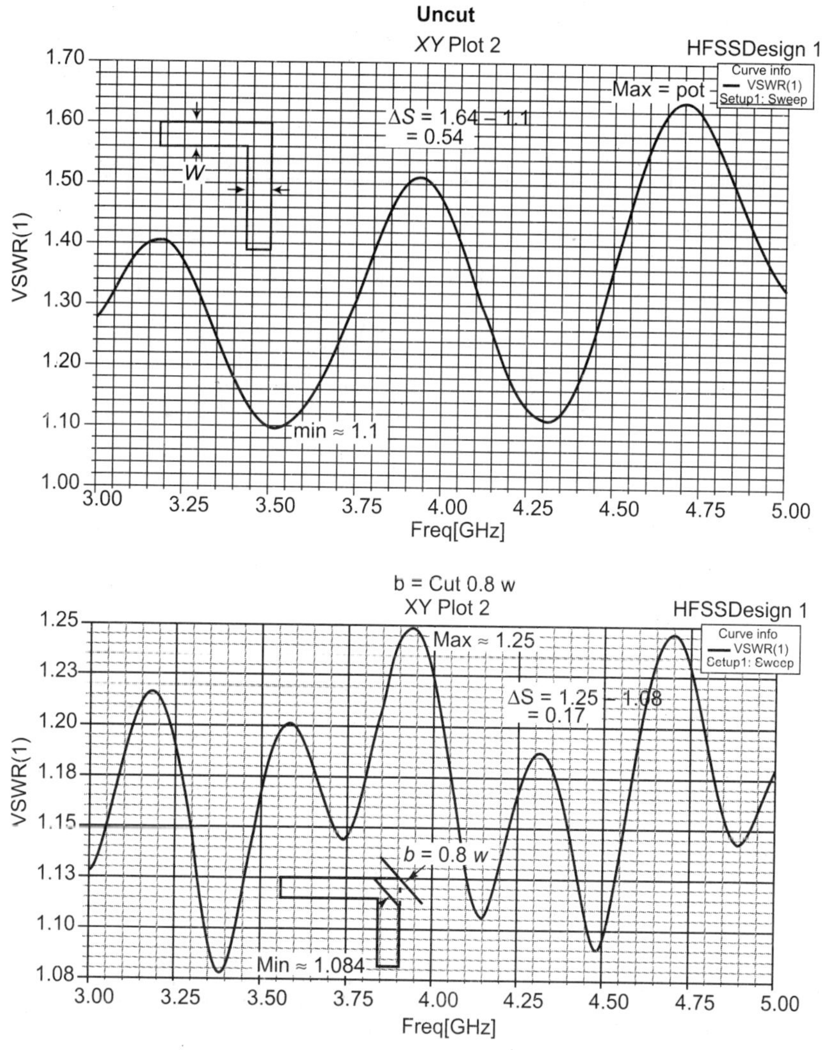

Fig. 6.30 (b) (Contd.)

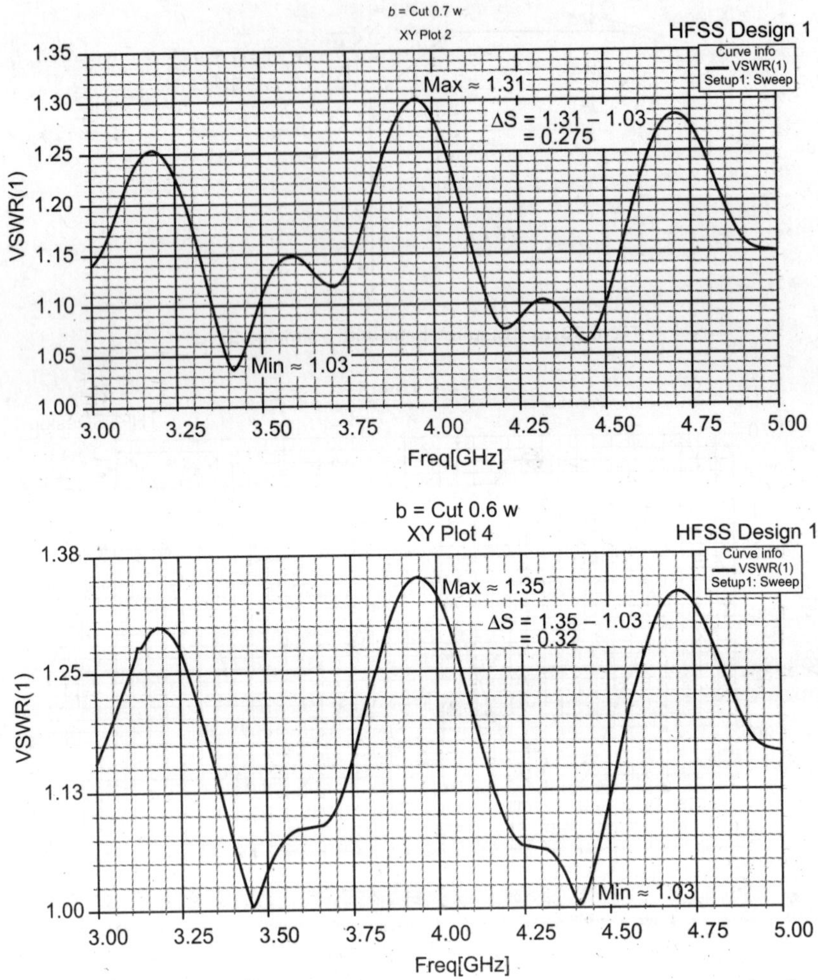

Fig. 6.30 (b) VSWR characteristics

It is seen that there is improvement is VSWR in a metered bend. Over the frequency range 1–2 GHz, VSWR ranges from 1.035–1.23 for uncut band and from 1.005–1.2 . . . for metered cut bend (b = 0.50 W, $\varepsilon_r = 4.4$).

Matched Terminations 6.4.7

Matched terminations are used in coaxial lines, strip lines and waveguides to absorb the incident power without appreciable reflection and radiation. A tapered lossy dielectric is placed at the end of a shorted line as shown in Fig. 6.31 to form a matched termination.

The length of the tapered section is kept about one to two guide wavelengths at the lowest frequency of operation for effective absorption of power. To increase the power dissipation, aquadag-coated sand is used as lossy material. High power (> 1 W) terminations use outer cooling fins for heat dissipation. Practical VSWR of these loads is in the range of 1.02–1.05 over a frequency bandwidth of the order of 20–30 % of the centre frequency.

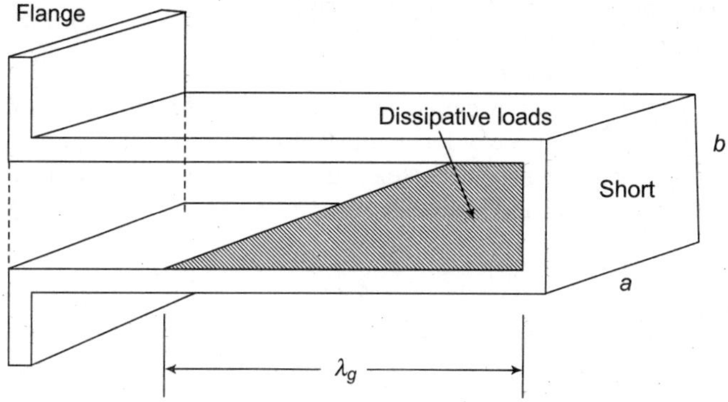

Fig. 6.31 Waveguide matched termination

A matched load is a single-port device having its ideal parameters:

$$Z_{\text{in}} = Z_0 = 50 \text{ ohm or } 75 \text{ ohm}$$

$$\Gamma = 0 = S_{11}$$

$$\text{VSWR} = 1.0$$

The load is located at the centre of the Smith chart as the reference point.

Example 6.16 A waveguide termination having VSWR of 1.1 is used to dissipate 100 watts of power. Find the reflected power.

Solution

$$|\Gamma| = \frac{S-1}{S+1} = \frac{1.1-1}{1.1+1} = \frac{0.1}{2.1} = 0.04762$$

$$P_{\text{ref}} = |\Gamma|^2 P_i = (0.04762)^2 \times 100 = 0.2268 \text{ W}$$

Short-Circuit Plunger 6.4.8

A short-circuit plunger is a variable short circuit to provide an adjustable reactive load $Z_{\text{in}} = jZ_0 \tan \beta l$, ranging from $-j\infty$ to $j\infty$, depending on the physical length l of the line and hence reflects all the incident power. These terminations are used for measurements of impedance or scattering parameters of a microwave device or circuit element. The short is achieved by a sliding block of copper as shown in Fig. 6.32(a).

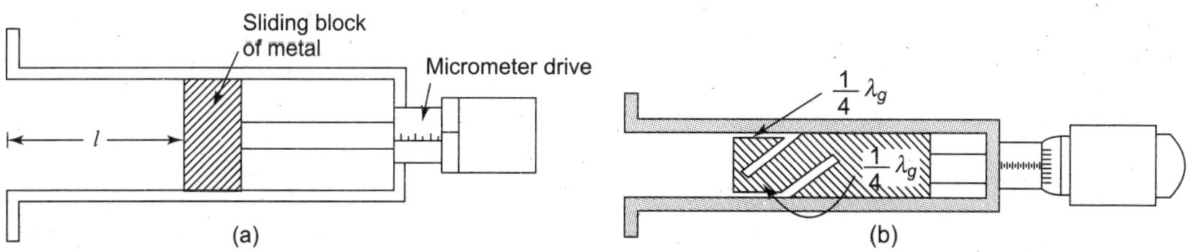

Fig. 6.32 Short-circuit plungers

The main requirements for the design of a short-circuit plunger are

1. Low contact resistance at the guide wall
2. Constant contact resistance for movement along the line

These requirements can be met better by employing a quarter-wavelength long spring contact assemblies to place the tips of spring contact at a node of longitudinal electric current as shown in Fig. 6.32(b).

Since there is difficulty in achieving an adequate contact throughout the entire perimeter of the plunger, choke plungers are often used as shown in Fig. 6.33.

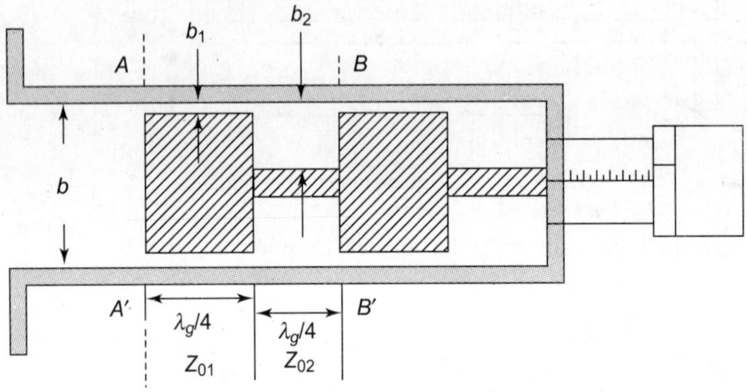

Fig. 6.33 Short-circuit choke plunger

A choke plunger uses the impedance transformation properties of a quarter-wave transformer. In the waveguide choke plunger, the front choke is a quarter-wave section. The second section is also a quarter guide wavelength long. The widths of the plungers are uniform and slightly less than the interior guide width of the broad wall. The height of the plungers are not uniform, the from one is $b - 2b_1$ and the second one is $b - 2b_2$, where b_1 is made as small as possible and b_2 is made as large as possible. The back section makes a sliding fit in the guide with almost zero gap. In this case, the contact resistance is in a current antinode and the input impedance is zero at the plane of reflection. If Z_{01} and Z_{02} are the characteristic impedances of the quarter wavelength coaxial sections $b - 2b_1$ and $b - 2b_2$, the impedance seen at the input plane AA' is

$$Z_{in} = (Z_{01}/Z_{02})^2 Z'_i = (b_1/b_2)^2 Z'_i \tag{6.71}$$

where Z'_i is the input impedance at the back section plane BB'. To make $Z_{in} \to 0$, $Z_{01} << Z_{02}$ and accordingly the gaps b_1 and b_2 are to be selected where $b_2 >> b_1$. If by good mechanical design, b_2 is made equal to $5b_1$, $Z_{in} = \dfrac{Z'_i}{25}$. Therefore, the shorting effect is improved 25 times as compared to a single non-choke type sliding short. For a coaxial line, shorting plungers are used to provide impedance-transforming properties which do not have physical contact with the line conductors.

The disadvantage of all non-contact plungers and choke plungers is the bandwidth limitation of 20 to 30% of the mid-frequency. In case of a circular waveguide in the TE_{01} mode, a short-circuit plunger does not require any choke connections since there is no longitudinal current. A simple sliding metal disc is adequate. The locus of reflection coefficient of this plunger is the outermost circle of the Smith chart ($\Gamma = 1\underline{|\pi}$).

Rectangular to Circular Waveguide Transition 6.4.9

This transition section is used to convert the dominant TE_{10} mode in the rectangular waveguide to TE_{11} dominant mode in circular waveguide and vice versa as shown in Fig. 6.34. The minimum length of the transition should be quarter wavelength to avoid abrupt dimensional changes and generation of higher-order modes. The usual length of complete section is kept $\geq 2\lambda_g$.

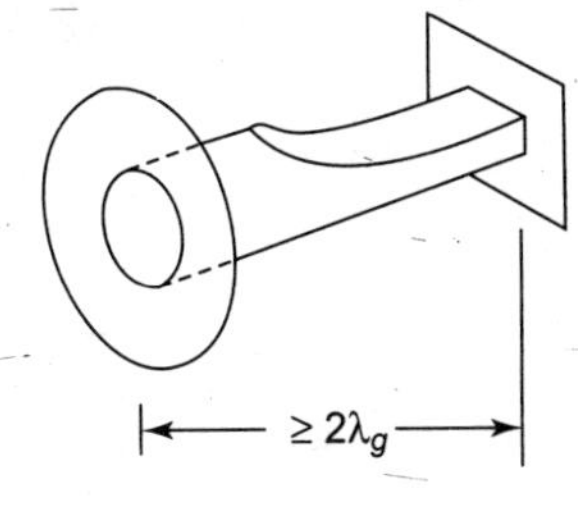

Fig. 6.34 Rectangular to circular waveguide transition

Waveguide Corners, Bends and Twists 6.4.10

The waveguide E and H plane corners and bends are used to alter the direction of the guide to any convenient angle, as shown in Fig. 6.35.

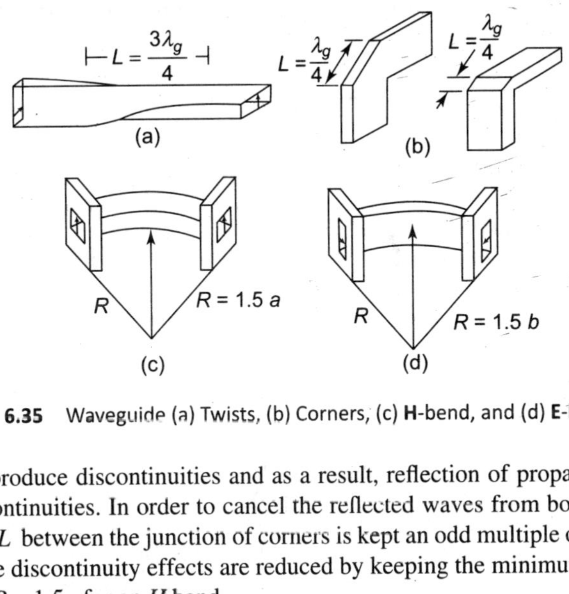

Fig. 6.35 Waveguide (a) Twists, (b) Corners, (c) **H**-bend, and (d) **E**-bend

Such arrangements produce discontinuities and as a result, reflection of propagating waves takes place at any point of the discontinuities. In order to cancel the reflected waves from both ends of the waveguide corner, the mean length L between the junction of corners is kept an odd multiple of the quarter wavelength. For waveguide bend, the discontinuity effects are reduced by keeping the minimum radius of curvature $R = 1.5b$ for an E bend and $R = 1.5a$ for an H bend.

A waveguide twist is used to change the polarization of the propagating wave by 90 degrees. The length of the twist is again kept equal to an odd multiple of the quarter guide wavelength.

Coaxial Line to Waveguide Adapters 6.4.11

In a rectangular waveguide, the electromagnetic fields are fed by means of a small coaxial line probe driven by a generator as shown in Fig. 6.36. For the dominant TE_{10} mode of excitation, the probe is inserted from the broad wall perpendicularly or parallel to the maximum **E** field. The depth d of the probe is kept small so that the *rf* current on the probe can be assumed constant which will radiate electromagnetic field inside the guide without any disturbance.

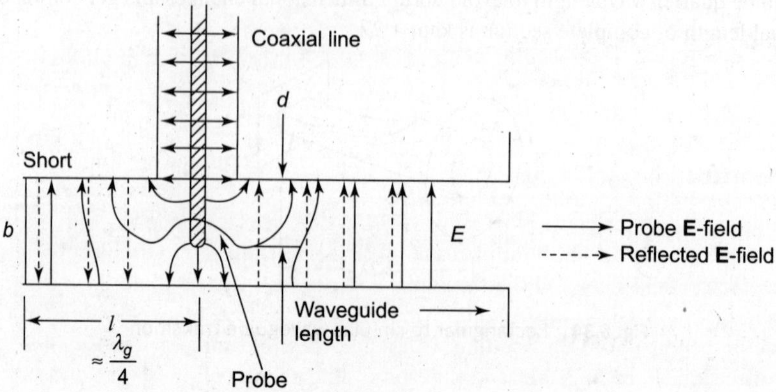

Fig. 6.36 Coaxial to rectangular ($a \times b$) waveguide adapter

In order to have radiation in one propagating direction only, a short circuit is placed at a distance l approximately equal to a quarter wavelength in the backward direction. By designing the values of l and d, the input impedance $Z_{in} = R_{in} + jX_{in}$ of the probe can be made pure resistive which is equal to the characteristic impedance Z_0 of the coaxial line feeding the signal to the guide:

$$Z_{in} = R = \frac{2\eta_0}{ab\beta_{10} k_0} \sin^2 \beta_{10} l \tan^2 (k_0 d/2) \tag{6.72}$$

where, $$\eta_0 = \sqrt{\mu_0 / \varepsilon_0}; \beta_{10} = k_0\sqrt{1-(\lambda_0/2a)^2}; k_0 = 2\pi/\lambda_0 \tag{6.73}$$

Since the electric field in the coaxial line for TEM mode and in the waveguide for TE_{10} mode in the vicinity of the probe are orthogonal to each other, higher-order modes (evanescent) are also excited near the probe which are highly attenuated within one wavelength distance. By making the probe diameter very small ($< 0.15\,a$) the reactance X_{in} of the probe resulting from higher order mode is made negligible small (or ideally zero) with proper choice of l and d. E-field radiated from the probe toward the waveguide propagation (*rhs*) becomes inphase with the E-field radiated back and reflected from the shorting end.

Coupling Loops 6.4.12

Electromagnetic fields can also be excited in waveguides by a coupling loop placed midway between the top and bottom walls of the guide with its plane transverse to the waveguide as shown in Fig. 6.37. Thus, maximum **H** field for TE_{10} mode is coupled to the loop. For propagation in one propagating direction only, a short-circuit plate is placed at a distance l on the reverse side. The input impedance of the probe is made pure resistive, equal to the characteristic impedance Z_0 of the feeder coaxial line.

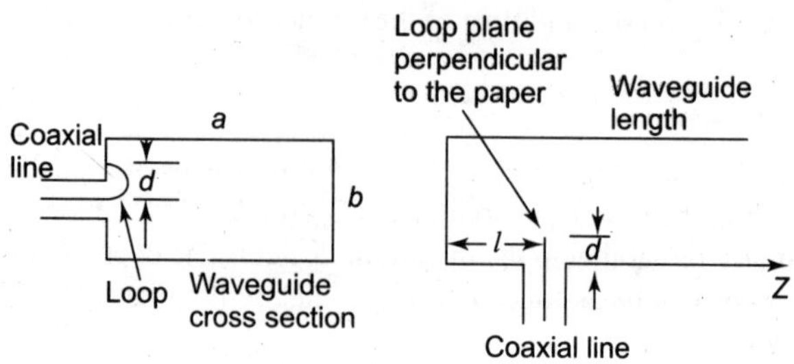

Fig. 6.37 Coupling loops

$$Z_{in} = Z_0 = R = \frac{k_0 \eta_0}{2ab\beta_{10}} (\pi/a)^2 (\pi d^2)^2 \sin^2 \beta_{10} l \tag{6.74}$$

Z_{in} is made equal to Z_0 by adjusting the diameter of the thin loop and l such that $d < \lambda_0/10$ and $\lambda_g/2 < l < \lambda_g/4$.

Coupling Aperture 6.4.13

The electromagnetic signal can be coupled into a waveguide from another guide by cutting a small aperture of radius $r_0 << \lambda_0$ in the common wall. If the aperture plane is perpendicular to the electric field, the normal component of the electric field lines and the tangential component magnetic field lines fringe through the aperture as shown in Fig. 6.38. Under this condition, the small aperture becomes equivalent to a combination of a electric dipole perpendicular to the aperture with its dipole moment being proportional to the normal component of the electric field in the guide, and a magnetic dipole in the plane of the aperture with its dipole moment proportional to the tangential component of the magnetic field. The constants of proportionality depend on the aperture size and shape. For **E** and **H** field couplings, the constants of proportionality are called the *electric* and *magnetic polarisabilities* α_e *and* α_m, respectively, given by

$$\alpha_e = -2/3\ r_0^3,\ \alpha_m = 4/3\ r_0^3 \tag{6.75}$$

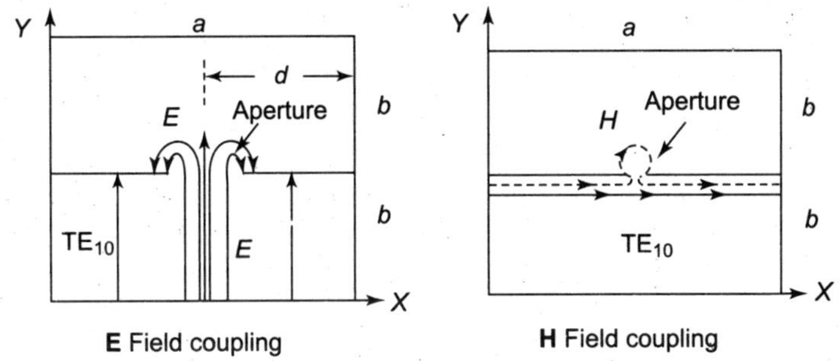

Fig. 6.38 Aperture coupling from lower to upper guide

When a small circular aperture of radius $r_0 << \lambda_0$ exists in a transverse wall in a rectangular waveguide operating in the TE_{10} mode, it offers an inductive susceptance B given by

$$B = \frac{-3ab}{8r_0^3 \beta_{10} Z_w};\quad Z_w = \text{guide wave impedance} \tag{6.76}$$

The transmission and reflection coefficients of the aperture are given by

$$T = \frac{j16 r_0^3 k_0 \eta_0}{3ab Z_w} \tag{6.77}$$

$$\Gamma = \frac{j16 r_0^3 \beta_{10}}{3ab} - 1 \tag{6.78}$$

A circular aperture of radius r_0 in the common broad wall between two rectangular waveguide couples the energy from one guide to the other. For the dominant TE_{10} mode of the input wave propagation, the aperture will be excited by a y-directed electric dipole and x-and z-directed magnetic dipoles. The normal (y) electric dipole and the axial (z) component of the magnetic dipole radiate in the upper guide symmetrically in both directions ($\pm z$). The transverse (x) component of the magnetic dipole radiates asymmetrically. The amplitude of the field in the upper guide can be controlled by adjusting the angle θ between the two waveguides [Fig. 6.60(a)] or by adjusting the aperture position d in transverse direction. When d is chosen to satisfy

$\sin(\pi d/a) = \dfrac{\lambda_0}{\sqrt{6}a}$, there will be zero power at Port 4 in the upper guide and input power at Port 1 will be coupled into Ports 2 and 3 only. The 4-port device is then called a *directional coupler* or *Bethe hole coupler* which is described in this chapter later.

Attenuators 6.4.14

Attenuators are passive devices used to control power levels in a microwave system by partially absorbing the transmitted signal wave. Both fixed and variable attenuators are designed using resistive films (aquadag coated dielectric sheet).

A coaxial fixed attenuator uses a film with losses on the centre conductor to absorb some of the power as shown in Fig. 6.39(a). The fixed waveguide type [Fig. 6.39(b)] consists of a thin dielectric strip coated with resistive film and placed at the centre of the waveguide parallel to the maximum **E** field. Induced current on the resistive film due to the incident wave results in power dissipation, leading to attenuation of microwave energy. The dielectric strip is tapered at both ends up to a length of more than half wavelength to reduce reflections. The resistive vane is supported by two dielectric rods separated by an odd multiple of quarter wavelength and perpendicular to the electric field [Fig. 6.39(b)].

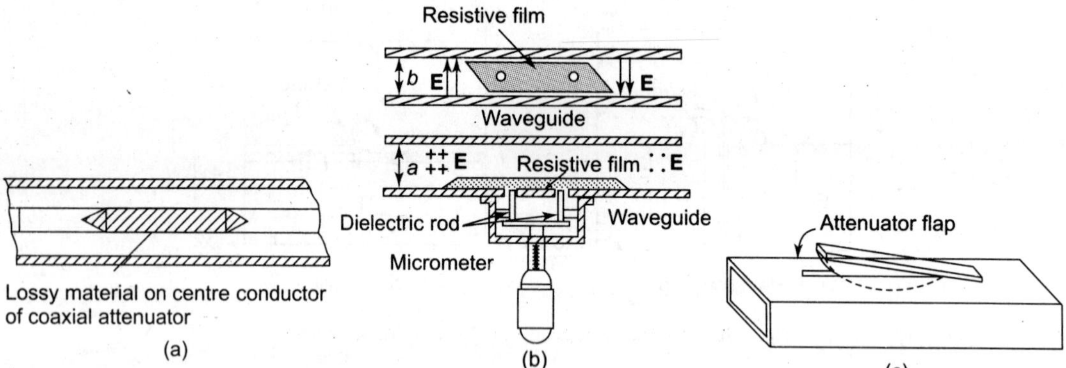

Fig. 6.39 Microwave attenuator: (a) coaxial line fixed attenuator (b) and (c) waveguide attenuators

A variable-type attenuator can be constructed by moving the resistive vane by means of micrometer screw from one side of the narrow wall to the centre where the **E** field is maximum [Fig. 6.39(b)] or by changing the depth of insertion of a resistive vane at an **E** field maximum through a longitudinal slot at the middle of the broad wall as shown in Fig. 6.39(c). A maximum of 90 dB attenuation is possible with VSWR of 1.05. The resistance card can be shaped to give a linear variation of attenuation with the depth of insertion.

Precision Variable Attenuator

A precision-type variable attenuator makes use of a circular waveguide section (C) containing a very thin tapered resistive card (R_2), to both sides of which are connected axisymmetric sections of circular-to-rectangular waveguide tapered transitions (RC_1 and RC_2) as shown in Fig. 6.40(a). The centre circular section with the resistive card can be precisely rotated by 360° with respect to the two fixed sections of circular to rectangular waveguide transitions. The induced current on the resistive card R_2 due to the incident signal is dissipated as heat and produces attenuation of the transmitted signal. The incident TE_{10} dominant wave in the rectangular waveguide is converted into a dominant TE_{11} mode in the circular waveguide. A very thin tapered resistive card is placed perpendicular to the **E** field at the circular end of each transition section so that it has a negligible effect on the field perpendicular to it but absorbs any component parallel to it. Therefore, a pure TE_{11} mode is excited in the middle section.

With reference to Fig. 6.40(b), if the resistive card in the centre section is kept at an angle θ relative to the **E** field direction of the TE_{11} mode, the component $E \cos \theta$ parallel to the card gets absorbed while the component $E \sin \theta$ is transmitted without attenuation. This later component finally appears as electric field component $E \sin^2 \theta$ in the rectangular output guide. Therefore, the attenuation of the incident wave is

$$\alpha = \frac{E}{E\sin^2\theta} = \frac{1}{\sin^2\theta} = \frac{1}{|S_{21}|}$$

or, $$\alpha\,(\text{dB}) = -40 \log (\sin \theta) = -20 \log |S_{21}| \tag{6.79}$$

Therefore, the precision rotary attenuator produces attenuation which depends only on the angle of rotation θ of the resistive card with respect to the incident wave polarization. Attenuators are normally matched reciprocal devices, so that

$$|S_{21}| = |S_{12}| \tag{6.80}$$

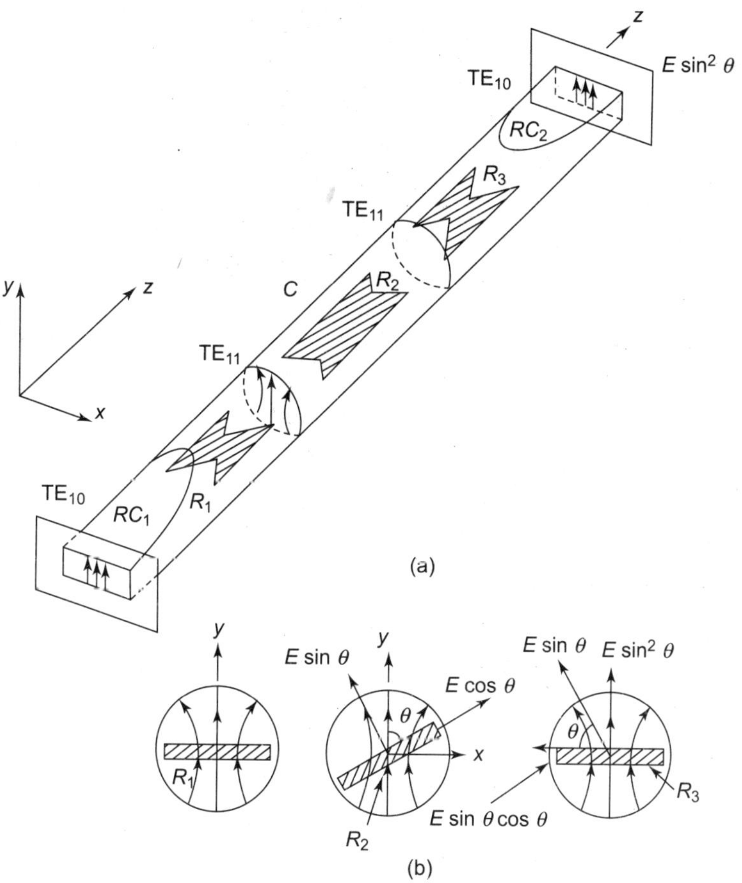

Fig. 6.40 Precision-type variable attenuator
R_1, R_2, R_3 : Tapered resistive cards
RC_1 and RC_2: Rectangular-to-circular waveguide transitions
C: Circular waveguide section

$$\text{and}\quad |S_{11}| \text{ or } |S_{22}| = \frac{\text{VSWR}-1}{\text{VSWR}+1} << 0.1 \tag{6.81}$$

where the VSWR is measured at the input or output port concerned. The *S*-matrix of an ideal precision rotary attenuator is

$$[S] = \begin{bmatrix} 0 & \sin^2\theta \\ \sin^2\theta & 0 \end{bmatrix} \tag{6.82}$$

Phase Shifters 6.4.15

A phase shifter is a device which provides variable insertion phase in a microwave signal path without altering the physical path length. Most phase shifters are two-port devices characterized by low insertion loss and low VSWR. The phase of a microwave signal can be changed by inserting a low loss dielectric material or a ferrite material in the wave propagation path. Accordingly, phase shifters are called ***dielectric phase shifters*** and ***ferrite phase shifters***, respectively. Dielectric phase shifters are reciprocal phase shifters. On the other hand, ferrite phase shifters can be reciprocal as well as non-reciprocal depending upon whether the variable differential phase shift through the device is a function of the direction of propagation. These are described below.

Dielectric Phase Shifters

A dielectric phase shifter can be realized by placing a lossless dielectric slab within a waveguide parallel to and at the position of maximum **E** field. A differential phase change is produced due to the change of wave velocity $u_p = \dfrac{1}{\sqrt{\mu_0\varepsilon_0\varepsilon_r}}\dfrac{1}{\sqrt{1-\left(\dfrac{f_c}{f}\right)^2}}$ through the dielectric slab compared to that through an empty waveguide in which $\varepsilon_r = 1$. Two ports are matched by reducing the reflections of the wave from the dielectric slab tapered at both ends as shown in Fig. 6.41.

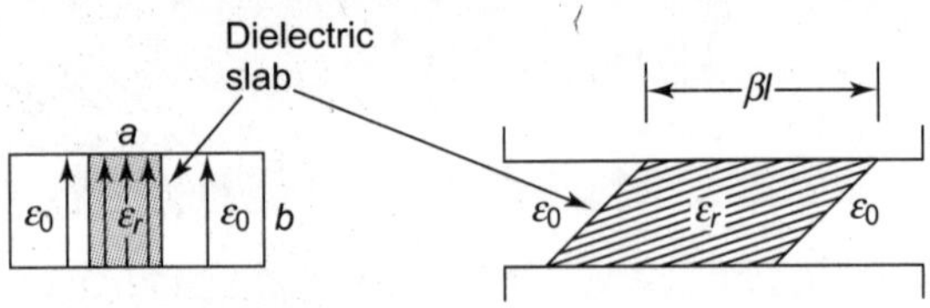

Fig. 6.41 Phase shifters

The propagation constant through a length l of a dielectric slab and of an empty guide excited in dominant mode TE_{10} are, respectively,

$$\beta_1 = \frac{2\pi}{\lambda_{g_1}},\quad \lambda_{g_1} = \frac{\dfrac{\lambda_0}{\sqrt{\varepsilon_r}}}{\sqrt{1-\left(\dfrac{\lambda_0}{2a\sqrt{\varepsilon_r}}\right)^2}} \tag{6.83}$$

$$\beta_2 = \frac{2\pi}{\lambda_{g_0}},\quad \lambda_{g_0} = \frac{\lambda_0}{\sqrt{1-\left(\dfrac{\lambda_0}{2a}\right)^2}} \tag{6.84}$$

Thus, the differential phase shift produced by the phase shifter is $\Delta\phi = (\beta_1 \sim \beta_0)l$. By adjusting the length l, different phase shifts can be produced.

The S-matrix of an ideal phase shifter can be expressed by

$$[S] = \begin{bmatrix} 0 & e^{-j\Delta\phi} \\ e^{-j\Delta\phi} & 0 \end{bmatrix} \tag{6.85}$$

Precision Dielectric Rotary Phase Shifter

A precision phase shifter can be designed as a rotary type as shown in Fig. 6.42. This uses a section of circular waveguide c containing a lossless dielectric plate of length $2l = \dfrac{\lambda_g}{2}$ called half-wave (180°) section. This section can be rotated over 360° precisely between two sections of circular to rectangular waveguide transitions A and B, each containing lossless dielectric plates of length $l = \dfrac{\lambda_g}{4}$ called quarter-wave (90°) sections oriented at an angle of 45° with respect to the broad wall of the rectangular waveguide ports at the input and output. The incident TE_{10} wave in the rectangular guide becomes a TE_{11} wave in the circular guide A. The half-wave section C produces a phase shift equal to twice its rotation angle θ with respect to the quarter-wave section. The dielectric plates are tapered through a length of quarter wavelength at both ends for reducing reflection due to discontinuity.

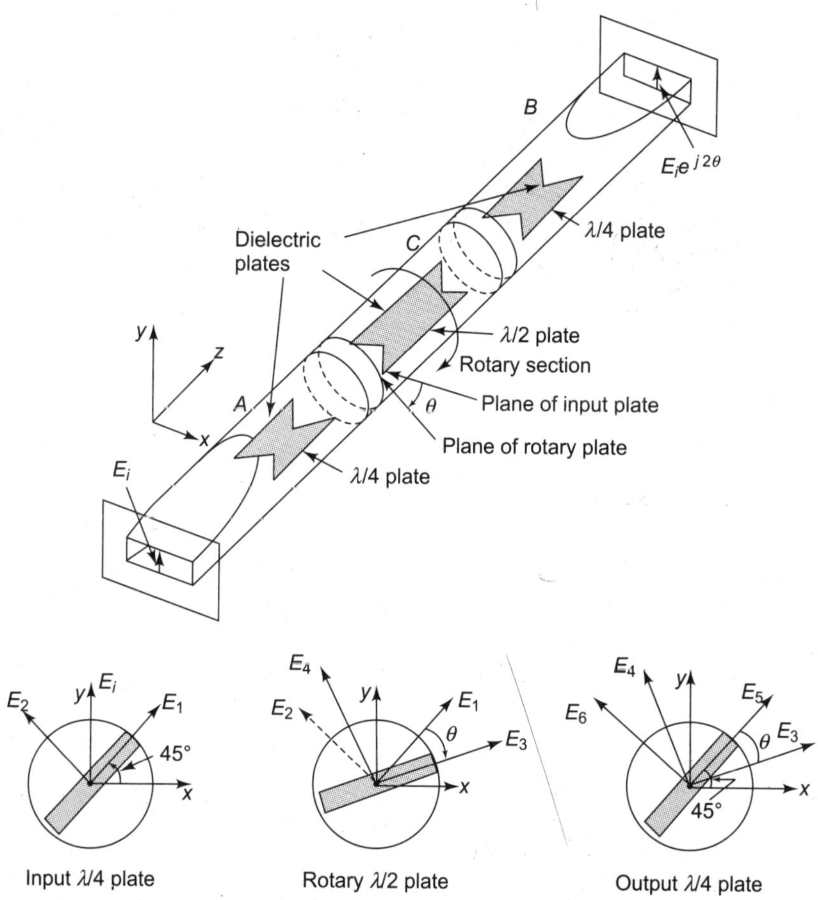

Fig. 6.42 Precision dielectric rotary phase shifter

The principle of operation of the rotary phase shifter can be explained as follows. The TE_{11} mode incident field E_i in the input quarter-wave section can be decomposed into two transverse components, one E_1, polarized parallel and other, E_2 perpendicular to quarter-wave plate. After propagation through the quarter-wave plate, these components becomes

$$E_1 = E_i \cos 45° e^{-j\beta_1 l} = E_0 e^{-j\beta_1 l} \qquad \cos 45° = \sin 45° = \frac{1}{\sqrt{2}} \tag{6.86}$$

$$E_2 = E_i \sin 45° e^{-j\beta_2 l} = E_0 e^{-j\beta_2 l} \tag{6.87}$$

where, $E_0 = E_i/\sqrt{2}$. The length l is adjusted such that these two components will have equal magnitude but a differential phase change of $(\beta_1 - \beta_2)\, l = 90°$. Therefore, after propagation through the quarter-wave plate, these field components become

$$E_1 = E_0\, e^{-j\beta_1 l} \tag{6.88}$$

$$E_2 = E_0\, e^{-j\beta_1 l} \cdot e^{j\pi/2} = E_0 e^{-j(\beta_1 l - \pi/2)} \tag{6.89}$$

Thus, the quarter-wave sections convert a linearly polarized TE_{11} wave to a circularly polarized wave and vice versa. Equations (6.88) and (6.89) are input components to half-wave section C.

After emergence from the half-wave section, the field components parallel and perpendicular to the half-wave plate can be represented as

$$E_3 = (E_1 \cos\theta - E_2 \sin\theta)\, e^{-j2\beta_2 l} = E_0\, e^{-j\theta}\, e^{-j3\beta_1 l} \tag{6.90}$$

$$E_4 = (E_2 \cos\theta + E_1 \sin\theta)\, e^{-j2\beta_2 l} = E_0\, e^{-j\theta}\, e^{-j3\beta_1 l}\, e^{-j\pi/2} \tag{6.91}$$

since, $2(\beta_1 - \beta_2)l = \pi$ or $-2\beta_2 l = \pi - 2\beta_1 l$ (6.92)

After emergence from the half-wave section, the field components E_3 and E_4 may again be decomposed into two TE_{11} modes, polarized parallel and perpendicular to the output quarter-wave plate B. At the output end of this quarter-wave plate, the field components parallel and perpendicular to the quarter-wave plate can be written as

$$E_5 = (E_3 \cos\theta + E_4 \sin\theta)\, e^{-j\beta_1 l} = E_0\, e^{-j2\theta}\, e^{-j4\beta_1 l}; \tag{6.93}$$

$$E_6 = (E_4 \cos\theta - E_3 \sin\theta)\, e^{-j\beta_2 l} = E_0\, e^{-j2\theta}\, e^{-j4\beta_1 l}; \tag{6.94}$$

Therefore, the parallel component E_5 and perpendicular component E_6 at the output end of the quarter-wave plate are equal in magnitude and in phase to produce a resultant field which is a linearly polarized TE_{11} wave

$$E_{out} = \sqrt{2}\, E_0\, e^{-i2\theta} e^{-j4\beta_1 l} = E_i\, e^{-j2\theta} e^{-j4\beta_1 l} = E_i e^{-j(2\theta + 4\beta_1 l)} \tag{6.95}$$

having the same direction of polarization as the incident field E_i with a phase change of $2\theta + 4\beta_1 l$. Since θ can be varied and $4\beta_1 l$ is fixed at a given frequency and structure, a phase shift of 2θ can be obtained by rotating the half-wave plate precisely through an angle of θ with respect to the quarter-wave plates. The rotation angle can be calibrated to obtain phase shift.

Ferrite Phase Shifters

It is known that most of the ferrite materials have relative dielectric constants in the range 9 to 16, and dielectric loss tangents are normally small (< 0.001). Unlike a dielectric substrate, the permeability of the ferrite depends on the RF magnetic field configuration in the medium relative to the static magnetization state. Consequently, the propagation constant in the ferrite media changes with permeability since $\beta = \omega\sqrt{\mu\varepsilon}$.

Phase shift in a ferrite device is generated by magnetizing the ferrite inside the wave guide by RF current such that the phase constant in the propagating line alters due to change in permeability with magnetization (Ref: para 2.18.3 in Chapter 2).

Ferrite phase shifters may be latching or non-latching, depending upon whether continuous holding current must be supplied to sustain the magnetic bias field. Figure 6.43 shows a schematic diagram of a twin-toroid phase shifter. The twin toroid is a latching, nonreciprocal device. It uses either a closed magnetic circuit or a magnetic circuit with very small air gaps. The relative permeability of the ferrite is controlled by adjusting the magnetic flux level existing in the closed magnetic circuit.

The dielectric spacer is used to concentrate the rf energy in the center of the waveguide. The walls of the ferrite toroids, which contact the dielectric spacer, are located in those regions of the waveguide which support a circularly polarized magnetic field. If β^+ is the propagation constant when a positive bias field saturates the ferrite and β^- be the same for negative bias field, the maximum amount of phase shift per unit length is $(\beta^+ - \beta^-)$. The variable phase shift less than this can be achieved by reducing the bias field level. When the the direction of propagation changes, the values of β^- and β^+ interchange.

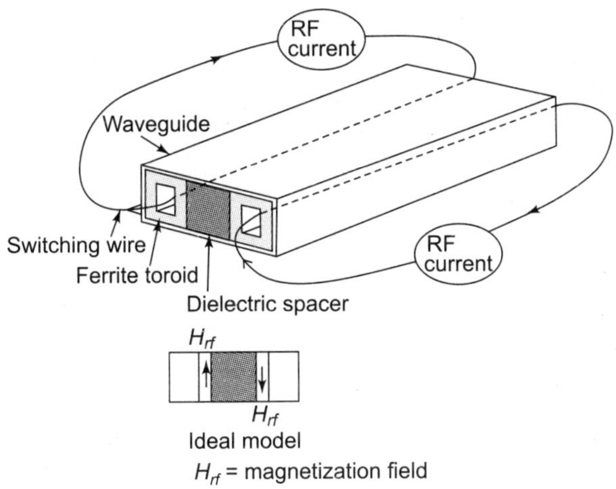

Fig. 6.43 Ferrite twin-toroid phase shifter

The device geometry and the saturation magnetization of the ferrite influence the frequencies for practical realization of these phase shifters. The conductive and dielectric losses are directly proportional to the length or inversely proportional to the saturation magnetization while the magnetic loss varies approximately directly with the saturation magnetization. The sum of these losses shows a minimum value in the frequency range given by

$$0.2 \le \frac{\omega_m}{\omega} \le 0.6$$

where $\omega_m = 2\pi\gamma(4\pi M_s)$, $\gamma = 2.8$ MHz/Oe, $4\pi M_s$ is the saturation magnetization and ω is the microwave radian frequency. The latching phase shifters have been realized from about 2 GHz to 94 GHz.

MIC Ferrite Phase Shifter

Passive MIC phase shifters are constructed with ferrite substrate using the property that the propagation constant of microwaves along the microstrip line on ferrite substrate can be changed by varying the magnitude and/or the direction of the static magnetic field H_0. Thus, the phase shift, for a fixed length of the microstrip line can be controlled electrically leading to low cost, small and highly reliable components.

The following tree shows the versions of ferrite MIC phase-shifters—analog and digital.

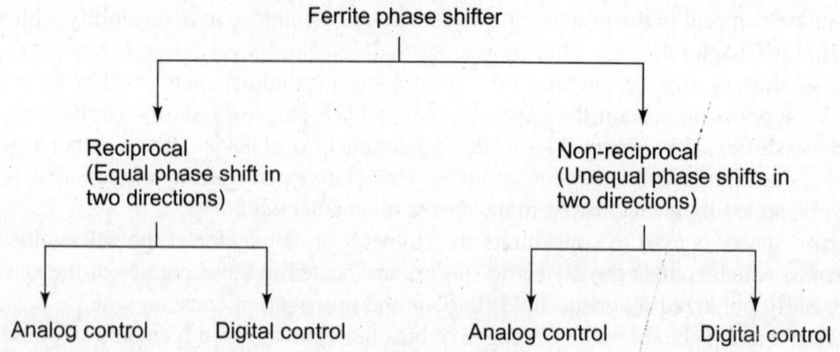

❑ *Analog Control* In analog phase shifter, the phase is controlled by varying either the magnitude or the direction of the magnetization vector **M** using applied static magnetic field $\mathbf{H}_0$.

❑ *Digital Control* In digital phase shifter, the phase is controlled by using high remanence ferrite materials and latching the substrate by a current pulse to change direction (±) of dc magnetic field digitally. The remanence of magnetic materials is defined in terms of its residual magnetic induction when an externally applied magnetizing current is reduced to zero as explained in the B–H curve in Fig. 6.44. A pulse current of sufficient amplitude, through a single turn wire threading the ferrite, is used to magnetize the ferrite substrate.

If n is the number of bits, then the least significant bit will have a ϕ-shift of $360°/2^n$. A digital phase shifter of n-bits will consist of n separately actuated sections giving $\frac{360°}{2^1}, \frac{360°}{2^2}, \ldots, \frac{360°}{2^n}$ of ϕ-shift. A series of different $\Delta\phi = 180°$, $90°$ and $45°$ can be cascaded in a single housing to form a 3-bit digital ϕ-shifter.

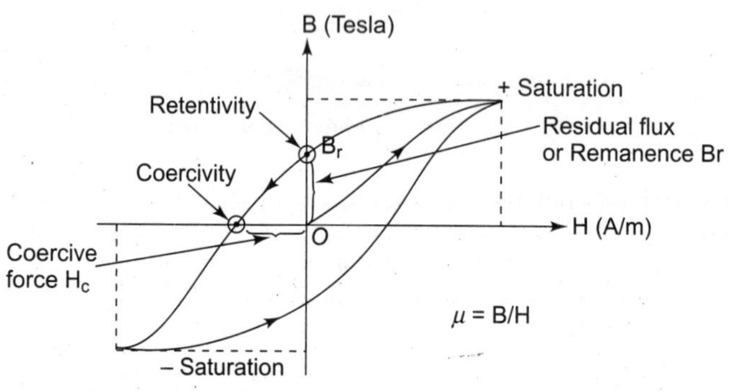

Fig. 6.44 Ferrite remanence

Reciprocal Phase Shifters

When a ferrite substrate is magnetized to saturation in a direction perpendicular to the microstrip line, as depicted by $\mathbf{H}_{dc}$ in Fig. 6.45, there is a very small interaction between the ferrite and the *RF* magnetic field. As a result, the effective *RF* relative permeability is approximately unity $\mu_{reff} \approx 1$. When the ferrite sample is magnetized to saturation in its plane parallel to the strip line (i.e., $\perp^r$ to the *RF* magnetic field), maximum interaction is obtained between the ferrite substrate and the *RF* magnetic field. Then the *RF* effective permeability becomes,

$$\mu_{\text{reff}} = \frac{\mu^2 - K^2}{\mu} \tag{6.96a}$$

where μ and K are the component of the permeability tensor ($\mu = B/H$, $K = \gamma 4\pi M/\omega$, M = magnetic dipole moment per unit volume)

When the applied magnetic field is small,

$$\mu_{\text{reff}} \approx 1 - (\omega_m/\omega)^2 \text{ for } \omega > \gamma 4\pi M_s = \omega_m \tag{6.96b}$$

where $4\pi M_s$ = ferrite magnetization, γ = gyromagnetic ratio (2.8 MHz/oersted). When $\omega \to \omega_m$, the magnetic loss cannot be neglected and μ_{reff} expression gets modified. The propagation constant of the TEM mode on the ferrite substrate is given by

$$\beta = \omega\sqrt{\varepsilon_0 \mu_0 \varepsilon_{\text{reff}} \mu_{\text{reff}}} \tag{6.96c}$$

The different values of μ_{reff} for the two mutually perpendicular directions of magnetizations yield two different values of the phase constant for wave propagation along the microstrip line on the ferrite substrate.

Writing in terms of a filling factor q as proposed by wheeler for dielectric substrate.

$$\varepsilon_{\text{eff}} = 1 + q_e(\varepsilon_r - 1);\ q_e \approx 0.6 \tag{6.96d}$$

$$\mu_{\text{eff}} = \frac{1}{1 + q_m\left(\dfrac{1}{\mu_r} - 1\right)} \tag{6.96e}$$

$$q_m = \frac{\mu_{\text{eff}}^{-1} - 1}{\mu^{-1} - 1} \tag{6.96f}$$

If magnetic filling factor $q_m = 0.5$, then for parallel magnetization

$$\beta_{11} \cong \omega\sqrt{\mu_0 \varepsilon_0 \varepsilon_{\text{reff}}}\ [\{2 - 2(\omega_m/\omega)^2\}/\{2 - (\omega_m/\omega)^2\}]^{1/2} \tag{6.96g}$$

and for perpendicular magnetization

$$\beta_{\perp} \approx \omega\sqrt{\mu_0 \varepsilon_0 \varepsilon_{\text{reff}}} \tag{6.96h}$$

When ferrite magnetization direction changed between two states, the fractional change in phase constant is

$$\frac{\Delta\phi}{\phi} = \frac{\Delta\beta}{\beta} = 1 - [\{2 - 2(\omega_m/\omega)^2\}/\{2 - (\omega_m/\omega)^2\}]^{1/2} \tag{6.96i}$$

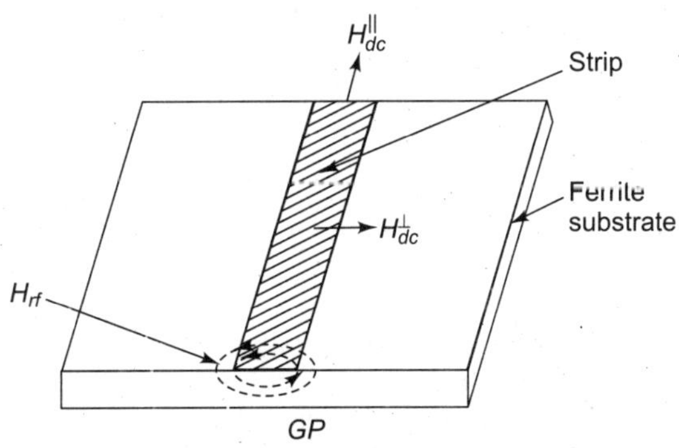

Fig. 6.45 Parallel and perpendicular magnetization

Both analog and digital controlled reciprocal phase shifters can be constructed using a loosely meander line microstrip conductor to minimise size as shown in Fig. 6.46. The adjacent lines of the meander scheme are spaced by distance > 5 times substrate height to minimize *RF* coupling and consequently to avoid non-reciprocal effects. DC current through perpendicular wires penetrating holes 1–1′ and 2–2′ generates magnetic fields for magnetization which can be varied, and also the direction can be changed as per requirement of phase change. In digital phase shifter two states of magnetization corresponding to two values of ϕ-shifts are obtained by sending dc currents through 1–1′ and 2–2′, respectively (2–2′ for parallel and 1–1′ for perpendicular magnetizations).

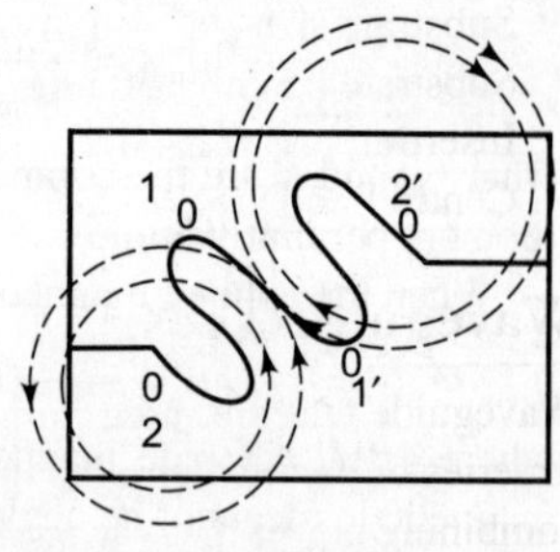

Fig. 6.46 Reciprocal ϕ-shifter schematic

Non-reciprocal Phase Shifter

In a non-reciprocal ϕ-shifters, a circularly polarized magnetic field region is obtained by using a meander microstrip configuration with light coupling between the lines as shown in Fig. 6.47. This is achieved by closely spacing the lines of length $\lambda/4$ at the centre frequency. Because of tight coupling, the two magnetic field components at a centre point *P* between two adjacent lines become spatially orthogonal. Since lines are $\lambda/4$ long, the *RF* current flowing at the centre of line 2 is delayed by 90° relative to the current flowing at the centre of line 1. Thus, the resulting *RF* magnetic field at *P* is circulary polarized. However, at points away from the centre of the lines, the polarization changes from circular to elliptical and then to linear at the ends.

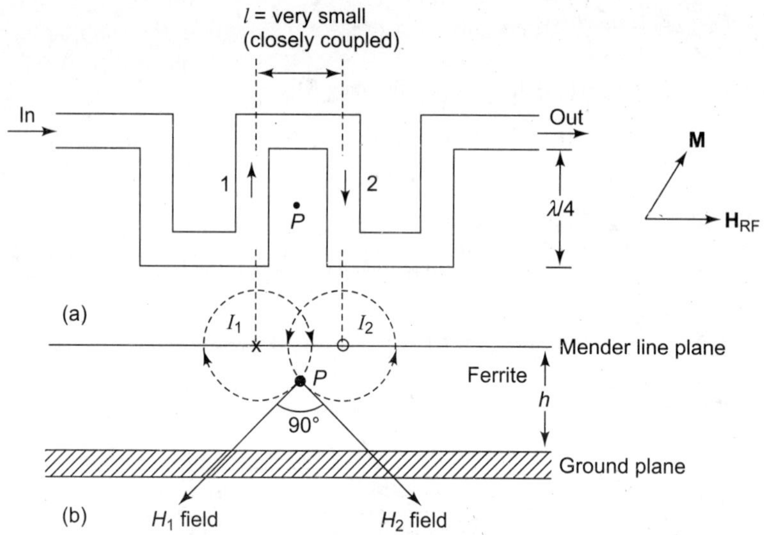

Fig. 6.47 Non-reciprocal phase shifter: (a) Top view of strip (b) Side view from the plane of the strip

By magnetizing the ferrite along the direction of the lines and reversing the direction of the magnetization, a non-reciprocal differential interaction is obtained resulting in a non-reciprocal differential phase shift. Digital/latching phase shifter can be obtained by using sufficiently thin ferrite and small wires through holes in the substrate for dc currents I_1 and I_2. The magnitude and direction of magnetization can be altered by current pulses. Typical parameters for such a phase shifter are the following:

Phase shift—1 bit (360° + 10°)

Substrate size—1″ × 1″
Substrate height = 50 mil; 1 mil = 0.254 mm
Insertion loss = 2.4 dB over 10% bandwidth
Centre frequency = 5.75 GHz

Waveguide Tees 6.4.16

Waveguide tees are three-port components. They are used to connect a branch or section of the waveguide in series or parallel with the main waveguide transmission line for providing means of splitting, and also of combining power in a waveguide system. The two basic types, *E*-plane (series) *T* and *H*-plane (shunt) *T*, are constructed as shown in Fig. 6.48. These are named according to the axis of the side arm which is parallel to the **E** field or the **H** field in the collinear arms, respectively.

Because of the junction, waveguide tees are poorly matched devices. Adjustable matching reactance can be introduced by means of a tuning screw at the centre. Because of symmetry and absence of non-linear elements in the junction, the *S*-matrix is symmetric: $S_{ij} = S_{ji}$; $i = 1, 2; j = 1, 2$. The general *S*-matrix for a tee junction is

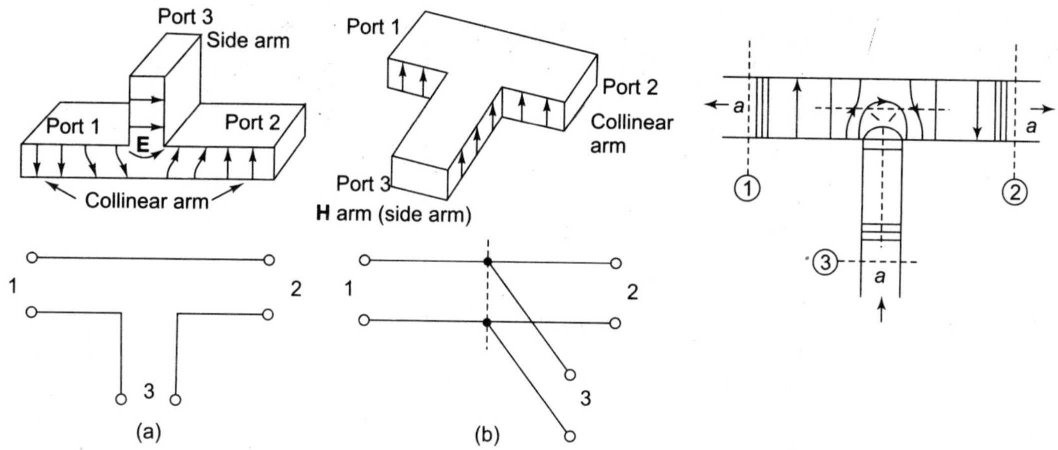

Fig. 6.48 Waveguide tees: (a) E-Tee or Series-T (b) H-Tee or Shunt-T

$$[S] = \begin{bmatrix} S_{11} & S_{12} & S_{13} \\ S_{12} & S_{11} & S_{23} \\ S_{13} & S_{23} & S_{33} \end{bmatrix} \tag{6.97}$$

E-Plane Tee

From considerations of symmetry and the phase relationship of the waves in each of the arms, it can be seen that a wave incident at Port 3 will result in waves at Ports 1 and 2, which are equal in magnitude but opposite in phase, i.e., $S_{31} = S_{13} = -S_{23} = -S_{32}$, $S_{12} = S_{21}$. If two in-phase input waves are fed into Ports 1 and 2 of the collinear arm, the output waves at Port 3 will be opposite in phase and subtractive. Sometimes, this third port is called the *difference arm*. By analogy with the voltage relationship in the series circuit, *E*-plane junction is also called a *series junction*. All diagonal elements of the *S*-matrix of an *E*-plane *T*-junction cannot be zero simultaneously since the tee junction cannot be matched to all the three arms simultaneously. Considering the Port 3 as matched, the *S*-matrix of a *E*-plane *T* can be derived as follows.

Denoting the incident and outgoing signal variables as the ith port by a_i and b_i, respectively, for an input power at Port 3, the net input power to Port 3 is $\frac{1}{2}(|a_3|^2 - |b_3|^2) = \frac{1}{2}|a_3|^2(1 - |S_{33}|^2)$, and the output power is $\frac{1}{2}(|b_1|^2 + |b_2|^2) = |a_3|^2 \cdot |S_{13}|^2$, since $|S_{31}| = |S_{32}|$ by symmetry. For lossless junction, input power must be equal to the output power, i.e.,

$$\frac{1}{2}|a_3|^2(1 - |S_{33}|^2) = |a_3|^2 \cdot |S_{13}|^2 \quad \text{or,} \quad (1 - |S_{33}|^2) = 2|S_{13}|^2$$

By suitable matching elements, we can make $S_{33} = 0$, so that $|S_{13}| = 1/\sqrt{2}$. From the symmetry characteristics described above,

$$S_{13} = S_{31} = 1/\sqrt{2},\ S_{23} = S_{32} = -1/\sqrt{2}$$

After matching Port 3, if one attempts to match either Port 1 or 2 by a similar method, the matching elements, such as irises or tuning screws will interact with each other and matching at Port 3 would be disturbed since no port is isolated. Based on power consideration, it can also be shown that $S_{11} = S_{22} = 1/2$ and $S_{12} = S_{21} = 1/2$ for $S_{33} = 0$. Therefore, with matching at Port 3, the S-matrix of an E-plane T can be expressed by real values with proper choice of reference plane:

$$[S] = \begin{bmatrix} 1/2 & 1/2 & 1/\sqrt{2} \\ 1/2 & 1/2 & 1/\sqrt{2} \\ 1/\sqrt{2} & -1/\sqrt{2} & 0 \end{bmatrix} = \frac{1}{2}\begin{bmatrix} 1 & 1 & \sqrt{2} \\ 1 & 1 & -\sqrt{2} \\ \sqrt{2} & -\sqrt{2} & 0 \end{bmatrix} \tag{6.98}$$

There are three cases to describe below:

❑ *Case 1* Input at port 3 and no inputs at port 1 and 2, $a_3 \neq 0$, $a_1 = a_2 = 0$.

$$b_1 = \frac{1}{\sqrt{2}} a_3$$

$$b_2 = -\frac{1}{\sqrt{2}} a_3$$

$$b_3 = 0$$

❑ *Case 2* Input at port 1 and Port 2, and no input at port 3. $a_3 = 0$, $a_1 = a_2 = a$.

$$b_1 = \frac{a}{2} + \frac{a}{2} = a$$

$$b_2 = \frac{a}{2} + \frac{a}{2} = a$$

$$b_3 = 0$$

❑ *Case 3* Input at port 1 and Port 2, and no input at port 3. $a_1 \neq 0$, $a_3 = a_2 = 0$.

$$b_1 = \frac{a_1}{2}$$

$$b_2 = \frac{a_1}{2}$$

$$b_3 = -\frac{a_1}{\sqrt{2}}$$

Similarly, we have all combinations of input and output.

H-Plane Tee

In an H-plane tee, if two in-phase input waves are fed into ports 1 and 2 of the collinear arm, the output waves at Port 3 will be in-phase and additive. Because of this, the third port is called the *sum arm*. Conversly, an input wave at Port 3 will be equally divided into ports 1 and 2 in phase. Because the magnetic field loops get divided into two arms 1 and 2 in a manner similar to currents between branches in the parallel circuit, an H-plane junction is also called a shunt junction. For a symmetrical and lossless junction, in absence of non-linear elements at the H-plane junction, the S-parameters are obtained in a similar manner as in the case of E-plane junction:

Since it is a three-port junction, the scattering matrix can be derived as follows:

[S] Matrix of order 3 × 3.

$$[S] = \begin{bmatrix} S_{11} & S_{12} & S_{13} \\ S_{21} & S_{22} & S_{23} \\ S_{31} & S_{32} & S_{33} \end{bmatrix} \tag{6.99a}$$

Because of plane of symmetry of the junction, the Scattering coefficients are

$$S_{23} = S_{13}$$

As the waves coming out of port 1 and 2 of the collinear arm will be opposite in phase and in same magnitude. The negative sign indicates phase difference.

If Port 3 is perfectly matched to the junction, $S_{33} = 0$

For symmetric property, $S_{ij} = S_{ji}$

$$\therefore \quad S_{12} = S_{21}: \quad S_{13} = S_{31}: \quad S_{23} = S_{32} = S_{13} \tag{6.99b}$$

With the above properties. [S] becomes,

$$[S] = \begin{bmatrix} S_{11} & S_{12} & S_{13} \\ S_{12} & S_{22} & S_{13} \\ S_{13} & S_{13} & 0 \end{bmatrix} \tag{6.99c}$$

For the symmetry and lossless properties,

$$[S]_t = [S] \text{ and } [S]^* [S]_t = [S]^* [S] = [U] \tag{6.99d}$$

or,

$$|S_{11}|^2 + |S_{12}|^2 + S_{13}|^2 = 1$$

$$|S_{12}|^2 + |S_{22}|^2 + S_{13}|^2 = 1$$

$$0 + |S_{13}|^2 + |S_{13}|^2 = 1$$

$$S_{13} \cdot S_{11}^* + S_{13} \cdot S_{12}^* = 0 \tag{6.99e}$$

From first two equations of (6.99e), we get

$$S_{11} = S_{22} \tag{6.99f}$$

From third equation in (6.99c),

$$S_{13} = \frac{1}{\sqrt{2}} = S_{23} = S_{32} \tag{6.99g}$$

From last equations of (6.99e), we get

$$S_{13}(S_{11}^* + S_{12}^*) = 0$$

but

$$S_{13} \neq 0 \quad \therefore (S_{11}^* + S_{12}^*) = 0$$

$$\therefore \qquad S_{12} = -S_{11}$$

Substituting these values in the first equation of (6.99e),

$$|S_{11}|^2 + |S_{11}|^2 + \frac{1}{2} = 1$$

or,
$$S_{11} = \frac{1}{2} = S_{22}$$

$$S_{12} = -\frac{1}{2}$$

Substituting these values in (6.99c), we get S-matrix of the H-plane Tee

$$[S] = \frac{1}{2}\begin{bmatrix} 1 & -1 & \sqrt{2} \\ -1 & 1 & \sqrt{2} \\ \sqrt{2} & \sqrt{2} & 0 \end{bmatrix} \qquad ((6.99h)$$

when Port 3 is match terminated. Two cases can be described:

- *Case 1* Input is given at Port 3 and no inputs at Ports 1 and 2, $a_3 \neq 0$, $a_1 = a_2 = 0$

$$b_1 = \frac{1}{\sqrt{2}} a_3$$

$$b_2 = \frac{1}{\sqrt{2}} a_3$$

$$b_3 = 0$$

Hence, the power coming out of ports 1 or port 2 is 3 dB down with respect to input power at Port 3, hence the H-plane Tee is called 3 dB splitter.

- *Case 2* Input is gives at Port 1 and Port 2, and no input at Port 3, $a_3 = 0$, $a_1 = a_2 = a$.

$$b_1 = \frac{a}{2} - \frac{a}{2} = 0$$

$$b_2 = \frac{a}{2} - \frac{a}{2} = 0$$

$$b_3 = \frac{a}{\sqrt{2}} + \frac{a}{\sqrt{2}} = \sqrt{2}a$$

Output at Port 3 is the addition of the two inputs at Port 1 and Port 2 and these are added in phase.

Example 6.17 Consider a lossless two-port network. (a) If the network is reciprocal, show that $|S_{21}|^2 = 1 - |S_{11}|^2$. (b) If the network is nonreciprocal, show that it is impossible to have unidirectional transmission, where $S_{12} = 0$ and $S_{21} \neq 0$.

(a)
$$[S] = \begin{bmatrix} S_{11} & S_{21} \\ S_{21} & S_{22} \end{bmatrix} \qquad S_{12} = S_{21} \text{ since reciprocal}$$

$[S]$ is unitary if lossless, so

First row: $|S_{11}|^2 + |S_{21}|^2 = 1$ (or 1st col)

$$|S_{21}|^2 = 1 - |S_{11}|^2$$

(b) $$[S] = \begin{bmatrix} S_{11} & S_{21} \\ 0 & S_{22} \end{bmatrix} \quad S_{12} \neq S_{21} \text{ since nonreciprocal}$$

First row: $|S_{11}|^2 + |S_{21}|^2 = 1$

First Column: $|S_{11}|^2 = 1$

$\therefore$ $|S_{21}| = 0$

Example 6.18 A 20 mW signal is fed into one of the collinear (port 1) of a lossless H-plane T-junction. Calculate the power delivered through each port when other ports are terminated in matched load.

Solution Since Ports 2 and 3 are matched terminated, $a_2 = a_3 = 0$, $|S_{11}| = 1/2$. The total effective power input to Port 1 is

$$P_1 = \frac{1}{2}|a_1|^2 (1 - |S_{11}|^2)$$
$$= 20\,(1 - 0.5^2) = 15 \text{ mW}$$

The power transmitted to Port 3 is

$$P_3 = \frac{1}{2}|a_1|^2 |S_{31}|^2$$
$$= 20 \times (1/\sqrt{2})^2 = 10 \text{ mW}$$

The power transmitted to Port 2 is

$$P_2 = \frac{1}{2}|a_1|^2 |S_{21}|^2$$
$$= 20 \times (1/2)^2 = 5 \text{ mW.}$$

Therefore, $P_1 = P_3 + P_2$

Example 6.19 In an H-plane T-junction, compute power delivered to the loads of 40 ohms and 60 ohms connected to arms 1 and 2 when a 10 mW power is delivered to the matched port 3.

Solution With Port 3 matched, the scattering matrix for H-plane T is

$$[S] = 1/2 \begin{bmatrix} 1 & -1 & \sqrt{2} \\ -1 & 1 & \sqrt{2} \\ \sqrt{2} & \sqrt{2} & 0 \end{bmatrix}$$

Therefore, input power at Port 3 is equally divided in arms 1 and 2. Since input at Port 3 = 10 mW = 0.01 W, power towards ports 1 and 2 = 0.005 W = $(1/2)\,|b_1|^2 = (1/2)\,|b_2|^2$. Considering first-order reflection, reflected power from ports 1 and 2 are

$$(1/2)\,|\Gamma_1 b_1|^2 \quad \text{and} \quad (1/2)\,|\Gamma_2 b_2|^2$$

Therefore, power delivered to load $Z_1 = 40$ ohms and $Z_2 = 60$ ohms are

$$P_1 = (1/2)\,|b_1|^2 - (1/2)\,|\Gamma_1 b_1|^2 = (1/2)\,|b_1|^2\,(1 - |\Gamma_1|^2)$$

and
$$P_2 = (1/2)\,|b_2|^2 - (1/2)\,|\Gamma_2 b_2|^2 = (1/2)\,|b_2|^2\,(1 - |\Gamma_2|^2)$$

Now taking the characteristic impedance of the line = 50 ohms

$$|\Gamma_1| = |40 - 50| / |40 + 50| = 1/9\ ;\ |\Gamma_1|^2 = 0.01234$$

$$|\Gamma_1| = |60 - 50| / |60 + 50| = 1/11\ ;\ |\Gamma_2|^2 = 8.2694 \times 10^{-3}$$

Therefore,
$$P_1 = 0.005\,(1 - 0.01234) = 4.938 \times 10^{-3} = 4.9383 \text{ mW}$$

$$P_2 = 0.005\,(1 - 8.2694 \times 10^{-3}) = 4.9586 \times 10^{-3} \text{ W}$$

$$= 4.9586 \text{ mW}$$

Hybrid or Magic-T

A hybrid tee is formed with the combination of the E-plane and H-plane tees and is called a *magic-T*. It has four ports as shown in Fig. 6.49(a) and 6.49(b).

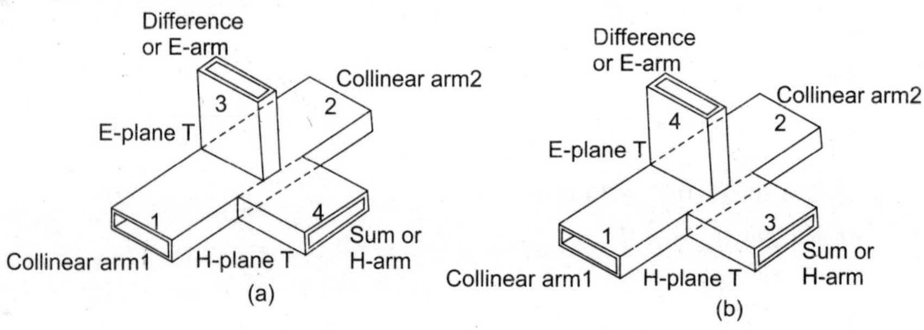

Fig. 6.49 Magic-T s with ports 3 and 4 interchanged

The magic-T has the following characteristics when all the ports are terminated with a matched loads. Let us consider the structure (a) with the ports as indicated.

1. If two waves of equal magnitude and equal phase are fed into ports 1 and 2, the output at Port 3 is subtractive and becomes zero and total output will appear additively at the port 4. Hence, Port 3 is called the difference or E-arm and 4, the sum or H-arm.
2. A wave incident at Port 3 (E-arm) divides equally between ports 1 and 2 but is opposite in phase with no coupling to Port 4 (H-arm). Thus,
$$S_{13} = -S_{23},\ S_{43} = 0 \tag{6.100}$$
3. A wave incident at Port 4 (H-arm) divides equally between ports 1 and 2 in phase with no coupling to port 3 (E-arm). Thus,
$$S_{14} = S_{41} = 1/\sqrt{2} = S_{24} = S_{42} \text{ and } S_{34} = 0 \tag{6.101a}$$
4. A wave fed into one collinear port, 1 or 2, will not appear in the other collinear Ports, 2 or 1, respectively. Hence, two collinear ports 1 and 2 are isolated from each other, making
$$S_{12} = S_{21} = 0 \tag{6.101b}$$

A magic-T can be matched by putting tuning screws suitably in the E and H-arms without destroying the symmetry of the junctions. Therefore, for an ideal lossless magic-T matched at ports 3 and 4, $S_{33} = S_{44} = 0$. The procedure of derivation of the *S*-matrix considers the symmetry property at the junction for which S_{14}

$= S_{41} = S_{24} = S_{42}$, $S_{31} = S_{13} = -S_{23} = -S_{32}$, $S_{34} = S_{43} = 0$, $S_{12} = S_{21} = 0$. Therefore, the S-matrix for a magic-T, matched at ports 3 and 4 given by

$$[S] = \begin{bmatrix} S_{11} & S_{12} & S_{13} & S_{14} \\ S_{12} & S_{22} & -S_{13} & S_{14} \\ S_{13} & -S_{13} & 0 & 0 \\ S_{14} & S_{14} & 0 & 0 \end{bmatrix} \tag{6.102}$$

From the unitary property applied to rows 1 and 2, we get

$$|S_{11}|^2 + |S_{12}|^2 + |S_{13}|^2 + |S_{14}|^2 = 1 \tag{6.103}$$

$$|S_{12}|^2 + |S_{22}|^2 + |S_{13}|^2 + |S_{14}|^2 = 1 \tag{6.104}$$

Subtracting these two equations:

$$|S_{11}|^2 - |S_{22}|^2 = 0 \quad \text{or,} \quad |S_{11}| = |S_{22}| \tag{6.105}$$

Form the unitary property applied to rows 3 and 4,

$$2\,|S_{13}|^2 = 1, \quad \text{or} \quad |S_{13}| = 1/\sqrt{2} \tag{6.106}$$

$$2\,|S_{14}|^2 = 1, \quad \text{or} \quad |S_{14}| = 1/\sqrt{2} \tag{6.107}$$

Substituting these values in Eq. (6.103),

$$|S_{11}|^2 + |S_{12}|^2 + 1/2 + 1/2 = 1 \quad \text{or,} \quad |S_{11}|^2 + |S_{12}|^2 = 0 \tag{6.108}$$

which is valid if

$$S_{11} = S_{12} = 0 \tag{6.109}$$

From Eqs (6.105) and (6.109), $S_{22} = 0$ (6.110)

Therefore,

$$[S] = \begin{bmatrix} 0 & 0 & S_{13} & S_{13} \\ 0 & 0 & -S_{13} & S_{13} \\ S_{13} & -S_{13} & 0 & 0 \\ S_{13} & S_{13} & 0 & 0 \end{bmatrix} \tag{6.111}$$

where $|S_{13}| = 1/\sqrt{2} = |S_{14}|$

By proper choice of reference planes in arms 3 and 4, it is possible to make both S_{13} and S_{14} real, resulting in the final form of S-matrix of magic-T.

$$[S] = 1/\sqrt{2}\begin{bmatrix} 0 & 0 & 1 & 1 \\ 0 & 0 & -1 & 1 \\ 1 & -1 & 0 & 0 \\ 1 & 1 & 0 & 0 \end{bmatrix} \tag{6.112a}$$

For the structure (b) where ports 3 and 4 are interchanged, the S-matrix becomes

$$[S] = 1/\sqrt{2}\begin{bmatrix} 0 & 0 & 1 & 1 \\ 0 & 0 & 1 & -1 \\ 1 & 1 & 0 & 0 \\ 1 & -1 & 0 & 0 \end{bmatrix} \tag{6.112b}$$

Example 6.20 A magic-T is terminated at collinear ports 1 and 2 and difference port 4 by impedances of reflection coefficients $\Gamma_1 = 0.5$, $\Gamma_2 = 0.6$ and $\Gamma_4 = 0.8$, respectively. If 1W power is fed at the sum port 3, calculate the power reflected at Port 3 and power transmitted to the other three ports.

Solution S-matrix for a matched magic-T with collinear ports 1 and 2 and sum and difference ports 3 and 4, respectively, is given by

$$[S] = 1/\sqrt{2}\begin{bmatrix} 0 & 0 & 1 & 1 \\ 0 & 0 & 1 & -1 \\ 1 & 1 & 0 & 0 \\ 1 & -1 & 0 & 0 \end{bmatrix}$$

If a_1, a_2, a_3 and a_4 be the normalized input voltages and b_1, b_2, b_3, and b_4 are the corresponding output voltage at ports 1, 2, 3, and 4 respectively, then

$a_1 = \Gamma_1 b_1$, $a_2 = \Gamma_2 b_2$, a_3 = input applied voltage at port 3, and

$a_4 = \Gamma_4 b_4$.

Now, $P_i = |a_3|^2/2 = 1\text{W}$, or, $a_3 = \sqrt{2}$ V

Therefore,

$$\begin{bmatrix} b_1 \\ b_2 \\ b_3 \\ b_4 \end{bmatrix} = 1/\sqrt{2}\begin{bmatrix} 0 & 0 & 1 & 1 \\ 0 & 0 & 1 & -1 \\ 1 & 1 & 0 & 0 \\ 1 & -1 & 0 & 0 \end{bmatrix}\begin{bmatrix} .5b_1 \\ .6b_2 \\ \sqrt{2} \\ .8b_4 \end{bmatrix}$$

or,

$$b_1 + 0 + 0 - 0.8\, b_4/\sqrt{2} = 1$$
$$0 + b_2 + 0 + 0.8\, b_4/\sqrt{2} = 1$$
$$-0.5b_1/\sqrt{2} - 0.6\, b_2/\sqrt{2} + b_3 + 0 = 0$$
$$-0.5b_1/\sqrt{2} + 0.6\, b_2/\sqrt{2} + 0 + b_4 = 0$$

The unknown quantities b's may be solved by Cramer's rule.

$$b_1 = \frac{\dfrac{1}{\sqrt{2}}\begin{bmatrix} \sqrt{2} & 0 & 0 & -0.8 \\ \sqrt{2} & \sqrt{2} & 0 & 0.8 \\ 0 & -0.6 & \sqrt{2} & 0 \\ 0 & 0.6 & 0 & \sqrt{2} \end{bmatrix}}{\dfrac{1}{\sqrt{2}}\begin{bmatrix} \sqrt{2} & 0 & 0 & -0.8 \\ 0 & \sqrt{2} & 0 & 0.8 \\ -0.5 & -0.6 & \sqrt{2} & 0 \\ -0.5 & 0.6 & 0 & \sqrt{2} \end{bmatrix}} = \frac{2.0782}{2.2381} = 0.928 \text{ V}$$

Similarly, $$b_2 = \frac{2.3981}{2.2381} = 1.07 \text{ V}$$

$$b_3 = \frac{1.7524}{2.2381} = 0.78 \text{ V}$$

$$b_4 = \frac{-0.2827}{2.2381} = -0.126 \text{ V}$$

Therefore,

$$\text{Power transmitted at Port 1} = \frac{1}{2}\,|b_1|^2 = 0.4309 \text{ W}$$

$$\text{Power transmitted at Port 2} = \frac{1}{2}\,|b_2|^2 = 0.5738 \text{ W}$$

$$\text{Power transmitted at Port 4} = \frac{1}{2}\,|b_4|^2 = 0.00797 \text{ W}$$

$$\text{Power reflected at Port 3} = \frac{1}{2}\,|b_3|^2 = 0.3065 \text{ W}$$

Note: Power absorbed at Port $i = 1/2\ (|b_i|^2 - |a_i|^2)$; $i = 1, 2, 4$. Total power absorbed by the system $= (|a_3|^2 - |b_3|^2)\dfrac{1}{2}$.

Application of Magic-T

The magic-*T* has a number of applications in various microwave circuits, such as the *E-H* tuner for impedance matching, balanced mixer in a microwave superheterodyne to balance out the local oscillator noise at the IF amplifier input, power combiner and duplexer.

❑ *E-H Tuner* In an *E-H* tuner (Fig. 6.50), both the *E* and *H*-arms are terminated by movable shorts which act as *E*-plane and *H*-plane stubs. The position of the shorts can be adjusted so that a wide range of load impedance may be matched to reduce the VSWR of a waveguide system connected through the collinear arms.

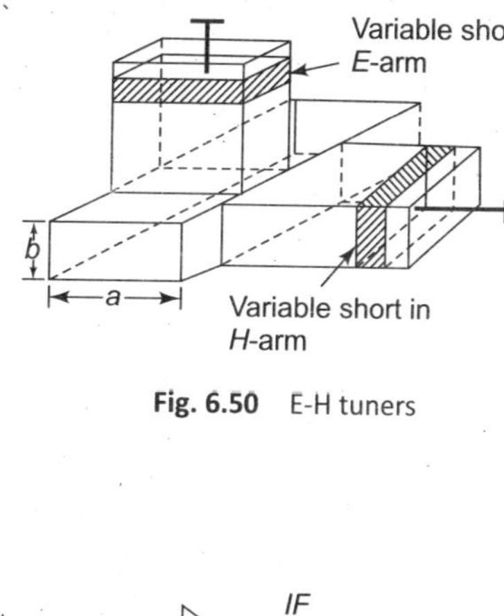

Fig. 6.50 E-H tuners

❑ *Balanced Mixer* In a balanced microwave mixer configuration, an incoming signal is fed to the *E*-arm and a local oscillator signal is fed to the *H* arm as shown in Fig. 6.51. When these two signals enter the collinear arms, the crystal diodes placed in these arms produce the IF signal or difference signal in the following manner.

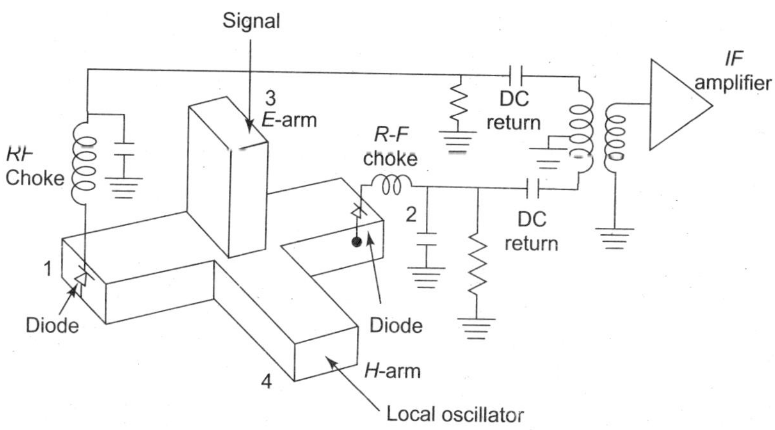

Fig. 6.51 Microwave mixer

The local oscillator signal from the H-arm will arrive at the diodes in collinear arms in-phase, whereas the incoming signal from the E-arm will arrive at the diodes out-of-phase. These signals are mixed in the non-linear diodes and produce IF signals in the collinear arms which are out-of-phase by 180°. Since local oscillator noise will be in-phase at the diodes, this gets cancelled at the balanced IF input to IF amplifier whereas, the IF signals are added up in phase for amplification in IF amplifier. LO and RF signals are uncoupled due to magic-T properties of E and H-arms. For equal power inputs at isolated ports 3 and 4, $P_3 = P_4 = P = \frac{1}{2}|a|^2$, where $a_3 = a_4 = a$. If Port 1 is matched, $a_1 = 0$. Therefore,

$$b_2 = (0 + 0 - a + a)/\sqrt{2} = 0.$$

and hence $a_2 = 0$. Consequently $b_3 = b_4 = 0$. Thus

$$b_1 = S_{13}\,a + S_{14}\,a = \sqrt{2}\,a$$

Therefore, output power at Port 1 is $P_1 = \frac{1}{2}|b_1|^2 = |a|^2 = 2P$ = sum of two equal input powers.

In a duplexer circuit of a radar system, a common antenna is connected to Port 1 while the transmitter and receiver are connected to isolated E and H-arms and a dummy load is connected at Port 2. Half of the power transmitted is coupled to the antenna and 50 % of the received power gets into the receiver, remaining 50 % power is absorbed in dummy load.

Ferromagnetic Insert and Component 6.4.17

Ferromagnetic materials (ferrite: Mg + Mn, Ni + Zn alloys) when placed in dc magnetic field, electromagnetic wave propagation becomes non-reciprocal. This property is used for construction of circulators and isolators.

Circulators

A circulator is a multiport junction in which the wave can travel from one port to the next immediate port in one direction only as shown in Fig. 6.52(a). Commonly used circulators are three-port or four-port passive devices although more number of ports is possible.

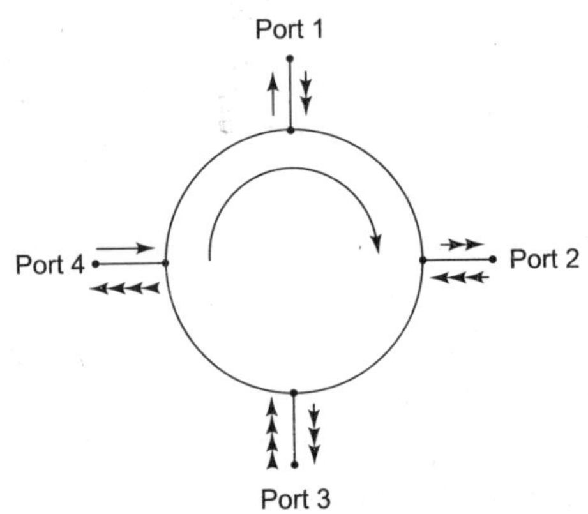

Fig. 6.52(a) Schematic diagram of a four-port circulator

❑ ***Four-port Circulator*** A four-port circulator can be constructed from two magic-T's and a non-reciprocal 180° phase shifter or a combination of two 3 dB side hole directional couplers with two non-reciprocal phase shifters as shown in Figs 6.52 (b) and (c).

In Fig. 6.52 (b), an input signal at Port 1 is split into two in-phase and equal amplitude waves in the collinear arms b and d of the magic-tee, T_1 and added up to emerge from Port 2 in the magic tee, T_2. On the other hand a signal at Port 2 will split into two equal amplitude and equiphase waves in the collinear arms of the magic-tee, T_2 and appears at point b and d out of phase due to presence of the non-reciprocal 180° phase shifter. These out-of-phase waves add up and appear from Port 3 in the magic-tee, T_1. In a similar manner, an input signal at Port 3 will emerge from 4, an input at Port 4 will appear at Port 1. Thus, the circulator property is exhibited.

In Fig. 6.52(c), each of the two 3 dB couplers introduces a 90° phase shift. An input signal at Port 1 is split into two components by the coupler 1 and the coupled signals are again split into two components by the coupler 2 with a 90° phase shift in each. Each of the two phase shifters produces additional phase shift so that the signal components at Port 2 are in phase, and at Port 4, they are out of phase. Since Port 3 is the decoupled port for the directional coupler, the input signal at Port 1 appears in Port 2. Similarly, signals from Port 2 to Port 3, from Port 3 to Port 4 and from Port 4 to Port 1.

A perfectly matched, lossless, and non-reciprocal four-port circulator has S-matrix:

$$[S] = \begin{bmatrix} 0 & 0 & 0 & 1 \\ 1 & 0 & 0 & 0 \\ 0 & 1 & 0 & 0 \\ 0 & 0 & 1 & 0 \end{bmatrix} \tag{6.113}$$

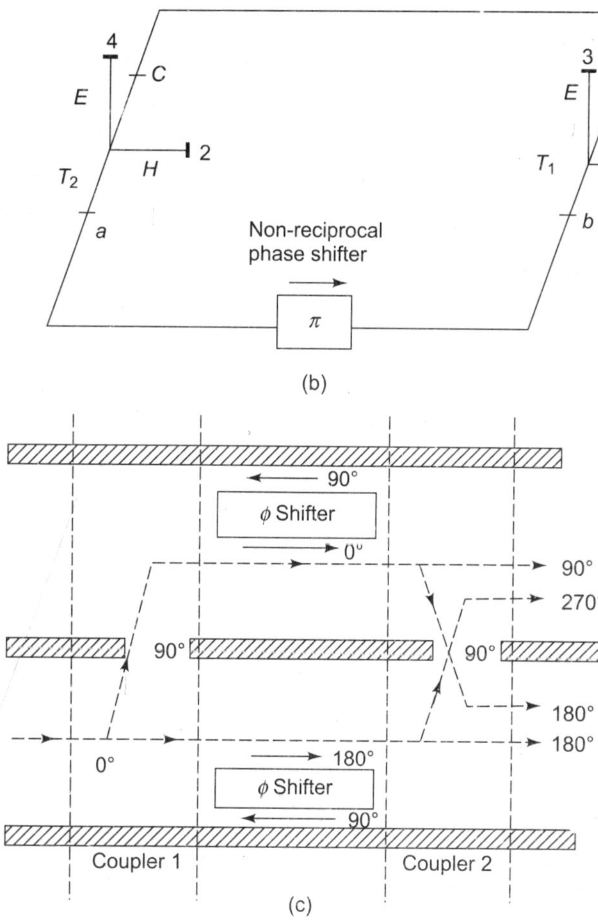

Fig. 6.52(b–c) Four-port circulators

❑ *Three-port Circulator* A three-port circulator is formed by a 120° H-plane waveguide or strip line symmetrical Y-junction with a central ferrite post or disc. A steady magnetic field $\mathbf{H}_0$ is applied along the axis of the disc as shown in Fig. 6.53. Depending on the polarization of the incident wave and the direction of $\mathbf{H}_0$, the microwave signal travels from one port to the immediate next one only.

For a perfectly matched, lossless, non-reciprocal three-port circulator, the S-matrix is

$$[S] = \begin{bmatrix} 0 & 0 & S_{13} \\ S_{21} & 0 & 0 \\ 0 & S_{32} & 0 \end{bmatrix} \tag{6.114}$$

If the terminal planes are properly chosen to make the phase angles of S_{13}, S_{21} and S_{32} zero,

$$S_{13} = S_{21} = S_{32} = 1$$

so that

$$[S] = \begin{bmatrix} 0 & 0 & 1 \\ 1 & 0 & 0 \\ 0 & 1 & 0 \end{bmatrix} \tag{6.115}$$

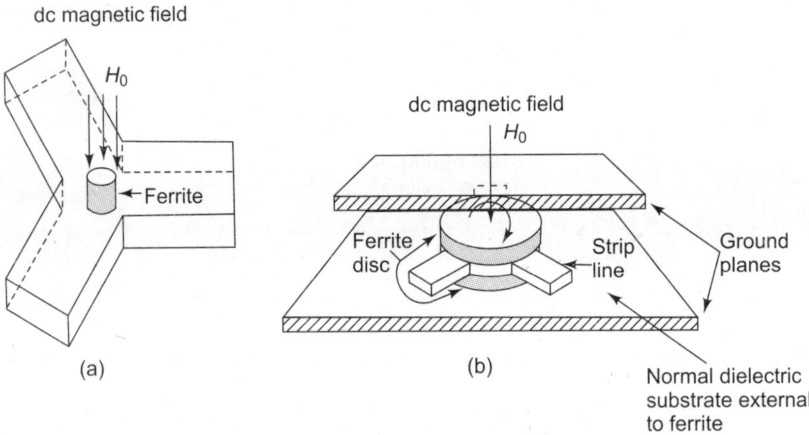

Fig. 6.53 Three-port circulator: (a) Waveguide type (b) Stripline type

The matching of the junction can be achieved by placing suitable tuning elements in each arm.

Since, in practice, losses are always present, the performance is limited by finite isolation and non-zero insertion loss. Typical characteristics can be represented by

Insertion loss < 1 dB
Isolation ≈ 30–40 dB
VSWR < 1.5

Example 6.21 Prove that it is impossible to construct a perfectly matched, lossless, reciprocal three-port junction.

Solution A perfectly matched three-port junction has a symmetric scattering matrix with zero diagonal elements:

$$[S] = \begin{bmatrix} 0 & S_{12} & S_{13} \\ S_{12} & 0 & S_{23} \\ S_{13} & S_{23} & 0 \end{bmatrix}$$

For a lossless junction, S-matrix is unitary, so that

$$S_{12}S_{12}^* + S_{13}S_{13}^* = 1 \quad \text{(1st row)}$$
$$S_{12}S_{12}^* + S_{23}S_{23}^* = 1 \quad \text{(2nd row)}$$
$$S_{13}S_{13}^* + S_{23}S_{23}^* = 1 \quad \text{(3rd row)}$$

and $S_{13}S_{23}^* = S_{12}S_{23}^* = S_{12}S_{13}^* = 0$

If S_{12} is not equal to zero, the fourth equation from above gives $S_{13} = 0 = S_{23}$. But this does not satisfy the third equation. Therefore, it is not possible to construct a perfectly matched, lossless, reciprocal three-port network.

Example 6.22 A three-port circulator has an insertion loss of 1 dB, isolation-30 dB and VSWR = 1.5. Find the S-matrix.

Solution The S-matrix of a three-port circulator is

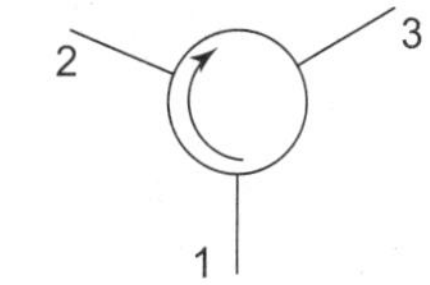

Fig. 6.54 Three-port circulator for Example 6.22

$$[S] = \begin{bmatrix} S_{11} & S_{12} & S_{13} \\ S_{21} & S_{22} & S_{23} \\ S_{31} & S_{32} & S_{33} \end{bmatrix}$$

Insertion loss = 1 dB = – 20 log $|S_{21}|$

or, $$|S_{21}| = 10^{-1/20} = 0.89$$

For the same insertion loss between ports 1 and 2, 2 and 3, 3 and 1, $|S_{21}| = |S_{32}| = |S_{13}| = 0.89$.

The isolation between the ports is 30 dB = – 20 log $|S_{31}|$

or, $$|S_{31}| = 10^{-30/20} = 10^{-1.5} = 0.032$$

$$= |S_{23}| = |S_{12}|$$

Since VSWR S = 1.5, reflection coefficient

$$|\Gamma| = \frac{S-1}{S+1} = \frac{1.5-1}{1.5+1} = 0.2 = |S_{11}|$$

$$= |S_{22}| = |S_{33}|$$

By placing reference planes suitably to make the phase of S-parameters zero, the S-matrix becomes

$$[S] = \begin{bmatrix} 0.200 & 0.032 & 0.890 \\ 0.890 & 0.200 & 0.032 \\ 0.032 & 0.890 & 0.200 \end{bmatrix}$$

❑ ***Microstrip Circulator*** A microstrip circulator is constructed in the same form as strip line one without having the top ground plane. The entire circulator can be made by using the whole ferrite substrate. The basic design criteria are

- Selection of radius R of ferrite disc
- Calculation of radius R of magnet cylinder for the whole ferrite substrate.

The value of R is given by

$$R = \frac{1.84}{\omega_0\sqrt{(\varepsilon_0\varepsilon_r\,\mu_0\mu_{\text{eff}})}} \quad (6.116a)$$

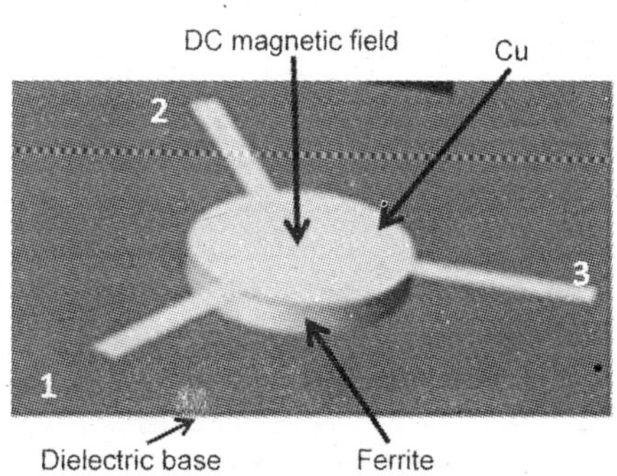

Fig. 6.55 Microstrip circulator

where ω_0 = Centre angular frequency
ε_0 = Free-space permittivity
ε_r = Relative permittivity of the ferrite
μ_0 = Free-space permeability
$\mu_{eff} = (\mu^2 - K^2)/\mu$ = scalar effective permeability
$\mu = 1 - p\sigma/(1 - \sigma^2)$

$$K = \frac{p}{1-\sigma^2}$$

$p = |\gamma|M_s/\omega$ = Normalized saturation magnetization
$= M_s/H_0$
$\sigma = |\gamma|H_0/\omega$ = Normalized biasing magnetic field
$= H_{dc}/H_0$

$H_0 = \dfrac{\omega}{|\gamma|}$ = Field required for gyromagnetic resonance in the infinite ferrite medium

For low magnetic loss, ferrite material selection is done such that

$$\gamma 4\pi M_s/\omega = 0.6 \tag{6.116b}$$

The insertion loss between coupled ports is determined by
1. Copper loss of the strip and ground plane
2. Dielectric loss of the input/output strips
3. Magnetic loss of the ferrite disc

The isolation between the ports is primarily dependent on the mismatching between the ports and the junction. Tuning screws, $\lambda/4$ transformer or tapered strips can be used for matching and this can be the parts of the mask used in photo-etching process. The magnetic disc can be located above or below the substrate.

Using YIG substrate of thickness $h = 0.055''$ one can get Isolation > 20 dB, VSWR < 1.2, insertion loss 1 L < 0.8 dB, power handling capability of 60 W over the frequency range 8.5–9.9 GHz.

The power-handling capability of such a device can be increased by
1. Lowering the impedance or increasing intrinsic line width
2. Increasing substrate thickness h
3. Decreasing $4\pi M_s$ of the material by substituting Al ions in the YIG material.

Isolators

An isolator is a two-port, non-reciprocal device which produces a minimum attenuation to wave propagation in one direction and very high attenuation in the opposite direction. Thus, when inserted between a signal source and load, almost all the signal power can be transmitted to the load and any reflected power from the load is not fed back to the generator output port. This eliminates variations of source power output and frequency pulling due to changing loads.

An isolator can be constructed in a rectangular waveguide ($a \times b$) operating in dominant mode as shown in Fig. 6.56(a). The non-reciprocal characteristics are obtained by establishing a steady magnetic field $\mathbf{H}_0$ in the y direction and placing a ferrite slab at any of the longitudinal planes $x = x_1$ near and parallel to the narrow waveguide wall, where the magnetic field exhibits circular polarization. This occurs at $x_1 = a/4$ or, $3a/4$.

For the propagation of waves in $+z$ direction, direction of rotation of $\mathbf{H}$ in the planes at $x_1 = a/4$ and $3a/4$ are opposite to each other. The non-reciprocal characteristic is achieved by placing a ferrite slab at any one of these two planes. The required steady state magnetic field $\mathbf{H}_0$ in the y-direction is established by placing permanent magnetic poles between the two broad walls.

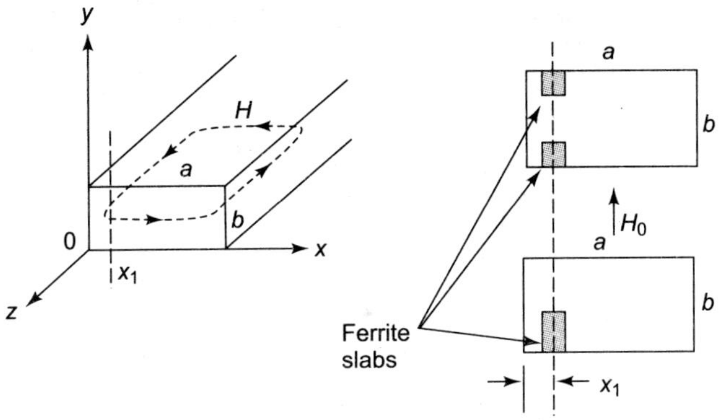

Fig. 6.56(a) Waveguide isolator

It is known that the attenuation in ferrite for negative/clockwise circular polarization is very small whereas for positive/counter clockwise circular polarization is very large at and near the resonance frequency $f \approx f_0$. Therefore, the ferrite slab is placed in such a way that while transmission it encounters negative circular polarization in the reverse direction. The steady magnetic field is set to be equal to the resonant value. The isolation of the order of 20 – 30 dB in the backward direction and a transmission loss of 0.5 dB in the forward direction can be achieved with a VSWR of the order of 1.1.

Since the reverse power is absorbed in the ferrite and dissipated as heat, the maximum power handling capability of an isolator is limited. To increase the capacity of heat dissipation, two ferrite slabs of smaller heights are used instead of one with a larger height.

The requirement of very high steady magnetic field (10,000 oersteds at $\lambda = 1$ cm) is the main drawback of waveguide resonance isolator at higher frequencies.

❑ ***Faraday Rotation Isolator*** A Faraday rotation isolator is a circular waveguide section axially loaded with a ferrite rod of smaller diameter as shown in Fig. 6.56(b).

The ferrite rod is subjected to a steady axial magnetic field $\mathbf{H}_0$ of strength much smaller than the resonant intensity so that dissipative loss in the ferrite is neglected. The dominant TE_{11} mode in the circular section can be decomposed into two oppositely rotating circularly polarized waves of equal magnitude. These waves encounter different permeabilities μ'_+ and μ'_- for the clockwise and anticlockwise directions of field rotation and exhibit changes in the phase velocities. This will result in a change in the plane of polarization of the main mode TE_{11} which will experience gradual rotation θ during propagation to the other end. The rotation angle θ is proportional to the length of the ferrite rod. In this case for the reverse wave the direction of rotation remains the same confirming the non-reciprocal characteristics of the ferrite.

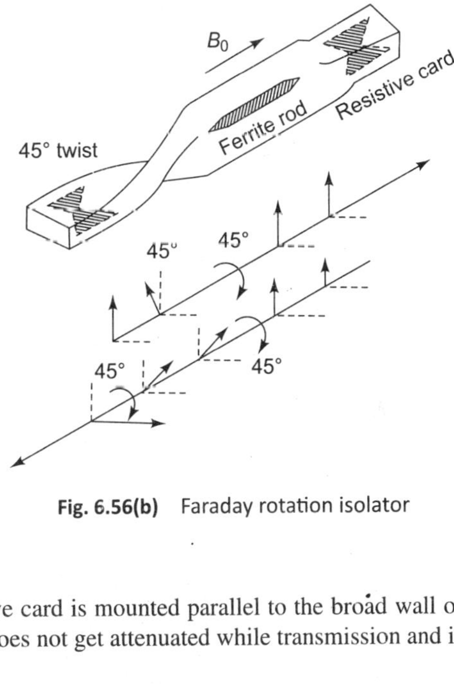

Fig. 6.56(b) Faraday rotation isolator

The isolator input is a 45° twist where a tapered resistive card is mounted parallel to the broad wall of the rectangular waveguide part. The dominant TE_{10} mode does not get attenuated while transmission and is rotated at 45° at the twist output and enters the circular waveguide through rectangular to circular waveguide transition as the TE_{11} mode. The length of the ferrite rod is selected so as to obtain Faraday rotation $\theta = 45°$ at the output and regain its original polarization. The plane of polarization of the reflected wave from the load is again rotated by the same angle $\theta = 45°$ and at the emergence through the 45° twist becomes aligned with the surface of the absorbing plate and gets absorbed. Thus, non-reciprocal isolation action lakes place. Typical insertion loss and isolation are approximately 1dB and 20~30 dB, respectively, for these isolators. Isolators are also available in the coaxial and microstrip forms.

For an ideal lossless, matched isolator

$$|S_{21}| = 1, |S_{12}| = |S_{11}| = |S_{22}| = 0 \tag{6.117a}$$

i.e.,

$$[S] = \begin{bmatrix} 0 & 0 \\ 1 & 0 \end{bmatrix} \tag{6.117b}$$

Example 6.23 A matched isolator has insertion loss of 0.5 dB and an isolation of 25 dB. Find the scattering coefficients.

Solution

Insertion loss = 0.5 dB = $-20 \log |S_{21}|$

or, $S_{21} = 10^{-.5/20} = 10^{-0.025} = 0.9441$

The isolation is 25 dB = $-20 \log |S_{12}|$

or, $S_{12} = 10^{-25/20} = 10^{-1.2} = 0.0631$

Since there is no reflection, $S_{11} = S_{22} = 0$. Therefore, the S-matrix for the isolator is

$$[S] = \begin{bmatrix} 0 & 10^{-1.2} \\ 10^{-0.025} & 0 \end{bmatrix} = \begin{bmatrix} 0 & 0.0631 \\ 0.9441 & 0 \end{bmatrix}$$

❑ ***Microstrip Isolator*** A microstrip isolator can be formed by match terminating one port of a 3-port circulator. Several termination methods are described with the help of Fig. 6.57(a–d).

(a) *Method 1* A chromium layer is deposited on ferrite initially in all circuits for adherence. A tapered Cr film of 0.6″ long with a surface impedance of 270-ohm termination gives VSWR < 1.2 over the frequency range 5.5–11.0 GHz.

(b) *Method 2* Initial Cr adherence is done as Method 1. Then the Cr film acts as a lumped 50-ohm resistor which is terminated in a quarter wavelength open circuit impedance. This arrangement provides VSWR < 1.2 over the frequency range 4.5–6.0 GHz.

(c) *Method 3* A 50-ohm lumped resistor shunt mounted across the strip line and terminated by a short at the end.

(d) *Method 4* An isolator in MIC form on ferrite substrate can be designed with a triangular patch having a narrow transverse slot and corner cuts as shown in Fig. 6.57(d). The dc magnetic field is applied perpendicular to the substrate plane. For X-band operation, the approximate size of the isolator may be of the order of 10 mm × 8 mm × 1/.8" and slot size of approximately 6 mm × 0.8 mm for best matching at 11 GHz. The insertion loss of nearly 2 dB and isolation of 18 dB could be achieved. The metallization on the Ni–Zn ferrite is deposited by sputtering a seed layer of titanium tungsten (TiW) prior to the final metallization.

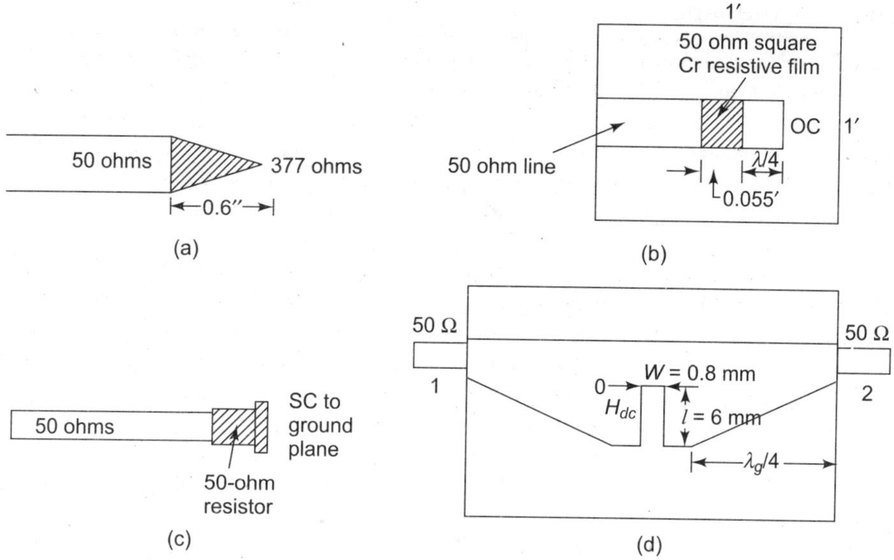

Fig. 6.57 Microstrip isolator

YIG Filters and Oscillators

Yttrium Iron Garnet (YIG) is a complex solid having ferromagnetic properties and is represented by $Me_3Fe_5O_{12}$ or $5Fe_2O_3\ 3Me_2O_3$ or $Me_3Fe_2(FeO_4)_3$, where Me is a rare-earth metal. It has much smaller losses than ordinary ferrites. A small sphere (40 mil diameter) of YIG usually has a very small loss and can be used as a resonator in microwave filters and oscillators with resonant frequency at $f_0 = 2.8\ H_0$, where f_0 is in MH_z and H_0 is steady external field in oersteds. Thus, the resonant frequency can be controlled electronically by changing H_0. The low frequency limit (about 3.5 GHz) of the YIG resonator is controlled by the minimum μ_0 applied for saturation of magnetization. This frequency can be brought down by doping pure YIG with gadolinium (Gd).

A small YIG sphere is placed at the centre of two mutually perpendicular loops in *yz* and *xz* planes (Fig. 6.58). A dc magnetic field $\mathbf{H}_0$ along *z*-axis is created perpendicular to the two loops. Input RF current is fed to the input loop and the output loop is connected to the load. In the steady $\mathbf{H}_0$ field, the YIG has a net magnetization in the $\mathbf{H}_0$ direction. The resulting magnetic field component H_x makes the magnetization vector precess about the *z*-axis resulting in a *y*-component of magnetization which in turn induces a voltage at the output loop. The off-frequency components of *RF* input couples a small power to the output loop. Thus, the system acts like a tuned band-pass filter at frequency f_0.

In a waveguide, the YIG sphere is placed in the coupling hole of two rectangular waveguides having dominant mode magnetic fields perpendicular to each other for filtering action.

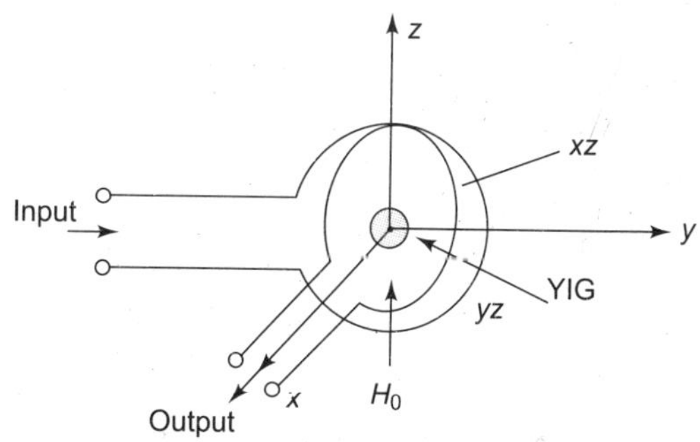

Fig. 6.58 YIG filters and oscillators

Directional Couplers 6.4.18

A directional coupler is a four-port passive device commonly used for coupling a known fraction of the microwave power to a port (coupled port) in the auxiliary line while flowing from the input port to the output port in the main line. The remaining port is an ideally isolated port and matched terminated. There are three basic types of directional couplers. One is a multiple aperture waveguide type as shown in Fig. 6.59(a), the second is a coupled coaxial, or strip or microstrip line as shown schematically in Fig. 6.59(b), and the third is a branch line couplers as shown in Fig. 6.59(c).

The theory of the waveguide coupler was first established by 'Bethe', using a single hole in the common broad wall of two rectangular waveguides, commonly known as *Bethe-hole* directional coupler. Practical waveguide directional couplers are multihole couplers in which the desired coupling response vs frequency can be achieved by proper selection of the number of holes and size of the holes. *Waveguide couplers are forward couplers since the coupled power in the ancillary guide flows in the same direction as the input power in the main guide, whereas the coaxial, strip and microstrip couplers are backward couplers because the coupled energy travels in the opposite direction of the input energy flow.*

The performance of a directional coupler is measured in terms of four basic parameters, i.e., coupling (C), transmission loss (T), directivity (D), and the return loss (R) when all the ports are matched. These are defined as

$$C\,(\text{dB}) = 10 \log (\text{Input power/Coupled output power}) = 10 \log(P_1/P_4) \tag{6.118}$$

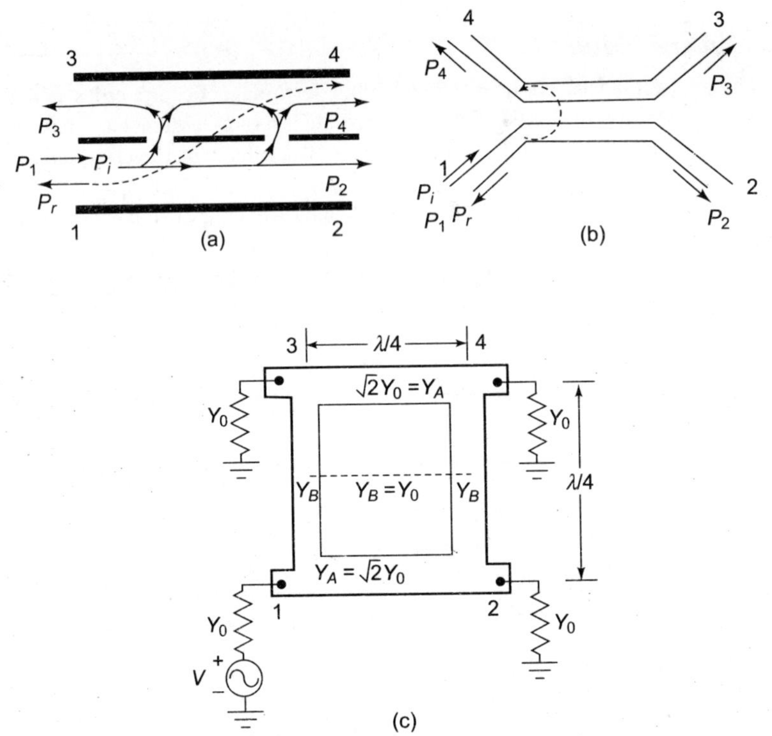

Fig. 6.59 Directional couplers: (a) Waveguide forward coupler (b) Strip and microstrip line backward coupler, and (c) Branch line coupler

$$D\text{ (dB)} = 10\log\text{ (Isolated port power/Coupled output power)}$$
$$=10\log(P_4/P_3) \tag{6.119}$$
$$T\text{ (dB)} = 10\log\text{ (Input power/Direct-line output power)}$$
$$=10\log(P_1/P_2) \tag{6.120a}$$
$$R\text{ (dB)} = 10\log\text{ (Input power/Reflected output power at input)}$$
$$=10\log(P_i/P_r) \tag{6.120b}$$

It may be noted that the coupled and isolated port numbering is interchange in Fig. 6.59(a) and 6.58(b) so that definitions (6.118)–(6.120b) are valid for both the couplers.

Waveguide Directional Coupler

A waveguide directional coupler commonly consists of two waveguides coupled together through one (Bethe-hole) or a number of small openings (multihole) in a common broad wall as shown in Fig. 6.60.

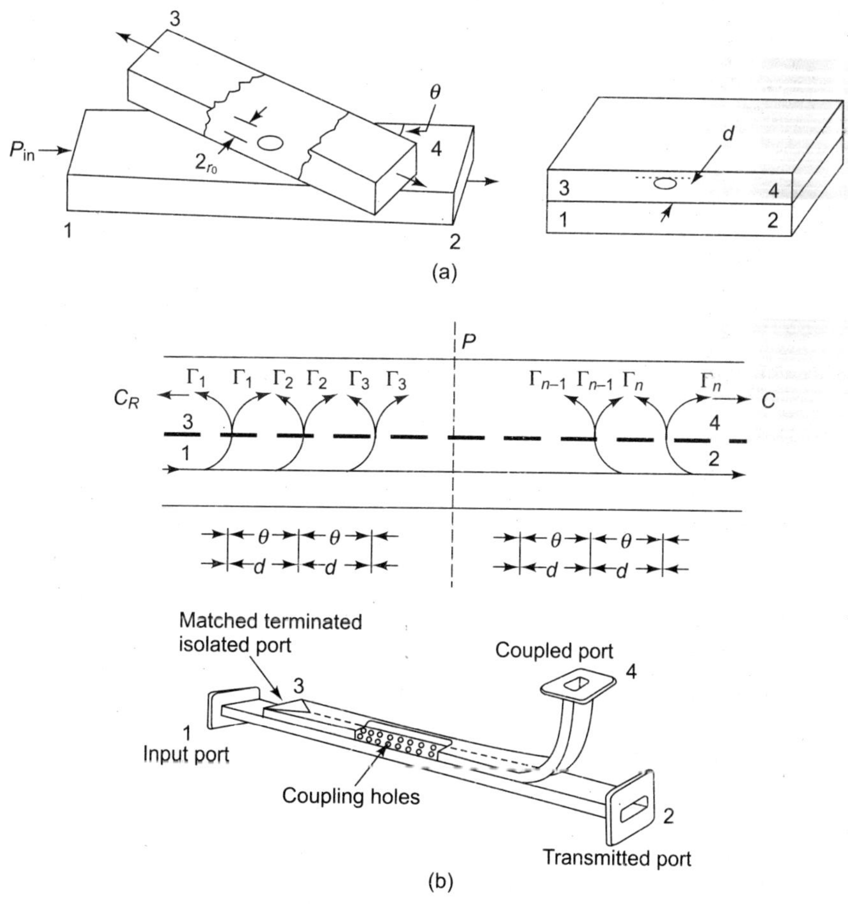

Fig. 6.60 Waveguide directional couplers: (a) Bethe-hole coupler (b) Multihole coupler

Bethe-hole Coupler

The Bethe-hole coupler is a single hole broad wall aperture coupled waveguide coupler, where the hole is located at the centre of a common broad wall of two waveguides inclined at an angle θ, or at an offset position d of two parallel waveguides as shown in Fig. 6.60(a). If the aperture is small compared to the wavelength, it can be considered as an electric dipole normal to the aperture with dipole moment proportional to the normal component of the electric field in the main guide at the aperture, plus a magnetic dipole in the plane of the aperture with dipole moment proportional to the tangential component of the exciting magnetic field at the aperture. The coupling to the auxiliary guide is due to radiation from these dipoles. The electric dipole radiates equally in both the directions longitudinally. But the magnetic dipole radiates asymmetrically in longitudinal directions. By varying the angle θ between the two waveguides, or by adjusting the distance d of the hole from the broad wall edge, powers in Port 3 and Port 4 can be controlled. Ideally, power at Port 3 can be zero (isolated) whereas that at Port 4 is maximum to achieve directional coupling.

Bethe Centre Hole For the centre-coupling hole, since practically $P_3 \neq 0$, the directivity is finite. Coupling and directivity of this Bethe-hole coupler are given by

$$C = -20 \log \left[\frac{4}{3} \frac{\beta r_o^3}{ab} \left(\cos\theta + \frac{\lambda_g^2}{2\lambda_0^2} \right) \right] \text{ dB} \tag{6.121}$$

$$D = 20 \log \frac{2\beta^2 \cos\theta + k_0^{\,2}}{2\beta^2 \cos\theta - k_0^{\,2}} \text{ dB} \tag{6.122}$$

where r_0 = Radius of hole

$a \times b$ = Waveguide cross-sectional dimension

$\beta = 2\pi/\lambda_g$

For w/g excitation in dominant mode TE_{10}

$$\lambda_g = \frac{\lambda_0}{\sqrt{\left[1 - \left(\frac{\lambda_0}{2a}\right)^2\right]}}$$

$$k_0 = 2\pi/\lambda_0$$

Here, the guide wall thickness is assumed negligibly small. The optimum angle θ (for $P_3 = 0$) is obtained from the relation $\theta = \cos^{-1}\left(\frac{\lambda_g^2}{2\lambda_0^2}\right)$.

Bethe Off-set Hole For the off-set hole, the optimum forward coupling ($P_3 = 0$) is obtained for the off-set value d given by

$$\sin(\pi d/a) = \frac{\lambda_0}{\sqrt{6}a} \quad \text{and} \quad \theta = \cos^{-1}\left(\frac{\lambda_g^2}{2\lambda_0^2}\right) \tag{6.123}$$

The maximum coupling and directivity are given by

$$C = 20 \log [(1 + X^2)/X] \text{ dB} \tag{6.124}$$

$$D = 20 \log (1/X) \text{ dB} \tag{6.125}$$

where $$X = \frac{16\pi r_0^3}{3ab\lambda_g}\sin^2\left(\frac{\pi d}{a}\right)$$

For non-zero finite wall thickness ($t \approx 0.51$ mm), the coupling decreases and will be 1 to 2 dB less compared to the zero thickness value.

Example 6.24 Design a centre-hole Bethe-hole directional coupler with air-filled rectangular waveguide of dimensions 0.9 in × 0.4 in at 9.8 GHz for 20 dB coupling and 40 dB directivity, respectively.

Solution

$$a = 0.9 \text{ in} = 2.286 \text{ cm},\ b = 0.4 \text{ in} = 1.016 \text{ cm},\ f_0 = 9.8 \text{ GHz}$$

$$\lambda_0 = c/f = \frac{30}{9.8} = 3.06 \text{ cm},\ k_0 = \frac{2\pi}{\lambda_0} = \frac{2\times 3.1415}{3.06} = 2.0533 \text{ rad/cm}$$

$$\lambda_g/\lambda_0 = \frac{1}{\sqrt{\left[1-(\lambda_0/2a)^2\right]}} = \frac{1}{\sqrt{\left[1-(3.06/2\times 2.286)^2\right]}} = 1.3458$$

$$\lambda_g = 1.3458 \times 3.06 = 4.118 \text{ cm},$$

$$\beta = 2\pi/\lambda_g = 1.525 \text{ rad/cm}$$

$$\cos\theta = 1/2\ (\lambda_g/\lambda_0)^2 = 1/2 \times (1.3458)^2 = 0.9055$$

$$\theta = \cos^{-1} 0.9055 = 25.09°$$

Therefore, $$C = 20 \text{ dB} = -20\log\left[\left(\frac{4}{3}\times\frac{1.525}{2.286\times 1.016}\times 2\times 0.9055\right)r_0^3\right]$$

$$= -20\log(1.5855\, r_0^3)$$

or, $\log(1.5855\, r_0^3) = -1$

or, $1.8555\, r_0^3 = 10^{-1} = 0.1$

or, $r_0^3 = 0.1/1.5855 = 0.0631$

Therefore,

$$r_0 = (0.0631)^{1/3} = 0.398 \text{ cm}$$

$$d = a/2 = 1.143 \text{ cm}$$

$$D = 20\log\frac{2\beta^2\cos\theta + (2\pi/\lambda_0)^2}{2\beta^2\cos\theta - (2\pi/\lambda_0)^2}$$

At the band edge, f_1

$$40 = 20\log\frac{2\beta_1^2\times 0.9055 + (2\pi/\lambda_0)^2}{2\beta_1^2\times 0.9055 - (2\pi/\lambda_0)^2}$$

or, $$\frac{2\beta_1^2\times 0.9055 + (2\pi/\lambda_0)^2}{2\beta_1^2\times 0.9055 - (2\pi/\lambda_0)^2} = 10^2 = 100$$

or, $$1.811 \times \left(\frac{2\pi}{\lambda_{g1}}\right)^2 + \left(\frac{2\pi}{\lambda_{01}}\right)^2 = 100\left[1.811\left(\frac{2\pi}{\lambda_{g1}}\right)^2 - \left(\frac{2\pi}{\lambda_{01}}\right)^2\right]$$

or, $$\frac{1.811}{\lambda_{g1}^2} + \frac{1}{\lambda_{01}^2} = \frac{181.1}{\lambda_{g1}^2} - \frac{100}{\lambda_{01}^2}$$

or, $$\frac{1}{\lambda_{g1}^2}[1.811 - 181.1] = -\frac{1}{\lambda_{01}^2}(100 + 1)$$

or, $$\frac{179.89}{\lambda_{g1}^2} = \frac{101}{\lambda_{01}^2}$$

or, $$\frac{\lambda_{g1}^2}{\lambda_{01}^2} = \frac{179.89}{101} = 1.775$$

or, $$\frac{1}{\sqrt{\left[1-[\lambda_{01}/(2\times 2.286)]^2\right]}} = 1.775$$

or, $$1 - \left(\frac{\lambda_{01}}{4.572}\right)^2 = \left(\frac{1}{1.775}\right)^2$$

or, $$\lambda_{01}^2 = \left[1-\left(\frac{1}{1.775}\right)^2\right](4.575)^2 = 9.1276$$

or, $$\lambda_{01} = 3.02 \text{ cm}$$

Therefore, $$f_1 = \frac{c}{\lambda_{01}} = \frac{30}{3.02} = 9.929 \text{ GHz}$$

$$f_2 = 2f_0 - f_1 = 2 \times 9.8 - 9.929 = 9.671 \text{ GHz}$$

Therefore, bandwidth for $D > 40$ dB is $\Delta f = f_1 \sim f_2 = 258$ MHz

Multihole Waveguide Coupler

The structure of a multihole coupler is symmetrical with respect to a transverse plane. For an input at Port 1 of the main guide, a fraction of input power is coupled into the auxiliary guide in a forward direction at Port 4. The successive openings are spaced quarter wavelength apart at a designed frequency so that the coupled waves travelling back towards Port 3 will be out of phase and cancelled; whereas the forward-coupled waves are in phase and thus, reinforce each other. Thus, Port 3 is an ideally isolated port and matched terminated for absorbing any coupled power flows in practical situation. Such a coupler is called a *forward coupler*. For an input to Port 2, coupled power gets absorbed at Port 3 and Port 4 is isolated.

Since in practice, due to non-ideal mechanical configurations, there is some coupled power available in the isolated port, the matched directional coupler is designed from specifications of two basic parameters—coupling and finite directivity.

For the reciprocal matched coupler, the scattering matrix of a directional coupler can be derived from the following general S-matrix:

$$[S] = \begin{bmatrix} 0 & S_{12} & 0 & S_{14} \\ S_{12} & 0 & S_{23} & 0 \\ 0 & S_{23} & 0 & S_{34} \\ S_{14} & 0 & S_{34} & 0 \end{bmatrix} \tag{6.126}$$

Since for a loss network, the S-matrix is unitary,

$$S_{12} S_{12}^* + S_{14} S_{14}^* = 1 \tag{6.127}$$

$$S_{12} S_{12}^* + S_{23} S_{23}^* = 1 \tag{6.128}$$

$$S_{23} S_{23}^* + S_{34} S_{34}^* = 1 \tag{6.129}$$

$$S_{14} S_{14}^* + S_{34} S_{34}^* = 1 \tag{6.130}$$

$$S_{12} S_{23}^* + S_{14} S_{34}^* = 0 \tag{6.131}$$

From the first two equations, we get

$$|S_{14}| = |S_{23}| \tag{6.132}$$

and from the second and third equations, we get

$$|S_{12}| = |S_{34}| \tag{6.133}$$

Further, by choosing reference plane of Port 1 with respect to that of Port 2 and the reference plane of Port 3 with respect to that of 4, we can make S-parameters real

$$S_{12} = S_{34} = \alpha, \text{ say} \tag{6.134}$$

where α is a positive real number. Then from the fifth Eq. (6.131),

$$\alpha (S_{23}^* + S_{14}) = 0$$

Since $\alpha \neq 0$, $S_{23}^* + S_{14} = 0$ (6.135)

Further, selecting the reference plane of Port 4 with respect to Port 1, we can make S_{14} real, so that

$$-S^*_{23} = S_{14} = \beta, \text{ say real} \tag{6.136}$$

$$\therefore \quad S_{23} = S_{23}^* = -\beta$$

Therefore, $$[S] = \begin{bmatrix} 0 & \alpha & 0 & \beta \\ \alpha & 0 & -\beta & 0 \\ 0 & -\beta & 0 & \alpha \\ \beta & 0 & \alpha & 0 \end{bmatrix} \tag{6.137}$$

where $\alpha^2 + \beta^2 = 1$ for conservation of energy. Here, α is called the transmission factor and β is the coupling factor.

From above, the coupling, directivity and transmission loss for a matched directional coupler can be expressed in terms of S-parameters as

$$C = 10 \log \frac{P_1}{P_4} = -20 \log |S_{41}| \tag{6.138}$$

$$D = 10 \log \frac{P_4}{P_3} = 20 \log \frac{|S_{41}|}{|S_{31}|} = 20 \log \frac{|S_{41}|}{|S_{42}|} \tag{6.139}$$

$$T = 10 \log P_1/P_2 = 20 \log |S_{21}| \tag{6.140}$$

where $|S_{31}| = |S_{42}|$ for symmetry.

The amount of coupling (3 dB, 6 dB, 10 dB, 20 dB, 30 dB, etc.) and the directivity (30–40 dB) depend upon the sizes and locations of the holes in the common wall.

Since the phase cancellation in the reverse direction between successive holes can occur only at the designed frequency which satisfy $d = \lambda_{g0}/4$, the bandwidth and the frequency response of the coupler are determined by designing the coupler following a suitable reverse voltage response function, such as, maximally flat or binomial response, Chebyshev response, etc., as shown in Fig. 6.61.

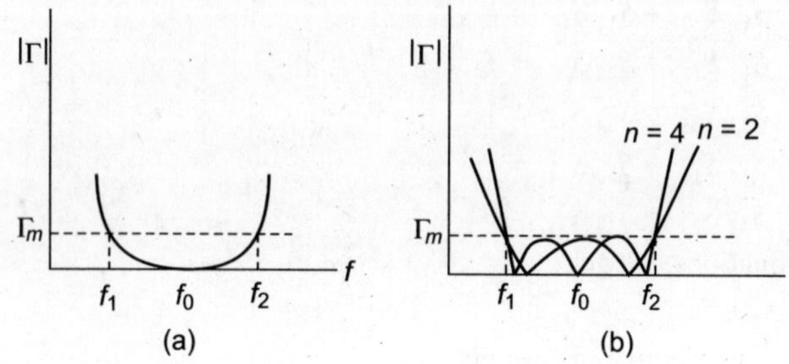

Fig. 6.61 Coupling characteristics of directional couplers: (a) Binomial (b) Chebyshev

For simplicity, we assume that the total power coupled is a small fraction of the incident power so that the incident wave amplitude can be considered to be same at each coupling aperture and only the phase will change at different holes. The total reverse and forward voltages at any frequency f can be represented for an unit input voltage at Port 1, respectively by,

$$\Gamma' = \Gamma_1 + \Gamma_2 e^{-j2\theta} + \Gamma_3 e^{-j4\theta} + \cdots + \Gamma_n e^{-j2(n-1)\theta}$$

$$= \sum_{k=1}^{n} \Gamma_k e^{-j2(k-1)\theta} \tag{6.141}$$

and

$$C' = e^{-j(n-1)\theta} \sum_{k=1}^{n} \Gamma_k \tag{6.142}$$

Here, $\theta = \dfrac{2\pi d}{\lambda_g}$, with $\theta = \dfrac{\pi}{2}$ at the design frequency f_0. Coupling holes are assumed to be non-directional due to small sizes. Normally, the design is made keeping the structure symmetrical with respect to a centre plane P as shown in Fig. 6.60(b) so that the size of the k_{th} hole is the same as that of the $(n+1-k)_{th}$ one. Under this condition, $\Gamma_1 = \Gamma_n, \Gamma_2 = \Gamma_{n-1}$, etc. Therefore, by changing the reference plane to P, from (6.141) and (6.142):

❑ *The Reverse Wave*

$$\Gamma = \Gamma' e^{j(n-1)\theta} = 2[\Gamma_1 \cos(n-1)\theta + \Gamma_2 \cos(n-3)\theta + \ldots + \Gamma_{n/2}\cos\theta]$$

$$= 2\sum_{k=1}^{n/2} \Gamma_k \cos\{n-(2k-1)\}\theta; \quad \text{for } n = \text{even number} \tag{6.143a}$$

and

$$\Gamma = \Gamma' e^{j(n-1)\theta} = 2\left[\Gamma_1 \cos(n-1)\theta + \Gamma_2 \cos(n-3)\theta + \ldots + \Gamma_{\frac{n+1}{2}}\right]$$

$$= 2\sum_{k=1}^{(n-1)/2} \Gamma_k \cos\{n-(2k-1)\}\theta; \qquad \text{for } n = \text{odd number} \tag{6.143b}$$

since $e^{jn\theta} + e^{-jn\theta} = 2\cos n\theta$.

❑ *The Forward or the Couple Wave*

$$C = e^{-j\frac{n-1}{2}\theta} 2\left[\Gamma_1 + \Gamma_2 + \cdots + \Gamma_{n/2}\right];\ n \text{ even}$$

or,
$$C = e^{-j\frac{n-1}{2}\theta} 2\sum_{k=1}^{n/2} \Gamma_k;\ n \text{ even} \tag{6.144a}$$

and,
$$C = e^{-j\left(\frac{n-1}{2}\right)\theta} 2\left[\Gamma_1 + \Gamma_2 + \cdots + \Gamma_{\frac{n-1}{2}} + \Gamma_{\frac{n+1}{2}}\right];\ n \text{ odd}$$

or,
$$C = e^{-j\frac{n-1}{2}\theta}\left[2\sum_{k=1}^{\frac{n-1}{2}} \Gamma_k + \Gamma_{\frac{n+1}{2}}\right];\ n \text{ odd} \tag{6.144b}$$

The coupling and directivity of the directional coupler can be defined from (6.143), and (6.144) as

$$\text{Coupling C (dB)} = -20\log |C| \tag{6.145}$$

$$\text{Directivity D(dB)} = -20\log\left|\frac{\Gamma}{C}\right| \tag{6.146}$$

The multihole directional couplers are designed for a desired coupling and directivity from the type of distribution by keeping the reverse voltage magnitude $|\Gamma|$ under a specified value Γ_m over the frequency band between the edges f_1 and f_2 (Fig. 6.61).

◆ *Binomial Coupler* Binomial couplers are designed by choosing the coupling coefficient Γ_k at a hole k proportional to the corresponding binomial coefficient of $(a + b)^{n-1}$ for total n holes. For a unit input voltage, the total reverse coupled voltage $\Gamma = 0$ at $f = f_0$ and increases slowly at either side of f_0 with a maximally flat realisable response as shown in Fig. 6.61(a).

Here,
$$\Gamma_k = {}^{n-1}C_{k-1}\Gamma_1 \tag{6.147}$$

where
$${}^{n-1}C_{k-1} = \frac{(n-1)!}{(k-1)!\,(n-k)!} \text{ binomial coefficient} \tag{6.148}$$

viz., for n holes $\Gamma_1 = {}^{n-1}C_0, \Gamma_2 = {}^{n-1}C_1, \Gamma_3 = {}^{n-1}C_2 \ldots$

Γ_k can be adjusted by varying the aperture size while maintaining its shape and position fixed. For the reverse wave for $f_1 \le f \le f_2$,

$$|\Gamma| = \left|2\Gamma_1\left[{}^{n-1}C_0 \cos(n-1)\theta + {}^{n-1}C_1 \cos(n-3)\theta + \cdots + {}^{n-1}C_{n/2}\right]\right|$$

$$= \left|2\Gamma_1 \sum_{k=1}^{n/2} {}^{n-1}C_{k-1} \cos(n - \overline{2k-1})\theta\right| \le \Gamma_m;\ n \text{ even} \tag{6.149}$$

and $$|\Gamma| = \left|2\Gamma_1\left[{}^{n-1}C_0\cos(n-1)\theta + {}^{n-1}C_1\cos(n-3)\theta + \cdots + {}^{n-1}C_{\frac{n-1}{2}}\right]\right|$$

$$= \left|2\Gamma_1\sum_{k=1}^{(n-1)/2} {}^{n-1}C_{k-1}\cos(n-2k-1)\theta + {}^{n-1}C_{(n-1)/2}\right| \le \Gamma_m;\ n \text{ odd} \tag{6.150}$$

Expanding cos $m\theta$ in each term of the above equation, in terms of cos θ, only one term containing $\cos^{n-1}\theta$ with coefficient A_{n-1} does not cancel, so that

$$\Gamma = 2\,\Gamma_1\,A_{n-1}\cos^{n-1}\theta \tag{6.151}$$

where A_{n-1} is a constant determined from the expansion of cos $(n-1)\theta$. Since all the derivatives up to the $(n-2)$th terms of Γ at mid-frequency $f = f_0$ are zero, the binomial distribution results in a maximally flat response without any ripple. Since the maximum value of Γ at the band edges is specified to be Γ_m,

$$\Gamma_m = 2\Gamma_1 A_{n-1}\cos^{n-1}\theta;\quad \theta_1 \text{ for } f_1, \text{ and } \theta_2 \text{ for } f_2$$

or

$$\Gamma_1\big|_{max} \le \frac{\Gamma_m}{2A_{n-1}\cos^{n-1}\theta};\ \theta_1 \le \theta < \theta_2 \tag{6.152}$$

since $|\Gamma| = \Gamma_m$ at f_1 and f_2 $\theta_2 = \pi - \theta_1$.

If the band-edge frequency ratio is defined by,

$$r = \lambda_{g1}/\lambda_{g2} = \theta_2/\theta_1, \text{ and } d = \lambda_{g0}/4,$$

$$\theta_1 = \pi/(1 + r) = 2\pi d/\lambda_{g1}$$

$$\theta_2 = \pi\, r/(1 + r) = 2\pi d/\lambda_{g2}$$

The coupling coefficients for a binomial coupler having specified number of holes are given as follows.

No. of holes n	Coefficients	
2	$a_1 = {}^{2-1}C_0 = 1 = a_2$,	
3	$a_1 = {}^{3-1}C_0 = 1 = a_3$,	$a_2 = {}^{3-1}C_1 = 2$
4	$a_1 = {}^{4-1}C_0 = 1 = a_4$,	$a_2 = {}^{4-1}C_1 = 3 = a_3$
5	$a_1 = {}^{5-1}C_0 = 1 = a_5$,	$a_2 = {}^{5-1}C_1 = 4 = a_4$,
		$a_3 = {}^{5-1}C_2 = 6$

The coefficient for other number of holes can be determined in a similar manner. Γ_1 can be computed from the desired coupling C for *a* given number of holes.

A major practical disadvantage of a binomial coupler is the wide variation between the coupling coefficients of different holes of the coupler, especially for a large number of holes. This leads to fabrication difficulties while maintaining equal-hole spacings of quarter wavelength.

Example 6.25 Design a maximally flat 20 dB directional coupler so that $D \ge 40$ dB in the band $r = 2$.

Solution Directivity = 20 log C/Γ = 20 log C – 20 log Γ

or, |Directivity| + |Coupling| = –20 log Γ

Here in the band, D + C ≥ 60 dB = –20 log Γ_m

Therefore, the maximum tolerance in reverse voltage $\Gamma_m = 10^{-3}$

Let n = number of holes to be selected. If $\Gamma_1, \Gamma_2, \Gamma_3, \ldots$, are the coupled wave at corresponding holes

$\Gamma_1 2^{n-1} \cos^{n-1} \theta_1 \le \Gamma_m;\ \theta_1 = \pi/(1+2) = \pi/3$ (A)

$$C = 20 \text{ dB} = -20 \log C$$

Therefore, $C = 10^{-1} = 0.1 = \Gamma_1 + \Gamma_2 + \Gamma_3 + \ldots + \Gamma_n$ (B)

For symmetrical coupler with binomial type,

$$\Gamma_1 = \Gamma_n$$

$$\Gamma_2 = (n-1)\Gamma_1 = \Gamma_{n-1}$$

$$\Gamma_3 = [(n-1)(n-2)/2]\,\Gamma_1 = \Gamma_{n-2}, \text{ etc.}$$

Selecting $n = 3$, $\Gamma_1 = \Gamma_3$

$$\Gamma_2 = (n-1)\Gamma_1 = 2\Gamma_1$$

Therefore, (B) reduces to

$$C = 0.1 = \Gamma_1 + 2\Gamma_1 + \Gamma_1 = 4\Gamma_1$$

$$\Gamma_1 = 0.1/4 = 0.025$$

Therefore, the left-hand side of (A) is

$$\frac{0.1}{4} \times 2^2 \cos^2\left(\frac{\pi}{3}\right) = \frac{0.1}{4} > \Gamma_m$$

Therefore, $n > 3$

Selecting $n = 5$, $\Gamma_1 = \Gamma_5$

$$\Gamma_2 = (n-1)\Gamma_1 = 4\Gamma_1$$

$$\Gamma_3 = [(n-1)(n-2)/2]\Gamma_1 = 6\Gamma_1,$$

Therefore, $C = 0.1 = 2(\Gamma_1 + \Gamma_2) + \Gamma_3 = 2(\Gamma_1 + 4\Gamma_1) + 6\Gamma_1 = 16\Gamma_1$

or, $\Gamma_1 = 0.1/16$

Therefore, the left-hand side of (A) is

$$\frac{0.1}{16} \times 2^4 \cos^4\left(\frac{\pi}{3}\right) = 6.25 \times 10^{-3} > \Gamma_m$$

Therefore, $n > 5$

Selecting $n = 6$, $\Gamma_1 = \Gamma_6$

$$\Gamma_2 = (n-1)\Gamma_1 = 5\Gamma_1$$

$$\Gamma_3 = [(n-1)(n-2)/2]\Gamma_1 = 10\Gamma_1,$$

Therefore, $C = 0.1 = 2\,(\Gamma_1 + \Gamma_2 + \Gamma_3) = 2\,(\Gamma_1 + 5\Gamma_1 + 10\Gamma_1) = 32\Gamma_1$

or, $\Gamma_1 = 0.1/32$

Therefore, the left-hand side of (A) is

$$\frac{0.1}{32} \times 2^5 \cos^5\left(\frac{\pi}{3}\right) = 3.125 \times 10^{-3} > \Gamma_m$$

Therefore, $n > 6$

Selecting $n = 7$, similarly, the left-hand side of (A) is

$$0.1\cos^6\left(\frac{\pi}{3}\right) = 1.5625 \times 10^{-3} > \Gamma_m$$

Selecting $n = 8$, in a similar way, the left-hand side of (A) is

$$0.1\cos^7\left(\frac{\pi}{3}\right) = 7.8125 \times 10^{-4} < \Gamma_m$$

Therefore, since the condition (A) is satisfied for $n = 8$, the total number of holes are 8 with the coupling coefficients:

$$\Gamma_1 = 0.1/2^7,\ \Gamma_2 = 7\Gamma_1,\ \Gamma_3 = 21\,\Gamma_1,\ \Gamma_4 = 35\,\Gamma_1$$

Example 6.26 A rectangular waveguide binomial coupler coupled by 5 circular holes in a common side/narrow wall to produce 30 dB coupling at 10 GHz. The guide width $a = 2.5$ cm and the height $b = 1.2$ cm. The dominant input mode of unit amplitude radiates a field of amplitude $-j\frac{4}{3}r_0^{\,3}(\pi/a)^2\frac{1}{ab\beta}$ in both directions in the other guide. Find the required hole radii and the frequency ratio for which the directivity is greater than 50 dB.

Solution Given,

$$|C| = \frac{4}{3}r_0^{\,3}\left(\frac{\pi}{a}\right)^2\frac{1}{ab\beta}, \text{ for narrow-wall waveguide coupler with number of holes } n = 5$$

Binomial coupling C = 30 dB at $f = f_0 = 10$ GHz. Waveguide dimension $a = 2.5$ cm and $b = 1.2$ cm

(a) To find C

Assuming symmetrical binomial-type coupler

$$C = \Gamma_1 + \Gamma_2 + \Gamma_3 + \Gamma_4 + \Gamma_5 = 2\,(\Gamma_1 + \Gamma_2) + \Gamma_3$$

where $\Gamma_1 = \Gamma_5$

$$\Gamma_2 = \Gamma_4 = (\mathrm{n} - 1)\,\Gamma_1 = 4\Gamma_1$$

$$\Gamma_3 = [(n-1)(n-2)/2]\,\Gamma_1 = 6\Gamma_1$$

Therefore, $C = 2\,(\Gamma_1 + \Gamma_2) + \Gamma_3 = 16\,\Gamma_1$

Given, $30 \text{ dB} = -20\log C$

$$C = 10^{-30/20} = 10^{-1.5} = 0.0316$$

Therefore, $\Gamma_1 = C/16 = 1.976 \times 10^{-3} = \Gamma_5$

$$\Gamma_2 = \Gamma_4 = 4\Gamma_1 = 7.905 \times 10^{-3}$$

$$\Gamma_3 = 6\Gamma_1 = 11.858 \times 10^{-3}$$

(b) To find hole spacing

$$\lambda_0 = c/f = 30/10 = 3 \text{ cm}$$

$$\lambda_{g0} = \frac{\lambda_0}{\sqrt{[1-(\lambda_0/2a)^2]}} = \frac{3}{\sqrt{[1-\{3/(2\times 2.5)\}^2]}} = 3.75 \text{ cm}$$

Therefore, hole spacing

$$d = \lambda_{g0}/4 = 0.9375 \text{ cms}$$

(c) To find hole radius r_0

$$\beta = 2\pi/\lambda_{g0} = \frac{2 \times 3.14}{3.75} = 1.6755 \text{ rad/cm.}$$

Therefore, $|C| = r_0^{\,3} \times \dfrac{4}{3} \times \dfrac{\pi^2}{(2.5)^2} \times \dfrac{1}{2.5 \times 1.2 \times 1.6755} = r_0^{\,3} \times 0.4188$

or, $$r_0 = \left(\frac{|C|}{0.4188}\right)^{1/3}$$

Thus radii of the holes are

$$r_{01} = \left(\frac{|\Gamma_1|}{0.4188}\right)^{1/3} = 0.1677 \text{ cm} = r_{05}$$

$$r_{02} = \left(\frac{|\Gamma_2|}{0.4188}\right)^{1/3} = 0.26625 \text{ cm} = r_{04}$$

$$r_{03} = \left(\frac{|\Gamma_3|}{0.4188}\right)^{1/3} = 0.3047 \text{ cm}$$

(d) Frequency range $r = \theta_2/\theta_1$

$$D = 50 \text{ dB} = 20 \log C/\Gamma_m;\ \Gamma_m = \text{total tolerable back wave amplitude.}$$

or, $20 \log \Gamma_m = -(\text{Directivity} + \text{Coupling}) = -80 \text{ dB},\ \Gamma_m = 10^{-4}$.

If θ_1 and θ_2 corresponds to band edges f_1 and f_2

$$\Gamma_1\, 2^{n-1} \cos^{n-1}\theta_1 = \Gamma_m$$

or, $\dfrac{10^{-1.5}}{16} \times 2^4 \times \cos^4\theta_1 = \Gamma_m = 10^{-4};$

or, $\cos^4\theta_1 = 10^{-4}/10^{-1.5} = 3.16227 \times 10^{-3}$

or, $\cos\theta_1 = 0.056$

Therefore, $\theta_1 = 86.77° = 180°/(1 + r)$

Therefore, $r = \dfrac{180°}{\theta_1} - 1 = 1.074$

❑ *Chebyshev Coupler* The Chebyshev coupler response is an equal ripple characteristic where the reverse coupling Γ is made proportional to a Chebyshev polynomial of order $n - 1$ (for number of holes = n)

$$T_p(x) = \cos[p \cos^{-1}(x)];\ -1 \le x \le +1,\ p = n - 1$$
$$= \cosh[p \cosh^{-1}(x)];\ x < -1,\ x > +1,\ p = n - 1 \quad (6.153)$$

where x is the parameter for change of variable and is a function of frequency:

$$x = \cos\theta/\cos\theta_1 = \cos\theta/|\cos\theta_2| \quad (6.154)$$

The polynomials have the following properties as shown in Chapter 5.

1. All polynomials pass through the point (1,1).
2. $-1 \le T_p(x) \le +1$, for $-1 \le x \le 1$.

3. All roots occur within $-1 \le x \le 1$, and all maxima and minima have values of +1 and –1, respectively.

Thus, we have within the band

$$|\Gamma| = \left|2\left[\Gamma_1 \cos(n-1)\theta + \Gamma_2 \cos(n-3)\theta + \cdots + \Gamma_{n/2}\cos\theta\right]\right|$$

$$= 2\sum_{k=1}^{n/2} \Gamma_k \cos\,(n-\overline{2k-1})\,\theta;$$

$$= \Gamma_m T_{n-1}(x) \qquad n \text{ even} \qquad (6.155a)$$

and

$$|\Gamma| = \left|2\left[\Gamma_1 \cos(n-1)\theta + \Gamma_2 \cos(n-3)\theta + \cdots + \Gamma_{\frac{n+1}{2}}\right]\right|$$

$$= \sum_{k=1}^{(n-1)/2} \Gamma_k \cos\,[n-(2k-1)]^{\theta} + \Gamma_{(n+1)/2};$$

$$= \Gamma_m T_{n-1}(x) \qquad n \text{ odd} \qquad (6.155b)$$

The Γ_k's are obtained by expanding the left-hand side in term of powers of $\cos\theta$ and substituting $x\cos\theta_1$ for $\cos\theta$ and equating the coefficients of like powers of x with those on the right-hand side.

A Chebyshev coupler gives the largest bandwidth for a specified Γ_m and n. Alternately, for a given bandwidth it gives the lowest ripple Γ_m.

❑ ***Coupler Design*** The design procedure of the multihole directional couplers follows the steps given below:

1. Select the type of distribution (binomial or Chebyshev)
2. $\Sigma\Gamma_k$ is calculated from the type of distribution and Γ_1 is computed from the coupling C for a chosen number of holes.
3. Test $|\Gamma| \le \Gamma_m$ for this n.
4. Select n to satisfy this condition over the body by repeating the procedure 2 to 3.
5. From the designed values of Γ_k's, the hole dimensions are determined from the empirical formulas given in Eq. (6.121).

Example 6.27 Design a five-hole 30 dB directional coupler with Chebyshev distribution for wavelength ratio of 2 at the band edges.

Solution For $n = 5$ and $r = 2$, the reverse voltage

$$\Gamma = 2\,[\Gamma_1 \cos\,(5-1)\theta + \Gamma_2 \cos\,(5-3)] + \Gamma_3$$
$$= 2\,[\Gamma_1 \cos 4\theta + \Gamma_2 \cos 2\theta] + \Gamma_3$$
$$= 2\,[\Gamma_1(8\cos^4\theta - 8\cos^2\theta + 1) + \Gamma_2\,(2\cos^2\theta - 1)] + \Gamma_3$$

At band edge $f = f_1$ and $\theta = \theta_1 = \dfrac{\pi}{1+r} = \pi/3$

A change of variable according to $\cos\theta = x\cos\theta_1$ gives $\cos\theta = x\cos(\pi/3) = x/2$. Therefore,

$$\Gamma = 2\,[\Gamma_1\,(x^4/2 - 2x^2 + 1) + \Gamma_2\,(x^2/2 - 1)] + \Gamma_3$$
$$= \Gamma_1 x^4 + (\Gamma_2 - 4\Gamma_1)\,x^2 + (\Gamma_3 + 2\Gamma_1 - 2\Gamma_2)$$

$$= \Gamma_m T_{5-1}(x)$$
$$= \Gamma_m (8x^4 - 8x^2 + 1)$$

Equating the coefficients

$$\Gamma_1 = 8\,\Gamma_m$$
$$\Gamma_2 = 4\,\Gamma_1 - 8\,\Gamma_m = 32\,\Gamma_m - 8\,\Gamma_m = 24\,\Gamma_m$$
$$\Gamma_3 + 2\,\Gamma_1 - 2\,\Gamma_2 = \Gamma_m \quad \text{or,} \quad \Gamma_3 = \Gamma_m - 16\,\Gamma_m + 48\,\Gamma_m = 33\,\Gamma_m$$

Now coupling

$$C = 2(\Gamma_1 + \Gamma_2) + \Gamma_3 = 30 \text{ dB} = 10^{-30/20} = 0.0316$$

or, $$2(8 + 24)\,\Gamma_m + 33\,\Gamma_m = 0.0316.$$

or, $$97\,\Gamma_m = 0.0316 \quad \text{or, } \Gamma_m = 3.26 \times 10^{-4}$$

Therefore, coupling values of the holes are

$$\Gamma_1 = 8\,\Gamma_m = 26.08 \times 10^{-4}$$
$$\Gamma_2 = 24\,\Gamma_m = 78.24 \times 10^{-4}$$
$$\Gamma_3 = 33\,\Gamma_m = 107.58 \times 10^{-4}$$

Directivity $D = -C + 20 \log (1/\Gamma_m) = 39.7$ dB

In case the coupling coefficients make the hole size large, holes may overlap to make the centre-to-centre distance equal to the quarter wavelength. Otherwise, also for large holes, the coupling coefficients become more frequency sensitive. Under such a situation, the number of holes are increased to meet the desired specifications of coupling and directivity.

Under such a situation, the number of holes is increased to meet the desired specifications of coupling and directivity.

Coupled Transmission Line Coupler

❑ *Coupled Co-axial Lines* The electromagnetic field from port 1 of one coaxial transmission line can be coupled to a second adjacent line when a narrow longitudinal slot is cut between the lines on the common outer conductor joint as shown in Fig. 6.62. The electric field in the input or primary line induces an equal and opposite charge on the centre conductor of the two lines.

This results in an electric field in the coupled or secondary line directed oppositely to that in the primary guide. But the magnetic field follows the same direction in the both the lines. Hence, the directions of power density flow **P** = **E** × **H** in the two guides are opposite. Therefore, the coupled power flows in the backward direction compared to forward power flow in the primary guide. For this reason, this coupler is called the *backward coupler.* The forward power in the coupled line at Port 4 is ideally zero and this port is called the *isolated port.* The coupling slot is made equal to the quarter wavelength in length for achieving maximum coupling as can be seen in the analysis section given below.

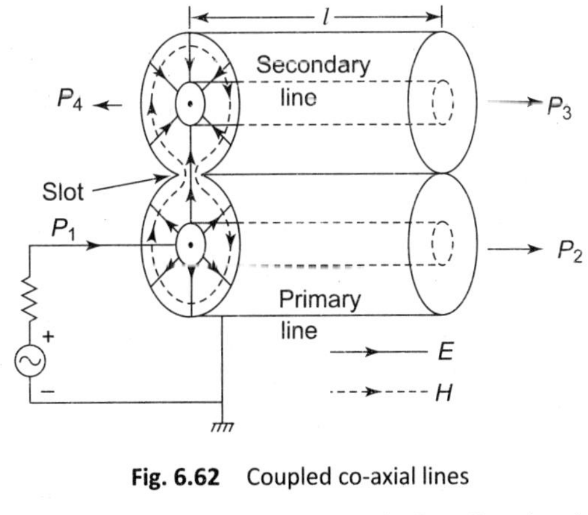

Fig. 6.62 Coupled co-axial lines

❑ *Microstrip and Strip Line Coupler* Of the various types of strip line and microstrip line directional couplers, the edge-coupled parallel conductor configuration is most extensively used in practical circuits.

Detailed analysis and design data on the characteristic impedances and effective dielectric constants of such structures are derived by S.B. Cohn using the conformal mapping method and even and odd mode method of analysis. A typical planer coupled transmission linear coupler is shown schematically in Fig. 6.63. The lines are coupled over a length l with spacing S between the adjacent edges. The width w of the lines are designed for desired impedance (typically 50 ohms). The lines are supported by dielectric substrate with a bottom ground plane.

Since the spacing S is usually very small, it would be difficult to install miniature co-axial line signal launching connectors at the end (port) of each line [Fig. 6.63(a)]. Each connector's outer diameter could be larger than the spacing S. Hence signals are fed /taken out of the ports through extra lengths of uncoupled feed lines which are at right angled to the coupled length l as shown in Fig. 6.63(b).The right angled bends are metered by a length d, nearly equal to the line width w to reduce impedance mismatch and charge accumulation due to sharp discontinuities. For 50 ohm strip line, $d = 1.131\ w$, and for a 50-ohm microstrip, $d = 1.194\ w$. As in a coaxial system, these planar couplers are also backward couplers.

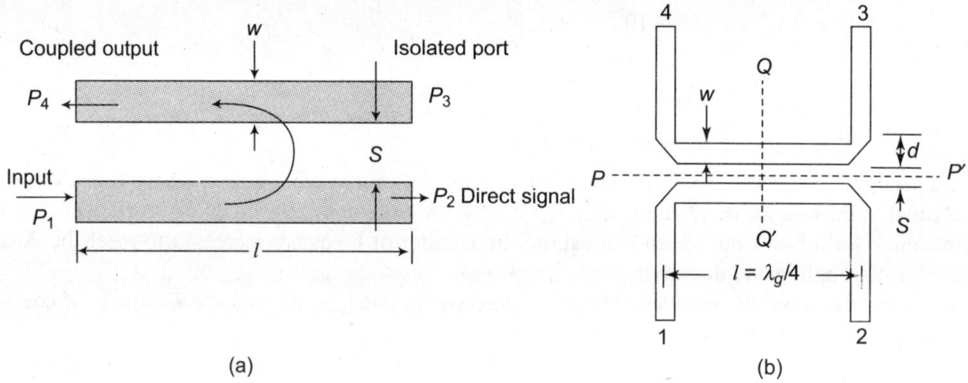

Fig. 6.63 Planar configuration of Transmission Line Coupler

❑ ***Even and Odd Mode Analysis of the Transmission Line Coupler*** A pair of coupled lines forms a 4-port device with two orthogonal planes of reflection symmetry—PP' and QQ' as shown in Fig. 6.63. As a result the scattering matrix of this four port symmetric device is

$$[S] = \begin{bmatrix} S_{11} & S_{12} & S_{13} & S_{14} \\ S_{12} & S_{22} & S_{23} & S_{24} \\ S_{13} & S_{23} & S_{33} & S_{34} \\ S_{14} & S_{24} & S_{34} & S_{44} \end{bmatrix} \tag{6.156a}$$

Because of symmetry, the attenuation in through in through port, coupling, isolation and directivity, can be defined in dB as follows:

$$\left.\begin{array}{ll} \text{Attenuation (between through ports)} & = -20\log_{10}\alpha \\ \text{Coupling between ports 1-4, 4-1, 2-3, 3-2} & = -20\log_{10}\beta \\ \text{Isolation}_{\text{ between ports 1-3, 3-1,}} & = -20\log_{10}|S_{31}| \\ \text{Isolation}_{\text{ between 2-4, 4-2}} & = -20\log_{10}|S_{24}| \\ \text{Directivity}_{\text{ Power at isolated port/Power at coupled port}} & = -20\log_{10}|S_{31}|/|S_{41}| \\ \text{Directivity} & = -20\log_{10}|S_{24}|/|S_{23}| \end{array}\right\} \tag{6.156b}$$

where

$$\left.\begin{aligned} \alpha &= |S_{12}| = |S_{21}| = |S_{34}| = |S_{43}| \text{ (between through ports)} \\ \beta &= |S_{14}| = |S_{41}| = |S_{23}| = |S_{32}| \text{ (between coupled ports 1-4, 4-1, 2-3, 3-2)} \\ \gamma &= |S_{31}| = |S_{13}| = |S_{24}| = |S_{42}| \end{aligned}\right\} \quad (6.156c)$$

To determine these S-parameters, we can apply a source to port 1 and then terminate all other ports with matched loads as shown in Fig. 6.64(a). The main properties of the parallel line coupler can be analyzed by decomposing the actual excitation into individual even and odd symmetry modes with reference to the plane of symmetry PP'. Response for actual excitation can be obtained by superimposing the responses of isolated even and odd modes for a symmetrical structure as shown in Fig. 6.64(b) and 6.64(c). In general, the characteristic impedances of these modes are not equal due to conditions that the capacitive coupling associated with these modes are different (i.e., $Z_0^e \neq Z_0^o$). Here superscript 'e' will be used for even mode and 'o' for odd mode.

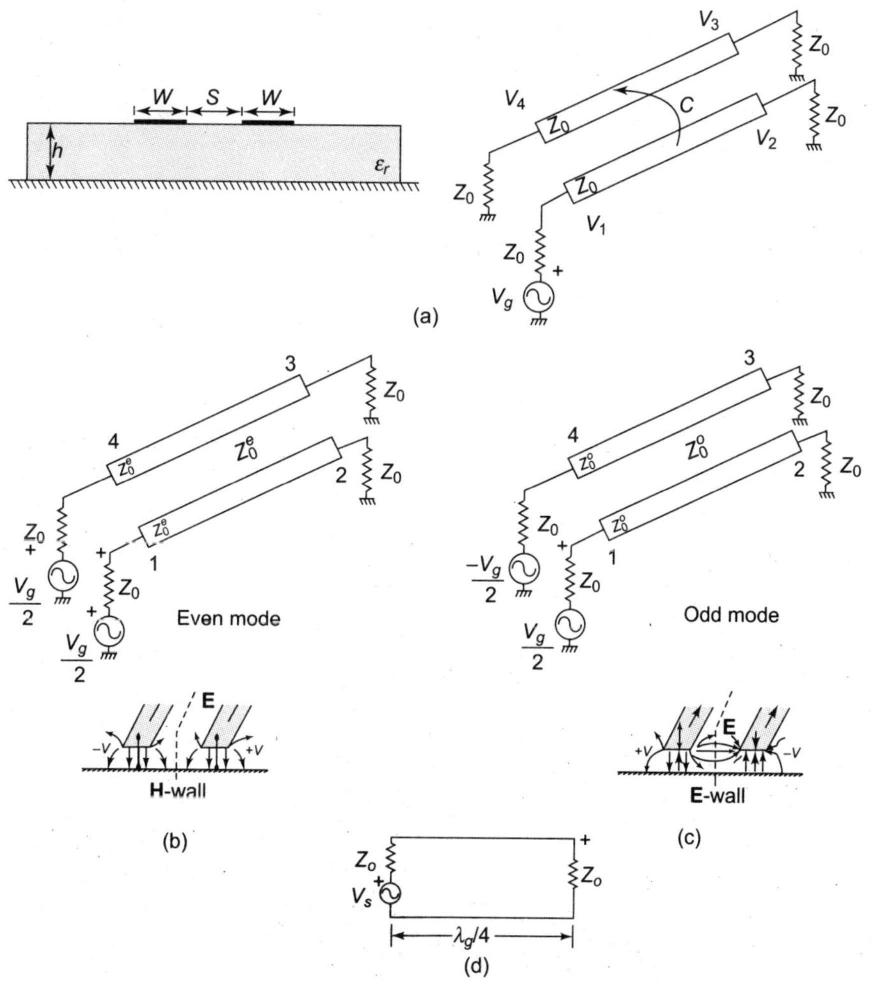

Fig. 6.64 Transmission-line couplers: (a) coupled transmission line (b) even-mode symmetric (c) odd-mode symmetry (d) single equivalent circuit

For odd mode of excitation, the instantaneous voltages at two strips are out of phase by 180° so that a majority of the electric field lines start from one conductor at positive potential to the other at negative potential. On the plane of symmetry *PP′*, **E** is normal with zero tangential component, and **H** is tangential with zero normal component. Therefore, *PP′* can be replaced by an electric wall ($\mathbf{E}_t = 0$).

For even mode of excitation, both the conductors are in phase of potential (positive, say) so that the **E** field lines originate from both of them and terminate on the ground planes/plane. On the plane of symmetry *PP′*, the electric field lines are completely tangential and magnetic field lines are completely normal. Therefore, PP' can be replaced by a magnetic wall ($\mathbf{H}_t = 0$).

These impedances are the major design parameters for any parallel coupled lines and are obtained in terms of coupling C and the single-line terminating characteristic impedance Z_0 in the following manner.

Let V^+ and V^- represent input voltage towards a port and output voltage from that port, respectively, having characteristic impedance Z_0.

The *S*-matrix equation of the coupler is

$$[b] = [S][a] \tag{6.156d}$$

where normalized voltages

$$a_i = V_i^+ / \sqrt{Z_0}$$

$$b_i = V_i^- / \sqrt{Z_0}\ ;\ i = 1, 2, 3, 4 \text{ for four ports} \tag{6.157}$$

For even-mode excitation,

$$a_1^e = a_4^e = a/2 \tag{6.158a}$$

and voltage maximum will occur at symmetry plane *PP′*. Thus, *PP′* is considered as an open circuit. Therefore,

$$b_1^e = b_4^e = S_{11}^e\, a/2; \qquad S_{11}^e = S_{44}^e \tag{6.158b}$$

$$b_2^e = b_4^e = S_{12}^e\, a/2 \qquad S_{12}^e = S_{43}^e \tag{6.158c}$$

For odd-mode excitation,

$$a_1^0 = -a_4^0 = a/2 \tag{6.159a}$$

and voltage null will occur at *PP′* for lossless network. Thus *PP′* is considered as a short-circuit plane. Therefore,

$$b_1^0 = -b_4^0 = S_{11}^0\, a/2 \tag{6.159b}$$

$$b_2^0 = -b_4^0 = S_{12}^0\, a/2 \tag{6.159c}$$

When both *e* and 0 modes are excited together, total solutions for input and output waves are

$$a_1 = a_1^e + a_1^0 = a; \qquad a_2 = 0 \text{ when matched}$$

$$a_3 = 0$$

$$a_4 = 0 \tag{6.160a}$$

$$b_1 = b_1^e + b_1^0 = (S_{11}^e + S_{11}^0)\, a/2 = S_{11}\, a; \quad a_1 = a \tag{6.160b}$$

$$b_2 = b_2^e + b_2^0 = (S_{12}^e + S_{12}^0)\, a/2 = S_{12}\, a \tag{6.160c}$$

$$b_3 = b_3^e + b_3^0 = b_2^e - b_2^0 = (S_{12}^e - S_{12}^0)\, a/2 = S_{13}\, a; \tag{6.160d}$$

$$b_4 = b_4^e + b_4^0 = b_1^e - b_1^0 = (S_{11}^e - S_{11}^0)\, a/2 = S_{14}\, a \tag{6.160e}$$

The input impedance of the coupler can be obtained from

$$Z_{in} = \frac{V_1^{+e} + V_1^{+0}}{I_1^{+e} + I_1^{+0}} \tag{6.161a}$$

where $$V_1^{+e} = \left(\frac{V_g}{2}\right)\frac{Z_1^e}{Z_1^e + Z_0} \tag{6.161b}$$

$$V_1^{+0} = \left(\frac{V_g}{2}\right)\frac{Z_1^0}{Z_1^0 + Z_0} \tag{6.161c}$$

$$I_1^{+e} = \left(\frac{V_g}{2}\right)\frac{1}{Z_1^e + Z_0} \tag{6.161d}$$

$$I_1^{+0} = \left(\frac{V_g}{2}\right)\frac{1}{Z_1^0 + Z_0} \tag{6.161e}$$

$$Z_1^e = Z_0^e\,\frac{Z_0 + jZ_0^e \tan\theta}{Z_0^e + jZ_0 \tan\theta} \tag{6.161f}$$

$$Z_1^0 = Z_0^0\,\frac{Z_0 + jZ_0^0 \tan\theta}{Z_0^0 + jZ_0 \tan\theta} \tag{6.161g}$$

where $\theta = \beta l$, the electrical length of the parallel coupled section of length l of a lossless ($\alpha = 0$) line in which it is assumed that $\theta^e = \theta^0 = \theta$.

For a perfect match,

$$Z_{in} = Z_0 = \frac{\left(\frac{V_g}{2}\right)\frac{Z_1^e}{Z_1^e + Z_0} + \left(\frac{V_g}{2}\right)\frac{Z_1^0}{Z_1^0 + Z_0}}{\left(\frac{V_g}{2}\right)\frac{1}{Z_1^e + Z_0} + \left(\frac{V_g}{2}\right)\frac{1}{Z_1^0 + Z_0}} = \frac{2Z_1^e Z_1^0 + Z_0(Z_1^e + Z_1^0)}{Z_1^e + Z_1^0 + 2Z_0} \tag{6.162a}$$

or, $$Z_0 = \sqrt{Z_1^e Z_1^0} \quad \text{for all } \theta,\ Z_0 = \sqrt{Z_0^e Z_0^0} \tag{6.162b}$$

Since the input impedance is equal to the characteristics impedance of the perfectly matched line.

Coupling and Directivity

The problem to find the coupling and directivity reduces to find the even and odd mode terminal voltages in a coupled section of electrical length $\beta l = \theta$, terminated at both ends by an impedance $\sqrt{Z_0^e Z_0^0}$, and excited by a voltage $V_g/2$. The transmission and the reflection coefficients are related to *ABCD* parameters for matched terminated symmetrical reciprocal network:

$$\Gamma = S_{11} = S_{22} = \frac{A + B/Z_0 - CZ_0 - D}{A + B/Z_0 + CZ_0 + D} \tag{6.163a}$$

$$T = S_{21} = S_{12} = \frac{2}{A + B/Z_0 + CZ_0 + D} \tag{6.163b}$$

where

$$\begin{bmatrix} A_i & B_i \\ C_i & D_i \end{bmatrix} = \begin{bmatrix} \cos\theta_i & jZ_{0i}\sin\theta_i \\ \frac{j\sin\theta_i}{Z_{0i}} & \cos\theta_i \end{bmatrix}; \quad \begin{matrix} i \to e & \text{for even mode} \\ \to 0 & \text{for odd mode} \end{matrix} \tag{6.164a}$$

Under the assumption that $\theta^e \approx \theta^0 = \theta$ and matching condition $Z_0 = \sqrt{Z_0^e Z_0^0}$, above equation leads to

$$\Gamma^e = -\Gamma^0 = \frac{j\left[\sqrt{\dfrac{Z_0^e}{Z_0^0}} - \sqrt{\dfrac{Z_0^0}{Z_0^e}}\right]\sin\theta}{2\cos\theta + j\left[\sqrt{\dfrac{Z_0^e}{Z_0^0}} - \sqrt{\dfrac{Z_0^0}{Z_0^e}}\right]\sin\theta} \tag{6.164b}$$

$$T^e = T^0 = \frac{2}{2\cos\theta + j\left[\sqrt{\dfrac{Z_0^e}{Z_0^0}} + \sqrt{\dfrac{Z_0^0}{Z_0^e}}\right]\sin\theta} \tag{6.164c}$$

Then $b_1 = b_3 = 0$ (6.164d)

$$b_2/a_1 = T = S_{21} = \frac{2}{2\cos\theta + j\left[\sqrt{\dfrac{Z_0^e}{Z_0^0}} + \sqrt{\dfrac{Z_0^0}{Z_0^e}}\right]\sin\theta} \tag{6.164e}$$

$$C = \frac{b_4}{a_1} = S_{41} = \frac{j\left[\sqrt{\dfrac{Z_0^e}{Z_0^0}} - \sqrt{\dfrac{Z_0^0}{Z_0^e}}\right]\sin\theta}{2\cos\theta + j\left[\sqrt{\dfrac{Z_0^e}{Z_0^0}} + \sqrt{\dfrac{Z_0^0}{Z_0^e}}\right]\sin\theta} \tag{6.165}$$

These equations show that

(1) the coupled voltages at Ports 2 and 4 are out of phase by 90°;

(2) coupling is a function of frequency or θ; and

(3) coupled voltage at Port 4 is maximum, when θ = odd multiple of $\pi/2$.
This corresponds to a minimum coupling length of $l = \lambda_{g0}/4$ at the midband of frequency f_0.

If C_0 is the midband coupling coefficient (for $\theta = \pi/2$, $l = \lambda_{g0}/4$)

$$C_0 = \frac{b_4}{a_1} = \frac{Z_0^e - Z_0^0}{Z_0^e + Z_0^0} \tag{6.166}$$

Therefore, the impedances required are

$$Z_0^e = Z_0^2/Z_0^0 = Z_0\sqrt{[(1 + C_0)/(1 - C_0)]} \tag{6.167}$$

and

$$Z_0^0 = Z_0^2/Z_0^e = Z_0\sqrt{[(1 - C_0)/(1 + C_0)]} \tag{6.168}$$

The coupling coefficient, transmission coefficient and directivity at any frequency can be written as

$$C = b_4/a_1 = \frac{j\,C_0\sin\theta}{\sqrt{[(1 - C_0^2)\cos\theta + j\sin\theta]}} \tag{6.169}$$

$$T = b_2/a_1 = \frac{\sqrt{(1-C_0^2)}}{\sqrt{[(1-C_0^2)\cos\theta + j\sin\theta]}} \tag{6.170}$$

$$D = b_3/a_4 = 0 \tag{6.171}$$

Since T is 90° out of phase with both a_1 and a_4, parallel coupled line is called 90° hybrid or quadrature coupler. $D = 0$ ideally, but in practice, θ is not same for even and odd modes. Therefore, the relation $Z_0 = \sqrt{Z_0^e Z_0^0}$ is only approximate. Moreover, due to imperfection in matching, directivity becomes non-zero. For loose coupling (more than 10 dB) the present approximation is sufficiently good. The frequency response of this coupler is obtained by writing

$$\theta = \frac{2\pi l}{\lambda_g} = \left(\frac{2\pi l}{u}\right) f \tag{6.172}$$

Since $l = \lambda_{g0}/4$, $\theta = \dfrac{\pi\lambda_{g0}}{2u}$, when u is the velocity of propagation. Assuming u a constant, θ is proportional to f. The ideal frequency response of this coupler is shown in Fig. 6.65.

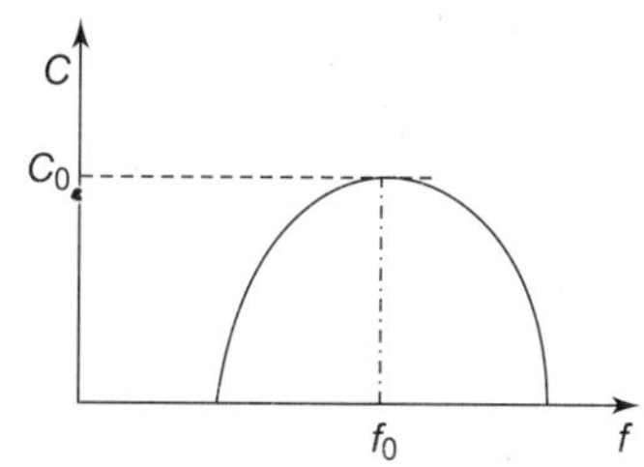

Fig. 6.65 Typical frequency response of a coupler

The closed form expression for the even and odd mode characteristic impedances Z_0^e and Z_0^0, respectively, of edge-coupled homogeneous symmetric strip-line coupler with negligible thickness strip conductors are given by

$$Z_0^e\sqrt{\varepsilon_r} = \frac{30\pi K(k_1^e)}{K(k^e)} \tag{6.173}$$

$$Z_0^0\sqrt{\varepsilon_r} = \frac{30\pi K(k_1^0)}{K(k^0)} \tag{6.174}$$

where
$$k^e = \tanh\left(\frac{\pi w}{2b}\right)\tanh\left(\frac{\pi}{2}\cdot\frac{w+S}{b}\right) \tag{6.175}$$

$$k^0 = \tanh\left(\frac{\pi w}{2b}\right)\coth\left(\frac{\pi}{2}\cdot\frac{w+S}{b}\right) \tag{6.176}$$

$$k_1^e = \sqrt{1-(k^e)^2}\ ;\quad k_1^0 = \sqrt{1-(k^0)^2} \tag{6.177}$$

Here, $K(k)$ and $K(k_1)$ are the complete elliptic integrals of the first kind. The Variation of $Z_0^e\sqrt{\varepsilon_r}$ and $Z_0^0\sqrt{\varepsilon_r}$ as a function of the dimensional parameters w/b and S/b is shown in Fig. 6.66.

The values of Z_0^e and Z_0^0 can be obtained from the coupling factor. The design data of w/b and S/b are obtained from the graph in Fig. 6.66 for a given substrate (ε_r).

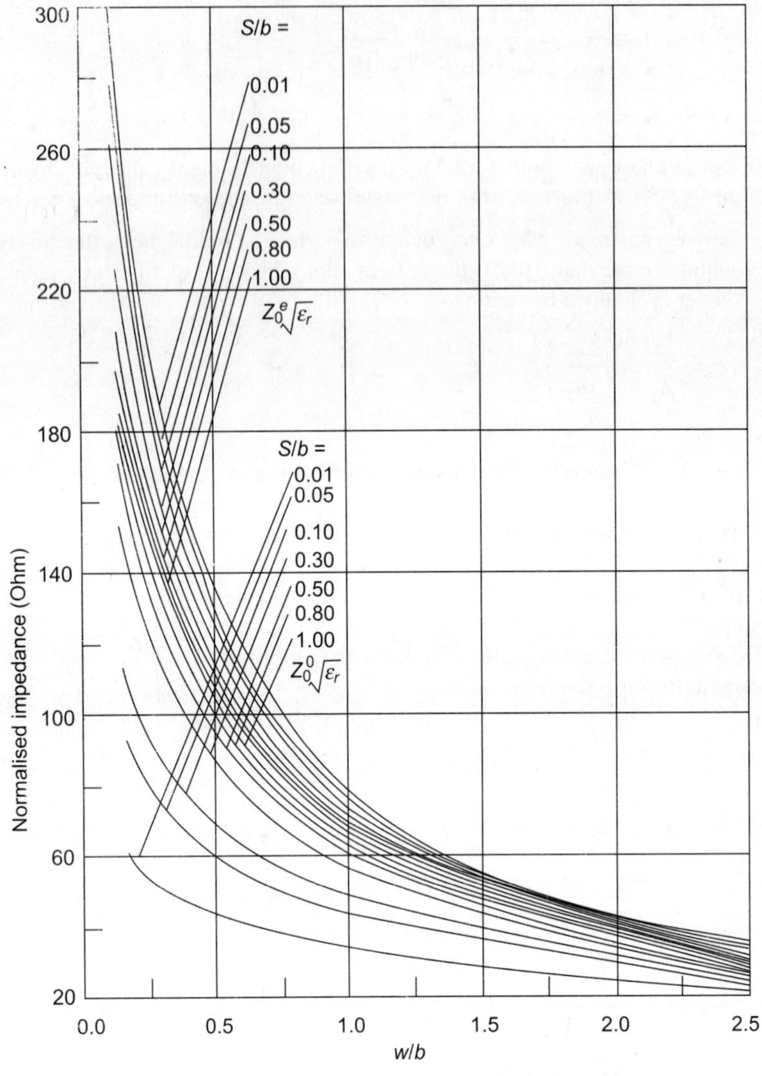

Fig. 6.66 $Z_0^e\sqrt{\varepsilon_r}$ and $Z_0^0\sqrt{\varepsilon_r}$ with w/b and S/b as parameters f or strip-line coupler

Example 6.28 Design a 10 dB stripline coupler at 5 GHz midband frequency with single feedline characteristic impedance Z_0 = 50 ohms, substrate permittivity ε_r = 9, substrate thickness b = 1 mm.

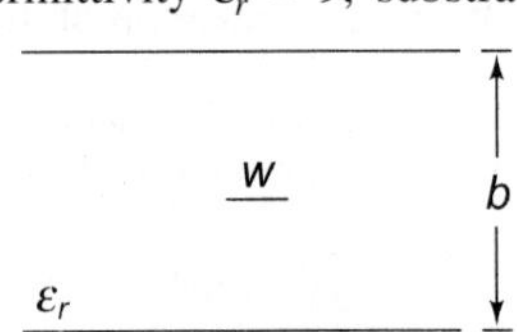

Fig. 6.67 Single feedline for Example 6.28

1. *Length of coupled section*

$$\frac{c}{f_0} = \frac{30}{5} = 6 \text{ cm}$$

$$\therefore \quad l = \frac{\lambda_{g0}}{4} = \frac{\lambda_0}{4\sqrt{\varepsilon_r}} = \frac{\lambda_0}{4\sqrt{9}} = 0.5 \text{ cm}$$

2. *Even and odd mode impedances*

$$-10\text{ dB} = 20\log C_0 \quad \text{or,} \quad C_0 = 10^{-1/2} = 0.3162,\ Z_0 = 50\text{ ohms}$$

$$Z_0^e = Z_0 \frac{\sqrt{(1+C_0)}}{\sqrt{(1-C_0)}} = 69.5\text{ ohms}$$

$$Z_0^0 = Z_0 \frac{\sqrt{(1-C_0)}}{\sqrt{(1+C_0)}} = 36\text{ ohms}$$

which satisfy $Z_0 \approx \sqrt{Z_0^e Z_0^0} = 50.02$ ohms; $Z_0\sqrt{\varepsilon_r} = 150$ ohms $\rightarrow \frac{w}{b} = 0.25$ for $S/b = 0.85$

3. *Line spacing S and width w*

S and w are obtained from impedance vs line dimensions graphs from Fig. 6.66. Hence, $S = 0.85$ mm, $w = 0.25$ mm.

Multisection Symmetrical Coupled Line Coupler

The bandwidth of coupled line coupler can be increased by cascading multiple sections of coupled lines as explained in Fig. 6.68. For perfect matching and increased directivity, the even- and odd-mode impedances of various sections satisfy the following relation

$$\sqrt{Z_{01}^e - Z_{01}^0} = \sqrt{Z_{02}^e - Z_{02}^0} = \cdots = \sqrt{Z_{0n}^e - Z_{0n}^0} = Z_0 \tag{6.178}$$

The midband coupling coefficient of the kth section is given by

$$C_{0k} = \frac{(Z_{0k}^e)^2 - 1}{(Z_{0k}^0)^2 + 1} \tag{6.179}$$

The coupling coefficients can be determined from the binomial or Chebyshev distributions in a similar manner as done in multisection quarterwave transformer or multihole directional couplers for a given coupling.

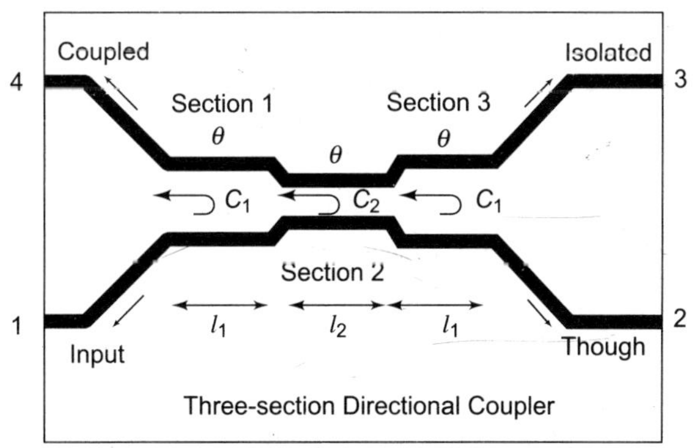

Fig. 6.68 Multisection coupled line coupler

Design of Microstrip Parallel Coupled Line Directional Coupler

The analysis of microstrip parallel coupled line directional coupler is same as described above at coupled transmission line coupler. The design of the coupler needs to determine the line widths, spacing and coupled line length for a given substrate. Figure 6.69 shows the schematic diagram of the coupler along with even and odd mode excitation schemes.

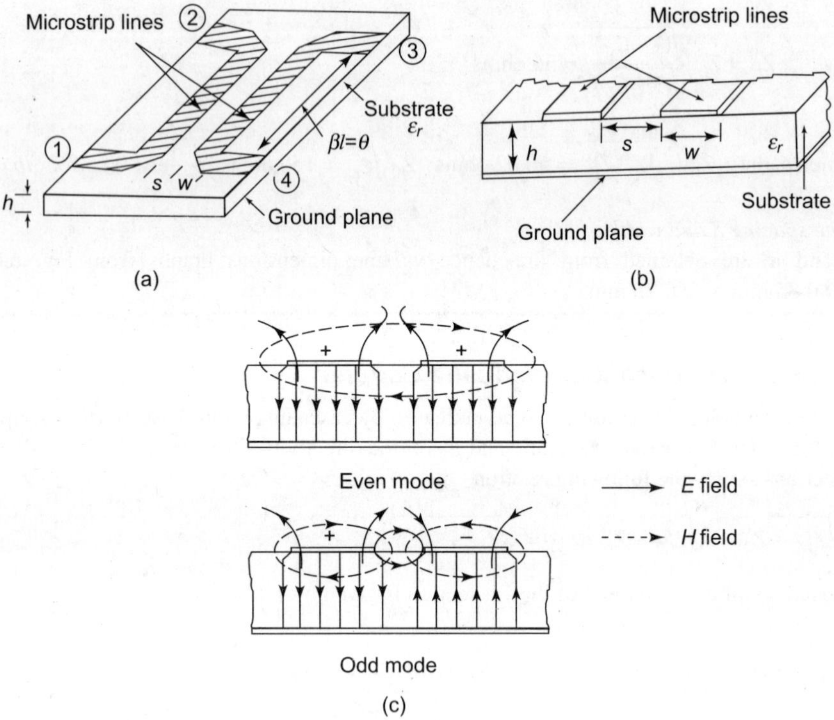

Fig. 6.69 Parallel coupled Microstrip lines: (a) coupler, (b) line dimensions (c) even and odd mode fields

❑ ***Coupler Line Width and Spacing*** We see that in the design of coupled lines, the following parameters are considered: w/h, S/h, ε_r, ε_{eff}, Z_0^e, Z_0^e, Z_0 and coupling. By using even- and odd-mode excitation as in the figure, we find even and odd-mode impedances Z_0^e and Z_0^0. Analysis yielding Z_0^e and Z_0^0 as functions of the shape ratio of w/h, S/h, and ε_r has been carried out by several workers.

In design of coupled transmission lines we use the following approximate synthesis technique while finding Z_0^e and Z_0^0:

- Firstly, with given single line characteristic impedance $Z_0 \approx \sqrt{Z_0^e\, Z_0^0}$ relative dielectric constant ε_r of the substrate and coupling coefficient C of the coupled lines, the shape ratios w/h are determined for equivalent single microstrip lines.
- Secondly, the shape ratio w/h and the spacing ratio S/h for the desired coupled microstrip are determined by using the single line shape ratios found in first step.

From C(in dB) and Z_0 , Z_0^e and Z_0^0 are determined as follows:

$$C = 10^{-C(dB)/20} \tag{6.180a}$$

Now, Z_0^0 and Z_0^e are expressed in terms of coupling coefficient:

$$Z_0^e = Z_0\sqrt{\frac{1+C}{1-C}} \text{ and } Z_0^0 = Z_0\sqrt{\frac{1-C}{1+C}} \tag{6.180b}$$

Therefore, Z_0^e and Z_0^0 for the coupled lines can be calculated from the above equations when absolute value of coupling C is known.

If we consider Z_0^e and Z_0^0 for single line as Z_0^{es} and Z_0^{0s};

$$Z_0^{es} = \frac{Z_0^e}{2} \text{ and } Z_0^{0S} = \frac{Z_0^0}{2} \tag{6.180c}$$

To find $\left(\frac{w}{h}\right)^{es}$ and $\left(\frac{w}{h}\right)^{os}$ from Z_0^{es} and Z_0^{0s}, let us use single line equations: given in Chapter 3 for Z_0 and ε_{eff} or the following:

$$\frac{w}{h} = \begin{cases} \dfrac{8e^A}{e^{2A}-2} & \text{for } w/h < 2 \\ \dfrac{2}{\pi}\left[B - 1 - \ln(2B-1) + \dfrac{\varepsilon_r - 1}{2\varepsilon_r}\left\{\ln(B-1) + 0.39 - \dfrac{0.61}{\varepsilon_r}\right\}\right] & \text{for } w/h > 2 \end{cases} \tag{6.180d}$$

where $$A = \frac{Z_0}{60}\sqrt{\frac{\varepsilon_r+1}{2}} + \frac{\varepsilon_r - 1}{\varepsilon_r + 1}\left(0.23 + \frac{0.11}{\varepsilon_r}\right) \tag{6.180e}$$

$$B = \frac{377\pi}{2Z_0\sqrt{\varepsilon_r}} \tag{6.180f}$$

Here, $Z_0 = Z_0^{es}$ and $w/h = (w/h)^{es}$ for even mode and $Z_0 = Z_0^{0s}$, $w/h = w/h)^{0s}$ for odd mode in equation (6.180d) – (6.180f)

The w/h and S/h for the desired coupled microstrip line are now determined using a family of approximate equations given Edward Terry as follows:

$$(w/h)^{es} = \frac{2}{\pi}\cosh^{-1}\left(\frac{2d - g + 1}{g+1}\right) \tag{6.180g}$$

$$(w/h)^{0s} = \frac{2}{\pi}\cosh^{-1}\left(\frac{2d-g-1}{g-1}\right) + \frac{4}{\pi(1+\varepsilon_r/2)}\cosh^{-1}\left(1 + 2\frac{w/h}{S/h}\right);\ \varepsilon_r \le 6 \tag{6.180h}$$

$$= \frac{2}{\pi}\cosh^{-1}\left(\frac{2d-g-1}{g-1}\right) + \frac{1}{\pi}\cosh^{-1}\left(1 + 2\frac{w/h}{S/h}\right);\ \varepsilon_r \ge 6 \tag{6.180i}$$

where $$g = \cosh\left(\frac{\pi S}{2h}\right) \tag{6.180j}$$

and $$d = \cosh\left(\pi w/h + \frac{\pi S}{2h}\right) \tag{6.180k}$$

As an approximation, neglecting second term in expression for $(w/h)^{0s}$ we get

$$S/h = \frac{2}{\pi}\cosh^{-1}\left[\frac{\cosh\{(\pi/2)(w/h)^{es}\} + \cosh\{(\pi/2)(w/h)^{os}\} - 2}{\cosh\{(\pi/2)(w/h)^{os}\} - \cosh\{(\pi/2)(w/h)^{es}\}}\right] \qquad (6.180l)$$

Considering $Z_0 = \sqrt{Z_0^e Z_0^0}$, w/h is obtained from $Z_0\sqrt{\varepsilon_0}$ vs w/h curves for different ε_r.

❑ ***Coupled Region Length*** Up to this stage the cross-sectional parameters of the coupler are determined. The effective permittivity needs to be determined to find the length of the coupling region.

In order to find the coupling region lengths, consideration is made that neither phase velocities nor the effective permittivities are equal for even and odd modes due to unequal distributions of field and capacitances for each mode for microstrip quasi-TEM configuration as shown in Fig. 6.70 specially for light coupling.

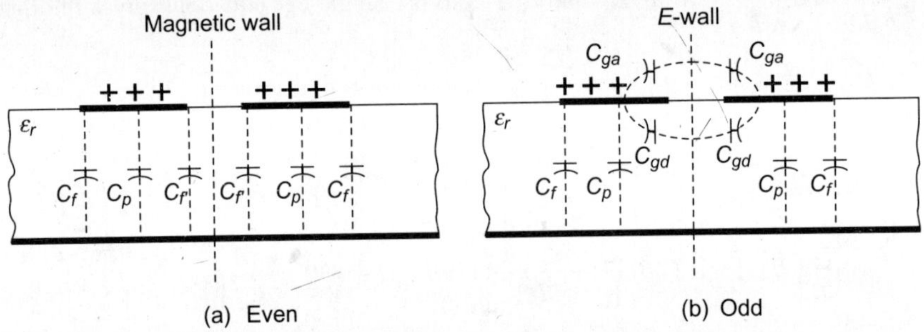

Fig. 6.70 Capacitances for microstrip-coupled lines: (a) Even mode (b) Odd mode

Here C_f, C_f' → Fringing capacitances
C_p → Parallel plate capacitance
C_{ga}, C_{gd} → Coupling gap capacitances in air and in dielectric, respectively.

For coupled lines,

$$Z_0 \approx \sqrt{Z_0^e Z_0^0} \text{ per strip} = \frac{1}{c\,C_{\text{air}}\sqrt{\varepsilon_{\text{eff}}}}; \qquad (6.180m)$$

For a single microstrip line of given width w and hight h

$$Z_0 = \sqrt{\frac{L}{C}} = vL = \frac{1}{vC}; \qquad (6.180n)$$

v = velocity of light in composite substrate media, c = velocity of light in air.

$$\varepsilon_{\text{eff}} \approx \frac{1+\varepsilon_r}{2}, \quad \mu = \mu_0$$

If substrate is air, $\varepsilon_{\text{eff}} \simeq \varepsilon_r = 1$,

$$Z_{0\text{air}} = \sqrt{\frac{L}{C_{\text{air}}}};$$

L remains unaltered for non-magnetic substrate $\mu = \mu_0$ (6.180o)

or, $$Z_{0\text{air}} = cL = \frac{1}{c\,C_{\text{air}}}; \; c = \text{velocity of light in air}$$

Combining $Z_0 = \sqrt{\dfrac{L}{C}} = \sqrt{\dfrac{\dfrac{1}{c^2 C_{air}}}{C}}$

or, $$Z_0 = \frac{1}{c\sqrt{C C_{air}}} \tag{6.180p}$$

$$v = \frac{1}{\sqrt{LC}}\,;\; C = \sqrt{\frac{1}{LC_{air}}}\,;\; v = \frac{c}{\sqrt{\varepsilon_{eff}}} \tag{6.180q}$$

or, $\sqrt{\varepsilon_{eff}} = \dfrac{c}{v}$

$$\therefore \quad \frac{C}{C_{air}} = \left(\frac{c}{v}\right)^2 = \varepsilon_{eff} = \left(\frac{\lambda_0}{\lambda_g}\right)^2; \tag{6.180r}$$

$$\therefore \quad Z_0 = \frac{Z_{0air}}{\sqrt{\varepsilon_{eff}}}; \tag{6.180s}$$

and $$v = \frac{c}{\sqrt{\varepsilon_{eff}}} \tag{6.180t}$$

$$\lambda_g = \lambda_0 / \sqrt{\varepsilon_{eff}} = \frac{300}{f\,(\text{GHz})\sqrt{\varepsilon_{eff}}}\ \text{mm}; \tag{6.180u}$$

For wide line $\varepsilon_{eff} \approx \varepsilon_r$, all E-field almost confined with in dielectric substrate.

For very thin line $\varepsilon_{eff} \approx \dfrac{1+\varepsilon_r}{2}$, field is almost equally shared by air and dielectric substrate.

For medium width, $\varepsilon_{eff} = 1 + q\,(\varepsilon_r - 1)$; $\dfrac{\varepsilon_r + 1}{2} \le \varepsilon_{eff} \le \varepsilon_r$

where q = filling factor ; $0.5 \le q \le 1$.

For coupled lines

$$Z_0\ \sqrt{Z_0^e\, Z_0^0}\ \text{per strip} = \frac{1}{c\, C_{air}\sqrt{\varepsilon_{eff}}} \tag{6.180v}$$

and for even and odd mode coupled lines

$$Z_0^e = \frac{1}{cC^e_{air}\sqrt{\varepsilon^e_{eff}}} = \frac{1}{c\sqrt{C^e\, C^e_{air}}}\,;\; \varepsilon_{eff} = \frac{C^e}{C^e_{air}} \tag{6.180w}$$

$$Z_0{}^0 = \frac{1}{cC^0_{air}\sqrt{\varepsilon^0_{eff}}} = \frac{1}{c\sqrt{C^0\, C^0_{air}}}\,;\; \varepsilon_{eff} = \frac{C^0}{C^0_{air}} \tag{6.180x}$$

C^e_{air} and C^0_{air} can be determined from the above equations for Z_0^e and Z_0^0 and determining C^e and C^0 from

$$\varepsilon^e_{eff} = (cC^e Z_0^e)^2 = \frac{C^e}{C^e_{air}}$$

$$\varepsilon^0_{eff} = (cC^0 Z_0^0)^2 = \frac{C^0}{C^0_{air}}$$

$$\therefore \qquad \sqrt{\varepsilon_{\text{eff}}} = \frac{\sqrt{\varepsilon_{\text{eff}}^e} + \sqrt{\varepsilon_{\text{eff}}^0}}{2} = \frac{cC^eZ_0^e + cC^0Z_0^0}{2} \tag{6.180y}$$

It can be shown also that,

$$\lambda_g^e = \frac{300}{f\,(\text{GHz})} \cdot \frac{Z_0^e}{Z_{0\text{air}}^e}\,(\text{mm}) = \frac{300}{f(\text{GH}_z)} \sqrt{\frac{C_{\text{air}}^e}{C^e}} \tag{6.181a}$$

$$\lambda_g^0 = \frac{300}{f\,(\text{GHz})} \cdot \frac{Z_0^0}{Z_{0\text{air}}^0}\,(\text{mm}) = \frac{300}{f\,(\text{GHz})} \sqrt{\frac{C_{\text{air}}^0}{C^0}} \tag{6.181b}$$

Length of coupled region

$$l = (2n-1)\,\lambda_{gm}/4;\ n = 0 \tag{6.181c}$$

where
$$\lambda_{gm} = \frac{\lambda_g^e + \lambda_g^0}{2} \text{ at midband frequency } f = f_0 \tag{6.181d}$$

$$= \frac{c}{4f_0\sqrt{\varepsilon_{\text{eff}}}}$$

The capacitances are determined from the following procedure [Edward Terry].
The total capacitances for each mode;

$$C^e = C_p + C_f + C'_f \tag{6.181e}$$

$$C^0 = C_p + C_f + C_{ga} + C_{gd}$$

The capacitance C_p simply relates to the parallel-plate line value and is given by

$$C_p = \varepsilon_0\varepsilon_r\,\frac{w}{h} \tag{6.181f}$$

Because of each microstrip taken alone as if for a single strip;

$$2C_f = \frac{\sqrt{\varepsilon_{\text{eff}}}}{cZ_0} - C_p \tag{6.181g}$$

$$C'_f = \frac{C_f}{1 + A\,(h/S)\tanh(8S/h)} \sqrt{\frac{\varepsilon_r}{\varepsilon_{\text{eff}}}} \tag{6.181h}$$

where
$$A = \exp\{-0.1\exp(2.333 - 2.53\,w/h)\} \tag{6.181i}$$

Here, c is the free space velocity. C_{ga}, is obtained by using an equivalent coplanar strip geometry,

$$C_{ga} = \varepsilon_0\frac{K(k')}{K(k)} = \varepsilon_0\frac{1}{\pi}\ln\left[2\frac{1+\sqrt{k'}}{1-\sqrt{k'}}\right];\ 0 \le k^2 \le 0.5; \tag{6.181j}$$

$$= \varepsilon_0\frac{\pi}{\ln\{2(1+\sqrt{k})/(1-\sqrt{k})\}};\ 0.5 \le k^2 \le 1;$$

where

$$k = \frac{S/h}{S/h + 2w/h} \tag{6.181k}$$

$$k' = \sqrt{1-k^2} \tag{6.181l}$$

The modified capacitance for coupled strip is given by

$$C_{gd} = \frac{\varepsilon_0 \varepsilon_r}{\pi} \ln\left\{\coth h\left(\frac{\pi}{4}\frac{S}{h}\right)\right\} + 0.65 C_f \left(\frac{0.02}{S/h}\sqrt{\varepsilon_r} + 1 - \varepsilon_r^{-2}\right); \tag{6.181m}$$

Combining all capacitances in Eq. [6.180y], the solution for ε_{eff} is obtained.

Branch-line Directional Couplers

Branch-line couplers are direct-coupled transmission line structures in which the main line is directly bridged to the secondary line by means of two shunt branches as shown in Fig. 6.71. These couplers can provide tight coupling and can handle high power.

The branch lengths and spacings are quarter wavelength long at the midband frequency f_0. The characteristic admittances of the series and shunt branches are Y_A and Y_B, respectively, for a symmetrical coupler and all output and input lines have the same characteristic admittance Y_0.

The analysis of a single section branch coupler can be carried out from even and odd symmetry circuits of Fig. 6.71 (b) and (c).

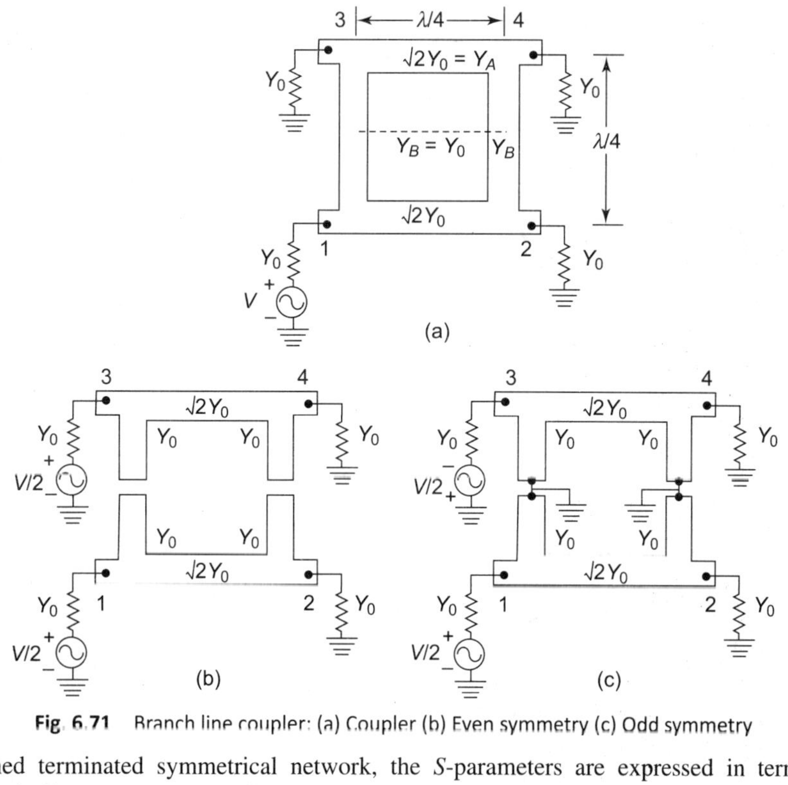

Fig. 6.71 Branch line coupler: (a) Coupler (b) Even symmetry (c) Odd symmetry

For matched terminated symmetrical network, the S-parameters are expressed in terms of $ABCD$ parameters in similar manner as described in 6.4.18 of coupled parallel coupler. The even and odd mode $[ABCD]$ for cascaded shunt, series and shunt sections can be expressed as

$$[ABCD]^e = \underset{\substack{\text{(shunt } \lambda/8 \text{ oc)} \\ \text{line}}}{\begin{bmatrix} 1 & 0 \\ jY_B & 1 \end{bmatrix}} \underset{\substack{\text{(series } \lambda/4\text{)} \\ \text{line}}}{\begin{bmatrix} 0 & j/Y_A \\ jY_A & 0 \end{bmatrix}} \underset{\substack{\text{(shunt } \lambda/8 \text{ oc)} \\ \text{line}}}{\begin{bmatrix} 1 & 0 \\ jY_B & 1 \end{bmatrix}}$$

$$= \begin{bmatrix} -Y_B/Y_A & j/Y_A \\ j(Y_A - Y_B^2/Y_A) & -Y_B/Y_A \end{bmatrix} \tag{6.182a}$$

$$\text{Similarly,} [ABCD]^0 = \begin{bmatrix} 1 & 0 \\ -jY_B & 1 \end{bmatrix} \begin{bmatrix} 0 & j/Y_A \\ jY_A & 0 \end{bmatrix} \begin{bmatrix} 1 & 0 \\ -jY_B & 1 \end{bmatrix}$$

$$\begin{pmatrix} \lambda/8\, S.c. \\ \text{Shunt} \end{pmatrix} \begin{pmatrix} \text{Series} \\ \lambda/4 \text{ line} \end{pmatrix} \begin{pmatrix} \lambda/8\, S.c. \\ \text{Shunt} \end{pmatrix}$$

$$= \begin{bmatrix} Y_B/Y_A & j/Y_A \\ j\left(Y_A - Y_B^2/Y_A\right) & Y_B/Y_A \end{bmatrix} \tag{6.182b}$$

Under matched condition, $S_{11} = S_{22} = 0$

$$\therefore \quad B/Z_0 = C\,Z_0$$

$$\therefore \quad S_{12} = S_{21} = \frac{1}{A + B/Z_0} = \frac{1}{A + BY_0} \tag{6.182c}$$

Therefore, from $[ABCD)]^e$ and $[ABCD]^0$

$$\frac{Y_0}{Y_A} = (Y_A - Y_B^2/Y_A)\frac{1}{Y_0} \quad \text{or,} \quad Y_A^2 - Y_B^2 = Y_0^2 \tag{6.182d}$$

Now, coupling coefficient

$$S_{14} = \frac{b_4}{a} = \frac{S_{12}^e - S_{12}^0}{2}$$

$$= \frac{1}{2}\left[\left(\frac{1}{A + BY_0}\right)^e - \left(\frac{1}{A + BY_0}\right)^0\right] \tag{6.182e}$$

$$\left(\frac{1}{A + BY_0}\right)^e = \frac{1}{-Y_B/Y_A + jY_0/Y_A} = \frac{Y_A}{-Y_B + jY_0}; \tag{6.182f}$$

$$\left(\frac{1}{A + BY_0}\right)^0 = \frac{1}{Y_B/Y_A + jY_0/Y_A} = \frac{Y_A}{Y_B + jY_0}; \tag{6.182g}$$

$$\therefore \quad S_{14} = \frac{1}{2}\left[\frac{Y_A}{-Y_B + jY_0} - \frac{Y_A}{Y_B + jY_0}\right]$$

$$= \frac{1}{2}\left[\frac{Y_AY_B + jY_AY_0 + Y_AY_B - jY_AY_0}{Y_B^2 + Y_0^2}\right] = \frac{1}{2}\,\frac{2Y_AY_B}{-Y_A^2}$$

$$\text{or} \quad S_{14} = \frac{-Y_B}{Y_A} \tag{6.182h}$$

Therefore, the coupling can be varied by varying characteristic admittances of series and shunt arms Y_A and Y_B, respectively. If $Y_A = \sqrt{2}\ Y_0$, $Y_B = Y_0$, then $S_{14} = -1/\sqrt{2}$. Therefore, under this condition the branch line is a 3 dB forward coupler with $\angle 180°$ phase between ports 4 and 1.

The through coefficient

$$S_{12} = \frac{S_{12}^e + S_{12}^0}{2} = \frac{1}{2}\left[\left(\frac{1}{A+BY_0}\right)^e + \left(\frac{1}{A+BY_0}\right)^0\right]$$

$$= \frac{1}{2}\left[\frac{Y_A}{-Y_B + jY_0} + \frac{Y_A}{Y_B + jY_0}\right] = \frac{2jY_AY_0}{-2Y_A^2} = j\frac{Y_0}{-Y_A}\ ; \tag{6.183}$$

If $Y_A = \sqrt{2}\ Y_0$, $S_{12} = \dfrac{-j}{\sqrt{2}}$. So that the branch line coupler is a 3 dB forward coupler with quadrature phase w.r.t. Port 2 and 1.

$$S_{13} = \frac{S_{11}^e - S_{11}^0}{2} = 0, \text{ port 3 is isolated.} \tag{6.184a}$$

For matched port and symmetry, the S-matrix of branch line coupler is

$$[S] = \begin{bmatrix} 0 & -j/\sqrt{2} & 0 & -1/\sqrt{2} \\ -j/\sqrt{2} & 0 & -1/\sqrt{2} & 0 \\ 0 & -1/\sqrt{2} & 0 & -j/\sqrt{2} \\ -1/\sqrt{2} & 0 & -j/\sqrt{2} & 0 \end{bmatrix} = \frac{1}{\sqrt{2}}\begin{bmatrix} 0 & -j & 0 & -1 \\ -j & 0 & -1 & 0 \\ 0 & -1 & 1 & -j \\ -1 & 0 & -j & 0 \end{bmatrix} \tag{6.184b}$$

The coupler is called 3 dB, 90° hybrid coupler in general.

Branching Synthesis of Couplers and Hybrids

It can be shown that any device, whose admittance matrix is such that diagonal elements are zero and off-diagonal elements are pure imaginary numbers, may be synthesized by means of $\lambda/4$ or $3\lambda/4$ TEM elements of the appropriate characteristic admittance. Consider a two-port network consisting of a $\lambda/4$ line of characteristic admittance Y_{0A} inserted in a transmission line of characteristic admittance Y_0. The system is operating in TEM mode (coaxial, strip or two-conductor line) so that the characteristic admittance is defined on a real voltage–current basis.

The admittance matrix of a section of line of length z is given by

$$[Y] = \frac{jY_0}{\sin\gamma z}\begin{bmatrix} -\cos\gamma z & 1 \\ 1 & -\cos\gamma z \end{bmatrix} \tag{6.185}$$

The admittance matrix of a $\lambda/4$ line of admittance Y_{0A} normalized on a Y_0 basis is

$$[Y] = \frac{Y_{0A}}{Y_0}\begin{bmatrix} 0 & +j \\ +j & 0 \end{bmatrix} \tag{6.186}$$

For $\quad l = 3\lambda/4$

$$[Y] = \frac{Y_{0A}}{Y_0}\begin{bmatrix} 0 & -j \\ -j & 0 \end{bmatrix} \tag{6.187}$$

From Eqs (6.186) and (6.187) it follows that any device, whose admittance matrix is such that diagonal elements are zero and off-diagonal elements are pure imaginary numbers, can be synthesized by means of $\lambda/4$ or $3\lambda/4$ TEM elements of the appropriate characteristic admittance. A few examples for hybrid or directional coupler are given in the following paragraphs. The basic method of synthesis is to preselect the admittance matrix of a hybrid or a directional coupler and manufacture such a matrix by branching quarter wavelength elements to obtain prototypes.

❑ *Synthesis of rat-Race Hybrid Coupler from Magic-tee* The S-matrix for a matched magic-T with collinear Ports 1 and 2, E and H Ports 4 and 3, respectively, is

$$[S] = 1/\sqrt{2}\begin{bmatrix} 0 & 0 & 1 & 1 \\ 0 & 0 & 1 & -1 \\ 1 & 1 & 0 & 0 \\ 1 & -1 & 0 & 0 \end{bmatrix} \tag{6.188}$$

When all the reference planes are moved away from the junction by $\theta = \pi/4$ without altering the isolation property between 1, 2, 3, and 4, now S-parameter is

$$S' = S_{12}e^{-j(\theta 1 + \theta 2)} = S_{12}e^{-j(\pi/d + \pi/4)}$$

$$= S_{12}e^{-j\pi/2} = -j\, S_{12}, \text{ etc.} \tag{6.189}$$

The new S-matrix becomes

$$[S] = -\, j/\sqrt{2}\begin{bmatrix} 0 & 0 & 1 & 1 \\ 0 & 0 & 1 & -1 \\ 1 & 1 & 0 & 0 \\ 1 & -1 & 0 & 0 \end{bmatrix} \tag{6.190}$$

Here, $[S]^* = -\,[S]$, $[S]_t = [S]$ (symmetry property). For a lossless network, the unitary property of $[S]$ gives:

$$[S]^*\,[S]_t = [U] \tag{6.191}$$

or, $$[S]^2 = -\,[U] \tag{6.192}$$

Now, $$[Y] = Y_0\,[U - S]\,[U + S]^{-1}; \tag{6.193}$$

By associative property of matrix operations

$$[Y] = Y_0[U - S]\,[U + S]\,[U - S]^{-1}\,[U + S]^{-1}$$

$$= [U - 2S + S^2]\,[U - S^2]^{-1} \tag{6.194}$$

So that $[Y] = -Y_0\,[S]$; since $[S]^2 = [U]$

or, $$[Y] = +\,Y_0\, j/\sqrt{2}\begin{bmatrix} 0 & 0 & 1 & 1 \\ 0 & 0 & 1 & -1 \\ 1 & 1 & 0 & 0 \\ 1 & -1 & 0 & 0 \end{bmatrix} \tag{6.195}$$

Thus, admittance matrix (6.195) may be synthesized by $\lambda/4$ lines for the positive coefficients and $3\lambda/4$ lines for the negative coefficients of admittance:

$$Y_{0A}/Y_0 = 1/\sqrt{2} \tag{6.196}$$

Here, the self-admittances parameters y_{11}, y_{22}, y_{33} and $y_{44} = 0$. Since $y_{12} = y_{34} = 0$, and $y_{21} = y_{43} = 0$, there is no direct element between Ports 1 and 2 and between Ports 3 and 4. The following elements are required to be inserted:

(i) Between Ports 1 and 3, a $\lambda/4$ element of normalized characteristic admittance $1/\sqrt{2}$

(ii) Between 1 and 4, a $\lambda/4$ element of normalized characteristic admittance $1/\sqrt{2}$

(iii) Between 2 and 3, a $\lambda/4$ element of normalized characteristic admittance $1/\sqrt{2}$

(iv) Between 2 and 4, a $3\lambda/4$ element of normalized characteristic admittance $1/\sqrt{2}$

A schematic circuit and its practical realization in coaxial line form is shown in Fig. 6.72. This circuit is called a *rat-race hybrid*.

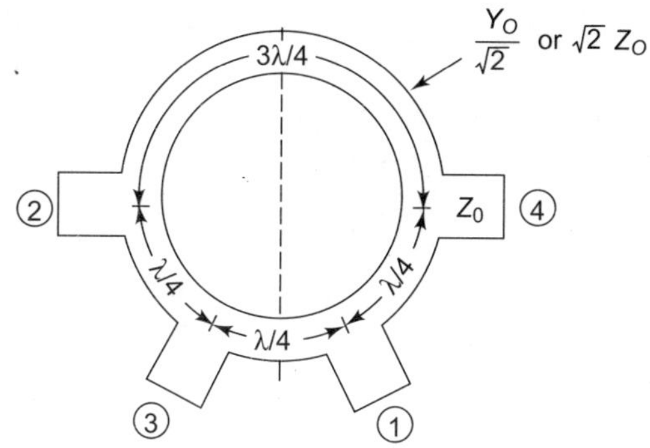

Fig. 6.72 Rat-race

Example 6.29 For the rat-race hybrid shown in Fig. 6.72, (a) calculate input admittance at Port 1 when all others are matched, (b) when input is fed at Port 1, find signal distributions in all other ports, (c) if input is fed to Port 3 and all others are matched, calculate power distribution in all ports.

Solution

(a) When all ports are matched, the admittance matrix equation is

$$\begin{bmatrix} I_1 \\ I_2 \\ I_3 \\ I_4 \end{bmatrix} = j/\sqrt{2}\begin{bmatrix} 0 & 0 & 1 & 1 \\ 0 & 0 & 1 & -1 \\ 1 & 1 & 0 & 0 \\ 1 & -1 & 0 & 0 \end{bmatrix}\begin{bmatrix} V_1 \\ V_2 \\ V_3 \\ V_4 \end{bmatrix}$$

When all ports except Port 1 are match, terminated by Z_0,

$$V_2 = -I_2$$
$$V_3 = -I_3$$
$$V_4 = -I_4$$

$$\therefore \quad \begin{bmatrix} I_1 \\ I_2 \\ I_3 \\ I_4 \end{bmatrix} = j/\sqrt{2}\begin{bmatrix} 0 & 0 & 1 & 1 \\ 0 & 0 & 1 & -1 \\ 1 & 1 & 0 & 0 \\ 1 & -1 & 0 & 0 \end{bmatrix}\begin{bmatrix} V_1 \\ -I_2 \\ -I_3 \\ -I_4 \end{bmatrix}$$

or,

$$I_1 = -(j/\sqrt{2})(I_3 + I_4)$$
$$I_2 = +(j/\sqrt{2})(-I_3 + I_4)$$
$$I_3 = -(j/\sqrt{2})(V_1 - I_2)$$
$$I_4 = +(j/\sqrt{2})(V_1 + I_2)$$

Adding, $I_3 + I_4 = j/\sqrt{2}\ (2V_1) = j\sqrt{2}\ V_1$

Substituting in the first equation above,

$$I_1 = -\frac{j}{\sqrt{2}}\cdot\frac{j}{\sqrt{2}}\cdot 2V_1 = V_1$$

Therefore, $I_1/V_1 = 1$ = normalized input admittance at Port 1. Thus, if all ports are matched, Port 1 also matched.

(b) When input is fed at Port 1, the S-matrix equation is

$$\begin{bmatrix} b_1 \\ b_2 \\ b_3 \\ b_4 \end{bmatrix} = -j/\sqrt{2}\begin{bmatrix} 0 & 0 & 1 & 1 \\ 0 & 0 & 1 & -1 \\ 1 & 1 & 0 & 0 \\ 1 & -1 & 0 & 0 \end{bmatrix}\begin{bmatrix} a_1 \\ 0 \\ 0 \\ 0 \end{bmatrix}$$

Therefore, $b_1 = b_2 = 0;\ P_1 = P_2 = 0$

$$b_3 = -ja_1/\sqrt{2}\ ;\ P_3 = \frac{1}{2}|b_3|^2 = |a_1|^2/4$$

$$b_4 = -ja_1/\sqrt{2}\ ;\ P_4 = \frac{1}{2}|b_4|^2 = |a_1|^2/4$$

Thus, the input to Port 1 is split equally and in phase at Ports 3 and 4, with no power reflected back to Port 1 and no output at Port 2. If there is mismatch at Port 4 it will not affect Port 3 and vice-versa since $S_{34} = 0$.

(c) When input is fed to Port 3 and all others are matched, the S-matrix equation is

$$\begin{bmatrix} b_1 \\ b_2 \\ b_3 \\ b_4 \end{bmatrix} = -j/\sqrt{2}\begin{bmatrix} 0 & 0 & 1 & 1 \\ 0 & 0 & 1 & -1 \\ 1 & 1 & 0 & 0 \\ 1 & -1 & 0 & 0 \end{bmatrix}\begin{bmatrix} 0 \\ 0 \\ a_3 \\ 0 \end{bmatrix}$$

$$b_1 = -ja_3/\sqrt{2}, \qquad P_1 = |a_3|^2/4$$

$$b_2 = -ja_3/\sqrt{2}, \qquad P_2 = |a_3|^2/4$$

$$b_3 = b_4 = 0, \qquad P_3 = P_4 = 0$$

Thus, input at Port 3 is equally split with the same phase at Ports 1 and 2, with no power reflected back to Port 3 and no output at Port 4.

A practical application of rat-race is shown in Fig. 6.73 as a hybrid transformer.

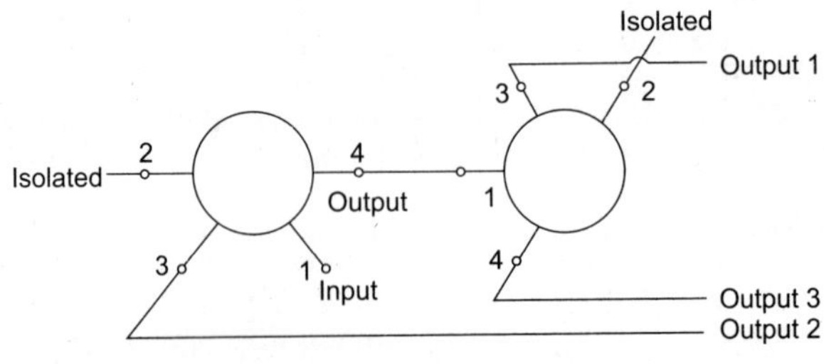

Fig. 6.73 Hybrid transformer

❑ *Synthesis of 3 dB, 90° hybrid from magic-T* By choosing the shift of reference planes 1 and 3 as $\theta_1 = \theta_3 = \pi/2$; and those 2 and 4 as $\theta_2 = \theta_4 = \pi$, the S-matrix of the magic-T reduces to

$$[S] = -1/\sqrt{2}\begin{bmatrix} 0 & 0 & 1 & j \\ 0 & 0 & j & 1 \\ 1 & j & 0 & 0 \\ j & 1 & 0 & 0 \end{bmatrix} \tag{6.197}$$

such that isolation between 1 and 2, 3 and 4 are not violated. Here, new parameters are

$$S_{13} = \frac{b_1\, e^{-j\theta 1}}{a_3\, e^{j\theta 3}} = \frac{b_1}{a_3} e^{-j(\theta 1 + \theta 3)};$$

or,

$$S_{13} = e^{-j(\pi/2 + \pi/2)} = -1 \tag{6.198}$$

$$S_{14} = e^{-j(\pi/2 + \pi)} = -j \tag{6.199}$$

$$S_{23} = e^{-j(\pi/2 + \pi)} = -j \tag{6.200}$$

$$S_{24} = e^{-j(\pi + \pi)} = 1, \text{ etc.} \tag{6.201}$$

The corresponding normalized admittance matrix

$$[Y] = [U - S][U + S]^{-1}$$

$$= j\begin{bmatrix} 0 & 1 & 0 & \sqrt{2} \\ 1 & 0 & \sqrt{2} & 0 \\ 0 & \sqrt{2} & 0 & 1 \\ \sqrt{2} & 0 & 1 & 0 \end{bmatrix} \tag{6.202}$$

Thus, $Y_{11} = Y_{13} = Y_{22} = Y_{24} = Y_{31} = Y_{33} = Y_{42} = Y_{44} = 0$. The physical line elements to be inserted are as follows:

1. Between ports (3, 4) and (1, 2): a $\lambda/4$ element of characteristic admittance = Y_0.
2. Between (2, 3) and (1, 4): a $\lambda/4$ element of characteristic admittance = $\sqrt{2}\, Y_0$.

The implementation of this hybrid is shown in Fig. 6.74. If power is fed at Port 1 with all the other ports matched, power distribution in all ports can be found as follows:

$$\begin{bmatrix} b_1 \\ b_2 \\ b_3 \\ b_4 \end{bmatrix} = -1/\sqrt{2}\begin{bmatrix} 0 & 0 & 1 & j \\ 0 & 0 & j & 1 \\ 1 & j & 0 & 0 \\ j & 1 & 0 & 0 \end{bmatrix}\begin{bmatrix} a_1 \\ 0 \\ 0 \\ 0 \end{bmatrix} \tag{6.203}$$

$$b_1 = b_2 = 0,\ b_3 = (-1/\sqrt{2})a_1,\ b_4 = (-j/\sqrt{2})(a_1) \tag{6.204}$$

Therefore, b_3 and b_4 are 90° out-of-phase but the magnitudes of power are same. This coupler is called 3 dB, 90° hybrid coupler.

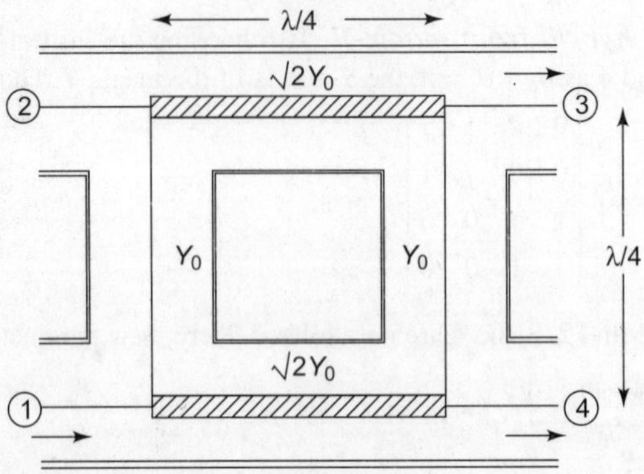

Fig. 6.74 3 dB, 90 degree hybrid

Example 6.30 Find the coupling to Port 2 when signal is fed to Port 1 with 3 and 4 connected to $\lambda/4$ short-circuited line sections.

Solution Power distribution can be found from the matrix equation $\{b\} = [S]\{a\}$.

or,
$$\begin{bmatrix} b_1 \\ b_2 \\ b_3 \\ b_4 \end{bmatrix} = -1/\sqrt{2} \begin{bmatrix} 0 & 0 & 1 & j \\ 0 & 0 & j & 1 \\ 1 & j & 0 & 0 \\ j & 1 & 0 & 0 \end{bmatrix} \begin{bmatrix} a_1 \\ 0 \\ b_3 \\ b_4 \end{bmatrix}$$

Since impedance of $\lambda/4$ short circuit line $= \infty$, Port 3 & 4 are opened, i.e. $\Gamma = 1$
Therefore,

$$b_1 = -\frac{1}{\sqrt{2}}(b_3 + j\,b_4)\,,\; b_2 = -\frac{1}{\sqrt{2}}(jb_3 + b_4)$$

$$b_3 = -\frac{1}{\sqrt{2}}a_1\,,\; b_4 = -\frac{1}{\sqrt{2}}j\,a_1$$

$$b_1 = -\frac{1}{\sqrt{2}}\left[-\frac{1}{\sqrt{2}}a_1 + j(-j/\sqrt{2})\,a_1\right] = -\frac{1}{\sqrt{2}}\left[-\frac{1}{\sqrt{2}}a_1 + \frac{a_1}{\sqrt{2}}\right] = 0$$

$$b_2 = -\frac{1}{\sqrt{2}}[jb_3 + b_4] = -\frac{1}{\sqrt{2}}\left[-\frac{j}{\sqrt{2}}a_1 - \frac{j}{\sqrt{2}}a_1\right]$$

$$= j\frac{1}{2}2a_1 = ja_1 = a_1e^{j\pi/2}$$

Now, in terms of current and voltage representation $V_2 = -I_2$ for matched termination.

Now $\{I\} = [Y]\{V\}$

or,
$$\begin{bmatrix} I_1 \\ I_2 \\ 0 \\ 0 \end{bmatrix} = j \begin{bmatrix} 0 & 1 & 0 & \sqrt{2} \\ 1 & 0 & \sqrt{2} & 0 \\ 0 & \sqrt{2} & 0 & 1 \\ \sqrt{2} & 0 & 1 & 0 \end{bmatrix} \begin{bmatrix} V_1 \\ -I_2 \\ V_3 \\ V_4 \end{bmatrix}$$

$$I_1 = j[-I_2 + \sqrt{2}\,V_4] = j[-jV_1 - j\sqrt{2}\,V_3 + \sqrt{2}\,V_4]$$

$$I_2 = j[V_1 + \sqrt{2}\,V_3]$$

$$0 = -V_2 I_2 + V_4$$

$$0 = +\sqrt{2}\,V_1 + V_3$$

Therefore, $$V_3 = -\sqrt{2}\,V_1$$

and $$I_2 = j(V_1 - \sqrt{2}\,V_2 V_1) = j(V_1 - 2V_1) = -jV_1$$

$$V_4 = \sqrt{2}\,I_2 = -j\sqrt{2}\,V_1$$

Therefore,

$$I_1 = j[-jV_1 - j\sqrt{2}(-\sqrt{2})V_1 + \sqrt{2}(-j\sqrt{2})V_1]$$

Normalized input admittance

$$Y_1 = I_1/V_1 = j[-j + 2j - 2j] = 1$$

Therefore, the port must be matched also. Actual strip-line configuration of the coupler is shown in Fig. 6.75.

If V_1 and V_2 are the input voltages at Ports 1 and 2, output voltage from Ports 3 and 4 are $-j(V_1 + V_2)/\sqrt{2}$ and $-j(V_1 - V_2)/\sqrt{2}$ or power $(V_1 + V_2)^2/2Z_0$ and $(V_1 - V_2)^2/Z_0$. Thus, sum and difference arms of magic-T are obtained.

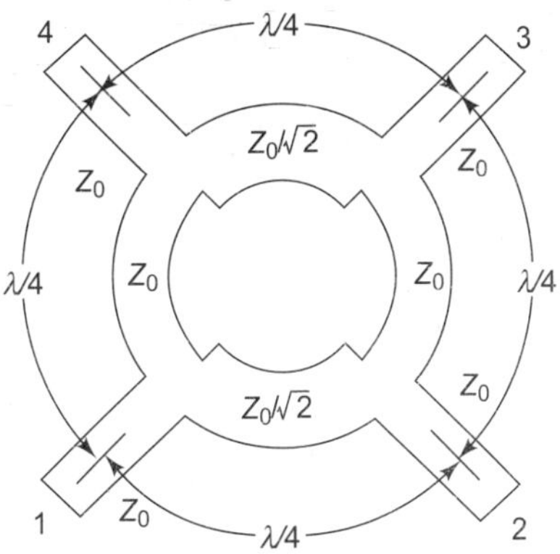

Fig. 6.75 Strip-branch line coupler

Power Dividers and Combiners 6.4.19

A power divider is a device to split the input power into a number of smaller amounts of power at multiple ports (N) to feed N number of branching circuits with isolation between the output ports. For simplicity, a two-way equal power divider is shown in Fig. 6.76 which is a three-port junction. For equal power division, the device consists of two quarter-wave sections with characteristic impedances Z_0 connected in parallel with the input line, which also has a characteristic impedance Z_0. A resistor $R = 2Z_0$ is connected between Ports 2 and 3 which are matched terminated. Since the input impedance at Port 1 is now $Z_0/2$, a quarter-wave matching transformer with characteristic impedance $\sqrt{2}\,Z_0$ is used to transform the Port 1 input impedance into the feeder line impedance Z_0. This also maintains zero current in the resistance R when Ports 2 and 3 are matched terminated.

With a help of even and odd mode analysis, it can be shown that the power applied to Port 1 divides equally between Ports 2 and 3 with zero loss in the balancing resistor R, and the voltage at either output port lags that at the input port by 90°. Thus, the device is a 3 dB, 90° power divider. It can be shown that the configuration also acts as a 3 dB power combiner when fed from the Ports 2 and 3 with the output taken at Port 1.

Turnstile Junction 6.4.20

A turnstile junction is a symmetrical six-port device as shown in Fig. 6.77. Ports 1, 2, 3 and 4 correspond to rectangular waveguide, operating in dominant TE_{10} mode. At the centre, a circular waveguide is placed with its axis perpendicular to the H-plane of the rectangular waveguide. Because of the rotational symmetry of the circular waveguide operating in TE_{11} mode, it can support two cross-polarized waves, each independent of the other, and thus exhibits two ports, 5 and 6, corresponding to each polarization. Therefore, the circular waveguide can deliver power through any one of these waves for which **E** is parallel to the polarization of the waves excited in the guide. If the polarization is somewhere in between, then both the ports 5 and 6 draw power.

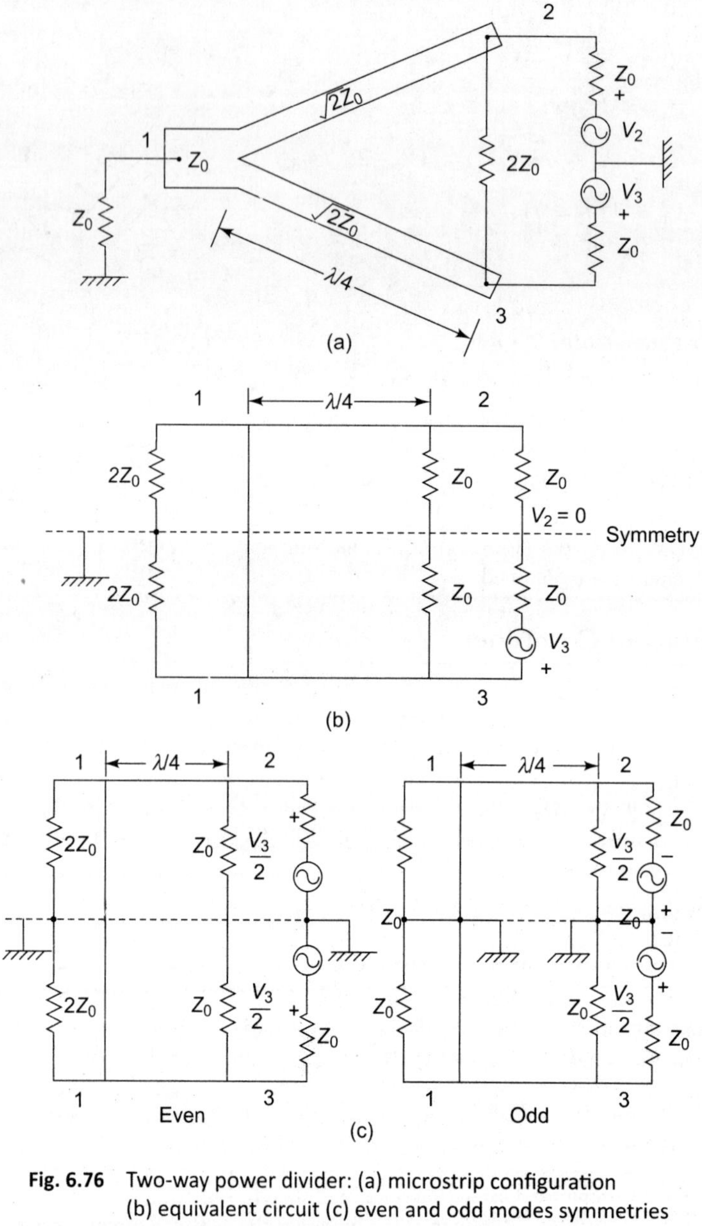

Fig. 6.76 Two-way power divider: (a) microstrip configuration (b) equivalent circuit (c) even and odd modes symmetries

There are four planes of symmetry P_1, P_2, P_3 and P_4 and one axis of symmetry A_{65}. Coupling between the circular and rectangular waveguides is achieved through a circular hole at the junction in the plane transverse to the circular guide. Plane P_1 interchanges the fields at Ports 2 and 4, reverses the fields at 6, but leaves fields at 1, 3 and 5 unchanged. The reference planes are set to make the scattering coefficients real for a matched turnstile junction:

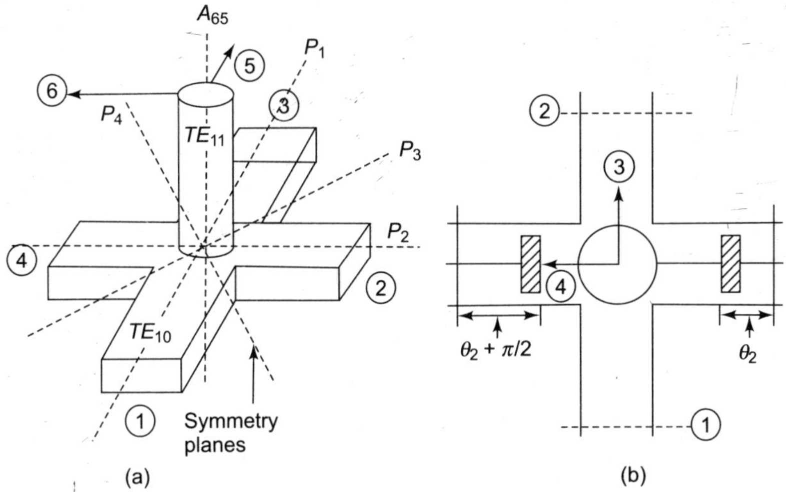

Fig. 6.77 Turnstile junction: (a) Isometric view (b) Top view of 4-port hybrid junction

$$[S] = 1/2\begin{bmatrix} 0 & 1 & 0 & 1 & \sqrt{2} & 0 \\ 1 & 0 & 1 & 0 & 0 & \sqrt{2} \\ 0 & 1 & 0 & 1 & -\sqrt{2} & 0 \\ 1 & 0 & 1 & 0 & 0 & -\sqrt{2} \\ \sqrt{2} & 0 & -\sqrt{2} & 0 & 0 & 0 \\ 0 & \sqrt{2} & 0 & -\sqrt{2} & 0 & 0 \end{bmatrix} \tag{6.205}$$

Ports 5 and 6 can be matched by tuning screws in the corresponding arms independently, since $S_{56} = 0$. Ports 1, 2, 3 and 4 are matched by inserting a tuning post along the axis A_{56} within the junction from the bottom.

Application of Turnstile Junction

There are many applications of a turnstile junction as described below.

❑ *Four port Hybrid Junction* By placing a short circuit at Port 2 at distance θ_2 from the reference plane and another at Port 4 at $\theta_4 = \theta_2 + \pi/2$, and with all other ports matched, the *S*-matrix equation for an input at Port 1 is

$$\begin{bmatrix} b_1 \\ b_2 \\ b_3 \\ b_4 \\ b_5 \\ b_6 \end{bmatrix} = [S]\begin{bmatrix} a_1 \\ -b_2 e^{j2\theta_2} \\ 0 \\ b_4 e^{j2\theta_2} \\ 0 \\ 0 \end{bmatrix} \tag{6.206}$$

This reduces the S matrix to

$$S = \frac{1}{\sqrt{2}}\begin{bmatrix} 0 & 0 & 1 & -e^{j2\theta_2} \\ 0 & 0 & -1 & -e^{j2\theta_2} \\ 1 & -1 & 0 & 0 \\ -e^{j2\theta_2} & -e^{j2\theta_2} & 0 & 0 \end{bmatrix} \tag{6.207}$$

If $\theta_2 = n\pi$, $n = 0, 1, 2, ...; e^{j\theta 2} = 1$, the $[S]$ becomes that of hybrid junction or magic-T with some phase change due to negative sign in the elements ($e^{j2n\pi} = 1$):

$$[S] = 1/\sqrt{2}\begin{bmatrix} 0 & 0 & 1 & -1 \\ 0 & 0 & -1 & -1 \\ 1 & -1 & 0 & 0 \\ -1 & -1 & 0 & 0 \end{bmatrix} \tag{6.208}$$

Here, ports 3 and 4 are contained in the same physical circular waveguide through cross polarization replacing the ports 5 and 6, respectively, of turnstile junction as shown in Fig. 6.62(b). An input signal at Port 1 produces equal and opposite signals at new ports 3 and 4. An input at port 2 produces equal and in-phase signals at 3 and 4. Ports 1 and 2, 3 and 4 are isolated to each other. This satisfies the properties of a hybrid four-port coupler.

❑ ***Matched Four-way Power Divider*** When an equal signal is fed into ports 5 and 6 in the turnstile junction with all other ports match terminated, the output signal can be expressed by

$$\begin{bmatrix} b_1 \\ b_2 \\ b_3 \\ b_4 \\ b_5 \\ b_6 \end{bmatrix} = [S]\begin{bmatrix} 0 \\ 0 \\ 0 \\ 0 \\ a \\ a \end{bmatrix} = 1/\sqrt{2}\begin{bmatrix} a \\ a \\ -a \\ -a \\ 0 \\ 0 \end{bmatrix} \tag{6.209}$$

This acts as a four-way (ports 1, 2, 3 and 4) equisignal distribution network. Here, the phase of signals b_1 and b_2 is opposite to that of b_3 and b_4.

❑ ***Matched Three-way Power Divider*** When a short circuit is placed in circular waveguide ports at a distance θ from the reference plane and input power is fed to Port 1, the output power distribution can be expressed by

$$\begin{bmatrix} b_1 \\ b_2 \\ b_3 \\ b_4 \\ b_5 \\ b_6 \end{bmatrix} = [S]\begin{bmatrix} a_1 \\ 0 \\ 0 \\ 0 \\ -b_5 e^{j2\theta} \\ -b_6 e^{j2\theta} \end{bmatrix} = 1/\sqrt{2}\begin{bmatrix} -\sqrt{2}\, b_5 e^{j2\theta} \\ a_1 - \sqrt{2}\, b_6 e^{j2\theta} \\ \sqrt{2}\, b_5 e^{j2\theta} \\ a_1 + \sqrt{2}\, b_6 e^{j2\theta} \\ \sqrt{2}\, a_1 \\ 0 \end{bmatrix} \tag{6.210}$$

Substituting the elements of the S-matrix and solving the above equation yields

$$b_1 = -a_1 e^{j2\theta}/2,\ b_2 = a_1/2,\ b_3 = a_1 e^{j2\theta}/2,\ b_4 = a_1/2,$$

$$b_5 = a_1/\sqrt{2} \text{ and } b_6 = 0$$

Since b_5 and b_6 are not used due to short-circuit terminations, and if $\theta = n\pi$, b_2, b_3 and b_4 are equiphase and equi-amplitudes always. Since b_1 is the reflected component at the input port, this configuration acts as a three-way power divider. Since the reflection coefficient at Port 1 is $e^{j2\theta}/2$,

$$\text{Input VSWR at Port 1} = \frac{|a_1| + |b_1|}{|a_1| - |b_1|} = 3.0$$

Because of high VSWR, an auxiliary matching structure is required at the input port. After matching, the power will be equally divided into the three rectangular waveguide Ports 2, 3 and 4.

REVIEW QUESTIONS

6.1 A shunt susceptance $j0.5$ mho is connected across a lossless transmission line with characteristic impedance 50 ohms, terminated by matched impedances. Find the S-matrix of the junction.

6.2 A series reactance j 40 ohm is connected between two lossless transmission lines of characteristic impedances 50 ohms and 75 ohms. Find the S-matrix of the junction. Assume matched terminated at both ends of these lines.

6.3 A reciprocal two-port microwave device has a VSWR of 1.5 and an insertion loss of 2 dB. Find the magnitudes of S-parameters for the device.

6.4 A waveguide load has a VSWR of 1.1 and is used to absorb an average power of 5 W. Find the reflected power and the return loss.

6.5 Find the scattering matrix for an ideal short-circuited section of waveguide with the reference ports chosen. Show that the coefficients are real numbers. How does the S-matrix change if the short-circuit position of Port 2 is variable?

6.6 A 10 dB attenuator having an input VSWR of 1.2 is terminated by matched load. Find the reflected power, the absorbed power and the transmitted power for 1W input.

6.7 Plot the dB attenuation versus angle for an ideal rotary attenuator.

6.8 An ideal rotary phase shifter is terminated with a matched load. Find the scattering matrix for the device. Express the transmitted voltage at the output in terms of the input for angles 0, 30, 45 and 90 degrees. Assume that the input voltage is real.

6.9 In a precision rotary phase shifter, the output quarter-wave plate, the transition and the rectangular waveguides are rotated by an angle of 30°. Show that phase of the transmitted wave will be changed by 30°.

6.10 The input power to the sum arm of an ideal matched magic-T is 1W. Find the output powers from the other arms when matched terminated.

6.11 Find the magnitude of scattering coefficient for a directional coupler having a coupling coefficient of 3 dB and directivity of 25 dB and VSWR of 1.2 for the main guide Port 1.

6.12 The input power to a lossless matched directional coupler is 100 mW. If the coupling coefficient is 20 dB and the directivity is 30 dB, find the output powers at other ports.

6.13 An ideal three-port circulator is fed at Port 1 with average power 100 W. If the power reflected by the antenna at the next port 2 is 100 μW, find the power outputs at all ports assuming they are match terminated.

6.14 A three-port circulator has an insertion loss of 1 dB, an isolation of 20 dB, and a VSWR of 1.2. Find the output power at ports 2 and 3 for an input power of 100 mW at Port 1.

6.15 An E-plane tee is matched terminated at all the ports with an input power of 5 mW fed at Port 2 (E-arm). Determine the power flow through the junction. What changes in the power distribution will occur if the power is fed at the collinear arm 1.

6.16 A directional coupler of 10 dB coupling and 40 dB directivity produces a transmission loss of 1 dB. For an input power of 10 mW at the input port of the main arm, determine the power at the other ports when matched terminated.

6.17 A microstrip edge-coupled directional coupler is designed for a coupling factor of 10 dB and a characteristic impedance of 50 ohms. Determine the even and odd mode characteristic impedances.

6.18 A Bethe-hole directional coupler with a centre circular aperture using the rectangular waveguide of size 0.9 in × 0.4 in is designed to operate at 10 GHz with 30 dB coupling. Find the aperture radius and the frequency band over which the directivity remains greater than 20 dB.

6.19 Design a three-hole Chebyshev directional coupler using centre apertures in the common broad wall between two rectangular waveguides of dimensions 0.9″ × 0.4″ to be operated at 9 GHz. Find the aperture radii, spacing and the bandwidth of the coupler for a coupling of 30 dB and directivity of 30 dB.

6.20 Design an air dielectric stripline edge-coupled 10 dB directional coupler for ground plane spacing of 1 cm at 5 GHz. Find the strip width and spacing for obtaining input and output line characteristic impedances equal to 50 ohms for air dielectric.

6.21 A branch-line 3 dB microstrip directional coupler at 5 GHz is designed with input and output line impedances of 50 ohms. Find the characteristic impedance of the through lines and the branch lines and their lengths in terms of the wavelength. The dielectric substrate is 20 mm thick and has a dielectric constant of 10. Find the width of the lines and the lengths of the through lines and branch lines.

6.22 A microstrip hybrid ring is constructed on 1 mm thick substrate with a dielectric constant of 2.5 at 3 GHz. Determine the widths of the transmission line and the ring as well as the radius of the ring for input and output line impedances of 50 ohms.

6.23 Show using *S*-matrix theory that a lossless non-reciprocal two-port microwave device cannot be constructed.

REFERENCES

1. Altman, J. L., *Microwave Circuits*, D. Van Nostrand Company, Inc. 1964.
2. A. J. Slobodnik, Jr. and R. T. Webster, "Experimental Validation of Microstrip Bend Discontinuity Models from 18 to 60 GHz," *IEEE Trans, Microwave Tech.*, Vol. 42, No. 10, October 1994, pp. 1872–1878.
3. Bahl, I. and P. Bhartia, *Microwave Circuit Design*, John Wiley and Sons, Inc., New York, 1988.
4. Bhat, Bharathi and Shiban K. Koul, *Stripline-like Transmission Lines for Microwave Integrated Circuits*, Wiley Eastern limited, New Delhi, 1989.
5. Bosma, H., On Stripline Y-Circulators at UHF, *IRE Trans.*, Vol. MTT-12, pp. 61–72, 1964.
6. Clarricoats, P., *Microwave Ferrites*, John Wiley and Sons, Inc., New York, 1961.
7. Cohn, S. B. and R. Levy, "History of microwave passive components with particular attention to directional couplers", *IEEE Trans.* MTT, Vol. 32 No. 9, 1984.
8. Collin, R. E., *Foundations for Microwave Engineering*, McGraw-Hill Inc., 1992.
9. Edwards, Terry, *Foundations for Microstrip Circuit Design*, John Wiley and Sons, New York, 1981.
10. Gupta, K. C. and Amarjit Singh, *Microwave Integrated Circuits*, John Wiley and Sons, New York, Inc., 1974.
11. Gupta, K. C., et al, *Microstrip Lines and Slotlines*, 2nd ed., Artech House, 1996.
12. Howe, J. R., *Stripline Circuit Design*, Dedham, Mass. Artech House, Inc. 1974.
13. Kirschning, M. et al, "Measuring and Computer Aided Modeling of Microstrip Discontinuities by an Improved Resonator Method," IEEE Microwave Theory and Techniques Symposium Digest, 1983, pp. 495–497.
14. Lance, A. L., *Introduction to Microwave Theory and Measurements*, New York, McGraw-Hill, 1964.
15. Lebedev, I., *Microwave Engineering*, MIR Publishers, Moscow, 1973.
16. Liao, S. Y., *Microwave Devices and Circuits*, Prentice-Hall of India Private Limited, New Delhi, 1985.
17. Marcuvitz, N., *Waveguide Handbook*, McGraw-Hill Book Company, New York, 1951.
18. Montogomery, C. G., R. H. Dicke, and E. M. Purcell, *Principles of Microwave Circuits*, McGraw-Hill Book Company, New York, 1948.
19. R. J. P. Douville and D. S. James, "Experimental Study of Symmetric Microstrip Bends and Their Compensation," *IEEE Trans. Microwave Theory Tech.*, Vol. 26, No. 3, March 1978, pp. 175–181.

20. Roddy, D., *Microwave Technology*, Prentice-Hall, Inc., 1986.

21. Seeger, John, A., *Microwave Theory Components and Devices*, Prentice-Hall, Englewood Cliffs, New Jersey, 1988.

22. Sooboo, R. F., *Theory and Application of Ferrites*, Prentice-Hall, Inc., Englewood Cliffs, N.J., 1960.

23. Suhl, H. and R. L. Walker, "Topics in guided wave propagation through gyromagnetic media," BSTJ, Vol. 33, Sept. 1954.

24. *"The Microwave Engineers Handbook and Buyers", Guide, 1963 Ed.*, pp. T-107 to T-111, Horizon House-Microwave, Inc., Brookline, Mass.

25. Veley, V. F., *Modern Microwave Technology*, Prentice-Hall, Inc., 1987.

26. Von Aulock, H. Wilhelm, and E. F. Clifford, *Linear Ferrite Devices for Microwave Applications*, New York, Academic Press, Inc, 1968.

27. Wheeler, G. J., *Introduction to Microwaves*, Englewood Cliffs, N.J., Prentice-Hall, 1963.

28. Wheeler, H. A., "Tx-line properties of parallel strips separated by a dielectric sheet," *IEEE Trans.* MTT, MTT-13, pp. 172–185, March 1965.

29. Wilkinson, E., "An N-way Hybrid Power Divider", *IEEE Trans*. MTT, pp. 116–118, 1960.

30. Wu, Y. S. and F. J. Rosenbaum, "Wide-Band Operations of Microstrip Circulators", *IEEE Trans*., Vol. MTT-13, pp. 15–27, 1965.

31. Young, Leo, *Advances in microwaves*, Vol. 1, New York, Academic Press, Inc., 1966.

7 MICROWAVE RESONATORS

INTRODUCTION 7.1

Microwave resonators are tunable circuits used in microwave oscillators, amplifiers, wavemeters and filters. At the tuned frequency, the circuit resonates, where the average energies stored in the electric field (or in the capacitor) is W_e, and the average energies stored in the magnetic field (or in the inductor) is W_m. At resonance, they are equal and the circuit impedance becomes purely real. The total energy is therefore twice the electric or magnetic energy stored in the resonator. The parameters which describe the performance of a resonator are

1. Resonant frequency f_r at which the energy in the cavity attains maximum value = $2W_e$ or $2W_m$
2. Quality factor Q which is a measure of the frequency selectivity of a cavity, defined by

$$Q = \frac{2\pi \times \text{Maximum energy stored } (W)}{\text{Average energy dissipated per cycle } (P)} = \frac{\omega_r W}{P}$$

$$= \frac{\text{Resonant frequency}}{\text{Half- power bandwidth}}$$

$$= \frac{f_r}{\Delta f}$$

3. Input impedance of the resonator specifying the matching with the input/output circuits

The Q-factor relates a resonant circuit's capacity for electromagnetic energy storage with its energy dissipation through heat. In order to obtain high Q, the circuits are made highly reactive and can be simulated by using short-circuited or open-circuited sections of low loss coaxial transmission lines in the frequency range of 100 to 1000 MHz. Typical values of Q up to about 10,000 can be achieved from such transmission sections.

At higher frequencies, transmission-line resonators do not give very high Q due to skin effect and radiation loss in braided cables. Therefore, most of the microwave resonators are constructed with rectangular or circular metallic cavities, which have a natural resonant frequency and behaves like a *LCR* circuit. The electric and magnetic energies stored in the volume of the cavity give rise to equivalent C and L, while the walls, having a finite conductivity, give rise to power loss which provides for the equivalent resistance R. The resonator bandwidth is inversely proportional to the Q-factor.

The following sections describe the analysis of microwave resonators, their performances and applications.

COAXIAL RESONATORS 7.2

There are three basic configurations of coaxial resonant cavities: (1) quarterwave coaxial cavity; (2) halfwave coaxial cavity; and (3) coaxial cavity with a shortening capacitance as shown in Fig. 7.1 along with their equivalent circuits.

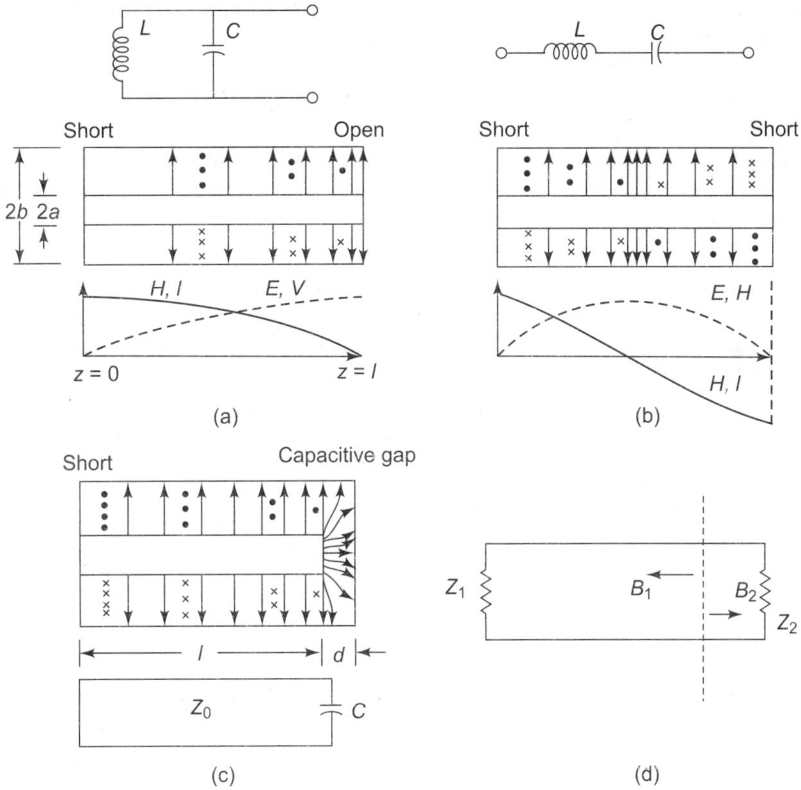

Fig. 7.1 Coaxial cavities and equivalent circuits: (a) quarterwave cavity (b) halfwave cavity (c) capacitive end cavity (d) resonance condition ×××···H, ——E

The general method to determine the resonant length of these cavities is based on that the total susceptance of the oscillatory system becomes zero at resonance, i.e., $B_1 + B_2 = 0$, where, B_1 and B_2 are the susceptances of the sections of the transmission lines looking towards left and right, respectively, from an arbitrary reference plane on the line. By taking the reference plane at right-hand side end of these cavities, with reference to Fig. 7.1 (d), the resonant length l is obtained as follows.

Quarterwave Coaxial Cavity 7.2.1

From the resonant condition,

$$jZ_0 \tan \beta l = \text{infinite} \quad \text{or,} \quad l = (2n - 1)\, \lambda/4 \tag{7.1a}$$

This cavity is essentially equivalent to a parallel resonant circuit.

Halfwave Coaxial Cavity 7.2.2

From the resonant condition

$$jZ_0 \tan \beta l = 0 \quad \text{or,} \quad l = (2n-1)\lambda/2 \tag{7.1b}$$

This is equivalent to series resonance.

Capacitive-End Coaxial Cavity 7.2.3

From the resonant condition,

$$jZ_0 \tan \beta l = -\frac{1}{j\omega_r C} \quad \text{or,} \quad l = \frac{\lambda}{2\pi}\tan^{-1}\left(\frac{1}{Z_0\omega_r C}\right) \tag{7.1c}$$

Here, $n = 1, 2, 3, \ldots$; Z_0 is the characteristic impedance of the coaxial line and C is the gap capacitance between the centre conductor and the shorting termination of Fig. 7.1(c). The dimensions a and b of all the coaxial cavities in TEM modes are restricted by the generation of the next higher order TE and TM mode, so that $\pi(b + a) < \lambda$.

Since the diameters of the coaxial cavities determine the power loss in the cavity, Q varies with b/a and attains a maximum value at $b/a = 3.6$. Due to microwave radiation from the open end of the quarterwave coaxial cavity, halfwave coaxial resonators are preferred over quarterwave sections and are used in microwave resonant wavemeters.

The coaxial cavity can be tuned either by changing the capacitive gap d, by means of a capacitive ridge at a fixed length l, or by changing length l by a variable shorting plunger at a fixed value of gap length d.

Analysis of Co-axial Cavity 7.2.4

A coaxial cavity resonator (Fig. 7.2) of length l with one shorting end is analyzed to find an expression of unloaded Q when operated in fundamental TEM mode.

For propagation in the + z direction, the fields inside the line are given by Eqs. 3.75a and 3.75b in Chapter 3. Since the line is shorted at $z = 0$, standing waves are generated with the field components

$$\mathbf{E} = 2j\frac{V_0}{\ln(b/a)}\frac{\hat{\boldsymbol{\rho}}}{\rho}\sin kz \tag{7.2a}$$

$$\mathbf{H} = 2\frac{V_0}{\eta\ln(b/a)}\frac{\hat{\boldsymbol{\phi}}}{\rho}\cos kz \tag{7.2b}$$

Fig. 7.2 Coaxial cavity with one end short

where $k = \omega\sqrt{\mu\varepsilon}$. Here, a and b are the outer and inner radii of the inner and outer conductors, respectively, and V_0 is the peak potential difference between the conductors.

Stored Energy

The electric energy density stored in the cavity is given by

$$w_e = \frac{1}{4}\varepsilon\vec{E}.\vec{E}^* = \varepsilon\left[\frac{V_0}{\ln(b/a)}\right]^2\frac{1}{\rho^2}\sin^2 kz \tag{7.3a}$$

The total electric energy stored in the cavity is, therefore,

$$W_e = \int_v w_e dV = \varepsilon\left[\frac{V_0}{\ln(b/a)}\right]^2 \int_a^b \int_0^{2\pi} \int_{-l}^0 \frac{1}{\rho^2}\sin^2 kz\rho d\rho d\varphi dz$$

$$= \frac{\pi\varepsilon V_0^2 d}{\ln(b/a)}\left\{1 - \frac{\sin 2kd}{2kd}\right\}$$

$$= W_0\left(1 - \frac{\sin 2kd}{2kd}\right) \tag{7.3b}$$

where $$W_0 = \frac{\pi\varepsilon V_0^2 d}{\ln(b/a)} \tag{7.3c}$$

Similarly the magnetic energy stored in the cavity is given by

$$W_m = \int_V w_m dV = \int_V \frac{1}{4}\mu\vec{H}.\vec{H}^* dV$$

$$= \mu\left[\frac{V_0}{\eta ln(b/a)}\right]^2 \int_a^b \int_0^{2\pi} \int_{-l}^0 \frac{1}{\rho^2}\cos^2 kz\,\rho d\,\rho d\,\phi dz$$

$$= \frac{\pi\mu V_0^2 d}{\eta^2 \ln(b/a)}\left(1 + \frac{\sin 2kd}{2kd}\right) = \frac{\pi\varepsilon V_0^2 d}{\ln(b/a)}\left(1 + \frac{\sin 2kd}{2kd}\right) = W_0\left(1 + \frac{\sin 2kd}{2kd}\right) \tag{7.3d}$$

It can be seen from the above expressions that

$$W_m + W_e = 2W_0 \tag{7.3e}$$

and

$$\frac{W_m - W_e}{W_m + W_e} = \frac{\sin 2kd}{2kd} \tag{7.3f}$$

Since at resonance, $W_e = W_m$,

$$W_0\left(1 - \frac{\sin 2kd}{2kd}\right) = W_0\left(1 + \frac{\sin 2kd}{2kd}\right) \tag{7.3g}$$

or $$\frac{\sin 2kd}{2kd} = 0 \tag{7.3h}$$

From the above it is evident that $2kd = mx$, $m = 1, 2, 3, \ldots; m \neq 0$.

Further examination reveals two cases of interest

(1) $kd = (p, ½)\pi$, $p = 0, 1, \ldots$ for which $m = 1, 3, 5, \ldots$
For this case, the electric field is maximum at $z = -d$, whereas the magnetic field = 0 which is equivalent to a parallel resonant circuit possessing infinite impedance.

(2) $kd = px$, $p = 1, 2, 3, \ldots$ for which $m = 2, 4, 6, \ldots$, and the electric field = 0 at $z = -d$, whereas the magnetic field is maximum. This is equivalent to a series-resonant circuit, in which, at resonance, the impedance is zero.

Conductor Loss

Due to the skin effect, the signal currents flows through the outer surface of the inner conductor and returns through the inner surface of the outer conductor. These current densities are given, respectively , by

$$\mathbf{J}_{sa} = \hat{n} \times \mathbf{H} = \hat{\boldsymbol{\rho}} \times \mathbf{H}_i = \hat{z}\frac{2V_0 \cos kz}{\eta a \ln(b/a)} \tag{7.3i}$$

and

$$\mathbf{J}_{sb} = \hat{n} \times \mathbf{H} = \hat{\boldsymbol{\rho}} \times \mathbf{H}_t = \hat{z}\frac{2V_0 \cos kz}{\eta b \ln(b/a)} \tag{7.3j}$$

The loss contribution from the inner conductor is

$$P_{\text{loss},a} = \frac{1}{2}R_s \int_0^{2\pi} \int_{-d}^{0} \mathbf{J}_{ia} \cdot \mathbf{J}_{ia}^* \; a d\phi dz = \frac{2\pi R_s V_0^2 d}{[\eta \ln(b/a)]^2}\frac{1}{a}\left(1+\frac{\sin 2kd}{2kd}\right) = \frac{2R_s W_0}{\mu \ln(b-a)}\frac{1}{a}\left(1+\frac{\sin 2kd}{2kd}\right) \tag{7.3k}$$

Similarly, the loss contribution from the outer conductor is

$$P_{\text{loss},b} = \frac{1}{2}R_s \int_0^{2\pi} \int_{-d}^{0} \mathbf{J}_{ib} \cdot \mathbf{J}_{ib} * b d\phi dz = \frac{2\pi R_s V_0^2 d}{[\eta \ln(b/a)]^2}\frac{1}{b}\left(1+\frac{\sin 2kd}{2kd}\right) = \frac{2R_s W_0}{\mu \ln(b/a)}\frac{1}{b}\left(1+\frac{\sin 2kd}{2kd}\right) \tag{7.3l}$$

At the short-circuit end ($z = 0$), $\hat{\mathbf{n}} = -\hat{\mathbf{z}}$, and

$$\mathbf{H} = 2\frac{V_0}{\eta \ln(b/a)}\frac{\hat{\boldsymbol{\phi}}}{\rho} \tag{7.3m}$$

from which $\mathbf{J}_s = 2\dfrac{V_0}{\eta \ln(b/a)}$

The loss contribution of the shorting plate is

$$P_{\text{loss},-d} = R_s \int_a^{b} \int_0^{2\pi} \mathbf{J}_s \cdot \mathbf{J}_s *, \rho d\rho l\phi = \frac{4\pi R_s V_0^2}{\eta^2 \ln(b/a)} = \frac{4R_s W_0}{\mu d} \tag{7.3n}$$

The total loss

$$P_{\text{loss}} = P_{\text{loss},a} + P_{\text{loss},b} + P_{\text{loss},-d} = \frac{4R_s W_0}{\mu}\left[\frac{\left(\frac{1}{a}+\frac{1}{b}\right)}{2\ln(b/a)} + \frac{1}{d}\right] \tag{7.3p}$$

Since $d = \dfrac{\left(p+\frac{1}{2}\right)\lambda}{2}$,

$$P_{\text{loss}} = \frac{4R_s W_0}{\mu}\left[\frac{\left(\frac{1}{a}+\frac{1}{b}\right)}{2\ln(b/a)} + \frac{4}{(2p+1)\lambda}\right] \tag{7.3q}$$

The quality factor is

$$Q = \frac{\omega(W_m + W_e)}{P_{\text{loss}}} = \frac{2\omega W_0}{P_{\text{loss}}} = \frac{\omega\mu}{2R_s}\frac{\lambda}{\left[\frac{\frac{\lambda}{2}\left(\frac{1}{a}+\frac{1}{b}\right)}{\ln(b/a)} + \frac{4}{(2p+1)}\right]} = \frac{\frac{\lambda}{\delta_s}}{\left[\frac{\frac{\lambda}{2}\left(\frac{1}{a}+\frac{1}{b}\right)}{\ln(b/a)} + \frac{4}{(2p+1)}\right]} \tag{7.3r}$$

Here,

$$\frac{\omega\mu}{2R_s} = \frac{1}{\delta_s} \tag{7.3s}$$

Input Impedance

At resonance, the input impedance is resistive (real).

The input resistance at parallel resonance can be obtained from

$$P_{\text{loss}} = \frac{V_0^2}{2R} = \frac{4R_s W_0}{\mu}\left[\frac{\left(\frac{1}{a}+\frac{1}{b}\right)}{2\ln(b/a)}+\frac{1}{d}\right] = \frac{4\pi R_1 \varepsilon V_0^2 d}{\mu \ln(b/a)}\left[\frac{\left(\frac{1}{a}+\frac{1}{b}\right)}{2\ln(b/a)}+\frac{1}{d}\right]$$

$$= \frac{4\pi R_s V_0^2}{\eta^2 \ln(b/a)}\left[\frac{\left(\frac{d}{a}+\frac{d}{b}\right)}{2\ln(b/a)}+1\right] \tag{7.3t}$$

Therefore,

$$R = \frac{\frac{\eta^2 \ln(b/a)}{8\pi R_s}}{\left[\frac{\left(\frac{d}{a}+\frac{d}{b}\right)}{2\ln(b/a)}+1\right]} = \frac{\frac{\eta Z_e}{R_s}}{\left[\frac{2\left(\frac{d}{a}+\frac{d}{b}\right)}{\ln(b/a)}+4\right]} \tag{7.3u}$$

The characteristic impedance of the symmetric coaxial line for a non-magnetic dielectric is given by

$$Z_0 = \frac{1}{2\pi}\sqrt{\frac{\mu}{\varepsilon}}\ln\left(\frac{b}{a}\right) = \frac{60}{\sqrt{\varepsilon_r}}\ln\frac{b}{a}\text{ ohm} \tag{7.3v}$$

where $\varepsilon_r = \varepsilon\varepsilon_0$, the dielectric constant of the dielectric medium between the conductors. The wavelength inside this TEM line is

$$\lambda_g = \frac{\lambda_0}{\sqrt{\varepsilon_r}} \tag{7.3w}$$

For $a = 1$ cm, $b = 3$ cm, $p = 0$, $\lambda = 4d = 160$ cm, $f = 0.1875$ GHz, $\delta = 4.82$ μm $Q = 3279.42$.

$R_s = 1/(\sigma\delta_s) = 3.577$ mΩ $Z_0 = 60 \ln (3) = 65.92$ Ω, where $R = 68.81$ kΩ.

High values of Q and R are expected for parallel resonance circuits.

WAVEGUIDE CAVITY RESONATORS 7.3

Waveguide cavity resonators are formed by shorting two ends of a section of a waveguide as shown in Fig. 7.3. The field components inside the cavity can be computed from the wave equations which satisfy the boundary condition of the zero tangential E field at all conducting walls. Within the cavity, various TE_{mnp}

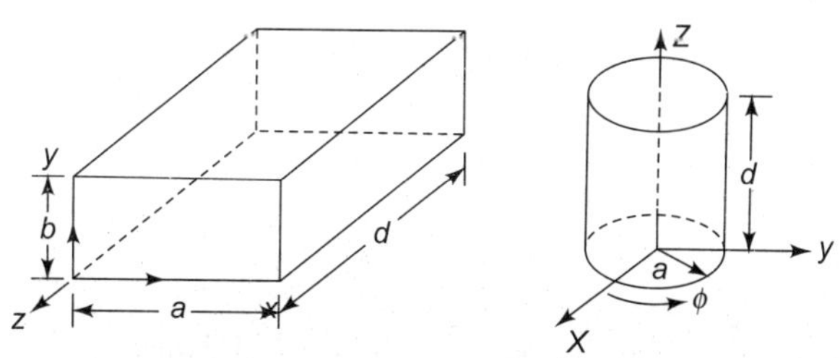

Fig. 7.3 Waveguide cavity resonators (a) Rectangular cavity (b) Circular cavity

and TM_{mnp} modes exist. Very high Q-factors and accompanying narrow bandwidths can be obtained with these resonators. External circuits are coupled to the cavity through transmission line probes or loops or apertures. The cavity fields can be very large at resonance compared to the input value because the stored energy oscillates between the electric and magnetic fields.

Rectangular Cavity 7.3.1

Because of the metallic surfaces at the boundaries, the field distributions possess standing wave patterns in all the three directions.

The following types of field components can exist in a rectangular cavity:

TE_{mnp} Mode Field

$$H_z = H_0 \cos(m\pi x/a)\cos(n\pi y/b)\sin(p\pi z/d); \tag{7.4}$$

$$H_y = \frac{1}{k_c^2}\frac{\partial^2 H_z}{\partial y\,\partial z}$$

$$= -(H_0/k_c^2)(p\pi/d)\left(\frac{n\pi}{b}\right)\cos(m\pi x/a)\sin(n\pi y/b)\cos(p\pi z/d) \tag{7.5}$$

$$H_x = \frac{1}{k_c^2}\frac{\partial^2 H_Z}{\partial x\,\partial z}$$

$$= -(H_0/k_c)^2(p\pi/d)\left(\frac{m\pi}{a}\right)\sin(m\pi x/a)\cos(n\pi y/b)\cos(p\pi z/d) \tag{7.6}$$

$$E_z = 0 \tag{7.7}$$

$$E_y = j\frac{\omega\mu}{k_c^2}\frac{\partial H_Z}{\partial x}$$

$$= -j\frac{\omega\mu H_0}{k_c^2}\left(\frac{m\pi}{a}\right)\sin\left(\frac{m\pi x}{a}\right)\cos\left(\frac{n\pi y}{b}\right)\sin\left(\frac{p\pi z}{d}\right) \tag{7.8}$$

$$E_x = -j\frac{\omega\mu}{k_c^2}\frac{\partial H_Z}{\partial y}$$

$$= j\frac{\omega\mu H_0}{k_c^2}\left(\frac{n\pi}{b}\right)\cos\left(\frac{m\pi x}{a}\right)\sin\left(\frac{n\pi y}{b}\right)\sin\left(\frac{p\pi z}{d}\right) \tag{7.9}$$

where, $m = 0, 1, 2, 3, \ldots$; $n = 0, 1, 2, 3, \ldots$; and $p = 1, 2, 3, 4, \ldots$; m and n simultaneously $\neq 0$.

Here, k_c is the cut-off wave number given by

$$k_c^2 = (m\pi/a)^2 + (n\pi/b)^2 \tag{7.10}$$

TM_{mnp} Mode Field

$$E_z = E_0 \sin(m\pi x/a)\sin(n\pi y/b)\cos(p\pi z/d) \tag{7.11}$$

$$E_y = \frac{1}{k_c^2}\frac{\partial^2 E_Z}{\partial y\partial z}$$

$$= \frac{-E_0}{k_c^2}\left(\frac{n\pi}{b}\right)\left(\frac{p\pi}{d}\right)\sin\left(\frac{m\pi x}{a}\right)\cos\left(\frac{n\pi y}{b}\right)\sin\left(\frac{p\pi z}{d}\right) \tag{7.12}$$

$$E_x = \frac{1}{k_c^2}\frac{\partial^2 E_Z}{\partial x \partial z}$$

$$= \frac{-E_0}{k_c^2}\left(\frac{m\pi}{a}\right)\left(\frac{p\pi}{d}\right)\cos\left(\frac{m\pi x}{a}\right)\sin\left(\frac{n\pi y}{b}\right)\sin\left(\frac{p\pi z}{d}\right) \quad (7.13)$$

$$H_z = 0$$

$$H_y = \frac{-j\omega\varepsilon}{k_c^2}\frac{\partial E_Z}{\partial x}$$

$$= \frac{-j\omega\varepsilon E_0}{k_c^2}\left(\frac{m\pi}{a}\right)\sin\left(\frac{m\pi x}{a}\right)\cos\left(\frac{n\pi y}{b}\right)\cos\left(\frac{p\pi z}{d}\right) \quad (7.14)$$

$$H_x = \frac{j\omega\varepsilon}{k_c^2}\frac{\partial E_Z}{\partial y}$$

$$= \frac{j\omega\varepsilon E_0}{k_c^2}\left(\frac{n\pi}{b}\right)\sin\left(\frac{m\pi x}{a}\right)\cos\left(\frac{n\pi y}{b}\right)\cos\left(\frac{p\pi z}{d}\right) \quad (7.15)$$

where, $m = 1, 2, 3, ...3$; $n = 1, 2, 3, ...$; and $p = 0, 1, 2, ...$; $m \neq 0, n \neq 0$

Here, k_c is given by the same Eq. 7.10.

For either TE_{mnp} or TM_{mnp} mode, the resonant frequency is given by

$$f_r = \frac{1}{2\sqrt{\mu\varepsilon}}\sqrt{(m/a)^2 + (n/b)^2 + (p/d)^2} \text{ Hz} \quad (7.16)$$

The dominant mode of such a resonator depends on the dimensions of the cavity. For $b < a < d$, the dominant mode is the TE_{101} mode. The field configurations of this dominant mode are shown in Fig. 7.4.

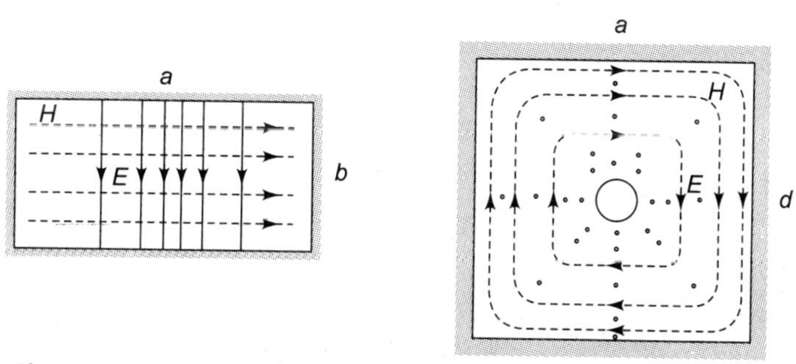

Fig. 7.4 Rectangular cavity dominant mode field configuration (TE_{101})

Example 7.1 Find the resonant frequencies of the first five lowest modes of an air-filled rectangular cavity of dimensions $x, y, z = 5$ cm, 4 cm, 2.5 cm. List them in ascending order.

Solution Given $a = 5$ cm, $b = 4$ cm, $d = 2.5$ cm, resonant frequencies of TE_{mnp} and TM_{mnp} modes can be calculated from the equation

$$f_r = \frac{1}{2\sqrt{(\mu\varepsilon)}}\sqrt{(m/a)^2 + (n/b)^2 + (p/d)^2} \text{ Hz}$$

$$= \frac{c}{2}\sqrt{(m/a)^2 + (n/b)^2 + (p/d)^2}\ \text{Hz}$$

Here, $c = 3 \times 10^{10}$ cm/s. Results are given below.

Resonant frequencies of rectangular cavity

TE_{mnp} modes

m	*n*	*p*	*GHz*
0	0	1	6
0	0	2	12
0	1	1	7.0755
0	1	2	12.5723
0	2	1	9.6047
0	2	2	14.1510
1	0	1	6.7082
1	0	2	12.3693
1	1	1	7.6852
1	1	2	12.9253
1	2	1	10.0623
1	2	2	14.4655
2	0	1	8.4853
2	0	2	13.4164
2	1	1	9.2770
2	1	2	13.9306
2	2	1	11.3248
2	2	2	15.3704

TM_{mnp} modes

m	*n*	*p*	*GHz*
1	1	0	4.8023
1	1	1	7.6852
1	1	2	12.9253
1	2	0	8.0777
1	2	1	10.0623
1	2	2	14.4655
2	1	0	7.0755
2	1	1	9.2770
2	1	2	13.9306
2	2	0	9.6047
2	2	1	11.3248
2	2	2	15.3704

Ascending order of modes are

Modes	TM_{110}	TE_{001}	TE_{101}	TE_{011} TM_{210}
Resonant GHz	4.8023	6.0000	6.7082	7.0755

Modes	TE_{111} TM_{111}	TM_{120}	TE_{201}
Resonant GHz	7.6852	8.0777	8.4853

It may be noted that here TM_{110} is dominant. For TE_{101} mode to be dominant, condition $d > a > b$ must be satisfied.

Circular Cavity 7.3.2

Electromagnetic field analysis shows that due to ϕ-symmetric structure of circular cylindrical cavity, field solutions possess harmonic solution in ϕ and standing waves in the radial and z-directions. The field components inside the cavity are described in terms of TE_{nmp} and TM_{nmp} modes as follows:

TE_{nmp} Mode Field

$$H_z = H_0 J_n (x'_{nm} \rho/a) \cos n\phi \sin (p\pi z/d) \tag{7.17}$$

$$H_\phi = -H_0 \left(\frac{p\pi}{d}\right)\left(\frac{n}{\rho}\right)\left(\frac{a}{x'_{nm}}\right)^2 J_n (x'_{nm} \rho/a) \sin n\phi \cos (p\pi z/d) \tag{7.18}$$

$$H_\rho = H_0 \frac{p\pi}{d}\left(\frac{a}{x'_{nm}}\right) J'_n (x'_{nm} \rho/a) \cos n\phi \cos (p\pi z/d) \tag{7.19}$$

$$E_z = 0 \tag{7.20}$$

$$E_\phi = jH_0\, \omega\mu\, (a/x'_{mn})\, J'_n (x'_{nm} \rho/a) \cos n\phi \sin (p\pi z/d) \tag{7.21}$$

$$E_\rho = jH_0\, \omega\mu\, (n/\rho)\, (a/x'_{nm})^2 J_n (x'_{nm}\rho/a) \sin n\phi \sin (p\pi z/d) \tag{7.22}$$

where, $n = 0, 1, 2,...$ is the number of periodicity in ϕ, $m = 1, 2, 3,...$ is the number of zeros of the field in ρ directions and $p = 1, 2, 3,...$ is the number of halfwaves in the z direction. Here, $k_\rho = x'_{nm}/a$ and x'_{nm} is the mth root of equation $J'_n(k_\rho a) = 0$ such that $x'_{11} = 1.841$, $x'_{21} = 3.054$, $x'_{01} = 3.832$, $x'_{12} = 5.331$, etc.

TM_{nmp} Mode Field

$$E_z = E_0 J_n (x_{nm} \rho/a) \cos n\phi \cos (p\pi z/d) \tag{7.23}$$

$$E_\phi = E_0 \frac{p\pi}{d} (n/\rho)(a/x_{nm})^2 J_n (x_{nm} \rho/a) \sin n\phi \sin n\phi \sin (p\pi z/d) \tag{7.24}$$

$$E_\rho = -E_0 \frac{p\pi}{d} (a/x_{nm}) J'_n (x_{nm} \rho/a) \cos n\phi \sin (p\pi z/d) \tag{7.25}$$

$$H_z = 0$$

$$H_\phi = -j\, E_0\, \omega\varepsilon\, (a/x_{nm})\, J'_n (x_{nm}\rho/a) \cos n\phi \cos (p\pi z/d) \tag{7.26}$$

$$H_\rho = -j\, \omega\varepsilon\, E_0\, (a/x_{nm})^2 J_n (x_{nm}\rho/a) \sin n\phi \cos (p\pi z/d) \tag{7.27}$$

where, $n = 0, 1, 2, ...$, $m = 1, 2, 3, ...$, and $p = 0, 1, 2, ...$ and $k_\rho = x_{nm}/a$. x_{nm} is the mth root of $J_n (k_\rho a) = 0$, so that $x_{01} = 2.405$, $x_{11} = 3.832$, $x_{21} = 5.135$, $x_{02} = 5.520$, $x_{12} = 7.016$, etc. The resonant frequencies for these modes are given by

$$TE_{nmp}: f_r = \frac{1}{2\pi\sqrt{(\mu\varepsilon)}} \sqrt{[(x'_{nm}/a)^2 + (p\pi/d)^2]}\ ;\ \text{Hz} \tag{7.28}$$

$$TM_{nmp}: f_r = \frac{1}{2\pi\sqrt{(\mu\varepsilon)}} \sqrt{[(x_{nm}/a)^2 + (p\pi/d)^2]}\ ;\ \text{Hz} \tag{7.29}$$

where x'_{nm} and x_{nm} are the mth roots of the equation $J_n'(x) = 0$ and $J_n(x) = 0$, respectively. The dominant mode in circular cavity will depend on the dimensions of the cavity. The smallest root out of x_{01} and x_{11}' generates the dominant mode.

We can write,
$$F = \frac{f_r(TM_{011})}{f_r(TE_{111})} = \frac{x_{01}}{\sqrt{[x'^2_{11} + (\pi a/d)^2]}} \tag{7.30}$$

For $d < 2a$, $F < 1$, the TM_{010} mode is dominant. For $d \geq 2a$, $F \geq 1$, the TE_{111} mode is dominant. Another mode TE_{011} is important for its high Q. The field configuration for these modes are shown in Fig. 7.5.

Here $$k_\rho^2 + k_z^2 = \left(\frac{x_{nm}}{a}\right)^2 + \left(\frac{p\pi}{d}\right)^2 = \omega_r\sqrt{\mu\varepsilon} = \omega_r/c \tag{7.31}$$

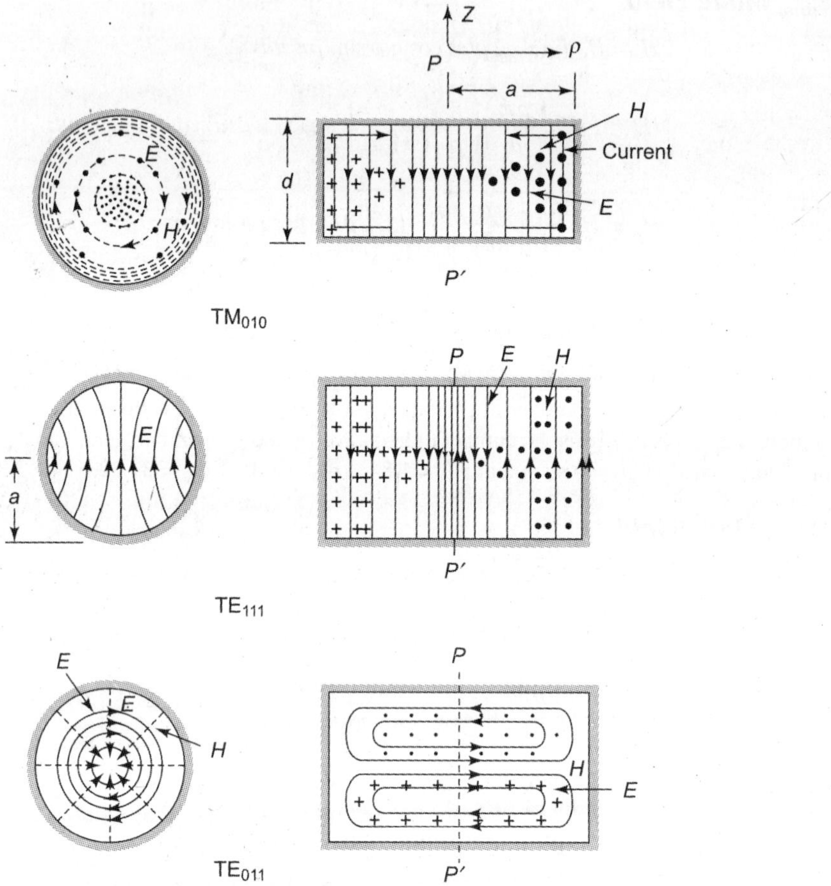

Fig. 7.5 Circular cavity field configurations (left-hand side is the cross-section through PP′)

Equations 7.28 and 7.29 show that TM and TE modes of the same order n, m, p have identical frequencies but have different field patterns. Such modes are known as degenerate modes. Equations 7.28 and 7.29 can also be written as

$$(2af_r)^2 = \left[\left(\frac{p\pi}{\sqrt{\mu\varepsilon}}\right)^2\left(\frac{a}{d}\right)^2\right]^2 + \left(\frac{\bar{x}_{nm}}{\sqrt{\mu\varepsilon}}\right)^2 \tag{7.32}$$

where $\bar{x}_{nm} = x_{nm}$ for TM_{nmp} mode and $\bar{x}_{nm} = x'_{nm}$ for TE_{nmp} mode. The above equation is used to draw a straight line mode chart as shown in Fig. 7.6 which is very useful in cavity design. It shows the range of frequencies and the ratio of diameter/length, in which only a single mode can be resonated if there is no degeneracy.

Example 7.2 Find the resonant frequencies of the five lowest modes of an air-filled cylindrical cavity of radius 1.905 cm and length 2.54 cm. List them in ascending order.

Solution Given $a = 1.905$ cm, $d = 2.54$ cm the resonant frequencies are calculated as follows

$$\text{TE}_{nmp}\text{: } f_r = \frac{1}{2\pi\sqrt{(\mu\varepsilon)}}\sqrt{[(x'_{nm}/a)^2 + (p\pi/d)^2]} \text{ ; Hz}$$

$$\text{TM}_{nmp}\text{: } f_r = \frac{1}{2\pi\sqrt{(\mu\varepsilon)}}\sqrt{[x_{nm}/a)^2 + (p\pi/d)^2]} \text{ ; Hz}$$

Modes	TM_{010}	TE_{111}	TM_{011}	TE_{211}	TM_{111} TE_{011}
f_r (GHz)	6.03	7.49	8.44	9.67	11.27

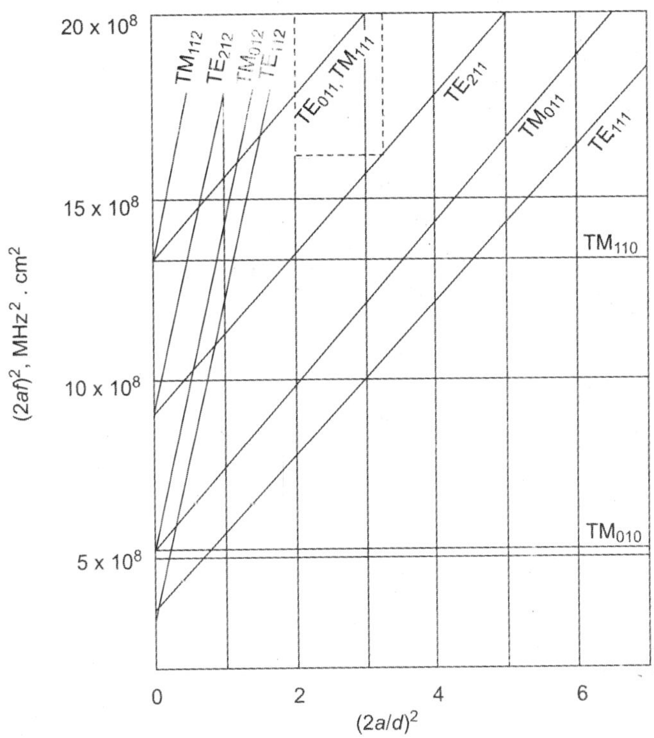

Fig. 7.6 Circular cavity mode chart [Source: Ref.1]

7.4 CAVITY EXCITATION AND TUNING

The power between the waveguide cavity resonator or a coaxial resonator, the signal source and the load can be coupled by means of a coaxial line whose centre conductor is extended inside the cavity in the form of a probe or a loop. The resonator cavity may also be coupled to a waveguide by means of a small aperture in the common wall. Various coupling configurations are shown in Fig. 7.7.

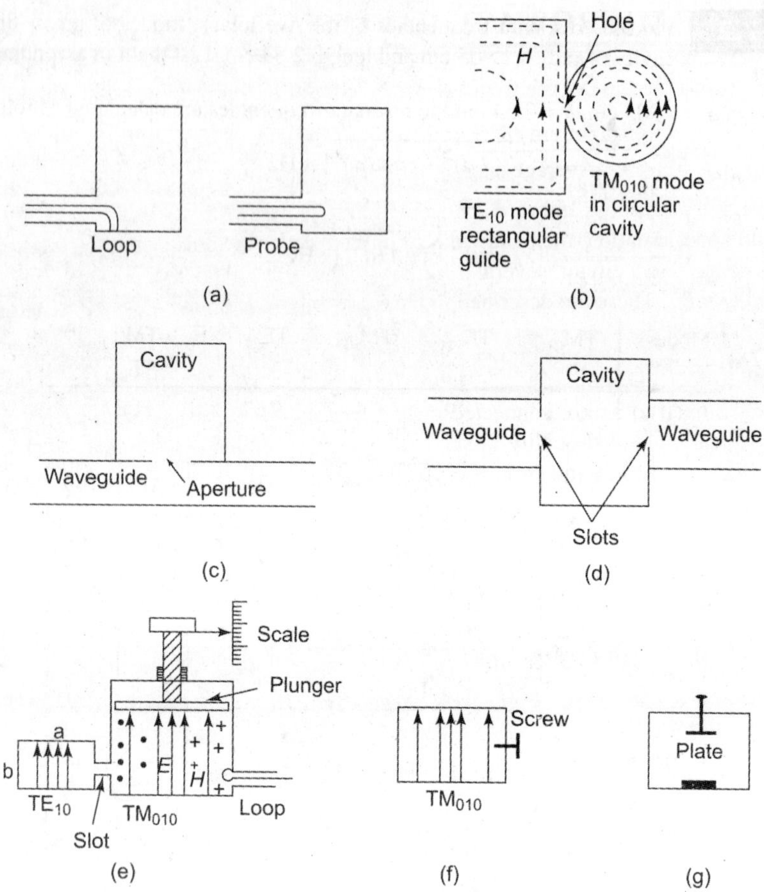

Fig. 7.7 Cavity excitations and tuning

A coaxial line probe parallel to **E**, placed at the position of maximum **E** field should be used to couple power between the cavity and the external circuit. The loop should be used for power coupling at the position of maximum magnetic field with the plane of the loop perpendicular to **H** field.

Microwave cavity resonators are used in a transmission system as reflection type or transmission type cavities. In reflection type, the cavity is used as load producing maximum impedance, while transmission type cavity is placed in series with the line producing maximum absorption of power.

There are a number of methods by which the cavity may be mechanically tuned to resonance:

1. A short-circuit plunger can be moved parallel to the axis to change the length of the cavity and thereby f_r as shown in Fig. 7.7(e).
2. A metallic screw may be inserted into the curved wall of the cavity (klystron tubes) to lower the cavity's distributed inductance and thereby increasing f_r as shown in Fig. 7.7(f).
3. A small plate may be placed at the centre of the cavity and its position is varied to control the variation in capacitance at the centre and thereby the f_r as shown in Fig. 7.7(g).

Q-FACTOR OF MICROWAVE CAVITIES 7.5

The quality factor Q is a measure of the frequency selectivity of a cavity resonator and is defined by

$$Q = 2\pi \frac{\text{Maximum energy stored } (W) \text{ during a cycle}}{\text{Average energy dissipated per cycle}} = \frac{\omega_r W}{P} \tag{7.33}$$

where P is the average power loss and $\omega_r = 2\pi$/oscillation time period, T.

To a very good approximation, the bandwidth Δf around the resonant frequency f_r is given by $Q = f_r/\Delta f$. When a cavity is coupled to a circuit, several different Q factors are defined: unloaded $Q = Q_0$, external $Q = Q_e$ and the loaded $Q = Q_L$. These are described in the following paragraphs.

Unloaded Q_0 7.5.1

When a cavity is assumed to be not connected to any external circuit or load, Q accounts for the internal losses and is called the unloaded quality factor Q_0. The Q_0 of a cavity is reduced when for all practical purposes the cavity is coupled to an external circuit or load. The equivalent circuits of an unloaded cavity, cavity coupled to a generator and cavity coupled to a load are shown in Fig. 7.8.

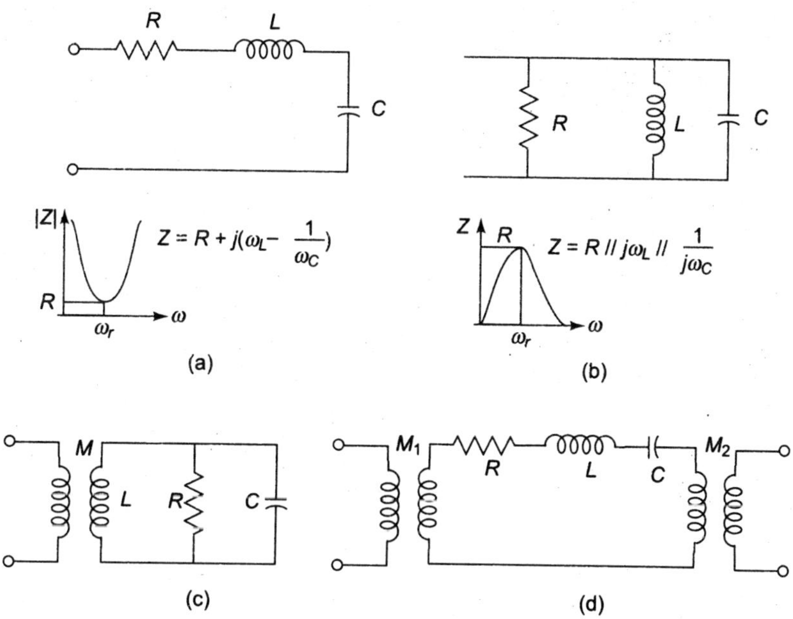

Fig. 7.8 Cavity equivalent circuits: (a) Series equivalent of an unloaded cavity (b) Parallel connection of an unloaded cavity (c) Cavity coupled to a generator (d) Cavity coupled to a load and a generator

The total energy stored in a cavity resonator is a constant and equal to the sum of the electric and magnetic energy stored. At resonant frequency, the electric and magnetic energies are equal and when the electric energy is maximum, the magnetic energy is zero and vice versa. The total maximum energy stored, therefore, can be written as

$$W = 2W_e = \frac{\varepsilon}{2}\int_V |\mathbf{E}|^2\, dV = 2W_m = \frac{\mu}{2}\int_V |\mathbf{H}|^2\, dV \tag{7.34}$$

where V is the resonant cavity volume. For a lossless dielectric inside the cavity, the energy dissipation is only due to joule loss in the cavity walls S and can be expressed by

$$P = \frac{R_s}{2}\int_S |\hat{\mathbf{n}} \times \mathbf{H}|^2 \cdot d\mathbf{S} \tag{7.35}$$

where $\hat{\mathbf{n}}$ is the unit outward normal vector to the inside surface S having current density $\mathbf{J} = \hat{\mathbf{n}} \times \mathbf{H}$. $\mathbf{J}$ results from the tangential component of $\mathbf{H}$.

Therefore, the unloaded Q factor can be computed from the field components using the following equation:

$$Q_0 = \frac{\omega\varepsilon \int_V |\mathbf{E}|^2\, dV}{R_s \int_S |\mathbf{H}_t|^2\, dS} \tag{7.36}$$

where suffix t stands for the tangential component. For highly conducting cavity walls (copper, $\sigma = 5.8 \times 10^7$ mho/m) at normal temperature, Q_0 has values of the order of 10^4–10^5. Even higher Q_0 up to $10^8 - 10^{10}$ can be obtained by using superconductive materials such as lead cooled to 2–5 K.

In practice, power loss by conductors P_c, dielectric fills P_d and radiation P_r from the openings can contribute to unloaded Q_0,

$$Q_0 = \frac{\omega_r W}{P_c + P_d + P_r} \tag{7.37}$$

or,

$$\frac{1}{Q_0} = \frac{1}{Q_c} + \frac{1}{Q_d} + \frac{1}{Q_r} \tag{7.38}$$

Here, conductor quality factor is defined by

$$Q_c = \omega_r W/P_c \tag{7.39}$$

The dielectric quality factor for a lossy dielectric fill

$$Q_d = \omega_r W/P_d = \frac{\omega_r\varepsilon \int_V |E|^2\, dV}{\sigma_d \int_V |E|^2\, dV} = \frac{\omega_r\varepsilon}{\sigma_d}; \quad \alpha_d = \frac{\omega\mu_d\sigma_d}{2\beta} = \frac{\sigma_d\eta_d}{2\beta}$$

$$\beta = k\sqrt{1-(f_c/f)^2}$$

$$= 1/\tan\delta \tag{7.40}$$

where the loss tangent for a dielectric of conductivity σ_d is given by

$$\tan\delta = \sigma_d/\omega_r\varepsilon \tag{7.41}$$

The radiation Q factor

$$Q_r = \omega_r W/P_r \tag{7.42}$$

For a completely closed cavity with highly conducting walls, radiation loss approaches zero and Q_r approaches infinity. For a cavity with apertures or transmission line or microstrip configuration, non-zero finite Q_r can be calculated. For the design of resonators unloaded Q is required to be determined as described in the following paragraphs.

Q-Factor of Transmission Line Resonator 7.5.2

The Q-factor of a transmission line resonator can be computed by taking into account the distributed nature of currents and voltages. The instantaneous voltage and current on the low loss short-circuited line are given by

$$v(x, t) = V_0 \sin \beta x \sin \omega t \tag{7.43}$$

$$i(x, t) = (V_0/Z_0) \cos \beta x \cos \omega t \tag{7.44}$$

where $x = 0$ at the right-hand side end (load). The stored electric and magnetic energies along the line of length l are given by

$$W_e(t) = (1/2)\, CV_0^2 \sin^2 \omega t \int_0^l \sin^2 \beta x dx$$

$$= (CV_0^2/4) \sin^2 \omega t \, [(\sin 2\beta l/2\beta) - l] \tag{7.45}$$

$$W_m(t) = \frac{LV_0^2}{2Z_0^2} \cos^2 \omega t \left[\frac{\sin 2\beta l}{2\beta} + 1\right] \tag{7.46}$$

Total stored energy

$$W = W_e(t) + W_m(t) = \frac{1}{4} C\, lV_0^2 \left[1 + \frac{\sin 2\beta l}{2\beta l} \cos 2\omega t\right] \tag{7.47}$$

The maximum stored energy ($\omega t = 0$)

$$W_{\max} = \frac{1}{4} C\, lV_0^2 \left[1 + \frac{\sin 2\beta l}{2\beta l}\right] \tag{7.48}$$

For both $\lambda/4$ and $\lambda/2$ lines, $\sin 2\beta l/2\beta l = 0$.
Therefore,

$$W_{\max} = 1/4\, (C\, lV_0^2) \tag{7.49}$$

Due to the conductor heating, average power loss is

$$P_c = (R_s/2) \int i^2\, dx = \frac{1}{4} \frac{RlV_0^2}{Z_0^2} \left[1 + \frac{\sin 2\beta l}{2\beta l}\right] \tag{7.50}$$

The insulation dielectric heating loss is

$$P_d = G/2 \int V^2 dx = \frac{GlV_0^2}{4} \left[1 - \frac{\sin 2\beta l}{2\beta l}\right] \tag{7.51}$$

For resonant lengths $l = \lambda/4$ or $\lambda/2$,

$$P_c = RlV_0^2/4Z_0^2 \tag{7.52}$$

$$P_d = \frac{GlV_0^2}{4} \tag{7.53}$$

The Q-factor of the short-circuited line cavity of resonant length is

$$Q = \frac{2\pi}{T} \frac{W}{P_c + P_d} = \frac{\omega_r LC}{RC + GL} \tag{7.54}$$

Thus, for small R and G, Q is very high and approaches infinity for lossless line ($R = 0 = G$). In a low loss case,

$$Z_0 = \sqrt{L/C} \tag{7.55}$$

$$\alpha = \frac{1}{2}\left(\frac{R}{Z_0} + GZ_0\right) \tag{7.56}$$

$$\beta_r = \omega_r \sqrt{LC} \tag{7.57}$$

Then for quarter and halfwave transmission line resonators.

$$Q = \beta_r/2\alpha \tag{7.58}$$

Because of the open structure, the skin effect loss and radiation loss at high frequencies lower the predicted Q.

Unloaded Q-Factor of Coaxial Cavity 7.5.3

Let us assume that a coaxial resonator is formed by short-circuiting two ends of an air-filled coaxial line of length d. The radii of the conductors, a and b are such that only TEM mode exists at resonance. The standing wave fields inside the resonator are

$$E_\rho = \frac{V_0}{\rho \ln b/a}(e^{-jk_z z} - e^{+jk_z z}) = \frac{2jV_0}{\rho \ln b/a} \sin k_z z \tag{7.59}$$

$$H_\phi = \frac{2V_0}{\eta\rho \ln b/a}(e^{-jk_z z} + e^{+jk_z z}) = \frac{2V_0}{\eta\rho \ln b/a} \cos k_z z \tag{7.60}$$

where $k_z = p\pi/d = \omega\sqrt{\mu\varepsilon}$, $p = 1, 2, 3, \ldots$ and V_0 = potential difference between the conductors. The resonant frequencies are given by

$$f_r = \frac{p}{2\sqrt{(\mu\varepsilon)}d} \tag{7.61}$$

It can be seen that both the electric and magnetic fields undergo p half-cycles of variation along the length of the resonator.

The time-average total energy stored is

$$W = 2W_e = \frac{\varepsilon}{2}\int_a^b\int_0^d\int_0^{2\pi} |E_\rho|^2 \rho d\, \rho d\phi\, dz$$

$$= \frac{\varepsilon}{2}\frac{4V_0^2}{\left(\ln\frac{b}{a}\right)^2}\int_0^{2\pi} d\phi \int_0^d \sin^2(p\pi z/d)\, dz \int_a^b \frac{d\rho}{\rho}$$

$$= \frac{\varepsilon}{2}\left[\frac{2V_0}{\ln(b/a)}\right]^2 \cdot 2\pi\frac{d}{2}\cdot\ln\frac{b}{a} = \frac{2\pi d\varepsilon V_0^2}{\ln b/a} \tag{7.62}$$

The time-average power loss on the cavity walls is given by

$$P = R_s/2\int_S |H_\phi|^2 ds = (R_s/2)\int_0^{2\pi}\left[\int_0^d \left(\frac{2V_0}{\eta a \ln(b/a)}\right)^2 \cos^2\frac{p\pi z}{d}\, dz\cdot a\right.$$

$$+ \int_0^d \left(\frac{2V_0}{\eta b \ln(b/a)}\right)^2 \cos^2\frac{p\pi z}{d}\, dz\cdot b$$

$$\left. + 2\int_a^b \left(\frac{2V_0}{\eta\rho \ln(b/a)}\right)^2 \cos^2\left[\left.\frac{p\pi z}{d}\right|_{z=0} \rho d\rho\right]\right] d\phi$$

$$= \frac{R_s}{2}\left(\frac{2V_0}{\eta \ln (b/a)}\right)^2 \left[2\pi a \frac{1}{a^2}\int_0^d \cos^2 \frac{p\pi z}{d}\, dz + 2\pi b \frac{1}{b^2}\int_0^d \cos^2 \frac{p\pi z}{d}\, dz + 4\pi \int_a^b \frac{d\rho}{\rho}\right]$$

$$= \frac{\pi R_s}{2}[2V_0/(\eta \ln b/a)]^2 \left[\frac{d}{a} + \frac{d}{b} + 4 \ln b/a\right] \quad (7.63)$$

Therefore, $Q = \omega_r W/P$

$$Q_{TEM} = \frac{p\pi\eta \ln (b/a)}{R_s\left[d\left(\frac{1}{a}+\frac{1}{b}\right) + 4\ln\left(\frac{b}{a}\right)\right]} \quad (7.64)$$

where $\omega_r \varepsilon = \frac{p\pi}{d}\sqrt{(\varepsilon/\mu)} = \frac{p\pi}{\eta d}$

$$\eta = \sqrt{(\mu/\varepsilon)} \quad (7.65)$$

Unloaded Q of Rectangular Cavity 7.5.4

Dominant TE_{101} Mode

The dominant mode in a rectangular cavity of dimensions $(a \times b \times d)$, $b < a < d$, is TE_{101}. The field distribution in such a cavity can be obtained from Eqs 7.4–7.9 and after normalizing by the factor

$$E_0 = \frac{j\omega\mu_0 H_0}{k_c^2}(\pi/a);\ \mu = \mu_0,\ \varepsilon = \varepsilon_0 \quad (7.66)$$

The field components become

$$E_y = E_0 \sin\left(\frac{\pi x}{a}\right)\sin\left(\frac{\pi z}{d}\right) \quad (7.67)$$

$$H_x = -j\,(E_0/\omega\mu_0)\,(\pi/d)\sin\left(\frac{\pi x}{a}\right)\cos\left(\frac{\pi z}{d}\right) \quad (7.68)$$

$$H_z = j\,(E_0/\omega\mu_0)\,(\pi/a)\cos\left(\frac{\pi x}{a}\right)\sin\left(\frac{\pi z}{d}\right) \quad (7.69)$$

where, $k_c = \pi/a$ (7.70)

Since at resonance, total energy is equal to the maximum energy stored in the E field, or that in the magnetic field, the resonant energy stored inside the cavity can be calculated from the maximum electric energy

$$W = (\varepsilon_0/2)\int_0^a\int_0^b\int_0^d |E_y|^2\, dx\, dy\, dz$$

$$= (\varepsilon_0 E_0^2/2)\int_0^a\int_0^b\int_0^d \sin^2(\pi x/a)\sin^2(\pi z/d)\, dx\, dy\, dz$$

$$= (\varepsilon_0/8)\, abd\, E_0^2 = \frac{1}{8}\varepsilon_0 E_0^2 * \text{cavity volume} \quad (7.71)$$

For a lossless dielectric the total power loss P in the cavity can be obtained from the ohmic losses in the six walls.

Therefore,

$$P = R_s/2\left\{2\int_0^a\int_0^b |H_x|^2_{z=0}\,dx\,dy + 2\int_0^b\int_0^d |H_z|^2_{x=0}\,dy\,dz + 2\int_0^a\int_0^d \left[|H_x|^2 + |H_z|^2\right]_{y=0} dx\,dz\right\} \tag{7.72}$$

Substituting the expressions for the field components and evaluating the integrals,

$$P = \frac{R_s\,\lambda^2\,E_0^2}{8\eta^2}\left[\frac{ab}{d^2} + \frac{bd}{a^2} + \frac{1}{2}\left(\frac{a}{d} + \frac{d}{a}\right)\right]; \tag{7.73}$$

where $\eta = \sqrt{\mu_0/\varepsilon_0}$, the intrinsic impedance of free space.

The Q-factor is then obtained from Eqs 7.71 and 7.73:

$$Q_0 = \frac{\omega_r W}{P} = \frac{\pi\eta}{4R_s}\,\frac{2b(a^2+d^2)^{3/2}}{ad(a^2+d^2)+2b(a^3+d^3)} = \frac{d}{\delta_s}\cdot\frac{1}{\frac{d}{b} + 2\left(1+\frac{d^3}{a^3}\right)\bigg/\left(1+\frac{d^2}{a^2}\right)} \tag{7.74}$$

where $\delta_s = \dfrac{1}{\sqrt{\pi f\mu\sigma}}$, the skin depth.

From the symmetry of Q_0 in a and d, it is seen that Q_0 is maximum for a square base cavity $a = d$, given by

$$Q_{0\,\max} = \frac{1.11\sqrt{(\mu/\varepsilon)}}{R_s\,(1+a/2b)} = \frac{d}{2\delta_s}\,\frac{1}{1+\frac{a}{2b}} \tag{7.75}$$

where $Q_{0\,\max}$ increases with decrease in a/b.

For an air-filled cubic cavity, $a = b = d$, so that

$$Q_0 = \frac{0.74\,\eta}{R_s} = \frac{279}{R_s} = \frac{d}{3\delta_s} = \frac{\lambda_g}{6\delta_s} \tag{7.76}$$

For an air dielectric, $\eta = 120\,\pi$ ohm, and for copper, $R_s = \sqrt{(\omega\mu_0/2\sigma)} = 0.0261$,ohm at 10 GHz. Then $Q \approx 10{,}690$ for a cubic cavity of dimensions $a = b = d$. The half power frequency bandwidth Δf is $\Delta f = f_r/Q$ = 10,000/10,690 MHz = 935.5 kHz.

TE_{mnp} and TM_{mnp} Modes

The general expressions for Q for the TE_{mnp} and TM_{mnp} modes can be obtained as

$$\text{TE: } Q_{mnp} = \frac{\eta\,abdk_c^2\,k^3}{4R_s[bd(k_c^4 + k_y^2 k_z^2) + ad(k_c^4 + k_x^2 k_z^2) + abk_c^2 k_z^2]} \tag{7.77}$$

$$\text{TM: } Q_{mnp} = \frac{\eta\,abdk_c^2\,k}{4R_s[b(a+d)k_x^2 + a(b+d)k_y^2]} \tag{7.78}$$

where, $k_x = m\pi/a,\ k_y = n\pi/b,\ k_z = p\pi/d,$

$$k = \sqrt{(k_x^2 + k_y^2 + k_z^2)};\ k_c = \sqrt{(k_x^2 + k_y^2)} \tag{7.79}$$

Closed rectangular cavities are used in resonant wavemeters, gun diode oscillators, etc. The cavities can be tuned with the aid of a short-circuited plunger at one end wall.

Unloaded Q of a Circular Cavity 7.5.5

For a circular cavity of radius a and length d, the unloaded Q-factor can be computed from the expressions of field components of a given mode. The three very important modes of practical interest are TM_{010}, TE_{111} and TE_{011}. For $d/2a < 1$, the TM_{010} is dominant, while for $d/2a \geq 1$ the TE_{111} mode is dominant. The ϕ symmetric TE_{011} mode is of particular interest because its Q is two to three times that of the dominant modes and there are no axial currents ($H_\phi = 0$). Thus the cavity tuning by changing the length d with a short-circuit plunger does not obstruct the electric currents on the wall. However, since TE_{011} mode is higher than the dominant mode, care must be taken to excite the cavity without the generation of other possible modes.

The resonant frequencies for various ratios of $d/2a$ are tabulated in Table 7.1 for the different modes.

Table 7.1 $(f_r)_{nmp}/(f_r)_{dominant}$

$d/2a$	TM_{010}	TE_{111}	TM_{110}	TM_{011}	TE_{211}	TM_{111} TE_{011}
0.5	1.0	1.5	1.59	1.63	1.80	2.05
1.0	1.0	1.0	1.59	1.19	1.42	1.72
1.5	1.13	1.0	1.80	1.24	1.52	1.87

Dominant TM_{010} Mode Q ($d < 2a$)

Here, $n = 0$, $m = 1$, $p = 0$. Therefore, non-zero cavity field components can be obtained from Eqs 7.23–7.27,

$$E_z = E_0 J_0 (x_{01} \rho/a) \tag{7.80}$$

$$H_\phi = -E_0 j\omega\varepsilon_0 \left(\frac{a}{x_{01}}\right) J_0' \left(\frac{x_{01}\rho}{a}\right)$$

$$= -jE_0\omega\varepsilon_0 \left(\frac{a}{x_{01}}\right) J_1 \left(\frac{x_{01}\rho}{a}\right)$$

$$= j\frac{E_0}{\eta} J_1 \left(\frac{x_{01}\rho}{a}\right) \tag{7.81}$$

where $k_\rho = x_{01}/a,\ x_{01} = 2.405,$

$$\omega\varepsilon_0 \left(\frac{a}{x_{01}}\right) = \frac{\omega\varepsilon_0}{k_p} = \sqrt{\varepsilon_0/\mu_0} = 1/\eta$$

and $J'_0 = -J_1$ (7.82)

The resonant frequency from Eq. 7.29 can be obtained as

$$f_r = \frac{x_{01}}{2\pi a \sqrt{(\varepsilon_0 \mu_0)}} \tag{7.83}$$

The quality factor

$$Q = \omega \frac{\text{Average stored energy}}{\text{Power dissipated}}$$

can be determined from the average stored energy and power dissipated in the cavity walls. At resonance the electric energy stored W_e = average magnetic energy stored W_m. Therefore, average stored energy is $W = 2W_e$ or,

$$2W_e = \varepsilon_0/2 \int_0^a \int_0^{2\pi} \int_0^d \mathbf{E}.\mathbf{E}^* \rho d\rho d\phi\, dz \tag{7.84}$$

Assuming no dielectric loss inside the cavity, the total power dissipation is due to ohmic loss in the conducting walls:

$$P = R_s/2 \int_{\text{walls}} |H_{\tan}|^2 ds \tag{7.85}$$

Now, $\mathbf{E}.\mathbf{E}^* = |E_z|^2 = E_0^2 J_0^2 (x_{01}\rho/a)$

Therefore, $W_e = (\varepsilon_0/4) \int_0^a \int_0^{2\pi} \int_0^d E_0^2 J_0^2 (x_{01}\rho/a)\, \rho d\rho d\phi\, dz$

$$= (\varepsilon_0 E_0^2/4)\, 2\pi . d. \int_0^a \rho J_0^2 (x_{01}\rho/a)\, d\rho$$

$$= (\varepsilon_0 E_0^2/4)\, 2\pi . d\, a^2/2\, J_1^2 (x_{01})$$

$$= (\varepsilon_0 E_0^2/4)\, \pi d\, a^2 J_1^2 (x_{01}) \tag{7.86}$$

$$\int_{\text{walls}} |H_{\tan}|^2 ds = \int_{\substack{\text{curved}\\ \text{walls}}} + \int_{\substack{\text{end}\\ \text{walls}}}$$

$$= \int_0^{2\pi} \int_0^d |H_\phi|^2_{\rho=a} a d\phi dz + 2\int_0^a \int_0^{2\pi} |H_\phi|^2_{z=0,d} \rho d\rho d\phi \tag{7.87}$$

$$\text{First integral} = \frac{E_0^2}{\eta^2} \int_0^{2\pi} \int_0^d [J_1^2 (x_{01}\rho/a)]_{\rho=a}\, a d\phi dz$$

$$= (E_0/\eta)^2\, 2\pi a d\, J_1^2(x_{01}) \tag{7.88}$$

$$\text{Second integral} = 2\frac{E_0^2}{\eta^2} \int_0^a \int_0^{2\pi} J_1^2 (x_{01}\rho/a)\, \rho d\rho d\phi$$

$$= 2\left(\frac{E_0}{\eta}\right)^2 \cdot 2\pi \int_0^a \rho J_1^2\left(\frac{x_{01}\rho}{a}\right) d\rho$$

$$= 4\pi\left(\frac{E_0}{\eta}\right)^2 \cdot \left(\frac{a^2}{2}\right) J_1^2 (x_{01}) \tag{7.89}$$

From Eqs 7.84, 7.87 and 7.88

$$P = R_s \frac{E_0^2}{\eta^2} \pi a J_1^2 (x_{01})(a+d) \tag{7.90}$$

Therefore, $$Q_0 = \omega_r 2 W_e/P = \omega_r \frac{\varepsilon_0 d a \eta^2}{2 R_s (a+d)} = \frac{x_{01} \eta}{2 R_s (1 + a/d)} \tag{7.91}$$

Since $x_{01}/a = \omega_r \sqrt{(\mu \varepsilon)}$ for $p = 0$, $x_{01} = 2.405$.

$$Q_{\text{TM010}} = \frac{1.202 \eta}{R_s (1 + a/d)} \tag{7.92}$$

Dominant TE$_{111}$ Mode Q ($d/2a \geq 1$)

Substituting the expressions for the field components from Eqs 7.17–7.22 for $n = 1$, $m = 1$ and $p = 1$ in integrals Eqs 7.84 and 7.85, the final result obtained for the Q is

$$Q_0 = \frac{\lambda_0}{2\pi\delta_s} \frac{[1-(1/x'_{11})^2][(x'_{11})^2 + (\pi a/d)^2]^{3/2}}{[(x'_{11})^2 + 2a/d\,(\pi a/d)^2 + (1-2a/d)(\pi a/x'_{11}d)^2]} \tag{7.93}$$

This shows that Q decreases as $1/\sqrt{f}$ for dominant TE$_{111}$ mode.

TE$_{011}$ Mode Q

Field distribution in circular cavity for TE$_{011}$ modes is ϕ independent ($n = 0$, $m = 1$, $p = 1$) and can be written as

$$H_z = H_0 J_0 (x'_{01}/a) \sin (\pi z/d) \tag{7.94}$$

$$H_\rho = H_0 (\pi a/d) x'_{01} J'_0 (x'_{01} \rho/a) \cos (\pi z/d) \tag{7.95}$$

$$E_\phi = j\omega\mu H_0 (a/x'_{01}) J'_0 (x'_{01} \rho/a) \sin (\pi z/d) \tag{7.96}$$

where, $$k'_\rho = x'_{01}/a,\; x'_{01} = 3.832 \tag{7.97}$$

The unloaded Q_0 can be obtained in a similar way

$$Q_0 = \frac{\eta}{2R_s} \frac{\left[\left(\frac{x'_{01}}{a}\right)^2 + \left(\frac{\pi}{d}\right)^2\right]^{3/2}}{\frac{1}{a}\left(\frac{x'_{01}}{a}\right)^2 + \frac{2}{d}\left(\frac{\pi}{d}\right)^2} \tag{7.98}$$

where $$\eta = \sqrt{(\mu/\varepsilon)},\; R_s - \sqrt{(\omega\mu/2\sigma)} = 1/\sigma\delta \tag{7.99}$$

For a cavity filled with a homogeneous dielectric of conductivity σ_d, the dielectric Q-factor is given by

$$Q_d = \omega\varepsilon/\sigma_d = 1/\text{loss tangent} \tag{7.100}$$

Example 7.3 A circular cylindrical air-filled cavity with a 3 cm radius and 10 cm length is excited in TE$_{111}$ mode. The 3 dB bandwidth is 2.5 MHz. Calculate the resonant frequency and the Q.

Solution Here, $n = m = p = 1$.

Therefore,

$$f_r = \frac{1}{2\sqrt{(\mu_0\,\varepsilon_0)}}\sqrt{[(x'_{11}/a)^2 + (\pi/d)^2]}$$

$$f_r = \frac{1}{2\sqrt{(4\pi\times10^{-7}\times 8.854\times10^{-12})}}[(1.841/0.03)^2 + (\pi/0.1)^2]^{1/2}$$

$$= 10.42 \text{ GHz}$$

$$Q_0 = f_r/\text{bandwidth} = \frac{10.42\times10^9}{2.5\times10^6} = 4168$$

Example 7.4 A circular air-filled copper cavity is excited in the TM_{010} at 9.375 GHz. The cavity has ratio length: radius = 1.5. Find the Q-factor.

Solution Surface resistance of Cu is $R_s = \sqrt{(\omega\mu/2\sigma)}$, $f = 9.375\times10^9$ Hz. $\eta = 377$ ohms. $d/a = 1.5$, $1 + a/d = 1 + 1/1.5 = 5/3$; $\sigma = 5.8\times10^7$ mho/m.

Therefore,

$$Q = 1.202\times377/(R_s\times5/3)$$

$$R_s = \sqrt{[(6.28\times9.375\times10^9\times4\pi\times10^{-7})/(2\times5.8\times10^7)]} = 2.5\times10^{-2}$$

$$\therefore \quad Q \approx 10875$$

Q-factors for TE_{nmp} and TM_{nmp} modes

The expression for the unloaded Q-factor for general TE_{nmp} and TM_{nmp} modes are

$$TE_{nmp}{:}Q_0 = \frac{\lambda_0}{2\pi\delta}\frac{[1-(n/x'_{nm})^2][(x'_{nm})^2 + (p\pi a/d)^2]^{3/2}}{[(x'_{nm})^2 + (2a/d)(p\pi a/d)^2 + (1-2a/d)\{np\pi a/(x'_{nm}d)\}^2]} \tag{7.101}$$

$$TM_{nmp}{:}Q_0 = \frac{\lambda_0}{2\pi\delta}\frac{[(x_{nm})^2 + (p\pi a/d)^2]^{1/2}}{(1+2a/d)}; \text{ for } p > 0 \tag{7.102}$$

$$= \frac{\lambda_0}{2\pi\delta}\frac{x_{nm}}{(1+a/d)}; \text{ for } p = 0 \tag{7.103}$$

The parameter $Q\,\delta/\lambda_0$ is independent of frequency but varies with the ratio diameter/length (= $2a/d$.) Therefore.

$$Q \propto \lambda_0/\delta \propto \lambda_0\sqrt{f}$$

$$\propto 1/\sqrt{f} \tag{7.104}$$

Figure 7.9 shows the variation of normalized value $Q_0\delta/\lambda_0$ with $2a/d$ for several modes. It is seen that Q_0 is higher for the TE_{011} mode as compared to the dominant mode TM_{010} for $2a/d > 1$ or TE_{10} for $2a/d \le 1$.

Circular cavity provides a very high Q over a wide frequency range. For this reason a circular cavity is preferably used for frequency meters when excited in the TE_{011} mode. The advantages of TE_{011} mode are: (1) Q_0 is 2 to 3 times that of TE_{111} mode, and (2) since $H_\phi = 0$, there is no axial current on the walls and consequently a short-circuit plunger at one end can move freely for tuning purpose without intersecting the current path. Thus, no significant leakage of signal takes place through the gap between the plunger plate and the circular walls.

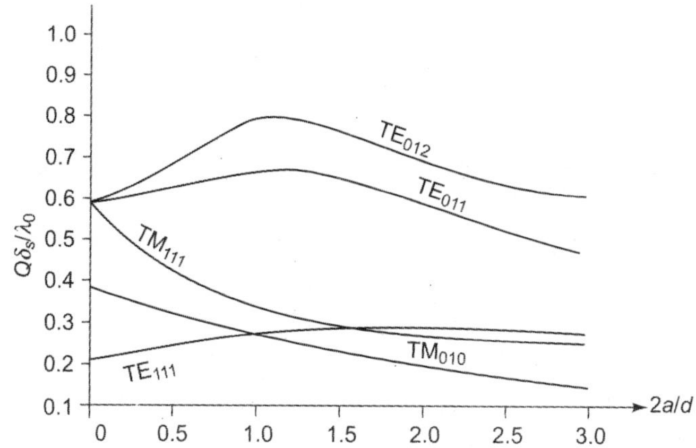

Fig. 7.9 Q_0 for circular cavity [Source: Ref. 1]

Example 7.5 Show that Q of a circular cavity operating in TM_{010} mode is greater than that of the square-based rectangular cavity with TE_{101} mode by 8.26%. The dimension of each cavity is such that the circular cavity is circumscribed by the square-based rectangular cavity.

Solution For TM_{010} mode in circular air-filled cavity of radius a and length d,

$$Q_1 = \frac{1.2025\eta}{R_s(1+a/d)}$$

For TE_{101} mode in a square base ($a' \times a'$) rectangular cavity ($a' \times b' \times a'$), $Q_2 = \eta\ 1.1107/[R_s(1 + a'/2b')]$. For a square-based rectangular cavity to circumscribe the circular cavity, the height and one of the sides of its base must be equal to the diameter $2a$, i.e., height $b' = d$ and base $a' = 2a$. Therefore,

$$\frac{Q_1 - Q_2}{Q_2} = \left[\frac{1.2025\eta}{R_s(1+a/d)} - \frac{1.1107\eta}{R_s(1+a'/2b')}\right] \Big/ Q_2$$

$$= \left[\frac{1.2025\eta}{R_s(1+a/d)} - \frac{1.1107\eta}{R_s(1+2a/2d)}\right] \Big/ Q_2$$

$$= \frac{1.2025 - 1.1107}{1.1107} = 0.0826$$

Therefore, Q of the circular cavity is greater than that of the square-based cavity by 8.26%. This is expected since $Q\ \alpha$ volume/surface area.

LOADED AND EXTERNAL Q 7.6

Since a cavity is always coupled to a generator and load through either a coaxial line probe or a slot in a common wall of a waveguide, energy dissipation should be calculated taking the dissipation in external load along with that in the cavity. The power loss P_e due to the presence of an external load in a cavity system results in an external quality factor $Q_e = \omega_r W/P_e$. We define loaded Q of the cavity as the total Q for the system at resonance expressed by

$$Q_L = 2\pi \frac{\text{Maximum energy stored } (W)}{\left(\begin{array}{l}\text{Energy dissipation in cavity per cycle, } P_0 + \text{energy} \\ \text{dissipation due to the presence of external load, } P_e \\ \text{per radian}\end{array}\right)} \tag{7.105}$$

or, $$Q_L = \frac{\omega_r W}{P_0 + P_e} \tag{7.106}$$

or, $$1/Q_L = 1/Q_0 + 1/Q_e \tag{7.107}$$

where, $$Q_0 = \omega_r W/P_0 = \text{unloaded } Q \text{ which is in dependent of coupling} \tag{7.108}$$

$$Q_e = \omega_r W/P_e = \text{external } Q \tag{7.109}$$

Q_L and Q_0 are related by the coupling coefficient k_e as

$$k_e = P_e/P_0 \tag{7.110}$$

$$Q_0 = (1 + k_e)Q_L = k_e Q_e \tag{7.111}$$

or, $$k_e = Q_0/Q_L - 1 \tag{7.112}$$

This shows that $Q_L < Q_0$, always. The following special cases are important for cavity resonators:

1. When $Q_e = Q_0$, the energy dissipation in the resonant cavity and that in the load are equal. Under this condition, the coupling is called critical and $Q_L = Q_0/2$.
2. When $Q_e < Q_0$, energy dissipation in the cavity is less than that in load, and the coupling is called overcoupling.
3. When $Q_e > Q_0$, energy dissipated in the cavity is greater than that in the load and the coupling is called undercoupling.
4. When Q_e = infinite, coupling = 0.

Q increases at a given frequency for higher-order modes, because of the fact that to support higher modes at a given frequency, the dimensions of the cavity increases and the ratio, volume/surface increases.

COUPLED CAVITIES 7.7

Different types of coupled cavities are used in microwave filter circuits and wavemeters. These are described below.

One-Port Reflection Cavity 7.7.1

When a cavity resonator is excited by a generator matched to the waveguide feeder line impedance Z_0, by means of a small centred hole in the transverse wall, the electrical equivalent is a parallel *R-L-C* circuit under steady-state conditions as shown in Fig. 7.10.

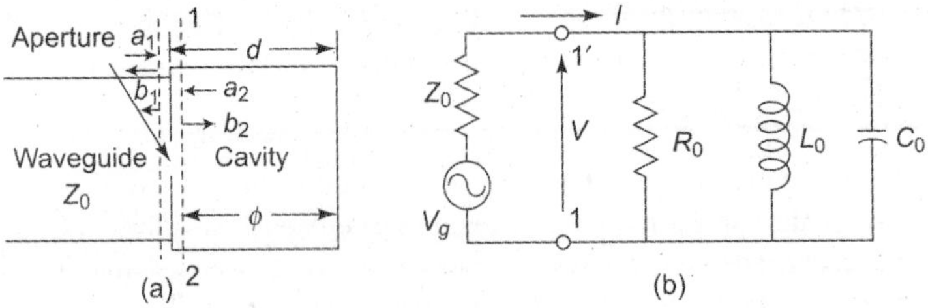

Fig. 7.10 (a) Aperture-coupled rectangular waveguide cavity (b) Its lumped equivalent circuit at resonance (ω_r)

The coupling mechanism may be represented by the s-matrix across the coupling ports 1 and 2 for a set of reference planes. The reference planes can be chosen to make both S_{11} and S_{22} real and negative.

$$[S] = \begin{bmatrix} S_{11} & S_{12} \\ S_{12} & S_{22} \end{bmatrix} \tag{7.113a}$$

Assuming that the coupling is lossless, by the unitary property of $[S]$,

$$S_{11} = S_{22} = -|S_{11}| \tag{7.113b}$$

$$S_{12} = S_{21} = \pm j\sqrt{1-|S_{11}|^2} \tag{7.113c}$$

$$|k| = \sqrt{1-|S_{11}|^2} = \pm jk \tag{7.113d}$$

Here, k is a real number (+ve or –ve) representing the coupling through the aperture. Then from (7.113a) – (7.113c),

$$[S] = \begin{bmatrix} -\sqrt{1-k^2} & jk \\ jk & -\sqrt{1-k^2} \end{bmatrix} \tag{7.113e}$$

When the electrical length $\phi = \beta d$ from the reference plane at the port 2 to the short end is an integral number of half-wavelengths, $|b_1|$ is minimum and $|b_2|$ is maximum, which yields minimum reflection and maximum coupling.

When βd = integral number of half wavelength + a quarter wavelength, $|b_1|$ = maximum, and $|b_2|$ = minimum. This is the antiresonance condition.

If α_T = the round trip attenuation from port 2 to the short circuit and back to port 2,

$$a_2 = -b_2\, e^{-(\alpha_T + j2\phi)}; \tag{7.114a}$$

Thus,

$$\begin{bmatrix} b_1 \\ b_2 \end{bmatrix} = \begin{bmatrix} -\sqrt{1-k^2} & jk \\ jk & -\sqrt{1-k^2} \end{bmatrix} \begin{bmatrix} a_1 \\ a_2 \end{bmatrix} \tag{7.114b}$$

from which

$$b_1 = \left(\sqrt{1-k^2} - \frac{k^2 e^{-(\alpha_T + j2\phi)}}{1-\sqrt{1-k^2}\, e^{-(\alpha_T + j2\phi)}} \right) a_1$$

$$b_2 = \frac{jk}{1-\sqrt{1-k^2}\, e^{-(\alpha_T \pm j2\phi)}}\, a_1 \tag{7.114c}$$

For given attenuation α_T, and coupling k

$|b_2|$ = maximum (cavity field)

$|b_1|$ = minimum (input reflection)

when $2\phi = 2n\pi;\ n = 0, 1, 2, \ldots$

or $\phi = n\pi$ (7.114d)

This is the resonance condition for which

$$b_1 = -\left(\frac{\sqrt{1-k^2} - e^{-\alpha_T}}{1-\sqrt{1-k^2}\, e^{-\alpha_T}} \right) a_1$$

$$b_2 = \frac{jk}{1-\sqrt{1-k^2}\, e^{-\alpha_T}} \cdot a_1 \tag{7.114e}$$

For a given α and k, the maximum value of b_2 is obtained from

$$\frac{d}{dk}|b_2| = 0 \frac{\sqrt{1-k^2} - e^{-\alpha_T}}{(1-\sqrt{1-k^2}e^{-\alpha_T})^2\sqrt{1-k^2}}|a_1|$$

or,
$$\sqrt{1-k^2} = e^{-\alpha_T}$$

or,
$$k = \pm\sqrt{1-e^{-2\alpha_T}} \tag{7.115a}$$

At resonance,

$$b_1 = 0$$

$$b_2 = \pm j\frac{1}{\sqrt{1-e^{-2\alpha_T}}}a_1 \tag{7.115b}$$

This is called critical coupling when all the incident power is absorbed in the cavity. α_T is kept very small for high Q.

It can be shown that there are three cases:

(i) $|k| = \pm\sqrt{1-e^{-2\alpha_T}} \approx \pm 1$ (7.116)

$b_1 \Rightarrow 0$

For critical coupling at resonance.

(ii) $|k| < \sqrt{1-e^{-2\alpha_T}} \approx 1$ (7.117)

$b_1 < 0$ at resonance when VSWR is minimum at the reference plane 1 for under coupling at resonance.

(iii) $|k| > \sqrt{1-e^{-2\alpha_T}} \approx 1$ (7.118)

$b_1 > 0$ at resonance when VSWR is maximum at the reference plane 1 for over coupling at resonance.

Under off-resonance condition or frequencies, VSWR is minimum at the reference plane 1 for all cases of coupling.

Electrical Equivalent Circuit of Reflection Cavity

The external behaviour of the reflection cavity can be determined at frequencies near resonance on the basis of an electrical equivalent parallel *R-L-C* circuit as shown in Fig. 7.9(b) for a particular cavity mode and a particular input waveguide mode.

The reference plane 1–1′ is called the ***plane*** of ***detuned short***, where the impedance is a minimum off-resonance (Γ = real and negative) and almost coincides with the plane of the small exciting iris hole.

Since the normalized impedance at any point is given by

$$\frac{Z(z)}{Z_0} = \frac{1+\Gamma}{1-\Gamma} \tag{7.119}$$

from the (7.113 h), the impedance at the reference plane 1–1′ at resonance is R_0 and is given by

$$\frac{R_0}{Z_0} = \left(\frac{1-\sqrt{1-k^2}}{1+\sqrt{1-k^2}}\right)\left(\frac{1+e^{-\alpha_T}}{1-e^{-\alpha_T}}\right) \tag{7.120}$$

= function of both coupling k and attenuation α_T.

The coupling is considered an ideal transformer of $n:1$ turns ratio and the losses by resistance R as shown in Fig. (7.11).

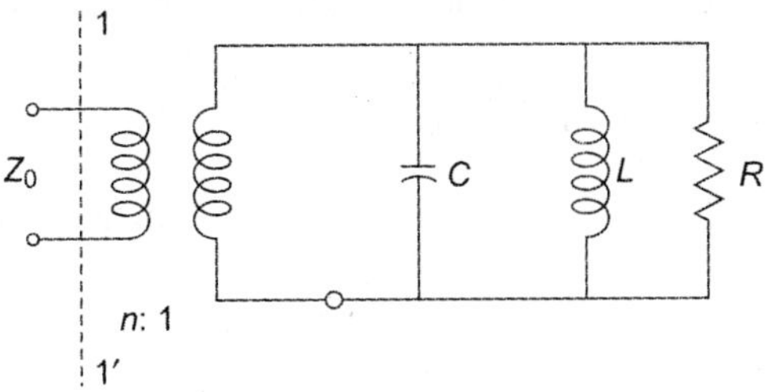

Fig. 7.11 Equivalent circuit with ideal transformer

Here, the following relationship can be written

$$R_0 = n^2 R$$
$$L_0 = n^2 L \tag{7.121}$$
$$C_0 = \frac{C}{n^2}$$

where,
$$n^2 = \frac{1-\sqrt{1-k^2}}{1+\sqrt{1-k^2}} \tag{7.122}$$

$$R = \frac{1+e^{-\alpha_T}}{1-e^{-\alpha_T}}$$

From (7.113 q), the normalized parameters for measurements are

$$\bar{R} = \frac{R_0}{Z_0} = \frac{n^2 R}{Z_0}$$
$$\bar{L} = \frac{L_0}{Z_0} = \frac{n^2 L}{Z_0} \tag{7.123}$$
$$\bar{C} = C_0 Z_0 = \frac{Z_0 C}{n^2}$$

The normalized admittance of the circuit is

$$\bar{Y} = \frac{1}{R} + j\left(\omega\bar{C} - \frac{1}{\omega\bar{L}}\right) = \bar{G} + j\bar{B}; \quad \bar{G} = \frac{1}{R}, \quad \bar{B} = \omega\bar{C} - \frac{1}{\omega\bar{L}} \tag{7.124}$$

At resonance ($\omega = \omega_r$),

$$\omega_r\bar{C} = \frac{1}{\omega_r\bar{L}} = 0$$

or,
$$\omega_r = \frac{1}{\sqrt{\bar{L}\bar{C}}} = \frac{1}{\sqrt{L_0 C_0}} \tag{7.125}$$

The normalized susceptance at any frequency ω can be expressed by

$$\bar{B} = \sqrt{\frac{\bar{C}}{\bar{L}}}\left(\frac{\omega}{\omega_r} - \frac{\omega_r}{\omega}\right) \tag{7.126}$$

Q of Reflection Cavity

As defined in section 7.6, following considerations are made of Q of reflection cavity.

❑ *Unloaded Q = Q_0*

From ac circuit theory, the energy stored in the circuit

$$\omega_0 = \frac{1}{Z}\bar{C}|V_0|^2 \tag{7.127}$$

where V_0 = peak voltage across across the tank circuit at resonance ω_r.

Power loss at resonance ω_r is

$$P_0 = \frac{|V_0|^2}{2\bar{R}} \tag{7.128}$$

By definition, the unloaded Q (independent of coupling) is

$$Q_0 = \frac{\omega_r \omega_0}{P_0} \tag{7.129a}$$

$$Q_0 = \bar{R}\,\bar{C}\,\omega_r = \sqrt{\frac{\bar{C}}{L}}\,\bar{R}$$

$$= R_0\sqrt{\frac{C_0}{L_0}} = R\frac{\sqrt{C}}{L} \tag{7.129b}$$

The unloaded Q can be expressed in terms of bandwidth of the cavity $\Delta\omega = \omega_2 \sim \omega_1$, where ω_1 and ω_2 are values of ω at half-power points w.r.t. the resonant frequency power. The power losses at half-power points can be expressed as

$$P_{1,2} = \frac{1}{2}P_0 = \frac{|V_{1,2}|^2}{2\bar{R}} = \frac{1}{2}\frac{|V_0|^2}{2\bar{R}} \tag{7.129c}$$

since $$|V_{1,2}| = \frac{|V_0|}{V_2}$$

For a constant current source,

$$|V_{1,2}| = i\,|\bar{z}_{1,2}| = i\frac{\bar{R}}{\sqrt{2}}$$

or, $$|\bar{z}_{1,2}| = \frac{\bar{R}}{\sqrt{2}} \tag{7.129d}$$

From (7.113v) and (7.129 d),

$$\bar{z}_{1,2} = \frac{1}{\bar{G} \pm j\bar{B}} \tag{7.129e}$$

Thus, for $\omega_2 > \omega_1$, $$\bar{B}_1 = \sqrt{\frac{\bar{C}}{L}}\left(\frac{\omega_1}{\omega_r} - \frac{\omega_r}{\omega_1}\right) = -\bar{G} \tag{7.129f}$$

$$\bar{B}_2 = \sqrt{\frac{\bar{C}}{L}}\left(\frac{\omega_2}{\omega_r} - \frac{\omega_r}{\omega_2}\right) = \bar{G}$$

For a small offset frequency from the resonance ($\Delta\omega \ll \omega_r$, $\Delta\omega_2 \ll \omega_r$)

$$\omega_1 = \omega_r - \frac{\Delta\omega_1}{2}$$

$$\omega_2 = \omega_r + \frac{\Delta\omega_2}{2} \tag{7.129g}$$

which yields

$$\bar{B}_1 = -\sqrt{\frac{\bar{C}}{\bar{L}}}\frac{\Delta\omega_1}{\omega_r} = -\bar{G}$$

$$\bar{B}_2 = \sqrt{\frac{\bar{C}}{\bar{L}}}\frac{\Delta\omega_2}{\omega_r} = \bar{G} \tag{7.129h}$$

When ω_1 and ω_2 are close to ω_r (for high Q cavity),

$$\Delta\omega_1 = \Delta\omega = \omega_2 - \omega_1$$

where $\Delta\omega$ = halfpower bandwidth of the cavity.

From the first equation of (7.129b),

$$\bar{G} = \frac{\sqrt{\bar{C}/\bar{L}}}{Q_0} \tag{7.129i}$$

From (7.129h) and (7.129i),

$$Q_0 = \frac{\omega_r}{\omega_2 - \omega_1} = \frac{\omega_r}{\Delta\omega} = \frac{f_r}{\Delta f} \tag{7.129j}$$

❑ *External* $Q = Q_e$

In similar derivation of Q_0, it can be shown that

$$Q_e = \omega_r C_0 = \frac{1}{\omega_r \bar{L}} = \sqrt{\frac{\bar{C}}{\bar{l}}} = \frac{Z_0}{n^2}\sqrt{\frac{C}{L}} \tag{7.129k}$$

❑ *Loaded* $Q = Q_L$

Since the circuit external to the cavity has normalized impedance,

$$\bar{z} = \frac{Z_0}{Z_0} = 1$$

the equivalent circuit for Q_L can be shown as in Fig. 7.12 for a constant current source i

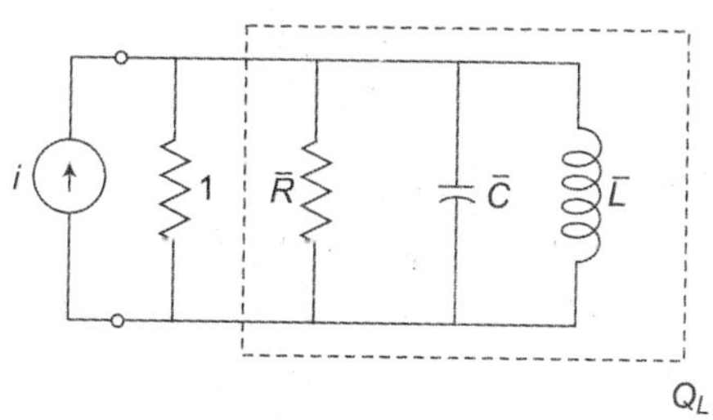

Fig. 7.12 Equivalent circuit for Q_L

Following the derivation of (7.129) for Q_0, with $\bar{R}$ replaced by $\frac{\bar{R}}{\bar{R}+1}$, the parallel combination of $\bar{R}$ and 1 and $\bar{G}$ by $\bar{G}+1$,

$$Q_L = \omega_r \bar{C}\,\frac{\bar{R}}{\bar{R}+1} = \sqrt{\frac{\bar{C}}{\bar{L}}}\cdot\frac{\bar{R}}{\bar{R}+1}$$

$$= \sqrt{\frac{\bar{C}}{\bar{L}}}\cdot\frac{1}{1+\bar{G}}$$

$$= \frac{\omega_r}{\Delta\omega_L} \tag{7.129l}$$

where $\Delta\omega_L = \omega_{2L} - \omega_{1L}$, such that

$$(\bar{B}_{1,2})_L = \sqrt{\frac{\bar{C}}{\bar{L}}}\left(\frac{\omega_{1,2}|_L}{\omega_r} - \frac{\omega_r}{\omega_{1,2}|_L}\right)$$

$$= \pm(1+\bar{G}) \tag{7.129m}$$

Therefore, from para (a), (b) and (c),

$$\left.\begin{aligned} Q_0 &= \bar{R}\sqrt{\bar{C}/\bar{L}} \\ Q_e &= \sqrt{\bar{C}/\bar{L}} \\ Q_L &= \sqrt{\bar{C}/\bar{L}}\left(\frac{\bar{R}}{1+\bar{R}}\right) \end{aligned}\right\} \tag{7.129n}$$

and

$$\frac{1}{Q_L} = \frac{1}{Q_0} + \frac{1}{Q_e} \tag{7.129p}$$

Coupling parameter,

$$k = Q_0/Q_e$$

$$Q_0 = (1+k)Q_L \tag{7.129q}$$

Two-Port Transmission Cavity 7.7.2

A transmission-type cavity is coupled to both a generator and load. Therefore, two apertures are involved, one on each side as shown in Fig. 7.13. The equivalent circuit permits one to devise many ways of measuring Q. The coupling of a cavity mode to a waveguide is represented by ideal transformers of turn ratios $n_1 : 1$ and $n_2 : 1$ at the generator and the load sides, respectively. Finally a combined equivalent circuit transforming load and generator sides into a single mesh at resonance facilitates the mesh analysis.

The following main parameters are important:

1. Resonant frequency f_r and corresponding VSWR, S_r.
2. Input and output coupling parameters k_1 and k_2, which measure the coupling efficiency (= External power loss/Internal power loss).
3. Transmission loss $T(\omega_r)$ at resonance (= Power output/Power input).
4. Q factors: Q_0, Q_L and Q_e.

By definition

$$Q_0 = \frac{\frac{1}{2}\omega_r L I^2}{\frac{1}{2} R I^2} = \frac{\omega_r L}{R} \tag{7.130}$$

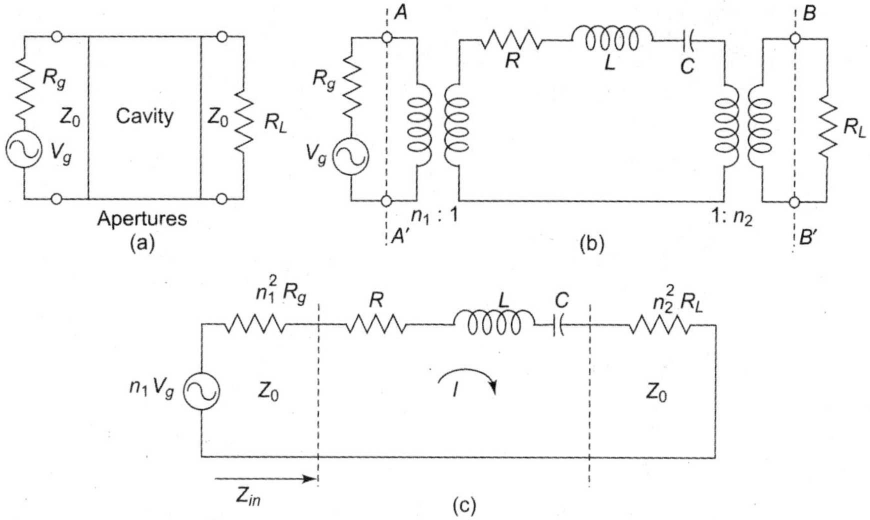

Fig. 7.13 Transmission-type cavity resonator and its equivalent circuits

$$Q_L = \frac{\frac{1}{2}\omega_r L I^2}{\frac{1}{2}(R + n_1^2 R_g + n_2^2 R_L)I^2} = \frac{\omega_r L}{(R + n_1^2 R_g + n_2^2 R_L)} \tag{7.131}$$

$$Q_e = \frac{\frac{1}{2}\omega_r L I^2}{\frac{1}{2}Z_0 I^2} = \frac{\omega_r L}{Z_0} \tag{7.132}$$

where $$1/Q_L = 1/Q_0 + 1/Q_e \tag{7.133a}$$

$$1/Q_e = \frac{Q_0 - Q_L}{Q_L Q_0} \tag{7.133b}$$

$$Q_e = \frac{Q_L Q_0}{Q_0 - Q_L} \tag{7.133c}$$

$$Q_e = Q_0/k_e \tag{7.133d}$$

$$k_1 = \frac{\frac{1}{2}n_1^2 R_g I^2}{\frac{1}{2}RI^2} = \frac{n_1^2 R_g}{R} \tag{7.134}$$

$$k_2 = n_2^2 R_L/R \tag{7.135}$$

$k = 1$ critical coupling

$k < 1$ undercoupling

$k > 1$ overcoupling

From Eqs 7.130–7.135,

$$Q_0 = Q_L(1 + k_1 + k_2) \tag{7.136}$$

The impedance of equivalent circuit can be expressed by

$$Z = R + n_1^2 R_g + n_2^2 R_L + j(\omega L - 1/\omega C)$$
$$= R[1 + k_1 + k_2 + jQ_0(\omega/\omega_r - \omega_r/\omega)] \quad (7.137)$$

and $I = n_1^2 V_g^2/Z^2$ (7.138)

In terms of matched generator and load, $R_g = Z_0 = R_L$. The load power

$$P_L = n_2^2 Z_0 |I|^2 = k_2 R |I|^2$$
$$= \frac{k_1 k_2 V_g^2/Z_0}{(1+k_1+k_2)^2 + Q_0^2(\omega/\omega_r - \omega_r/\omega)^2} \quad (7.139)$$

Generator power available under matched condition

$$P_0 = \frac{1}{4}\frac{V_g^2}{Z_0} \quad (7.140)$$

Therefore, the transmission loss of the cavity

$$T(\omega) = \frac{P_L}{P_0} = \frac{4k_1 k_2}{(1+k_1+k_2)^2 + Q_0^2(\omega/\omega_r - \omega_r/\omega)^2} \quad (7.141)$$

At resonance $\omega = \omega_r$

$$T(\omega_r) = \frac{4k_1 k_2}{(1+k_1+k_2)^2} \quad (7.142)$$

Therefore, $T(\omega) = \dfrac{T(\omega_r)}{1+Q_L^2(\Delta\omega/\omega_r)^2}$; for $\omega = \omega_r + \Delta\,\omega/2$ (7.143)

If $\Delta\,\omega$ = half-power bandwidth

$$T(\omega) = \frac{1}{2}T(\omega_r);\ \Delta\omega \ll \omega_r$$

or, $T(\omega)/T(\omega_r) = 1/2 = \dfrac{1}{1+Q_L^2(\Delta\omega/\omega_r)^2}$ (7.144)

Therefore, $Q_L = \omega_r/\Delta\omega$ when $\Delta\omega \ll \omega_r$ for high Q (7.145)

An equivalent resistance, seen by the generator is R_{eg} for maximum power transfer and can be expressed by

$$n_1^2 R_{eg} = R + n_2^2 R_L \quad \text{or,} \quad R_{eg} = \frac{R + n_2^2 R_L}{n_1^2} \quad (7.146)$$

The normalized value

$$R_{eg}/Z_0 = \frac{R + n_2^2 R_L}{n_1^2 Z_0} \quad (7.147)$$

At resonance, for resistive impedance, the cavity VSWR

$$S_r \approx R_{eg}/Z_0 = \frac{R + n_2^2 R_L}{n_1^2 Z_0} = \frac{1+k_2}{k_1} \quad (7.148)$$

Therefore, $k_1 = \dfrac{4}{4S_r - T(\omega_r)(1+S_r)^2}$ (7.149)

$$k_2 = k_1 S_r - 1 \quad (7.150)$$

The above analysis provides working formulae for experimental measurements of Q:

$$Q_L = f/\Delta f \tag{7.151}$$

$$Q_0 = Q_L(1 + k_1 + k_2) \tag{7.152}$$

$$k_1 = \frac{4}{4S_r - T(\omega_r)(1 + S_r)^2} \tag{7.153}$$

$$k_2 = k_1 S_r - 1 \tag{7.154}$$

In the series resonant circuit

$$T = 1 - \Gamma$$

$$S_{21} = 1 - S_{11}$$

The measurement parameters at resonance are f_r, S_r, $T(\omega_r)$ and 3 dB bandwidth Δf.

Loop-coupled Cavity 7.7.3

In a loop-coupled cavity, since the loop is very small, the current in the loop can be considered to be constant. The plane of the loop is placed perpendicular to the magnetic flux lines of a given resonant mode. The equivalent circuit of the loop coupled cavity is shown in Fig. 7.14.

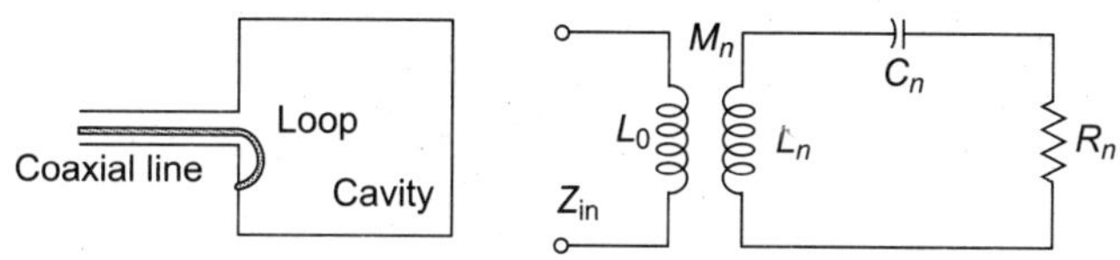

Fig. 7.14 Equivalent circuit of single-port loop-coupled cavity

The input impedance of the circuit in the vicinity of the nth resonance can be expressed by

$$Z_{in} = j\omega L_0 + \frac{\omega_n^2 M_n^2}{2jL_n(\omega - \omega_n - j\omega_n/2Q_n)} \tag{7.155}$$

where L_0 is the self-inductance of the coupling loop, $L_n = Q_n R_n/\omega_n$, $C_n = Q_n/\omega_n R_n$. For maximum coupling, the loop is placed at a position of maximum magnetic field of a given mode.

RE-ENTRANT CAVITY 7.8

In microwave tubes it is essential that for efficient energy transfer from an electron beam to high Q cavity resonators, the electron transit time across the cavity field region must be very small. Consequently, the cavity grids need to be spaced very closely to form re-entrant structure as shown in Fig. 7.15. The **E** field is concentrated in the small gap g. This capacitance region can be used to control flow of electrons passing through this gap when it is grided (for a klystron tube). The tuning of the cavity can be accomplished by means of short-circuit plungers.

The re-entrant cavity of length d with a gap thickness $g << d$ may be considered as a coaxial line with radii of the inner and outer conductors as a and b, respectively. The gap capacitance C and the coaxial line below the gap provide equal and opposite reactances at the plane of the capacitance at resonance such that

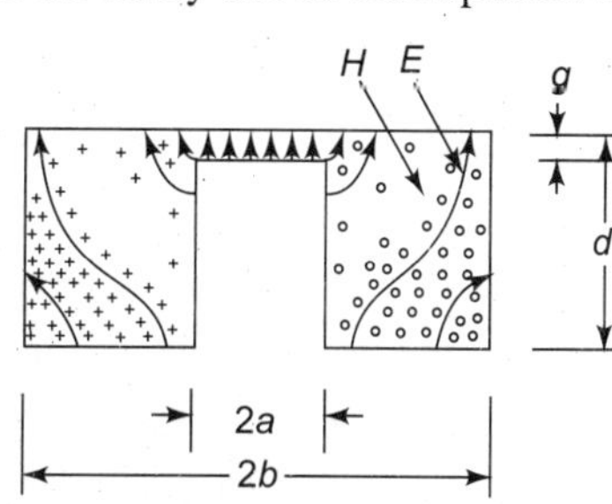

Fig. 7.15 Re-entrant cavity

$$1/\omega_r C = Z_0 \tan(2\pi d/\lambda_g)$$

or,
$$d = (\lambda_g/2\pi)\tan^{-1}(1/Z_0\omega_r C) \tag{7.156}$$

where
$$C = \varepsilon_0 \pi a^2/g, \tag{7.157}$$

and $$Z_0 = 60 \ln b/a \tag{7.158}$$

for air dielectric. For a small capacitance C, $Z_0\, \omega_r C << 1$, $\beta d \approx \pi/2$, i.e., the line is practically a quarterwave long. For large C or small gap, the length d should be shortened. Thus, for a given gap width or C, the resonant lengths d of the cavity can be varied from multiple of quarter wavelength to a smaller value satisfying the resonant condition in Eq. 7.156.

By increasing C, the electric energy stored in the cavity increases. A corresponding increase in the magnetic energy ($W_e = W_m$ at resonance) has to be provided by a larger microwave current in the cavity walls, which results in higher dissipative losses due to finite conductivity of the cavity walls. Consequently the unloaded Q_0 of coxial cavity decreases.

The cavity mode in these configurations is TEM type when $d > (b - a)$; $b < \lambda_g/4$. If $d < (b - a)$ and $b \approx \lambda_g/4$, the cavity mode is TM type with the electric field directed mainly along the cavity axis. The TM mode cavity can be easily tuned by adjusting the capacitive gap g, provided that the resonant mode is TM_{010}.

HOLE-AND-SLOT CAVITY 7.9

Microwave resonant cavity of the hole-and-slot type, shown in Fig. 7.16, is used in multicavity magnetron oscillators. The resonant frequency of the cavities can be determined from the relation

$$\omega_r = 1/\sqrt{LC} \tag{7.159}$$

The function of the lumped inductance L is carried out by the surface of the hole which is equivalent to a loop of metallic band. The lumped capacitance C is formed by the slot cut through the copper block.

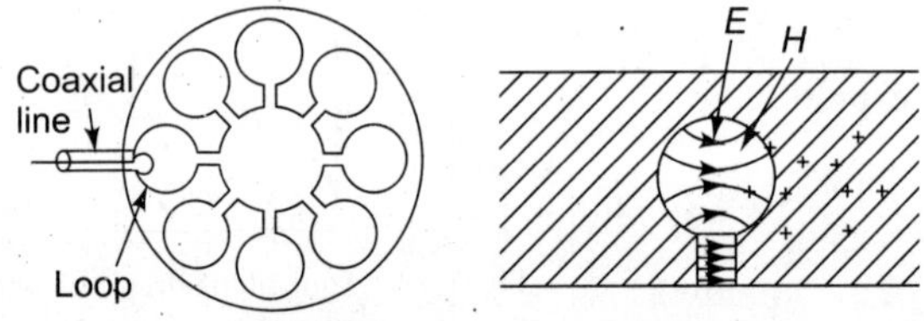

Fig. 7.16 Hole-and-slot cavity

MICROSTRIP RESONATORS 7.10

In microwave integrated circuits, resonators of planar configurations are used for fixed tuned circuits, determining the effective permittivity ε_{eff} at a given frequency and various discontinuity parameters. A microstrip resonator can be formed by a closed loop on a substrate or a line left open or a line short-circuited at both ends simply by a metallic patch of rectangular or circular geometry. By loosely coupling an r.f. test signal to such a resonator, frequency pulling effects are avoided, and the reflected energy and resonator characteristic can be analyzed.

There are three phenomena associated with an open circuit:

(a) The fringing fields extending beyond the physical end of the strip

(b) Radiation loss from the open end decreases by using high ε_{r0} substrate

Phenomena (b), require equivalent shunt conductance at the open end of the line. In most cases this will not be taken into consideration. But their minimization may be carried out. The phenomenon (a) dominates and is accounted for by an equivalent fringe capacitance C_f connected at the open end. A variety of resonator configurations are described below:

Microstrip Line Gap-Coupled Resonators 7.10.1

The microstrip line gap-coupled resonators is formed by sections of microstrip lines coupled through gaps as shown in Fig. 7.17. The resonator consists of a half wavelength open-circuited microstrip line that is capacitively coupled to an input microstrip line along the length or from the side as shown in

Figs 7.17 (a) and (b), respectively. The open-circuited end contains fringing electric field, resulting in accumulation of electric charge. This in turns results in an equivalent fringing capacitance C_{f0} as shown in Fig. 7.18(a). The capacitance C_{f0} is compensated by an additional inductance which arises due to increases in the length d beyond one wavelength by Δd_0. Δd_0 is the equivalent extra length of microstrip line, having all the propagation characteristics of those applicable to the main line, i.e., having same Z_0 and hence same W/h and ε_{eff}. Equating the reactances of C_{f0} and that of electrically open-circuited extra length Δd_0,

$$\frac{1}{\omega C_{f0}} = Z_0 \cot(\beta\ \Delta d_0) = \frac{Z_0}{\tan(\beta \Delta d_0)} \tag{7.160}$$

For $\quad \Delta d_0 << \lambda_g, \tan(\beta\ \Delta d_0) \cong \beta \Delta d_0 = \omega C_{f0} Z_0$

or, $\quad \Delta d_0 = cZ_0 C_{f0}/\sqrt{\varepsilon_{eff}};\ c = \dfrac{1}{\sqrt{\mu_0 \varepsilon_0}} = 3 \times 10^8$ m/s

It can be shown (TE) that

$$\Delta d_0 = 0.412\,h \left[\frac{\varepsilon_{eff} + 0.3}{\varepsilon_{eff} - 0.258}\right]\left[\frac{W/h + 0.262}{W/h + 0.813}\right] \tag{7.161}$$

Similarly, at the gap region also is end effect length Δd_g. Therefore, the resonant condition can be expressed by

$$d + \Delta d_g + \Delta d_0 = \lambda_g/2 \tag{7.162}$$

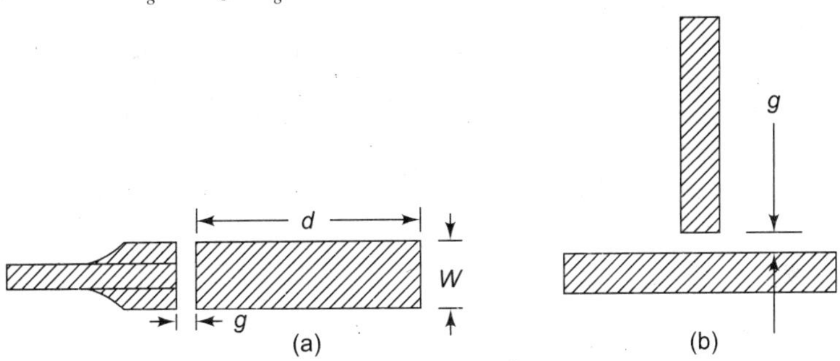

Fig. 7.17 Microstrip resonators

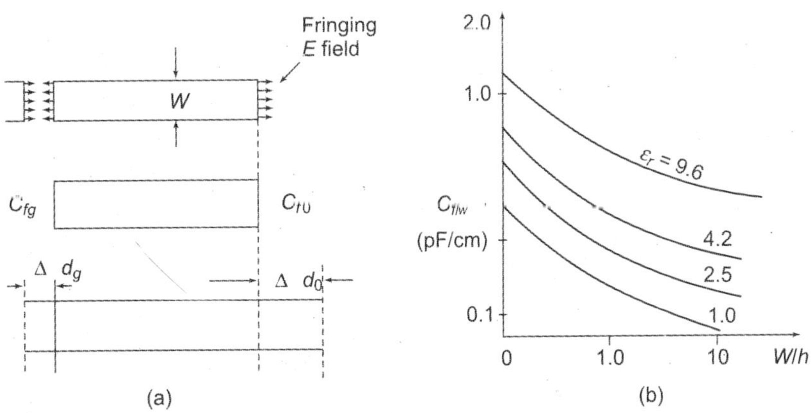

Fig. 7.18 Fringing capacitance as a function of the line width W and the substrate thickness h (Source: Ref. 21)

The amount of fringing capacitance can be calculated as a function of line width W and the substrate thickness h using the assumption of static potential fields [1, 22, 23]. Typical values of the open-circuited fringing capacitance are shown in Fig. 7.18(b). The quasi-TEM mode analysis of microstrip lines can be used with reasonable accuracy for frequencies up to 40 GHz and for substrate thickness down to 5 mil. At higher frequencies, the electric field is more confined between the resonator line patch and ground plane resulting in an increase in the effective dielectric constant as well as increased dielectric loss and conductor loss. This frequency dispersion of the effective dielectric constant helps to determine the effective dielectric constant of a substrate. Taking a pair of resonators of lengths d_1 and d_2 having identical widths, gaps and substrate, dispersion results can be expressed for closely spaced resonant frequencies f_1 and f_2 as [Bendek and Silvester].

$$\varepsilon_{\text{eff}} = \left[\frac{hc\,(2f_1 - f_2)}{2f_1 f_2\,(d_2 - d_1)}\right]^2 \tag{7.163}$$

where h is the substrate thickness and C is the velocity of light of light in free space. Total end effect length can be determined as [R.E. Collin]

$$\Delta d = \Delta d_g + \Delta d_0 = \frac{f_2 d_2 - 2 f_1 d_1}{2 f_1 - f_2} \tag{7.164}$$

Series Gap Capacitance

The series gap capacitance can be represented by an equivalent π-network as shown in Fig. 7.19(a). The significant capacitance that exists across the gap is denoted by C_2, whereas, C_1 represents the fringing capacitance between the line and the ground at each end of the line at a symmetrical gap. The values of the gap capacitance can be determined from the graphs of Fig. 7.19(b) as a function of the gap spacing/substrate thickness. Resonators of this type are commonly used in microstrip filters. Garg and Bahl [21] have given expressions for the gap capacitances as

$$C_1 = \frac{C_e(\varepsilon_r)}{2}\ ;\ 0.5 \le W/h \le 2.0 \tag{7.165}$$

$$C_2 = \frac{C_0(\varepsilon_r) - C_1(\varepsilon_r)}{2}\ ;\ 2.5 \le \varepsilon_r \le 15 \tag{7.166}$$

where, $C_0(\varepsilon_r) = C_0(9.6)\,(\varepsilon_r/9.6)^{0.8}$

$C_e(\varepsilon_r) = C_e(9.6)\,(\varepsilon_r/9.6)^{0.9}$

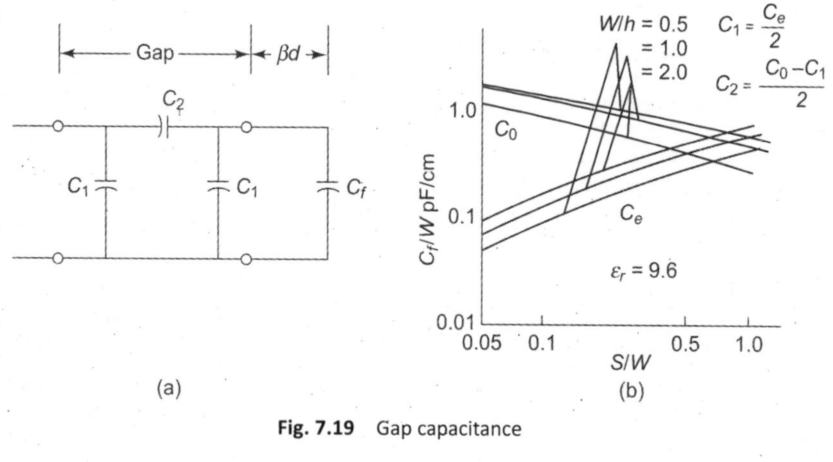

Fig. 7.19 Gap capacitance

$C_0\,(9.6)/W = (g/W)^{m}{}_{0}\,\exp\,(k_0)$ pF/m

$C_e\,(9.6)/W = (g/W)^{m}{}_{e}\,\exp\,(k_e)$ pF/m

$$\begin{array}{ll|l}
m_0 & = W/h\,(0.619\log(W/h) - 0.3855 & \\
k_0 & = 4.26 - 1.453\log(W/h) & 0.1 \le W \le 1.0 \\
m_e & = 0.8675 & \\
k_e & = 2.043\,(W/h)^{0.12} & 0.1 \le g/W \le 0.3 \\
m_e & = 1.565/(W/h)^{0.16} - 1 & \\
k_e & = 1.97 - 0.03/(W/h) & 0.3 \le g/W \le 1.0
\end{array}$$

Rectangular Microstrip Disk Resonator 7.10.2

A microstrip rectangular patch on a dielectric substrate over a ground plane (Fig. 7.20) forms a dielectric loaded cavity with the ground plane and the patch as electric walls and magnetic side walls at the periphery where $\mathbf{H}_{tan} = 0$ (simulates an open circuit). The cavity exhibits higher-order mode resonances. The fields inside the cavity can be found accurately by assuming that the substrate thickness $h << \lambda$, so that only one **E** field component E_z (constant in z) exists which satisfies the scalar *Helmholtz equation*

$$(\nabla^2 + k^2)\,E_z\,(x, y) = 0 \tag{7.167}$$

where $k = \omega\sqrt{\mu_0\,\varepsilon}$. Hence the possible modes are TM_{mn0}. The z-variation is zero due to small thickness of the substrate ($h << \lambda$). If the length and width of the patch are l and W respectively, and the origin of the coordinates is at one corner such that $0 \le z \le h$, $0 \le x \le l$, $0 \le y \le W$, the following boundary conditions must be satisfied.

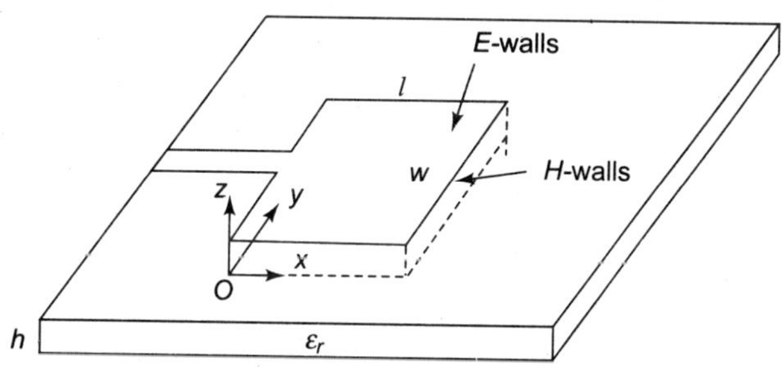

Fig. 7.20 Microstrip rectangular disk resonator

$$H_x\,(y = 0) = H_x\,(y = W) = 0$$

or

$$\left.\frac{\partial E_z}{\partial y}\right|_{y=0} = \left.\frac{\partial E_z}{\partial y}\right|_{y=W} = 0$$

$$H_y\,(x = 0) = H_y\,(x = l) = 0 \quad \text{or,} \quad \left.\frac{\partial E_z}{\partial x}\right|_{x=0} = \left.\frac{\partial E_z}{\partial x}\right|_{x=l} = 0 \tag{7.168}$$

to yield the field solutions of TM_{mn0} modes

$$E_z = A_{mn}\cos\,(m\pi x/l)\cos\,(n\pi\,y/W) \tag{7.169}$$

$$H_x = \frac{j}{\omega\mu}\frac{\partial E_z}{\partial y} = \frac{-n\pi\,A_{mn}}{W\,\omega\mu}\cos\left(\frac{m\pi x}{l}\right)\sin\left(\frac{n\pi y}{W}\right) \tag{7.170}$$

$$H_y = \frac{-j}{\omega\mu}\frac{\partial E_z}{\partial x} = \frac{+m\pi A_{mn}}{\omega\mu l}\sin\left(\frac{m\pi x}{l}\right)\cos\left(\frac{n\pi y}{W}\right) \tag{7.171}$$

where $k_x^2 + k_y^2 = (m\pi/l)^2 + (n\pi/W)^2 = k_r^2 = \omega_r^2\mu_0\varepsilon$ and A_{mn} is an *amplitude constant* that depends on the excitation. Thus, the resonant frequency of the cavity in TM_{mn0} mode to z is

$$(f_r)_{mn} = \frac{1}{2\pi\sqrt{(\mu_0\,\varepsilon)}}[(m\pi/l)^2 + (n\pi/W)^2]^{1/2}; \quad \begin{matrix} m = 0, 1, 2, \ldots \\ n = 0, 1, 2, \ldots \end{matrix} \tag{7.172}$$

when m and n are not both zero simultaneously.

Resonant frequencies occur when l or W is equal to an integral multiple of half wavelength in the substrate. For $l > W$, the lowest-order resonant frequency occurs when $m = 1$ and $n - 0$, i.e., for TM_{100} mode:

$$(f_r)_{100} = \frac{1}{2l\sqrt{(\mu_0\,\varepsilon)}} \tag{7.173}$$

The Q-factor of the resonator is determined in the following manner. The total stored energy in the resonator for dominant mode TM_{100} is

$$W_s = 2\,W_e = \frac{\varepsilon'}{2}\int_0^h\int_0^W\int_0^l |E_z|^2\,dx\,dy\,dz$$

$$= \frac{\varepsilon'}{2}\int_0^h\int_0^W\int_0^l A^2{}_{10}\cos^2(\pi x/l)\,dx\,dy\,dz$$

$$= \frac{\varepsilon' h A_{10}^2 W}{2}\int_0^l \cos^2\left(\frac{m\pi x}{l}\right)dx = \frac{\varepsilon' h A_{10}^2\cdot hWl}{4} \tag{7.174}$$

The power loss in the dielectric

$$P_d = \frac{\omega\varepsilon''}{2}\int_0^h\int_0^W\int_0^l |E_z|^2 dx\,dy\,dz \frac{\omega\varepsilon''}{2}\frac{2W}{\varepsilon'} = \frac{\omega\varepsilon'' W}{\varepsilon'}$$

$$= \frac{\omega\,A_{10}^2\varepsilon'' hWl}{4} \tag{7.175}$$

The conductor power loss for the bottom and top conducting surfaces:

$$P_c = 2\frac{R_s}{2}\int_0^l\int_0^W \left(|H_x|^2 + |H_y|^2\right)dx\,dy = R_s\int_0^l\int_0^W |H_y|^2\,dx\,dy;\quad H_x = 0 \text{ for } n = 0$$

$$= R_s\int_0^l\int_0^W \left(\frac{\pi A_{10}}{\omega\mu l}\right)\sin^2\left(\frac{\pi x}{l}\right)dx\,dy = R_s\left(\frac{\pi A_{10}}{\omega\mu l}\right)^2\frac{Wl}{2} \tag{7.176}$$

The Q-factor

$$Q_0 = \frac{\omega W_s}{P_c + P_d} = \frac{W\varepsilon' A_{10}^2\,hWl/4}{W\varepsilon' A_{10}^2\,Whl/4 + R_s\left(\frac{\pi A_{10}}{\omega\mu l}\right)^2\frac{Wl}{2}}$$

$$= \frac{\omega\varepsilon'h}{\omega\varepsilon''h + 2R_s(\pi/\omega\mu l)^2} = \frac{W\varepsilon'h}{\omega\varepsilon''h + \left(\frac{\sigma\delta}{2}\right)\left(\frac{\pi\delta}{l}\right)^2} \tag{7.177}$$

The above expression for Q neglects the radiation loss due to fringing fields at the edges of the patch.

Microstrip Circular Disk Resonator 7.10.3

A circular Cu disk of radius a on the substrate of a high dielectric constant forms a cavity with the patch and the ground plane as electric walls on which $\hat{n} \times \mathbf{E} = 0$ and cylindrical magnetic walls surrounding the patch on which $\hat{n} \times \mathbf{H} = 0$ as shown in Fig. 7.21. The patch can be excited by a coaxial line probe from the ground plane or an input microstrip line on the plane of the patch. It will be assumed that the height (h) of the substrate of relative dielectric constant (ε_r) is small compared to the free space wavelength (λ_0). The resonator is excited by a line current (I_0) along the z-axis through a probe of negligible diameter which forms the centre conductor of a coaxial line fed from the ground plane at $\rho = \rho_0$, $\phi = 0°$, where $0 < \rho_0 < a$.

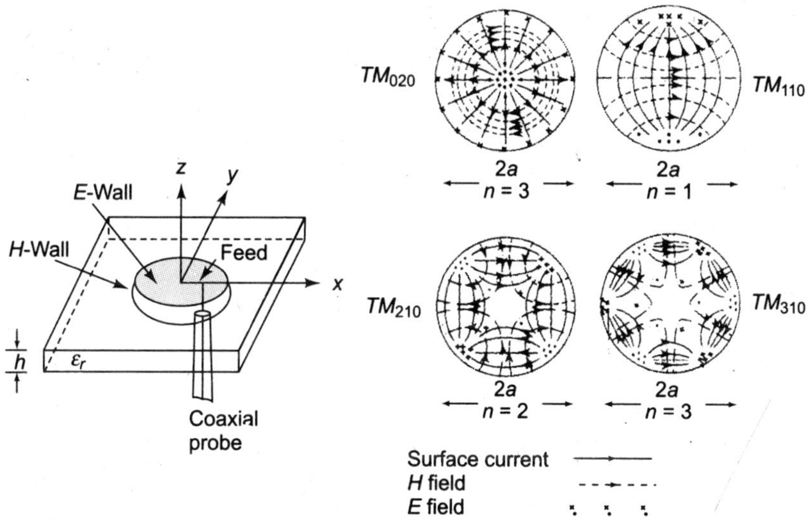

Fig. 7.21 Circular disk microstrip resonator

The cavity model of the disk structure will be based on the following assumptions

(a) Because of the close proximity between the disk and the ground plane, the source current will be assumed uniform along z and therefore the modal fields will have no z-variations in the frequency band of interest, and there is no charge accumulation in the region between the disk and the ground plane.

(b) The above implies that only non-zero field components are E_z, H_ρ and H_ϕ in the region bounded by the microstrip and the ground plane.

(c) The electric current in the microstrip must have no components normal to the edge at any point on the edge, implying zero tangential component of $\mathbf{H}$ along the edges: $H_\phi = 0$ at $\rho = a$ (magnetic wall).

The region between the disk and the ground plane will therefore be considered as a cavity with magnetic walls at the edges and electric walls at the top and bottom. It will be assumed that such a cavity would radiate no power and have a purely reactive input impedance of either zero or infinite value at resonance.

The resonant modes are denoted by TM_{nm0}, where n and m denote variations in the azimuthal and radial directions, respectively, with zero variation in the z-direction.

The components of modal fields in the resonator can be obtained by solving the wave equation in the cylindrical coordinate system with delta function excitation. However, without much loss of generality, the cavity fields can be obtained from the source free wave equation

$$\nabla_t^2 E_z + k_\rho^2 E_z = 0 \tag{7.178}$$

Following the method of separation of variables, we seek to find solutions of the form

$$E_z = R(\rho)\,\Phi(\phi) \tag{7.179}$$

The ϕ-symmetric ring structure yields harmonic solution

$$\Phi(\phi) = \cos n\phi \tag{7.180}$$

The solution $R(\rho)$ corresponds to Bessel's equation of order $n.k_\rho$ represents the wave number in the radial direction and is always real. Therefore, in the disk structure there will be formation of cylindrical standing waves due to two magnetic side walls. This implies that the solution of R should be represented by the Bessel function of the first kind $\mathbf{J}_n(k_\rho \rho)$,

$$R(\rho) = \mathbf{J}_n(k_\rho \rho) \tag{7.181}$$

Since the fringing fields in the radial direction extend into free space, wave number k_ρ is determined by the effective dielectric constant ε_{eff}; $k_\rho = k_0\sqrt{\varepsilon_{\text{eff}}}$. For a very thin substrate ($h << a$), $\varepsilon_{\text{eff}} \approx \varepsilon_r$ and $k_\rho = k_0 \sqrt{\varepsilon_r}$. The components of the modal fields for TM_{nm0} modes are, therefore

$$E_z = \mathbf{J}_n(k_\rho \rho)\cos n\phi \tag{7.182}$$

$$H_\rho = \frac{j}{\omega\mu_0}\frac{\partial E_z}{\partial \phi} = (-jn/\omega\mu_0)\,\mathbf{J}_n(k_\rho \rho)\sin n\phi \tag{7.183}$$

$$H_\phi = \frac{j}{\omega\mu_0}\frac{\partial E_z}{\partial \rho} = (-jk_\rho/\omega\mu_0)\,\mathbf{J}'_n(k_\rho \rho)\cos n\phi \tag{7.184}$$

From the boundary conditions

$$H_\phi = 0 \quad \text{at} \quad \rho = a \tag{7.185}$$

or,

$$\frac{\partial E_z}{\partial \rho} = 0 = k_\rho \mathbf{J}'_n(k_\rho a) \tag{7.186}$$

where prime indicates differentiation with respect to the argument. The resonant frequencies are obtained, from the roots x'_{nm} of

$$\mathbf{J}'_n(k_\rho a) = 0 \quad \text{or} \quad k_\rho = x'_{nm}/a \quad \text{or} \quad f_r = \frac{x'_{nm}}{2\pi a\sqrt{\mu\varepsilon a}} \tag{7.187}$$

The dominant mode is designated as the TM_{110} mode for which x'_{nm} is minimum $= x'_{11} = 1.841$. The total stored energy in the resonator for the dominant mode is given by

$$W = 2W_e = \frac{\varepsilon'_r \varepsilon_0}{2}\int_0^h\int_0^{2\pi}\int_0^a \mathrm{J}_1^2(k_\rho \rho)\cos^2\phi\,\rho\,d\rho\,d\phi\,dz$$

$$= \frac{\pi\varepsilon'_r\varepsilon_0 h a^2}{4}\left(1 - \frac{1}{k_\rho^2 a^2}\right)\mathbf{J}_1^2(k_\rho a) \tag{7.188}$$

since $\mathbf{J}_1'(k_\rho a) = 0$

The power loss in the dielectric

$$P_d = \frac{\omega \varepsilon_r''}{2} \int_0^h \int_0^{2\pi} \int_0^a |E_z|^2 \rho \, d\rho \, d\phi \, dz \frac{2\,\omega \varepsilon_r''}{\varepsilon_r'} W_e \tag{7.189}$$

The conductor power loss

$$P_c = 2\frac{R_s}{2} \int_0^{2\pi} \int_0^a \left[|H_\phi|^2 + |H_\rho|^2 \right] \rho d\, \rho \, d\phi$$

$$= \frac{\pi R_s}{k_0^2 \eta^2} \int_0^a \left[k_\rho^2 \mathbf{J}_1^2 (k_\rho \rho) + \frac{\mathbf{J}^2 (k_\rho \rho)}{\rho^2} \right] \rho \, d\rho$$

$$= \frac{\pi R_s k_\rho^2 a^2}{2 k_0^2 \eta^2} \left(1 - \frac{1}{k_\rho^2 a^2} \right) \mathbf{J}_1^2 (k_\rho a) \tag{7.190}$$

The Q-factor $$Q = \frac{\omega W}{P_d + P_c} = \frac{k_0 h}{2 R_s/\eta + \varepsilon_r'' k_0 h/\varepsilon_r'} = \frac{1}{\delta/h + \varepsilon_r''/\varepsilon_r'} \tag{7.191}$$

where $$\delta = \frac{1}{\sqrt{\pi f \mu \sigma}}$$

This expression neglects the radiation loss due to fringing field and conductor loss for currents on the top of the patch. These effects reduce the practical Q achievable.

Example 7.6 Find Q of a microstrip disk resonator at dominant TM_{110} mode operating at 5 GHz. The copper patch is etched on alumina substrate with $\varepsilon'_r = 9.6$, $\varepsilon_r'' = 0.0002$ and $h =$ 1/16 inch.

Solution

$$Q = \frac{k_0 h}{2 R_s/\eta + \varepsilon_r'' k_0 h/\varepsilon_r'}$$

Here, $\varepsilon_r' = 9.6$, $\varepsilon_r'' = 0.0002$, h = 1/16 inch = 1.5875 mm

$$\lambda_0 = 30/5 = 6 \text{ cm}$$

$$k_0 = 2\,\pi/6 = \pi/3 \text{ rad/cm}$$

$$k_0 h = \pi/3 \times 0.15875 = 0.1663 \text{ radian}$$

$$R_s = \sqrt{(\omega \mu_0 / 2\sigma)}$$

$$= \left[\frac{2\pi \times 5 \times 10^9 \times 4\pi \times 10^{-7}}{2 \times 5.8 \times 10^7} \right] = 1.8 \times 10^{-2} \quad \text{ohm/m}^2$$

$$Q = \frac{0.1663}{2 \times 1.8 \times 10^{-2}/120\pi + 0.0002 \times 0.1663/9.6} = 1680.50$$

Microstrip Ring Resonator 7.10.4

Figure 7.22 shows a circular microstrip ring resonator with inner and outer radii a and b, respectively. Ring resonators have found applications in circulators, hybrid junctions and filters. They can also be used to measure propagation constants in microstrips, based on the principle that the ring is resonant when its mean circumference equals an integral number of half wavelengths, i.e., $(a + b)/2 = nc/(2 f_r \sqrt{\varepsilon_{\text{eff}}})$. A simple approach to find the resonant frequencies will be described from a cavity model similar to the case of a disk. The region between the ring and the ground plane will therefore be considered as a cavity with magnetic walls at the edges and electric walls at the top and bottom. The resonant modes are denoted by TM_{nm0}, where n and m denote variations in the azimuthal and radial directions, respectively, with zero variation in the z-direction.

The components of modal fields in the ring structure can be obtained by solving the wave equation in the cylindrical coordinate system in a similar way as in the case of disk. Solution R should be represented by the Bessel function of the first kind $J_n(k_\rho \rho)$ and Bessel function of the second kind $N_n(k_\rho \rho)$

$$R(\rho) = J_n(k_\rho \rho) - A_n\, N_n(k_\rho \rho) \tag{7.192}$$

to exclude origin, where the modal field is zero. Here, A_n is a constant dependent on frequencies and to be determined from the boundary conditions. Since the fringing fields in the radial direction extend into free space, wave number k_ρ is determined by the effective dielectric constant ε_{eff}, $k_\rho = k_0 \sqrt{\varepsilon_{\text{eff}}}$. For a very thin substrate $b - a >> h$, $\varepsilon_{\text{eff}} \approx \varepsilon_r$ and $k_\rho = k_0 \sqrt{\varepsilon_r}$. The components of the modal fields for TM_{nm0} modes are therefore

$$E_z = R(\rho) \cos n\phi \tag{7.193}$$

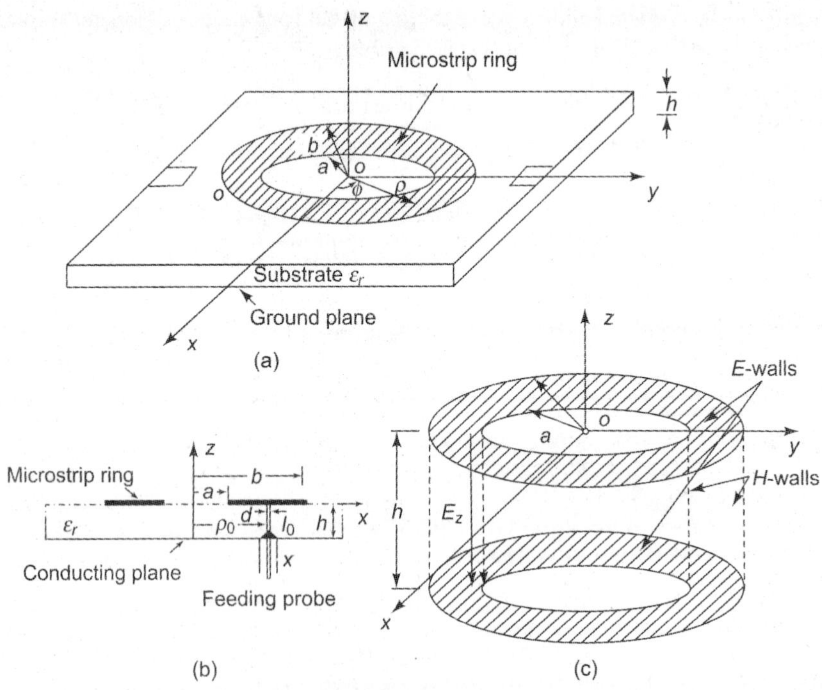

Fig. 7.22 Microstrip ring resonator: (a) Isometric view (b) Excitation by probe (c) Cavity model

$$H_\rho = \frac{j}{\omega\mu_0}\frac{\partial E_z}{\partial \phi} \tag{7.194}$$

$$H_\phi = \frac{j}{\omega\mu_0}\frac{\partial E_z}{\partial \rho} \tag{7.195}$$

From the boundary conditions,

$$H_\phi = 0 \quad \text{at} \quad \rho = a, b \tag{7.196}$$

or,

$$\frac{\partial E_z}{\partial \rho} = 0 = k_\rho\,[\mathbf{J}_n'(k_\rho a) - A_n N_n'(k_\rho a)$$

$$= k_\rho\,[\mathbf{J}_n'(k_\rho b) - A_n N_n'(k_\rho b) \tag{7.197}$$

where prime indicates differentiation with respect to the argument. This leads to the eigen value equation of the ring resonator for the resonances

$$\frac{\mathbf{J}_n'(k_\rho a)}{N_n'(k_\rho a)} = \frac{\mathbf{J}_n'(k_\rho b)}{N_n'(k_\rho b)} = A_n \tag{7.198}$$

The eigen value solutions k_ρ give the resonant frequency f_r, with

$$k_\rho = \frac{2\pi f_r}{c}\sqrt{\varepsilon_r},\ c = \text{speed of light in vacuum} \tag{7.199}$$

The field components in the ring resonator are

$$E_z = \left[\mathbf{J}_n(k_\rho \rho) - \frac{\mathbf{J}_n'(k_\rho a)}{N_n'(k_\rho a)} N_n(k_\rho \rho)\right]\cos n\phi \tag{7.200}$$

$$H_\rho = (-jn/\omega\mu_0)\left[\mathbf{J}_n(k_\rho \rho) - \frac{\mathbf{J}_n'(k_\rho a)}{N_n'(k_\rho a)} N_n(k_\rho \rho)\right]\sin n\phi \tag{7.201}$$

$$H_\phi = (-jk/\omega\mu_0)\left[\mathbf{J}_n(k_\rho \rho) - \frac{\mathbf{J}_n'(k_\rho a)}{N_n'(k_\rho a)} N_n(k_\rho \rho)\right]\cos n\phi \tag{7.202}$$

These equations determine the modal field patterns of TM_{nm0} modes inside microstrip ring resonator. The surface electric current on the inner surface of the patch is given by

$$\mathbf{J}_S = -\hat{z}\times\mathbf{H} = \hat{\boldsymbol{\rho}}H_\phi - \hat{\boldsymbol{\phi}}H_\rho$$

$$= -\hat{\boldsymbol{\rho}}(jk/\omega\mu_0)\left[\mathbf{J}_n(k_\rho \rho) - \frac{\mathbf{J}_n'(k_\rho a)}{N_n'(k_\rho a)} N_n(k_\rho \rho)\right]\cos n\phi$$

$$+\hat{\boldsymbol{\phi}}(jn/\omega\mu_0)\left[\mathbf{J}_n(k_\rho \rho) - \frac{\mathbf{J}_n'(k_\rho a)}{N_n'(k_\rho a)} N_n(k_\rho \rho)\right] \tag{7.203}$$

The mode field patterns and current distributions on the microstrip ring can be obtained from Eqs 7.200–7.203 and are shown in Fig. 7.23.

The Q-factor of the microstrip ring resonator can be computed in the same way as in the case of a disk.

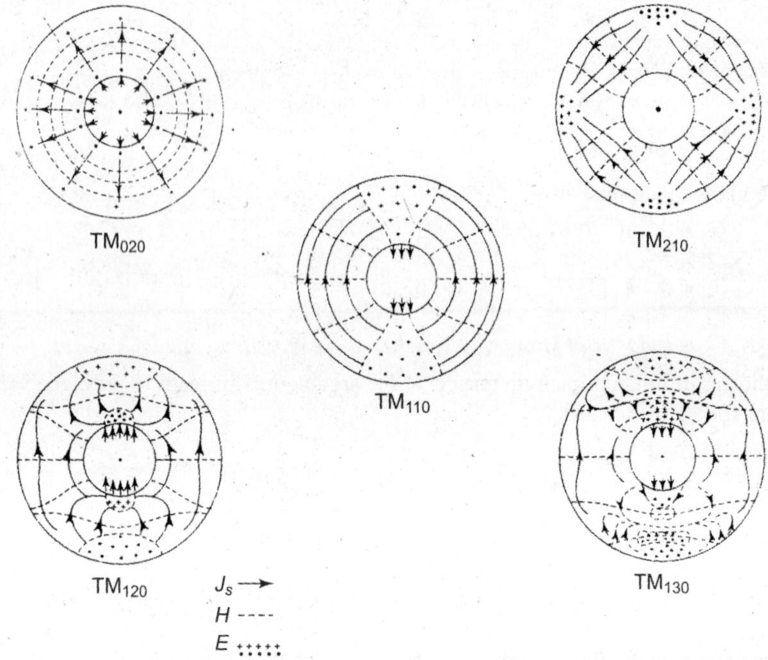

Fig. 7.23 Mode field patterns and current distributions in a microstrip ring resonator

REVIEW QUESTIONS

7.1 A short-circuited coaxial line is made of copper and air dielectric. The inner and outer conductor diameters are 1 cm and 3 cm, respectively and the length is 40 cm. Find the antiresonant frequency, the Q, and the input resistance at resonance.

7.2 An air-filled 50 ohm transmission line is shorted at one end and terminated at the other end in a lumped capacitor of 4 pF. Calculate the length of the transmission line for resonance at 1 GHz.

7.3 Find the smallest possible size of a cubical cavity and the modes which resonate at 5 GHz.

7.4 A rectangular waveguide resonator of cross-sectional dimensions 2.3 cm × 1 cm is filled with air. What should be the length of the resonator for TE_{101} mode resonance at 10 GHz? What is the next higher mode of resonance and the corresponding resonant frequency?

7.5 Calculate the resonant frequency for the TE_{101} mode of a rectangular cavity formed by shorting the ends of an X-band waveguide with $a = 0.9''$. $b = 0.4''$ and $d = 2.4''$.

7.6 Design a rectangular cavity to have a dominant resonant mode frequency of $f_r = 9.8$ GHz having dimension $a = d$ and $b = a/2$.

7.7 Find the resonant frequency and Q of a copper cubic rectangular cavity of 1000 cubic cm dimension for the TE_{101} mode.

7.8 A rectangular copper cavity of height $b = 1.2$ cm, width $a = 2.5$ cm resonates at 10 GHz in dominant mode. The cavity is critically coupled to a rectangular guide of dimensions $a \times b$ by a small circular hole in the transverse wall. Determine the cavity length d. Find the unloaded and loaded Q's for air dielectric inside the cavity.

7.9 An air-filled circular cylindrical cavity resonates at 3 GHz in TM_{010} mode. The resonator is now filled with a lossless dielectric material with dielectric constant 2.2. Calculate the new resonance frequency.

7.10 An air-filled cylindrical cavity has a radius of 3 cm. The cavity is tuned by means of a plunger that allows the length to be varied from 5 to 7 cm. Determine the dominant resonant modes and the range of resonant frequencies.

7.11 The resonant frequency of a cavity is 8.8 GHz. It is critically coupled to an external circuit. The measured bandwidth with loading is 4.0 MHz. Calculate the loaded and unloaded Q's.

7.12 A cylindrical cavity of radius $a = 2$ cm and length 5 cm is filled with a dielectric with permittivity $\varepsilon = (2.5 - j0.0001)\varepsilon_0$. The cavity is made of copper. Find the resonant frequency and Q for the dominant mode.

7.13 Design a microstrip circular disk resonator operating in the TM $_{110}$ mode at 2 GHz with the substrate of relative dielectric constant $6 - j0.005$ and is 1/16″ thick. The disk is made of copper. Determine the required radius of the disk and Q of the resonator.

REFERENCES

1. Collin, R .E., *Foundation of Microwave Engineering*, Second edition, McGraw-Hill, Inc. 1992.
2. Montgomery, C. G., R. H. Dicke and E. M. Purcell, *Principles of microwave circuits*, McGraw-Hill Book Company, New York, 1948.
3. Ragan, G. L., (Ed.) *Microwave Transmission Circuits*, McGraw-Hill Book Company, New York, 1948.
4. Slater, J. C., *Microwave Electronics*, D. Van Nostrand Company, Inc., Princeton, N.J., 1950.
5. Goubau, G., *Electromagnetic Waveguides and Cavities*, Chap. 2, Pergamon Press, New York, 1961.
6. Kurokawa, K., "The expansions of electromagnetics fields in cavities", *IRE Trans.*, Vol. MTT-6, pp. 178–187, April, 1958.
7. Kaifez, D. and P. Guilon, *Electron Resonators*, Artech House Books, Dedham, Mass., 1986.
8. Van Bladel, J., "The Excitation of Dielectric Resonators of Very High Permittivity", *IEEE Trans.*, Vol. MTT-23, pp. 208–217, 1975.
9. Chow, K. K., "On the solution and field pattern of cylindrical dielectric resonators", *IEEE Trans.*, Vol. MTT-14, p. 439, 1966.
10. Kraus, J. D., *Electromagnetics*, McGraw-Hill Book Company, Inc., 1953.
11. Slater, P. C. and N. H. Frank, *Electromagnetism*, McGraw-Hill Book Company, 1947.
12. Plonsey, R. and R. E. Collin, *Electromagnetic Fields*, McGraw-Hill Book Company, Inc., 1961.
13. Javod, M. and P. M. Brown, *Field Analysis and Electromagnetics*, McGraw-Hill Book Company, 1963.
14. Argence, E. and T. Kahan, *Theory of Waveguides and Cavity Resonators*, Blackie and Son Limited, London, 1967.
15. Marcavitz, N., *Waveguide Handbook*, M.I.T Radiation laboratory series, Vol. 10; McGraw-Hill, New York, 1951.
16. Ramo, S., J. R. Whinnery and T. Van Duzer, *Fields and Waves in Communication Electronics*, 2nd ed. New York, John Wiley, 1984.
17. Waldron, R. A., *Theory of Guided Electromagnetic Waves*, London, Van Nostrand Reinhold, 1970.
18. Wu, Y. S. and Rosenbourn, "Mode chart for microstrip ring resonators", *IEEE Trans. on Microwave Theory and Techniques*, Vol. MTT-21, July, 1973, pp. 487–489.
19. Ali, S.M., W. C. Chew, and J. A. Kong, "Vector hankel transform analysis of annular ring microstrip antenna", *IEEE Trans. on Antennas Propagation*, Vol. AP-30, July, 1982, pp. 637–644.
20. Terry Edwards, *Foundations for Microstrip Circuit Design*, Second Edition, John Wiley and Sons, 1992.
21. Garg, Ramesh and I. J. Bahl, "Microstrip discontinuities", *Int. J. Electronics*, 45, No. 1, 1978, pp. 81–87.
22. Silvester, P. and Benedek, "Equivalent Capacitances of Microstrip Open Circuits", *IEEE Trans.* Vol. MTT-20, pp. 511–516, 1972.
23. Benedek, P. and P. Silvester, "Equivalent Capacitances for Microstrip Gap and Steps", *IEEE Trans.* Vol. MTT-20, pp. 729–733, 1972.
24. Das, A., S. K. Das and S. P. Mathur, "Radiation characteristics of higher order modes in annular ring microstrip antenna", *Proc. IEE, Pt. H.*, October, 1984.

8 MICROWAVE FILTERS

INTRODUCTION 8.1

Microwave filters are two-port, reciprocal, passive, linear devices which attenuate heavily the unwanted signal frequencies while permitting transmission of wanted frequencies. There are three types of construction of filters—a *reflective filter* which consists of capacitive and inductive elements producing ideally zero reflection loss in the pass band and very high attenuation in the stop band, *absorptive filters* which dissipate the unwanted signal internally and pass the wanted signal, and a *lossy filter* which uses lossy material in the filter to heavily attenuate the rejected signal but permit passage with low loss to the wanted signal. This chapter describes mainly the reflective microwave filters constructed from purely reactive elements. The practical filters have a small non-zero attenuation in the pass band and a small signal output in the attenuation or stop band due to the presence of resistive losses in reactive elements and propagating medium. A microwave filter is designed to operate between resistive source and load impedances of 50 ohms in most microwave systems. Mostly, reactive elements L and C are used and the filter is assumed to be lossless with its nominal characteristic *resistive impedance* R_0. For basic filter theory, the readers may refer fundamental books on filters. Here, some special features and design are described.

FILTER PARAMETERS 8.2

In designing a filter, the following important parameters are generally considered.

1. Pass-band width
2. Stop-band attenuation and frequencies
3. Input and output impedances
4. Return loss
5. Insertion loss
6. Group delay

The most important parameters among the above is the amplitude response given in terms of the insertion loss vs frequency characteristics. Figure 8.1 shows a basic block diagram of a filter fed by a generator and terminated by a load.

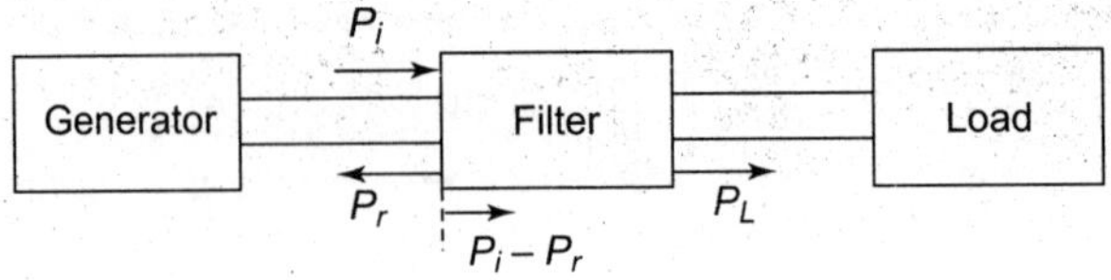

Fig. 8.1 A filter between a generator and a load

Let P_i be the incident power at the filter input, P_r is the reflected power, P_L is the power passed on to the load. The insertion loss of the filter is then defined by, under load match condition ($R_L = R_0$),

$$IL(\text{dB}) = 10 \log \frac{P_i}{P_L} = 10 \log \frac{P_i}{P_i - P_r} = 10 \log \frac{1}{1-|\Gamma|^2} \tag{8.1}$$

where $P_L = P_i - P_r$ if the filter is lossless and Γ is the voltage reflection coefficient given by $|\Gamma|^2 = P_r/P_i$.

The return loss of the filter is defined by

$$RL\ (\text{dB}) = 10 \log \frac{P_i}{P_r} = 10 \log \frac{1}{|\Gamma|^2} \tag{8.2}$$

which quantifies the amount of impedance matching at the input port.

The group delay is important for the multi-frequency or pulsed signals to determine the frequency dispersion or deviation from constant group delay over a given frequency band and is defined by

$$T_d = \frac{1}{2\pi} \times \frac{d\phi_t}{d f} \tag{8.3}$$

where ϕ_t is the *transmission phase.*

The most commonly used microwave filters are the low-pass, high-pass, band-pass and band-stop filters. The electrical equivalent circuits of these filters along with typical insertion loss response characteristics are shown in Fig. 8.2.

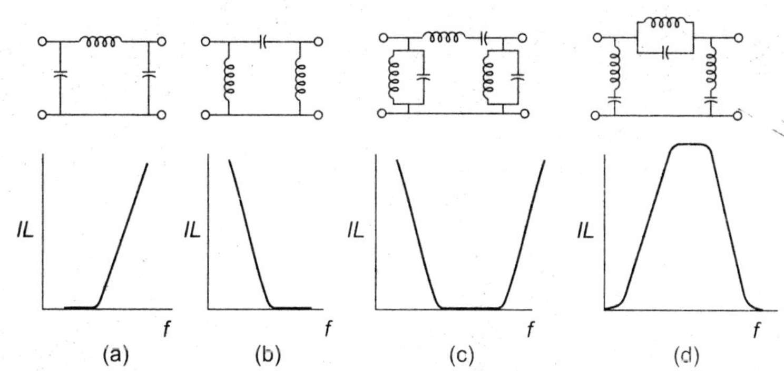

Fig. 8.2 Electrical equivalent circuits of filters and their responses: (a) low-pass (b) high-pass (c) bandpass (d) band stop

MISMATCH EFFECTS 8.3

A filter has a nominal characteristic resistive impedance R_0 such that when it is terminated by this impedance at both ends there is no reflection at either of the ports. If the filter is not matched terminated, there are mismatched effects as described below.

Consider that a lossless filter network with characteristic resistive impedance R_0 is terminated by a resistive load R_L and fed by a generator of voltage V_g and resistive impedance R_g as shown in Fig. 8.3, where $R_g \neq R_L \neq R_0$. The maximum power P_1 delivered to the load without the presence of the filter is given by

$$P_1 = i_L^2 R_L = \frac{V_g^2 R_L}{(R_g + R_L)^2} \tag{8.4}$$

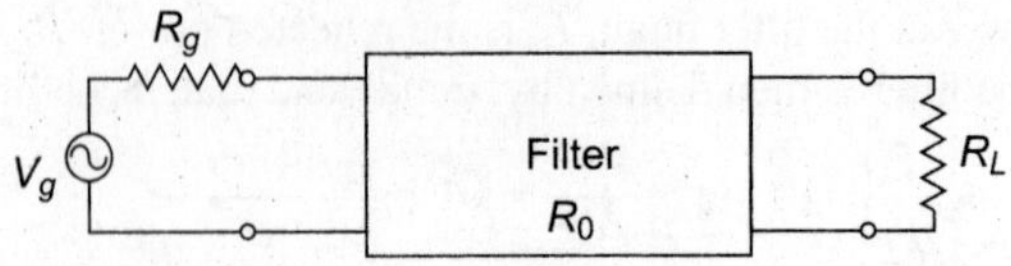

Fig. 8.3 Filter network with mismatch terminations

The power delivered to the load when the filter is inserted is expressed by

$$P_2 = \frac{|V_L|^2}{R_L} \tag{8.5}$$

The insertion loss of the filter is given by

$$IL_1\,(\text{dB}) = 10\log\frac{P_1}{P_2} = \alpha = 20\log\left(\frac{R_L}{R_g + R_L}\right) + 20\log\left(\frac{V_g}{V_L}\right) \tag{8.6}$$

If $$R_g = R_L \neq R_0,\; P_1 = \frac{V_g^2}{4R_L},\; P_2 = \frac{V_L^2}{R_L} \tag{8.7}$$

$$IL_2\,(\text{dB}) = 20\log\left(\frac{1}{2}\frac{V_g}{V_L}\right) = \alpha_0,\ \text{say} \tag{8.8}$$

$$\therefore\quad 20\log\left(\frac{V_g}{V_L}\right) = \alpha_0 + 6.021,\ \text{dB}$$

If the filter is matched to the line, $R_g = R_0 = R_L$, $V_L = V_g/2$ and

$$IL_3(\text{dB}) = 20\log\left(\frac{1}{2}\left|\frac{V_g}{V_L}\right|\right) = 0 \tag{8.9}$$

Therefore, the mismatched insertion loss is, in general,

$$IL\,(\text{dB}) = \alpha = 20\log\left(\frac{V_g}{V_L}\right) - 20\log\left(1 + \frac{R_g}{R_L}\right);$$

or, $$\alpha = \alpha_0 + 6 - 20\log\left(1 + \frac{R_g}{R_L}\right); \tag{8.10}$$

This shows that, if $R_L > R_g$, $\alpha > \alpha_0$. If $R_L < R_g$, $\alpha < \alpha_0$, the insertion loss becomes insertion gain, i.e., the output level becomes higher than input level in the pass-band.

MICROWAVE REALIZATION OF THE FILTER ELEMENTS 8.4

In microwave filters, lumped elements of the filter circuit are simulated by means of sections of waveguides, coaxial lines, strip or microstrip lines, cavity resonators, and resonant irises, etc. A few examples of these circuits are shown in Fig. 8.4 (a)-(h).

Figure 8.4 (a) shows that a shunt inductance is realized by means of a short-circuit stub of length less than the quarter wavelength formed in the narrow wall of a rectangular waveguide or by a step-in conductor

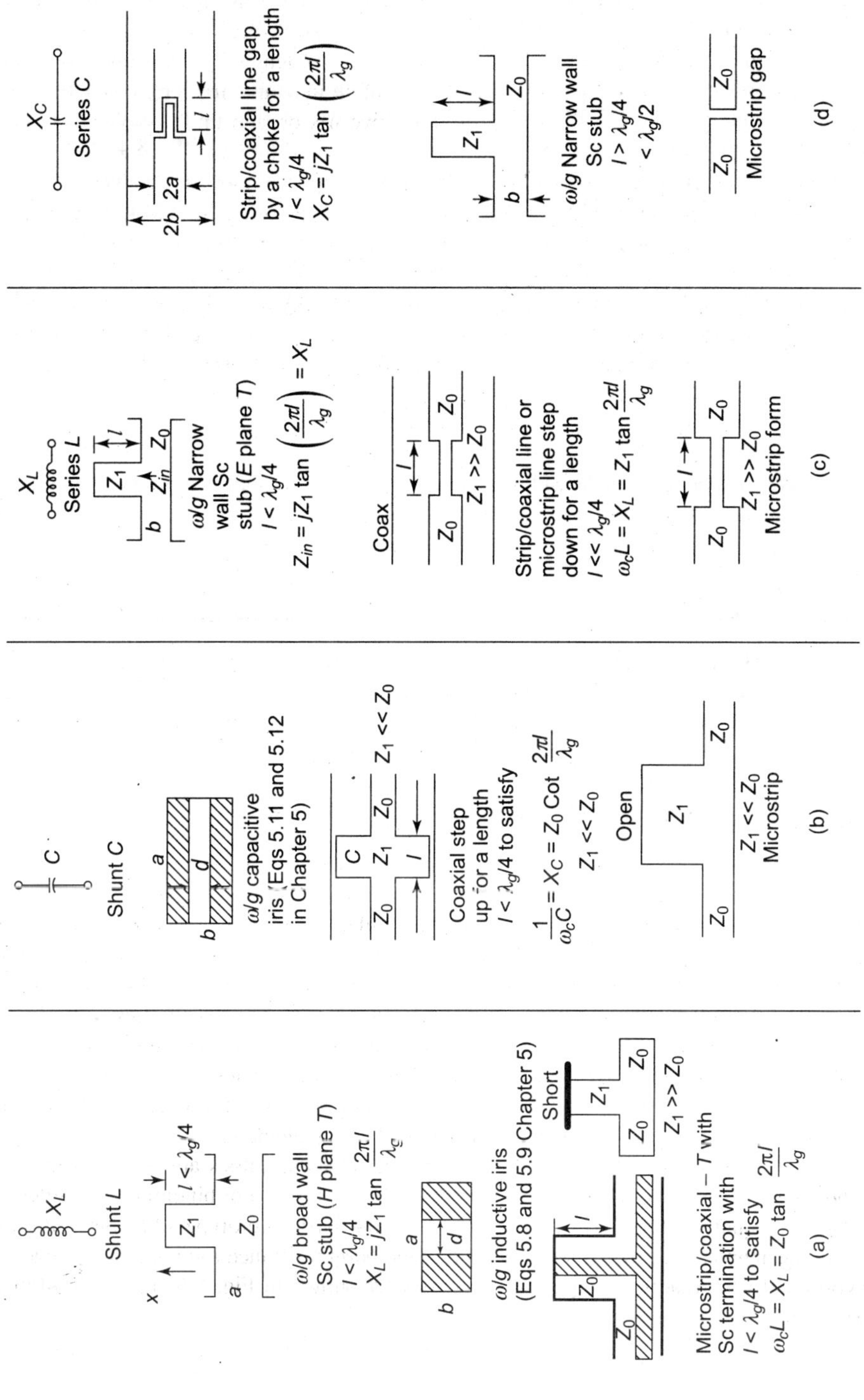

Fig. 8.4 Realization of lumped elements in microwaves

of coaxial and microstrip lines. A shunt capacitance can be realized by a similar stub of length greater than quarter wavelength but less than half wavelength or by a coaxial line gap formed by choke of length l less than a quarter wavelength or by a gap in microstrip lines as shown in Fig. 8.4 (b). Figure 8.4 (c) shows simulation of a series inductance by a short-circuit waveguide stub of length less than quarter wavelength in the plane parallel to the waveguide broadwall, or by means of an inductive iris, or a coaxial- microstrip-T. A shunt capacitance can be simulated by means of capacitive iris or step up is coaxial/microstrip lines. Similarly, LP and HP configurations are realized using this concept as shown in Fig. 8.4 (d).

The equivalent lumped element values of the microwave components are themselves functions of the frequency and have in general, an infinite number of poles or zeros, or both. Whereas, according to the *Foster's Reactance Theorem*, input impedance of lumped element filter circuits possess finite number of poles or zeros or both. Hence, the established synthesis procedure for the design of lumped element filters can be applied to design microwave filters only for a narrow frequency band. For physical realizability of a passive filter network, the magnitude of reflection coefficient $|\Gamma(\omega)| \le 1$ can be expressed as the ratio of two polynomials to result in the insertion loss expression in terms of even polynomials $M(\omega^2)$ and $N^2(\omega)$:

$$\text{Insertion loss} = 1 + \frac{M(\omega^2)}{N^2(\omega)} \tag{8.11}$$

Suitable forms for the polynomials M and N are chosen to obtain desired response characteristics in the filter design which are described below.

FILTER DESIGN 8.5

The two-filter synthesis techniques are popular. These are the *image parameter method,* and the *insertion loss method.* Out of these two methods, only the insertion loss method gives complete specifications of a physically realizable frequency characteristic over the entire pass and the stop bands from which the microwave filters are synthesized or designed most preferably.

Prototype Low-pass Design by Insertion Loss Method 8.5.1

Basic design of microwave filters of types low-pass, band-pass and band-stop, operating at arbitrary frequency bands and between arbitrary resistive loads, are made from a prototype low-pass design through some frequency transformation, element normalization and the simulation of these elements by means of sections of microwave transmission line. In this method, a physically realizable network is synthesized that will give the desired insertion loss vs frequency characteristics. This method consists of the following steps:

1. Design of a prototype low-pass filter with the desired pass band characteristics.
2. Transformation of this prototype network to the required type (low-pass, high-pass, band-pass, or band-stop) filter with the specified centre and band-edge frequencies.
3. Realization of the network in microwave form by using sections of microwave transmission lines whose reactances correspond to those of distributed circuit elements.

Figure 8.5 shows a prototype low-pass circuit where the element values are designed from standard low-pass response approximations such as ***Butterworth*** (or Maximally flat or binomial) *pass band response* and ***Chebyshev*** (or equal ripple) *pass-band response.* In the basic prototype filter derived from these approximations, the angular cut-off frequency ω'_c and termination resistance r are both normalized to unity. The Butterworth and Chebyshev response characteristics are shown in Fig. 8.6 and are described in the following paragraphs.

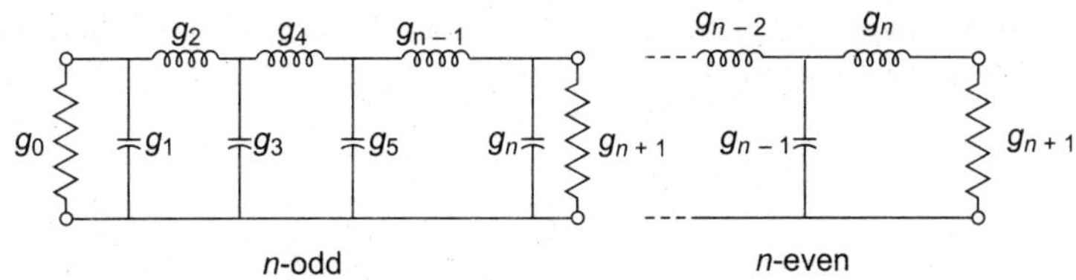

Fig. 8.5 Prototype low-pass filter

Butterworth Response

Insertion loss approximation for a low-pass Butterworth filter prototype is expressed by

$$IL = 1 + a_m^2\, \omega'^{2n}; \; \omega' = \omega/\omega_c \tag{8.12}$$

where the pass band ranges from $\omega = 0$ to $\omega = \omega_c$. Butterworth approximation exhibits a flat response in the pass-band and a monotonically increasing attenuation in the stop band. The maximum insertion loss in the pass band is 3 dB at ω_c so that $a_m^2 = 1$. The rate of increase of the insertion loss for $\omega > \omega_c$ depends on the exponent $2n$, which in turn is related to the number of filter sections used in a filter network of n reactive elements, L_x represents an insertion loss at a given frequency $\omega'_x = \omega_x/\omega_c$ in the stop band.

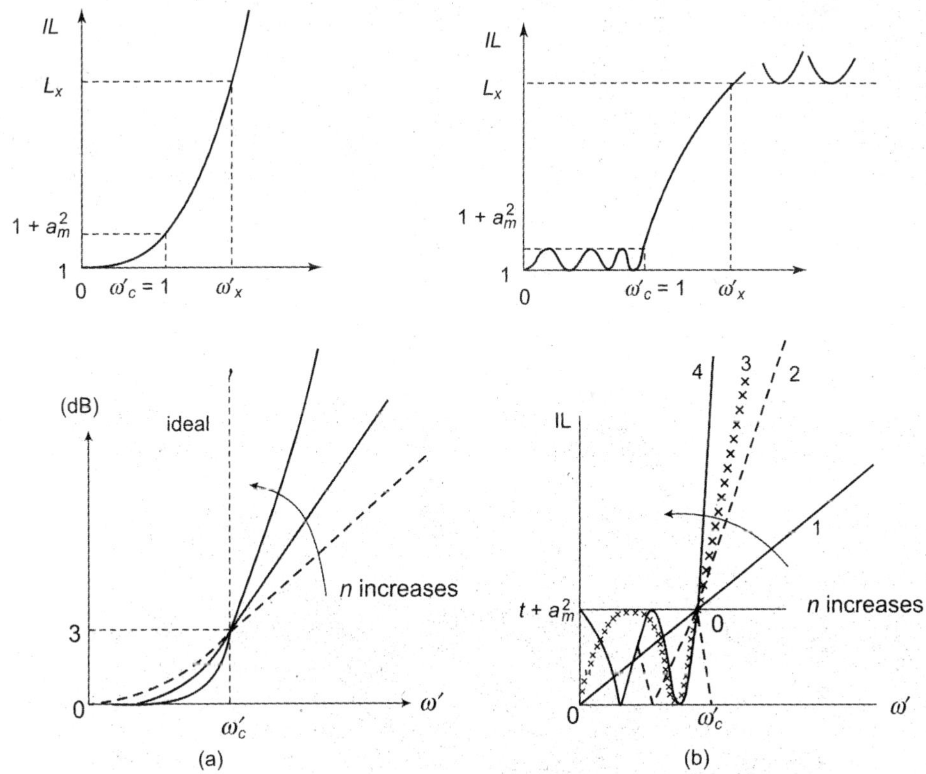

Fig. 8.6 IL responses of filters: (a) Butterworth (b) Chebyshev

Chebyshev Response

The approximation for a low-pass Chebyshev prototype is expressed by

$$IL = 1 + a_m^2\, T_n^2(\omega'); \; \omega' = \omega/\omega_c \tag{8.13}$$

where n denotes the degree of approximation (i.e., number of reactive elements) and a_m is the ripple factor. $T_n(\omega')$ is the Chebyshev polynomial of degree n given by

$$\begin{aligned} T_1(x) &= x, \\ T_2(x) &= 2x^2 - 1, \\ T_3(x) &= 4x^3 - 3x, \\ T_4(x) &= 8x^4 - 8x^2 + 1 \\ T_n(x) &= 2xT_{n-1}(x) - T_{n-2}(x) \\ &= \cos(n \cos^{-1}x) \text{ for } x \le 1 \text{ ports and} \\ &= \cosh(n \cosh^{-1}x) \text{ for } x > 1 \text{ stop band} \end{aligned} \tag{8.14}$$

where $x = \omega'$. The insertion loss oscillates between 1 and $1 + a_m^2$ in the pass-band, becomes $1 + a_m{}^2$ at the cut-off frequency ω_c and increases monotonically beyond cut-off (stop band $\omega' > 1$) at a much faster rate as compared with the Butterworth filter. L_x represents insertion loss at a frequency $\omega'_x = \omega_x/\omega_c$ in the stop band.

The element values of low-pass ladder network derived from Butterworth and Chebyshev responses for the normalized values of cut-off angular frequency $\omega'_c = 1$, and load resistance r = unity, can be calculated from the following equations:

Butterworth prototype element values

$$g_o = 1 = g_{n+1} \text{ for all } n$$

$$g_k = 2 \sin[(2k-1)\pi/2n],\ k = 1, 2, \ldots, n \tag{8.15}$$

Chebyshev prototype element values

$$g_o = 1 \text{ for all } n$$

$$g_1 = 2p_1/\sinh(\beta/2n)$$

$$g_k = \frac{4p_{k-i} \cdot p_k}{q_{k-i} \cdot g_{k-j}};\ k = 2, 3, 4, \ldots, n$$

$$\begin{aligned} g_{n+1} &= 1 \text{ for } n \text{ odd} \\ &= \coth^2(\beta/4) \text{ for even } n \end{aligned} \tag{8.16}$$

where,

$$p_k = \sin\frac{(2k-1)\pi}{2n},\ k = 1, 2, 3, \ldots, n$$

$$q_k = \sinh^2(\beta/2n) + \sin^2(k\pi/n),\ k = 1, 2, \ldots, n$$

$$\beta = \ln[\coth(A_m/17.37)];\ A_m \text{ in dB} \tag{8.17}$$

Here, A_m is the ripple height = a_m^2. The values of g_k can be obtained from tables of g_k as given below.

Table 8.1 *Values of g_k for Butterworth filter (IL = 3 dB at $\omega' = 1$)*

k	n = 1	2	3	4	5	6	7
1	2.0	1.414	1.0	0.7654	0.618	0.5176	0.445
2	1.0	1.414	2.0	1.8480	1.618	1.4140	1.247
3		1.000	1.0	1.8480	2.000	1.9320	1.802
4			1.0	0.7654	1.618	1.9320	2.000
5				1.0000	0.618	1.4140	1.802
6					1.000	0.5176	1.247
7						1.0000	0.445

Table 8.2 *Values of g_k for Chebyshev filter 0.1 dB ripple*

k	n = 1	2	3	4	5	6	7
1	0.3052	0.8430	1.0315	1.1468	1.1468	1.1681	1.1811
2	1.0000	0.6220	1.1474	1.3061	1.3712	1.4039	1.4228
3		1.3554	1.0315	1.7703	1.9750	2.0562	2.0966
4			1.0000	0.8180	1.3712	1.5170	1.5733
5				1.3554	1.1468	1.9029	2.0966
6					1.0000	0.8618	1.4228
7						1.3554	1.1811

0.2 dB ripple

k	n = 1	2	3	4	5	6	7
1	0.4342	1.0378	1.2275	1.3028	1.3394	1.3598	1.3722
2	1.0000	0.6745	1.1525	1.2844	1.3370	1.3632	1.3781
3		1.5386	1.2275	2.9761	2.1660	2.2934	2.2756
4			1.0000	0.8468	1.3370	1.4555	1.5001
5				1.5386	1.3394	2.0974	2.2756
6					1.0000	0.8838	1.3761
7						1.5386	1.3722

0.5 dB ripple

k	n = 1	2	3	4	5	6	7
1	0.6986	1.4029	1.5963	1.6703	1.7058	1.7254	1.7372
2	1.0000	0.7071	1.0969	1.1926	1.2296	1.2479	1.2583
3		1.9841	1.5963	2.3661	2.5408	2.6064	2.6381
4			1.0000	0.8419	1.2296	1.3137	1.3444
5				1.9841	1.7058	2.4758	2.6381
6					1.0000	0.8696	1.2583
7						1.9841	1.7372

Filter Transformations from Prototype 8.5.2

In order to design actual low-pass, high-pass, band-pass and band-stop filters, the transformations of the low-pass prototype filters with normalized cut-off frequency $\omega_c' = 1$ and having the source and load resistance of 1 ohm are made into the desired type with required source and load impedances using frequency and impedance transformations. These procedures are described below.

Low-pass Filters

For the design of a loss-pass filter from the low-pass prototype response parameters, frequency scaling is required to change the normalized cut-off frequency to absolute cut-off frequency ω_c by dividing all inductances and capacitances by ω_c without altering the resistances. Impedance scaling is done by changing the source and load resistances from 1 to R_L (with $R_L = R_G$), multiplying all resistances and inductances by R_L, and dividing all capacitances by R_L. The above two scalings are done simultaneously. Thus, the actual low-pass filter element values are

$$C_k = \frac{g_k}{\omega_c R_L} \text{ farad} \tag{8.18}$$

$$L_k = \frac{g_k R_L}{\omega_c} \text{ henry} \tag{8.19}$$

High-pass Filters

The low-pass prototype network is transformed into a high-pass filter by transforming series inductances into series capacitances and shunt capacitances into shunt inductances using the frequency transformation

$$\omega' = \frac{\omega_c}{\omega} \tag{8.20}$$

where ω_c and ω are the band edge and variable angular frequencies of the highpass filter. The element values are

$$C_k = \left(\frac{1}{g_k \omega_c R_L}\right) \tag{8.21}$$

$$L_k = \left(\frac{R_L}{g_k \omega_c}\right) \tag{8.22}$$

Band-pass Filter

To map the low-pass prototype to a band-pass filter, the following frequency transformation is used:

$$\omega' = \frac{\omega_0}{\omega_2 - \omega_1}\left(\frac{\omega}{\omega_0} - \frac{\omega_0}{\omega}\right); \quad \omega_0 = \sqrt{\omega_1 \omega_2} \tag{8.23}$$

where, ω_0, ω_1, and $\omega_2 - \omega_1$ are centre angular frequency, variable angular frequency and angular frequency bandwidth, respectively, and ω_1 and ω_2 are the angular frequency band limits.

Applying the frequency transformation to series inductances and shunt capacitances of the low-pass prototype gives

Series-tuned Series Elements

$$L_k = \frac{g_k R_L}{\omega_2 - \omega_1}; \quad C_k = \frac{\omega_2 - \omega_1}{\omega_0^2 g_k R_L} \tag{8.24}$$

Shunt-tuned Shunt Elements

$$L_k = \frac{R_L}{g_k}\left(\frac{\omega_2 - \omega_1}{\omega_0^2}\right); \quad C_k = \frac{g_k}{R_L}\left(\frac{1}{\omega_2 - \omega_1}\right) \tag{8.25}$$

where,

$$\omega_0^2 = \frac{1}{L_k C_k} \tag{8.26}$$

Band-stop Filters

The transformation from low-pass prototype to band-stop is given by

$$-\frac{1}{\omega'} = \frac{\omega_0}{\omega_2 - \omega_1}\left(\frac{\omega}{\omega_0} - \frac{\omega_0}{\omega}\right) \tag{8.27}$$

where all the quantities used in Eq. 8.27 are defined as in a band-pass filter. Here series inductance is mapped into a shunt-tuned circuit in series with the generator and load with element values

$$\omega_0 C_k = \frac{1}{\omega_0 L_k}; \quad C_k = \frac{1}{(\omega_2 - \omega_1) g_k R_L} \tag{8.28}$$

and shunt capacitance into a series-tuned circuit with element values

$$\omega_0 L_k = \frac{1}{\omega_0 C_k}; \quad L_k = \frac{R_L}{g_k}\left(\frac{1}{\omega_2 - \omega_1}\right) \tag{8.29}$$

The above results are summarized in Table 8.3.

Table 8.3 *Filter element values*

Proto-type element	*Low-pass filter elements*	*High-pass filter elements*	*Band-pass filter elements*	*Band-stop filter elements*
Series arm g_k	$L_k = \frac{g_k Z_L}{\omega_c}$	$C_k = \frac{1}{g_k Z_L \omega_c}$	$L_k = \frac{g_k R_L}{(\omega_2 - \omega_1)}$ $C_k = \frac{\omega_2 - \omega_1}{\omega_0^2 g_k R_L}$	$L_k = \frac{g_k Z_L(\omega_2 - \omega_1)}{\omega_0^2}$ $C_k = \frac{1}{g_k R_L (\omega_2 - \omega_1)}$
Shunt arm g_k	$C_k = \frac{g_k}{Z_L \omega_c}$	$L_k = \frac{Z_L}{g_k \omega_c}$	$L_k = \frac{R_L}{g_k}\left(\frac{\omega_2 - \omega_1}{\omega_0^2}\right)$ $C_k = \frac{g_k}{R_L}\left(\frac{1}{\omega_2 - \omega_1}\right)$	$C_k = \frac{g_k(\omega_2 - \omega_1)}{Z_L \omega_0^2}$ $L_k = \frac{R_L}{g_k(\omega_2 - \omega_1)}$

MICROWAVE LOW-PASS FILTERS 8.6

Low-pass filter elements can be designed from the low-pass prototype by using the scaled Eqs 8.18 and 8.19. Low-pass filters are conveniently constructed using TEM structures such as coaxial lines, strip lines and microstrip lines. Microstrip filters are used in satellite, airborne communication, and EW systems for their small size, light weight, and low cost. Coaxial line, strip line and microstrip filters can be realized using such line sections having appropriate impedances to simulate reactive elements. The design is a good approximation of an idealized lumped-element circuit. In coaxial-line and micro strip-line low-pass filters, the shunt capacitors can be realized by very short sections ($\ll \lambda_g/4$) of relatively low characteristic impedance line, and the series inductors by short sections ($< \lambda_g/4$) of relatively high characteristic impedance. Figure 8.7 shows realization of some filter elements.

From the basic transmission-line theory, the length of the predominantly inductive line and the predominatly capacitive line are obtained from the input reactance and susceptances formulae of the filter section, respectively.

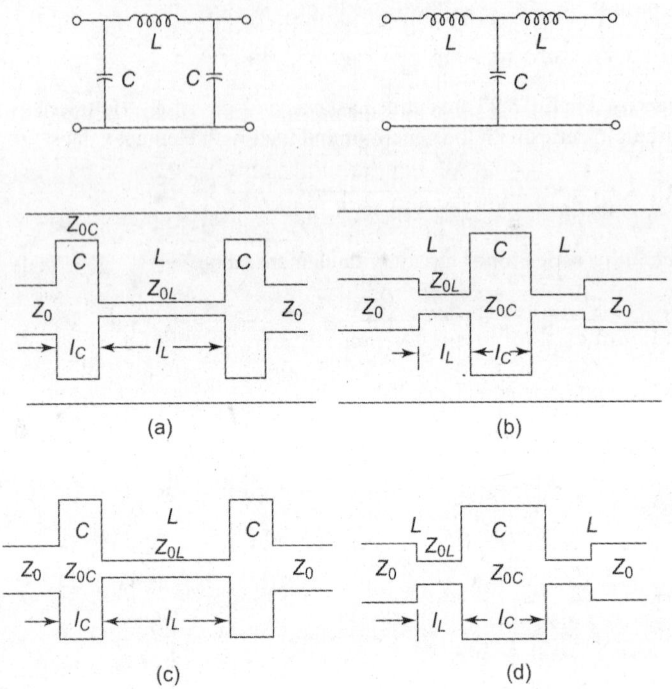

Fig. 8.7 Microwave low-pass filters: (a) Coaxial π-section (b) Coaxial T-section (c) Microstrip π-section (d) Microstrip T-section

Inductive Length

For inductive reactance X_L of short length l_L of a lossless line with short-circuit termination,

$$\omega_c L = Z_{0L} \tan \beta l_L, \quad l_L < l_g/4 \tag{8.30}$$

or

$$l_L = \frac{\lambda_{gL}}{2\pi} \tan^{-1}\left(\frac{\omega_c L}{Z_{0L}}\right) \approx f_c \lambda_{gL} L/Z_{0L} \tag{8.31}$$

Capacitive Length

For capacitive reactance X_c of a short length l_c of lossless line with open-circuit termination,

$$\frac{1}{\omega_c C} = Z_{0C} \cot \beta l_C \tag{8.32}$$

or

$$l_C = \frac{\lambda_{gC}}{2\pi} \cot^{-1}(\omega_c Z_{0C} C) \approx f_c \lambda_{gc} Z_{oc} C \tag{8.33}$$

where λ_g represents the corresponding guide wavelength, L and C are the inductance and capacitance required, and Z_0 represents the characteristic impedance of the corresponding section.

In practice, however, fringe capacitance C_f at the ends of the inductive line must be taken into account, and must be added to C where C_f is given by

$$C_f = \frac{1}{\omega Z_{0L}}\left(\tan\frac{\pi l_L}{\lambda_{gL}}\right) \qquad l_c \approx \frac{\lambda_{gC}}{2\pi}\cot^{-1}\left[\omega_c Z_{0C}(C+C_f)\right] \tag{8.34}$$

Waveguide form of low-pass filters cannot be realized since waveguides are basically high-pass lines. A five-section low-pass microstrip filter configuration is shown in Fig. 8.8.

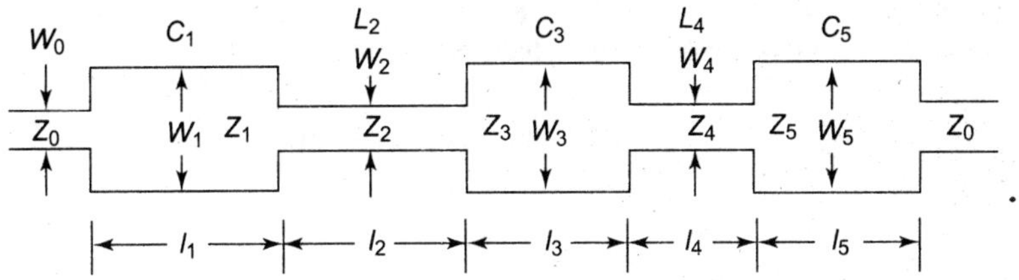

Fig. 8.8 Five-section low-pass microstrip filter

Example 8.1 Design a microstrip low-pass filter with 2 GHz cut-off frequency, 30 dB attenuation at 3.5 GHz frequency for Chebyshev attenuation response with 0.2 dB ripple. Use alumina substrate of thickness 0.63 mm.

Solution Given, $f_c = 2$ GHz, 30 dB attenuation frequency $f_x = 3.5$ GHz and $1 + a_m^2 = 0.2$ dB. For low-pass prototype attenuation,

$$A = 1 + a_m^2 T_n^2(\omega'); \ \omega' = \omega/\omega_c$$

For the given ripple magnitude

$$0.2 \text{ dB} = 10 \log (a_m^2 + 1) \quad \text{or,} \quad a_m^2 = 10^{0.2/10} - 1 = 0.047$$

Stop-band attenuation at f_x'.

$$30 \text{ dB} = 10 \log [1 + a_m^2 \cos h^2 (n \cos h^{-1} \omega_x')$$

or, $\omega_x' = \omega_x/\omega_c = 3.5/2 = 1.75$

Therefore, $1 + 0.047 \cos h^2 [n \cos h^{-1} (1.75)] = 10^{30/10} = 1000$

or, $\cos h^2 [n \cos h^{-1} (1.75)] = (1000 - 1)/0.047$

or, $n \approx 5$

Therefore, a five-section LP filter is required.

The prototype elements for $n = 5$, ripple = 0.2 dB, are $g_0 = 1 = g_6$, $g_1 = g_5 = 1.3394$, $g_2 = g_4 = 1.337$, $g_3 = 2.166$.

The lumped element values for cascaded π-sections for 50-ohm terminations are obtained as

$$C_k = \frac{g_k}{R_L \omega_C}$$

$$L_k = \frac{g_k R_L}{\omega_C} \quad \text{or,} \quad C_1 = C_5 = \frac{1.3394}{50 \times 2\pi \times 2 \times 10^9} \text{ F} = 2.132 \text{ pF}$$

$$C_3 = \frac{2.166}{50 \times 2\pi \times 2 \times 10^9} \text{ F} = 3.447 \text{ pF}$$

$$L_2 = L_4 = \frac{1.337 \times 50}{2\pi \times 2 \times 10^9} = 5.32 \text{ nH}$$

These lumped elements can be realized by sections of microstrip lines having length and width with corresponding suffices as shown in Fig. 8.8 where the capacitor and inductor line impedances are assumed $Z_{0c} = 20$ ohms $\ll Z_0$ and $Z_{0L} = 100$ ohms $\gg Z_0$. $Z_0 = 50$ ohms for input/output lines.

These elements are realized as follows:

For alumina substrate at 2 GHz, $\varepsilon_r = 9.9$. The effective permittivity

$$\varepsilon_{\text{eff}} = \frac{\varepsilon_r + 1}{2} + \frac{\varepsilon_r - 1}{2}\left[\left(1 + \frac{12\,h}{W}\right)^{-1/2} + 0.04(1 - W/h)^2\right]; \frac{W}{h} \le 1$$

$$= \frac{\varepsilon_r + 1}{2} + \frac{\varepsilon_r - 1}{2}(1 + 12h/W)^{-1/2}; \; W/h \ge 1$$

$$= \frac{9.9 + 1}{2} + \frac{9.9 - 1}{2}\left(1 + \frac{12h}{W}\right)^{-1/2} = 5.45 + 4.45\left(1 + \frac{12h}{W}\right)^{-1/2}$$

$$Z_0 = \frac{120\pi}{2\pi\sqrt{\varepsilon_{\text{eff}}}} \ln\left[\frac{8h}{W} + \frac{0.25h}{W}\right]; \; W/h \le 1$$

$$= \frac{120\pi}{\sqrt{\varepsilon_{\text{eff}}}}\left[\frac{W}{h} + 1.393 + 0.667 \ln\left(\frac{W}{h} + 1.444\right)\right]^{-1} \quad W/h \ge 1$$

$$\lambda_g = \frac{\lambda_0}{\sqrt{\varepsilon_{\text{eff}}}}$$

Design parameters at 2 GHz

$$\lambda_0 = 30/2 = 15 \text{ cm.} \quad h = 0.63 \text{ mm}$$

From the above equations for Z_0, ε_{eff} and λ_{g0} the following are found.

For $Z_0 = 50$ ohms, $W_0 = 0.6$ mm, $\varepsilon_{\text{eff}} = 6.66$, $\lambda_{g0} = 5.81$ cm

For $Z_{0c} = 20$ ohms, $W_{0C} = 2.64$ mm, $\varepsilon_{\text{eff}} = 7.71$, $\lambda_{gc} = 5.40$ cm

For $Z_{0L} = 100$ ohms, $W_{0L} = 0.075$ mm, $\varepsilon_{\text{eff}} = 6.03$, $\lambda_{gL} = 6.11$ cm

and

$$l_{c1} = l_{c5} = 4.59 \text{ mm}$$

$$L_{L1} = l_{L4} = 7.25 \text{ mm}$$

$$l_{c3} = 7.43 \text{ mm}$$

MICROWAVE HIGH-PASS FILTERS 8.7

High-pass microwave filters can be designed from the low-pass prototype by using the transformation (8.20)–(8.22) and the response characteristics of Fig. 8.9. In coaxial and microstrip forms, short ($< \lambda_g/4$) lengths of relatively high characteristic impedance tee-connected to the main line, approximates the shunt inductors. The series capacitors are obtained by very small ($\ll \lambda_g/4$) gaps in the line as shown in Fig. 8.9.

In waveguide form, the change in broadwall dimension may be utilized to simulate the lump equivalent circuit. The shunt inductor can be realized by an inductive iris or by a H-plane T-junction.

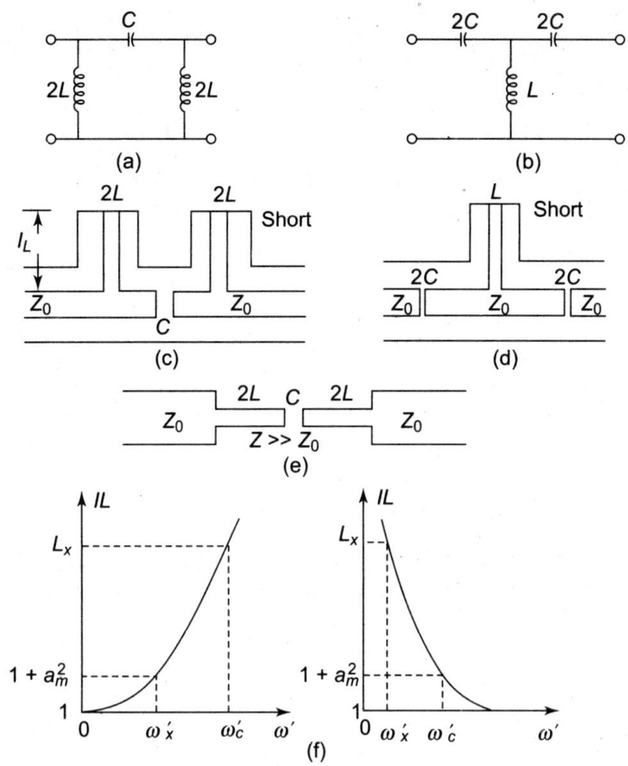

Fig. 8.9 High-pass filters: (a) π-section (b) T-section (c) Coaxial π-section (d) Coaxial T-section (e) Microstrip π-section (f) Butterworth response of prototype low-pass and high-pass

MICROWAVE BANDPASS FILTERS 8.8

Band-pass microwave filters are narrowband filters and can be designed from the low-pass prototype using the transformations (8.23)–(8.26). The low-pass prototype circuit in Fig. 8.5 can be transformed to band-pass filter circuit of Fig. 8.10 (a). The response characteristics of low-pass and its transformation to band-pass is shown in Fig. 8.10 (b). The resonator circuits of these filters are realized by the use of a cascaded strip line, coaxial line or cavity resonators of suitable configurations as described below.

Quarter-wave Coupled Cavity Bandpass Filters 8.8.1

These filters are realized by waveguide cavities coupled through irises as shown in Fig. 8.11 (a). The equivalent circuit for any kth section loaded with two identical inductive irises with normalized susceptances $-jb_k$ is also shown in Fig. 8.11 (b). Each pair of irises separated by distance l_k forms a cavity k. Mumford and Collin have given detailed analysis of such filters. For high Q, narrowband filter $b_k >> 1$ and is given by

$$b_k = 2(Q_k^2 - 1)^{1/2} \tag{8.35}$$

where for the kth resonator Q

$$Q_k = \frac{g_k}{2} \frac{\beta_0}{\beta_2 - \beta_1} \tag{8.36}$$

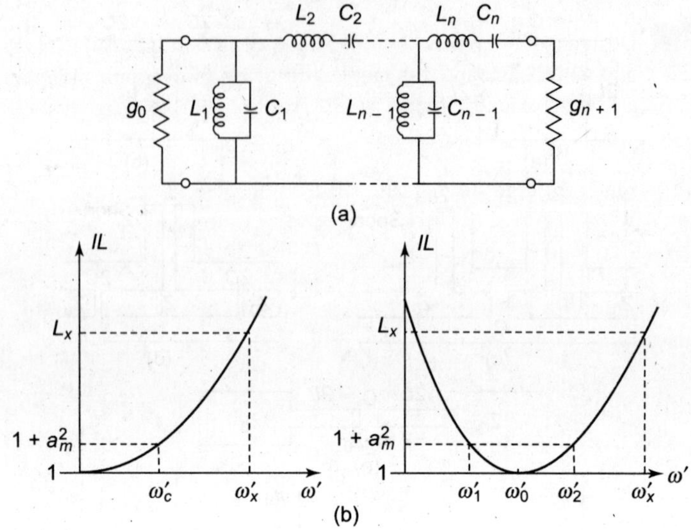

Fig. 8.10 (a) A bandpass filter circuit derived from the low prototype, (b) Bandpass response transformed from a low-pass prototype

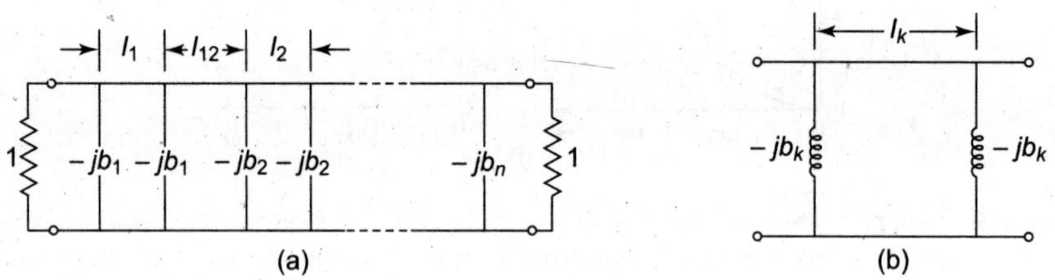

Fig. 8.11 (a) Quarter-wave coupled waveguide cavity filter (b) Single section equivalent circuit

and the length of the cavity

$$l_k = \frac{1}{\beta_0}\left[\tan^{-1}\left(\frac{-2}{b_k}\right)\right] \tag{8.37}$$

Here, g_k is the normalized value of the element of prototype filter, The physical length of the quarter-wave coupling line between the kth and $k + 1$ th cavities will be

$$l_{k,\,k+1} = \frac{l_k + l_{k+1}}{2} - \frac{\lambda_{g0}}{4} \tag{8.38}$$

The required iris dimensions can be found from Chapter 5. The insertion loss of the Chebyshev bandpass filter is given by

$$IL = 1 + a_m^2 T_n^2\left[\frac{\beta_0}{\beta_2 - \beta_1}\left(\frac{\beta}{\beta_0} - \frac{\beta_0}{\beta}\right)\right] \tag{8.39}$$

where $$\beta_0 = \sqrt{(\beta_1 \beta_2)} \tag{8.40}$$

Example 8.2 Design a quarter-wave coupled three-cavity filter having Chebyshev response with maximum passband ripple of 0.1 dB using a waveguide of dimensions $a = 2.286$ cm and $b = 1.016$ cm. The passband extends from $f_1 = 10$ GHz to $f_2 = 10.4$ GHz. Inductive diaphragm with circular holes are to be used.

Solution For narrowband and large susceptance values of the irises,

$$b_k = 2\,(Q_k^2 - 1)^{1/2}$$

$$Q_k = \frac{\beta_0}{\beta_2 - \beta_1} \times \frac{g_k}{2}$$

At centre frequency

$$\beta_0 = 2\pi/\lambda_{g0},\ \tan \beta_0 l_k = -2/b_k$$

Actual distance between cavity k and $k + 1$ is

$$l_{k,\,k+1} = \frac{l_k + l_{k+1}}{2} - \frac{\lambda_{g0}}{4}$$

For a three-cavity Chebyshev filter with $n = 3$, and ripple 0.1 dB, $a_m^2 = 0.0233$; $g_1 = 1.0315$, $g_2 = 1.1474$, $g_3 = 1.0315$.

Now, at band edges the propagation constants k_0 in free space and β in a rectangular waveguide ($a \times b$) excited in dominant TE_{10} mode are given by

$$k_{01} = \frac{2\pi \times 10 \times 10^9}{3 \times 10^{10}} = 2.0944$$

$$k_{02} = \frac{2\pi \times 10 \cdot 4 \times 10^9}{3 \times 10^{10}} = 2.1782$$

$$\beta_1 = [k_{01}^2 - (\pi/a)^2]^{1/2} = 1.5805$$

$$\beta_2 = [k_{02}^2 - (\pi/a)^2]^{1/2} = 1.6899$$

Centre of the band occurs at

$$\beta_0 = \sqrt{(\beta_1\, \beta_2)} = 1.6343$$

$$f_0 = \sqrt{(f_1\, f_2)} = 10.198 \text{ GHz}$$

$$\lambda_0 = \frac{30}{f_0} = 2.942 \text{ cm},\ \lambda_{g0} = \frac{2\pi}{\beta_0} = \frac{2 \times 3.1415}{1.6343} = 3.845 \text{ cm}$$

Now,

$$\frac{\beta_0}{\beta_2 - \beta_1} = \frac{1.6343}{1.6899 - 1.58011} = 14.9386$$

Therefore,

$$Q_1 = 14.9386 \times \frac{1.0315}{2} = 7.7046$$

$$Q_2 = 14.9386 \times \frac{1.1474}{2} = 8.5703$$

$$Q_3 = 14.9386 \times \frac{1.0315}{2} = 7.7046$$

and,

$$b_1 = 2 \times (7.7046^2 - 1)^{1/2} = 15.2788$$

$$b_2 = 2 \times (8.5703^2 - 1)^{1/2} = 17.0235$$
$$b_3 = b_1 = 15.2788$$

Calculation of cavity length l_k

$$\tan \beta_0 l_k = \frac{-2}{b_k} \quad \text{or} \quad \tan\left(\beta_0 \times \frac{180°}{\pi}\right) l_k = -\frac{2}{b_k}$$

$$\text{or,} \ \tan\left(\frac{1.6343 \times 180°}{3.1415}\right) l_k = -\frac{2}{b_k};$$

For $k = 1$,

$$\tan 93.64\, l_k = -2/b_k = -2/15.2788 = -0.1309 = -\tan 7.4576$$
$$= \tan(180° - 7.4576°) = \tan 172.5424°$$

Therefore, $l_1 = 172.5424/93.64 = 1.843$ cm

$$\tan 93.64\, l_2 = -2/b_2 = -2/17.0235 = -\tan 6.7006$$
$$= \tan(180° - 6.7006)$$
$$= \tan 173.299$$

Therefore, $l_2 = \dfrac{173.299}{93.64} = 1.851$ cm

$$l_3 = l_1 = 1.843 \text{ cm}$$

Distance between cavities

$$l_{k,\,k+1} = \frac{l_k + l_{k+1}}{2} - \frac{\lambda_{g0}}{4} = \frac{l_k + l_{k+1}}{2} - \frac{3.845}{4}$$

$$= \frac{l_k + l_{k+1}}{2} - \frac{1}{4} \times \frac{2\pi}{\beta_0} = \frac{l_k + l_{k+1}}{2} - 0.96114$$

$$l_{1,2} = \frac{l_1 + l_2}{2} - 0.96114 = \frac{1.843 + 1.851}{2} - 0.96114 = 0.885 \text{ cm}$$

$$l_{2,3} = \frac{l_2 + l_3}{2} - 0.96114 = \frac{1.851 + 1.843}{2} - 0.96114 = 0.885 \text{ cm}$$

Radii of diaphragm holes Assuming TE_{10} mode of excitation, radius of the kth hole is given by

$$r_k^3 = 3ab/(8\beta_0 b_k)$$

Therefore, $r_1 = \left[\dfrac{3 \times 2.286 \times 1.016}{8 \times 1.6343 \times 15.2788}\right]^{1/3} = 0.327$ cm

$$r_2 = \left[\frac{3 \times 2.286 \times 1.016}{8 \times 17.0235 \times 1.5343}\right]^{1/3} = 0.3152 \text{ cm}$$

$$r_3 = 0.3267 \text{ cm}$$

Example 8.3 Design a symmetrical three-section, maximally flat BP quarter-wave coupled filter such that the centre frequency in the passband and the bandwidth are 10,000 and 100 MHz, respectively.

Compare the insertion loss of this filter with that of a four-section filter in the same pass and attenuation bands in the frequency range 9800 to 10,200 MHz.

Solution

$$\text{Fractional bandwidth of the filter} = \frac{\text{bandwidth}}{\text{centre frequency}} = \frac{100}{10{,}000} = \frac{1}{100} = 1\%$$

In our case, $n = 3$ (odd) and we assume maximally flat response with 3 dB insertion loss at the band edges. Hence elements of *LP* prototype are

$$g_0 = 1,\ g_1 = 1,\ g_2 = 2,\ g_3 = 1$$

Given, $f_0 = 10{,}000 \text{ MHz} = \sqrt{(f_1 f_2)} = 10^4$, and

$$f_2 - f_1 = 100 \text{ MHz}$$

Therefore, $f_2 = 10.05$ GHz

$$f_1 = 9.95 \text{ GHz}$$

X-band waveguide dimensions

$$a = 0.9'' = 2.286 \text{ GHz}$$

$$b = 0.4'' = 1.016 \text{ GHz}$$

$$\beta_0 = 2\pi\sqrt{\left[\left(\frac{f_0}{c}\right)^2 - \left(\frac{1}{2}a\right)^2\right]} = 2\pi\sqrt{\left[\left(\frac{10}{30}\right)^2 - \left(\frac{1}{2\times 2.286}\right)^2\right]}$$

$$= 1.5805 \text{ rad/cm}$$

$$\lambda_{g0} = \frac{2\pi}{\beta_0} = \frac{2\pi}{1.5805} = 3.975 \text{ cm}$$

$$\beta_1 = 2\pi\sqrt{\left[\left(\frac{f_1}{c}\right)^2 - \left(\frac{1}{2}a\right)^2\right]} = 2\pi\sqrt{\left[\left(\frac{9.950}{30}\right)^2 - \left(\frac{1}{2\times 2.286}\right)^2\right]}$$

$$= 1.5665 \text{ rad/cm}$$

$$\beta_2 = 2\pi\sqrt{\left[\left(\frac{f_2}{c}\right)^2 - \left(\frac{1}{2}a\right)^2\right]} = 2\pi\sqrt{\left[\left(\frac{10.05}{30}\right)^2 - \left(\frac{1}{2\times 2.286}\right)^2\right]}$$

$$= 1.5943 \text{ rad/cm}$$

$$\frac{\beta_0}{\beta_2 - \beta_1} = \frac{1.5805}{1.5943 - 1.5665} = 56.8225$$

Loaded Q of the sections are

$$Q_1 = \frac{\beta_0}{\beta_2 - \beta_1}\frac{g_1}{2} = 28.4263$$

$$Q_2 = \frac{\beta_0}{\beta_2 - \beta_1}\frac{g_2}{2} = 56.8525$$

$$Q_3 = \frac{\beta_0}{\beta_2 - \beta_1}\frac{g_3}{2} = 28.4263$$

Therefore, for hole sincefitances

$$b_1 = 2\,(Q_1^2 - 1)^{1/2} = 56.8173$$

$$b_2 = 2\,(Q_2^2 - 1)^{1/2} = 113.6874$$

$$b_3 = 2\,(Q_3^2 - 1)^{1/2} = 56.8173$$

Calculation of l_k at $f = f_0$

$$\tan \beta_0 l_k = -2/b_k$$

or, $\tan\,(90.555\, l_1)^\circ = -2/b_1 = \tan\,(177.98^\circ)$

or, $(90.555\, l_1)^\circ = 177.98^\circ$

or, $l_1 = 177.98^\circ/90.555 = 1.965$ cm

$\tan\,(90.555\, l_2)^\circ = -2/b_2 = \tan\,(178.992^\circ)$

or, $(90.555\, l_2)^\circ = 178.992^\circ$

$l_2 = 178.992/90.555 = 1.976$ cm

Since $b_1 = b_3,\; l_3 = l_1 = 1.965$ cm

Calculation of $l_{k,k+1}$ at f_0

$$l_{1,2} = \frac{l_1 + l_2}{2} - \frac{\lambda_{g0}}{4} = 0.9767 \text{ cm}$$

$$l_{2,3} = \frac{l_2 + l_3}{2} - \frac{\lambda_{g0}}{4} = 0.9767 \text{ cm}$$

Radius 'r' of diaphragm

Assuming TE_{10} mode is excited:

$$b_k = \frac{3\,ab}{8\,\beta_0\, r_k^3}$$

Therefore,

$$r_k = \left[\frac{3\,ab}{8\,\beta_0\, b_k}\right]^{1/3}$$

$$r_3 = r_1 = \left[\frac{3 \times 2.286 \times 1.016}{8 \times 1.5805 \times 56.8173}\right]^{1/3} = 0.2136 \text{ cm}$$

$$r_2 = \left[\frac{3 \times 2.286 \times 1.016}{8 \times 1.5805 \times 113.6874}\right]^{1/3} = 0.1693 \text{ cm}$$

Insertion loss vs frequency

$$IL = 1 + (\omega')^{2n} = 1 + \left[\frac{f_0}{f_2 - f_1}\left(\frac{f}{f_0} - \frac{f_0}{f}\right)\right]^{2n}$$

$$= \left[1 + 10^{12}\left(\frac{f}{10} - \frac{10}{f}\right)^6\right]; \text{ Three-cavity filter}$$

$$= \left[1 + 10^{16}\left(\frac{f}{10} - \frac{10}{f}\right)^8\right]; \text{ Four-cavity filter}$$

The results are tabulated below,

f(GHz)	*Band*	*IL(dB) (3 cavity)*	*IL(dB) (4 cavity)*
9.8	Attenuation	36.4	48.5
9.9	Attenuation	18.2	24.3
9.95	Bandedge	3.0	3.05
10.0	Centre	0.0	0.0
10.05	Bandedge	3.0	3.0
10.1	Attenuation	17.0	24.0
10.2	Attenuation	35.9	47.8

Therefore, insertion loss characteristic is more sharp in a four-cavity filter.

Direct Coupled Cavity Band-pass Filters 8.8.2

Direct coupled cavity filters consist of a number of cavity resonators directly coupled by inductive irises as shown in Fig. 8.12. Due to zero separation between the cavities these filters are more compact than the corresponding quarter-wave coupled cavity filters. The design formulae are given below based on *S.B. Cohn's* [6] *design method.*

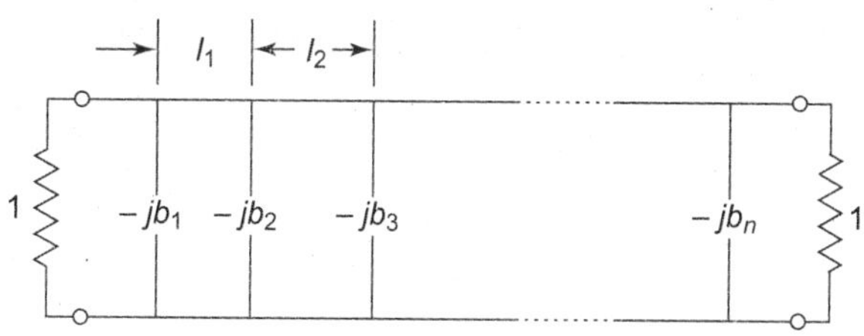

Fig. 8.12 Direct coupled cavity band-pass filter

The normalized susceptances of the irises are

$$b_1 = \frac{1 - \omega/g_1}{\sqrt{(\omega/g_1)}} \tag{8.41}$$

$$b_k = \frac{1}{\omega}\left(1 - \frac{\omega^2}{g_k g_{k-1}}\right)\sqrt{g_k g_{k-1}} \tag{8.42}$$

$$b_n = \frac{1 - \omega r/g_n}{\sqrt{(\omega r/g_n)}} \tag{8.43}$$

where
$$\omega = \frac{\pi}{2}\frac{\beta_2 - \beta_1}{\beta_0} \tag{8.44}$$

Here, $r = 1$ for odd n in Chebyshev filter and for all integer values of n for Butterworth filter.

For Chebyshev filter with even n

$$r = 2a_m^2 + 1 - 2a_m\sqrt{(1 + a_m^2)} \tag{8.45}$$

$$\omega' = \frac{\beta_0}{\beta_2 - \beta_1}\left(\frac{\beta}{\beta_0} - \frac{\beta_0}{\beta}\right) \tag{8.46}$$

and g_k are the element values from the low-pass prototype filter. The length of the kth cavity at centre frequency f_0 is

$$l_k = \frac{\lambda_{g0}}{2} - \frac{\lambda_{g0}}{4\pi}\left[\tan^{-1}\frac{2}{b_{k+1}} + \tan^{-1}\frac{2}{b_k}\right] \tag{8.47}$$

Example 8.4 Design a four-cavity direct coupled microwave filter of Chebyshev response where the passband is to extend from f_1 = 9500 MHz to f_2 = 10,500 MHz with 0.1 dB ripple.

Solution The values of $k_0 = \omega/c$ at band edge frequencies f_1 and f_2 are

$$k_{01} = \left[\frac{2 \times 3.1416 \times 0.95 \times 10^{10}}{3 \times 10^{10}}\right] = 1.9897$$

$$k_{01} = \left[\frac{2 \times 3.1416 \times 1.05 \times 10^{10}}{3 \times 10^{10}}\right] = 2.1991$$

The corresponding values of β_1 and β_2 for a rectangular waveguide ($a \times b$) are

$$\beta_1 = [k_{01}^2 - (\pi/a)^2]^{1/2} = 1.4468$$

$$\beta_2 = [k_{02}^2 - (\pi/a)^2]^{1/2} = 1.7235$$

where $a \approx 2.3$ cm and $b \approx 1$ cm.

The centre of the band occurs at

$$\beta_0 = (\beta_1\beta_2)^{1/2} = 1.5791$$

or, $$f_0 = \sqrt{(f_1 f_2)} = 9987.5 \text{ MHz}$$

From the table of values of g_k for the Chebyshev filter with 0.1 dB ripple and $n = 4$,

$$g_1 = 1.1088$$
$$g_2 = 1.3061$$
$$g_3 = 1.7703$$
$$g_4 = 0.8180$$

The value of ω is calculated from the formula

$$\omega = \frac{\pi}{2}\cdot\frac{\beta_2 \sim \beta_1}{\beta_0} = 0.2752$$

Ripple 0.1 dB = $10 \log (1 + a_m^2)$

or, $$a_m^2 = 0.0233.$$

Because n is even and the response required is Chebyshev, the source impedance $r \neq 1$ but is given by the relation

$$r = 2a_m^2 + 1 - 2a_m\sqrt{(1 + a_m^2)}$$

$$= 2 \times 0.0233 + 1 - 2\sqrt{(0.0233)}\sqrt{(1+0.0233)} = 0.7377$$

Calculation of b_k

$$b_1 = \frac{1-\omega/g_1}{\sqrt{(\omega/g_1)}} = \frac{1-0.45036/1.1088}{\sqrt{0.4536/1.1088}} = 0.9318$$

$$b_2 = \frac{1}{\omega}\left(1-\frac{\omega^2}{g_1 g_2}\right)\sqrt{(g_2 g_1)}$$

$$= \frac{1}{0.45036}\left(1-\frac{(0.45036)^2}{1.1088\times 1.3061}\right)\sqrt{(1.1088\times 1.3061)} = 2.298$$

$$b_3 = \frac{1}{\omega}\left(1-\frac{\omega^2}{g_3 g_2}\right)\sqrt{(g_3 g_2)} = 3.080209$$

$$b_4 = \frac{1-\omega r/g_4}{\sqrt{\omega r/g_1}} = \frac{1-0.45036\times 0.7377/0.8180}{\sqrt{[0.45036\times 0.7377/0.8180]}} = 0.9318$$

The length of the kth cavity at $\beta = \beta_0$ is

$$l_k = \frac{\lambda_{g0}}{2} - \frac{\lambda_{g0}}{4\pi}\left(\tan^{-1}\frac{2}{b_{k+1}} + \tan^{-1}\frac{2}{b_k}\right)$$

$$\lambda_0 = \frac{2\pi}{[\beta_0^2 + (\pi/a)^2]^{1/2}} = 3.001\text{cm}$$

and $\quad f_0 = 9.6721 \times 10^9$ Hz

Now $\quad \lambda_{g0} = \dfrac{\lambda_0}{\sqrt{[1-(\lambda_0/2a)^2]}} = \dfrac{3.001}{\sqrt{[1-(3.001/2\times 2.3)^2]}} = 4.2218$ cm

$$l_1 = \frac{\lambda_{g0}}{2} - \frac{\lambda_{g0}}{4\pi}\left[\tan^{-1}\frac{2}{2.2988} + \tan^{-1}\frac{2}{0.9318}\right] = 0.1575 \text{ cm}$$

$$l_2 = \frac{\lambda_{g0}}{2} - \frac{\lambda_{g0}}{4\pi}\left[\tan^{-1}\frac{2}{3.08021} + \tan^{-1}\frac{2}{2.298}\right] = 0.7471 \text{ cm}$$

$$l_3 = \frac{\lambda_{g0}}{2} - \frac{\lambda_{g0}}{4\pi}\left[\tan^{-1}\frac{2}{0.9318} + \tan^{-1}\frac{2}{3.08021}\right] = 0.8718 \text{ cm}$$

To calculate diaphragm dimensions, we use circular irises for which

$$r_0^3 = 3ab/(8b_k\beta_0)$$

For $\quad a = 0.9''$ and $b = 0.4''$

$$r_{01} = \left[\frac{3\times 0.9\times 0.4\times (2.54)^2}{8\times 0.9318\times 1.48826}\right]^{1/3} = 0.85635 \text{ cm.}$$

$$r_{02} = \left[\frac{3\times 0.9\times 0.4\times (2.54)^2}{8\times 2.2988\times 1.48826}\right]^{1/3} = 0.63378 \text{ cm.}$$

$$r_{03} = \left[\frac{3\times0.9\times0.4\times(2.54)^2}{8\times3.08021\times1.48826}\right]^{1/3} = 0.57489 \text{ cm.}$$

$$r_{04} = 0.70341$$

Thus, the radii and the spacings are

$r_{01} = 0.85635$ cm $\qquad l_1 = 0.1575$ cm

$r_{02} = 0.63378$ cm $\qquad l_2 = 0.7471$ cm

$r_{03} = 0.57489$ cm $\qquad l_3 = 0.8718$ cm

$r_{04} = 0.70341$ cm

This is the final form of the filter.

Microstrip Bandpass Filters 8.8.3

Bandpass filters can be constructed in microstrip forms by using a combination of low and high impedance lines with different terminations as shown in Fig. 8.13 for simulation of series and parallel resonance circuits.

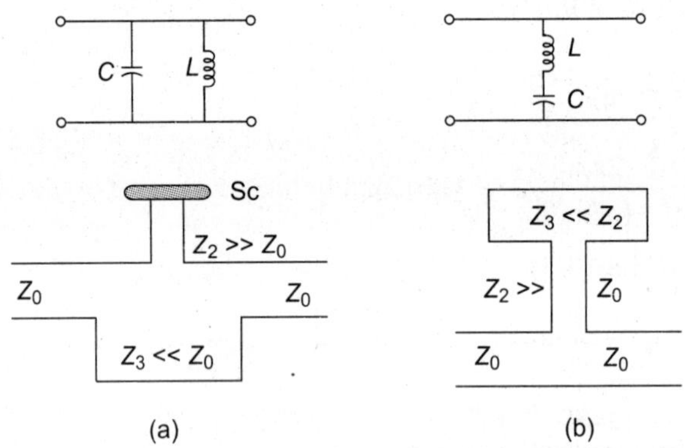

Fig. 8.13 Bandpass simulation

Other realization with end coupled, parallel coupled, and interdigital configurations are shown in Fig. 8.14 (a) and (b) and 8.18. Parallel coupling is much stronger than end coupling so that the realizable bandwidths are greater. The most compact configuration is the interdigital one. The design procedure of these filters are described below.

End-coupled Microstrip Bandpass Filters

Figure 8.14(a) shows the general layout of an end-coupled microstrip bandpass filter.

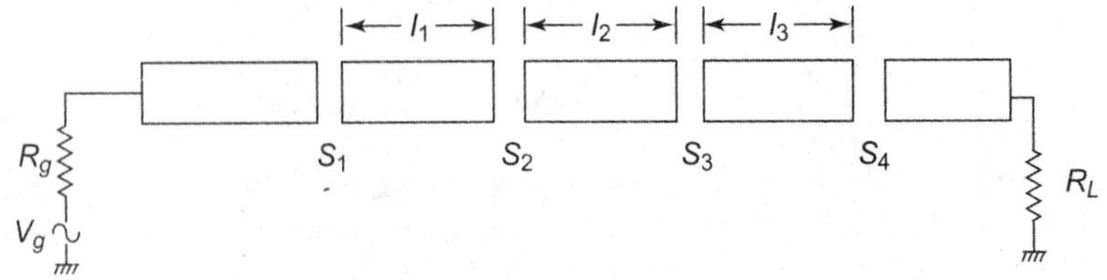

Fig. 8.14 (a) End-coupled microstrip bandpass filter

Here, the tuned circuits are realized from open or short-circuited transmission line sections of one quarter or half wavelength long. The resonators are coupled by means of gap capacitances between the resonator sections. The resonator lengths l and the coupling gaps S between successive resonators are important design parameters. To reduce insertion loss in the pass-band, the gaps are usually much smaller than the substrate height to enable tight coupling. The resonator lengths depend on the guide wavelength, coupling reactance and the gap capacitance. This configuration provides relatively narrow bandwidth. Since this structure is large, it is not a much preferred configuration.

Parallel-coupled Bandpass Filter

To achieve a more compact structure than an end-coupled filter, parallel-coupled microstrip configuration as shown in Fig. 8.14 (b) is preferred. As discussed in the theory of couplers in Chapter 6, since the maximum coupling occurs over a quarter-wave long coupling region, to achieve resonance, each resonator element is nearly half-wavelength long at the centre frequency and kept open at both ends, where $l_1, l_2 \ldots\ldots l_n \cong \lambda_g/4$.

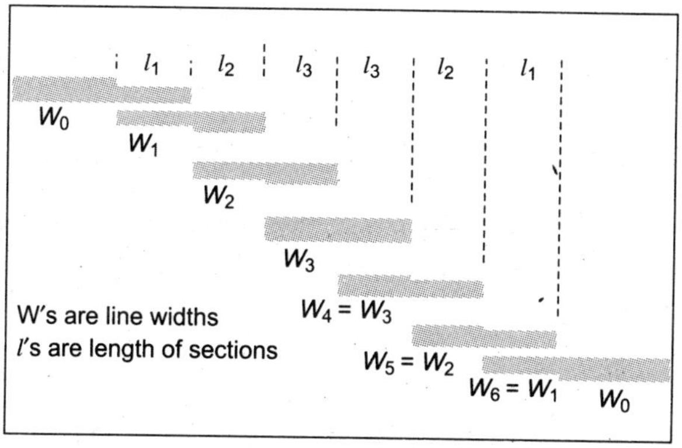

Fig. 8.14(b) Parallel-coupled bandpass filter

Detailed derivation of the coupling and impedance of parallel coupled filters have been given by Matthaei [15] using even and odd mode excitations which are summarized below for Chebyshev response of n section band-pass filter.

Let the filter be terminated in input and output lines with characteristic impedance Z_0. Each resonator has an electrical length of π at the centre frequency f_0. The maximum effective length l_e of each resonator is calculated after correcting for the capacitive end loading at each open-circuited end, such that

$$l_e = \frac{\lambda_g}{4} - l_{e0} \tag{8.48}$$

where l_{e0} is the hypothetical extension of the line to make up the end fringing given by [20],

$$l_{e0} = 4.12\, h \left(\frac{\varepsilon_{\text{eff}} + 0.3}{\varepsilon_{\text{eff}} - 0.258} \right) \left(\frac{W/h + 0.262}{W/h - 0.813} \right) \tag{8.49}$$

and due to Cohn [7],

$$\left(\frac{l_{e0}}{h} \right)_{\max} = \frac{2}{\pi} \ln 2 \approx 0.441 \tag{8.50}$$

The bandpass filter circuit consisting of parallel resonant circuit in the shunt arm and series resonant circuit in the series arm can be converted to either cascade of all parallel resonator networks together with impedance inverters K or to cascade of all series resonator networks together with admittance inverters J as shown in Fig. 8.15, where the inverters are basically quarter-wave transformers to transform any load impedance by the quantity Z'^2, where Z' is the characteristic impedance of the quarter-wave transforming line. In this process, (1) the resonant frequency must remain constant : $\omega_0^2 = 1/LC$ and, (2) impedance at similar planes in each network must be equal.

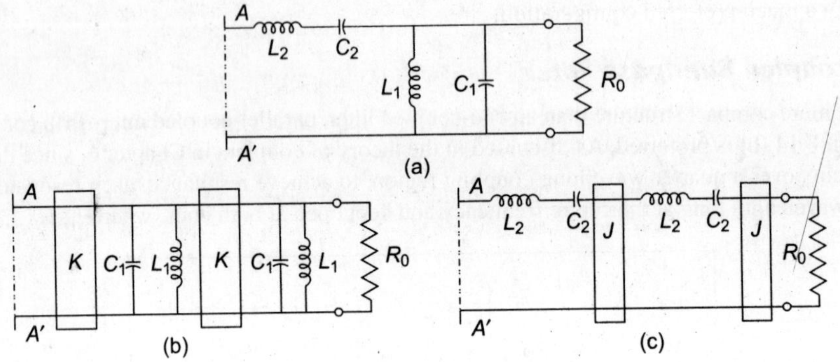

Fig. 8.15 Impedance and admittance inverters: (a) A section of band-pass filter (b) K-inverter (c) J-inverter

The inverter concept can be explained by considering a quarter-wave lossless line of characteristic impedance Z_0 terminated in an impedance Z_L. The input impedance in the line

$$Z_i = \frac{Z'^2}{Z_L} \tag{8.51}$$

Normalizing by Z',

$$\frac{Z_i}{Z'} = \frac{1}{Z_L/Z'} \quad \text{or,} \quad z_i = 1/z_L \tag{8.52}$$

where normalized impedances are $z_i = Z_i/Z'$, $z_L = Z_L/Z'$. This concept is represented by the following equivalent circuit of a simple L-section as shown in Fig. 8.16, where the impedances shown are normalized with respect to characteristic impedance of the quarter-wave section.

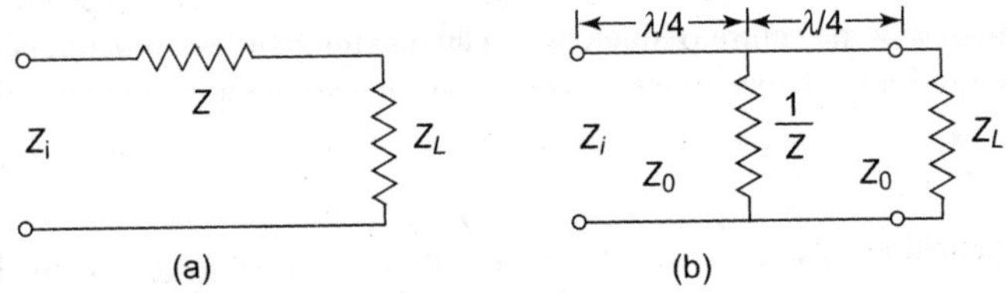

Fig. 8.16 Quarter-wave coupling network: (a) Lumped element circuit (b) Quarter-wave coupling equivalent

The details of impedance and admittance inverter formulae are given by Matthaei et al [15] and Collin [8] which are summarized below without any derivation.

For all series network configurations with admittance inverters J, the admittances of the inverters are as follows.

For the first coupling section,

$$\frac{J_{1,0}}{Y_0} = \left(\frac{\pi F}{2 g_0 g_1}\right)^{1/2} \tag{8.53}$$

For the final coupling section,

$$\frac{J_{n+1.n}}{Y_0} = \left(\frac{\pi F}{2g_n\, g_{n+1}}\right)^{1/2} \tag{8.54}$$

For the intermediate coupling section,

$$\frac{J_{k+1.k}}{Y_0} = \left(\frac{\pi F}{2\omega_c'\sqrt{g_k g_{k+1}}}\right), \quad k = 1 \text{ to } n-1 \tag{8.55}$$

where $F = (f_2 - f_1)/f_0$ the fractional bandwidth.

The frequency transformation from the low-pass prototype to the bandpass is then

$$\omega' = \frac{\omega_0}{\omega_2 - \omega_1}\left(\frac{\omega}{\omega_o} - \frac{\omega_o}{\omega}\right).$$

$$\omega_0 = \sqrt{\omega_1\omega_2}$$

In the microstrip filter, the line dimensions are determined from the knowledge of even- and odd-mode coupled line impedances which are given in terms of admittances of the inverters

$$(Z_0^e)_{k+1,\,k} = Z_0(1 + J_{k+1,\,k}\,Z_0 + J^2_{k+1\,k,}\,Z_0^{\,2}) \tag{8.56a}$$

and $$(Z_0^o)_{k+1,\,k} = Z_0(1 - J_{k+1,\,k}\,Z_0 + J^2_{k+1\,k,}\,Z_0^{\,2}) \tag{8.56b}$$

where Z_0 is the characteristic impedance of the input/output lines of the filter.

Parallel-coupled Bandpass Filter Simulation

In parallel-coupled microstrip filter design, it will be assumed that the odd- and even-mode phase velocities are equal and the characteristic impedance $Z_0 \approx \sqrt{Z_0^e Z_0^o}$. Several workers described the variation of Z_0^o and Z_0^e as functions of *W/h*, *S/h* and ε_r. Bryant and Weiss [4] have given a rigorous solution of the electrostatic problem for coupled microstrip lines and results are shown graphically for $\varepsilon_r = 1.0$ and 9.0 as shown in Fig. 8.17.

Akhtarzad et al [1] described a method to find *W/h* and *S/h* for the coupled structure which are independent of the substrate ε_r for a considerable range.

In this method, the following stages are involved:

1. Determination of $(W/h)_s$ of equivalent single microstrip line.
2. Determination of *W/h* and *S/h* for the coupled microstrip lines using $(W/h)_s$ formed at stage 1.

For the stage 1, even and odd mode impedance for a single microstrip is given by

$$Z_{0s}^e = Z_0^e/2 \tag{8.57a}$$

$$Z_{0s}^o = Z_0^o/2 \tag{8.57b}$$

For these values of Z_{0s}^e and Z_{0s}^o, $(W/h)_s^e$ and $(W/h)_s^o$ are determined from graphs or from closed form expressions given in Chapter 3 (Z_0 vs *W/h*). The *W/h* and *S/h* ratios for the coupled lines then can be determined from Akhtarzad et al. [1] expressions:

$$(W/h)_s^e = \frac{2}{\pi}\cosh^{-1}\left(\frac{2d - g + 1}{g+1}\right) \tag{8.57c}$$

$$(W/h)_s^o = \frac{2}{\pi}\cosh^{-1}\left(\frac{2d - g - 1}{g - 1}\right) \tag{8.57d}$$

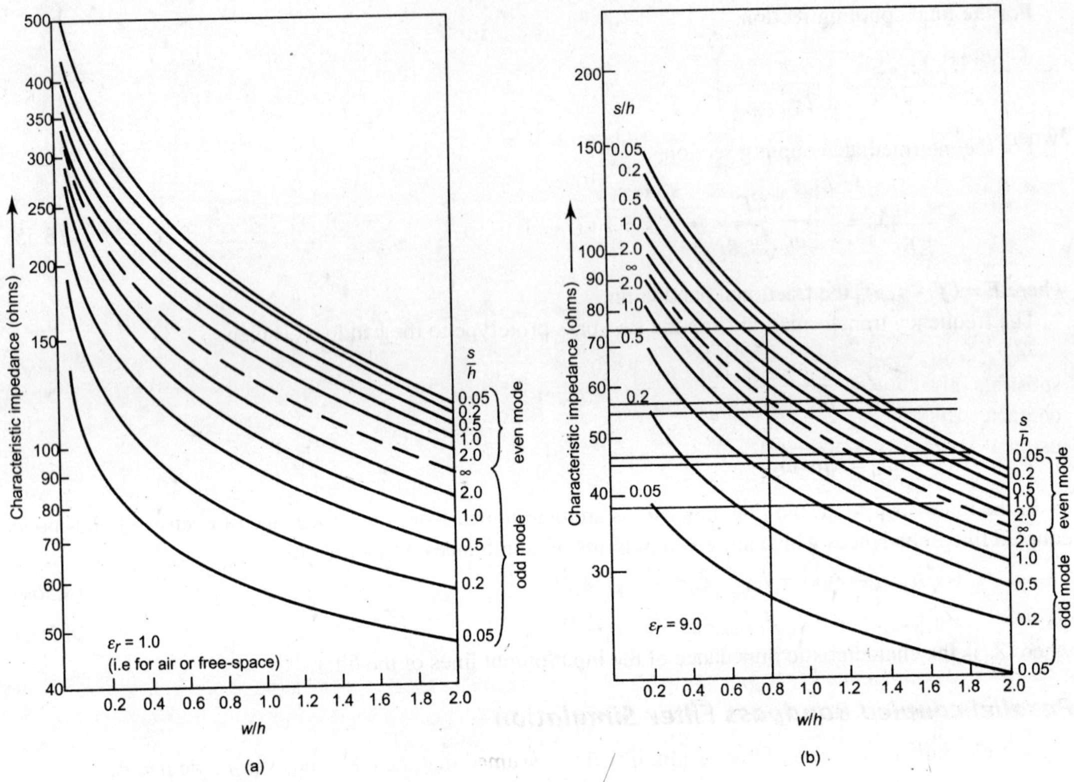

Fig. 8.17 Numerical results for the even- and odd-mode characteristic impedances of parallel-coupled microstrip lines: Source [4] 1968, IEEE

where
$$g = \cosh \cosh\left(\frac{\pi}{2}\cdot\frac{S}{h}\right) \tag{8.57e}$$

$$d = \cosh\left(\frac{\pi W}{h} + \frac{\pi}{2}\cdot\frac{S}{h}\right) \tag{8.57f}$$

Combining the above two equations, an explicit formula for couple line S/h is obtained:

$$\frac{S}{h} = \frac{2}{\pi}\cosh^{-1}\left[\frac{\cosh\left\{\frac{\pi}{2}\left(\frac{W}{h}\right)_s^e\right\} + \cosh\left\{\frac{\pi}{2}\left(\frac{W}{h}\right)_s^o\right\} - 2}{\cosh\left\{\frac{\pi}{2}\left(\frac{W}{h}\right)_s^o\right\} - \cosh\left\{\frac{\pi}{2}\left(\frac{W}{h}\right)_s^e\right\}}\right] \tag{8.57g}$$

The length of the coupled region is made $\lambda_g/4 = \dfrac{\lambda_0}{4\sqrt{\varepsilon_{\text{eff}}}}$ for maximum coupling at midband frequencies.

Therefore, the physical lengths of the resonators and the coupled regions are half and quarter wavelengths, respectively. Since W/h is different for different resonators, ε_{eff} of each resonator will have different values. Even- and odd-mode values are again used to find the length of each section.

Coupled region length $l = \dfrac{\lambda_{gm}}{4}$

where $\lambda_{gm} = \dfrac{\lambda_g^o + \lambda_g^e}{4}$ = Midband wavelength

λ_g^o = Odd-mode guide wavelength (mm) = $\dfrac{300}{f(\text{GHz})}\dfrac{Z_0^o}{Z_{0\text{air}}^o}$

λ_g^e = Even-mode guide wavelength (mm) = $\dfrac{300}{f(\text{GHz})}\dfrac{Z_0^e}{Z_{0\text{air}}^e}$

$Z_{0\text{ air}}^e$, $Z_{0\text{ air}}^o$ = even- and odd-mode characteristic impedances of the air-spaced lines without the actual substrate present for the specified designed values of *W*/*h* and *S*/*h*. Air-spaced even- and odd-mode characteristic impedances $Z_{0\text{ air}}^e$ and $Z_{0\text{air}}^o$ can be obtained from the single-line parameters in the same manner using analytic expressions of Chapter 3 for Z_0 vs *W*/*h* ($\varepsilon_r = 1$):

$$Z_{0\text{ air}}^e = 2\,Z_{0s\text{air}}^e$$

$$Z_{0\text{ air}}^o = 2\,Z_{0s\text{air}}^o$$

Subsequently, guide wavelengths and couple line lengths are determined.

The design steps for the parallel-coupled bandpass filters are summarized here:

1. Determine the number of sections from the specified attenuation characteristics.
2. Determine the values of the prototype elements to realize the specifications.
3. Determine the one-type resonator network (cascaded series resonators, say) and determine the inverter values.
4. Obtain the even- and odd-mode coupled-line characteristic impedances Z_0^e and Z_0^o.
5. Determine the microstrip widths and separations (*W*, *S*) of the parallel-coupled resonant lines from the graphs of Z_0^e, Z_0^o vs *W*/*h* and *S*/*h* or synthesis technique due to Akhtarzad et al. [1].
6. Calculate the coupled section length l_k which is slightly less than quarter wavelength at centre frequency to account for the end fringing.

Example 8.5 Design a parallel-coupled microstrip bandpass filter having Chebyshev response characteristic with passband ripple of 0.1 dB between the band edge frequencies of 9.98 GHz and 11.03 GHz. The 20 dB minimum attenuation occurs at 11.33 GHz.

Solution Centre frequency $f_0 = \sqrt{(9.98 \times 11.03)} = 10.49$ GHz

For ripple 0.1 dB = 10 log (a_m^2 + 1)

$$a_m^2 = (10^{0.1/10}) - 1 = 0.0233$$

(i) *Number of resonators* Fractional bandwidth

$$F = (f_2 - f_1)/f_0 = 0.1001$$

For an *n*-resonator filter stopband attenuation

$$20 \text{ dB} = 10 \log [1 + 0.0233 \cosh^2 (n \cosh^{-1} \omega_x')]$$

where,

$$\omega_X' = \frac{\omega_0}{\omega_2 - \omega_1}\left(\frac{\omega_X}{\omega_0} - \frac{\omega_0}{\omega_X}\right)$$

$$= \frac{10.49}{11.03 - 9.98}\left(\frac{11.33}{10.49} - \frac{10.49}{11.33}\right) = 1.541$$

Therefore,

$$\cosh(n \cosh^{-1} 1.541) = [(10^2 - 1)/0.0233]^{1/2} = 65.18 = \cosh h\,(4.87°)$$

or,

$$n = \frac{4.87°}{0.998} = 4.88 \approx 5$$

Therefore, five resonator band-pass filter is needed.

(ii) *Prototype values*
For $n = 5$, ripple of 0.1 dB, the low-pass prototype elements are $g_0 = g_6 = 1.0$, $g_1 = g_5 = 1.1468$, $g_2 = g_4 = 1.3712$ and $g_3 = 1.9750$. Thus, the filter is symmetrical in structure.

(iii) *Admittance inverter admittances* For symmetry for the first and final coupling,

$$\frac{J_{1,0}}{Y_0} = \frac{J_{6,5}}{Y_0} = \sqrt{\left(\frac{\pi F}{2 g_0 g_1}\right)} = \sqrt{\left(\frac{\pi F}{2 g_5 g_6}\right)} = \sqrt{\left(\frac{\pi \times 0.1001}{2 \times 1.1468}\right)} = 0.3703$$

For the intermediate coupling section (k = 1 to 2),

$$\frac{J_{2,1}}{Y_0} = \frac{\pi F}{2\sqrt{(g_1 g_2)}} = \frac{J_{5,4}}{Y_0} = \frac{\pi \times 0.1001}{2\sqrt{(1.1468 \times 1.3712)}} = 0.1254$$

$$\frac{J_{3,2}}{Y_0} = \frac{\pi F}{2\sqrt{(g_2 g_3)}} = \frac{J_{4,3}}{Y_0} = \frac{\pi \times 0.1001}{2\sqrt{(1.3712 \times 1.9750)}} = 0.09555$$

(iv) *Even- and odd-mode coupled impedances* For symmetry, only k = 0 to 2 are required to be computed:

$k = 0$ $\quad (Z_0^e)_{10} = 50\,(1 + 0.3703 + 0.1371)$; $(Z_0^o)_{10} = 38.54$ ohms
$\quad = 75.37$ ohms

$k = 1$ $\quad (Z_0^e)_{21} = 50\,(1 + 0.1254 + 0.0157)$; $(Z_0^o)_{21} = 44.52$ ohms
$\quad = 57.06$ ohms

$k = 2$ $\quad (Z_0^e)_{32} = 50\,(1 + 0.09555 + 0.00913)$; $(Z_0^o)_{32} = 45.68$ ohms
$\quad = 55.23$ ohms

(v) *Microstrip line dimensions for $\varepsilon_r = 9.0$*
Using Akhtarzad et al. synthesis

$k = 0$ $\quad Z_{0s}^e = Z_0^e/2 = 75.37/2 = 37.68\ \Omega$
$\quad Z_{0s}^o = Z_0^o/2 = 38.54/2 = 19.27\ \Omega$

From the graph or closed form expression of Z_0 vs W/h in Chapter 3 (redrawn in Fig. 8.18),

$(W/h)_s^e = 1.8 \quad (W/h)_s^o = 4.5$

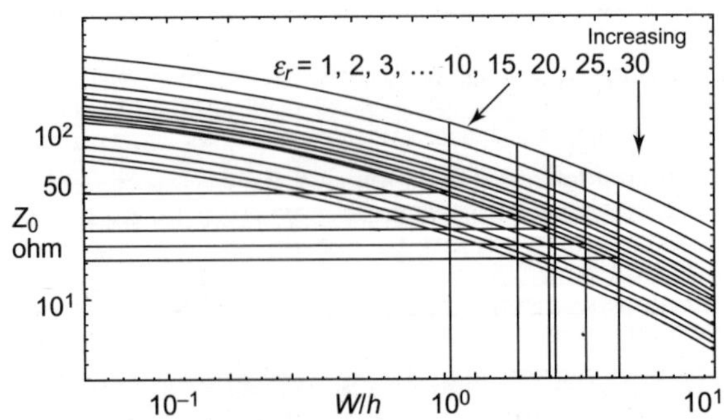

Fig. 8.18 Z_0 vs W/h for microstrip

From Eq. 8.57 (c-g), values of couple line dimensions:

$$S/h = 0.144 \quad W/h = 0.83$$

$k = 1$
$$Z_{0s}^{e} = Z_0^e/2 = \frac{57.06}{2} = 28.53\ \Omega$$

$$Z_{0s}^{o} = Z_0^o/2 = \frac{44.52}{2} = 22.26\ \Omega$$

$$(W/h)_s^e = 2.3 \quad (W/h)_s^o = 3.3$$

$$S/h = 0.61 \quad W/h = 0.92$$

$k = 2$
$$Z_{0s}^{e} = Z_0^e/2 = \frac{55.23}{2} = 27.615\ \Omega$$

$$Z_{0s}^{o} = Z_0^e/2 = \frac{45.68}{2} = 22.84\ \Omega$$

$$(W/h)_s^e = 2.5 \quad (W/h)_s^o = 3.2$$

$$S/h = 0.82 \quad W/h = 0.96$$

For 50-ohm input and output lines, the characteristic impedance is given by the analytical formula in Chapter 3,

For $\varepsilon_r = 9$,
$W/h = 1.1$ for $Z_0 = 50$ ohm.

(vi) *Resonator length*

The physical lengths of the resonators and the coupled regions are half-wavelengths and full wavelengths, respectively.

$$\text{Coupled region length } l = \frac{\lambda_{gm}}{4}$$

where
$$\lambda_{gm} = \frac{\lambda_g^o + \lambda_g^e}{2} = \text{Midband wavelength}$$

$$\lambda_g^o = \text{Odd-mode guide wavelength} = \frac{300}{f\,(\text{GHz})}\frac{Z_0^o}{Z_{0\,\text{air}}^o}\ (\text{mm})$$

$$\lambda_g^e = \text{Even-mode guide wavelength} = \frac{300}{f\,(\text{GHz})}\frac{Z_0^e}{Z_{0\text{air}}^e}\ (\text{mm})$$

$Z_{0\,\text{air}}^e$, $Z_{0\,\text{air}}^o$ = even- and odd-mode characteristics impedances of the air-spaced lines

For the specified designed values of W/h and S/h,

	$Z_{0\,\text{air}}^e$ (ohm)	$Z_{0\,\text{air}}^o$ (ohm)	λ_g^e (mm)	λ_g^o (mm)	λ_{gm} (mm)	$l = (\lambda_{gm}/4)$ (mm)
$k = 0$	200	110	10.8	10.1	10.45	2.6
$k = 1$	160	140	10.23	9.13	9.68	2.4
$k = 2$	140	150	11.3	8.74	10.02	2.5

Final design values for $\varepsilon_r = 9.0$, 50 ohm input/output lines. $W/h = 1.1$

Coupled lines	W/h	S/h	$l \approx (\lambda_{gm}/4)$
$k = 0$	0.83	0.14	2.6
$k = 1$	0.92	0.61	2.4
$k = 2$	0.96	0.82	2.5
$k = 3$	0.92	0.61	2.4
$k = 4$	0.83	0.14	2.6

The filter configuration is shown in Fig. 8.14 (b).

Interdigital Bandpass Filters

The interdigital configuration is the most compact filter where the resonators are placed side by side with one end short-circuited and the other end open-circuited alternatively as shown in Fig. 8.19.

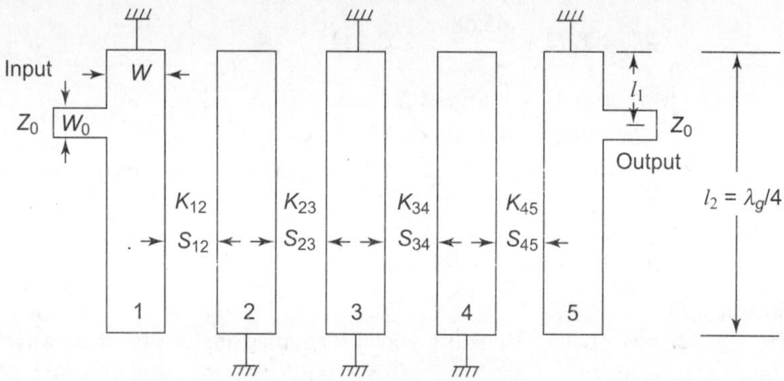

Fig. 8.19 Interdigital bandpass filter

The normalized coupling coefficient of a pair of resonators is given by [3]

$$K_{n,\,n+1} = \frac{f_2 - f_1}{f_0\sqrt{(g_n\, g_{n+1})}} \tag{8.58}$$

where $f_0 = (f_1 + f_2)/2$, the centre frequency, g_n's are low-pass prototype element values normalized to $\omega_c' = 1$. The loaded Q for the standalone single-tuned filter is given by

$$Q_L = \frac{f_0\, g_1}{(f_2 - f_1)} = \frac{f_0\, g_{n+1}}{(f_2 - f_1)} \tag{8.59}$$

The position of the input and output line points l w.r.t short-circuit location can be calculated from the following:

$$\frac{Q_L}{Z_0/Z_{01}} = \frac{\pi}{[4\sin^2(\pi l_1/2l_2)} \tag{8.60}$$

where Z_0 is the characteristic impedance of input/output lines and Z_{01} is filter internal impedance .

The dimensions of the filter are found from K vs S/h experiment [3, 19] graph and $Q/(Z_0/Z_{01})$ vs. l_1/l_2 graph of Fig. 8.20.

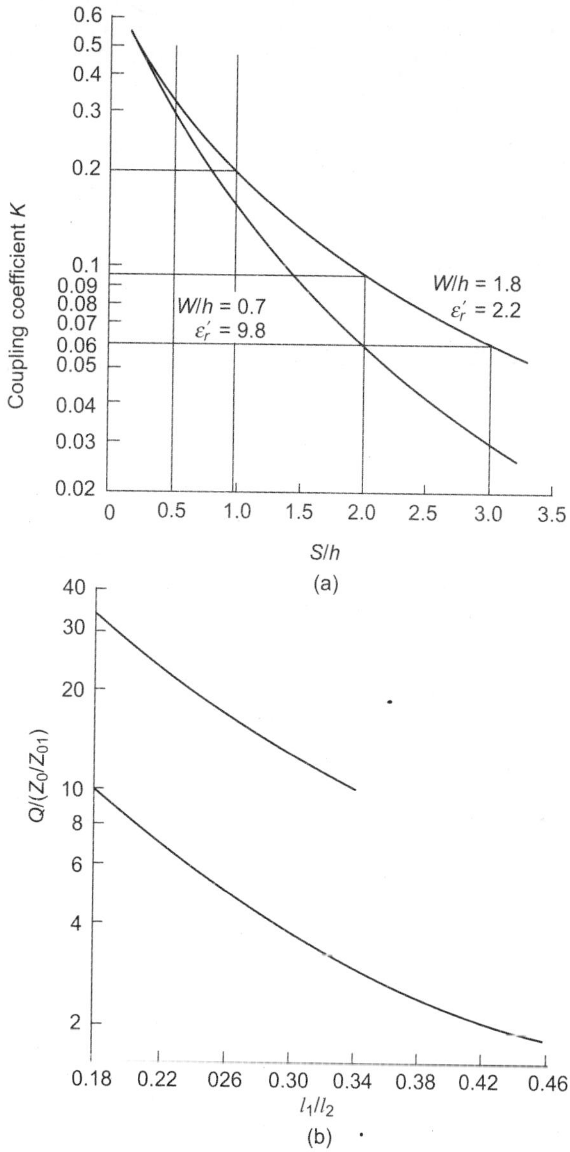

Fig. 8.20 Coupling coefficient and Q of interdigital filter [JS Wong[19] and Bahl et al[3]] (a) K vs S/h (b) $Q_L/(Z_0/Z_{01})$ vs l_1/l_2

Example 8.6 Design a microstrip bandpass interdigital filter having centre frequency $f_0 = 4$ GHz, bandwidth $\Delta f = 0.4$ GHz, 35 dB attenuation points at $f_x = 4 \pm 0.4$ GHz. Attenuation response is Chebyshev type with 0.2 dB ripple. Use substrate dielectric constant $\varepsilon_r = 9.8$ and height $h = 1.27$ mm.

Solution For the low-pass prototype Chebyshev response, the attenuation for normalized cut-off frequency $\omega'_c = 1$ is given by

$$A = 1 + a_m^2 T_n^2(\omega') \ ; \ \omega' = \omega/\omega_c$$

where a_m^2 is the ripple height, n is the order of the filter, ω'_c = bandwidth over which attenuation has maximum ripple a_m^2.

Given ripple magnitude

$$0.2 \text{ dB} = 10 \log (a_m^2 + 1) \quad \text{or,} \quad a_m^2 = 10^{0.2/10} - 1$$

Hence Chebyshev attenuation response is

$$\text{Pass-band: } A = 1 + (10^{0.2/10} - 1) \cos^2 (n \cos^{-1} \omega'); \ \omega' \leq 1$$

$$\text{Stop-band: } A = 1 + (10^{0.2/10} - 1) \cosh^2 (n \cos^{-1} \omega'); \ \omega' \geq 1$$

For low-pass to band-pass frequency transformation

$$\omega' = \frac{\omega_0}{\omega_2 - \omega_1}\left(\frac{\omega}{\omega_0} - \frac{\omega_0}{\omega}\right) \quad \text{or,} \quad \omega'_x = \frac{f_0}{f_2 - f_1}\left(\frac{f_x}{f_0} - \frac{f_0}{f_x}\right)$$

$$= \frac{4}{0.4}\left(\frac{4.4}{4} - \frac{4}{4.4}\right) = 1.9091$$

For stop-band 35 dB attenuation point at $\omega'_x = 4 \pm 0.4$ GHz.

Therefore,

$$35 = 10 \log [1 + (10^{0.2/10} - 1) \cosh^2 (n \cosh^{-1} \omega'_x)$$

or, $$\frac{35}{10} = \log [1 + 0.047 \cosh^2 (n \cosh^{-1} (1.9091))]$$

or, $$[1 + 0.047 \cosh^2 (n \cosh^{-1} (1.9091))] = 10^{3.5} = 3162.3$$

or, $$[1 + 0.047 \cosh^2 (n \times 1.26281)] = 3162.3$$

or, $$\cosh^2 (1.26281\, n) = 3161.3/0.047 = 67261.7$$

or, $$\cosh (1.26281\, n) = 259.35$$

or, $$n = \frac{\cosh^{-1}(259.35)}{1.26281} = 4.95$$

$$n \cong 5$$

Therefore, five resonators are required (odd number).

The prototype low-pass values obtained from Table 8.2 for $n = 5$ and ripple = 0.2 dB are

$$g_1 = 1.0 = g_6$$

$$g_1 = g_5 = 1.3394$$

$$g_2 = g_4 = 1.337$$

$$g_3 = 2.166$$

In microwave form, the interdigital band-pass filter can be realized by selecting input and output line characteristic impedance $Z_0 = 50$ ohms and the filter internal impedance $Z_{01} = 58$ ohms (say).

The loaded Q of the filter resonator

$$Q_L = f_0 g_1/\Delta f = 4 \times 3394/0.4 = 13.394$$

Normalized coupling coefficients between n and n+1 th resonator is

$$K_{n,\, n+1} = \frac{f}{f_0 \sqrt{(g_n\, g_{n+1})}}$$

Therefore, $$K_{12} = K_{45} = \frac{0.4}{4\sqrt{1.3394 \times 1.337}} = 0.0747$$

$$K_{23} = K_{34} = \frac{0.4}{4\sqrt{1.337 \times 2.166}} = 0.0588$$

Taking $W/h = 0.7$, $\varepsilon_r = 9.8$

$$W = 0.889 \text{ mm}, h = 1.27 \text{ mm}, W/h < 1$$

From Fig. 8.19,

$$S_{12} = S_{45} = 2.2 \text{ mm}$$

$$S_{23} = S_{34} = 2.52 \text{ mm}$$

$$l_1/l_2 = 0.145,$$

$$l_2 = \lambda_0/(4\sqrt{\varepsilon_{\text{eff}}}); \lambda_0 = c/f_0 = 30/4 = 7.5 \text{ cm}$$

$$\varepsilon_{\text{eff}} = \frac{\varepsilon_r + 1}{2} + \frac{\varepsilon_r - 1}{2}\left[\left(1 + \frac{12h}{W}\right)^{-1/2} + 0.04\left(1 - \frac{W}{h}\right)^2\right]$$

for $W/h \leq 1$

or,
$$\varepsilon_{\text{eff}} = \frac{9.8 + 1}{2} + \frac{9.8 - 1}{2}\left[\left(1 + \frac{12 \times 1.27}{0.889}\right)^{-1/2} + 0.04\left(1 - \frac{0.889}{1.27}\right)^2\right]$$

$$= 5.4 + 4.4 \times (0.2348 + 3.6 \times 10^{-3}) = 6.44$$

$$l_2 = 7.4 \text{ mm}, l_1 = 1.07 \text{ mm}$$

To find input/output line width W_0 from Z_0 vs W/h curves in Chapter 3,

$$W_0/h = 1.0$$

MICROWAVE BAND-STOP FILTERS 8.9

The band-stop microwave filters are narrow-band filters and can be designed from the low-pass prototype by using the frequency transformation

$$\frac{1}{\omega'} = \frac{\omega_0}{\omega_2 - \omega_1}\left(\frac{\omega}{\omega_0} - \frac{\omega_0}{\omega}\right) \tag{8.61}$$

The low-pass prototype circuit, band-stop filter circuit and their attenuation response are shown in Fig. 8.21, where ω_x represents a frequency for high attenuation in the stop band.

Realization by Quarter-wave Couplings 8.9.1

To realize a band-stop filter in transmission line form, it is more convenient to use only resonant series or shunt branches linked by quarter-wave sections of transmission lines, i.e., by the impedance or admittance inverters. Therefore, a lumped element ladder network consisting of reactive elements can be replaced by the quarter-wave coupling network of Fig. 8.22.

The characteristic impedances of the quarter-wave coupling sections are Z' for n even and Z_0 for n odd, where n is the number of resonators.

Therefore,

For n odd, $Z'/Z_0 = 1$ (8.62)

$$\text{For } n \text{ even, } Z'/Z_0 = \left(\frac{1}{g_0\, g_{n+1}}\right)^{1/2} \tag{8.63}$$

The filter can be designed by defining reactance slope parameters of the resonators as follows:

$$\frac{X_1}{Z_0} = \frac{\omega_0}{g_0\, g_1(\omega_2 - \omega_1)} \tag{8.64}$$

$$\frac{X_k}{Z_0} = \left(\frac{Z'}{Z_0}\right)^2 \frac{g_0\, \omega_0}{g_k\,(\omega_2 - \omega_1)};\ k = \text{even} \tag{8.65}$$

$$= \frac{\omega_0}{g_0\, g_k(\omega_2 - \omega_1)};\ k = \text{odd} \neq 1 \tag{8.66}$$

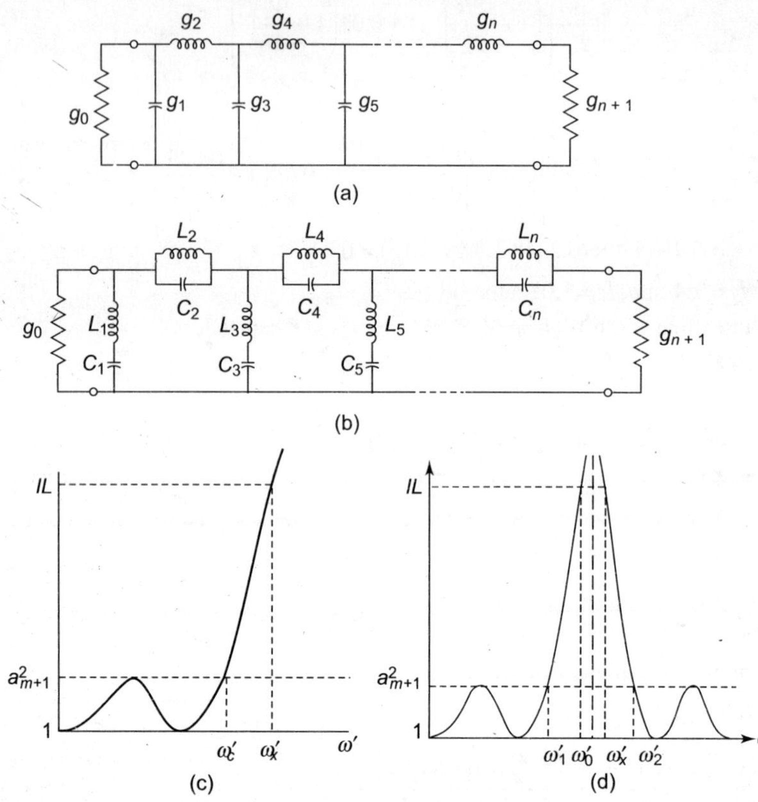

Fig. 8.21 Prototype and band-stop filters: (a) Prototype circuit (b) Equivalent band-stop circuit (c) Prototype low-pass response (d) Band-stop response

The dual realizåtion circuit is a quarter-wave coupled network consisting of parallel resonant circuit in series as shown in Fig. 8.23. The slope parameters are obtained by replacing Z_0, Z' by Y_0, Y' and x_k by b_k keeping all other conditions same.

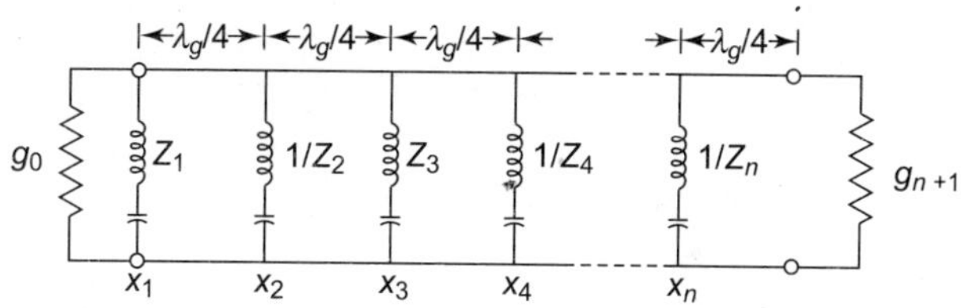

Fig. 8.22 Microwave equivalent circuit of a ladder bandstop network

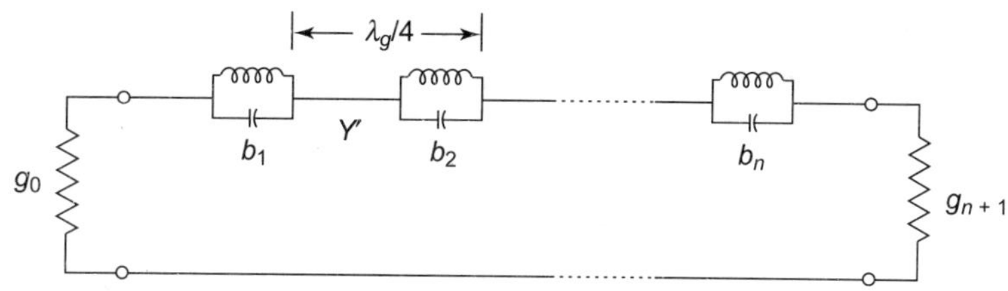

Fig. 8.23 Quarter-wave coupled bandstop filter

Example 8.7 Design a Chebyshev bandpass filter between band edge frequencies $f_1 = 3.5$ GHz, $f_2 = 4.5$ GHz, with high attenuation of 20 dB at 4.75 GHz and ripple of 0.2 dB. Use substrate of $\in_r = 4.4$, thickness $h = 1.6$ mm.

Solution

MATLAB Programme of Chebyshev BP Filter Design

```
h = 1.6 mm, epr = 4.4

fl=3.5%GHz band edge fre.
f2=4.5%G.Hz band edge fre.
fc=sqrt(fl*f2)%GHz centre fre.
fx=4.75%GHz
Zo=50;%ohm
dbripple=0.2%dB
dbx=20%dB
EPR=4.4;
am2=(10.^(dbripple./10))-1;% am^2

%Number of resonators n
F=(f2-f1)/fc;
wy=(fc/(f2-f1))*((fx/fc)-(fc/fx));
n=(acosh(sqrt((( 10^(dbx/10))-1)/((10^(dbripple/10))-1 ))))/(acosh(wy))

% n=5;% live section BPF
%prototype low pass elements

g0=1.0;
g1=1.3394;
g2=1.3370;
g3=2.1660;
g4=g2;
g5=g1;
```

```
g6=g0;

% Filter elements
% Admittance inverter admittances For symmetry

M1=sqrt((pi*F)/(2*g0*g1));%M=JZ0
M2=(pi*F)/(2*(sqrt(g1*g2)));
M3=(pi*F)/(2*(sqrt(g2*g3)));
M4=(pi*F)/(2*(sqrt(g3*g4)));
M5=(pi*F)/(2*(sqrt(g4*g5)));
M6=sqrt((pi*F)/(2*g5*g6));

%Even and odd-mode coupled impedances
%For symmetry only k-0 to 2 are require to be computed
%for even mode.odd mode when k=0
Zoe1=Zo*(1+M1+M1^2);
Zoo1=Zo*(1-M1+M1^2);

%For even mode,odd mode when k=1
Zoe2=Zo*(1+M2+M2^2);
Zoo2=Zo*(1-M2+M2^2);

%for even mode.odd mode when k=2
Zoe3=Zo*(1+M4+M4^2);
Zoo3=Zo*(1-M4+M4^2);

%for even mode,odd mode when k=2
Zoe3=Zo*(1+M4+M4^2);
Zoo3=Zo*(1-M4+M4^2);
% using Akhtarzad et al systhesis
%k=0
Zoes=Zoe1/2;
Zoos=Zoo1/2;
Z01=sqrt(Zoe1*Zoo1);
%k=1
Zoes1=Zoe2/2;
Zoos1=Zoo2/2;
Z02=sqrt(Zoe2*Zoo2);
%when k=2
Zoes2=Zoe3/2;
Zoos2=Zoo3/2;
Z03=sqrt(Zoe3*Zoo3);

EPRair=1;
EPRaireff=1;

%Theory shows Zoe>>Zoo;thus W/h<2 for Zoe and W/h>2 for Zoo expected
%values of couple line dimessions, S=s/h for k=0,1,2
%k=0
Z0air=Z01;
A=(Zoes./60).*sqrt((EPR+1)./2)+((EPR-1)./(EPR+1)).*(0.23+0.11/EPR);
```

```
wbyhes=8.*exp(A)./((exp(2.*A))-2);% for W/h<2

B=377.*pi./(2.*Zoos.*sqrt(EPR));
wbyhos=(2./pi).*(B-1-log(2.*B-1)+((EPR-1)./(2.*EPR)).*(log(B-1)...
   +0.39-0.61 ./EPR));%for W/h>2

sbyh1=(2/pi)*acosh(((cosh((pi./2)*wbyhes)+cosh((pi./2)*wbyhos)-2))/(cosh...
((pi/2)*wbyhos)-cosh((pi./2)*wbyhes)))%k=0
%k=1
A1=(Zoes1./60).*sqrt((EPR+1)./2)+((EPR-1)./(EPR+1)).*(0.23+0.11/EPR);
wbyhes1=8.*exp(A1)./((exp(2.*A1))-2);% for W/h<2

B1=377.*pi./(2.*Zoos1.*sqrt(EPR));
wbyhos1=(2./pi).*(B1-1-log(2.*B1-1)+((EPR-1)./(2.*EPR)).*(log(B1-1)...
   +0.39-0.61./EPR));%for W/h>2

sbyh2=(2/pi)*acosh(((cosh((pi./2)*wbyhes1)+cosh((pi./2)*wbyhos1)-2))/(cosh...
((pi/2)*wbyhos1)-cosh((pi./2)*wbyhes1)))%k=1
%k-2
A2=(Zoes2./60).*sqrt((EPR+1)./2)+((EPR-1)./(EPR+1)).*(0.23+0.11/EPR);
wbyhes2=8.*exp(A2)./((exp(2.*A2))-2);% for W/h<2
%w/h for k=0,1,2

A11=(Z01./60).*sqrt((EPR+1)./2)+((EPR-1)./(EPR+1)).*(0.23+0.11/EPR);
wbyh1=8.*exp(A11)./((exp(2.*A11))-2)% for W/h<2 & k=0
A22=(Z02./60).*sqrt((EPR+1)./2)+((EPR-1)./(EPR+1)).*(0.23+0.11/EPR);
wbyh2=8.*exp (A22)./((exp(2.*A22))-2)% for W/h<2 & k=1
A33=(Z03./60).*sqrt((EPR+1)./2)+((EPR-1)./(EPR+1)).*(0.23+0.11/EPR);
wbyh3=8.*exp(A33)./((exp(2.*A33))-2)% for W/h<2 & k=2

%Input/Output 50 Ohm lines

Z0=50.00;
A50=(Z0./60).*sqrt((EPR+1)./2)+((EPR-1)./(EPR+1)).*(0.23+0.11/EPR);
Wbyh50ohm=8.*exp(A50)./((exp(2.*A50))-2)% for W/h<2

c=3*10^8;%velocity of light m/S
EPR0=10^ (-9)./(36.*pi);%F/m

% e and o modes Capacitances to find λg

%k=0. even mode Ce1 calculation to find λge  1
Cp1=EPRO*EPR*wbyh1;
EPReff1=((EPR+1)/2)+((((EPR-1)/2)*((1+(12/wbyh1)^(-1/2))+(0.04*...
(1-wbyh1)^2))));
Cf 1 =((sqrt(EPReff1 )/(c* Z01 ))-Cp 1 )/2;
D1=exp((-0.1)*exp(2.333-2.53*wbyh1));
Cfdash1=(Cf1/(1+D1*(1/sbyh1)*tanh(8*sbyh1)))*sqrt(EPR/EPReff1);
Ce1=Cp1+Cf1+Cfdash1;
EPRair=1;
EPReffair=1;
Cpair1=EPRO*EPRair*wbyh1;
```

```
Cfair 1 =((sqrt(EPReffair)/(c*Z01))-Cpair1)/2;
Cfdashair1=(Cfair1/(1+D1*(1/sbyh1)*tanh(8*sbyh1)));

%calculation of λge 1
lamge1=(300./fc)*sqrt(Ceair1/Ce1);

%k=0, odd mode Co1 calculation to find λgo 1
k1=(sbyh1/(sbyh1+(2.*wbyh1)));
k1dash=sqrt(1-k1^2);
ifj(k1^2<=0.5);
Cga1=(EPR0*pi)/(log((2*(1+sqrt(k1)))/(1-sqrt(k1))));
else
   Cga1=(EPR0/pi)*(log((2*(1+sqrt(k1dash)))/(1-sqrt(k1dash))));
end;
Cgd1=(EPR0*EPR/pi)*log(coth((pi/4)*sbyh1))+(0.65*Cf1)*((0.02/sbyh1)*.
   sqrt(EPR)+1-(EPR)A(-2));
Co1=Cp1+Cf1+Cga1+Cgd1;

%k=0. odd mode Co1 calculation to find λgo 1
k1=(sbyh1/(sbyh1+(2.*wbyh1)));
k1dash=sqrt(1-k1^2);
if(k1^2<=0.5);
Cga1=(EPR0*pi)/(log((2*(1+sqrt(k1)))/(1-sqrt(k1))));
else
   Cgal=(EPR0/pi)*(1og((2*(1+sqrt(k1dash)))/(1-sqrt(k1dash))));
end;
Cgd1=(EPR0*EPR/pi)*1og(coth((pi/4)*sbyh1))+(0.65*Cf1)*((0.02/sbyh1)*...
   sqrt(EPR)+1-(EPR)^(-2));
Co1=Cp1+Cf1+Cga1+Cgd1;

%k=0, Coair1 calculation to find λgo 1

CgaAir1=EPR0*pi/(log((2*(1+sqrt(k1)))/(1-sqrt(k1dash))));
CgdAir1=(EPR0*EPRair/pi)*log(coth((pi/4)*sbyh1))+(0.65*Cfair1)*((0.02/...
   sbyh1 )*sqrt(EPRair)+1 -(EPRair)^(-2));
Coair 1 =Cpair 1 +Cfair 1 +CgaAir 1 +Cgd Air 1;

%calculation of λgo 1

lamgo1=(300./fc)*sqrt(Coair1/Co1);
lamg 1 by4=(lamge 1 +lamgo 1 )/8

%k=1, Ce2 calculation to find out λge 2

Cp2=EPR0*EPR*wbyh2;
EPReff2=((EPR+1)/2)+((((EPR-1)/2)*((1+(12/wbyh2)^(-1/2))+(0.04*...
   (1-wbyh2)^2))));
Cf2=((sqrt(EPReff2)/(c*Z02))-Cp2)/2;
D2=exp((-0.1)*exp(2.333-2.53*wbyh2));
Cfdash2=(Cf2/(1+D2*(1/sbyh2)*tanh(8*sbyh2)))*sqrt(EPR/EPReff2);
Ce2=Cp2+Cf2+Cfdash2;
```

```
%Cear2 calculation to find out λge 2
Cpair2=EPR0*EPRair*wbyh2;
Cfair2=((sqrt(EPReffair)/(c*Z02))-Cpair2)/2;
Cfdashair2=(Cfair2./(1+D2*(1./sbyh2)*tanh(8.*sbyh2)));
Ceair2=Cpair2+Cfair2+Cfdashair2;

%calculation of λge 2
lamge2=(300./fc)*sqrt(Ceair2./Ce2);

%k=1, Co2 calculation to find out λgo 2

k2-(sbyh2/(sbyh2+(2*wbyh2)));
k2dash=sqrt(1-k2^2);
if(k2^2<=0.5);
Cga2=(EPR0*pi)/(log((2*(1+sqrt(k2)))/(1-sqrt(k2))));
else

Cga2=(EPR0/pi)*(log((2*(1+sqrt(k2dash)))/(1-sqrt(k2dash))));
end;
Cgd2=(EPR0*EPR/pi)*log(coth((pi/4)*sbyh2))+(0.65*Cf2)*((0.02/sbyh2)*…
    sqrt(EPR)+1-(EPR)^(-2));
Co2=Cp2+Cf2+Cga2+Cgd2;

%k=1, Coair2 calculation to find out λgo 2

CgaAir2=EPR0*pi/(log((2*(1+sqrt(k2)))/(1-sqrt(k2dash))));
CgdAir2=(EPR0*EPRair/pi)*log(coth((pi/4)*sbyh2))+(0.65*Cfair2)*((0.02/…
    sbyh2)*sqrt(EPRair)+1-(EPRair)^(-2));
Coair2=Cpair2+Cfair2+CgaAir2+CgdAir2;

%calculation of λgo 2

lamgo2=(300./fc)*sqrt(Coair2./Co2);
lamg2by4=(lamge2+lamgo2)/8

%k=2. Ce3 calculation to find out λge 3

Cp3=EPR0*EPR*wbyh3;
EPReff3=((EPR+1)/2)+((((EPR-1)/2)*((1+(12/wbyh3)^(-1/2))+(0.04*...
    (1-wbyh3)^2))));
Cf3=((sqrt(EPReff3)/(c*Z03))-Cp3)/2;
D3=exp((-0.1)*exp(2.333-2.53*wbyh3));
Cfdash3=(Cf3/(1+D3*(1/sbyh3)*tanh(8*sbyh3)))*sqrt(EPR/EPReff3);
Ce3=Cp3+Cf3+Cfdash3;

%Ceair1 calculation to find out 1 λge 3
Cpair3=EPR0*EPRair*wbyh3;
Cfair3=((sqrt(EPReffair)/(c*Z03))-Cpair3)/2;
Cfdashair3=(Cfair3/(1+D3*(1/sbyh3)*tanh(8*sbyh3)));
Ceair3=Cpair3+Cfair3+Cfdashair3;
```

```
%calculation of lamge3
lamge3=(300./fc)*sqrt(Ceair3/Ce3);

%k=2, Co3 calculation to find out λgo 3
k3=(sbyh3/(sbyh3+(2*wbyh3)));
k3dash=sqrt(1-k3^2);
if(k3^2<=0.5);
Cga3=(EPR0*pi)/(log((2*(H-sqrt(k3)))/(1-sqrt(k3))));
else
   Cga3=(EPR0/pi)*(log((2*(1+sqrt(k3dash)))/(1-sqrt(k3dash))));
end;
Cgd3=(EPR0*EPR/pi)*log(coth((pi/4)*sbyh3))+(0.65*CD)*((0.02/sbyh3)*…
   sqrt(EPR)+1-(EPR)^(-2));
Co3=Cp3+CD+Cga3+Cgd3;

%k=2, Coair3 calculation to find out λgo 3
CgaAir3=EPR0*pi/(log((2*(1+Sqrt(k3)))/(1-sqrt(k3))));
CgdAir3=(EPR0*EPRair/pi)*log(coth((pi/4)*sbyh3))+(0.65*Cfair3)*((0.02/...
   sbyh3)*sqrt(EPRair)+1-(EPRair)^(-2));
Coair3=Cpair3+Cfair3+CgaAir3+CgdAir3;

%calculation of λgo 3

lamgo3=(300./fc)*sqrt(Coair3/Co3);
lamg3by4=(lamge3+lamgo3)/8
h=1.6;%mm
s1mm=sbyh1*h
s2mm=sbyh2*h
s3mm=sbyh3*h
s4mm=s2mm
s5mm=s1mm
w1mm=wbyh1*h
w2mm=wbyh2*h
w3mm=wbyh3*h
w4mm=w2mm
w5mm=w1mm
w50mm=Wbyh50ohm
```

%RESULTS OF CHEBYSHEV BP FILTER

Design parameters

fl=3.5GHz, f2=4.5GHz, fc=3.9686GHz, ripple= 0.2 dB, Attn =20 dB fx=4.75GHz, ε_r=4.4

Designed values Section=n=5

S/hl =0.0192, S/h2 =0.0721, S/h3 =0.1250

W/hl =1.4453, W/h2 =1.7720, W/h3 =1.8256/ W/h50ohm =1.9119

Length=$\lambda_g/4$ in mm

$\lambda_{g1}/4$=10. 6976, $\lambda_{g2}/4$=10.7243, $\lambda_{g3}/4$=11.7578, $\lambda_{g4}/4$= 10.7243, $\lambda_{g5}/4$=10.6976

Spacing

```
S1mm =0.0307, S2mm = 0.1154, S3mm =0.2000, S4mm = 0.1154, S5mm = 0.0307
```

Line widths

```
W1mm =2.3126, W2mm = 2.8352, W3mm =2.9210, W4mm =2.8352, W5mm = 2.3126, W50mm =1. 9119
```

Practical Realization in Microwave Form 8.9.2

The series and parallel resonant circuits can be approximately realized by short-circuited or open-circuited, resonant stubs with a series gap capacitance in coaxial, strip and microstrip forms and with a shunt inductive irises in waveguide form. These are summarized in Table 8.4 along with the approximate design formulae. The details are described by Matthaei [15].

Table 8.4 *Band-stop filter element realization*

Resonator	Approximate configuration		Design formulae
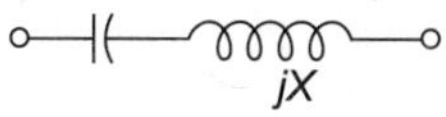	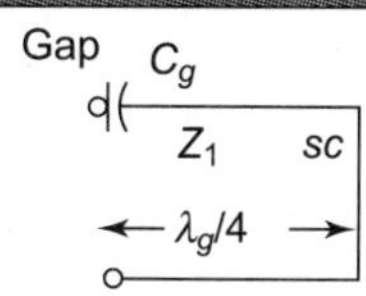	Coaxial strip line	$F(\phi_0) = 2x/Z_1$
	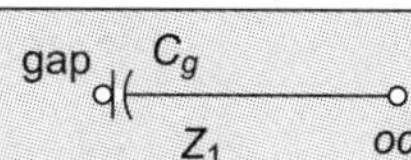	Microstrip line	$\omega_0 C_g = 1/(Z_1 \tan \phi_0)$
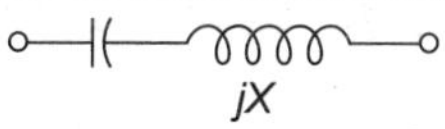	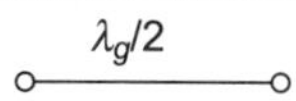	Coaxial strip line Microstrip line	$F(\phi_0) \approx x/Z_1$ $\omega_0 C_g = 1/(Z_1 \tan \phi_0)$
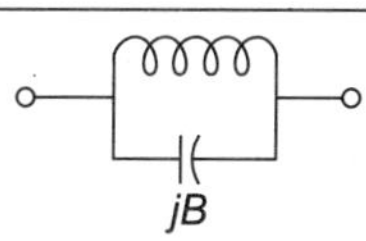	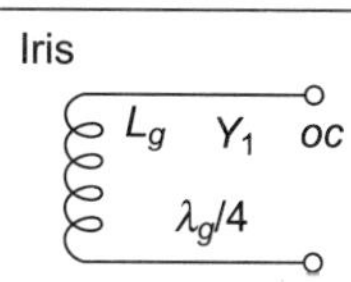	Waveguide	$F(\phi_0) = 2b/Y_1$ $\omega_0 C_g = 1/(Y_1 \tan \phi_0)$
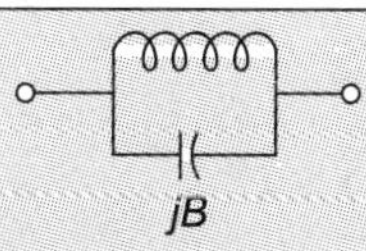	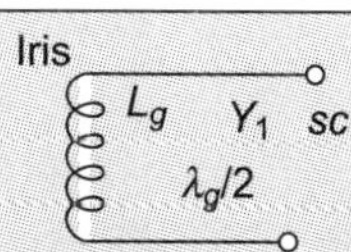	Waveguide	$F(\phi_0) = b/Y_1$ $\omega_0 L_g = 1/(Y_1 \tan \phi_0)$
$F(\phi_0) = \phi_0 \sec^2\phi_0 + \tan \phi_0$			

The length of the stub is determined by

$$l_k = \frac{\lambda_g \phi_0}{2\pi} \tag{8.67}$$

The line width can be determined from the knowledge of the characteristic impedance. The gaped length for the capacitance C_g can be approximated by

$$d_k = \frac{\lambda_g}{4} \quad \text{or} \quad \frac{\lambda_g}{2} - l_k \tag{8.68}$$

as the case may be depending on the short-circuit and open-circuit conditions. The inductive irises can be designed by calculating the value of L_g from Table 8.4. Values of $F(\phi_0)$ vs ϕ_0 is shown in Table 8.5.

Table 8.5 F(ϕ) Vs. ϕ °

ϕ	$F(\phi)$	ϕ	$F(\phi)$	ϕ	$F(\phi)$	ϕ	$F(\phi)$	ϕ	$F(\phi)$	ϕ	$F(\phi)$	ϕ	$F(\phi)$
20.00	0.76	68.00	10.96	75.50	24.99	80.55	58.56	83.30	116.48	86.00	328.42	88.70	3226.52
21.00	0.80	68.50	11.47	75.60	25.34	80.60	59.18	83.35	118.24	86.05	336.85	88.75	3497.78
22.00	0.85	69.00	12.01	75.70	25.69	80.65	59.81	83.40	120.40	86.10	345.61	88.80	3804.74
23.00	0.90	69.50	12.60	75.80	26.80	80.70	60.46	83.45	121.89	86.15	354.72	88.85	4153.96
24.00	0.95	70.00	13.23	75.90	26.42	80.75	61.11	83.50	123.77	86.20	364.20	88.90	4553.58
25.00	1.00	70.50	13.91	76.00	26.79	80.80	61.78	83.55	125.70	86.25	374.06	88.95	5013.77
26.00	1.06	70.60	14.06	76.10	27.17	80.86	62.45	83.60	127.68	86.30	384.33	89.00	5547.42
27.00	1.10	70.70	14.19	76.20	27.57	80.90	63.14	83.65	129.70	86.35	395.02	89.05	6171.05
28.00	1.16	70.80	14.34	76.30	27.97	80.95	63.84	83.70	131.77	86.40	406.17	89.10	6906.09
29.00	1.22	70.90	14.49	76.40	28.38	81.00	64.55	83.75	133.90	86.45	417.80	89.15	7780.72
30.00	1.28	71.00	14.64	76.50	28.80	81.05	65.27	83.80	136.07	86.50	429.93	89.20	8832.70
31.00	1.34	71.10	14.79	76.60	29.22	81.10	66.00	83.85	138.30	86.55	442.61	89.25	10113.40
32.00	1.40	71.20	14.95	76.70	29.66	81.15	66.75	83.90	140.58	86.60	455.85	89.30	11694.29
33.00	1.47	71.30	15.11	76.80	30.11	81.20	67.51	83.95	142.92	86.65	469.69	89.35	13677.07
34.00	1.54	71.40	15.27	76.90	30.57	81.25	68.28	84.00	145.32	86.70	484.18	89.40	16210.46
35.00	1.61	71.50	15.43	77.00	31.04	81.30	69.07	84.05	147.78	86.75	499.35	89.45	19518.95
36.00	1.69	71.60	15.60	77.10	31.52	81.35	69.87	84.10	150.31	86.80	515.24	89.50	23954.27
37.00	1.77	71.70	15.77	77.20	32.01	81.40	70.68	84.15	152.90	86.85	531.91	89.55	30092.89
38.00	1.85	71.80	15.94	77.30	32.51	81.45	71.51	84.20	155.55	86.90	549.39	89.60	38933.26
39.00	1.94	71.90	16.11	77.40	33.02	81.50	72.35	84.25	158.28	86.95	567.76	89.65	52329.04
40.00	2.03	72.00	16.29	77.50	33.55	91.55	73.21	84.30	161.08	87.00	587.06	89.70	74046.28
41.00	2.13	72.10	16.47	77.60	34.09	81.60	74.08	84.35	163.95	87.05	607.36	89.75	112731.26
42.00	2.23	72.20	16.65	77.70	36.64	81.65	74.97	84.40	166.91	87.10	628.74	89.80	192063.13
43.00	2.34	72.30	16.84	77.80	35.21	81.70	75.88	84.45	169.94	87.15	651.27	89.85	397933.81
44.00	2.45	72.40	17.03	77.90	35.79	81.75	76.80	84.50	173.06	87.20	675.03	89.90	1266238.98
45.00	2.57	72.50	17.22	78.00	36.39	81.80	77.74	84.55	176.26	87.25	700.11	89.95	27096342.23
46.00	2.70	72.60	17.42	78.10	37.00	81.85	78.70	84.60	179.55	87.30	726.63		

(Contd.)

Table 8.5 (*Contd.*)

ϕ	$F(\phi)$	ϕ	$F(\phi)$	ϕ	$F(\phi)$	ϕ	$F(\phi)$	ϕ	$F(\phi)$	ϕ	$F(\phi)$	ϕ	$F(\phi)$
47.00	2.84	72.70	17.62	78.20	37.62	81.90	79.67	84.65	182.94	87.35	754.67		
48.00	2.98	72.80	17.82	78.30	38.27	81.95	80.66	84.70	186.42	87.40	784.38		
49.00	3.14	72.90	18.03	78.40	38.92	82.00	81.67	84.75	190.01	87.45	815.87		
50.00	3.31	73.00	18.24	78.50	39.60	82.05	82.70	84.80	193.70	87.50	849.30		
51.00	3.49	73.10	18.45	78.60	40.29	82.10	83.75	83.75	197.49	87.55	884.83		
52.00	3.68	73.20	18.67	78.70	41.01	82.15	84.82	84.90	201.40	87.60	922.64		
53.00	3.89	73.30	18.80	78.80	41.74	82.20	85.92	84.95	205.43	87.65	962.92		
54.00	4.11	73.40	19.12	78.90	42.49	82.25	87.03	85.00	209.58	87.70	1005.91		
55.00	4.35	73.50	19.35	79.00	43.26	82.30	88.16	85.05	213.86	87.75	1051.84		
56.00	4.61	73.60	19.58	79.10	44.06	82.35	89.32	85.10	218.27	87.80	1100.99		
57.00	4.90	73.70	19.82	79.20	44.87	82.40	90.50	85.15	222.82	87.85	1153.67		
58.00	5.21	73.80	20.06	79.30	45.71	82.45	91.70	85.20	227.51	87.90	1210.22		
59.00	5.55	73.90	20.31	79.40	46.58	82.50	92.93	85.25	232.36	87.95	1271.04		
60.00	5.93	74.00	20.56	79.50	47.46	82.55	94.18	85.30	237.36	88.00	1336.56		
61.00	6.34	74.10	20.82	79.60	48.83	82.60	95.46	85.35	242.52	88.05	1407.28		
61.50	6.57	74.20	21.08	79.70	49.32	82.65	96.76	85.40	247.85	88.10	1483.78		
62.00	6.80	74.30	21.35	79.80	50.29	82.70	98.09	85.45	253.36	88.15	1566.68		
62.50	7.05	74.40	21.62	79.90	51.28	82.75	99.45	85.50	259.06	88.20	1656.74		
63.00	7.31	74 50	21.90	80.00	52.31	82.80	100.84	85.55	264.95	88.25	1754.79		
63.50	7.59	74 60	22.18	80.10	53.37	82.85	102.26	85.60	271.05	88.30	1861.81		
64.00	7.88	74 70	22.47	80.15	53.91	82.90	103.70	85.65	277.36	88.35	1978.94		
64.50	8.19	74.80	22.76	80.20	54.46	82.95	105.18	85.70	283.89	88.40	2107.48		
65.00	8.51	74.90	23.06	80.25	55.02	83.00	106.69	85.75	290.66	88.45	2248.96		
65.50	8.86	75.00	23.37	80.30	55.59	83.05	108.24	85.80	297.67	88.50	2405.19		
66.00	9.23	75.10	23.68	80.35	56.16	83.10	109.81	85.85	304.94	88.55	2578.29		
66.50	9.62	75.20	24.00	80.40	56.75	83.15	111.42	85.90	312.48	88.60	2770.77		
67.00	10.04	75.30	24.32	80.45	57.34	83.20	113.07	85.95	320.30	88.65	2985.64		
67.50	10.48	75.40	24.65	80.50	57.94	83.25	114.76						

The design procedure follows the following steps.

1. Number of resonators n is determined from the maximum attenuation at band-edge and the value of given attenuation in the stop band at a given frequency ω_x using the frequency transformation.
2. From n and band-edge attenuation, the prototype element values are determined.
3. Slope parameters are determined for a selected realizable network.
4. Selecting the stub impedances, the gap capacitances or iris inductances and the electrical length of the stub are determined at resonance frequency.
5. Stub lengths and the dimensions of the capacitance gaps or the irises are found.

A few examples of the microwave form of bandstop filters are shown in Fig. 8.23.

Example 8.8 Design a stripline Chebyshev narrow bandstop filter having frequency of infinite attenuation at 4 GHz, fractional bandwidth 0.05 pass-band ripple 0.5 dB and 26 dB minimum attenuation occurs at the 2% of the centre frequency. Use dielectric substrate or $\varepsilon_r = 2.56$ and height $h = 1.8$ mm. Filter's nominal characteristics impedance is 50 ohms.

Solution

(i) *Number of resonators = n*

The 2% of $f_0 = 4 \times 0.02 = 80$ GHz.

The frequency at 26 dB attenuation is

$$f_x = 4\text{ GHz} + \frac{80\text{ MHz}}{2} = 4.04\text{ GHz}.$$

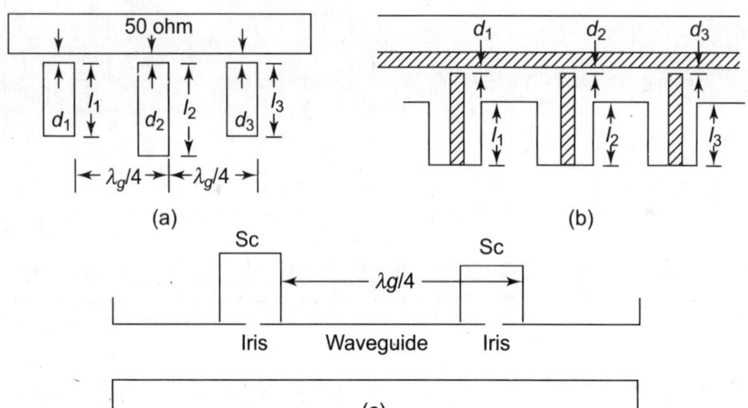

Fig. 8.24 Bandstop filter: (a) Strip line (b) Coaxial line (c) Waveguide

Therefore, $f_x/f_0 = 4.04/4 = 1.01$

$\Delta f/f_0 = 0.05$

From frequency transformation at 26 dB attenuation point

$$\frac{1}{\omega'_x} = \frac{f_0}{\Delta f}\left(\frac{f_x}{f_0} - \frac{f_0}{f_x}\right) = \frac{1}{0.05}\,(1.01 - 0.99)$$

or, $\omega'_x = 2.5$

Ripple $0.5\text{ dB} = 10 \log (a_m^2 + 1)$ or, $a_m^2 = 0.122$

At 26 dB attenuation point,

$$26 = 10 \log [1 + a^2_m \cosh^2 (n \cosh^{-1} \omega'_x)]$$

or,

$$[1 + 0.122 \cosh^2(n \cosh^{-1} 2.5)] = 10^{2.6}$$

$$[\cosh^2 (n \cosh^{-1} 2.5)] = \frac{10^{2.6}}{0.122}$$

or, $$[\cosh (n \cosh^{-1} 2.5)] = 57.1 = \cosh 4.74°$$

Therefore, $$n = \frac{4.74}{\cosh^{-1} 2.5} = 3$$

(ii) *LP prototype element values*

For $n = 3$, ripple = 0.5 dB, from Table 8.2

$$g_0 = g_4 = 1.0$$

$$g_1 = g_3 = 1.5963$$

$$g_2 = 1.0969$$

$$\omega'_c = 1$$

(iii) *Microwave realization*

The three-resonator band-stop filter, its transformed configuration using quarter wavelength coupling lines and strip-line realization with open-ended resonant half-wavelength stubs and gap capacitance are shown below.

Since $n = 3$ (odd), the impedance of the quarter wavelength coupling lines are $Z' = Z_0$. The reactance slope parameters of the series resonators are

$$\frac{x_1}{R_0} = \frac{\omega_0}{g_0 g_1(\omega_2 - \omega_1)} = \frac{x_3}{R_0} = \frac{1}{1.5963 \times 0.05} = 12.53$$

$$\frac{x_2}{R_0} = \left(\frac{Z_1}{R_0}\right)^2 = \frac{g_0 \omega_0}{g_2(\omega_2 - \omega_1)} = \frac{1}{1.0969 \times 0.05} = 18.23$$

Therefore, $x_1 = x_3 = 12.53 \times 50 = 626.5$ ohm

$x_2 = 18.23 \times 50 = 911.6$ ohm

The gap capacitance reactance at resonant

$$\omega_0 C_{gk} = \frac{1}{Z_k \tan \phi_0}$$

where, $\phi_0 \sec^2\phi_0 + \tan \phi_0 = \dfrac{x_k}{Z_k} = F(\phi_0)$

(iv) *Guide Wavelength*

Strip-line substrate height $h = 1/8'' = 3.175$ cm,

$$\varepsilon_r = 2.56$$

$$\lambda_{g0} = \frac{\lambda_0}{\sqrt{\varepsilon_r}} = \frac{c/f_0}{\sqrt{\varepsilon_r}} = \frac{30/4}{\sqrt{2.56}} = 4.68 \text{ cm}$$

$$\lambda_{g0}/4 = \frac{4.68}{4} = 1.17 \text{ cm}$$

(v) *Line widths*

Assuming resonator line impedances $Z_{b1} = Z_{b2} = Z_{b3} = 60\ \Omega$. ($t = 0$) line widths are calculated as follows from graph in Chapter 3:

For $Z_0 =$ 50 ohm line $\sqrt{\varepsilon_r}\ Z_0 = \sqrt{2.56} \times 50 = 80$ ohms

For $Z_0 =$ 60 ohm line, $\sqrt{\varepsilon_r}\ Z_0 = \sqrt{2.56} \times 60 = 96$ ohms

From $\sqrt{\varepsilon_r}\ Z_0$ vs W/h graphs in Chapter 3,

$$Z_0 = 50 \text{ ohms},\quad W_0/h = 0.74,\quad W_0 = 0.74 \times 3.175 = 2.35 \text{ mm}$$

$$Z_0 = 60 \text{ ohms},\quad W_1/h = 0.55,\quad W_1 = 0.55 \times 3.175 = 1.75 \text{ mm}$$

$$W_2 = W_3$$

Stub length

$$F(\phi_{01}) = F(\phi_{03}) = \phi_{01}\sec^2\phi_{01} + \tan\phi_{01} = \frac{x_1}{Z_{b_1}} = \frac{626.5}{60} = 10.442$$

$$F(\phi_{03}) = \frac{911.6}{60} = 15.19$$

From the table of ϕ_0 vs $F(\phi_0)$, we find

$$\phi_{01} = \phi_{03} = 67.5° = 1.1781 \text{ radians}$$

$$\phi_{02} = 71.35° = 1.2453 \text{ radians}$$

The stub lengths are given by

$$l_1 = l_3 = \frac{\lambda_{g0}\phi_{01}}{2\pi} = \frac{4.68 \times 1.1781}{2\pi} = 0.8775 \text{ cm}$$

$$l_2 = \frac{\lambda_{g0}\phi_{01}}{2\pi} = \frac{4.68 \times 1.2453}{2\pi} = 0.9275 \text{ cm}$$

Capacitance gap

$$d_k = \frac{\lambda_{g0}}{4} - l_k$$

$$d_1 = 1.17 - 0.8775 = 0.2925 \text{ cm} = d_3$$

$$d_2 = 1.17 - 0.9275 = 0.2425 \text{ cm}$$

REVIEW QUESTIONS

8.1 Design a maximally flat low-pass 50 ohm microstrip filter having cut-off frequency 2 GHz, and insertion loss of 30 dB at 4 GHz. Use lossless dielectric substrate of $\varepsilon_r = 9.6$ and thickness = 0.635 mm.

8.2 Design a low-pass 50 ohm Chebyshev coaxial line filter with 0.1 dB pass-band ripple and number of sections = 3. The band edge frequency is 2 GHz.

8.3 Design a low-pass stripline Chebyshev filter with $f_0 = 2$ GHz, having a ripple of 0.1 dB and stop-band attenuation of 30 dB at 3 GHz. The filter may be realized using high-impedance and low-impedance sections. The input and output impedances are 50 ohms.

8.4 Design coaxial line maximally flat 50 ohm high-pass filter when the number of sections is 3 and the band edge frequency is 1 GHz.

8.5 It is required to design a symmetrical three-section, band-pass, quarter-wave coupled filter such that the central frequencies in the pass-band and the bandwidth are 9375 and 125 MHz, respectively.

(a) Determine the loaded Q_L per section.

(b) Plot the estimated insertion loss due to the filter in the frequency range of 9000 and 10,000 MHz.

8.6 Design a three-cavity quarter-wave coupled Chebyshev band-pass filter with following specifications.
Waveguide dimensions $a = 0.9''$, $b = 0.4''$.
Band-edges at $f_2 = 10{,}400$ MHz, $f_1 = 10{,}000$ MHz
Pass-band tolerance $a_m^2 = 0.0233$
Inductive diaphragms with circular holes are to be used. Determine the hole radii and diaphragm spacings.

8.7 Find even- and odd-mode impedances of a 6-section parallel coupled Chebyshev microstrip band-pass filter at 4 GHz with a 100 MHz bandwidth. The microstrip parameters are $\varepsilon_r = 9.8$ and $h = 0.5$ mm. The input and output impedances are 50 ohms.

REFERENCES

1. Akhtarzad, S., et al, "The Design of Coupled Microstrip Lines", *IEEE Trans.* MTT–23, No. 6, June, 1975, pp. 486–492.
2. Altman, J. L., *Microwave Circuits*, D. Van Nostrand Company, Inc., 1964.
3. Bahl, I. J. and P. Bhartia, *Microwave Solid State Circuit Design*, John Wiley and Sons, 1998.
4. Bryant, T. G. and J.A. Weiss, "Parameter of Microstrip Transmission Lines and of Coupled Pairs of Microstrip Lines", *IEEE Trans.* MTT-16, No. 12, Dec. 1968, pp. 1021–1027.
5. Cohn, S. B., "Direct Coupled Resonator Filters," *Proc. IRE*, Vol. 45, pp. 187–196, February 1957.
6. Cohn, S. B., "Microwave band-pass filter containing high-Q dielectric resonators," *IEEE Trans.*, Vol. MTT-16, pp. 218–227, 1968.
7. Cohn, S. B., "Problems in Strip Transmission Lines", *IRE Trans.*, PG MTT-3, No. 2, March, 1955, pp. 119–126.
8. Collin, R. E., *Foundations for Microwave Engineering*, McGraw-Hill Inc., International Editions, 1992.
9. Fox, A. G., "Waveguide Filters and Transformers", *Bell Telephone System Lab*, Rept. MM-41-160-25, 1941.
10. Ghose, R. N., *Microwave Circuit Theory and Analysis*, McGraw-Hill book Company, Inc., 1963.
11. Hammerstad, E. O. and F. Bekkadal, "A Microstrip Handbook", *ELAB Report*, S T F 44 A74169, N7034, University of Trondheim-NTH, Norway, 1975.
12. Howe, H., Jr., "Nomographs Aid and Filter Designer", *Microwave and RF*, Vol. 24, Oct. 1985, pp. 103 107.
13. Malherbe, J. A. G., *Microwave Transmission Line Filters*, Aptech house books, Dedham, Mass, 1979.
14. Matthaei, G. L., "Design of Wide-band (and narrow-band) band-pass Microwave filters on the Insertion Loss Basis," *IRE, Trans.* Vol. MTT-8, pp. 580–593, 1960.
15. Matthaei, G. L., L. Young, and E. M. T. Johns, *Microwave Filters, Impedance Matching Networks and Coupling Structures*, McGraw-Hill, New York, 1964.
16. Mumford, W. W., "Maximally Flat Filters in Waveguides", *BSTJ*, Vol. 27, pp. 684–714, Oct. 1948.
17. Rhodes, J. D., *Theory of Electrical Filter*, John Wiley and Sons, Inc., New York, 1976.
18. Terry Edwards, *Foundation for Microstrip Circuit Design*, Sec. Edition, John Wiley and Sons, 1981.
19. Wong, J. S., "Microstrip Tapped-Line Filter Design", *IEEE Trans.* MTT., Vol. MTT-27, Jan. 1979, pp. 44–50.
20. Zverev, A. L., *Handbook of Filter Synthesis*, Wiley, New York, 1967.

9 MICROWAVE VACUUM TUBE DEVICES

INTRODUCTION 9.1

The principle of operation of a vacuum tube at microwave frequencies (above 1 GHz) is different from electronic vacuum tubes, such as triodes, tetrodes, and pentodes. These conventional electronic vacuum tubes fail to operate above 1 GHz because several factors contribute to rapidly decrease tube efficiency. Three parameters of ordinary vacuum tubes become increasingly important as frequency rises (Fig. 9.1): inter-electrode capacitance, lead inductance, and electron transit time.

With the increase in input signal frequency, the effective grid-to-cathode impedance of the tube decreases due to the interelectrode capacitance (C_{gk}). Beyond 100 MHz, the reactance of the grid-to-cathode capacitance becomes so small that much of the signal is short-circuited within the tube. Since the interelectrode capacitances are effectively in parallel with the external tuned circuits, they will also affect the resonance frequency of the tuned circuits.

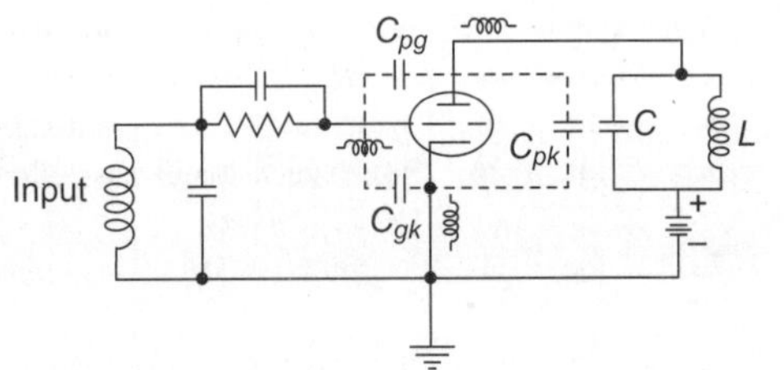

Fig. 9.1 Interelectrode capacitances in a vacuum tube

Since the lead inductances within a tube are effectively in parallel with the interelectrode capacitance, the net result is to raise the frequency limit. However, the inductance of the cathode lead is common to both the grid and plate circuits. This provides a path for degenerative feedback, resulting in decrease in overall circuit efficiency.

The transit time required for electrons to travel from the cathode to the plate is insignificant at low frequencies. However, at high frequencies, transit time becomes an appreciable portion of a signal cycle. Transit times in excess of 0.1 cycle cause a significant phase shift between plate current and grid voltage resulting in a significant decrease in tube efficiency.

For efficient operation of the tube, the plate current must be in phase with the grid-signal voltage and 180 degrees out of phase with the plate voltage. Due to appreciable transit time of electrons in the cathode grid space, the grid potential may attain a negative half cycle by the time the electrons, which started from the cathode during the positive half cycle of the grid, reach the grid. This makes the electrons oscillate back and forth in the cathode-grid space or return to the cathode resulting in reduction of efficiency. The current arriving at the plate now lags the grid-signal voltage. As a result, the power output of the tube decreases and the plate power dissipation increases.

Several methods are available to reduce the limitations of conventional tubes. But none of them works well when frequency increases beyond 1 GHz. Interelectrode capacitance can be reduced by moving the electrodes further apart or by reducing the size of the tube and its electrodes. Moving the electrodes apart increases the transit time, and reducing the size of the tube lowers the power-handling capability. The net effect is an upper frequency limit of approximately 1 GHz, beyond which the use of conventional tubes are not practical.

Efficient microwave tubes usually operate on the theory of electron velocity modulation concept to avoid the problems encountered in conventional tubes. The electron transit time is used in the conversion of dc power to *RF* power. An electron can be accelerated or decelerated by an electrostatic field. The negatively charged electron will be attracted to the positively charged electrode, travel in a direction opposite to electric field and will increase in velocity by absorbing dc energy. An electron travelling in the same direction of the electric field, the negatively charged electrode will repel the electrons and cause it to decrease in velocity and consequently giving up energy to the field. This velocity modulation of electrons is explained in Fig. 9.2.

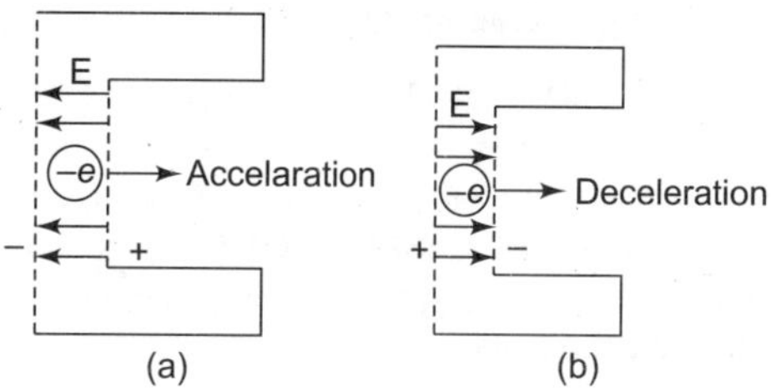

Fig. 9.2 Velocity modulation of electrons

KLYSTRONS 9.2

A kylstrons is a specialized linear beam vacuum tube, invented by Russel and Sigurd Variun (American) in 1937. There are two basic configurations of klystron tubes, one is called *reflex klystron,* used as a low-power microwave oscillator and another is called *multicavity klystron,* used as a low-power microwave amplifier. These are described below.

Reflex Klystron Oscillator 9.2.1

The schematic configuration of a reflex klystron tube is shown in Fig. 9.3, which uses only a single re-entrant microwave cavity as resonator. The electron beam emitted from the cathode *K* is accelerated by the grid *G* and passes through the cavity anode *AB* to the repeller space between the cavity anode and the repeller electrode *R*. The feedback required to maintain oscillations within the cavity is obtained by reversing the electron beam emitted from *K* towards *R* and sending it back through the cavity. The electrons in the beam are velocity modulated before the beam passes through the cavity second time and will give up the energy to the cavity to maintain oscillations. This type of a klystron oscillator is called a reflex klystron because of the reflex action of the electron beam.

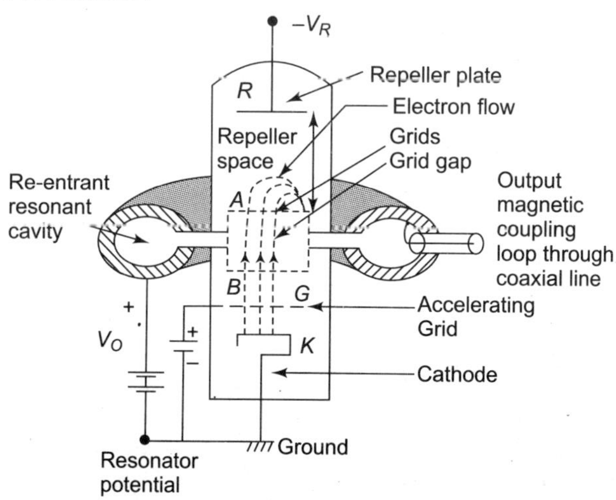

Fig. 9.3 Functional diagram of a reflex klystron

The physical design of the tube controls the number of modes possible in practical applications. Normally a range of four modes of operation are available. The choice of mode is determined by the power available from the mode and the band of frequencies over which the circuit can be tuned.

a. Mechanism of Oscillation

Due to dc voltage in the cavity circuit, *RF* noise is generated in the cavity. This electromagnetic noise field in the cavity becomes pronounced at cavity resonant frequency. As explained in Fig. 9.4, the electrons passing through the cavity gap *d* experience this *RF* field and are velocity modulated in the following manner.

The electrons *a* shown in Fig. 9.5 which encountered the positive half cycle of the *RF* field in the cavity gap *d* will be accelerated, those (reference electrons) *b* which encountered zero *RF* field will pass with unchanged original velocity, and the electrons *c* which encountered the negative half cycle will be retarded on entering the repeller space.

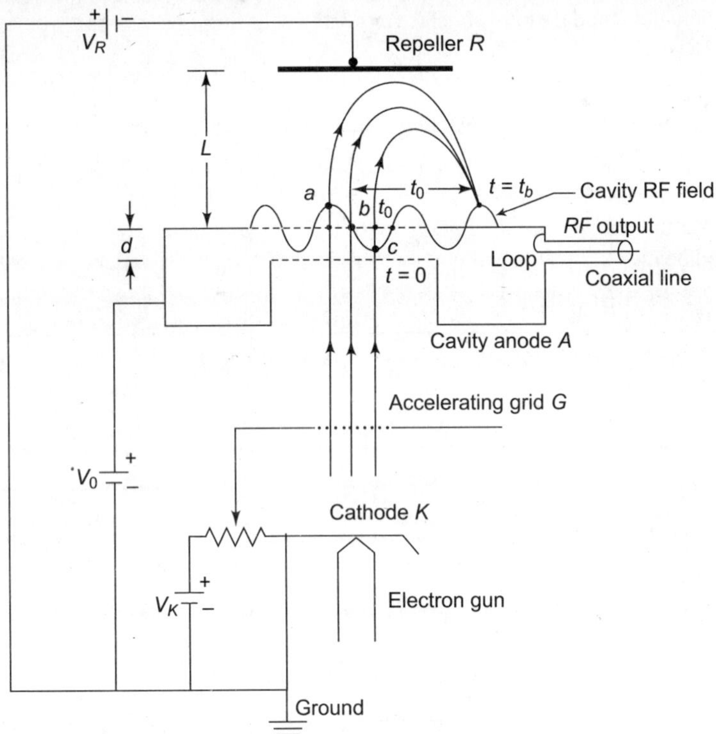

Fig. 9.4 Reflex klystron operation

All these velocity-modulated electrons will be repelled back to the cavity by the repeller due to its negative potential. The repeller distance *L* and the voltages can be adjusted to receive all the voltage-modulated electrons at a same time on the positive peak of the cavity *RF* cycle. Thus the velocity modulated electrons are bunched together and lose their kinetic energy when they encounter the positive cycle of the cavity *RF* field. This loss of energy is thus transferred to the cavity to conserve the total power. If the power delivered by the bunched electrons to the cavity is greater than the power loss in the cavity, the electromagnetic field amplitude at the resonant frequency of the cavity will increase to produce microwave oscillations. The *RF* power is coupled to the output load by means of a small loop which forms the centre conductor of the coaxial line. When the power delivered to the cavity by the electrons becomes equal to the total power loss in the cavity system, a steady microwave oscillation is generated and maintained at resonant frequency of the cavity.

b. Mode of Oscillation

The bunched electrons in a reflex klystron can deliver maximum power to the cavity at any instant which corresponds to the positive peak of the *RF* cycle of the cavity oscillation. If T is the time period at the resonant frequency, t_0 is the time taken by the reference electron to travel in the repeller space between entering the repeller space at b and the returning to the cavity at positive peak voltage on formation of the bunch, then

$$t_0 = (n + 3/4)\,T = NT \tag{9.1}$$

where $N = n + 3/4$, $n = 0, 1, 2, 3, \ldots$, as shown in Fig. 9.5.

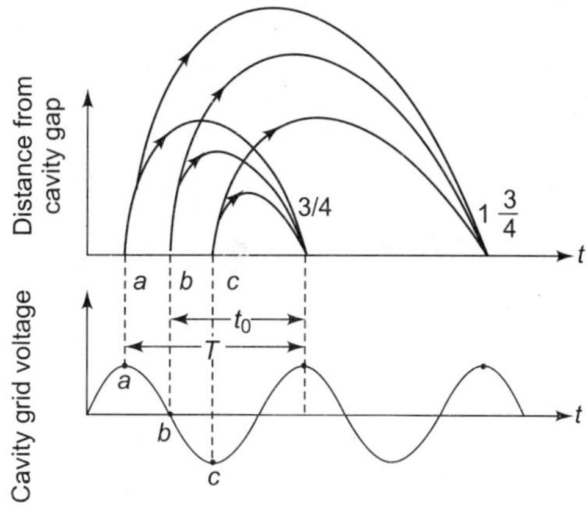

Fig. 9.5 Reflex klystron modes

Thus by adjusting repeller voltage for a given dimension of the reflex klystron, the bunching can be made to occur at $N = n + 3/4$ positive half cycle. Accordingly the mode of oscillation is named as $N = \frac{3}{4}, 1\frac{3}{4}, 2\frac{3}{4}$, etc., for modes $n = 0, 1, 2, 3, \ldots$, respectively. It is obvious that the lowest-order mode 3/4 occurs for a maximum value of repeller voltage when the transit time t_0 of the electrons in the repeller space is minimum. Higher modes occurs at lower repeller voltages. Since at the highest repeller voltage, the acceleration of the bunched electrons on return is maximum, the power output of the lowest mode is maximum.

c. Power Output and Efficiency

The power output of a reflex klystron is maximum if the bunched electrons on return cross the cavity gap when the gap field is positive maximum. This power can be calculated from various parameters as described below with reference to Figs 9.4 and 9.5.

❑ *Velocity Modulation* For calculation of *RF* power, we will assume the following:

1. Cavity grids and repeller are plane parallel and very large in extent.
2. No *RF* field is excited in the repeller space.
3. Electrons are not intercepted by the cavity anode grid.
4. No debunching takes place in the repeller space.
5. The cavity *RF* gap voltage amplitude V_1 is small compared to the dc beam voltage V_0 : $V_1 << V_0$.

The electron velocity u attained due to the dc beam voltage V_0 while entering the cavity gap at $t = 0$ is uniform, given by

$$u = u_0 = \sqrt{(2eV_0/m)} = 5.93 \times 10^5 \sqrt{V_0} \text{ m/sec.} \tag{9.2}$$

where V_0 is in volt and $u = 0$ at cathode. The instantaneous cavity *RF* voltage can be written as

$$V(t) = V_1 \sin \omega t \tag{9.3}$$

where $V_1 << V_0$. The average transit time through the cavity gap d and the transit angle are given by, respectively,

$$t_g = d/u_0,\ \theta_g = \omega t_g = \omega d/u_0 \tag{9.4}$$

The average microwave voltage in the cavity gap can be written as

$$V_{av} = (1/t_g)\int_0^{t_g} V_1 \sin \omega t\, dt = \frac{V_1[1-\cos \omega t_g]}{\omega t_g}$$

$$= \frac{V_1 \sin^2 (\theta_g/2)}{\theta_g/2} = V_1\, \beta_1 \sin \theta_g/2 \tag{9.5}$$

where
$$\beta_1 = \frac{\sin (\theta_g/2)}{\theta_g/2} = \frac{\sin\left(\dfrac{\omega d}{2u_0}\right)}{\dfrac{\omega d}{2u_0}} \tag{9.6}$$

is called the *beam-coupling coefficient* of the cavity gap. Therefore, the coupling between the electron beam and the cavity varies with the cavity gap d is $\dfrac{\sin X}{X}$ form. Here, $\beta_1 = 1$ for $d \to 0$.

The exit velocity from the cavity gap after velocity modulation is given by

$$u(t_g) = \sqrt{[2e(V_0 + V_{av})/m]u} = \sqrt{\frac{2eV_0}{m}\left[1 + (V_1\beta_1/V_0)\sin(\theta_g/2)\right]} \tag{9.7}$$

where $\beta_1 V_1/V_0$ is called the *depth of modulation.* If the modulation amplitude is small (<<1), Eq. 9.7 for velocity modulation becomes,

$$u(t_g) = u_0\left[1 + \frac{\beta_1 V_1}{2V_0}\sin(\theta_g/2)\right]$$

$$= u_0\left[1 + \frac{\beta_1 V_1}{2V_0}\sin(\omega t_g - \theta_g/2)\right] \tag{9.8}$$

❑ ***Transit Time*** The round trip transit time in the repeller space is given by

$$t_r = \frac{2\,\text{velocity}\,(u)}{\text{acceleration}\,(a)} \tag{9.9}$$

The factor 2 in the numerator arises because of the to-and-fro journey. The electron acceleration is given by

$$a = \frac{eE}{m} = \frac{e}{m}\frac{V_0 + V_R + V_1 \sin \omega t}{L} \approx \frac{e}{m}\frac{V_0 + V_R}{L} \tag{9.10}$$

Therefore,
$$t_r = \frac{2u(t_g)}{a} = \frac{2u_0\, mL\left[1 + \dfrac{\beta_1 V_1}{2V_0}\sin(\omega t_g - \theta_g/2)\right]}{e(V_0 + V_R)} \tag{9.11}$$

Since the reference electron does not undergo any velocity modulation, its transit time in repeller space

$$t_0 = \frac{2u_0}{a} = \frac{2u_0\, mL}{e(V_0 + V_R)} = NT \tag{9.12}$$

$$= \frac{2\pi N}{\omega} \tag{9.13}$$

From Eqs 9.11 and 9.13

$$t_r = t_0\left[1+\frac{\beta_1 V_1}{2V_0}\sin\left(\omega t_g - \frac{\theta_g}{2}\right)\right] \tag{9.14}$$

❑ ***Density Modulation and Beam Current*** The time of arrival of an electron to the cavity gap can be expressed by

$$t_b = t_g + t_r = t_g + t_0\left[1+\frac{\beta_1 V_1}{2V_0}\sin\left(\omega t_g - \frac{\theta_g}{2}\right)\right]$$

$$= t_g + \frac{2\pi N}{\omega} + \frac{\pi N}{\omega}\frac{\beta_1 V_1}{V_0}\sin\left(\omega t_g - \frac{\theta_g}{2}\right)$$

$$= t_g + \frac{2\pi N}{\omega} + \frac{X}{\omega}\sin\left(\omega t_g - \frac{\theta_g}{2}\right) \tag{9.15}$$

where $X = \pi N\beta_1 V_1/V_0$ is called the *bunching parameter* of the reflex klystron. (9.16)

From Eq. 9.15,

$$\frac{dt_b}{dt_g} = 1 + X\cos(\omega t_g - \theta_g/2) \tag{9.17}$$

The bunched electrons on return constitute the bunched beam current $i_b\left(\frac{dQ}{dt}\right)$ such that the conservation of charge gives

$$I_0|dt_g| = i_b|dt_b| \tag{9.18}$$

where I_0 is dc the beam current. From Eqs 9.17 and 9.18, we get

$$i_b = \frac{I_0}{\dfrac{dt_b}{dt_g}} = I_0[1 + X\cos(\omega t_g - \theta_g/2)]^{-1} \tag{9.19}$$

From Eq. 9.15, since $V_1 << V_0$, $X << 1$, $t_b = t_g + 2\pi N/\omega$, so that

$$i_b = I_0[1 + X\cos(\omega t_b - 2\pi N - \theta_g/2)]^{-1} \tag{9.20}$$

By Fourier expansion, the beam current of a reflex klystron oscillator can be expressed by

$$i_b = I_0 + 2I_0\sum_{n=1}^{\infty} J_n(nX)\cos n\left(\omega t_b - 2\pi N - \frac{\theta_g}{2}\right)$$

$$= I_0 + 2I_0 J_1(X)\cos\omega\left(t_b - t_0 - \frac{t_g}{2}\right)$$

$$+ \sum_{n=2}^{\infty} 2I_0 J_n(nX)\cos n\omega\left(t_b - t_0 - \frac{t_g}{2}\right); \tag{9.21}$$

The fundamental component of the *RF* induced current in the cavity is, therefore,

$$i_{RF} = \beta_1 2I_0 J_1(X)\cos\omega\left(t_b - t_0 - \frac{t_g}{2}\right)$$

or, $$i_{RF}(t_b) \approx 2I_0\beta_1 J_1(X)\cos(\omega t_b - 2\pi N) \tag{9.22}$$
where $t_g/2 << 2\pi N$, is neglected for smallness.

❑ *Power Output* The magnitude of the fundamental *RF* current in the cavity is given by
$$|i_{RF}| = 2I_0\,\beta_1\,J_1(X) \tag{9.23}$$
The rms *RF* power delivered to the cavity is
$$P_{RF} = V_1\, i_{RF}/2 = V_1\, I_0\, \beta_1\, J_1(X) \tag{9.24}$$
Since, from Eq. 9.16
$$\frac{V_1}{V_0} = \frac{2X}{\beta_1 2\pi N} = \frac{X}{\beta_1 \pi N}$$
Therefore, Eq. 9.24 reduces to
$$P_{RF} = \frac{V_0 I_0 X J_1(X)}{\pi N} \tag{9.25}$$
From Eqs 9.2 and 9.13
$$\frac{2\sqrt{(2eV_0/m)}\, mL}{e(V_0+V_R)} = \frac{2\pi N}{\omega} \quad \text{or,} \quad \pi N = \sqrt{(2V_0 m/e)}\cdot \omega L/(V_0+V_R)$$
Therefore, $$P_{RF} = \frac{V_0 I_0 X J_1(X)(V_0+V_R)}{2\pi f L}\cdot\sqrt{(e/2mV_0)} \tag{9.26}$$

❑ *Efficiency* The dc power supplied by the beam voltage V_0 is
$$P_{dc} = V_0 I_0 \tag{9.27}$$
Therefore, the electronic efficiency of a reflex klystron oscillator is
$$\eta = \frac{P_{RF}}{P_{dc}} = \frac{X J_1(X)}{\pi N} \tag{9.28}$$
where $X = \pi N \beta_1 V_1/V_0$, $N = n + 3/4$, the mode number ($n = 0, 1, 2, 3, \ldots$), $\beta_1 = \dfrac{\sin\theta_g/2}{\theta_g/2}$ and $u_0 = \sqrt{(2eV_0/m)}$.

It can be shown from Bessel's function table, $X J_1(X)$ attains a maximum value of 1.252 at $X = 2.408$. Thus the optimum values are
$$P_{RF} = 0.3986\, I_0\, V_0/N \tag{9.29}$$
$$\eta = 0.3986/N \tag{9.30}$$
Although these equations show that the lowest is N, the highest are P_{RF} and η, it has been observed that it is not possible to get 3/4 mode in reflex klystron, so that $N = 1\frac{3}{4}$ mode leads maximum *RF* power output and efficiency:
$$P_{RF\max} = 0.227\, V_0\, I_0 \tag{9.31}$$
$$\eta_{\max} = 22.7\ \% \tag{9.32}$$

d. Mode Curve

Since the output power and frequency can be electronically controlled by varying the repeller voltage, expansions for these parameters in terms of repeller voltage are important to draw mode curves.

From Eqs 9.26 and 9.29,

$$P_{RF} = \frac{0.3986\, V_0 I_0 (V_0 + V_R)}{2 f L} \sqrt{\frac{e}{2 m V_0}} \tag{9.33}$$

Operation frequency

$$f_{\text{MHz}} = \frac{(V_0 + V_R)\, N}{L_{\text{cm}} \sqrt{V_0} \times 6.47 \times 10^{-2}} \tag{9.34}$$

or $$fL = N \sqrt{\frac{2eV_0}{m}} \left(1 + \frac{V_R}{V_0}\right)$$

and $$|V_R| = \sqrt{(8\, m/e)} \cdot (fL/N) \cdot \sqrt{V_0} - V_0 \tag{9.35}$$

$$= 6.74375 \times 10^{-6} f (L/N) \sqrt{V_0} - V_0 \tag{9.36}$$

where f is in Hz and L in metres.

From Eq. 9.36,

$$d\,|V_R|/df_{\text{Hz}} = 6.74375 \times 10^{-6}\, L\, \sqrt{V_0}/N;$$

$$N = n + 3/4,\ n = 0, 1, 2, 3, \ldots \tag{9.37}$$

The nature of variation of output power and frequency with repeller voltage for different modes are shown in Fig. 9.6.

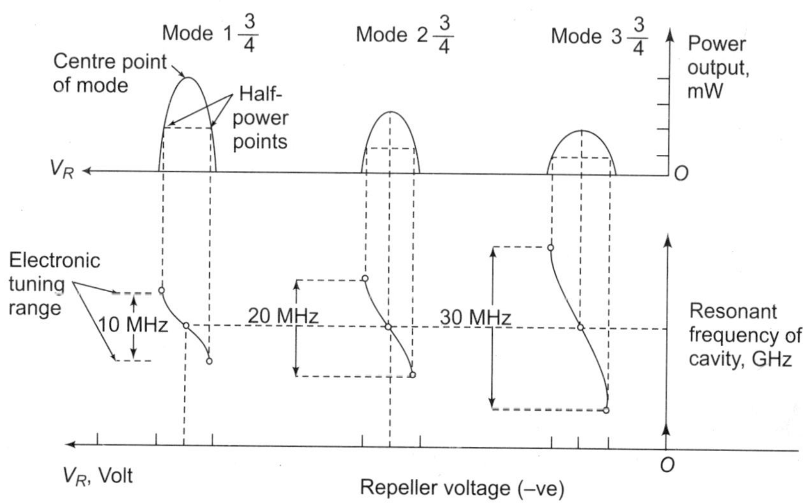

Fig. 9.6 Electronic tuning and output mode power of a reflex klystron

The variation in output power for different modes of operation can be explained by the debunching process of the electrons in the repeller field space, that is spreading out of the bunched electrons before they reach electrostatic fields across the cavity grid. The lower concentration of electrons in the returning bunch delivers less power to the oscillating cavity, which in turn reduces the amplitude of the cavity oscillations and causes a decrease in output power. In higher modes of operation, the electron bunches are formed more slowly. The long drift time in the higher modes allows more time for this electron interaction and, as a result, the effects of debunching are more severe.

Each mode has a centre frequency determined by the physical size of the cavity. The output power increases as the repeller voltage is made more negative. This results in decrease in the transit time of the electron bunches. Electronic tuning cannot change the centre frequency of the cavity, but vary the frequency within the mode of operation.

From Eqs 9.35 and 9.37, it is seen that once the cavity of a reflex klystron has been tuned to the correct frequency f_0, both the beam voltage V_0 and repeller voltage V_R are adjusted to get maximum of power output for a given mode N. The tuning frequency f_0 can be varied mechanically by a short circuit plunger as shown in Fig. 9.7 and adjusting the beam voltage V_0 with klystron power supply.

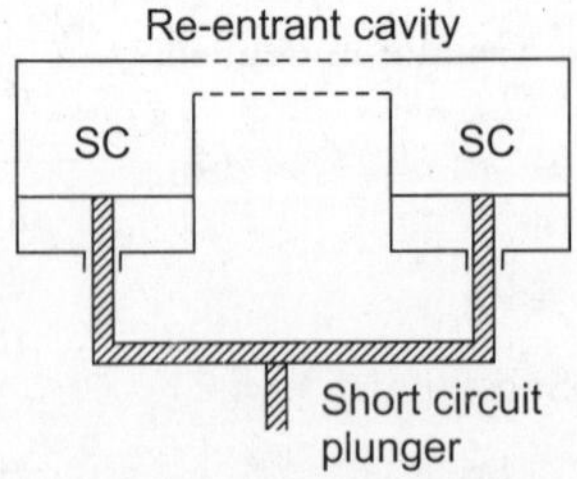

Fig. 9.7 Reflex klystron tuning

The performance of a reflex klystron is specified as
Frequency range : 2 – 200 GHz
Bandwidth : ± 30 MHz for $\Delta V_R = \pm 10$ volts
Power output : 10 mW – 2.5 W

Reflex klystrons are widely used as microwave sources in the laboratory, and in microwave transmitters, and as LO in superheterodyne radar receivers.

Example 9.1 A reflex klystron is to be operated at frequency of 10 GHz, with dc beam voltage 300 V, repeller space 0.1 cm for $1\frac{3}{4}$ mode. Calculate $P_{RF\max}$ and corresponding repeller voltage for a beam current of 20 mA.

Solution

$$P_{RF\max} = \frac{0.398\, V_0 I_0}{N} = \frac{0.398 \times 300 \times 20 \times 10^{-3}}{1\frac{3}{4}} = 1.365 \text{ Watts}$$

$$|V_R| = 6.74 \times 10^{-6} f_{(\text{Hz})} L_{(\text{m})} \sqrt{V_0}/N - V_0$$

$$L(m) = 0.1 \times 10^{-2} \text{ m} = 10^{-3} \text{ m}$$

$$N = 1\frac{3}{4} = 1.75$$

$$|V_R| = 6.74 \times 10^{-6} \times 10 \times 10^9 \times 10^{-3} \times \sqrt{300}/1.75 - 300$$

Therefore, $V_R = -\,367.08$ volts

Example 9.2 A reflex klystron is operated at 5 GHz with dc beam voltage 350 V, repeller spacing 0.5 cm for $N = 3\frac{3}{4}$ mode. Calculate bandwidth over $\Delta V_R = 1$ V.

Solution

$$N = 3\frac{3}{4} = 15/4$$

$$\Delta V_R = 6.7438 \times 10^{-6} \times L_m \times \Delta f_{\text{Hz}} \sqrt{V_0}/N$$

or, $1 = 6.74 \times 10^{-6} \times 0.5/100 \times \Delta f_{\text{Hz}} \times \sqrt{350} \times 4/15)$

or, $\Delta f = 5.948$ MHz

e. Electronic Admittance

The electronic admittance Y_e of the reflex klystron is defined by the ratio of induced bunch beam current i_b of Eq. 9.22 and cavity gap voltage at the time of bunching t_b of Eq. 9.3.

$$Y_e = i_{RF}(t_b)/V(t_b) = \frac{2I_0 \beta_1 J_1(X)\cos(\omega t_b - \omega t_0)}{V_1 \sin \omega t_b}$$

$$= \frac{2I_0 \beta_1 J_1(X)}{V_1} \frac{e^{j\omega(t_b - t_0)}}{e^{j(\omega t_b - \pi/2)}} = \frac{2I_0 \beta_1 J_1(X)}{V_1} e^{j(\pi/2 - \omega t_0)} \tag{9.38}$$

Substituting $V_1 = XV_0/\pi N\beta_1$ from (9.16) in the above Eq. 9.38.

$$Y_e = \frac{\pi N\beta_1 2I_0 \beta_1 J_1(X)}{XV_0} e^{j(\pi/2 - \omega t_0)} \tag{9.39}$$

$$= G_e + jB_e$$

Thus the electronic admittance is a function of the dc beam admittance I_0/V_0, the dc transit angle $2\pi N$, the transit angle through the cavity gap θ_g and the signal voltage V_1.

The equivalent circuit for the reflex klystron is shown in Fig. 9.8 where L and C represent the magnetic and electric storage of energy in the cavity, G_c the cavity loss conductance, G_L the load conductance. If the bunched electrons return to the cavity gap a little before reference transit time $t_0 = (n + 3/4)T$, the ac beam current lags behind the field and an inductive reactance appears in the circuit for B_e and if the bunched electrons return to the gap a little after t_0, the ac current leads the field and a capacitive reactance is presented to the circuit for B_e. The condition for oscillations is satisfied when G_e becomes negative and the total conductance in the circuit is negative.

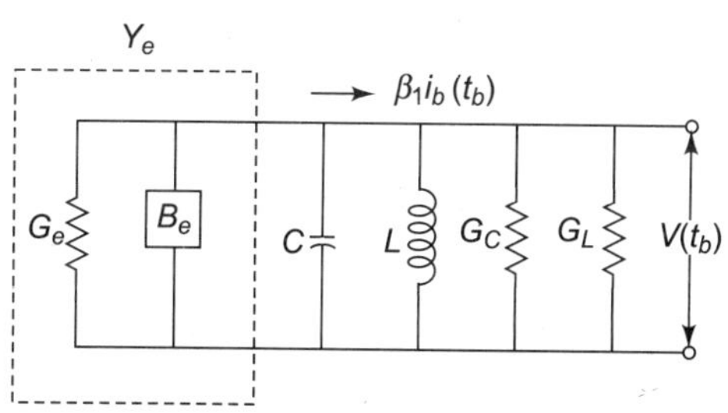

Fig. 9.8 Equivalent circuit of a reflex klystron

$$|-G_e| \geq G_c + G_L = \frac{1}{R_{sh}}; \; R_{sh} = \text{effective shunt resistance} \tag{9.40}$$

Variation of G_e and B_e in the complex Y_e plane is schematically shown in Fig. 9.9 which forms a spiral. $|V_R| = \infty$ corresponds to the origin. a decrease in $|V_R|$ moves outwards along the spiral. The oscillation will be there for the value of $2\pi N$ for which the spiral lies in the area to the left of the line $-(G_c + G_L)$ when Eq. 9.40 is satisfied for

$$\omega t_0 = 2\pi(n + 3/4) = 2\pi N, \; n = 0, 1, 2, \ldots$$

If $Y_e = \pm jB_e$, oscillations take place at a frequency lower or higher than the cavity resonance frequency as shown in Figs 9.6 and 9.9.

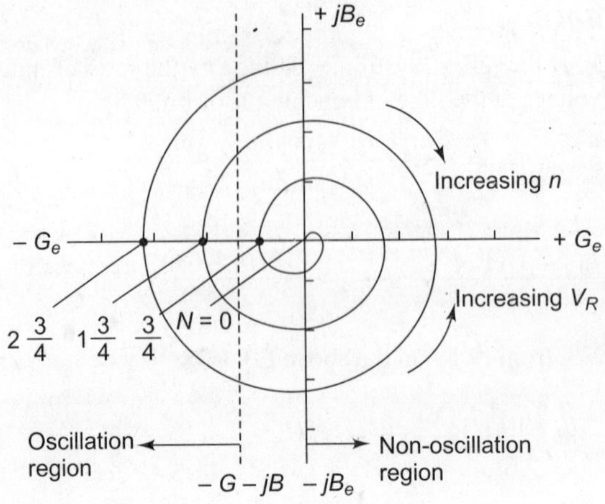

Fig. 9.9 Electronic admittance of a reflex klystron

f. Modulation of a Reflex Klystron

A reflex klystron is widely used as a microwave source for laboratory experiments. To find the characteristics of the line or a load, the most common measurable parameters are VSWR and the position of a voltage standing wave minimum. To avoid the use of the costly microwave receivers or power meters, microwave signals are low-frequency modulated and this modulated signal is probed and detected by a crystal detector. The detected microwave signals that carry the original amplitude and phase variation information of the microwave signal are measured using a low-frequency receiver to obtain the desired parameters.

Two basic modulations are used for this purpose: (1) amplitude modulation by square waves, and (2) frequency modulation by sawtooth waves, as shown in Figs 9.10 and 9.11.

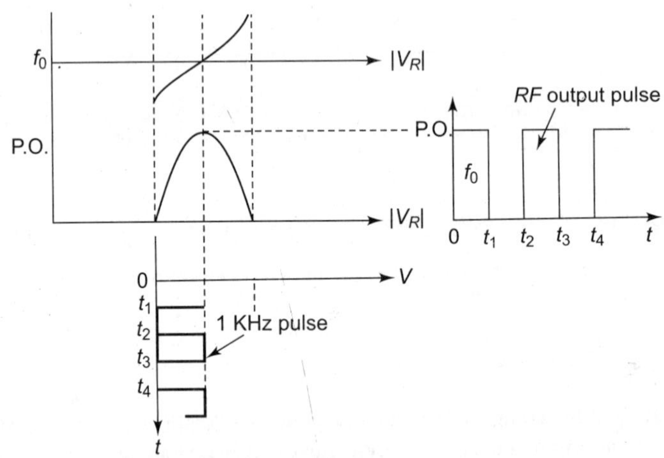

Fig. 9.10 Amplitude modulation in reflex klystron

In amplitude modulation, the dc repeller voltage is adjusted at the left edge of the mode power curve and a low-frequency (usually 1 KHz) square voltage wave is superimposed on the dc repeller voltage. The amplitude of the square pulse is adjusted to attain the power maximum point. Due to square-wave variation of effective repeller voltage, the output is a pulsed signal at a constant frequency f_0. Here, care must be taken such that the negative half cycle of the modulating amplitude should not enter into the higher mode or part of the same mode to avoid oscillation at two or more frequencies simultaneously.

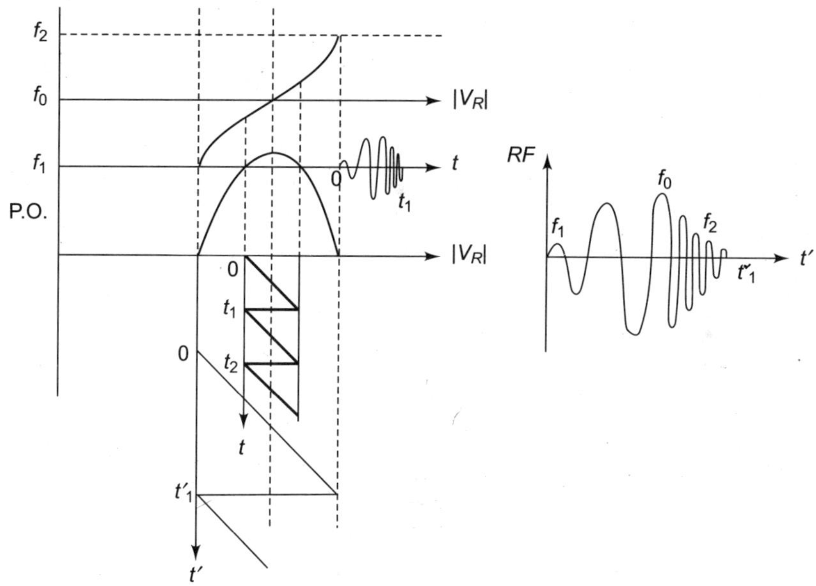

Fig. 9.11 Frequency modulation in reflex klystron

Frequency modulation is useful for frequency sweeping over the band of frequency for a given mode by using a superimposed sawtooth voltage at the repeller plate. Here, the positioning of the dc repeller voltage and the amplitude of the sawtooth wave are so adjusted that the frequency sweeping taken place over a nearly linear part of the frequency variation and at the same time over a small portion of the power output curve maintain almost constant power output and linear frequency sweeping for the response characteristics obtainable in a oscilloscope, where the detected output is fed to the y plate and the modulating signal to the x plate for triggering.

Example 9.3 A reflex klystron is operated at 9 GHz with a dc beam voltage of 600 V for $1\frac{3}{4}$ mode, repeller space length of 1 mm, and dc beam current of 10 mA. The beam-coupling coefficient is assumed to be 1. Calculate the repeller voltage, electronic efficiency and output power.

Solution

Given

$f_0 = 9$ GHz, $\quad V_R = ?$

$N = 1\frac{3}{4}$, $\quad P_{RF} = ?$

$V_0 = 600$ Volt, $\quad \eta = ?$

$L = 1$ mm

$\beta_1 = 1$, $\quad I_0 = 10$ mA.

From Eq. 9.36

$$|V_R| = 6.74 \times 10^{-6} \times 1 \times 10^{-3} \times 9 \times 10^9 \times \frac{\sqrt{600}}{1\frac{3}{4}} - 600 = 249 \text{ V}$$

$$P_{RF\max} = \frac{0.398V_0 I_0}{N} = \frac{0.398\times600\times10\times10^{-3}}{1\frac{3}{4}}$$

$$= 0.2274 \times 600 \times 10 \times 10^{-3} = 1.3644 \text{ Watts}$$

$$\eta_{\max} = \frac{XJ_1(X)}{\pi N} = \frac{0.398}{N} = \frac{0.398}{1\frac{3}{4}} = 22.74\ \%$$

Two-cavity Klystron Amplifier 9.2.2

The two-cavity klystron tubes are widely used for microwave amplification as shown by the schematic circuit diagram in Fig. 9.12.

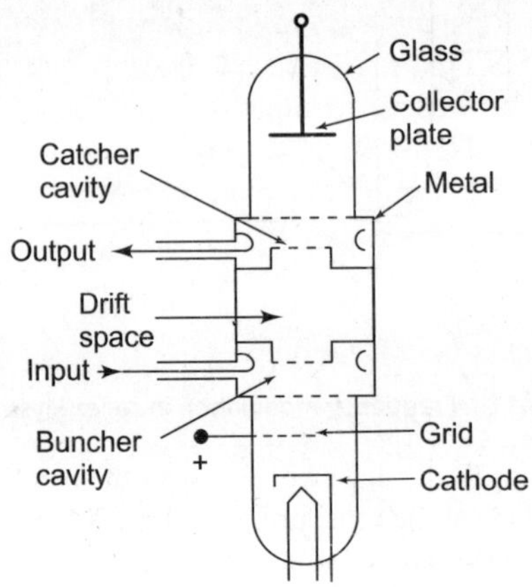

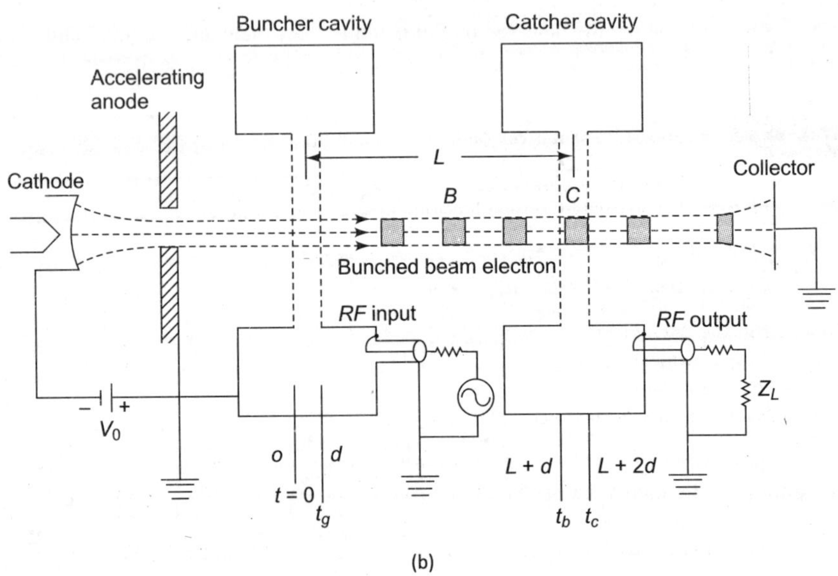

Fig. 9.12 Two-cavity klystron amplifier: (a) Schematic diagram (b) Functional diagram

a. Mechanism of Operation

The basic principles of two-cavity klystron operation can be described in the following manner. A high velocity electron beam produced by the accelerating anode is successively passed through an input re-entrant cavity resonator (buncher cavity) grid, a field free drift space of length L, an output re-entrant cavity resonator (catcher cavity) grid and finally collected by a collector electrode. The electron beam is focussed to travel axially without spreading during transit by applying an axial magnetic field produced by an external coil current. The input *RF* signal to be amplified excites the buncher cavity with a coupling loop. The combination of the anode voltage V_0 and the cavity gap width d are such that the transit of electrons through each cavity gap is less than the quarter of the time period T of the input signal cycle. The electrons passing through the buncher grids are accelerated/retarded/passed through with unchanged initial dc velocity depending upon when they encounter the *RF* signal field at the buncher cavity gap at positive/negative/zero crossing phase of the cycle, respectively, as shown by the distance–time plot in Fig. 9.13. This is called the *applegate diagram.* Thus, the electron beam is velocity modulated to form bunches or undergoes density modulation in accordance with the input *RF* signal cycle. While passing through the catcher cavity grid, this density modulated electron beam induces *RF* current in the output cavity and thereby excites the *RF* field in the output cavity at input signal cycle. The phase of field in the output cavity is opposite to that of the input cavity so that the bunched electrons are retarded by the output gap voltage. The loss of kinetic energy of the electrons on retardation process transfers *RF* energy to the output cavity continuously at signal cycle. The amplitude of the signal at output cavity attains a steady large value when the loss of kinetic energy of the bunched electrons compensates the output cavity circuit losses. The amplified signal is coupled out from the catcher cavity through a current loop to the load.

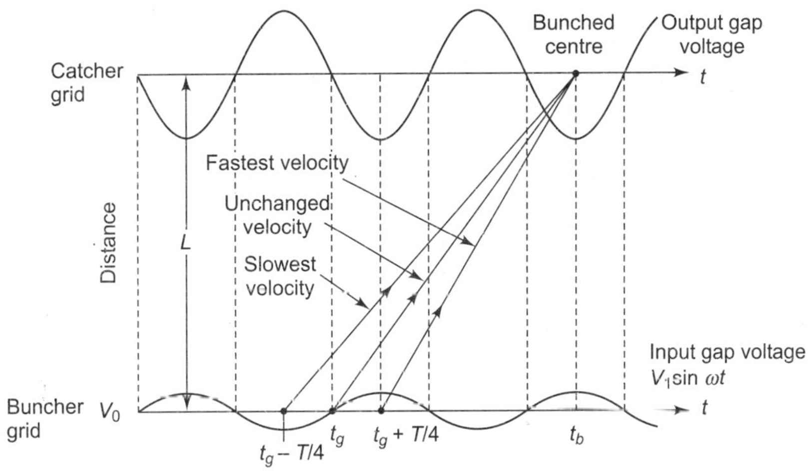

Fig. 9.13 Applegate diagram

The current induced in the catcher cavity is rich in harmonics up to say, 15. The cavity is tuned to the fundamental or any harmonic as desired.

b. Analysis of a Two-cavity Klystron Amplifier

The analysis for *RF* amplification by a two-cavity klystron amplifier is based on the following assumptions:

1. The transit time in the cavity gap is very small compared to the period of the input *RF* signal cycle.
2. The input *RF* signal amplitude V_1 is very small compared to the dc beam voltage V_0 = The anode potential with respect to the cathode potential.

3. The cathode, anode, cavity grids and collector are all parallel and the cavity grids do not intercept any electron while passing.
4. No space charge or debunching take place at the bunch point.
5. The *RF* fields are totally confined in the cavity gaps so that field is zero in the drift space L.
6. The electrons leave the cathode with zero initial velocity.

❑ ***Velocity Modulation*** Due to potential difference V_0 between the anode and cathode, the electrons form a high current density beam with the axial velocity u_0 which is obtained by equating kinetic energy and potential energy:

$$u_0 = \sqrt{(2eV_0/m)} = 5.93 \times 10^5 \sqrt{V_0} \text{ m/s} \tag{9.41}$$

Let the signal voltage across the gap be $V_1 \sin \omega t$. The electron beam enters the buncher cavity at a time $t = t_1$ with velocity u_0 and passes out at $t = t_2$ as shown in Fig. 9.14.

The transit time and transit angle through the buncher gap d is

$$t_2 - t_1 = t_g = d/u_0,\ \theta_g = \omega t_g \tag{9.42}$$

respectively. Due to input *RF* signal in the buncher cavity, the average *RF* voltage in the buncher gap can be obtained as

$$V_{av} = (1/t_g) \int_{t_1}^{t_2} V_1 \sin \omega t \, dt = (V_1/\omega t_g)(\cos \omega t_1 - \cos \omega t_2)$$

or,
$$V_{av} = V_1 \beta_1 \sin (\omega t_1 + \theta_g/2) \tag{9.43}$$

where, $\beta_1 = (\sin \theta_g/2)/(\theta_g/2)$, is called the buncher cavity beam-coupling coefficient. As electrons pass through the buncher gap, their velocities are increased, decreased or unchanged, depending on the positive, negative, or zero *RF* voltage across the grids when they pass through. At time $t_{av} = (t_1 + t_2)/2$ the electron is midway across the gap with velocity say u_{av} say. Then

$$u_{av}/u_0 = \sqrt{[(V_0 + V_{av})/V_0]} \quad \text{or,} \quad u_{av} = u_0 \left[1 + \frac{V_1\beta_1}{V_0} \sin(\omega t_1 + \theta_g/2)\right]^{1/2} \tag{9.44}$$

Since $V_1 << V_0$, The binomial expansion of Eq. 9.44 gives

$$u_{av} \approx u_0 \left[1 + \frac{M}{2} \sin (\omega t_1 + \theta_g/2)\right] \tag{9.45}$$

Thus, the electrons in the beam are velocity modulated by the input *RF* signal with depth of velocity modulation $\Delta m = \beta_1 V_1/V_0$. These electrons exit from the buncher gap at time $t_1 + t_2$ to the field free drift space between the two cavities with velocity

$$u_a(t_2) \approx u_0 \left[1 + \frac{M}{2} \sin (\omega t_2 - \theta_g/2)\right] \tag{9.46}$$

❑ ***Transit Time in the Drift Space*** If t_3 is the time when the bunched electrons are at the catcher grid after travelling through the field free drift space L,

$$t_3 = t_2 + \frac{L}{u(t_2)} = t_2 + \frac{L}{u_0[1 + \frac{M}{2} \sin (\omega t_2 - \theta_g/2)]}$$

$$\approx t_2 + (L/u_0)\left[1 - \frac{M}{2} \sin (\omega t_2 - \theta_g/2\right] \tag{9.47}$$

Therefore, the transit time t_d in the drift space is given by

$$t_d \approx t_3 - t_2 = L/u_0 . \left[1 - \frac{M}{2} \sin (\omega t_2 - \theta_g/2\right] \tag{9.48}$$

❑ ***Density Modulation*** Because of the difference in velocities of the electrons in the velocity modulated beam, the electrons will form bunches, i.e., become density modulated, in accordance with the input signal cycle.

This bunching action can be illustrated graphically by plotting distance as a function of time for the drift space, what is known as *Applegate diagram* as shown in Fig. 9.10 for the velocities $u_{min} = u_0 (1 - M/2)$, $u_{max} = u_0 (1 + M/2)$ and $u_{unchange} = u_0$. A maximum degree of bunching takes place when the buncher and catcher cavities are spaced to satisfy the condition

$$t_d = t_3 - t_2 = L/u(t_2) = t_0[1 - (M/2) \sin (\omega t_2 - \theta_g/2] \qquad (9.49)$$

where $t_0 = L/u_0$, is the dc transit time in drift space. The corresponding transit angle in the drift space L is

$$\omega t_d = \omega(t_3 - t_2) = \omega t_0 - (M\omega t_0/2) \sin (\omega t_2 - \theta_g/2)$$
$$= \theta_0 - (M/2)\, \theta_0 \sin (\omega t_2 - \theta_g/2) \qquad (9.50)$$

where $\theta_0 = \omega t_0$, the dc transit angle. If N is the number of *RF* cycles elapsed during the transit time of reference electron with velocity u_0, at the point of bunching ωt_0's

$$t_0 = \frac{L}{u_0} = NT = \frac{N}{f} = \frac{2\pi N}{\omega}$$

Therefore, $\theta_0 = \omega t_0 = 2\pi N$ (9.51)

The catcher cavity is placed so that the electron bunching occurs within the catcher gap. During traversal through the gap the bunched electrons induce positive charges in the catcher grids by repelling conduction electrons in it as shown in Fig. 9.14.

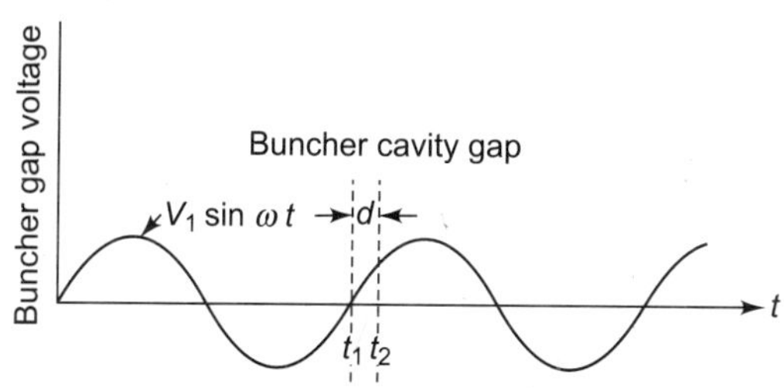

Fig. 9.14 The buncher gap voltage

This process results in a series of positive current pulses in the external circuit at the rate equal to the periodic time of the input signal and width equal to the transit time across the gap. The Fourier transform of this pulse waveform gives a large number of frequency components. The output cavity is tuned to the fundamental component equal to the frequency of the input cycle and provides a sinusoidal signal at the output port of the catcher cavity.

Let I_0 = dc beam current in the buncher, and i_b = bunched beam current at catcher = dQ/dt_3

Assuming there is no loss of electrons, total charge

$$Q = \Sigma dQ = \Sigma I_0\, dt_2 = \Sigma i_b\, dt_3 \quad \text{or,} \quad i_b = \frac{I_0}{\Sigma(dt_3/dt_2)} \qquad (9.52)$$

From Eq. 9.48,

$$t_3 = t_2 + \frac{2\pi N}{\omega}\left[1 - \frac{M}{2}\sin\left(\omega t_2 - \frac{\theta_g}{2}\right)\right]$$

$$= t_2 + \frac{2\pi N}{\omega} - \frac{M\pi N}{\omega}\sin\left(\omega t_2 - \frac{\theta_g}{2}\right)$$

$$\frac{dt_3}{dt_2} = 1 - M\,\pi N \cos\left(\omega t_2 - \frac{\theta_g}{2}\right) = 1 - X \cos\left(\omega t_2 - \frac{\theta_g}{2}\right) \tag{9.53}$$

where, $$X = \pi M N = \frac{\pi N \beta_1 V_1}{V_0} \tag{9.54}$$

and is defined as the *bunching parameter* of a two-cavity klystron.

From Eqs 9.52 and 9.53,

$$i_b = \frac{I_0}{\Sigma[1 - X\cos(\omega t_2 - \theta_g/2)]} \tag{9.55}$$

Equation 9.55 is the expression of the bunched beam current in the output cavity. Substituting $t_2 = t_3 - t_0 = t_3 - 2\,\pi N/\omega$ in Eq. 9.55.

$$i_b = I_0\,\Sigma[1 - X\cos(\omega t_3 - 2\,\pi N - \theta_g/2]^{-1} \tag{9.56}$$

Expanding as a Fourier series and finding the coefficients

$$i_b = I_0 + 2I_0 \sum_{n=1}^{\infty} J_n\,(n\,X)\cos n(\omega t_3 - \theta_g - \theta_0) \tag{9.57}$$

This is the generalized expression of a bunched beam current consisting of a dc component I_0 plus the fundamental ac component of amplitude $2\,I_0 J_1(X)$ and harmonics of amplitude $2I_0 J_n(nX)$, $n = 2, 3, 4, \ldots$

The klystron is generally tuned to fundamental ac component of current given by

$$i_f(t_3) = 2I_0 J_1\,(X)\cos(\omega t_3 - \theta_g - \theta_0) \tag{9.58}$$

The current can be made maximum when $J_1\,(X)$ is max = 0.582 at $X = 1.841$ by adjusting the dc beam voltage V_0.

From Eqs 9.51, 9.53 and 9.58, it can be shown that the fundamental component of current is maximum for an optimum drift space length,

$$L_{\text{op}} = \frac{3.682\,u_0\,V_0}{\omega\,\beta_1 V_1} \tag{9.59}$$

❑ ***Beam Spreading/Debunching*** At the point of bunching the electrostatic force of repulsion between the electrons does not allow electronic collision, but may cause beam spreading or undesirable debunching. This reduces the efficiency of the klystrons.

❑ ***Power Output*** The fundamental component of *RF* beam current passing through the output cavity gap induces a current in the catcher cavity

$$i_c = i_f\,\beta_2 \tag{9.60}$$

where β_2 is the beam-coupling coefficient of the catcher cavity gap. $\beta_2 = \beta_1$ when both buncher and catcher cavities are identical.

Therefore, from Eqs 9.58 and 9.60

$$\begin{aligned} i_c &= 2I_0\,\beta_2 J_1(X)\cos(\omega t_3 - \theta_g - \theta_0) \\ &= I_2\,\beta_2\cos(\omega t_3 - \theta_g - \theta_0);\ I_2 = 2I_0 J_1\,(X) \end{aligned} \tag{9.61}$$

The corresponding *RF* voltage across the catcher cavity is given by

$$V_c = V_2\cos(\omega t_3 - \theta_g - \theta_0 - \phi) = \beta_2 I_2 R_{\text{sh}} \tag{9.62}$$

where ϕ is the phase angle between i_c and V_c. The average power delivered to the output cavity is

$$P_0 = \frac{1}{2\pi}\int_0^{2\pi} i_c\ i_c\,V_c\,d\,(\omega t_3); = \beta_2 I_0\,V_2 J_1\,(X)\cos\phi \tag{9.63}$$

The coupling between the beam and *RF* field in the gap is better for higher values of $\beta \leq 1$.

For maximum power output $\cos\phi = 1$, $\phi = 0°$, $X = 1.841$ and $\beta_2 = 1$. Therefore,

$$P_{0\,max} = I_0 V_2 J_1(X) = 0.582\, I_0 V_2 \tag{9.64}$$

Since power output is a function of $X = \pi N \beta_1 V_1/V_0$, it can be varied by changing V_0 as shown in Fig. 9.15. From Eq. 9.51,

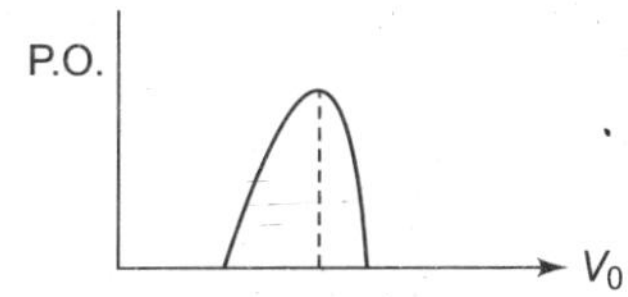

Fig. 9.15 Power output vs beam voltage

$$u_0 = \frac{Lf}{N}$$

or, $$\sqrt{(2eV_0/m} = \frac{Lf}{N}$$

or, $$V_0 = \frac{m}{2e}(Lf/N)^2 \tag{9.65}$$

❑ ***Efficiency*** Efficiency η of a two-cavity klystron can be defined as the ratio of *RF* output power to the dc beam power

$$\eta = \frac{P_{ac}}{P_{dc}} = \frac{\beta_2 I_0 V_2 J_1(X)}{I_0 V_0} = \frac{\beta_2 V_2 J_1(X)}{V_0} \tag{9.66}$$

The efficiency becomes maximum when $X = 1.841$

$$\eta_{max} = 0.582\, \beta_2 V_2/V_0 \tag{9.67}$$

If the coupling is perfect $\beta_2 = 1$

$$i_{c\,max} = 2 I_0 \times 0.582 \tag{9.68}$$

and $V_2 = V_0$. Then the $\eta_{max} = 58.2\%$

c. Equivalent Circuit

The output cavity can be represented by an equivalent circuit as shown in Fig. 9.16 where G_{sh} represents the effective shunt conductance which accounts for the catcher cavity wall resistance, beam-loading resistance and external load resistance. The electric and magnetic stored energies are represented by C and L, respectively, so that the resonant frequency of the cavity is

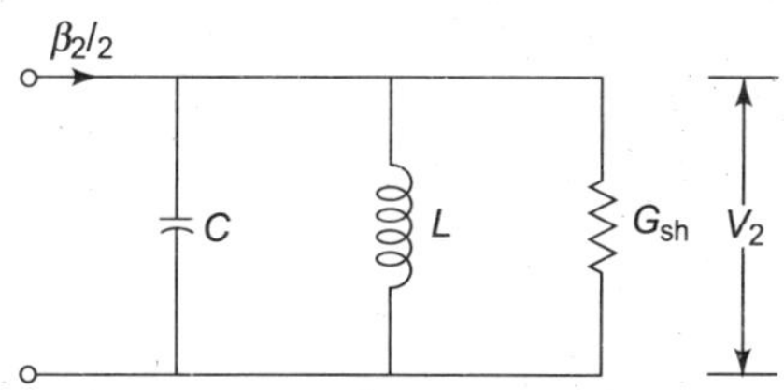

Fig. 9.16 Equivalent circuit of output cavity

$$f_0 = \frac{1}{2\pi\sqrt{LC}} \tag{9.69}$$

The power output is given by

$$P = |i_c|^2/(2G_{sh}) = 4 I_0^2 \beta_2^2 J_1^2(X)/(2G_{sh}) \tag{9.70}$$

d. Voltage Gain and Mutual Conductance

The voltage gain for the klystron is defined by

$$A_v = |V_2/V_1|$$

From the equivalent circuit

$$V_2 = \frac{\beta_2 I_2}{G_{sh}}$$

$$X = \frac{\pi N V_1 \beta_1}{V_0}$$

Therefore, $$V_1 = \frac{X V_0}{N \pi \beta_1} = \frac{2V_0}{\beta_2 \theta_0}$$

and $$A_v = \frac{\beta^2 \theta_0 I_0 J_1(X)}{X V_0 G_{sh}} \tag{9.71}$$

where $\beta_1 = \beta_2$ for identical cavities.

The dc beam conductance $G_0 = I_0/V_0$

The mutual conductance of the klystron amplifier is defined as

$$|G_m| = |i_c|/V_1 = 2\,\beta_2\, I_0\, J_1\,(X)/V_1$$

$$= \beta^2\, \theta_0\, G_0\, J_1\,(X)/X \tag{9.72}$$

Therefore, $$A_v = G_m/G_{sh} \tag{9.73}$$

G_m decreases with increase in X.

For maximum power output, $X = 1.841$, so that

$$|G_m| = 0.316\, G_0\, \beta^2\, \theta_0 \tag{9.74}$$

Two-cavity klystron amplifier is not a low-noise device. Therefore, it is normally used in the transmitter, not in the receiver. Some typical applications are, troposcatter transmitter, satellite communication ground stations and UHF TV transmitter power amplifier. The typical frequency range of applications is from C-band to about 60 GHz. The average cw power outputs is in the range 100 kW–250 kW for high power applications. A 30–60 dB power gain is possible with a bandwidth of 10–60 MHz and 30–40% efficiency in practice.

Example 9.4 A two-cavity klystron amplifier is tuned at 3 GHz. The drift space length is 2 cm and beam current is 25 mA. The catcher voltage is 0.3 times the beam voltage. It is assumed that the gap length of the cavity << the drift space so that the input and output voltages are in phase ($\beta = 1$). Compute (a) power output and efficiency for $N = 5\frac{1}{4}$, (b) beam voltage, input voltage and output voltage for maximum power output of $5\frac{1}{4}$ mode.

Solution Beam voltage

$$V_0 = \frac{m}{2e}\,(Lf/N)^2 = \frac{9.1\times10^{-31}}{2\times1.6\times10^{-19}}\left[\frac{2\times10^{-2}\times3\times10^9}{5+\frac{1}{4}}\right]^2 = 371.4 \text{ volts}$$

For maximum power output, the bunching parameter

$$X = 1.84 = \pi N V_1/V_0$$

Therefore, the buncher voltage magnitude

$$V_1 = \frac{X V_0}{\pi N} = \frac{1.84\times371.4}{3.1415\times5\frac{1}{4}} = 41.43 \text{ volts}$$

Catcher voltage $V_2 = 0.3\, V_0 = 111.4$ volts

Output power $P_0 = \beta_2\, I_0\, V_2\, J_1(X) \cos\phi$

$$P_{0max} = I_2 V_0 J_2(X)$$

or,
$$P_{0max} = 0.582\, I_0\, V_2$$
$$= 0.582 \times 25 \times 10^{-3} \times 111.4 \text{ watts} = 1.621 \text{ watts}$$
$$P_{dc} = I_0 V_0 = 25 \times 10^{-3} \times 371.4 \text{ watts} = 9.3 \text{ watts}$$

Efficiency $\eta_{max} = 0.582\, V_2/V_0 = 0.582 \times 111.4/371.4 = 17.46\ \%$

Example 9.5 A two-cavity klystron operates at 5 GHz with a dc beam voltage of 10 kV and a 2 mm cavity gap. For a given input *RF* voltage, the magnitude of the gap voltage is 100 volts. Calculate the transit time at the cavity gap, the transit angle, and the velocity of the electrons leaving the gap.

Solution

dc beam velocity
$$u_0 = 0.593 \times 10^6 \sqrt{V_0}$$
$$= 0.593 \times 10^6 \sqrt{(10 \times 10^3)} \text{ m/s}$$
$$= 0.593 \times 10^8 \text{ m/s}$$

Gap transmit time
$$t_g = \frac{d}{u_0} = \frac{2 \times 10^{-3}}{0.593 \times 10^8} = 33.7 \times 10^{-12} \text{ sec}$$

The gap transit angle
$$\theta_g = \omega t_g = 2\pi \times 5 \times 10^9 \times 33.7 \times 10^{-12} \text{ rad}$$
$$= 1.059 \text{ rad} = 60.7 \text{ deg}$$

The beam-coupling coefficient
$$\beta_1 = \frac{\sin(\theta_g/2)}{\theta_g/2} = 0.505/0.5295 = 0.9537$$

The velocity of electrons leaving the input cavity gap
$$u(t) = u_0 \left[1 + \frac{\beta_1 V_1}{2V_0} \sin(\omega t + \theta_g/2)\right]$$
$$= 0.593 \times 10^8\, [1 + (0.954 \times 100)/(2 \times 10 \times 10^3) \sin(\omega t + \theta_g/2)]$$
$$= 0.593 \times 10^8\, [1 + 0.00477 \sin(\omega t + 0.5295)]$$

is varying sinusoidally at the input cycle

The maximum velocity
$$u(t)_{max} = u_0\,(1 + M/2) = 0.593 \times 10^8\,(1 + 0.00477) = 0.5958 \times 10^8 \text{ m/s}$$

The minimum velocity
$$u(t)_{min} = u_0\,(1 - M/2) = 0.593 \times 10^8\,(1 - 0.00477) = 0.5902 \times 10^8 \text{ m/s}$$

Example 9.6 A two-cavity klystron operates at 10 GHz with $I_0 = 3.6$ mA, $V_0 = 10$ kV. The drift space length is 2 cm, the output cavity total shunt conductance is $G_{sh} = 20\ \mu$ mho and beam-coupling coefficient $\beta_2 = 0.92$. Find the maximum voltage gain.

Solution Maximum voltage gain
$$A_{max} = \frac{\beta^2 \theta_0 I_0 J_1(X)_{max}}{X V_0 G_{sh}}$$

dc beam velocity $u_0 = 0.593 \times 10^6 \sqrt{V_0} = 0.593 \times 10^6 \times \sqrt{10} \times 10^3$

$$= 0.593 \times 10^8 \text{ m/s}$$

Transit angle in drift space

$$\theta_0 = \frac{\omega L}{u_0} = \frac{2\pi \times 10 \times 10^6 \times 2 \times 10^{-2}}{0.593 \times 10^8} = 21.19 \text{ rad}$$

$$A_{max} = \frac{0.92 \times 0.92 \times 21.19 \times 3.6 \times 0.582}{1.841 \times 10 \times 10^3 \times 20 \times 10^{-6}} = 102.1$$

Example 9.7 An identical two-cavity klystron amplifier operates at 4 GHz with $V_0 = 1$ kV, $I_0 =$ 22 mA, cavity gap = 1 mm and drift space = 3 cm. If dc beam conductance and catcher cavity total equivalent conductance are 0.25×10^{-4} mhos and 0.3×10^{-4} mhos respectively, calculate,

(a) the beam-coupling coefficient, dc transit angle in the drift space and the input cavity voltage magnitude for maximum output voltage
(b) voltage gain and efficiency, neglecting the beam loading

Solution

(a) dc beam velocity $u_0 = 0.593 \times 10^6 \sqrt{V_0} = 0.59 \times 10^6 \sqrt{1} \times 10^3$
$= 1.88 \times 10^7$ m/s

Gap transit angle $\theta_g = \dfrac{\omega d}{u_0} = \dfrac{2\pi \times 4 \times 10^9 \times 1 \times 10^{-3}}{1.88 \times 10^7}$ rad = 1.337 rad = 76.6 deg

The beam-coupling coefficient

$$\beta_1 = \beta_2 = \frac{\sin\left(\dfrac{\theta_g}{2}\right)}{\dfrac{\theta_g}{2}} = \sin 38.3°/0.6685 = 0.927$$

dc transit angle in the drift space

$$\theta_0 = \frac{\omega L}{u_0} = \frac{2\pi \times 4 \times 10^9 \times 3 \times 10^{-2}}{1.88 \times 10^7} \text{ rad} = 40.11 \text{ rad}$$

For maximum output voltage $X = 1.84$, $J_1(X) = 0.582$, so that the input cavity gap voltage magnitude

$$V_1 = \frac{2V_0 X}{\beta_1 \theta_0} = \frac{2 \times 10^3 \times 1.841}{0.927 \times 40.11} \text{ volt} = 99 \text{ V}$$

(b) The voltage gain

$$A_v = \frac{\beta^2 \theta_0 J_1(X) I_0}{X V_0 G_{sh}} = \frac{0.927 \times 0.927 \times 40.11 \times 0.582 \times 22 \times 10^{-3}}{1.841 \times 0.55 \times 10^{-4} \times 10^3} = 4.36 = 12.8 \text{ dB}$$

Catcher voltage $V_2 = A_v \times V_1 = 4.36 \times 99 = 431.64$ V

Power efficiency

$$\eta = \frac{P_{RF}}{p_{dc}} = \frac{\beta_2 I_0 J_1(X) V_2}{I_0 V_0} = \frac{\beta_2 J_1(X) V_2}{2V_0} = \frac{0.927 \times 0.582 \times 431.64}{10^3}$$

$$= 23.28 \%$$

e. Beam Loading

It can be shown that when the buncher cavity gap is negligibly small, the average energy of the electron in the beam leaving the cavity over a cycle, is nearly equal to the energy with which they enter the cavity. However, when the buncher cavity gap is appreciable, the average energy of the electrons leaving the buncher gap is larger than that when they enter the gap because the electrons encounter the *RF* field for a longer duration. This excess energy must be supplied by the buncher cavity to the electron beam for bunching. Thus the electron beam is loaded by the cavity energy. This phenomenon is called "beam loading."

Two-cavity electron can be used as oscillator by providing a feedback path from the output cavity to the input cavity through a coaxial line.

The Multicavity Klystron 9.2.3

Klystron amplification, power output, frequency bandwidth and efficiency can be greatly improved by the addition of intermediate cavities between the input and output cavities of the basic klystron. Additional cavities are used to velocity-modulate the electron beam and produce increased power output. All intermediate cavities in a multicavity klystron operate in the same manner as described in two cavity case. As a representative, three-cavity klystron is shown in Fig. 9.17.

Unlike two-cavity klystron, the entire drift-tube assembly, the three cavities, and the collector plate of the three-cavity klystron are operated at ground potential for reasons of safety. The electron beam is accelerated toward the drift tube by a large negative pulse applied to the cathode. Magnetic focus coils are placed around the drift tube to minimise beam spreading. In high power klystrons the concave shape of the cathode aids the focus of the beam.

The electrons are velocity and density modulated by the cathode pulse. They are acted upon by *RF* fields developed across the input and middle cavities. While the electrons are travelling through the drift space between the cavities, bunches are formed. As a result, bunched electrons deliver energy to the output cavity from its kinetic energy that was originally absorbed from the cathode pulse.

Klystron amplifiers have been built with as many as five intermediate cavities to improve the electron bunching process and increase amplifier gain. The overall efficiency of the tube is also improved to a smaller extent. A multicavity klystron amplifier tube will produce high gain and a narrow bandwidth if all the cavities are tuned to the same frequency (called synchronous tuning). When each of the cavities are tuned to slightly different frequencies (called staggered tuning), the bandwidth will be appreciably increased at the cost of reduction in gain of the amplifier.

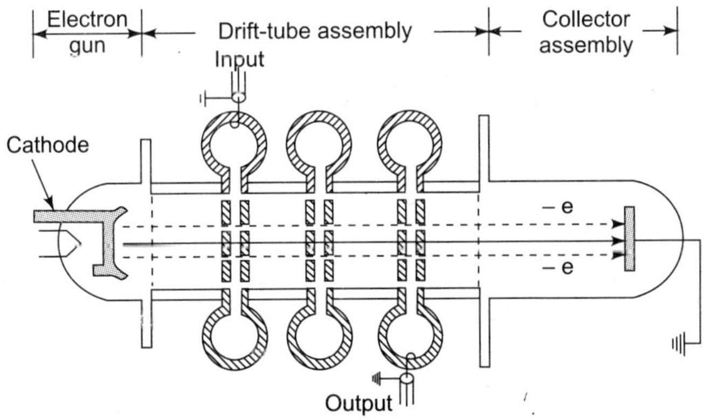

Fig. 9.17 Three-output cavity klystron

TRAVELLING-WAVE TUBE AMPLIFIER 9.3

The travelling-wave tube (TWT) forms a high-gain (> 40 dB), low-noise (NF < 10 dB), wide-band (> octave) microwave amplifier. Travelling-wave tubes are designed for the frequency range 0.3–50 GHz. The physical construction of a typical TWT is shown in Fig. 9.18. The TWT contains an electron gun, which produces and then accelerates an electron beam along the axis of the tube. The surrounding static magnet provides a magnetic field along the axis of the tube to focus the electrons into a tight beam. A longitudinal helix slow wave non-resonant guide is placed at the centre of the tube that provides a low-impedance transmission line for the *RF* energy within the tube. The *RF* input and output are coupled onto and removed from the helix by directional couplers that have no physical connection to the helix. The TWT is designed with helix delay structure to slow the travelling-wave down to or below the speed to the electrons in the beam.

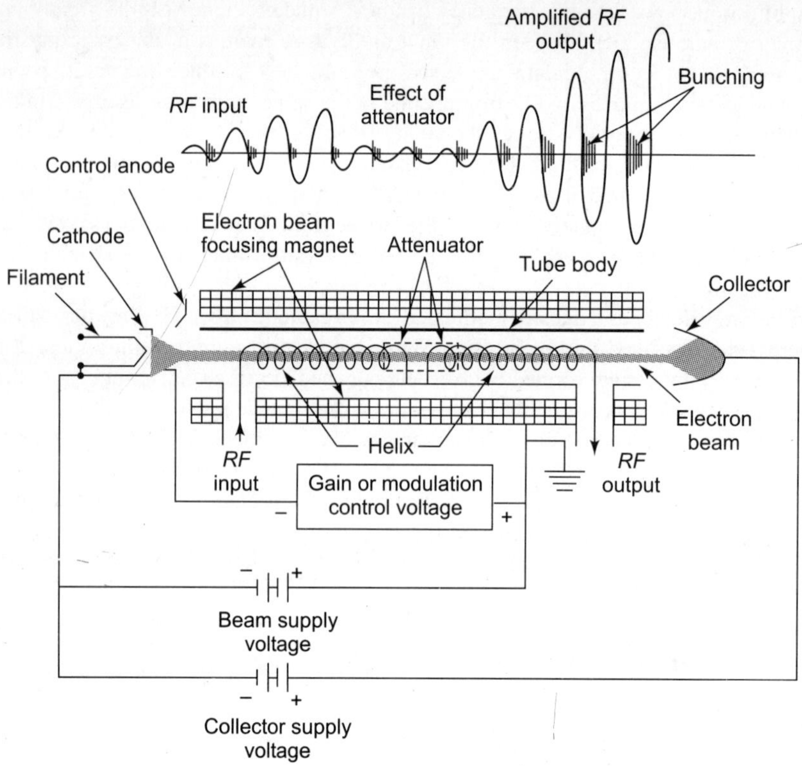

Fig. 9.18 TWT amplifier tube and circuit

The electrons of the beam are accelerated to travel faster than the waves travelling on the helix wire through velocity modulation caused by the interaction between the travelling-wave fields and the electron beam. This effect results in bunching and the electrons give up energy to the travelling-wave when the fields of the correct polarity slow down the bunches. The energy from the bunches increases the amplitude of the travelling-wave in a progressive action that takes place all along the length of the TWT.

The *RF* signal wave injected at the input end of the helix travels down the helix wire at the speed of light, but the coiled shape causes the wave to travel a much greater total distance than the electron beam. Changing the number of turns or the diameter of the turns in the helix wire, the speed at which the *RF* signal wave travels in the form of axial E field down the tube, can be varied. DC beam voltage is adjusted so that beam velocity is slightly greater than that of the axial field. The helical delay structure has the added advantage of causing a large proportion of electric fields that are parallel to the electron beam provides maximum interaction between the fields and the electron beam to form bunching.

As the electron bunches release energy to the signal on the helix, amplification begins. The initial amplified signal causes a denser electron bunch which in turn, amplifies the signal even more. This process continues as the *RF* wave and the electron beam travel down the length of the tube. When the loss in the system is compensated by this energy transfer, a steady amplification of the microwave signal appears at the output end.

An attenuator is placed over a part of the helix on midway to attenuate any reflected waves generated due to impedance mismatch that could be fed back to the input to cause oscillations. The attenuator is placed after sufficient length of the interaction region so that the attenuation of the amplified signal is insignificant compared to the amplification. TWTAs are commercially available for application at frequencies above 1 GHz, with a power gain up to 60 dB.

The internal attenuator reduces the gain of the tube. The TWT also produces heat which must be dissipated by either airconditioning or liquid-cooling systems. All of these factors reduce the overall efficiency of the TWT, but the advantages of high gain and wide bandwidth are usually enough to overcome the disadvantages.

Analysis of TWTA 9.3.1

a. Wave Propagation in Helix

In TWTA, an axial microwave E field component is being created inside a helix slow-wave guide by injecting the input singal into it. This field will decelerate the electrons in the axial beam at some sections, and at other sections it will accelerate them resulting in density modulation in the beam in accordance with the input singal. The helix reducing the phase velocity v_p of the wave propagation from the velocity of light to that of the beam can be maintained. The singal, however, travels along the helix conductor profile at approximately the velocity of light c.

If d is the diameter of the helix and p is the helix pitch, the time taken by the signal along the wire must be equal to that taken by the axial wave, so that

$$T = \frac{p}{v_p} = \frac{\sqrt{p^2 + (\pi d)^2}}{c};$$

($2\pi r = \pi d$ = circumference to make a straight line)

or,
$$v_p = \frac{cp}{\sqrt{p^2 + (\pi d)^2}} \approx \frac{cp}{\pi d} = \frac{\omega}{\beta} \tag{9.75}$$

when $p << \pi d$. Equation 9.75 is useful in designing a helix slow wave structure.

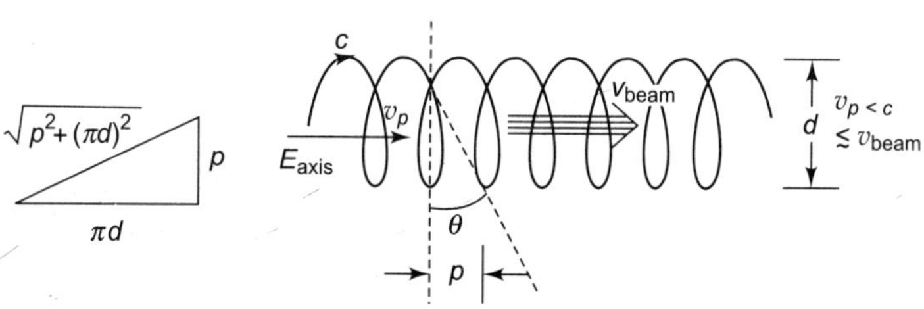

Fig. 9.19 Helical line

❑ *Beam and Field Interaction* The electromagnetic wave travelling along the helix axis, has a longitudinal component of electric field as shown in Fig. 9.20. These fields exert force **F** = –*e***E** on the electrons.

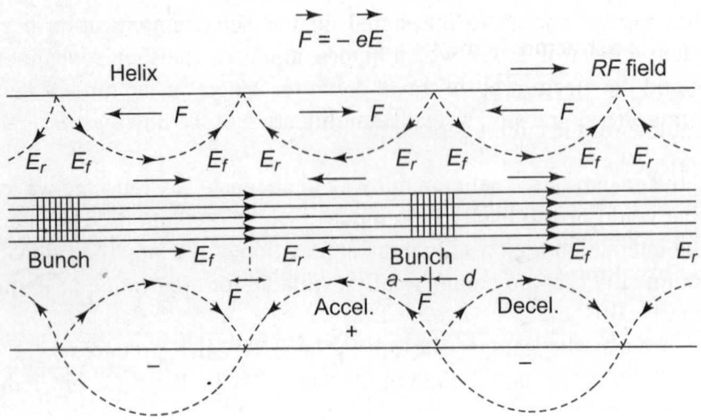

Fig. 9.20 Bunch formation in a TWT due to beam and field interaction

An electric field directed against the electron flow accelerates the electrons and that directed along the electron flow decelerates the electrons. Thus, a density modulation in accordance with the input signal frequency will occur to form bunches.

As the dc velocity of the beam is maintained slightly greater than the phase velocity of the travelling wave, more electrons face the retarding field than the accelerating field, and a great amount of kinetic energy is transferred from the beam to the electromagnetic field. Thus the field amplitude increases forming a more compact bunch and a larger amplification of the signal voltage appears at the output of the helix.

❑ *Focussing* The entire electron beam moves axially without spreading by the influence of an axial magnetic field generated by means of a magnet around the helix. A permanent magnet is used in low-power tubes and a current-carrying solenoid for high-power tubes. The disadvantages of the solenoid are that it is relatively bulky and it consumes power. These arrangements are suitable for high-power tubes where *RF* power output is more than a few kW. For satellite communications and other low-power requirements where weight and power consumption must be minimised, a permanent magnet is used for focussing.

❑ *Reflection of Waves* An attenuator placed midway along the helix attenuates reflected waves from any load mismatches at the output and prevents them from reaching the input and causing oscillations.

The attenuator will attenuate both the forward and reflected waves on the helix without affecting the electron beam. Thus, the bunched electrons after exit from the attenuator reinduce the forward wave on the helix with the same frequency and the corresponding field induces a new amplified microwave signal on the helix.

b. Gain Characteristics

The analysis of a TWT amplifier involves complicated mathematical techniques. A detailed description of theory and design was developed by Pierce. The output power gain is defined as

$$A_p = 10 \log \left| \frac{\text{output voltage}}{\text{input voltage}} \right|^2 = -9.54 + 47.3\, NC \text{ dB} \tag{9.76}$$

The first term – 9.54 dB represents a loss due to the fact that the input wave divides into three waves of equal magnitude and only one (axial) of these waves is amplified. N is the length of the interaction region in wavelengths,

$$N = l/\lambda_e \tag{9.77}$$

where l is the length of the slow-wave structure in metres and

$$\lambda_e = 2\pi\, u_0/\omega,\ u_0 = \sqrt{[(2e/m)\, V_0} \tag{9.78}$$

The factor C is the gain parameter of the circuit defined by

$$C = (I_0 Z_0/4V_0)^{1/3} \tag{9.79}$$

where I_0 is the dc beam current, V_0 is the dc beam voltage, and Z_0 the characteristic impedance of the helix.

The peak power output of a single-helix-type tube is limited to about 3 kW because of the difficulty in removing heat due to ohmic loss from the helix conductor. For low inputs, the small signal (or linear) gain is almost constant. As the *RF* power inputs is increased, the *RF* output power does not increase in proportion, but instead attains a maximum and finally starts to decrease. The point at which the output power is maximum is termed as the saturation point, and the gain at this point is the saturation gain as shown in Fig. 9.21. There will be a range of input for which the output remains in saturation, and this range defines the overdrive capability of the TWT.

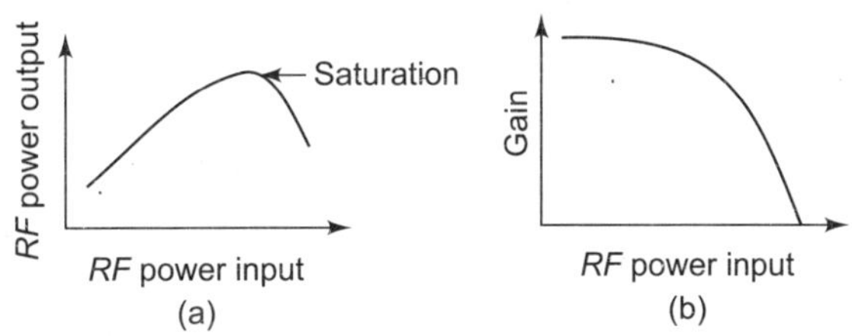

Fig. 9.21 RF power output and gain characteristics

❑ *Distortion* As the operating level approaches saturation, the non-linearity of the transfer characteristic gives rise to intermodulation distortion.

Another form of distortion occurs in a TWTA as a result of phase non-linearities. The absolute time delay between input and output of a TWT at a fixed input level is generally not significant. However, at higher input levels, where more of the energy in the beam is converted to output power, the average beam velocity is reduced, and therefore, the delay, this results in a phase shift at the output relative to the phase shift at saturation as shown in Fig. 9.22. Such a phase shift converts phase modulation at the output for an amplitude modulation on the input. This distortion can be troublesome where unwanted amplitude variations at the inputs could appear as a modulated signal.

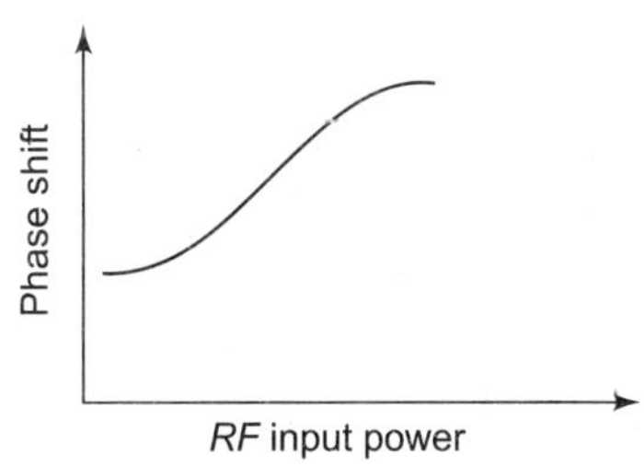

Fig. 9.22 Phase-shift characteristic

When TWT is operated near or at saturation, harmonic distortion occurs due to more intense electron bunching which produces sharp current peaks rich in harmonics. These harmonics must be filtered at the output.

Fluctuations in cathode emission and in electron velocity contribute to the current noise in TWT. For low-noise operation, the TWTs have noise factors of less than 10 dB. Medium power TWTs have noise factors in the range 18 to 30 dB, and high power TWTs have noise factors in the range 24 to 27 dB.

c. Comparison between TWTA and Klystron Amplifier

The performance characteristics of TWTA and a klystron amplifier are compared in Table 9.1.

Table 9.1 *Comparison of TWTA and klystron amplifier*

Klystron amplifier	*TWTA*
1. Linear beam or '*O*' type device	Linear beam or '*O*' type device
2. Uses cavities for inputs and output circuits	Uses non-resonant wave circuit
3. Narrow-band device due to use of resonant cavities	Wideband device because of use of non-resonant wave circuit

d. Applications of TWT

TWT amplifiers are used in medium-power satellite and higher-power satellite transponder output.

Example 9.8 A helix travelling-wave tube operates at 4 GHz under a beam voltage of 10 kV and beam current of 500 mA. If the helix impedance is 25 ohms and the interaction length is 20 cm, find the output power gain in dB.

Solution Given $V_0 = 10$ kV, $I_0 = 500$ mA, $Z_0 = 25$ ohms, $f = 4$ GHz and $l = 20$ cm.

$$u_0 = 0.593 \times 10^6 \sqrt{V_0} = 0.593 \times 10^6 (10 \times 10^3)^{1/2} = 0.593 \times 10^8 \text{ m/sec}$$

$$N = l/\lambda_e = l\omega/2\pi u_0 = \frac{0.2 \times 2\pi \times 4 \times 10^9}{2\pi \times 0.593 \times 10^8} = 13.49$$

$$C = (I_0 Z_0/4V_0)^{1/3} = \left[\frac{500 \times 10^{-3} \times 25}{4 \times 10 \times 10^3}\right]^{1/3} = 0.068$$

Therefore, $A_p = -9.45 + 47.3 \times 13.49 \times 0.068 = 33.85$ dB

MAGNETRON OSCILLATOR 9.4

Construction and Operation of Basic Magnetron 9.4.1

A magnetron oscillator is used to generate high microwave power required in radar and communication systems. Magnetrons are crossed field tubes (M-type) in which the dc magnetic field and the dc electric field are perpendicular to each other. A high power microwave oscillator uses a travelling-wave cylindrical magnetron tube as shown in the schematic cross-sectional diagram in Figs 9.23 and 9.24.

Magnetron consists of a cylindrical cathode K of finite-length-radius a at the centre surrounded by a cylindrical anode A of radius b. The anode is a slow-wave structure consisting of several re-entrant cavities equispaced around the circumference and coupled together through the anode–cathode space by means of slots. Radial electric field is established by dc voltage V_0 in between the cathode and the anode and an axial dc magnetic flux denoted by B_0 is maintained in the positive z-direction by means of a permanent magnet or an electromagnet. Magnetron theory of operation is based on the motion of electrons under the influence of combined electric and magnetic fields. The electrons emitted from the cathode try to travel to the anode. But with the influence of crossed fields **E** and **H** in the space between the anode and the cathode, it experiences resultant force $\mathbf{F} = -e\mathbf{E} - e(\mathbf{v} \times \mathbf{B})$, where **v** is the velocity vector of the electron considered and takes a curved trajectory. Due to excitation of the anode cavities by *RF* noise voltage in the biasing circuit, the *RF* field lines are fringed out of the cavity slot to the space between the anode and cathode. The accelerated electrons in the trajectory, when retarded by this *RF*, field, transfer energy from the electron to the cavities to grow *RF* oscillations. When the system *RF* losses balance the *RF* oscillation energy, a stable oscillation is achieved. Output power is extracted through an external line coupled to the cavity.

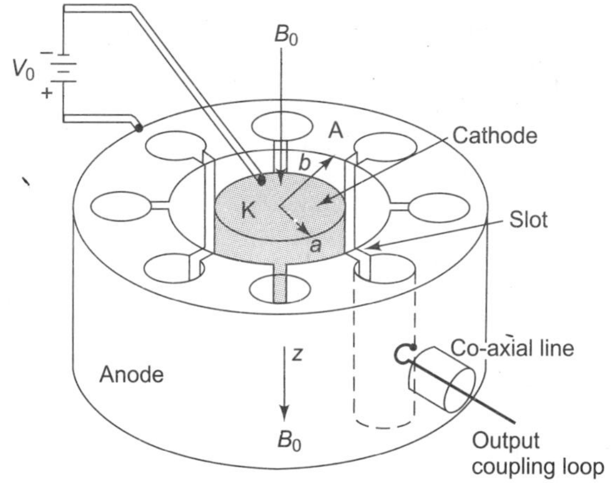

Fig. 9.23 Basic magnetron oscillator

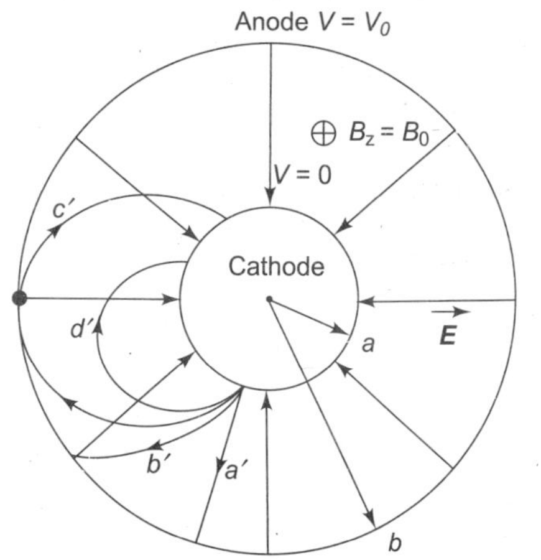

Fig. 9.24 Electron trajectory

Equations of Electron Trajectory 9.4.2

After emergence from the cathode with zero velocity (say), the electrons will acquire a velocity **v** having a tangential as well as a radial velocity components due to force **F** exerted by the crossed fields **E** and **H**;

$$\mathbf{F} = -e\mathbf{E} - e(\mathbf{v} \times \mathbf{H}) \tag{9.80}$$

In Fig. 9.24, trajectories a', b', c', and d' of the electrons are shown for different magnetic field strengths. At zero magnetic field, the electrons take the straight path a', by the influence of electric field only and collected by the anode. For a given V_0 if the magnetic field is increased, the electrons take the curved path b' due to the above force **F** to reach the anode. At a critical value of magnetic field B_c, say, the electrons just graze the anode surface at radius b and take the path c' to return to the cathode for a given voltage V_0. This value B_c is called the cut-off magnetic flux density. If the magnetic field is greater than B_c, all the electrons return to the cathode as shown by a typical path d' without reaching the anode. Therefore, when the magnetic field increases from zero to maximum, the anode current decreases from a maximum value to zero. The average velocity of the electron in the z-direction is constant, and is given by

$$v_z = E_0/B_0 \tag{9.81}$$

where $E_0 = V_0/(b - a)$ is the dc electric field, B_0 = dc magnetic flux density.

In general, an electron in the trajectory of the radius of curvature r and at velocity v experiences three radial forces: $-e\mathbf{E}$, $-e(\mathbf{v} \times \mathbf{B})$ and the centrifugal force mv^2/r such that for equilibrium.

$$\frac{mv^2}{r} + eE = evB \tag{9.82}$$

where the electric field is in the radial direction only and is given by

$$E(r) = \frac{-V_0}{r \ln \dfrac{b}{a}} \tag{9.83}$$

In the absence of an electric field, the electrons move in a circular path and return to the cathode, when

$$mv^2/r = evB \tag{9.84}$$

or, $v/r = eB/m = \omega$, called the *cyclotron angular frequency*.

The equations of motion for electrons in crossed electric and magnetic fields can be given as

$$m\,(dv/dt) = \mathbf{F} = -e\,\mathbf{E} - ev \times \mathbf{B} \tag{9.85}$$

Since the electron is emitted from the cathode in the direction opposite to **E** the equation of motion for electrons in cylindrical coordinates, are

$$\frac{d^2r}{dt^2} - r(d\phi/dt)^2 = +(e/m)\left[E_r - rB_z\frac{d\phi}{dt}\right] \tag{9.86}$$

and $$\frac{1}{r}\frac{d}{dt}\left(r^2\frac{d\phi}{dt}\right) = \frac{eB_z}{m}\frac{dr}{dt}$$

where $B_0 = B_z$ is assumed in the $+z$-direction.

Cut-off Magnetic Field and Voltage

From the second equation,

$$\frac{d}{dt}(r^2 d\phi/dt) = \frac{eB_z}{m}\frac{rdr}{dt} = \frac{\omega}{2}\frac{d}{dt}(r^2) \quad \text{or,} \quad r^2\,d\phi/dt = \omega r^2/2 + K \tag{9.87}$$

where K is an integration constant. At $r = a$, $d\phi/dt = 0$, and $K = -\omega a^2/2$. Therefore, the angular velocity of the electrons is

$$d\phi/dt = \omega/2 \cdot (1 - a^2/r^2) \tag{9.88}$$

Since the electrons move in direction perpendicular to the magnetic field, the kinetic energy of the electrons is given by the electric field only

$$eV = 1/2 \cdot m \cdot [(dr/dt)^2 + (r\,d\phi/dt)^2] \tag{9.89}$$

At $r = b$, $V = V_0$, and $dr/dt = 0$ for the electrons to just graze the anode, so that

$$(b\,d\phi/dt)^2 = 2(e/m)\,V_0 \quad \text{or,} \quad b^2[\omega/2.(1 - a^2/b^2)]^2 = (2e/m)\,V_0 \tag{9.90}$$

Substituting $\omega = eB_c/m$ at grazing.

$$b^2[eB_c/2m\,(1 - a^2/b^2)]^2 = 2eV_0/m \quad \text{or,} \quad B_c = \frac{(8V_0\,m/e)^{1/2}}{b(1 - a^2/b^2)} \tag{9.91}$$

Thus, if $B_0 > B_c$ for a given V_0, the electrons will not reach the anode. For a given B_0, the cut-off voltage is given by

$$V_c = \frac{e}{8m}b^2\,(1 - a^2/b^2)^2\,B_0^2 \tag{9.92}$$

Here, if $V_0 < V_c$, for a given B_0, the electrons will not reach the anode. Equations 9.91 and 9.92 for B_c and V_c are called the *Hull cut-off* magnetic and voltage equations, respectively.

Resonant Modes in a Magnetron 9.4.3

The nature of field distribution in the magnetron cavities is such that the alternating *RF* magnetic flux lines pass through the cavities parallel to the cathode axis, and the alternating *RF* electric fields are concentrated across the slot and fringe out to the interaction space between the anode and the cathode, in the transverse direction. For N resonant coupled cavities of the anode, there exist N resonant frequencies or modes. The equivalent resonant circuit is shown in Fig. 9.25.

Since the slow-wave structure is closed on itself, the total phase shift around the internal periphery must be an integral multiple of 2π for possible oscillations. The phase shift between two adjacent cavities is given by

$$\phi_n = \frac{2\pi n}{N} \tag{9.93}$$

where $n = 0, \pm 1, \pm 2, \pm 3, \ldots, \pm N/2$, indicates the nth mode of oscillation; $n \neq 0$ because this would indicate zero RF fringing fields in the interaction region.

For continuous interaction between the electrons and the RF fields for transfer of energy, the anode dc voltage V_0 is adjusted to coincide the average rotational velocity of the electrons with the phase velocity of the RF field in the interaction space.

Since having opposite phase in successive cavities, excitation is maximum in the cavities, $\phi_n = \pi$ or π-mode is commonly used for magnetrons oscillators, where $n = N/2$.

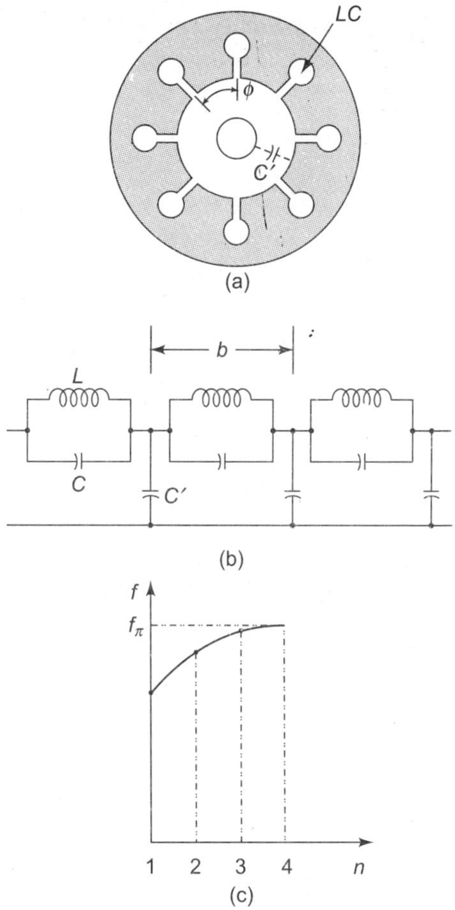

Fig. 9.25 Equivalent resonant circuit of magnetron: (a) Approximate equivalent (b) Band-pass equivalent circuit schematic (c) Mode frequencies

Mode Separation

Since each mode corresponds to a different frequency, the various modes are detuned differently from the fundamental resonant frequency of the cavities but differs very little in frequency from each other as shown in Fig. 9.25. This makes it difficult to separate the π-mode from the next immediate mode for maximum excitation of the cavities. Hence it is possible for the frequency to jump from one mode to another, which is highly undesirable. The separation of π-mode frequency from other modes is commonly done by the strapping method as shown in Fig. 9.26.

In this method, two metallic rings are arranged with one ring connected to the even-numbered anode and the other to the odd-numbered anode poles. Thus for π-mode, each ring is at the same potential and no π-mode current flows in the straps and the strap inductance has no effect. But the two rings having opposite potentials provide a capacitive loading in parallel to the capacitance C at each slot of the resonant cavities and lowers the frequency of π-mode. For the other modes each ring experiences a phase difference between the successive connection points and the resulting current flow gives rise to an inductive field with reduced capacitive effect. The inductance is in parallel with the slot, thus raising the unwanted mode frequency. Therefore, the strapping method increases the frequency separation between the π-mode and the higher adjacent modes.

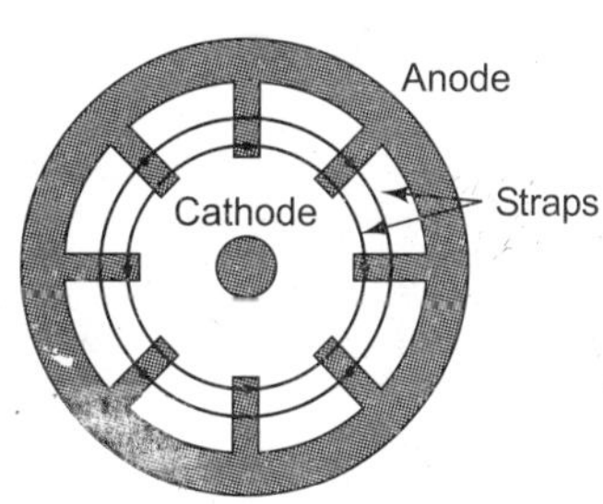

Fig. 9.26 Strapping method

❑ *Average Drift Velocity* Although the electrons move in both the ϕ and r directions, they possess average drift velocity in the ϕ direction given by

$$v_\phi = \frac{E_r}{B_z} \tag{9.94}$$

for $v_\phi < \frac{E_r}{B_z}$, the electrons tend to be deflected towards the anode and be collected. For $v_\phi > \frac{E_r}{B_z}$, the electrons will be deflected towards the cathode.

❑ *Hartree Voltage* Magnetrons are usually designed to operate in the π-mode where the phase difference between adjacent resonators is 180°. For strong interaction between the wave on the anode structure and the electron beam, the phase velocity of the wave should be nearly equal to the drift velocity v_ϕ and the oscillations for π-mode start at beam voltage

$$V_{0h} = \frac{2\pi f}{N}(b^2 - a^2)B_0 \tag{9.95}$$

which is known as the *Hartree* voltage. Here, f is the operating frequency and N is number of the resonators. A plot of Hull cut-off voltage and the Hartree voltage vs B_0 is shown in Fig. 9.27 where the region of oscillations is indicated, by the shaded area.

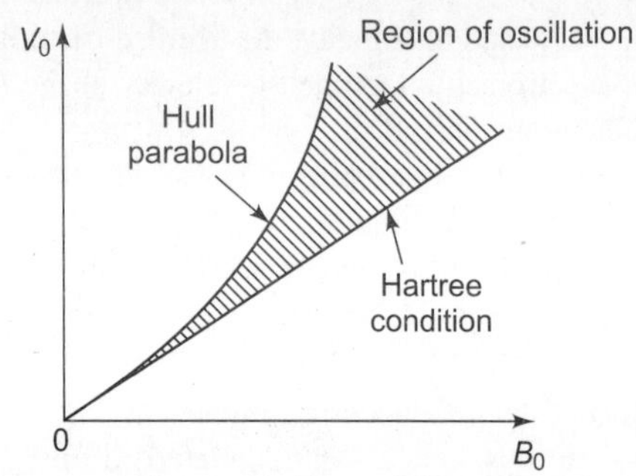

Fig. 9.27 Region of oscillation

Mechanism of Oscillations 9.4.4

The electrons emitted from the cathode try to travel towards the anode with the influence of dc electric field **E** but take a curved trajectory with the influence of dc magnetic field **B**. the total force on an electron is $\mathbf{F} = -e(\mathbf{E} + \mathbf{v} \times \mathbf{B})$, where **v** is the velocity of the electron in the space between the cathode and the anode. One can postulate the existence of *RF* oscillations in the resonant structure due to noise voltages in the dc biasing circuit. The *RF* electric field at the resonant frequency of the π-mode structure is in the slot of the cavities and fringes out in the cathode–anode space as shown in Fig. 9.28.

When the dc magnetic field exceeds the cut-off value, the electrons try to return back to the cathode in absence of the *RF* field. Due to fringing *RF* electric field, the electron *a* experiences a retarding electric field and electron *b* experiences an accelerating electric field. The retarded electron *a* experiences reduced magnetic force ($-e\mathbf{v} \times \mathbf{B}$) due to its reduced velocity and moves towards the anode. By adjusting the dc anode voltage and dc magnetic field, the circumferential velocity component of the electron can be made such that the electron *a* takes approximately one half cycle of the *RF* oscillation to travel from one slot position to the next. This makes electron *a* experience a retarding field and ultimately reaches the anode surface after continuously delivering energy to the *RF* oscillations.

On the other hand, the electron *b* is accelerated by the *RF* field to return quickly to the cathode. Thus, since electron *b* remains in the interaction space for a much shorter duration compared to the electron *a*, the energy absorbed by *b* is much smaller than the energy delivered by *a* to the system to sustain oscillations.

The electron *b* on bombardment to the cathode causes heating loss in the cathode (~5% of the anode power). The oscillation amplitude grows due to retarded electrons and a steady state is reached when the losses in the system are compensated by the *RF* oscillation continuously.

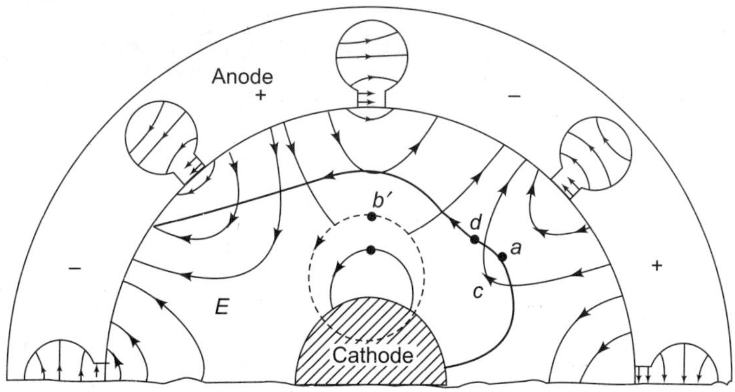

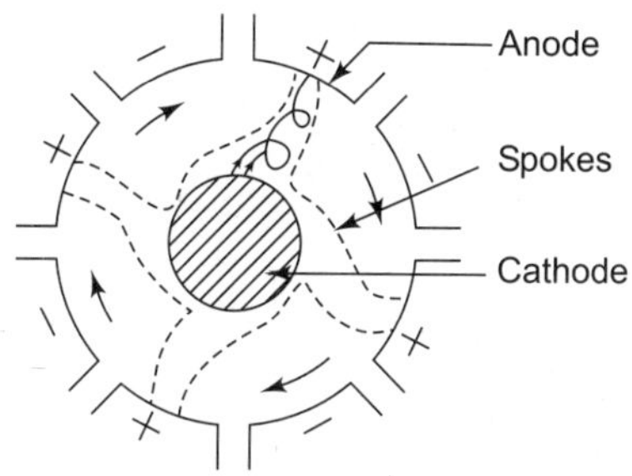

Fig. 9.28 Mechanism of oscillation

Phase Focussing

The electrons around *a*, such as *c* and *d* are acted upon both radial and tangential component of the *RF* field in such a way that the electron *c* moves faster than *a* and that *d* moves slower than *a* to form a bunch around *a*. Then these electrons from *c* to *d* are confined to spokes or focussed and terminated to the alternate anode. This is called phase focussing. For π-mode these spokes have angular velocity equal to two anode poles per cycle and the electrons within the spokes deliver energy to the oscillations before they are collected by the anode.

Tuning

The fine tuning of the magnetron oscillator can be done by changing (adding) the capacitance between the ring-strap by placing a tuning ring (*C*-ring) as shown in Fig. 9.29. This reduces the resonant frequency of the π-mode by an amount depending upon the position of the *C*-ring.

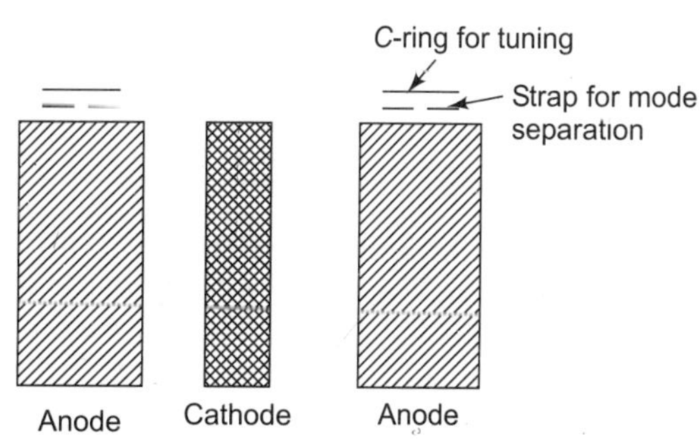

Fig. 9.29 Tuning of a magnetron

Power Output and Efficiency 9.4.5

A magnetron can deliver a peak power output of up to 40 MW with the dc voltage of 50 kV at 10 GHz. The average power output is of the order of 800 kW. The magnetron possesses a very high efficiency ranging from 40 to 70%. Magnetrons are commercially available for peak power output from 3 kW and higher.

Applications 9.4.6

Magnetrons are widely used on radar transmitters, industrial heating, and microwave ovens. A microwave oven requires a standard power of 600 W and frequencies of 915 MHz or 2450 MHz. For industrial heating magnetrons generate power of the order of kW at MHz frequencies. For radar applications, a magnetron generates peak power of the order of MW in GHz frequency range.

Example 9.9 A pulsed cylindrical magnetron is operated with the following parameters

Anode voltage = 25 kV

Beam current = 25 A

Magnetic density = 0.34 Wb/m^2

Radius of cathode cylinder = 5 cm

Radius of anode cylinder = 10 cm

Calculate (a) the angular frequency, (b) the cut-off voltage, (c) the cut-off magnetic flux density.

Solution

(a) Angular frequency $= eB_0/m = 1.759 \times 10^{11} \times 0.34$
$= 0.5981 \times 10^{11}$ radian

(b) The cut-off voltage $= (eB_0^2b^2/8m)\,(1 - a^2/b^2)^2$
$= 1/8 \times 1.759 \times 10^{11} \times 0.34^2 \times (10 \times 10^{-2})^2 \times (1 - 5^2/10^2)^2$
$= 142.97$ kV

(c) The cut-off magnetic flux density $= \dfrac{(8V_0m/e)^{1/2}}{b(1 - a^2/b^2)}$

$$= \left(\frac{8 \times 25 \times 10^3 \times 1}{1.759 \times 10^{11}}\right)^{1/2} [10 \times 10^{-2}\,(1 - 5^2/10^2)]^{-1}$$

$= 142.2$ mWb/m^2

THE BACKWARD-WAVE OSCILLATOR (BWO) 9.5

A **Backward Wave Oscillator (BWO)**, also called **carcinotron**, is a vacuum tube belonging to the traveling-wave tube family and is used as a powerful and stable microwave source. It is used to generate microwaves up to the terahertz (10^{12} Hz) frequency range. It can cover a wide electronic tuning range by varying the accelerating voltage. An electron beam, generated from the electron gun, is interacting with a slow-wave structure (Fig. 9.30). The BWO extracts energy from the electron beam by using a backward wave that travels from the collector toward the electron gun (cathode) and the oscillations are generated. The group velocity of the electromagnetic wave generated in the slow-wave circuit is directed oppositely to the direction of motion of the electrons. The output power is coupled out near the electron gun.

The typical BWO is constructed from a folded transmission line or waveguide that winds back and forth across the path of the electron beam, as shown in Fig. 9.30. The folded waveguide in the illustration serves the same purpose as the helix in a Traveling Wave Tube (TWT). The fixed spacing of the folded waveguide limits the bandwidth of the BWO. Since the frequency of a given waveguide is constant, the frequency of the BWO is controlled by the transit time of the electron beam. The transit time is controlled by the collector potential. Thus, the output frequency can be changed by varying the collector voltage, which is a definite

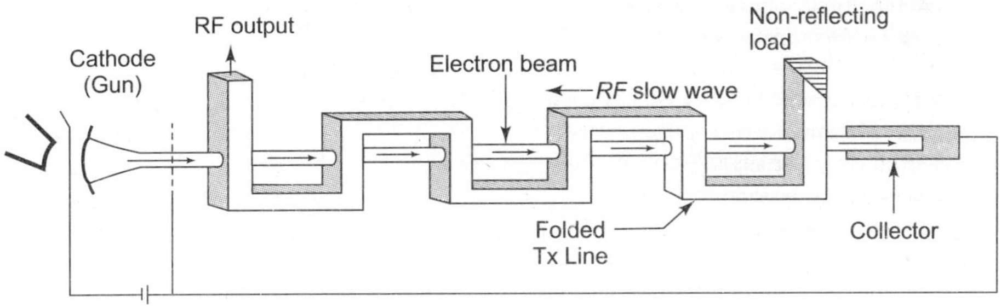

Fig. 9.30 Serpentine rectangular waveguide slow-wave circuit

advantage. As in the TWT, the electron beam in the BWO is focused by a magnet placed around the body of the tube.

The fixed spacing of the folded waveguide limits the bandwidth of the BWO. Since the frequency band of a given waveguide is constant, the frequency of the BWO is controlled by the transit time of the electron beam. The transit time is controlled by the collector potential. Therefore, the output frequency of a BWO can be changed by varying the collector voltage. This is a definite advantage.

There are two main subtypes of the BWO, the **M-type (M-BWO)**, the most powerful, and the **O-type (O-BWO)**. The O-type typically delivers power in the range of 1 mW at 1 THz to 50 mW at 200 GHz. The backward wave oscillators were first invented by Bernard Epsztein (**M-type**) and by Rudolf Kompfner (**O-type**) in 1951. The slow-wave structures must

(a) support a radio frequency (RF) electric field with a longitudinal component, and
(b) be periodic in the direction of the beam and behave like microwave filters with pass bands and stop bands.

Duc to the periodicity of the geometry, the EM fields inside are identical from cell to cell except for a constant phase shift, ϕ. As the magnitude of the space harmonics decreases rapidly when the value of n is large, the interaction can be significant only with the fundamental or the first space harmonic.

The **M-type backward wave oscillator**, uses crossed static electric field E and magnetic field B, similar to the magnetron, for focusing an electron sheet beam drifting perpendicularly to E and B, along a slow-wave circuit, with a velocity E/B. Strong interaction occurs when the phase velocity of one space harmonic of the wave is equal to the electron velocity. Both E_z and E_y components of the RF field are involved in the interaction (E_y parallel to the static E field). Electrons which are in a decelerating E_z electric field of the slow wave, lose the potential energy they have in the static electric field E and reach the circuit. The sole electrode is more negative than the cathode, in order to avoid collecting those electrons having gained energy while interacting with the slow-wave space harmonic.

The **O-type backward wave oscillator** uses an electron beam longitudinally focused by a magnetic field, and a slow-wave circuit interacting with the beam. A collector collects the beam at the end of the tube.

It has been shown that a signal-to-noise ratio of 120 dB per MHz can be achieved in a BWO.

REVIEW QUESTIONS

9.1 A reflex klystron is operated at 5 GHz with an anode voltage of 1000 V and the cavity gap of 2 mm. Calculate the gap transit angle. Find the optimum length of the drift region. Assume $N = 1\frac{3}{4}$, $V_R = -500$ V.

9.2 A reflex klystron operates at the peak mode of $n = 2$ with $V_0 = 280$ V, $I_0 = 22$ mA, and $V_1 = 30$ V. Determine (a) the input power, (b) the output power, (c) the efficiency.

9.3 A reflex klystron operates at 8 GHz at the peak of $n = 2$ mode with $V_0 = 300$ V, $R_{sh} = 20$ k-ohm, and $L = 1$ mm. If the gap transit time and beam loading are neglected, find the (a) repeller voltage, (b) beam current necessary to obtain a *RF* gap voltage of 200 V, and (c) the electronic efficiency.

9.4 A reflex klystron operates at the peak of $n = 1$ mode with a dc power input of 30 mW and $V_1/V_0 = 0.3$. Determine (a) the efficiency, and (b) total output power.

9.5 A two-cavity klystron amplifier is operated with a beam voltage of 3 kV. If the coupling coefficient is 0.9 and the magnitude of the signal voltage at the input cavity gap is 100 V, find the velocities of the electrons leaving the input gap.

9.6 A two-cavity klystron amplifier has a beam current of 400 mA, an output cavity coupling coefficient of 0.95 and the magnitude of the voltage across the output cavity gap is 75 V. Calculate the maximum magnitude of the fundamental induced current and the maximum power output.

9.7 A two-cavity klystron is operated at 10 GHz with $V_0 = 1200$ V, $I_0 = 30$ mA, $d = 1$ mm, $L = 4$ cm and $R_{sh} = 40$ k-ohm. Neglecting beam loading calculate (a) input *RF* voltage V_1 for a maximum output voltage, (b) voltage gain, (c) efficiency.

9.8 A helix travelling-wave tube is operated with a beam current of 300 mA, beam voltage of 5 kV, and characteristic impedance of 20 ohms. What length of the helix will be selected to give an output power gain of 50 dB at 10 GHz?

9.9 A TWT has $V_0 = 3$ kV, $I_0 = 3$ mA, $f = 10$ GHz, $Z_0 = 25$ ohms and normalised circuit length $N = 50$. Calculate (a) the gain parameter C, (b) dB power gain.

9.10 A cylindrical magnetron is operated at 5 GHz with $a = 3$ cm, $b = 5$ cm, $N = 16$, $V_0 = 20$ kV, $B_0 = 0.05$ T. Calculate the Hull cut-off voltage and cut-off magnetic flux density, and Hartree voltage. How do the cut-off voltage and Hartree voltage vary with B_0? Indicate the operating range.

9.11 An *X*-band pulsed cylindrical magnetron has $V_0 = 30$ kV, $I_0 = 80$ A, $B_0 = 0.01$ Wb/sq.m, $a = 4$ cm, and $b = 8$ cm. Calculate the cyclotron angular frequency, cut-off voltage and cut-off magnetic flux density.

9.12 For a magnetron, $a = 0.6$ m, $b = 0.8$ m, $N = 16$, $B = 0.06$ T, $f = 3$ GHz and $V_0 = 1.6$ kV. Calculate the average drift velocity for electrons in the region between the cathode and anode.

REFERENCES

1. Atwater, H. A., *Introduction to Microwave Theory*, Mcgraw-Hill Book Company, 1962.
2. Collin, R. E., *Foundations for Microwave Engineering*, McGraw-Hill Book Company, 1966.
3. Gewartowski, J.B. and H. A. Watson, *Principles of Electron Tubes,* Princeton, N. J., Van Nostrand D., 1966.
4. Gittins, J. F., *Power Travelling-wave Tubes,* New York, American Elsevier, 1965.
5. Hansen, W. James, *Eliminate Confusion in TWTA Specifications, Microwaves and RF,* July, 1984.
6. Hughes, T.W.T., and *TWTA Designer's Handbook,* Torrance, Calif., Hughes electron dynamics division.
7. Kleen, W. J., *Electronics of Microwave Tubes*, Academic Press, New York, 1958.
8. Liao, S. Y., *Microwave Electron Tubes*, Prentice-Hall, Inc., Englewood Cliffs, N. J., 1988.
9. Mendel, J. T., *Helix and Coupled-cavity Travelling-wave Tubes, Proc. IEEE*, 61, No. 3, 280–298, March, 1973.
10. Pierce, J. R., *Travelling Wave Tubes*, Princeton, N.J., Van Nostrand D, 1950.

11. Reich, H. J., *Microwave Principles*, Van Nostrand Reinhold Company, 1957 and Affiliated East-West Press Pvt, Ltd., New Delhi, 1972.

12. Sims, G. D. and I.M. Stephenson, *Microwave Tubes and Semiconductor Devices*, Interscience Publishers New York, 1963.

13. Terman, F. E., *Electronic and Radio Engineering*, McGraw-Hill Book Company, Inc. 1955.

10 MICROWAVE SOLID-STATE DEVICES AND CIRCUITS

INTRODUCTION 10.1

A wide range of microwave semiconductor devices have been developed since 1960s for detection, mixing, frequency multiplication, phase-shifting, attenuating, switching, limiting, amplification, and oscillation. In most of the low-power applications, solid-state devices have replaced electron beam or vacuum-tube devices because of the advantages of their small size, light weight, high reliability, low cost and capability of being incorporated into microwave integrated circuits. Some of the widely used devices are described in this chapter.

DIODES 10.2

Microwave diodes are classified as crystal diodes and Schottky diodes for mixing and detection, PIN diode for attenuation, modulation, switching, phase-shifting and limiting, Varactor diode for frequency multiplication, parametric amplification and tuning, Tunnel diode and Gunn diode for oscillation, and Read (IMPATT, TRAPATT and BARITT) diodes for amplification and oscillation. An important consideration of frequency and power limitations of these diodes is the requirement of the correct thickness of the semiconductor layer to appropriately tailor the transit time effect.

Point-Contact Diode 10.2.1

A typical point-contact diode is made with Si, the equivalent electrical circuit are illustrated in Fig. 10.1. The diode essentially consists of a pointed tungsten wire (~0.08 mm dia) made in the form of a spring that presses against the surface of a silicon (p-type) wafer (~1.6 mm square) suitably "doped" with impurities making a rectifying contact. The equivalent circuit parameters are described below.

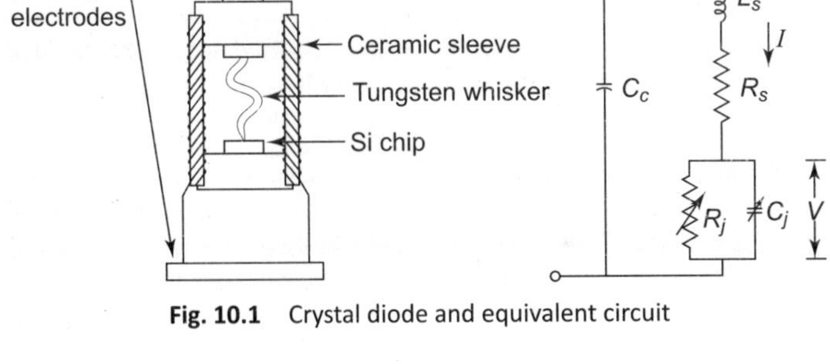

Fig. 10.1 Crystal diode and equivalent circuit

Here, R_s and L_s are the series lead resistance and inductance, respectively, C_c is the case capacitance, and R_j and C_j the effective resistance and capacitance for the junction. The value of R_j is small for forward bias and large for reverse bias. L_s and C_c can be tuned out by matching elements.

Schottky Diode 10.2.2

Schottky diodes are metal-semiconductor barrier diodes as shown in Fig. 10.2. The diode is constructed on a thin silicon (n^+-type) substrate by growing epitaxially in an n-type active layer of about 2-micron thickness.

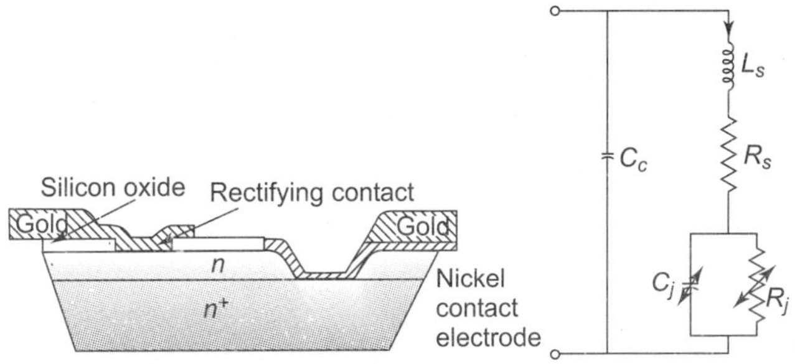

Fig. 10.2 Schottky diode and its equivalent circuit

A thin SiO_2 layer is grown thermally over this active layer. Metal-semiconductor junction is formed

R_j = resistance of metallic junction

C_j = barrier capacitance (0.3 – 0.5 pF)

R_s = bulk resistance of heavily doped Si substrate (4 – 6 ohm)

L_s = inductance of gold whisker wire (0.4 – 0.9 nH)

C_c = Case capacitance

by depositing metal over SiO_2. Schottky diodes also exhibit a square-law characteristic and have a higher burnout rating, lower $1/f$ noise and better reliability than point-contact diodes. When the device is forward biased, the major carriers (electrons) can be easily injected from the highly doped n-semiconductor material into the metal. When it is reverse-biased, the barrier height becomes too high for the electrons to cross and no conduction takes place.

RF power flow in the device is limited by power dissipation in R_s and is shorted across C_j. C_c and L_s produce *RF*-mismatch and can be matched by external circuit.

Diode Detector Circuit 10.2.3

A microwave diode can be used for detection of microwave signal. For input signal power (<10 W) the forward *I–V* characteristic is approximately parabolic and follows the square law : $I \alpha V^2$ as shown in Fig. 10.3(a). If the microwave signal voltage $v = V \cos \omega t$ is applied across the diode, the diode current is given by

$$i = I_0 (e^{aV} - 1) = I_0(e^{V/nVT} - 1) \approx I_0 e^{V/nVT} ; \tag{10.1}$$

where I_0 is the diode reverse saturation current, $a = 1/(nV_T)$, n is a constant, typically 1.1 for Schottky diodes and 1.4 for point contact diodes, V_T the thermal voltage whose value is 26 mV at room temperature << V. For a small signal amplitude, series expansion of this current gives

$$i = I_0[aV \cos \omega t + (a^2 V^2/4) (1 + \cos 2 \omega t)] \tag{10.2}$$

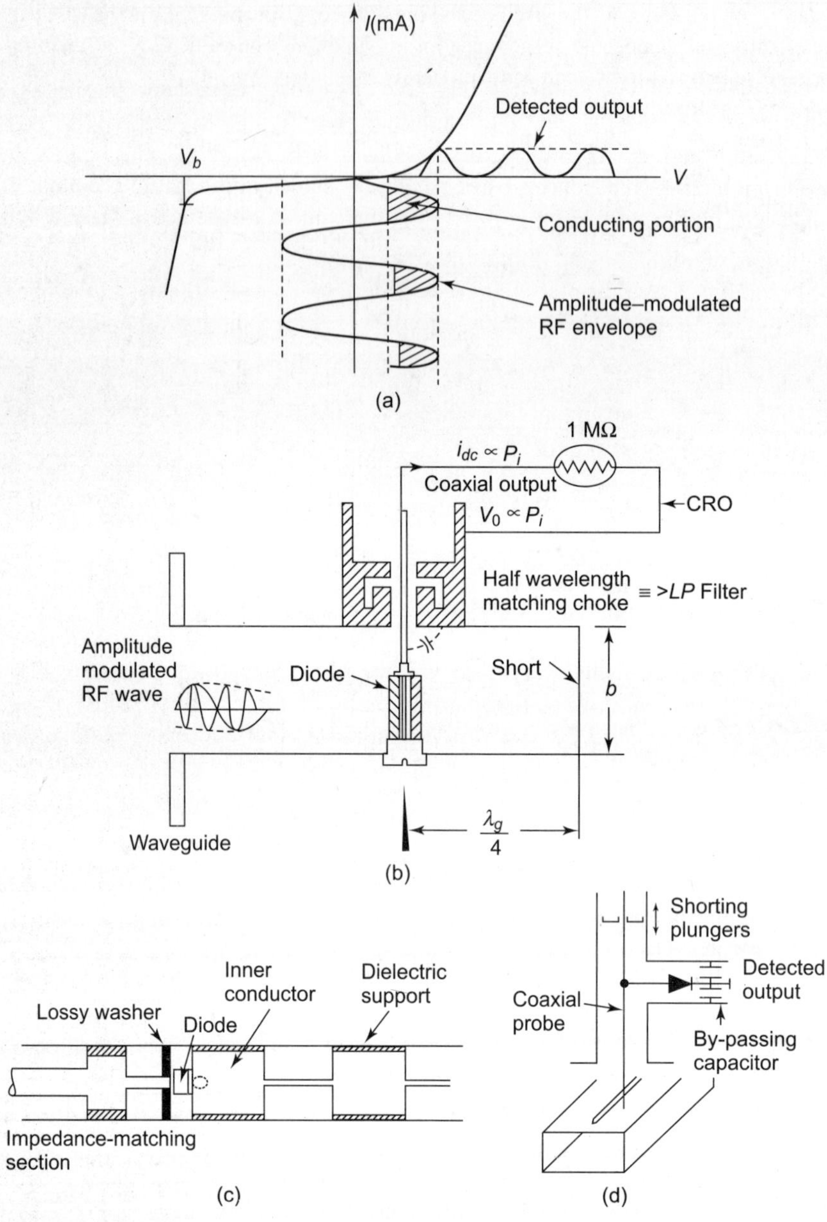

Fig. 10.3 Microwave detector mount: (a) Diode I-V and detection characteristics (b) Waveguide detector (c) Coaxial detector (d) Tunable probe detector

Thus the diode direct current $I = I_0 a^2 V^2/4$ is proportional to the microwave input power and hence the name—square law detector. The ac components are filtered out by the detector circuit. For a large power input (> 10 W) the *V–I* characteristic becomes more linear, so that input power needs to be attenuated for operation in the square-law region. Reverse biased output (current or power) of the detector is nearly zero.

The series resistance R_s and junction capacitance C_j limit the sensitivity of the detector circuit by reducing the signal power output. After tuning out L_s and C_c, if the rms signal voltage across the diode is V and signal current I, total power in the circuit is

$$P_t = I^2 R_s + V^2/R_j \tag{10.3}$$

Power absorbed in R_j is

$$P_a = V^2/R_j \tag{10.4}$$

which is also used for detection.

Hence, the power loss is the ratio of useful power input to the total power input

$$P_d/P_t = \frac{1}{1+(I^2/V^2)R_s R_j} = \frac{1}{1+|Y_j|^2 R_s R_j}$$

$$= \frac{1}{1+R_s/R_j + \omega^2 C_j^2 R_s R_j}; \; Y_j = \frac{1}{R_j} + j\omega C_j \tag{10.5}$$

The microwave detector diodes are sensitive and operate with RF signal without any dc bias. The diode is mounted in a waveguide or a coaxial line (Fig. 10.3), which contains matching elements, so that VSWR < 1.3 and the microwave power is absorbed without appreciable reflection. An RF bypass capacitance in the output circuit is constructed so that the microwave signal is contained in the detector without coupling to the measuring instrument such as VSWR meter which is a high gain, low-noise amplifier tuned to modulating frequency of 1 kHz. The output impedance of the detector circuit is designed either high (50–200 ohms) or low (2.5–10 ohms). The detector circuit is matched using short-circuit stub.

Example 10.1 A low-level point contact detector diode has $R_j = 2$ ohms, $R_s = 5$ ohms and $C_j = 0.5$ pf. Calculate the power loss in dB for operation at 5 GHz.

Solution Power loss in dB $= 10 \log (P_d/P_t)$

$$= 10 \log (1 + R_s/R_j + \omega^2 C_j^2 R_s R_j)^{-1}$$

$$= -10 \log(1 + 5/2 + (2\pi \times 5 \times 10^9 \times 0.5 \times 10^{-12})^2 \times 2 \times 5)$$

$$= -5.4 \text{ dB}$$

Diode Mixer 10.2.4

At frequencies above about 1 GHz a silicon crystal diode can be used satisfactorily as a mixer to produce a lower frequency *IF* signal. It has low conversion loss, low noise, and ability to stand momentary overloads. Figure 10.4 shows a schematic circuit of a coaxial line broadband mixer used up to frequencies of about 3 GHz. Single diode mixers are useful where signal levels are relatively large compared to noise. A mixer is a non-linear device since it is operated in the non-linear region of the *I-V* characteristics of the diode.

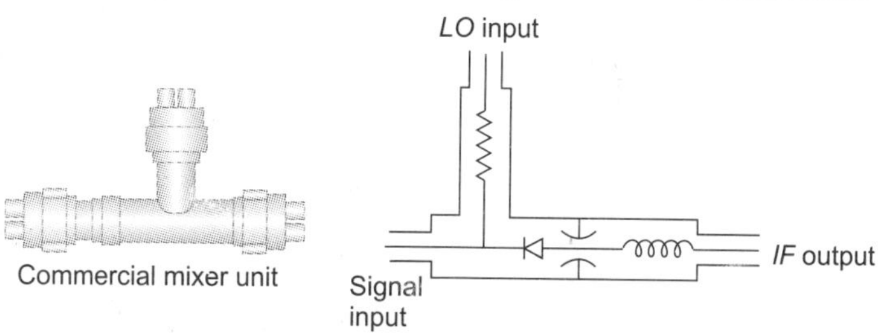

Fig. 10.4 Coaxial-line single diode mixer

For most communication applications where signals are not always large, balanced mixers are employed using two or four diodes in a hybrid ring or a magic-*T* configuration, respectively, as shown in Fig. 10.5,

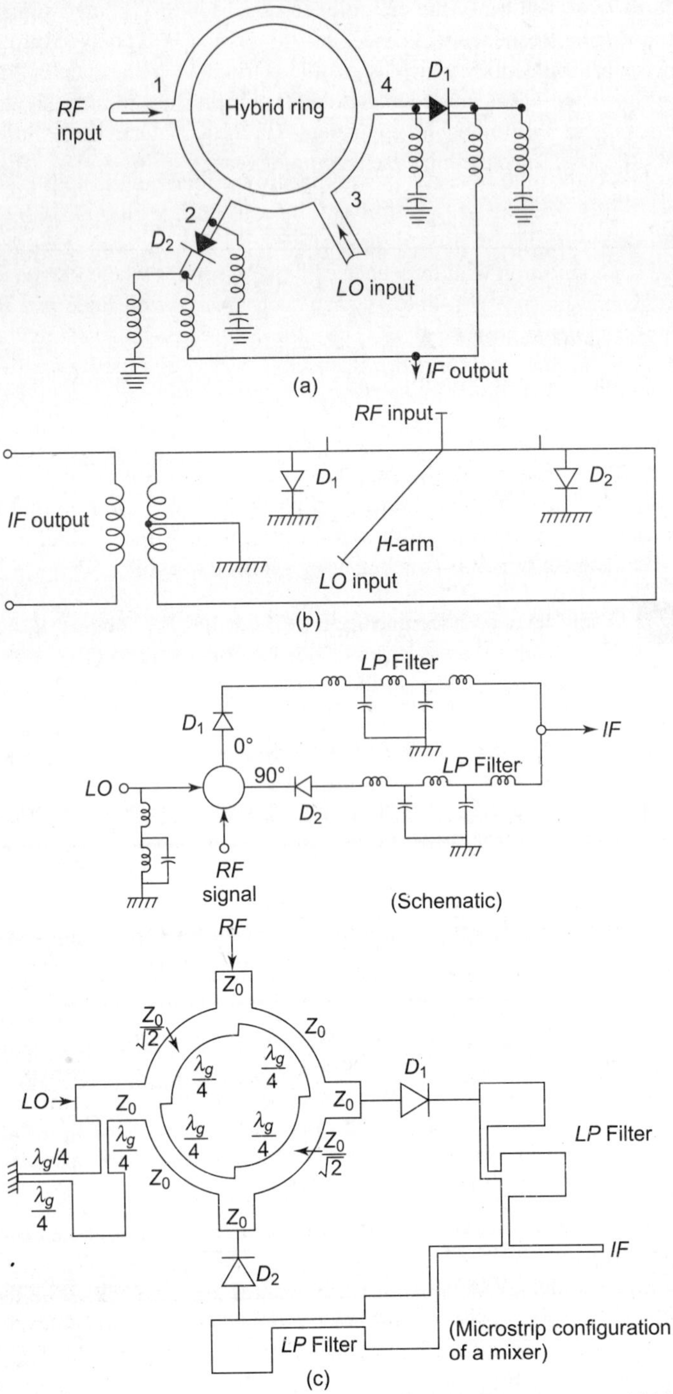

Fig. 10.5 Balanced mixers using (a) microstrip hybrid ring, (b) waveguide matgic- *T*, and (c) branch line coupler

where the *RF* signal port and the local oscillator port are decoupled. The *RF* signal splits at the input, appears at the remaining ports out of phase and recombines at these ports with in-phase local oscillator signals in the mixer diodes to produce an *IF* signal. Since the circuit is designed for 180° phased signals to produce *IF*, the oscillator noise close to signal frequency will cancel and the desired *IF* is coupled through the centre tap coil. Figure 10.5 (c) shows a balanced mixer which uses a microstrip branch line coupler. The required phase shifts in each path of the signal and *LO* at each schottky diode are met. The undesired signals are filtered out by the different *LP* filters. Due to 3 dB branch line coupler, both the signal and the *LO* powers are coupled to the mixer diodes with little loss. Thus the mixer sensitivity increases.

Harmonic Mixer 10.2.5

Harmonic mixers are used in connection with frequency calibration set-ups where a standard low frequency signal causes harmonic generation in a non-linear device such as a crystal diode and then heterodyned with another microwave test signal to produce an *IF*. The harmonic power generated by a crystal decreases rapidly with the order of the harmonic. Experience has shown that at about 10th and higher orders, the harmonic amplitude is inversely proportional to the harmonic number. The schematic circuit of a harmonic crystal mixer system is shown in Fig. 10.6. Crystals such as 1 N23 B, 1 N21 B are employed and mounted in waveguides or coaxial lines. A standard signal level of 1 V is sufficient in order to compromise between a large value that gives low conversion loss and a small value that gives minimum noise.

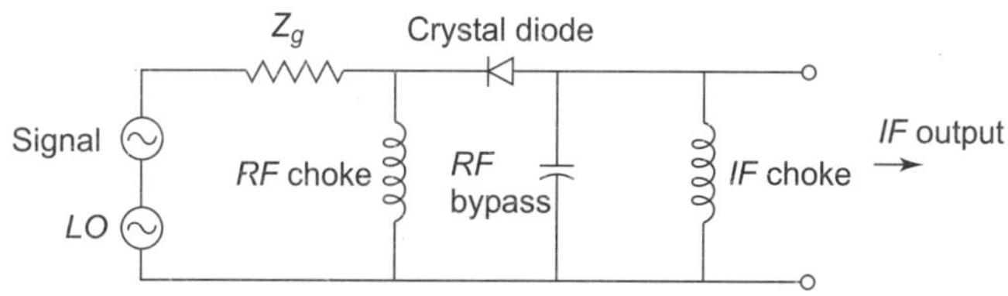

Fig. 10.6 Schematic circuit of a harmonic crystal mixer

The *RF* choke provides dc return for the crystal. The *RF* bypass prevents *RF* leakage through the *IF* terminals. The IF choke provides a dc return with negligible effect on the *IF* power. The local oscillator voltage $V_p \cos \omega_p t$ produces currents in the crystal that contain the fundamental frequency plus harmonics $n\omega_p$, $n = 1, 2, 3, ...$ An incoming signal $V_s \cos \omega_s t$ will beat with these frequencies and produce sum and difference frequency components $n\omega_p \pm \omega_s$. The difference frequencies can be tuned out through an *IF* amplifier. Thus harmonic operation can be used with low frequency local oscillator to generate *IF*.

A reasonably good impedance match should exist between the crystal unit and the source of signal power, and load impedance Z_{IF} for optimum *IF* signal level. Since the input and output impedances will depend upon the amplitude of the local oscillations, proper amplitude of local oscillator is to be adjusted. The degree to which input *RF* power is converted to *IF* power (at the difference frequency) is expressed in a quantity known as *conversion loss:*

$$\textbf{Conversion loss (dB)} = 10 \log_{10} \frac{IF \text{ output power available}}{RF \text{ input power available}} \quad (10.6)$$

The available *IF* power is the *IF* power delivered to a load whose impedance is the conjugate of the *IF* output impedance Z_{IF} of the mixer.

The conversion loss in crystal mixers ordinarily decreases with increasing amplitude of the local oscillator signal. Part of the signal power which does not reappear as *IF* output power is dissipated within the crystal. Typical conversion loss values range from 4.5 to 7 dB for the fundamental frequency. Conversion loss is higher for mixing with higher harmonics.

The output noise of mixer crystals has a distinct effect on the receiver operation. This is measured in terms of the output noise ratio:

$$\textbf{Output noise ratio, } N_r = \frac{\text{Available } IF \text{ output noise power}}{\text{Available thermal or Johnson noise power of an equivalent resistance}} = \frac{N_0}{kTB_{IF}} \tag{10.7}$$

The corresponding noise temperature

$$T_n = N_r T = \frac{N_0}{k\,B_{IF}} \tag{10.8}$$

A Design Example of a Harmonic Mixer

An *X*-band waveguide harmonic mixer is shown in Fig. 10.7. The *RF* and local oscillator signals are heterodyned in the crystal and the difference frequencies are obtained at a type BNC output jack. A reflecting choke (high series reactance) is used at the type *N* input to prevent loss of signal and the particular harmonic to be used in mixing. The waveguide is below cut-off for the fundamental and the lower harmonics. A tunable short is used for input impedance matching. An *RF* by-pass gap capacitor is used at the BNC port to prevent *RF* leakage through the *IF* terminal. The reflecting choke can be considered as a short-circuited radial guide of axial width $\delta < \lambda/2$. Under this condition, only TM_{on} modes will be supported by this guide. A high series reactance is obtained by keeping the depth *A* slightly less than the quarter wavelength at the signal or the harmonic frequency to be used.

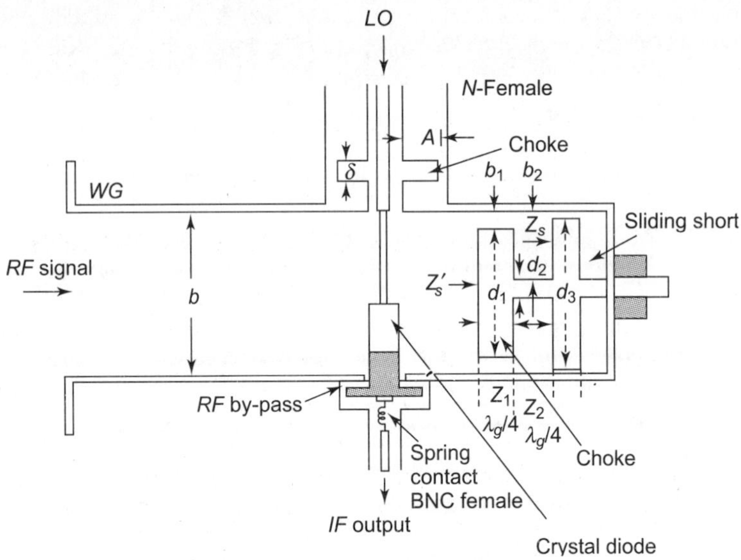

Fig. 10.7 Waveguide harmonic crystal mixer

Thus, the designed values are $f_0 = 10$ GHz, $\lambda_0 = 3$ cm, $\delta = 5$ mm, and $A = 7.2$ mm. For 50 Ω coaxial line with air dielectric ($\varepsilon_r = 1$), the characteristic impedance

$$Z_0 = 50 = \frac{138}{\sqrt{\varepsilon_r}} \log_{10} \text{(outer radius/inner radius)} \tag{10.9}$$

or outer radius/inner radius = 2.3.

The gap by-pass capacitance C_g is selected to pass the desired *IF* and by-pass the *RF*, local oscillator signal and its harmonics and also the higher frequency *IF*'s.

$$\text{Therefore, } \frac{1}{\omega C_g} << 50\ \Omega$$

is achieved by a half-millimetre gap with teflon filling. A good contact between the BNC connector and diode is obtained by means of a spring contact.

A choke-type short-circuit plunger is used for impedance tuning. The impedance seen at the input of the plunger is

$$Z'_s = (Z_1/Z_2)^2\, Z_s \tag{10.10}$$

where Z_s is the impedance of the sliding short and Z_1 and Z_2 are the characteristic impedances of the coaxial lines formed between the waveguide walls and the plunger. If $Z_2 >> Z_1$, Z'_s will approximate to a short circuit by a factor $(Z_1/Z_2)^2$ better than Z_s does. The width of the plunger is uniform and slightly less than the interior guide width. However, the height of the plunger is made non-uniform. The gap $b_1 << b_2$ is to make $Z_1 << Z_2$. The back section makes a sliding fit in the waveguide. The selection of the dimensions are made as

$$b_2 = 8b_1 \text{ with } b_1 = 0.5 \text{ mm, and } b_2 = 4 \text{ mm.}$$

For an *X*-band waveguide of 0.9″ × 0.4″ size,

$$b = 10.16 \text{ mm}$$

$$d_1 = b - 2b_1 = 9.16 \text{ mm}$$

$$d_2 = b - 2b_2 = 2.16 \text{ mm}$$

$$\lambda_g/4 = 9.938 \text{ mm at 10 GHz}$$

A bandwidth of 10 to 20% can be achieved with this design.

PIN Diode and Its Applications 10.2.6

A PIN diode consists of a high-resistivity intrinsic semiconductor layer between two highly doped p^+ and n^+ semiconductor (Si, GaAs) layers as shown in Fig. 10.8 along with its equivalent circuit. The device acts as an electrically variable resistor related to the *i* layer thickness.

The intrinsic layer has a very large resistance in reverse bias and it decreases in forward bias. When mobile carriers from *p* and *n* regions are injected into the *i* layer, carriers take time such that the diode ceases to act as a rectifier at microwave frequency and appears as a linear resistance. This property makes it usable as a variable attenuator at microwave frequencies.

R_j, C_j = Junction resistance, capacitance of *i* layer

R_s = Bulk semiconductor (p^+ and n^+) layer and contact resistance.

L_p, C_p = package inductance, capacitance

R_j is a variable element, and C_j is constant (≈ 0.2 pf) at frequencies > 1 GHz. The resistivity of the *i* layer is of the order of 1000 ohms-cm, and $R_s \approx 1$ ohm. L_p and C_p may be omitted when the diode is directly mounted in chip form into the microwave integrated circuit. Because of the large breakdown voltage (~ –500 V) compared to an ordinary *pn* junction diode, the PIN diode can be biased at a high negative potential so that large ac signal superimposed on dc cannot make the device forward biased. For the device to remain in the high impedance state, in the presence of a large microwave signal, the dc

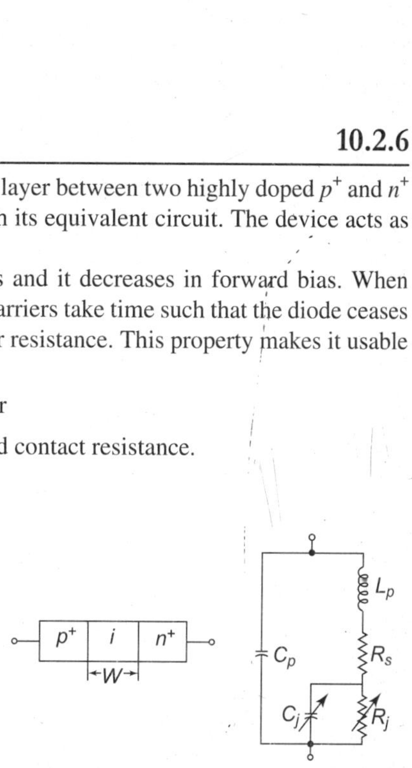

Fig. 10.8 PIN diode and equivalent circuit

reverse voltage must be high. At high power applications the i region width W should be wide but less than diffusion length L_d, otherwise the centre layer voltage drop will increase and the forward bias impedance will become high. The junction capacitance decreases with the increase in depletion region. To keep reverse bias capacitance low, the cross-section area of the device must be so small such that forward bias impedance does not become high.

Applications of PIN Diode

❑ PIN Switches

The PIN device is designed such that under reverse or zero biasing, R_j is extremely large and C_j(0.02–2 pf) plays the dominant role, whereas C_p, L_p, R_s are negligibly small. For forward biasing R_j is very small ($<<1/j\omega C_j$) so that R_f(0.1–2 ohms) = $R_j + R_s$ plays the dominant role. Therefore, for reverse bias, a high capacitive impedance is presented to the microwave signal and for forward bias, the diode represents a very low resistance. Thus by changing the bias it acts as a switch as described below.

- **Single-pole Switch** Figure 10.9 shows schematic circuits of a single PIN switch in shunt and series mounting configurations. AC blocking inductor is realized from a high impedance strip line section and dc blocking capacitor is realized from a gap in the line. For shunt configuration reverse biasing produces transmission ON due to high impedance shunt and forward biasing produces transmission OFF due to low impedance shunt. For series configuration, transmission is ON for forward bias and OFF for reverse bias. In practice for shunt mounting, due to non-zero forward bias resistance, isolation between input and output is not infinite. Similarly, for reverse bias, shunt capacitor is not infinite and a non-zero insertion loss results. The case is reversed for series mounting.

Because of the large breakdown voltage (~ – 500 V) compared to an ordinary diode, a PIN diode can be biased at high negative potential so that large ac signal superimposed on dc, cannot make the device forward biased. Referring to Fig. 10.8, where a microwave signal is superimposed on the dc negative bias voltage, it is apparent that for the device to remain in the "high impedance" state, in the presence of large microwave voltages, the dc reverse bias must be high. A high power PIN microwave switch is therefore, characterized by a wide i-region of width W. This width W must not be larger than the diffusion length L_d, otherwise, the centre layer voltage drop will increase and the forward bias impedance will become high. Also to keep reverse bias capacitance low, the cross-section area of the device must be small. But it must not be so small that the forward bias impedance increases. This switch is called **Single-Pole Single-Throw (SPST)** switch.

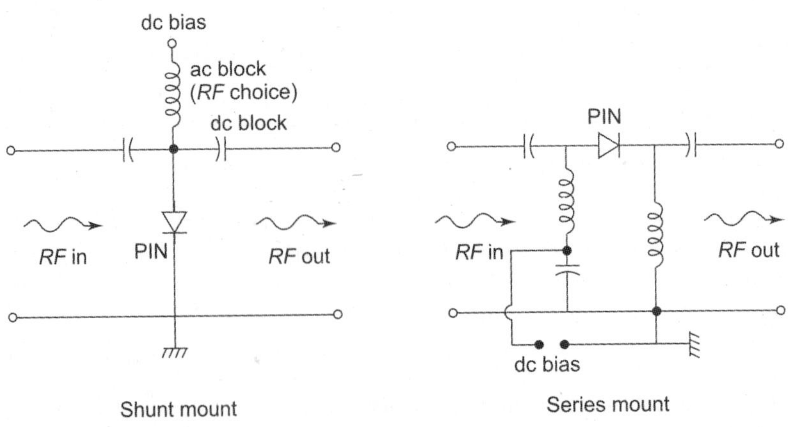

Fig. 10.9 Single PIN switch (SPST)

Example 10.2 A shunt-mounted PIN diode in a TEM transmission line having characteristic impedance $Z_0 = 50$ ohms can be represented by a shunt impedance $Z = R + jX$. Calculate the insertion loss and isolation at a frequency $f_0 = 2$ GHz. The forward resistance $R_f = 0.1\ \Omega$ and capacitance $C_j = 0.02$ pf.

Solution Let the line be terminated by characteristic impedance Z_0. Therefore, input impedance measured across Z at AA' towards the load is $Z_{in} = Z_0$.

Therefore, effective load at $AA' = Z_e = ZZ_0/(Z + Z_0)$

Reflection coefficient at AA' is

$$\Gamma = (Z_e - Z_0)/(Z_e + Z_0)$$

Therefore, the transmission coefficient

$$T = 1 + \Gamma = 1 + \frac{Z_e - Z_0}{Z_e + Z_0} = \frac{2Z_e}{Z_e + Z_0} = 2Z/(2Z + Z_0)$$

Now insertion loss

$$\alpha = \text{Incident power/Power delivered to load}$$

$$= \frac{|V_i|^2}{|V_t|^2} = \frac{|V_i|^2}{|TV_i|^2} = \frac{1}{|T|^2} = \left|\frac{2Z + Z_0}{2Z}\right|^2 = \left(1 + \frac{Z_0}{2Z}\right)^2$$

Assuming that there is no loss in the line and in the diode,

$$\alpha\,(\text{dB}) = 20 \log_{10}\left(1 + \frac{Z_0}{2Z}\right) \text{dB}$$

Now, $\quad Z \approx R_f = 0.1\ \Omega$ for forward bias

$$Z \approx \frac{1}{\omega C} = 1/[2\pi f \times 0.02 \times 10^{-12})\ \Omega \quad \text{for reverse bias}$$

$$= 1/[2\pi \times 2 \times 10^9 \times 0.02 \times 10^{-12})] \text{ reverse bias}$$

$$= 3.98\ \text{k}\Omega$$

Therefore,

$$\text{Isolation (forward bias)} = 20 \log\left(1 + \frac{50}{0.2}\right) = 48 \text{ dB}$$

$$\text{Insertion (Reverse bias)} = 20 \log\left(1 + \frac{50}{3978.9}\right) = 0.108 \text{ dB}$$

◆ **Double Pole Switch** The PIN diode double switch circuit using two diodes called **Single-Pole Double-Throw (SPDT)** is shown in Fig. 10.10. Diodes are biased through *RF* chokes to isolate the ac component from the dc source. The impedance matching between *RF* feeder line and the switch is obtained from the quarter-wave lines which provide zero impedance when the switch is OFF (open circuit) and infinite impedance when the switch is ON (short circuit).

Thus,

(i) when D_1 is forward biased, Z_{i2} = very large (ideally ∞)

(ii) when D_2 is reverse biased, Z_{i3} = very small (ideally zero)

Therefore, input *RF* power is directed to Port (3), while Port (2) remains isolated. Input power can be switched to Port (2), keeping Port (3) isolated by changing bias of D_1 and D_2, i.e., D_1 is reverse biased and D_2 is forward biased.

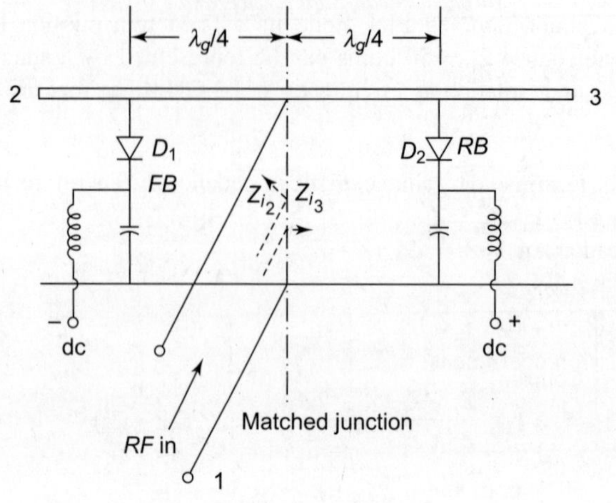

Fig. 10.10 SPDT double switch

Since in practice under forward bias condition a diode has non-zero finite resistance ($R_f \neq 0$), some power is transmitted to Port (2) in the first case giving rise to finite isolation. Similarly, under reverse bias condition the diode reactance X becomes finite and hence some non-zero finite insertion loss to Port (3) appears during transmission. For high power operation, Si is preferred to GaAs.

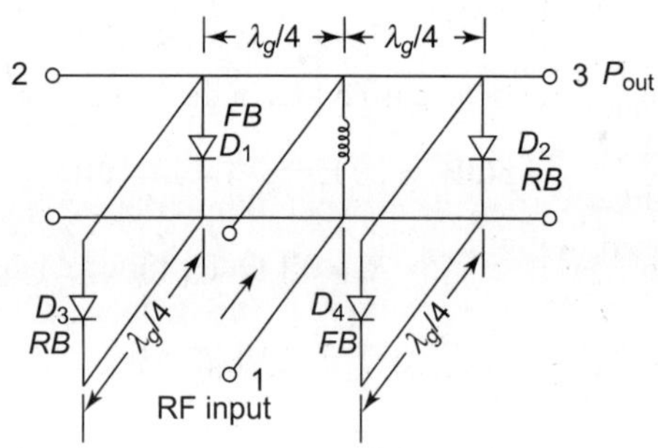

◆ **SPDT Switch Using 4 Pin Diodes** Insertion loss and isolation in double switch can be improved by putting two other diodes on both sides at a distance $\lambda/4$ from D_1 and D_2. The basic configuration of a SPDT switch using four PIN diodes is shown in Fig. 10.11, along with an MIC model. To have power in Port (3), the diode states will be as follows: D_1 forward biased, D_2 reverse biased, D_3 reverse biased and D_4 forward biased. The quarter-wave transformer transforms the short circuit impedance of any forward-biased diode to an open circuit at the *RF* port and eliminates any reactive loading on the *RF* port.

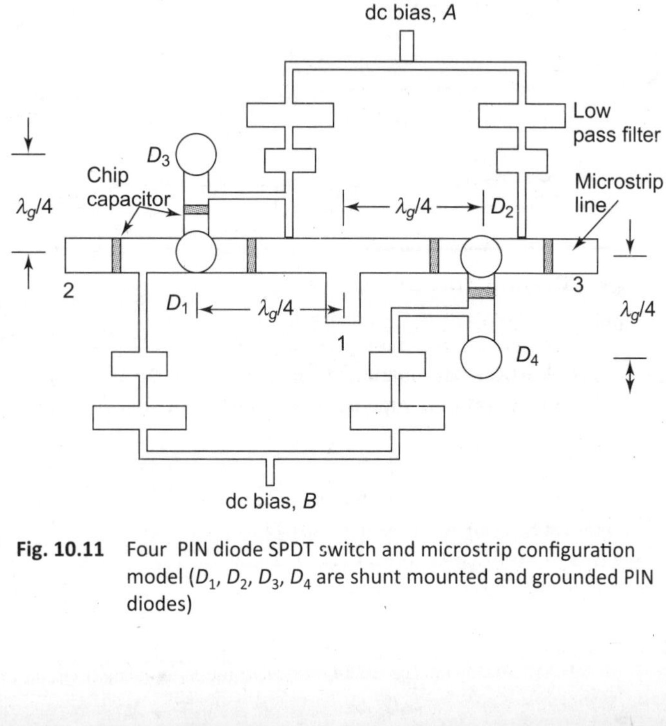

Fig. 10.11 Four PIN diode SPDT switch and microstrip configuration model (D_1, D_2, D_3, D_4 are shunt mounted and grounded PIN diodes)

Under this condition, Port (3) will be connected to Port (1) with Port (2) isolated. For power in Port (2), the logic of the diode states will be just opposite as compared to the above. In the MIC model as shown in the figure, the diodes are shunt mounted with the cathodes grounded.

❑ ***PIN Duplexer*** The basic PIN duplexer circuit is shown in Fig. 10.12, which uses two 3 dB, 90° hybrids and two PIN-diodes D_1 and D_2. For simplicity biasing circuits are not shown.

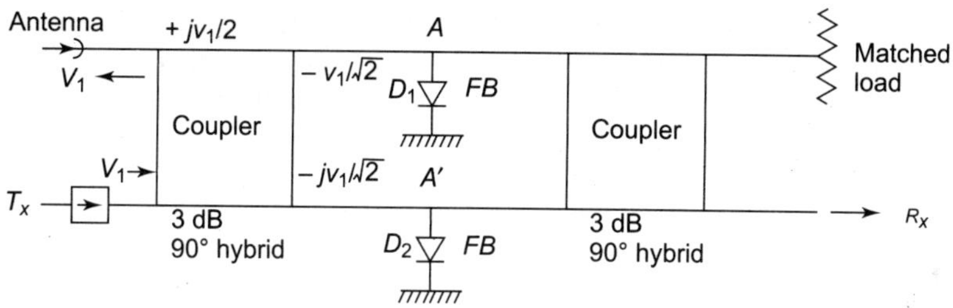

Fig. 10.12 PIN duplexer

◆ **Transmit Mode** For transmission of *RF* power from T_x, both the diodes are forward biased to act as short circuit at *AA′* as shown in Fig. 10.13. Almost all *RF* power is transmitted through the antenna. Total reflected power at T_x is zero. Because of non-zero finite forward resistance ($R_f \neq 0$) of the diode, some *RF* power is transmitted to the matched load and gets absorbed. Power in the R_x port is zero.

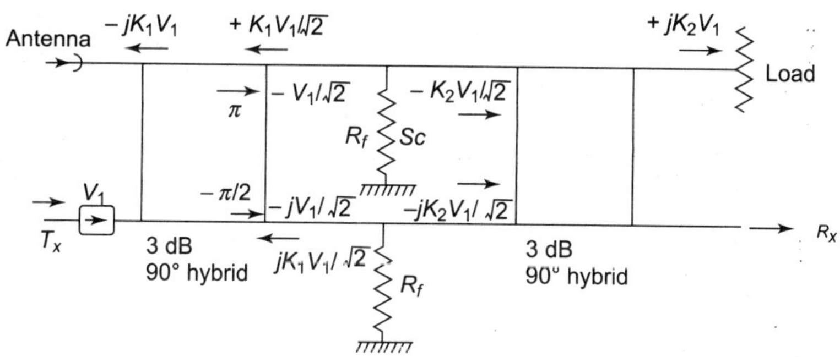

Fig. 10.13 Transmit mode equivalent circuit of a PIN duplexer ($k_1 = \Gamma_{sc} = -1$)

◆ **Receive Mode** For reception mode *RF* power enters R_x from the antenna and the isolator isolates any power entering T_x as explained in Fig. 10.14 when D_1 and D_2 are reverse biased and act as open circuit.

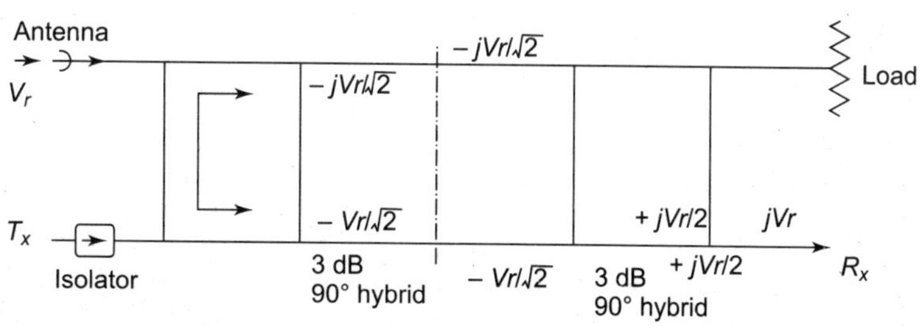

Fig. 10.14 Receive-mode equivalent circuit of PIN duplexer

◆ **PIN Phase-shifter** Electronic phase-shifters designed using PIN diodes are extensively used in phased array in quantities of thousands. Three types of variable phase-shifters using PIN diodes are described here

(i) Transmission type switched line ϕ-shifter
(ii) Reflection type hybrid coupled ϕ-shifter
(iii) Transmission type loaded line ϕ-shifter

Transmission-type Switched Line ϕ-shifter Phase shift is obtained by perturbing the parameters of the transmission line. Here D_1, D_2, D_3, and D_4 are identical PIN diodes. Operation is based on switching the diodes from one type (FB) biasing to another (RB) so that a differential phase change occurs at the output. This is explained in Fig. 10.15. Four different states are described below.

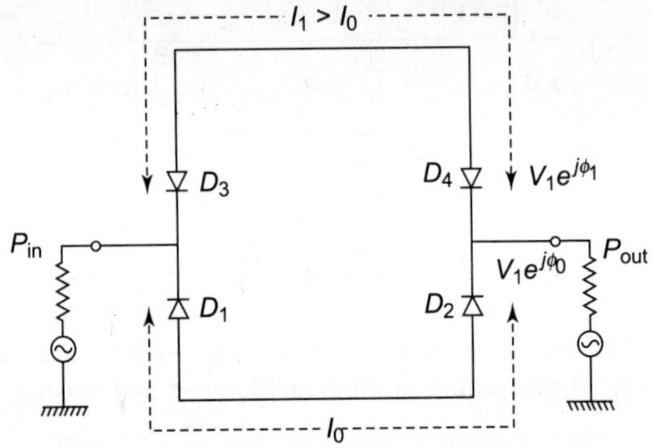

Fig. 10.15 Transmission-type switched-line ϕ-shifter

State 1

D_1 and D_2 are forward biased and offer short circuit.

D_3 and D_4 are reverse biased and offer open circuit.

Therefore, the signal passes through l_0 line. If the input voltage is V_0, output voltage becomes $V_0e^{j\phi 0}$, where $\phi_0 = \beta l_0$, the phase shift due to propagation through the line l_0.

State 2

D_1 and D_2 are reverse biased and offer open circuit.

D_3 and D_4 are forward biased and offer short circuit.

Therefore, the signal passes through l_1 line. If input voltage is V_0, output voltage becomes $V_0\, e^{j\phi_1}$, where $\phi_1 = \beta l_1$ is the amount of phase shift. Therefore, differential ϕ-shift $\Delta\phi = \phi_1 - \phi_0 = \beta\,(l_1 - l_0)$ becomes a fixed ϕ-shift. Such ϕ-shifter is called the 1-bit ϕ-shifter. For 4-bit ϕ-shifter, a cascade of 22.5°, 45°, 90° and 180° ϕ-shifters gives all 360° ϕ-shift in 22.5° steps.

If all upper line diodes are OFF, signal will pass through all lower lines with net ϕ-shift of 4 ϕ_0 which can be the reference. If lower diodes are not OFF, signal will pass through all upper diodes with net ϕ-shift $= 4\phi_1$. Therefore, the total phase shift is

$$\Delta\phi = 4\,(\phi_1 - \phi_0)$$

By tapping from different sections we can get phase shifts of $\Delta\phi = \phi_1 - \phi_0$, $2\phi_1 - \phi_0$, etc. The disadvantage of this model is that, leakage through the reverse bias diode and non-zero forward resistance decreases isolation between the ports and increases the insertion loss. Moreover, the number of diodes per bit of phase shift is large.

Reflection-type Hybrid ϕ-shifter As shown in Fig. 10.16, here ϕ-shift is obtained by changing the value or location of the terminating impedance. A hybrid coupler (or circulator) is used to separate input and output signals.

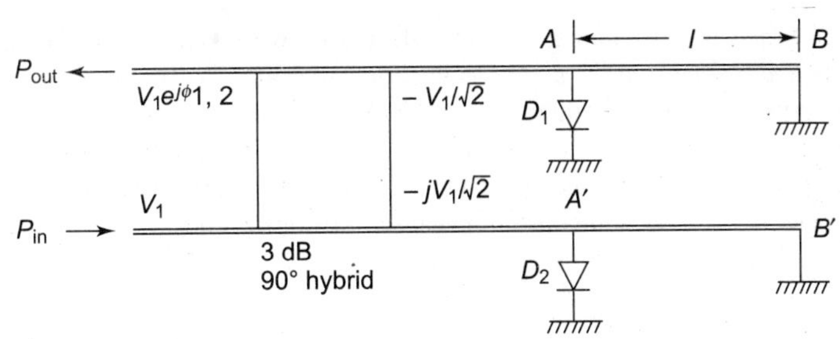

Fig. 10.16 Reflection-type hybrid ϕ-shifter

Two states are described below for phase shift.

State 1

D_1 and D_2 are forward biased and offer short circuit at AA' (at shorter length) to produce a phase shift $\phi_1 = \phi_0 = \beta l_0$.

State 2

D_1 and D_2 are reverse biased and offer open circuit at AA'. The physical short at BB' produces an effective short at a longer length to produce a phase shift $\phi_2 = \phi_0 + 2\beta l$

Therefore, phase change, $\Delta\phi = 2\beta l$. By changing the line length l or by changing the terminating impedance, the differential ϕ-shift can be obtained.

Transmission-type Loaded Line ϕ-shifter This is the simplest model of a PIN phase-shifter as shown in Fig. 10.17. Phase shift is obtained by loading the T_x-line between the input and output.

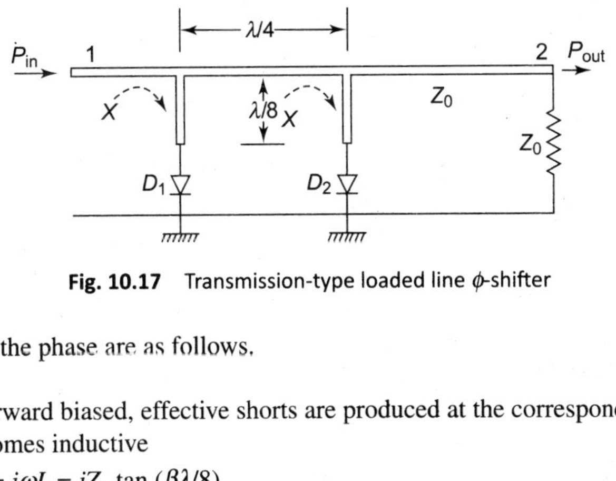

Fig. 10.17 Transmission-type loaded line ϕ-shifter

Two states contributing the phase are as follows.

State 1

When D_1 and D_2 are forward biased, effective shorts are produced at the corresponding diode locations and the input reactance becomes inductive

$$X = j\omega L = jZ_0 \tan(\beta\lambda/8)$$

This produces a phase ϕ_0. Therefore, for input voltage V_1, output becomes $V_1e^{j\phi 0}$.

State 2

When D_1 and D_2 are reverse biased, effective open circuits are produced at the corresponding diode locations and the input reactance becomes capacitive

$$X = 1/j\omega C = -jZ_0 \cot(\beta\lambda/8)$$

This produces a phase ϕ_1. Therefore, for an input voltage V_1, output becomes $V_1e^{j\phi 1}$. Therefore, phase change $\Delta\phi = \phi_1 - \phi_0$. This phase can be maximized by equating $1/\omega C = \omega L$. Here $\lambda/4$ line between the diodes provides $\lambda/2$ path for incidence and reflected signals to get cancelled at input side.

❑ *PIN Attenuator* Since the resistance of diode decreases with increase of forward bias, the diode acts as a variable attenuator when put in series or in shunt on a line. At frequencies < 10 MHz the diode shows rectifying property, whereas at frequencies >> 10 MHz it behaves as a variable resistor or attenuator. Figure 10.18 shows schematic circuit of series-and shunt-attenuator configurations. In a series circuit, the attenuation decreases with increase of biasing current with consequent decrease of *RF* resistance. In a shunt circuit, attenuation increases with biasing current because most of *RF* energy is absorbed in the diode.

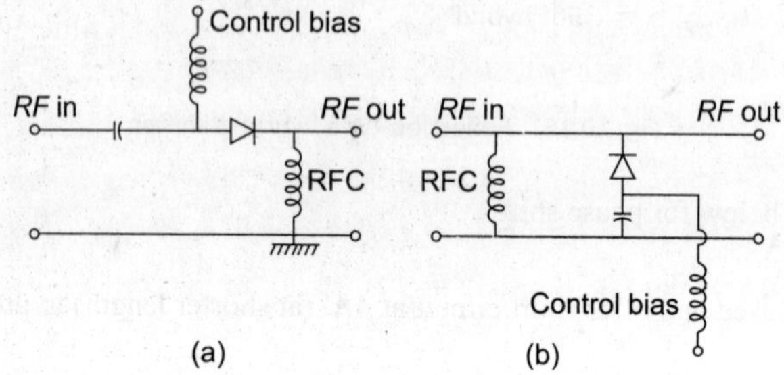

Fig. 10.18 PIN attenuator: (a) Series (b) Shunt

GUNN DIODES (TRANSFERRED ELECTRON DEVICES—TEDS) 10.3

Gunn diodes are known as **Transferred Electron Devices (TEDs)** and are low-noise devices at the microwave frequencies. Gunn diodes are named after JB Gunn, who observed in 1963, a periodic fluctuation of current passing through *n*-type gallium arsenide (n-GaAs) semiconductor sample when the terminal voltage difference exceeded a certain threshold value. Similar effects were also observed in *p*-type GaAs, Indium phosphate (InP) and in other semiconductors (CdTe, Tnsb, InAs, etc). Without having two different semiconductors forming a junction, the Gunn devices are named diodes due to two metallic end contacts as cathode and anode in a single bulk semiconductor such as *n*-GaAs or *p*-GaAs as shown in Fig .10.19(a).

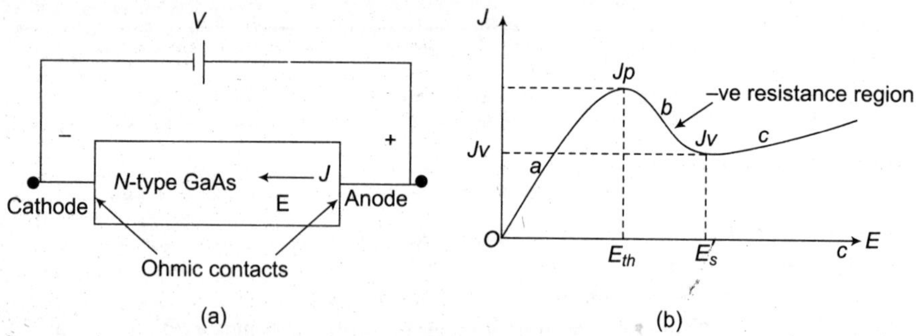

Fig. 10.19 Gunn diode biasing and *I-V* characteristics

I-V Characteristics and Basic Operation

When the GaAs sample is biased with a dc source of voltage *V*, an electric field is generated inside the sample with electric current *I*. The typical characteristic of current density *J* vs. electric field *E* in the semiconductor is shown in Fig. 10.19(b). It is seen that when the electric field in the sample is less than a **critical** value E_{th}, called the **threshold field**, the sample obeys Ohm's law. When *E* exceeds E_{th} but is less than a **sustaining**

value E_s, the slope of *J-E* curve is negative, i.e. the differential resistance $\left(\frac{dV}{dI} \text{or} \frac{dE}{dJ}\right)$ of the device is negative. This differential negative resistance (usually stated as the "negative resistance") of the device is utilized for making microwave Gunn oscillators. When the negative resistance of the device is compensated by the positive resistance of the external resonance circuit, sustained oscillations take place.

The value of the threshold field varies with the type of the semiconductor materials used. For *n*-type GaAs, the critical or the threshold field is of the order of 3.2 kV/cm and for InP, it is around 10.5 kV/cm. For an *n*-type GaAs, specimen of 25 μm length, the oscillation frequency is around 4.5 GHz.

Gunn diodes are normally used as low-power oscillator at microwave frequencies in transmitter and also as local oscillator in receiver front ends.

Constructional Details and Electrical Equivalent Circuit of a Gunn Diode

The basic structure of a Gunn diode is shown in Fig. 10.20 (a), which consists of *n*-type GaAs semiconductor with regions of high doping (n^+). Although there is no junction, this is called a diode with reference to the positive end (anode) and negative end (cathode) of the dc voltage applied across the device. Since GaAs is a poor conductor, considerable heat is generated in the diode. The diode should be well bonded into a heat sink (Cu-stud).

The electrical equivalent circuit of a Gunn diode is shown in Fig. 10.20 (b), where C_j and $-R_j$ are the diode capacitance and resistance, respectively, R_s includes the total resistance of lead, ohmic contacts, and bulk resistance of the diode, C_p and L_p are the package capacitance and inductance, respectively. The negative resistance has a value that typically lies in the range of –5 to –20 ohms.

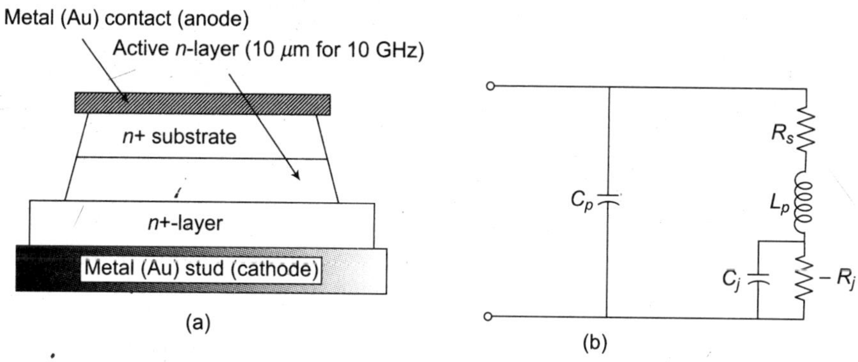

Fig. 10.20 Constructional details and the electrical equivalent circuit of a Gunn diode

Theory of Operation 10.3.1

The Gunn effect can be explained from the three- valley model of electron energy ε(eV) band vs electron momentum K diagram, viz ε-K diagram of certain compound semiconductors, e.g. GaAs, InP. The energy-band diagram of *n*-type GaAs is depicted in Fig. 10.21. One of the important characteristics of GaAs is its superior electron mobility. Gallium arsenide (GaAs) is a direct gap material with a maximum valence band and a minimum conduction band occurring at the same wave vector. The conduction band is separated from the valence band by band-gap energy of 1.42 eV. The conduction band has three sub-bands or valleys (Γ, L, and X) separated from the valence band by energy gaps as shown in Fig. 10.21.

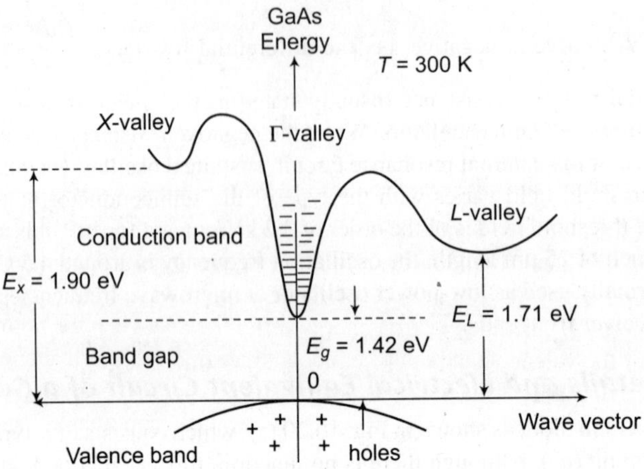

Fig. 10.21 Multi-valley conduction band energies of GaAs

The conductivity of the material is directly proportional to the electron mobility μ_n. The mobility of electrons in the conduction band depends upon

- concentration of impurity, N in the semiconductor,
- temperature, T Kelvin, and
- is inversely related to the electron effective mass, m_n.

The current density is related to the mobility by the following equation:

$$J = en\mu_n E = e(n_\Gamma \mu_\Gamma + n_L \mu_L + n_X \mu_X)E \tag{10.11}$$

Here, e = electronic charge = 1.6×10^{-19} C, n represents electron concentration, μ is the electron mobility, and E is the electric field in the semiconductor sample. Suffices Γ, L, X represent the values of n and μ in the Γ, L, and X valleys, respectively.

For GaAs, the effective mass of these electrons in the conduction band is 0.067 times the mass of free electrons. It can be seen that the lowest valley in the conduction band is narrowest and has highest slope which corresponds to the electrons with low effective mass and highest mobility, while upper valleys are wider with lower slopes and characterized by larger effective masses and lower electron mobility. The following explanation can be given for the J-E characteristic of GaAs.

At room temperature of 300 K, most of the electrons in the conduction band are in the lowest central valley Γ and much fewer in the upper valleys. Therefore, when Gunn diode is biased with a dc voltage, an electric field E is established across it. At low bias voltage or low field state, the high mobility electrons in lowest valley increase the conductivity of the material and hence, the current in the sample increases steadily with the applied field (region a in Fig. 10.19a). As the applied voltage increases, the electric field also increases further, and more and more electrons are transferred from higher mobility lower valley to the lower mobility upper valleys resulting in decrease of conductivity. When E-field exceeds a certain critical value E_{th}, majority of the conduction band electrons transferred to the lower mobility upper valley and the conductivity decreases with increase in field. This results in decrease of current in the sample (region b). This explains the negative resistance of the sample, i.e., $\frac{dJ}{dE}$ is negative and the device behaves as a differential negative resistance. This is called the transferred electron effect and the the Gunn diodes are also called **Transfer Electron Device (TED)**.

When $E = E_s$, called the sustaining field, almost all the conduction band electrons are in the upper valley. For $E > E_s$, the net mobility of the electrons in the conduction band is determined by the upper valley mobility which is substantially less than that in the lower valley. Thus, for $E > E_s$, the slope of the J-E curve is positive but less than that of the lower valley. This explains the slope difference between the region in the *J-E* diagram for $E < E_{th}$ and $E > E_s$.

Necessary Criteria for Exhibiting Gunn Effect

The band structure of a semiconductor must satisfy the following conditions in order to exhibit Gunn effect.

(i) There must be at least two sub-bands in the conduction band of the semiconductor. Electrons in the lower conduction sub-band should have sufficiently higher mobility than the mobility in the upper valley sub-band.

(ii) The band-gap energy E_g, i.e. the electron energy difference between the bottom of the conduction band to the top of the valence band must be much larger than the thermal energy $kT = 0.026$ eV at room temperature of 300 K. This means $E_g >> 0.026$ eV.

(iii) The energy separation ΔE between the two sub-bands in the conduction band must be much less than E_g but sufficiently higher than kT. This means that at low field most of the electrons are in the lower valley. Otherwise, the semiconductor will break down before the electron starts to transfer to the upper valley.

The most common semiconductors silicon and germanium do not meet all the above criteria. GaAs (gallium arsenide), InP (indium phosphate), and CdTe (cadmium telluride) satisfy the above conditions and hence satisfy the ***transferred electron effect.*** Table 10.1 describes some properties of Si, Ge and GaAs.

Table 10.1 Properties of Si. Ge. and GaAs at 300K

Properties	*Si*	*G.*	*GaAs*
Breakdown Field, V/m	approx. 3×10^5	approx. 1×10^5	approx. 4×10^5
Dielectric Constant	119	16.0	13 1
Effective Density of States in the Conduction Band, N/cm^3	2.8×10^{19}	1.04×10^{10}	4.7×10^{17}
Effective Density of States in the Valence Band, Nv/cm^3	1.04×10^{18}	6.0×10^{18}	7.0×10^{18}
Energy Gap at 300K (eV)	1.12	066	1.424
Intrinsic Carrier Concentration (cm^3)	1.45×10^{10}	2.4×10^{13}	1.79×10^6
Electron Mobility μ (cm^2/V.s)	1500	3900	8500
Hole Mobility μ_p (cm^2/V.s)	475	1900	400
Thermal Conductivity at 300 K (W/cm·°C)	15	0.6	0.46

Example 10.3 Calculate the conductivity of an n-GaAs gunn diode at room temperature. Given average electron density and mobility in the conduction band are 10^{17}/cm^3 and 8500 cm^2/Vs.

Solution The conductivity is given by

$$\sigma = en\mu = 1.6 \times 10^{-19} \times 10^{17} \times 8500 \text{ mho}$$

Modes of Operation

There are two principle modes of operation that result in microwave oscillations in a Gunn diode. These are the Gunn mode or the transit-time (TT) mode and the limited-space-charge (LSA) mode. Under special circumstances, two other modes—quenched domain mode and delayed mode—are possible as described below.

❑ *High Field Domain or Gunn Mode of Oscillations* In GaAs (*n* or *p* type) and other multivalley compound semiconductors, a decrease in mobility in the upper valley in the conduction band with increase in electric field leads to decrease in drift velocity of electrons. This yields the formation of a high field domain for microwave generation. This will now be explained with the help of Fig. 10.22.

Let us assume that at point *A* in the device, there exists an excess –ve charge due to random noise fluctuation on biasing or by non-uniformity in the doping as shown in Fig. 10.22 (a). Some electric field will be created by the accumulated charge. The field to the left (cathode side) of point *A* is lower than that to the right (anode side). If the diode is biased at the field point E_A on the *J-E* curve, the carriers (or the current) flowing into point *A* from cathode are larger than those flowing out of *A* towards anode since the velocity on the left hand side of *A* is larger than that on the right-hand side of *A*. This will increase the excess – ve charge at *A*. When the *RF* noise cycle reverses, the field in the left of *A* is lower than that before and the field to the right of *A* is greater than the original one, resulting in more space charge accumulation. This process continues until the low and high fields both reach values outside the differential –ve resistance region of *J-E* curve and settle at points *a* and *c* in Fig. 10.22(b), where the currents in the two field regions are equal.

Therefore, no further accumulation of charge at *A* occurs and a stable accumulation layer is formed. The analogous effect occurs in the event of a random deficiency of electrons at some point in the sample resulting in a depletion layer. When the enhancement (or the accumulation) and the depletion layers approach each other, they attract each other and together pass through the diode in the form of a domain.

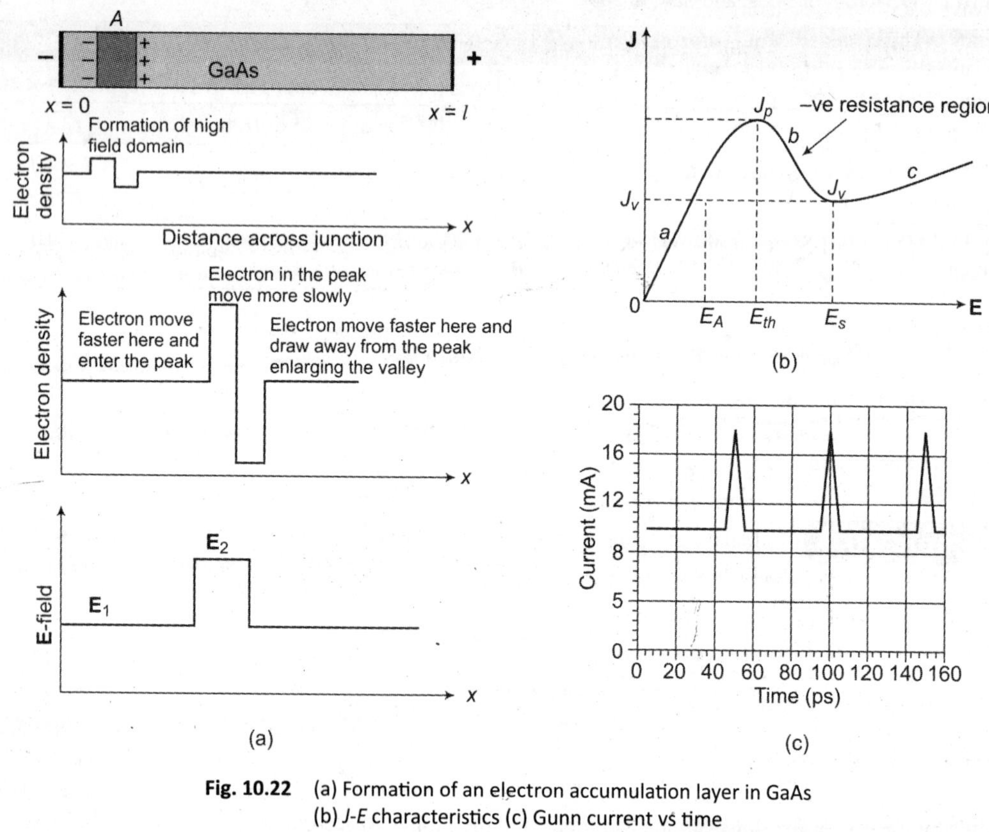

Fig. 10.22 (a) Formation of an electron accumulation layer in GaAs (b) *J-E* characteristics (c) Gunn current vs time

Since for a given bias, the terminal voltage $V = -\int E\,dx$ across the semiconductor remains constant for a constant bias voltage and the field in the interior of the domain is very high, the field outside falls below the threshold field E_{th}. Thus, no new domain can be found until the existing one disappears at the anode. Usually, the domain is formed near the cathode where the instability in the carrier distribution is more likely.

◆ **Microwave Generation** The high field domain moves through the sample with uniform velocity and gets collected at the anode. A new domain is again formed at the cathode and the process repeats itself. Thus, a pulse current output is obtained with the intrinsic period T equal to the transit time of the domain through the effective length l_{eff} of the sample:

$$T = \frac{l_{eff}}{v_d}$$

The intrinsic frequency of oscillation is given by

$$f = \frac{1}{T} = \frac{v_d}{l_{eff}}$$

where v_d is the domain velocity (approximately the drift velocity of electrons) and the effective length that the domain travels $l_{eff} \approx l$, the sample length. Since the drift velocity is very high and the semiconductor length is very small, the current fluctuation occurs at microwave frequencies to produce an output signal in the low impedance external RF circuit.

❑ *Other Modes of Oscillations* When the Gunn diode is connected to a resistive load, the frequency of microwave oscillations is determined by the time taken by the high field domain through the diode (transit frequency). This is called the **Gunn or Transit Time (TT) mode**. The efficiency of this intrinsic mode is low, usually below 10%. However, if the external load is a tuned circuit (as is the case in most TED oscillators), the resonator frequency can be 'imposed' upon the Gunn diode. Several operating modes are possible in this case, viz. **Limited Space Charge Accumulation (LSA) mode, quenched domain mode and delayed domain mode.**

◆ **Limited Space Charge Accumulation (LSA) Mode** When the Gunn diode is mounted as a part of a resonant circuit as shown in Fig. 10.23 and the resonant circuit is tuned to a frequency several times greater than that of the TT mode, the dipole domains do not have sufficient time to form. The circuit operates as a negative resistance oscillator when the dc voltage is adjusted to a value greater than the threshold voltage nearly at the midpoint of the negative resistance region. The resistance load R_L is

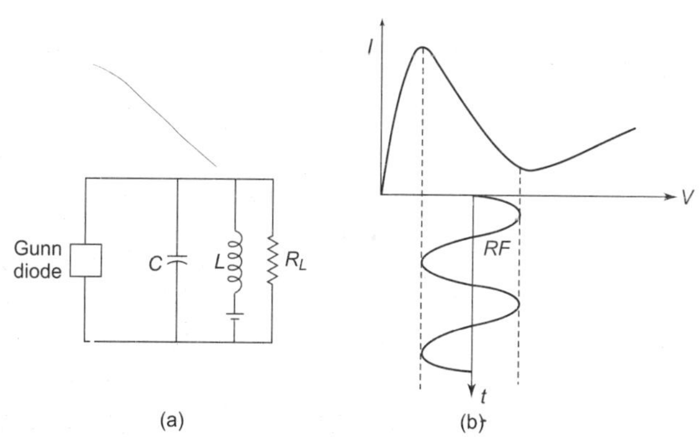

Fig. 10.23 Gunn oscillator operating in LSA mode and RF oscillating voltage

adjusted to a value of about 20% greater than the maximum negative resistance value of the device to enable oscillations to start and stay steady. The amplitude of the oscillations builds up and becomes steady when the average negative resistance of the Gunn diode becomes equal to the load resistance R_L. The peak-to-peak amplitude of the microwave oscillations is approximately equal to the voltage range in the negative resistance region.

Any accumulation of electrons near the cathode has time to collapse while the signal is below threshold. In this mode, the internal electric field would be uniform and proportional to the applied voltage. The current in the diode is then proportional to the drift velocity at this field level.

LSA mode of operation can produce several watts of power with efficiencies of 20%. The output power decreases with frequency, viz., 1W at 10 GHz and several mW at 100GHz.

The following condition must be satisfied for the LSA mode.

$$2 \times 10^4 < \frac{n_0}{f} < 2 \times 10^5$$

Also, the requirement for the LSA mode is that the period of oscillations, should be much greater than the dielectric relaxation time τ_d. Fig. 10.23(a) shows the LSA oscillation mode where time period $T = 3\tau_d$.

◆ **Quenched Domain Mode** If the resonant circuit is tuned to a value slightly above that of the TT mode, instantaneous voltage across the diode drops below the sustaining voltage V_s during the –ve half cycle as shown in Fig. 10.23(a). The domain collapses before it reaches the anode. When the bias voltage swings back above the threshold, a new domain is formed and the process repeats. The Gunn diode will operate mostly like Gunn mode. This mode of operation is called *quenched domain mode*. The oscillations occur at the frequency of the resonant circuit rather than at the transit time frequency. This mode occurs when, $f.l > 2 \times 10^7$ cm/s. Theoretically, the efficiency of the quenched domain mode is around 13%.

◆ **Delayed Domain Mode** If the external resonator is tuned below the transit-time frequency of the Gunn mode, the dipole domains will arrive at the anode well in time but the formation of a new dipole domain will be delayed until the oscillation voltage increases above the threshold value. This type of operating mode is called the delayed mode. Figure.10.23 (a) is the usual J-E or I-V characteristics of an *n*-GaAs material. Figure 10.23(b) depicts *RF* voltage superimposed over the dc bias voltage. At the moment when the domain reaches the anode, the instantaneous value of the diode voltage (dc bias voltage + *RF* voltage) is less than the threshold value V_{th}. The formation of the new domain is delayed till the voltage swings over the threshold value. In this case, therefore the oscillation period is greater than the transit time, i.e,

$$T_0 > \tau_i \quad \tau_i = \text{intrinsic domain transit time.}$$

This mode occurs when 10^6cm/s $< f.\ l < 10^7$ cm/s where l = length of the device.

The efficiency of this mode is around 20%.

Performance of Guun Diode Oscillators

Gunn diode oscillators have better noise performance than IMPATT oscillators but power output, both in continuous wave and pulse modes, are significantly lower than those of IMPATTS. Because of low noise, Gunn oscillators are used as local oscillators in receivers and as primary sources where continuous wave power of 100 mW or less is required. InP Gunn diodes posses higher power and efficiency than GaAs diodes. The state-of-the-art performance for GaAs and InP Gunn diodes are given below.

GaAs diodes

Power output ≈ 300 mW of CW power at 35 GHz

≈ 30 mW of CW power at 100 GHz

InP diodes

Power output ≈ 500 mW of CW power at 35 GHz

≈ 50 mW of CW power at 100 GHz

Since GaAs is a poor conductor, considerable heat is generated in the diode. The diode should be well bonded into a heat sink (Cu-stud).

Gunn Diode Oscillator 10.3.2

Gunn diode oscillators are commonly used in radars as *LO* and also as signal source in the laboratory. A Gunn diode oscillator can be designed by mounting the diode inside a waveguide cavity formed by a short-circuit termination at one end and by an iris at other end as shown in Fig. 10.24. The diode is mounted at the centre perpendicular to the broadwall where the electric field component is maximum under the dominant TE_{10} mode. The intrinsic frequency f_0 of oscillation depends on the electron drift velocity v_d due to high field domain through the effective length l.

$$f_0 = v_d/l \tag{10.12}$$

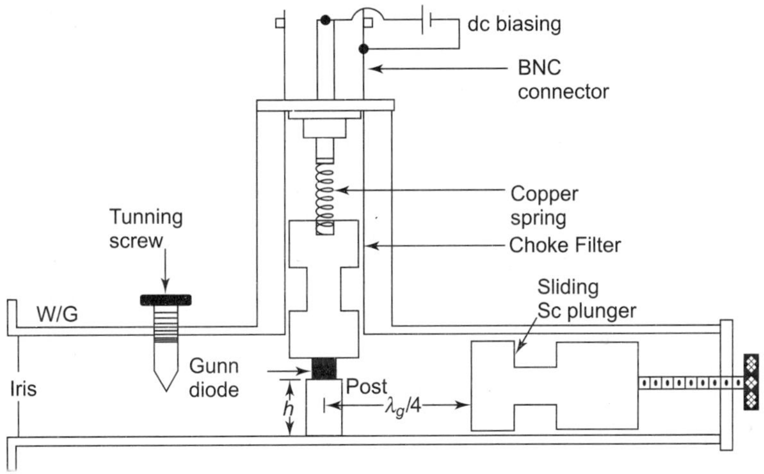

Fig. 10.24 Gunn diode oscillator circuit

For GaAs, $v_d \approx 10^7$ cm/s. Normally, the cavity is tuned to resonate at the intrinsic frequency f_0 by adjusting the short position of the sliding choke plunger. A tuning screw is inserted perpendicularly at the centre of the broadwall for fine frequency tuning.

The total resistive loading from the cavity and the external load should be around 20 per cent higher than the Gunn device resistance $-R_j$ so that the circuit resistance $-R_L R_j/(R_L - R_j)$ will be negative. The cavity has an impedance transforming property between the high impedance of the output waveguide and the required low value for the Gunn diode. The Gunn diode is placed on a metal post. The post diameter is nearly 5 mm at 10 GHz and the height of the post is slightly less than $\lambda/4$. Dc bias voltage is applied to the diode through BNC fitting. The degree of coupling to the external waveguide is adjusted by selecting the inductive iris dimension.

The power output of the Gunn diode oscillator is in the range of a few watts for CW operation at biasing values 10 V and 1A at 30–40 GHz. A frequency tuning range of nearly 2% can be achieved. For pulsed operation, peak powers are typically 100–200 W.

The power output of the Gunn diode is limited by the difficulty of heat dissipation from the small chip. The advantages are small size, ruggedness, and low cost.

Example 10.4 The drift velocity of electrons is 2×10^7 cm/s through the active region of length 10×10^{-4} cm. Calculate the natural frequency of the diode and the critical voltage.

Solution

The natural frequency $f_0 = \dfrac{v_d}{l} = \dfrac{2\times10^7}{10\times10^{-4}} = 20$ GHz

The critical voltage $V = l \times$ critical field for GaAs

$= 10 \times 10^{-4} \times 3.2$ kV/cm

$= 10 \times 10^{-4} \times 3.2 \times 10^3$ V $= 3.2$ volts

AVALANCHE TRANSIT-TIME DEVICES (ATTD) 10.4

Avalanche transit-time devices (WT Read, 1958) are semiconductor junction structures with highly doped *p* and *n* regions.' Reverse bias of sufficient strength produces *E*-field that is high enough to create an avalanche of carriers through impact ionization within adjacent, highly doped *p* and *n* layers. The generated carriers transit a drift space, which causes a delay in their collection by the anode. If the delay is sufficient to cause the current to be out of phase with the voltage across the device, then an apparent negative resistance appears across the structure's terminals. There are three types of these devices—(1) Impact Ionization Avalanche Transit Time (IMPATT), (2) Trapped Plasma Avalanche Triggered Transit (TRAPATT), and (3) Barrier Injected Transit Time (BARITT). The IMPATT diodes have efficiency of the order of 3% CW power and 60% pulsed power and can be operated from 500 MHz to 100 GHz. The power outputs lie between 1 W (CW) and over 400 W (pulsed). TRAPATT is suitable for low frequency (1–3 GHz) applications with pulsed power output of several hundred watts and efficiency 20–60%. BARITT has the advantage of low-noise figures (< 15 dB) but with low power and smaller bandwidth.

IMPATT Diodes 10.4.1

IMPATT (Impact ionization Avalanche Transit Time) diodes rely on the effect of carrier impact ionization breakdown across a reverse biased *p-n* junction and the drift of carrier in the high field region of a semiconductor to produce differential negative resistance at microwave frequencies. The device was first proposed by ***WT Read in 1958*** where he had shown the negative resistance behaviour of an n^+–p–i–p^+ junction structure. Such diodes can be manufactured from Si, GaAs, or InP. However, GaAs provides the highest efficiency, the highest operating frequency, and least noise figure. But the fabrication process is more difficult and is more expensive than Si. Silicon IMPATT diodes can operate up to 100 GHz with typical dc-to-*RF* efficiency of 5 to 10 %. IMPATTs offer highest CW and peak power and are being exploited in many civilian applications like missile seekers. IMPATT diodes have many forms, viz., Read device n^+pip^+ or p^+nin^+, abrupt junction p^+nn^+, and p^+in^+ diode. A typical package of p^+–n–n^+ diode is shown in Fig. 10.25. A metallized cathode and plated heat sink as anode are also shown.

IMPATT diodes are used as oscillators as well as amplifiers. These devices, although noisier compared to other solid state sources, can be injection locked to produce highly stable high power oscillators at microwave and mm-wave bands.

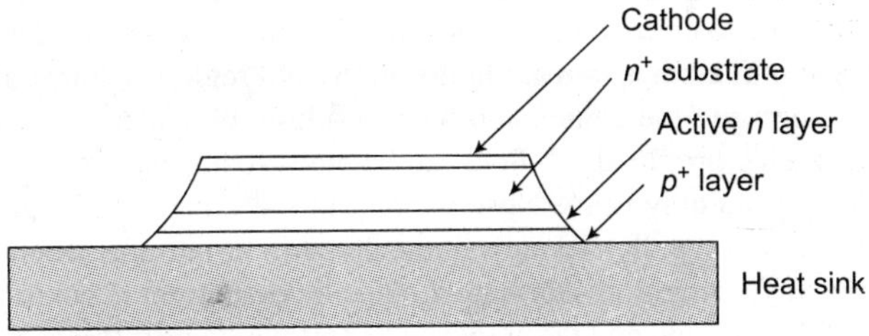

Fig. 10.25 A typical p^+-n-n^+ diode package

READ Diode

The semiconductor layer structure with the doping profile of the Read diode n^+pip^+ are shown in Fig. 10.26. The n^+ and p^+ are the heavily doped regions, p is moderately doped, and i-region is slightly n-type with doping concentration of about $10^{13}/cm^3$.

❑ ***Operating Principle*** Under reverse-biased condition, the characteristics of different regions of the n^+ $p-i-p^+$ Read diode, electric field distribution, and doping concentration are shown in Fig. 10.26. When the reverse bias voltage exceeds the breakdown voltage V_B, a maximum electric field of very high value (MV/m) appears at the n^+ p junction. The E-field in the intrinsic region (i-region) remains below breakdown. The p-region is very thin. The space between the n^+–p junction and the i-p^+ junction is called the space charge region.

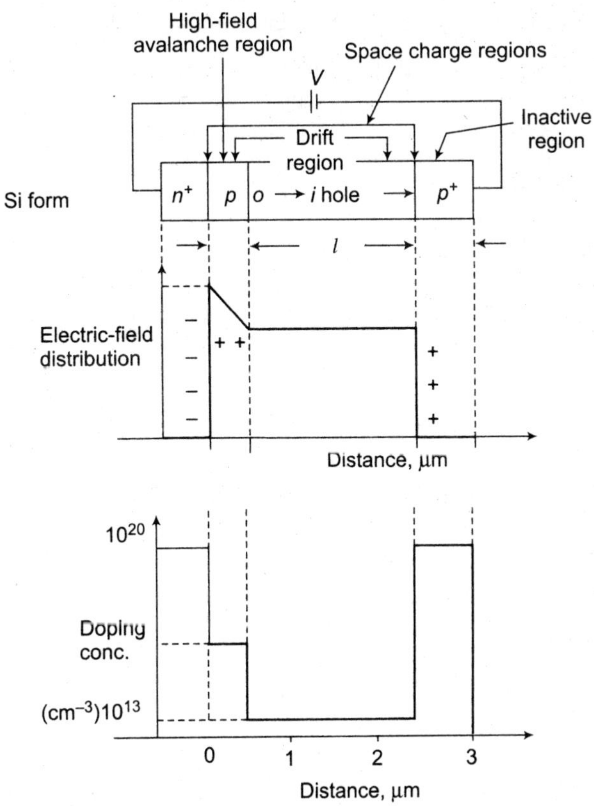

Fig. 10.26 Read diode profile

The generated electrons due to electric field, move through the *i*-region with saturated drift velocity and are collected at the p^+ region. The holes moving in this high-field region acquire sufficient energy to excite valence electrons of the atom into the conduction band resulting in avalanche multiplication of electron-hole pairs. By designing doping profile, the *E*-field can be made to have a very sharp peak, so that "impact avalanche" multiplication occurs only in the close vicinity of the junction. The process is cumulative and the carrier density increases very rapidly. To prevent the diode from burning, a constant current bias source is used to maintain the average current at safe limit I_0. The diode current is contributed by the conduction electrons which move to the n^+ region and the associated holes which drift through *i* space charge region to the p^+ region, under the influence of a lower but steady electric field. The drift time is given by

$$t_d = \frac{l}{v_d}$$

where v_d is the drift velocity of the holes ($\sim 10^5$ m/s for $E \sim 0.5$ MV/m), and l is the length of the drift region. If $l = 2$ mm, $t_d = 20$ ps.

The breakdown voltage for silicon $p^+ – n$ junction is given by

$$|V_B| = \frac{\rho_n \mu_n \varepsilon |E_{max}|^2}{2}$$

where ρ_n = resistivity of the semiconductor, μ_n = electron mobility , ε = permittivity of semiconductor, E_{max} = maximum breakdown electric field at the $n^+ – p$ junction. The breakdown voltage depends on the doping concentration. For Si, $V_B \sim 50$ V whehe doping concentration is 10^{16} /cm^3 and $V_B \sim 10$ V for doping of the order of 10^{17}/cm^3. For GaAs , V_B is higher than Si for the same doping level.

❑ ***Negative Resistance Property and Mechanism of Oscillations*** Let us consider a Read diode biased just below breakdown voltage (V_B) and mounted in a microwave resonator circuit. Noise that is present in the device is broadband but due to high *Q*-resonant, the noise in the circuit may be considered sinusoidal at the resonant frequency. The concept of noise and its effect on the circuit is important for understanding the mechanism of oscillation.

To understand the negative resistance property of IMPATT diode, let an ac voltage be superimposed on the dc bias voltage as shown in Fig. 10.27(c).Since the bias voltage is near V_B, the total voltage (dc + ac) across the terminals exceeds the breakdown voltage in the positive half of the ac voltage cycle, and at time $t = 0 +$ avalanche multiplication process starts. The avalanche multiplication grows exponentially with time while the field is above the critical breakdown value. The carriers generated at the junction due to avalanche or the injected junction current reaches its peak at $t = T/2$, where *T* is in the time period of the *AC* voltage. Beyond $t = T/2$, the generated carriers decay rapidly. The injected carrier current at $n^+ – p$ junction is, therefore, at short duration pulse delayed by $T/4$ (i.e by $\pi/2$ phase) as shown in Fig. 10.27(d).

As the injected holes traverse through the drift space (*i*-region) towards the –ve terminal, a pulse current $I_e(t)$ of width equal to the half period of the ac voltage, is induced in the external circuit. The generated holes move with constant saturated drift velocity through the drift zone and, hence, the current in the external circuit remains constant till the time the carriers reach the cathode contact. The transit time t_d through the drift zone is equal to l/v_d. If $t_d = T/2$, the duration of the external current is from $t = T/2$ to $t = T$. It is seen that since the external current exists over the entire –ve cycle of the AC voltage, the incremental terminal resistance of the IMPATT device is negative. The effective transit time of carriers must be equal to half the time period of the oscillator voltage, i.e.

$$t_d = T/2 \quad (T = \text{time period})$$

$$= \frac{1}{2f}$$

where frequency of oscillation is

$$f = \frac{1}{2t_d} = \frac{v_d}{2l} \tag{10.13}$$

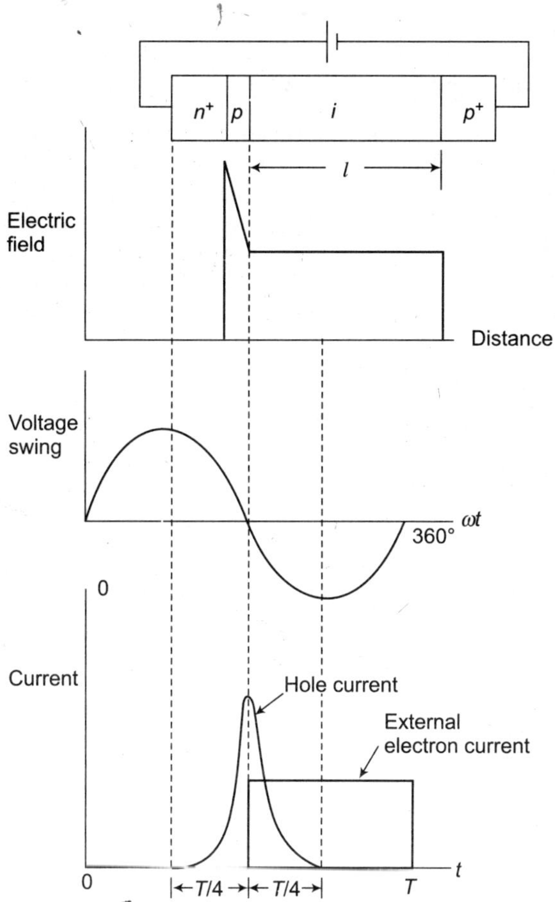

Fig. 10.27 IMPATT diode operation

This is the basic design equation for an IMPATT device.

Example 10.5 Design the width of the *i*-region for an Si Read diode for operation at 10 Hz.

Solution

The frequency of oscillation $f = \frac{v_d}{2l}$

Therefore, the intrinsic length = $\frac{v_d}{2f}$

For silicon, $v_d = 10^7$ cm/s.

Therefore, the length of the intrinsic region is

$$l = \frac{10^7}{2\times10\times10^9}\text{ cm}$$
$$= 5\ \mu\text{m}$$

Mounting and Equivalent Circuit of IMPATT

IMPATT diodes are mounted in coaxial lines, waveguides, or microstrip lines to form microwave circuits for oscillations and amplifications as shown in Fig. 10.28. A simplified equivalent circuit for IMPATT diode chip is shown in Fig. 10.29. Here, R_d is the diode negative resistance consisting of the series lead resistance R_s and the negative resistance $-R_j$ due to impact avalanche and transit time process. C_j is the junction capacitance at the break-down voltage, and L_p and C_p are the package lead inductance and capacitance, respectively. The diode mount is designed in such a way that the overall reactance of the circuit is tuned to zero at resonance by controlling L_p. Since the power extracted by the ac field from the dc field compensates for the power

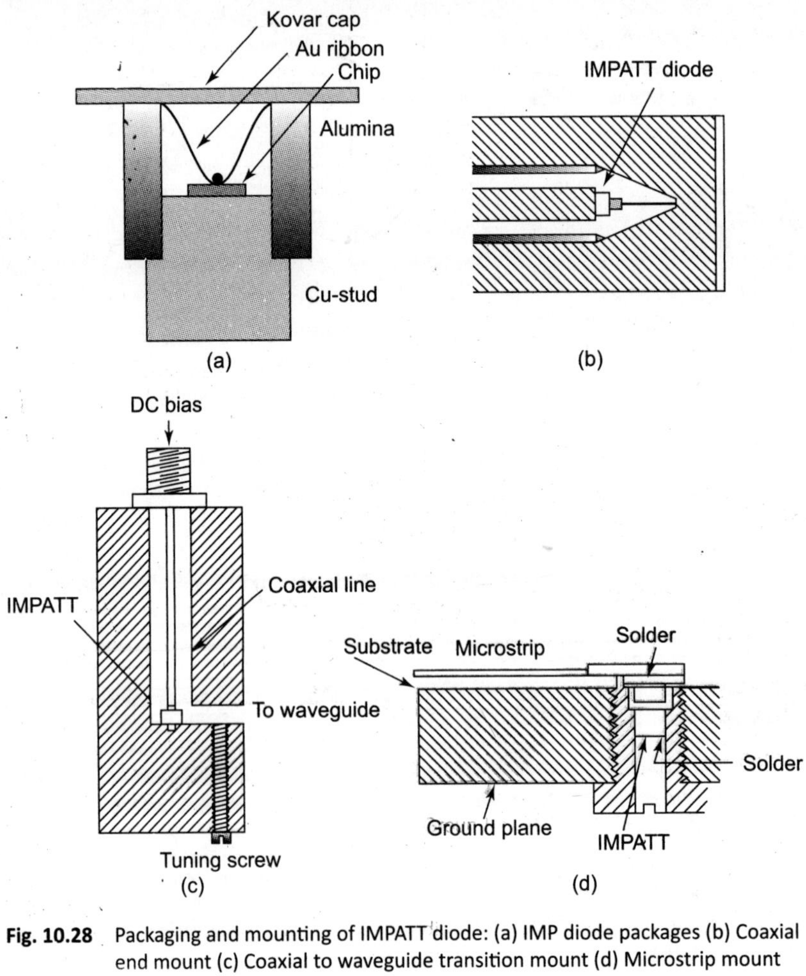

Fig. 10.28 Packaging and mounting of IMPATT diode: (a) IMP diode packages (b) Coaxial end mount (c) Coaxial to waveguide transition mount (d) Microstrip mount

dissipation in positive resistance of the circuit, the total resistance R_d must also be zero. Since the diode chip impedance is given by

$$Z_d = -|R_d| - \frac{j}{\omega C_j} \quad (10.14)$$

the load impedance must be $Z_L = -Z_d$, or

$$R_L = |R_d| \quad (10.15a)$$

$$X_L = \frac{1}{\omega C_j} \quad (10.15b)$$

Since R_d varies with both bias and signal currents, for a given load and bias, stable oscillations are obtained when $|R_d| = R_L$. The corresponding peak RF current determines the load power. Since total negative resistance in the circuit is low (= 10 ohms) for sustained oscillation, multisection quarter-wave transformers must be incorporated for impedance matching between the diode circuit and the load as shown in Fig. 10.29.

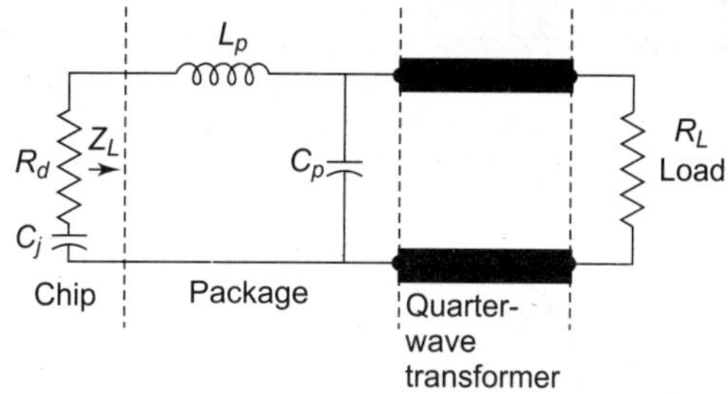

Fig. 10.29 Equivalent circuit of IMPATT diode

Power Output and Efficiency

The maximum power output of an IMPATT diode is limited by the properties of the semiconductor material and the impedance of the external microwave circuit. The maximum voltage that can be applied across the diode is given by

$$V_m = E_m \cdot l$$

Here, E_m is the maximum field and l is the depletion length. The maximum voltage is limited by the breakdown voltage V_B. The maximum current is given by

$$\begin{aligned} I_m &= J_m A \\ &= \sigma E_m A \\ &= (\varepsilon/t_d) E_m A \\ &= (v_d/l).\varepsilon E_m A \end{aligned}$$

where σ is in the conductivity, ε is the semiconductor permittivity, A is the junction area, t_d is the transit time, v_d the drift velocity and L is the drift length.

The upper limit of the input power is

$$P_m = I_m V_m = E_m^2\, \varepsilon\, v_d A$$

The capacitor of the device is defined as

$$C = (\varepsilon A)/\, l$$

Noting that $t_d = T/2 = \frac{1}{2f}$, above equations are combined to yield

$$P_m f^2 = \frac{E_m^2 c_d^2}{4\pi^2 X_C}$$

where $$X_C = \frac{1}{2\pi f C}.$$

It is seen that for a given circuit reactance X, the product $P_m f^2$ is constant for a given semiconductor material. Therefore, the maximum power obtained for a single IMPATT device varies as $1/f^2$. The higher the frequency of operation, the lower is the power.

The efficiency of the device is given by

$$\eta = P_{ac}/P_{dc} = (V_{ac}/V_{dc})\,(I_{ac}/I_{dc})$$

For an ideal Read IMPATT, the ratio (V_{ac}/V_{dc}) is about 0.5 and the ratio (I_{ac}/I_{dc}) is about $2/\pi$ and hence, the η would be around 30%. However, in practical IMPATT diodes, the efficiency is much lower due to several fundamental effects like space-charge effect , ionization-saturation effect, skin effect, etc.

Example 10.6 Calculate the oscillation frequency and power output of an IMPATT diode for the following given parameters:

Drift region length = 5 μm
Saturation drift velocity (v_d) = 2×10^7 cm/s
Breakdown voltage = 90 V
Maximum operating voltage = 100 V
Maximum operating current = 300 mA
Efficiency = 10%

Solution

l = 5 μm, $v_d = 2 \times 10^7$ cm/s, V_B = 90 V, V_m =100 V, I_m = 300 mA, and η = 10%

(i) The oscillation frequency is

$$f = v_d/(2l) = (2 \times 10^7)/(2 \times 5 \times 10^{-4}) = 20 \text{ GHz}$$

(ii) The power output is

$$P = \eta P_{dc} = \eta V_m I_m = 0.10 \times 100 \times 0.3 = 3 \text{ W}$$

Example 10.7 An IMPATT diode with nominal frequency 10 GHz, has C_j = 0.5 pf, L_p = 0.5 nH and C_p = 0.3 pF at breakdown bias of 80 V and bias current 80 mA. The *RF* peak current is 0.65 A for $R_d = -2$ ohms. Find (a) the resonant frequency of oscillation, and (b) the efficiency.

Solution

(a) For sustained oscillation, physical load $R_L = |Z_d| = 2$ ohm

The effective load impedance (refer Fig. 10.29).

$$Z_L = j\omega L_p + \frac{R_L}{1 + j\omega C_p R_L} = j\omega L_p + \frac{R_L}{\sqrt{[1 + (\omega C_p R_L)^2]}\angle \tan^{-1}(\omega C_p R_L)}$$

At nominal operating frequency of 10 GHz

$$\omega L_p = 2\pi \times 10 \times 10^9 \times 0.5 \times 10^{-9} \text{ ohm} = 31.42 \text{ ohm}$$

$$\omega C_p R_L = 2\pi \times 10 \times 10^9 \times 0.3 \times 10^{-12} \times 2 = 0.038,$$

$$\tan^{-1}(\omega C_p R_L) = 2.18°$$

Therefore, $(\omega C_p R_L)^2 << 1$ and $Z_L \approx R_L + j\omega L_p$

This shows that the role of C_p in circuit resonance is negligible. Therefore, the resonant frequency

$$f = \frac{1}{2\pi\sqrt{(C_j L_p)}} = \frac{1}{2\pi(0.5 \times 10^{-12} \times 0.5 \times 10^{-9})^{1/2}} = \frac{\sqrt{10^3}}{2\pi \times 0.5} = 10.06 \text{ GHz}$$

(b) The output power

$$P_L = 1/2\ (RF \text{ peak current})^2 \times R_L = \frac{1}{2} \times 0.65^2 \times 2$$
$$= 0.4225 \text{ W}$$

Average dc input power

$$P_{dc} = \text{Breakdown voltage} \times \text{dc bias current} = 80 \times 0.08 = 6.4 \text{ W}$$

Efficiency

$$\eta\ \% = \frac{P_L}{P_{dc}} \times 100 = \frac{0.4225}{6.4} \times 100 \approx 6.6\%$$

IMPATT Diode Power Amplifier 10.4.2

The IMPATT diode can be used as amplifier with the same basic circuit arrangement as oscillator, provided $R_L > |R_d|$ and a circulator is incorporated as shown in Fig. 10.30.

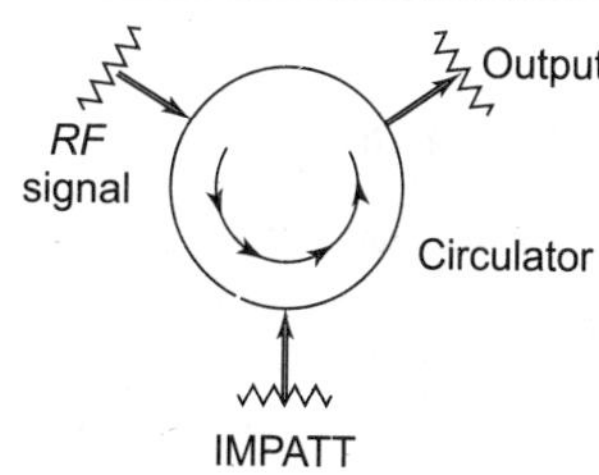

Fig. 10.30 IMPATT circulator-type amplifier

The negative resistance is used to terminate one port of the circulator and the actual load is connected to other port. The input *RF* power is fed from the remaining port. The negative resistance results in voltage reflection coefficient at the port greater than unity. Thus the average power from the source P_{av} circulates to the negative resistance port and the reflected power is greater than the incident power. The reflected power is delivered to the load. The power gain is

$$G_p = \frac{|\Gamma|^2 P_{av}}{P_{av}} = |\Gamma|^2 > 1 = \left|\frac{-|R_d| - R_L}{-|R_d| + R_L}\right|^2 \quad (10.16)$$

Therefore, if $R_d = -2$ ohm and $R_L = 3$ ohm, $G_p = |-5/1|^2 = 25$.

TRAPATT Diode 10.4.3

TRAPATT diodes are manufactured from Si, and have $p^+n\ n^+$ (or $n^+\ pp^+$) configurations as shown in Fig. 10.31. The *p-n* junction is reverse biased beyond the breakdown region, so that the current density is higher. This causes the electric field in the space charge region to be decreased and the carrier transit time is increased. Consequently, the frequency of operation becomes lower and is limited to below 10 GHz, although efficiency of the diode increases due to low power dissipation.

The TRAPATT diode is mounted inside a coaxial resonator at a position of maximum *RF* voltage swing. When avalanche occurs at dc reverse-bias plus *RF* swing beyond the threshold of breakdown, a plasma of holes and electrons are generated. This plasma density results in a high potential difference across the junction in opposition to the dc reverse bias voltage. At this relatively low voltage the plasma gets trapped. But since the external circuit current flow, the voltage rises and the trapped plasma is released producing current pulse across the drift space. The total transit time is the sum of the delay time in releasing the trapped plasma and the drift time. Since the transit time is longer due to low voltage, the operating frequency is limited below 10 GHz. Moreover, the current pulse is associated with low voltage. Therefore, the power dissipation is low and efficiency is higher.

The major disadvantages of TRAPATT are (1) high noise figure (60 dB) limits its use as an amplifier, (2) it generates strong harmonics due to the short duration current pulse.

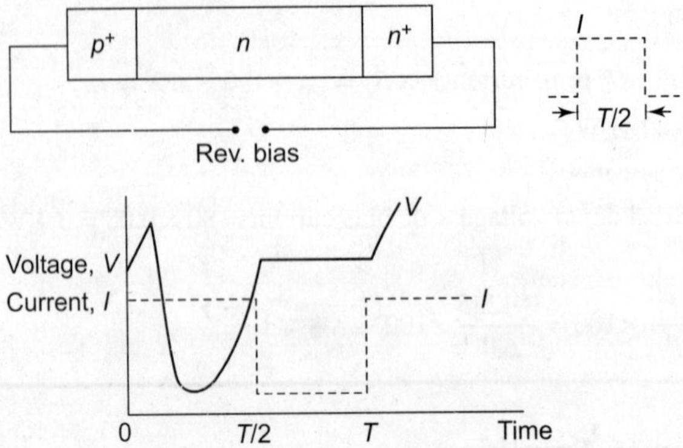

Fig. 10.31 TRAPATT diode

BARITT Diodes 10.4.4

BARITT diodes are formed by forward biased *p-n* junction with *p-n-p* or *p-n-i-p* or *p-n*-metal, or metal-*n*-metal configurations. Minority charge carries are injected into the drift region. The transit time through the drift region provides the required phase-shift between the current and voltage to give a negative resistance. When the diode is mounted in a resonator, a noise spike generates microwave voltage across the diode. During positive half cycle the total voltage produces a sharp pulse of minority carrier current in the drift region. During the drift time, a constant external current delivers energy to the resonator from the dc bias source to maintain a continuous oscillation. BARITT diodes are low power (mw), low efficiency, and less noisy devices. They are used as local oscillator at microwave frequencies (4–8 GHz). Table 10.2 shows a comparison between various TEDs and ATTDs.

Table 10.2 *Comparison between Gunn, IMPATT, TRAPATT and BARITT*

	Gunn	*IMPATT*	*TRAPATT*	*BARITT*
Operating	1–100 GHz	0.5–100 GHz	1–10 GHz	4–8 GHz
Bandwidth	2% of centre frequency	One-tenth of centre frequency	—	Narrow
Power output	A few watts (CW), 100–200W (pulsed)	1W (CW), 400 W pulsed	Several 100W (pulsed)	Low (mw)
Efficiency	—	3% CW, 60% pulsed	20–60% pulsed	Low (2%)
Noise figure	—	High, 30 dB	High, 60 dB	Less noisy than MPATT (< 15 dB)
Application	Oscillator	Oscillator, amplifier	Oscillator	Local oscillator
Construction	n^+nn^+ GaAs single crystal	*n+pip+* reverse-bias *p-n* junction	p^+nn^+ or $n^+ p p^+$ reverse-bias *p-n* junction	*p-n-p*,or *p-n-i-p*, or *p-n*-metal or metal - *n* - metal, forward - bias *p* - *n* junction
Basic semiconductors	GaAs InP	Si,Ge, GaAs or InP	Si	Si/metal
Harmonics	—	Less	Strong	Less
Size	Small	Small	Small	Small
Ruggedness	Yes	Yes	Yes	Yes

TUNNEL DIODES 10.5

Tunnel diodes are heavily doped *p-n* junction diodes that have a negative resistance over a portion of its *I-V* characteristic as shown in Fig. 10.32. These diodes are used as microwave amplifier or oscillators. Due to heavy doping, the width of the depletion region becomes very thin and an overlap occurs between the conduction-band level on the *n*-side and the valence-band level on the *p*-side. Initially under little forward bias condition the conduction-band electrons tunnel through the depletion layer resulting in increase of current with forward voltage. The current reaches maximum of I_p at voltage V_p.

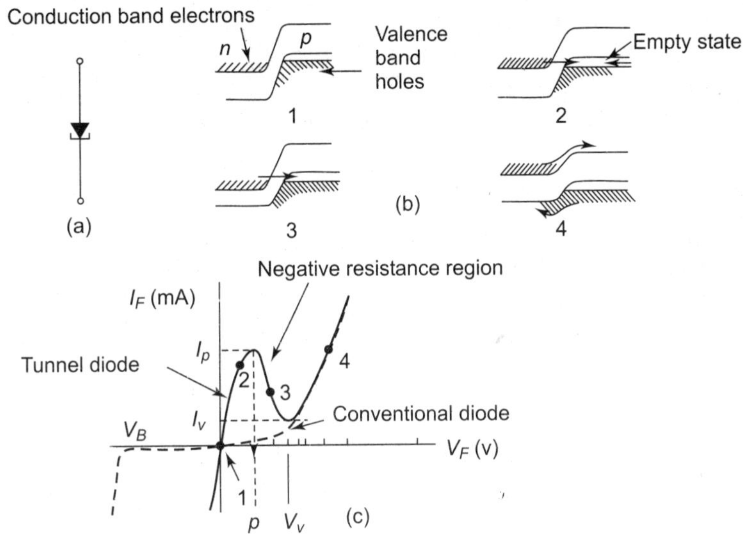

Fig. 10.32 Tunnel diodes and *I-V* characteristics: (a) Symbol (b) Energy-level diagram (c) *I-V* characteristics

For further increase of forward bias, the conduction band electron energy levels are raised above the available energy levels in the valence band and becomes equal to levels in forbidden band. No direct tunnelling occurs and current decreases with increases in forward voltage till V_v showing a negative resistance characteristics for use in amplifiers or oscillators. As the forward voltage is increased beyond V_v, current increases with forward voltage in the same manner as a semiconductor diode.

The difference between tunnel diode and normal *p-n* junction diode is given in Table 10.3.

Table 10.3 Comparison between tunnel diode and normal *p-n* junction diode

	Tunnel diode	*Normal p-n diode*
1.	Doping levels at *p* and *n* sides are very high.	Doping is normal in both *p* and *n* sides.
2.	Tunnelling current consists of majority carriers–electrons from *n*-side to the *p*-side.	Current consists of minority carriers—holes from *p*-side to the *n*-side.
3.	Majority carriers current responds much faster to voltage changes—suitable to microwaves.	Majority carrier current does not respond so fast to voltage changes—suitable for low-frequency applications only.

(*Contd.*)

Table 10.3 (*Contd.*)

	Tunnel diode	*Normal p-n diode*
4.	At a small value of reverse voltage a large current flows due to considerable overlap between conduction band and valence band—useful as frequency converter.	Current is extremely small (leakage current) up to considerable reverse bias voltage and then increases abruptly to extremely high at a particular voltage called breakdown voltage.
5.	Low-power device.	Lower power device.
6.	Shows negative resistance characteristics—useful for reflection amplifiers and oscillators.	Does not show negative resistance—used as detector and mixers.
7.	It is a low-noise device.	Moderate noise characteristics.
8.	Preferred semiconductors are Ge and GaAs.	Preferred semiconductors are Ge and Si.

Equivalent Circuit 10.5.1

The equivalent circuit of a tunnel diode is shown in Fig. 10.33. R_j and C_j are the diode junction resistance (100 ohms) and capacitance (pf), L_s(nH) is the lead inductance, R_s (few ohms) includes lead resistance and bulk resistance and bulk resistance of the semiconductor. The impedance of the circuit is expressed by

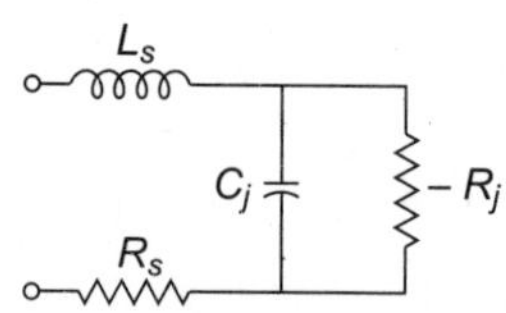

Fig. 10.33 Equivalent circuit of a tunnel diode

$$Z = R_s - \frac{R_j}{1+(\omega R_j C_j)^2} + j\omega R_j C \left[\frac{L_s}{R_j C_j} - \frac{R_j}{1+(\omega R_j C_j)^2} \right] \quad (10.17)$$

From the condition of oscillation that

$$\text{Re}(Z) = 0 \quad (10.18)$$

$$\text{Im}(Z) = 0 \quad (10.19)$$

two important frequencies can be obtained as

$$\text{Resistive cut-off frequency } f_r = \frac{1}{2\pi R_j C_j} \sqrt{\left[\frac{R_j}{R_s} - 1\right]} \quad (10.20)$$

$$\text{Self-resonant frequency } f_0 = \frac{1}{2\pi} \sqrt{\left[\frac{1}{L_s C_j} - \frac{1}{(R_j C_j)^2}\right]} \quad (10.21)$$

Tunnel-Diode Amplifiers 10.5.2

Tunnel diode can be used as reflection amplifier (TDRA) incorporating a circulator to isolate the source from the load (Fig. 10.34).

The characteristic impedance Z_0 of the circulator must be greater than that of the diode's negative resistance R_j. The voltage reflection coefficient at the diode port is

$$\Gamma = \frac{(Z_{in}/Z_0) - 1}{(Z_{in}/Z_0) + 1} \quad (10.22)$$

Neglecting C_j,

$$\Gamma = \frac{\left[(-R_j/Z_0)-1\right]}{\left[(-R_j/Z_0)+1\right]} \quad \text{or, } |\Gamma|^2 > 1 \text{ for } Z_0 > R_j \tag{10.23}$$

Therefore, load power is

$$P_L = |\Gamma|^2 P_{in} \quad \text{or,}$$

$$\text{Power gain} \quad G_p = |\Gamma|^2 = \left|\frac{R_j + Z_0}{R_j - Z_0}\right|^2 \tag{10.24}$$

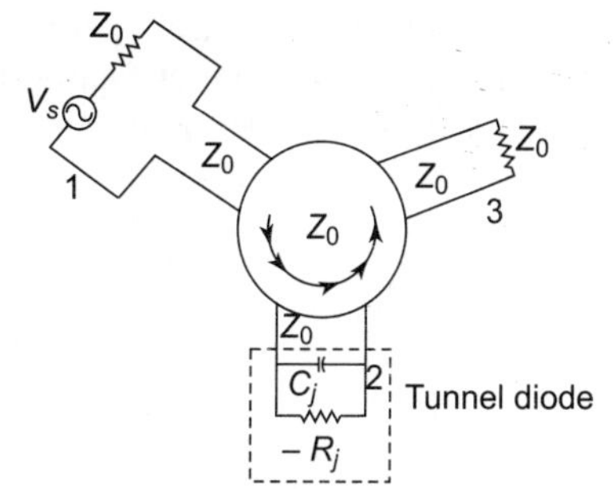

Fig. 10.34 Tunnel-diode reflection amplifier

Several advantages of tunnel diodes are as follows:

1. Require very simple dc power supply
2. Broadband operation possible
3. Low-noise figure (<5 dB at 10 GHz) due to low current levels
4. Immune to the natural radiation in the solar system and suitable for space communication

Tunnel Diode Oscillator (TDO) 10.5.3

A tunnel diode oscillator circuit consists of a tank circuit coupled with the diode by means of a capacitive divider (Fig. 10.35). When the power is switched on, a surge current produces oscillation in the tank circuit. The *R–C* values make the dc bias at the centre of the negative resistance characteristic of the diode. Sustained oscillation occurs if the magnitude of the negative resistance of the diode is equal or greater than the resistance of the tank circuit. The oscillator circuit can generate microwave signals up to frequencies of about 100 GHz.

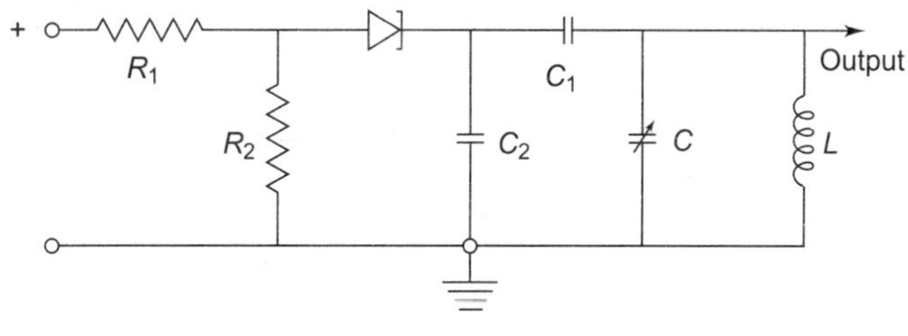

Fig. 10.35 Tunnel-diode oscillator

VARACTOR DIODES 10.6

Varactor diodes are *p-n* junction diodes which provide a voltage variable junction capacitance in microwave circuit when reverse biased.

$$C_j = C_0 (1 + V_R/V_B)^{\,n} \tag{10.25}$$

where C_0 is the junction capacitance for zero bias, V_R is the magnitude of reverse bias voltage below breakdown, V_B is the barrier potential and n is a constant whose values are 0.5 and 0.33 for abrupt and linear graded *p-n* junction respectively. In this operation as the magnitude of reverse bias voltage is increased, the depletion layer width W_d will increase and the junction capacitance C_j will decrease in accordance with

$$C_j \,\alpha\, \frac{1}{W_d}, \quad W_d \,\alpha\, |V_R| \tag{10.26}$$

This non-linear $C_j - V$ characteristics permits the varactor diode to be used for frequency multiplication and parametric amplification at microwave frequencies. Fig. 10.36 shows the typical structure and equivalent circuit of a varactor diode. Here C_j is the junction capacitance, R_j is the junction reverse resistance, and R_s is the diode bulk resistance. Since R_j is very large, the simplified model is a series combination of C_j and R_s. The diode has a wide capacitance variation and the cut-off frequency of operation at a given reverse voltage is defined by

$$f_c = \frac{1}{2\pi R_s C_j(\text{min})} \tag{10.27}$$

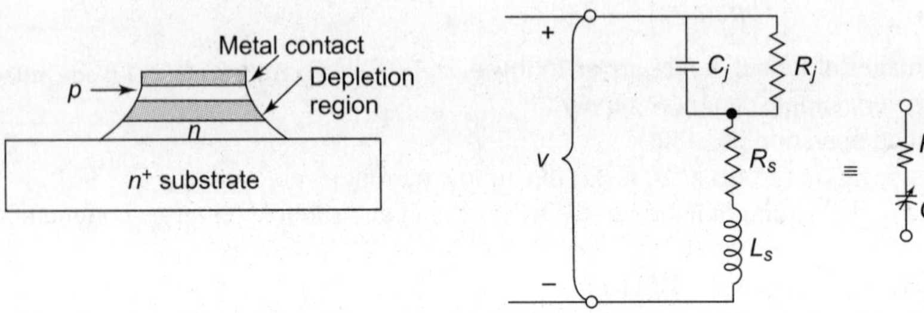

Fig. 10.36 Varactor diode and its equivalent circuit

Due to skin effect, R_s increases with increase in frequency, and therefore, the varactor is normally used at frequencies below $0.2 f_c$.

Varactor Frequency Multipliers

When an ac voltage (pumping voltage) $v_p = V_p \sin \omega_p t$ is applied to a reverse biased diode, the diode capacitance varies as (Fig. 10.37),

$$C_j(t) = C_0 [1 + (V_R + V_p \sin \omega_p t)/V_B]^{-n}; \tag{10.28}$$

Expanding in a hormonic series

$$C_j(t) = C_0 - C_1 \sin \omega_p t + C_2 \sin 2\,\omega_p t - C_3 \sin 3\,\omega_p t + \ldots \tag{10.29}$$

The diode ac current due to the above time-varying capacitance can be expressed as

$$i(t) = I_1 \cos \omega_p t + I_2 \cos 2\omega_p t + I_3 \cos 3\omega_p t + \ldots \tag{10.30}$$

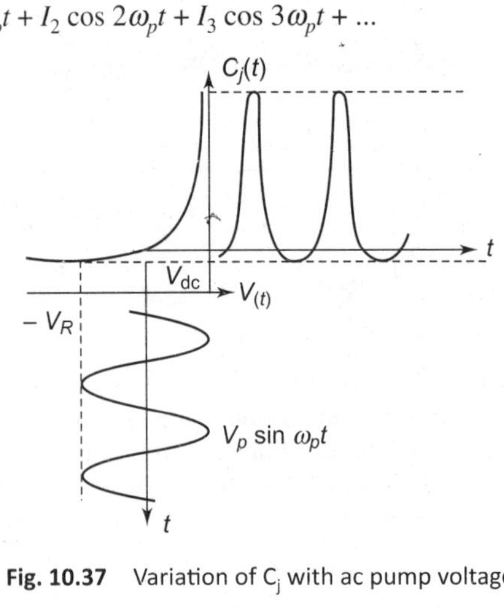

Fig. 10.37 Variation of C_j with ac pump voltage

Therefore, the varactor diode can be used as frequency multiplier or a harmonic generator where a resonant circuit tunes to the desired harmonic.

Figure 10.38 shows a 1–2–3 tripler. The input signal at frequency f is fed from a stable crystal-controlled low-frequency generator to the reverse biased varactor diode through a buffer amplifier. The input resonant circuit 1 prevents unwanted frequencies reaching the diode from the input's side and the input circuit from the diode circuit. The output resonator circuit 3 is tuned to the third harmonic to produce the final signal at $3f$ frequency. The intermediate resonant circuit 2 called the idler, eliminates the heterodying between the input and output frequencies. Since the harmonic current is capacitive, very little power loss occurs and the varactor multiplier operates at high efficiency without adding excessive noise.

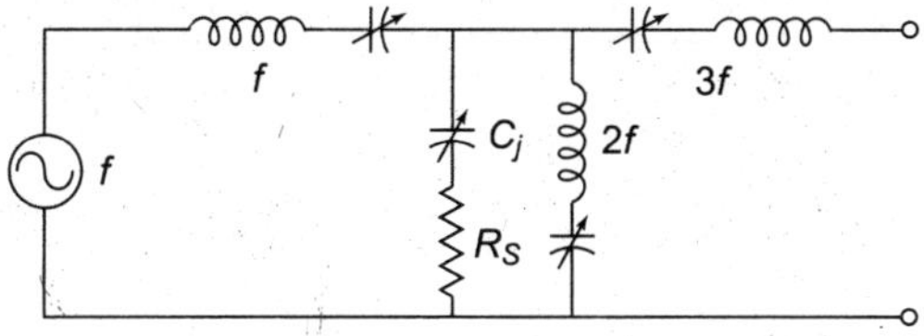

Fig. 10.38 Varactor diode multiplier circuit

PARAMETRIC AMPLIFIER 10.7

A parametric amplifier is named because of its operation due to the periodic variation of the device's parameters such as capacitance of the varactor diode, under the influence of a suitable pump signal. If a small input signal at a frequency f_g and the ac power source operated as a pumping signal at a frequency f_p, are applied together to the varactor diode, linear amplification of a small signal results due to time-varying capacitance of the diode. Pump signal provides the power required for amplification and the power output is either at the input frequency f_g or at the idler frequency $f_i = f_p - f_g$.

Manley-Rowe Relations

A set of power-conservation (Manley–Rowe) relations are described below which are useful in determining the maximum gain of the parametric amplifier. When two sinusoidal signals at frequencies f_g and f_p are applied across a lossless time varying non-linear capacitance $C_j(t)$, signals at the harmonics of f_g and f_p are generated which can be separated by band-pass filters and their power made to be dissipated in separate resistive loads (Fig. 10.39).

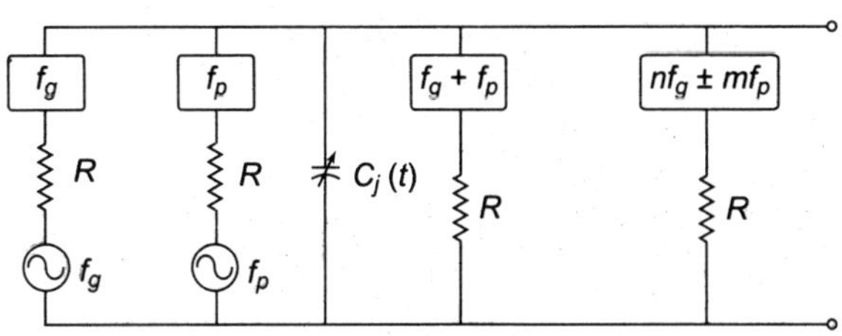

Fig. 10.39 Manley–Rowe relations

From the conservation of power,

$$\sum_{n=0}^{\infty} \sum_{m=-\infty}^{\infty} \frac{n P_{nm}}{n f_g + m f_p} = 0 = \sum_{n=-\infty}^{\infty} \sum_{m=0}^{\infty} \frac{m P_{nm}}{n f_g + m f_p} \tag{10.31}$$

where P_{nm} represents the average power at the output frequencies $\pm|n f_g + m f_p|$.

The above relations are called the *Manley–Rowe* or *power-conservation relations*. The sign convention for the power term P_{nm} is such that it is positive when the power is supplied by the two generators or power flowing into the non-linear capacitance, otherwise it is negative.

Example 10.8 A sinusoidal input signal at frequency f_g and pump signal at frequency f_p are applied across a time varying non-linear capacitance. If the output circuit is a band-pass filter with resistive series load at frequencies $f_g + f_p$, calculate the power gain.

Solution From Manley–Rowe relations,

$$\sum_{n=0}^{\infty}\sum_{m=-\infty}^{\infty}\frac{n\,P_{nm}}{n\,f_g + m\,f_p} = 0 = \sum_{n=-\infty}^{\infty}\sum_{m=0}^{\infty}\frac{m\,P_{nm}}{n\,f_g + m\,f_p} \tag{10.32}$$

Here, m and n vary from –1 through zero to +1, respectively. Therefore, Manley–Rowe relations are

$$\frac{P_{10}}{f_g} + \frac{P_{11}}{f_g + f_p} = 0 = \frac{P_{01}}{f_p} + \frac{P_{11}}{f_g + f_p} \tag{10.33}$$

Since f_g, f_p, $f_g + f_p > 0$, P_{10} and $P_{01} > 0$, the average power at the frequency $f_g + f_p$ is $P_{11} < 0$ to satisfy above relations, i.e., at the output frequency $f_0 = f_g + f_p$, the average power is negative in the output load. This means that power dissipation in the output load is negative. Therefore, power is delivered from the non-linear time varying capacitor $C_j(t)$ at the output frequency $f_0 = f_g + f_p$ to the load, where $f_0 > f_p > f_g$.

The power gain of the signal is

$$\frac{P_{11}}{P_{10}} = \frac{f_0}{f_g} = \frac{f_g + f_p}{f_g} = 1 + \frac{f_p}{f_g} \tag{10.34}$$

The Manley–Rowe relations give the maximum signal gain possible. However, due to presence of losses in practical cases, the gain is less than $1 + f_p/f_g$. Such parametric amplifiers are called *up-converters*. If the output frequency $f_0 = f_g - f_p$, the parametric device is called the parametric *down-converter* having power loss in place of gain.

Analysis of a Parametric Amplifier 10.7.1

When a pump voltage $v_p = V_p \sin \omega_p t$ and a small signal voltage $v_g = V_g \sin \omega_g t$ (where $V_g << V_p$ and $V_p <<$ dc reverse bias voltage magnitude $|V_R|$) are applied simultaneously to time varying non-linear capacitance $C_j(t)$ of a varactor diode, the current through the diode is

$$i = \frac{dQ(V)}{dt} = \frac{dQ(v_p + v_g)}{dt} u \tag{10.35}$$

Expanding $Q\,(v_p + v_g)$ in Taylor series about the small signal voltage $v_g = 0$,

$$Q\,(v_p + v_g) \approx Q(v_p) + \left.\frac{\partial Q}{\partial v}\right|_{v_g=0} .v_g;\ \text{for } v_g << v_p \tag{10.36}$$

Therefore,
$$i = \frac{d\,Q(v_p)}{dt} + \frac{d}{dt}\left[\left.\frac{\partial Q}{\partial v}\right|_{v_g=0} .v_g\right] \tag{10.37}$$

This shows that the diode current is given by

$$i = \frac{d\,Q(v_p)}{dt} + \frac{d}{dt}\,[C_j(t)\,v_g] \tag{10.38}$$

and the junction capacitance is

$$C_j(t) = C\left[1 + \frac{V_R + V_p \sin\omega_p t + V_g \sin\omega_g t}{V_B}\right]^{-n} \tag{10.39}$$

For a linearly graded junction, $n = 1/3$ and since $|V_g| << |V_p| << |V_R|$, $C_j(t)$ can be expanded in a harmonic series yielding

$$C_j(t) \approx C_0 - C_1 \sin \omega_p t \tag{10.40}$$

This time-varying capacitance is utilized for parametric amplifications.

Negative Resistance Amplifier 10.7.2

Figure 10.40 shows a parametric amplifier equivalent circuit where the varactor diode represents a negative resistance at signal frequency to provide amplification of the signal. It is assured that the parasitic elements of the equivalent circuit of the varactor diode is tuned out by the external tuning elements L-C so that the diode can be represented by the series combination of $C_j(t)$ and R_s. The non-linear capacitance results in small-signal currents at the harmonics of the input frequencies, f_g, f_p and the sum $f_g + f_p$ and difference $f_g \sim f_p$ components of the pumping and input frequencies. External tuned circuits and band-pass filters allow currents at desired frequencies, viz., i_g at f_g in the inputs loop, i_p at f_p in the pumping circuit and i_i at $f_i = f_p - f_g$ in the idler loop. The output is obtained across the load R_L in the idler circuit. In negative resistance amplifiers difference frequency component $f_g \sim f_p$ is utilized for the output signal and is called idler frequency $f_i = f_p - f_g$.

The presence of tuned circuits allow only the first harmonic of the signals to pass, so that

$$C_j(t) = C_0 - C_1 \sin \omega_p t \tag{10.41}$$

$$i_g = I_g \sin \omega_g t \tag{10.42}$$

$$i_i = I_i \sin \omega_i t \tag{10.43}$$

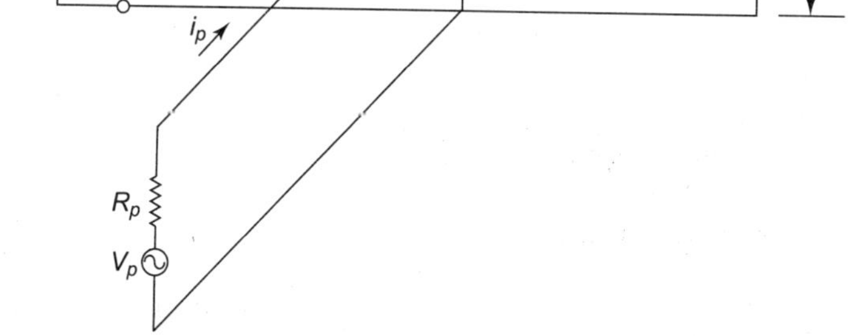

Fig. 10.40 Negative resistance parametric amplifier equivalent circuit

The diode current

$$i = i_g + i_i + i_p \tag{10.44}$$

The signal voltage across the diode is given by

$$V_d = Q/C = \int idt\,/\,C \tag{10.45}$$

Since i_p involves only pumping frequency, it does not contribute to the signal current. Hence,

$$V_d = -\left[\frac{I_g}{\omega_g}\cos\omega_g t + \frac{I_i}{\omega_i}\cos\omega_i t\right] \Big/ \left[C_0 - C_1 \sin\omega_p\right]$$

$$\approx \frac{-1}{C_0}\left[\frac{I_g}{\omega_g}\cos\omega_g t + \frac{I_i}{\omega_i}\cos\omega_i t\right]\left(1 + \frac{C_1}{C_0}\sin\omega_p t\right) \tag{10.46}$$

This voltage contains terms containing cos $\omega_g t$, cos $\omega_i t$, sin $\omega_p t$ cos $\omega_g t$ and sin $\omega_p t$ cos $\omega_i t$. The product terms result in signal voltage components at frequencies $\omega_p \pm \omega_g$ and $\omega_p \pm \omega_i$. Due to resonant circuits, only the difference frequency $\omega_p - \omega_g = \omega_i$ or $\omega_p - \omega_i = \omega_g$ will contribute to the signal voltage across the diode. Therefore,

$$V_d = \frac{1}{C_0}\left[\frac{I_g}{\omega_g}\cos\omega_g t + \frac{C_1}{C_0}\frac{I_i}{2\omega_i}\cos\omega_g t\right] \tag{10.47}$$

Factor 1/2 appears in the second term due to trigonometric identities sin $\omega_p t$ cos $\omega_i t$ = $\frac{1}{2}\left[\sin(\omega_p + \omega_i)t + \sin(\omega_p - \omega_i)t\right]$. The negative-resistance amplifier is an unstable device with narrow bandwidth and high gain. Since $|V_g| << |V_p|$, contribution of the first term is small compared to the second term and thus

$$V_d \approx -\frac{1}{C_0^2}\frac{C_1}{2\omega_i}I_i \sin\omega_g t = -\frac{1}{C_0^2}\frac{C_1}{2\omega_i}I_g \sin\omega_g t$$

$$= -K I_g \sin \omega_g t; \quad K = \frac{-C_1 I_i}{2C_0^2\omega_i I_g} \tag{10.48}$$

Thus, the voltage component is out of phase to the current showing that the diode behaves like a negative resistance R^- to the signal frequency.

$$R^- = \frac{C_1 I_i}{C_0^2 2\omega_i I_g} = K \tag{10.49}$$

The total input resistance of the input loop is

$$R_{\text{in}} = R_g + R^- \tag{10.50}$$

Negative resistance circulator type amplifier can be designed if $|R^-| > R_g$ as shown in Fig. 10.41.

The signal power circulates around to the negative resistance port where the magnitude of the voltage reflection coefficient > 1, so that the reflected power is larger than the incident power which is obtained from the next port. The power gain is $G = |\Gamma|^2$ where,

$$\Gamma = \frac{R_{\text{in}}/Z_0 - 1}{R_{\text{in}}/Z_0 + 1} \tag{10.51}$$

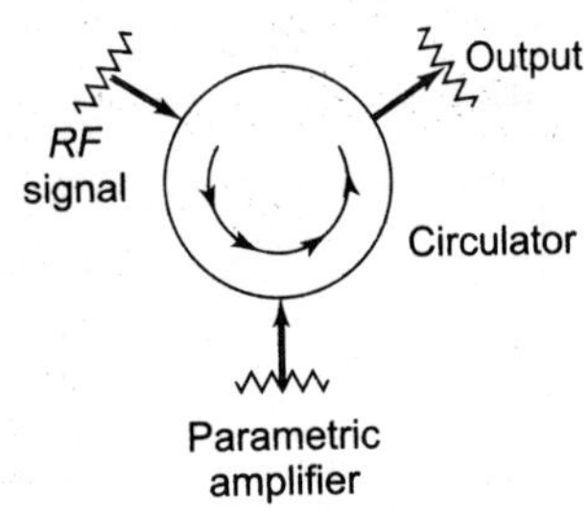

Fig. 10.41 A negative resistance circulator-type parametric amplifier

Parametric amplifiers are very low noise amplifiers.

Parametric Up-Converter 10.7.3

Figure 10.42 shows an equivalent circuit of a parametric up-converter in which a pump voltage at frequency f_p and a signal voltage at frequency f_g are applied through respective tuned circuits to the varactor diode. The output power is taken from the output idler circuit at the upper sideband frequency $f_i = f_p + f_g$. Physically, power is transferred from the pumping source to the idler circuit to achieve power gain with low noise. The series tuned circuits allow only currents with respective frequencies, f_g, f_p and f_i in each loop. The circuit losses are small compared to the loss in the diode resistance R_s and external loads R_g and R_L. For maximum gain, $R_L = R_g$ and maximum gain for a lossless ($R_s = 0$) diode is, $G_m = \omega_i/\omega_g$ as predicted by the Manley–Rowe relations.

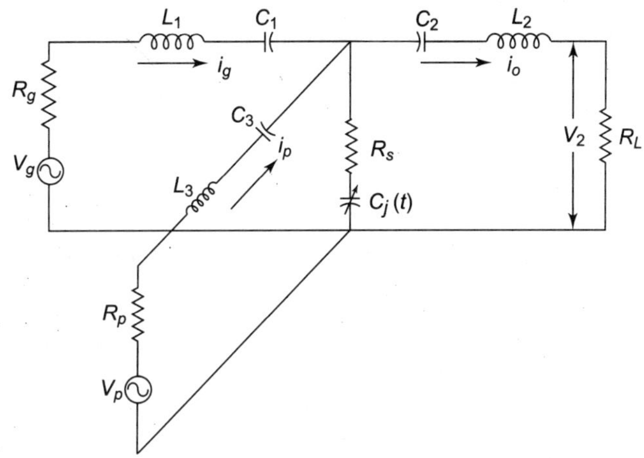

Fig. 10.42 Parametric up-converter

For high gain, an up-converter requires a large ratio of output to input frequency, but limited to operation at f_g = 1 GHz. The up-converter is a stable device with a wide bandwidth and low gain. The advantages of the up-converter over the negative resistance devices are as follows

1. Input impedance >> 0
2. More stable
3. Power gain independent of source impedance
4. No circulator is required
5. Larger bandwidth (~5%)

Advantages and Limitations of Parametric Amplifiers 10.7.4

Parametric amplifiers have the following advantages and limitations:

1. **Noise Figure** Because of minimum resistive elements, thermal noise in parametric amplifiers is very less compared to that of transistor amplifier.
   Typically, noise figure is in the range 1–2 dB.
2. **Frequency** The upper frequency limit (40–200 GHz) is set by the difficulty of obtaining a source power at the pump frequency and by the frequency at which the varactor capacitance can be pumped. The lower frequency limit is set by the cut-off frequency of the microwave components used in the circuit.
3. **Bandwidth** Bandwidth of the parametric amplifier is small due to the presence of tuned circuits. Bandwidth can be increased by stagger tuning.

4. **Gain** The gain is limited (20–80 dB) by the stabilities of the pump source and the time varying capacitance.

Because of its low noise, parametric amplifiers are used in space communication systems, radio telescopes and tropo-receivers.

MICROWAVE TRANSISTORS AND CIRCUITS 10.8

After long use of klystron, TWT, tunnel diodes and varactor diodes in microwave amplifiers, microwave transistors have taken an indispensable position in amplifier design. They are rugged, low-cost, reliable, and can be easily integrated in MICs. They can be used up to about 100 GHz with low-noise figure, broad bandwidth and medium power output. Similarly, after magnetrons, reflex klystrons, Gunn and IMPATT diodes, microwave transistors are mostly used in solid-state sources over a wide frequency range with medium power.

Ordinary transistors are *npn* or *pnp* junction transistors—bipolar and unipolar FET. High frequency operation of these devices is limited by (1) diffusion capacitance at base-emitter junction, (2) space charge capacitances at base-emitter and base-collection junction, (3) base spreading resistance, and (4) bulk resistance of the base, which in turn increase the transit time of the charge carriers in the device.

Microwave transistors are miniaturized designs to reduce device and package parasitic capacitances and inductances and to overcome the finite transit time of the charge carriers in the semiconductor materials. The most commonly used semiconductors are Si and GaAs. Since transit times are dependent on the electron mobility and saturation velocity in the semiconductor material, GaAs is better than Si for high frequency devices because of its higher electron mobility and saturation velocity compared with Si. Si-bipolar has advantage over GaAs due to in expensive, durable, higher gain and moderate noise characteristics. By means of molecular beam epitaxy techniques, the High Electron Mobility Transistors (HEMT) are developed presently which can operate at frequencies of the order of 100 GHz.

There are three categories of microwave transistors: (1) low-noise transistor which is employed at the front end, since this is the major determinant of the overall system noise, (2) low-level transistor which is used to drive power stage, and (3) power transistor which amplifies final power output.

Microwave transistor amplifiers are constructed either as hybrid Microwave Integrated Circuits (MIC) where the transmission lines and matching networks are realized by microstrip circuit elements and the discrete components such as chip capacitors, resistors, and transistors are soldered in place, or as Monolithic Microwave Integrated Circuits (MMIC) where all active devices and passive circuit elements are fabricated on a single semiconductor (GaAs) crystal.

There are two basic types of construction of microwave transistors, bipolar and unipolar. Bipolar is a three-semiconductor *n-p-n* region junction structure where charge carriers of both negative (electrons) and positive (holes) polarities are involved in transistor operation. Unipolar transistors are junction gate and insulated gate field-effect transistors (FETs). These are one or two semiconductor region structures where dominant carriers are of single polarity (electrons or holes). Most widely used transistors in microwave analog circuits are Si-bipolar for UHF-S band, and Si bipolar/ GaAs FET-used in analog and high-speed digital ICs in S–C band.

Heterojunction bipolar transistors (HBT) are developed for operation in the mm wave region.

Microwave Bipolar Transistors (< 5 GHz) 10.8.1

The microwave bipolar transistors are planar in form and mostly Si *n-p-n* type operating up to 5 GHz. GaAs also is used for performance improvements in the operating frequency, in high temperatures, and in high radiation field. Geometries of these devices are (a) inter-digitated, (b) overlay, and (c) matrix forms with wide emitter area to overcome transit time limitations. A typical bipolar construction is shown in

Fig. 10.43. Epitaxial *n* layer is formed by condensing a single crystal film of semiconductor material upon a low resistivity Si wafer of substrate n^+. Above this, a *p*-type diffused base and n^+-type diffused emitter are formed. Typically the emitter width *W* is 1 micron, base thickness is 2 microns and emitter length is 25 microns. The packaging is done (Fig. 10.43(c)) with minimum beam leads to reduce inductance.

For power amplification, class *C* operation is preferred to class *A*/*AB* mode to minimise collector-base transit time. In class *C*, both emitter-base and collector-base junctions are reverse biased so that no current flows in absence of a signal. When an *RF* voltage of sufficient magnitude is applied to an emitter-base junction, for a fraction of *RF* cycle the junction is forward biased, the electrons are injected to the base and the injected carriers transit the base by combined diffusion and drift flow processes. They are accelerated in the collector-base depletion region.

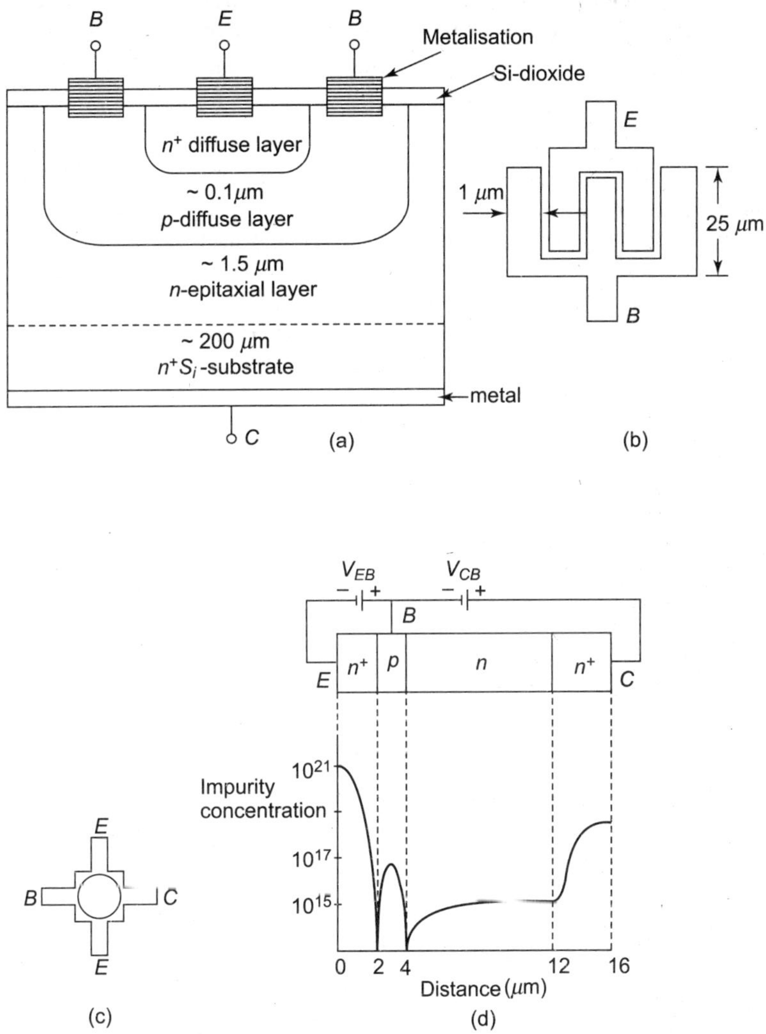

Fig. 10.43 Geometries of a microwave bipolar junction transistor (BJT): (a) Cross-section of Si-bipolar BJT, (b) Top view (c) Packaged transistor (d) Concentration profile

High-frequency Limitations of Bipolar Devices

The microwave bipolar transistor is a non-linear *p-n-p* or *n-p-n* device, and its principle of operation is similar to that of the low-frequency devices, but requirements for dimensions (<< λ), heat sinking and packaging are much more severe due to the following limitations.

1. At high frequencies, the reactances due to junction capacitances limit the gain. The values of these capacitances depend on the depletion layer's width or the bias voltage and it provides feedback paths. However, the feedback effect is reduced in *CB* mode of operation. The junction capacitances must be reduced as much as possible.
2. Signal loss in the lead inductances can be minimized by reducing the length in packaging.
3. The transit time taken by the majority of charge carriers to cross from the emitter to collector limits the high frequency. The total transit time is a contribution from

   T_{jeb} = Emitter-base junction capacitance charging time
   T_b = Base region transit time
   T_{bc} = Base collector-region depletion-layer transit time
   T_{jbc} = Base collector junction capacitance charging time

Since a forward-biased emitter base junction is assumed, the transit time associated with this depletion region can be ignored. In general, the transit time can be reduced by using narrow *p* and *n* regions, especially the base. Transit time can also be reduced by reducing the depletion layer width at the reverse-biased collector-base junction. This needs use of a higher collector voltage without avalanche breakdown. But corresponding increase in the power dissipation may damage the device.

The upper frequency limit of these devices is nearly 22 GHz. The cut-off frequency of the device is given by

$$f_T = 1/(2\pi T) \text{ at which current gain} \approx 1 \tag{10.52}$$

where

$$T = T_{jeb} + T_b + T_{bc} + T_{jbc}$$
$$\approx T_b + T_{bc} \tag{10.53}$$

since base and base-collector depletion layer transit time is much larger than the junction capacitance charging time. For $N_d = 5 \times 10^{15}$/cc, or 1 ohm-cm resistivity, $T_{bc} = 1.7 \times 10^{-11}$ s, $T_b = 8.3 \times 10^{-10}\, W_B^{\,2}$ s, where W_B is the basewidth, Therefore,

$$f_T \approx \frac{1}{2\pi(T_b + T_{bc})} \tag{10.54}$$

The equivalent circuits of *CB* and *CE* modes are shown in Fig. 10.44, where r_e = emitter resistance, r_b = base resistance, C_c = collector depletion layer capacitance and C_e = emitter base junction capacitance, α = $\alpha = \dfrac{\partial i_c}{\partial i_e} \le 1$, current gain in *CB* mode, $\beta = \partial i_c/\partial i_b = \dfrac{\alpha}{1-\alpha} \gg 1$, current gain in *CE* mode.

Power Frequency Limitations

Power capacity of the transistor is limited by the device cut-off frequency.

Let L = Emitter collector distance
v_s = Maximum saturated drift velocity of carriers in semiconductor
E_m = Maximum **E** field that can be sustained in a semiconductor without having dielectric breakdown
$V_m = E_m L$ = Maximum allowable applied voltage
$X_c = 1/(2\pi f_T C_c)$ = Reactance impedance of collector base capacitance C_c.
I_m = Maximum current of the device

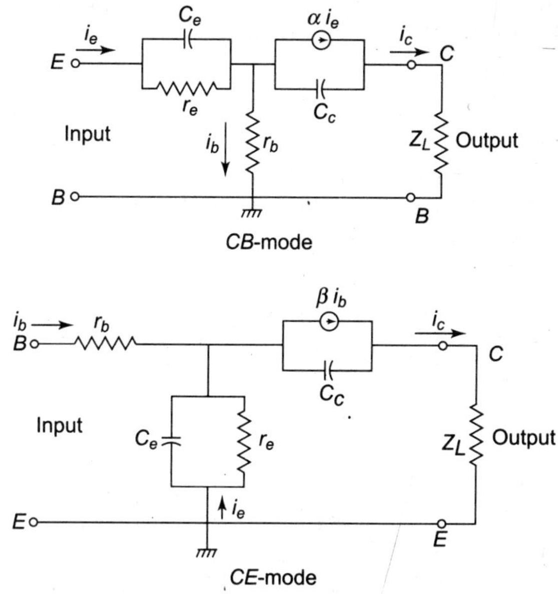

Fig. 10.44 Microwave equivalent circuits of a bipolar transistor

Then, $$V_m f_T = E_m L.\frac{1}{2\pi T} = \frac{E_m L}{2\pi}\frac{1}{L/v_s} = E_m v_s/2\pi \tag{10.55}$$

$$I_m f_T = \frac{V_m}{X_c}\frac{1}{2\pi T} = \frac{E_m v_s}{2\pi X_c} \tag{10.56}$$

and, $$V_m f_T . I_m f_T = V_m I_m f^2_T = P_m f^2_T \tag{10.57}$$

Therefore, $P_m f_T^2 = \dfrac{E_m v_s}{2\pi}\dfrac{E_m v_s}{2\pi X_c}$ or, $P_m X_c f^2_T = (E_m v_s/2\pi)^2$

or, $$(P_m X_c)^{1/2} f_T = E_m v_s/2\pi \tag{10.58}$$

Thus, for a given device impedance, the power capacity of the transistor decreases with increase in the device cut-off frequency as shown in Fig. 10.45. Typical values of the parameters are

f_T = 5 GHz; noise figure = 2 dB at 1 GHz, 4 dB at 2 GHz and 5 dB at 4 GHz.

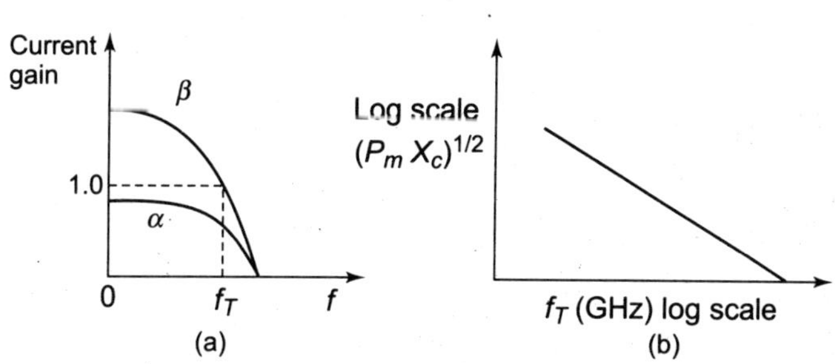

Fig. 10.45 Power frequency limitations: (a) Current gain (b) Power vs frequency

Power output = 50 W (0.5 GHz)

10 W (2.5 GHz)

5 W (4 GHz)

Example 10.9 An Si microwave transistor has reactance of 1 ohm, transit time cut-off frequency of 4 GHz, maximum **E**-field 1.6×10^5 V/m and saturation drift velocity 4×10^5 m/s. Determine the maximum allowable power.

Solution

Given $X_c = 1$ ohm, $E_m = 1.6 \times 10^5$ V/m

$f_T = 4$ GHz, $v_s = 4 \times 10^5$ m/s

$$P_m X_c = \left(\frac{E_m v_s}{2\pi f_T}\right)^2$$

or,

$$P_m = \left(\frac{E_m v_s}{2\pi f_T}\right)^2 \frac{1}{X_c} = \left(\frac{1.6\times10^5 \times 4\times10^5}{2\times3.14\times4\times10^9}\right)^2$$

$= 6.48$ watts

Biasing Circuits

Microwave transistor biasing circuits are designed for (1) a stable operating point against variations in the device parameters and temperature, and (2) isolation of the biasing circuit from the high frequency circuit. The first requirement is fulfilled by designing a dc feedback and the second one by connecting *RF* chokes in series with bias resistance and capacitive shunt across the dc circuit elements as shown in Fig. 10.46. When I_c increases, voltage drop across R_E increases. Consequently, I_B decreases to prevent increase in I_c.

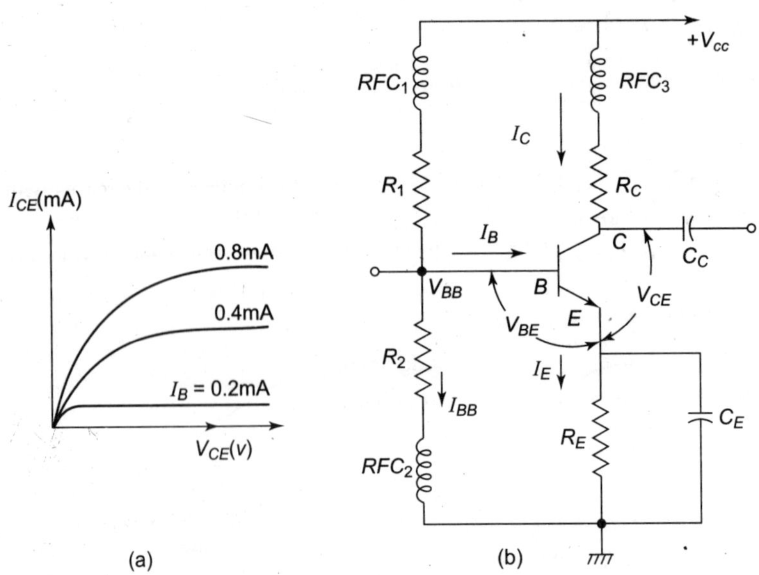

Fig. 10.46 Biasing circuit and *I–V* characteristic

Biasing Example

$$V_{cc} = 30 \text{ V}$$
$$V_{BE} = 1 \text{ V}$$
$$V_{BB} = 3 \text{ V}$$
$$I_{BB} = 1.5 \text{ mA}$$
$$I_{CO} = 0, \beta = 50$$
$$Q \to (10 \text{ V}, 15 \text{ mA})$$

$$\therefore \quad I_B = \frac{I_c}{\beta} = 0.3 \text{ mA}$$

$$R_B = \frac{V_{BB} - V_{BC}}{I_B} = 6.7 \text{ k}\Omega$$

$$R_1 = \frac{V_{CE} - V_{BB}}{I_{BB} + I_B} = 3.9 \text{ k}\Omega$$

$$R_2 = \frac{V_{BB}}{I_{BB}} = 2 \text{ k}\Omega$$

$$R_C = \frac{V_{CC} - V_{CE}}{I_C + I_{BB} + I_B} \quad 1.2 \text{ k}\Omega$$

$$R_E = \frac{V_{BB} - V_{BE}}{I_B} = 6.7 \text{ k}\Omega$$

$RFC \to$ 2–3 turns of No. 36 enameled wire on a 0.1 inch air core.

Microwave Unipolar Transistors 10.8.2

Microwave unipolar transistors are field-effect transistors manufactured from semiconductors incorporating **gallium** or **indium** which have better high-frequency performance due to the higher electron mobility compared to Si. Two very commonly used transistors are **Metal–Semiconductor Field Effect Transistor (MESFET) and High-Electron-Mobility Transistor (HEMT)** as described below. Microwave transistors have low noise and thus replaced the parametric amplifiers as the later are expensive and difficult to fabricate. Since only one type of carriers (*n* or *p* electrons) are responsible for the operation of the device, these devices are called unipolar. MESFET can operate in 2–20 GHz range and give a single gain of 10-15 dB at 2 GHz with noise figure less than 1 dB. HEMT can operate up to 100 GHz and give a single stage gain of 6–8 dB at 50 GHz over a bandwidth of 5 GHz and noise figure of 5 dB.

Metal-Semiconductor Field Effect Transistor (MESFET)

MESFET stands for **Metal–Semiconductor Field Effect Transistor with three terminals—source(S), gate (G), and drain (D)** as shown in Fig 10.47(a). A schematic top view is also shown in Fig. 10.47(b). A Schottky (metal-semiconductor) junction is used for a gate instead of normal *p*-*n* junction. MESFETs are usually constructed using GaAs and InP semiconductors. It consists of a conducting *n*-type semiconductor channel positioned between a source and drain contact region as shown in the Fig. 10.47(a). *p*-type channel is not used since hole mobility is much higher than that of electrons. The substrate is a semi-insulating material (GaAs doped with chromium) with resistivity about 10^8 ohm-cm. The carrier channel is a highly doped (*n*) GaAs. The material contacts are Au for S and D, and Al for the Schottky gate junction. A typical GaAs MESFET has the following structural parameters:

Epitaxial layer thickness = 0.2 to 0.4 μm

Substrate thickness = 100 to 200 μm

Channel doping concentration = 8×10^{16} to 2×10^{17} per cm^3

For high-power MESFET as many as 10 source and drain fingers adjacent to gate fingers are used to obtain large drain currents and hence large output power is possible in interdigitated construction. as shown in Fig. 10.47(c).

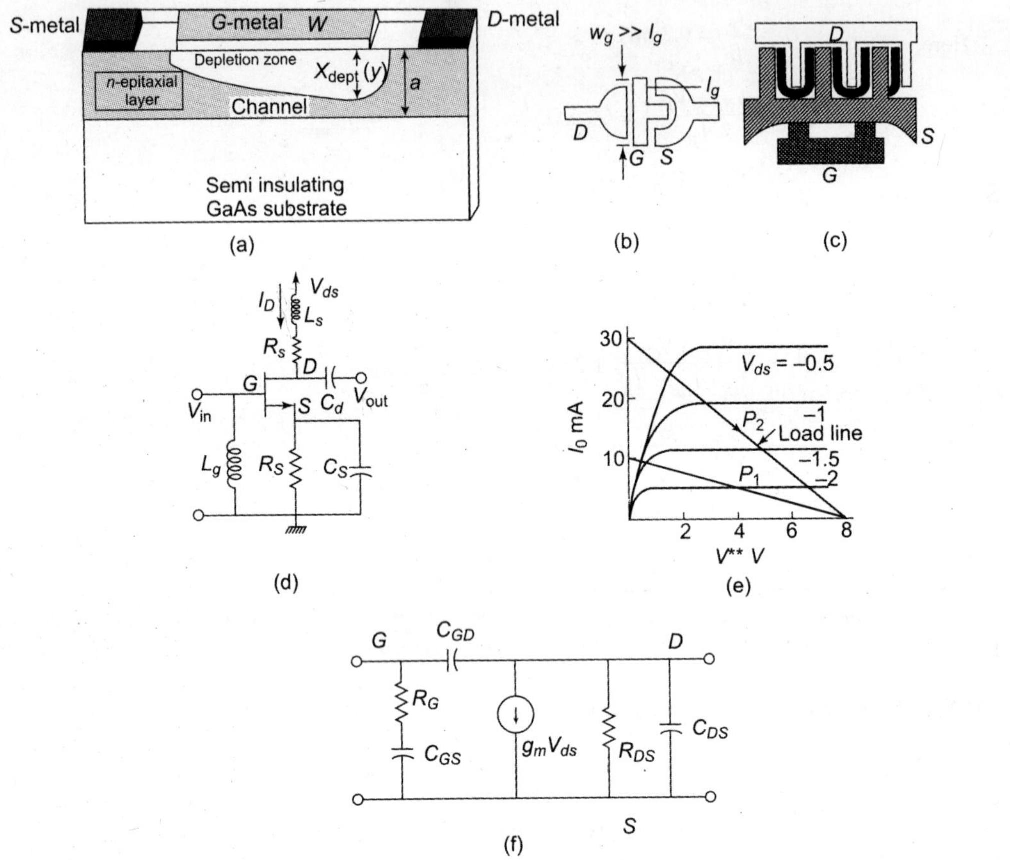

Fig. 10.47 (a) The schematic diagram of Metal-Semiconductor-Field-Effect-Transistor (MESFET) (b) Top view of MESFET (c) High power MESFET (d) MESFET biasing circuit (e) I_D vs – V_{ds} characteristics (f) Small-signal equivalent circuit of MESFET

❑ ***Operation*** The biasing circuit of a MESFET is shown in Fig. 10.47 (d). Other than the normal *R-C* biasing elements, small inductors , L_d and L_g are connected at the drain and gate terminals, respectively, for *rf* isolation. The drain voltage is +ve w.r.t. the source and the G-S junction is thus reverse biased. A depletion layer is therefore formed in the n-epitaxial layer along the gate length *L*. The drain is positively biased so that the carriers (electrons) flow from S to D through the channel. The current in the channel causes a voltage drop along the channel length so that the Schottky barrier gate becomes progressively more reverse biased and consequently the depletion layer thickness further increases. Ultimately, this may cause pinch-off of the channel toward the drain end. Since the voltage on the gate controls the depletion region width, current from the source to drain can be controlled by the gate voltage and the device can be used as a small-signal amplifier. The output drain current I_D vs drain-to-source voltage V_{ds} is shown in Fig. 10.47(e) as a function of gate-to-source voltage V_{gs} For maximum dynamic range, the dc voltage of the gate is made –ve w.r.t. the source by dc grounding the gate by means of an *RF* choke L_g.

The pinch-off voltage can be expressed as

$$V_P = \frac{eN_d a^2}{2\varepsilon_r}$$

Here,

e = Electronic charge
N_d = Doping concentration
ε_r = Dielectric constant of GaAs

Small Signal Equivalent Circuit

A simplified model of small-signal equivalent circuit of MESFET is shown in Fig. 10.47 (f). Here,

R_G = Bonding parasitic resistance of the gate
C_{GS} = Parasitic capacitance between gate and source
C_{GD} = Parasitic capacitance between gate and drain
R_{DS} = Drain to source resistance
C_{DS} = Parasitic capacitance between drain and source
g_m = Transconductance of the MESFET
V_{ds} = The supply voltage to drain

The operation of MESFET at high *RF* frequencies is limited by two factors:

- Transit time $\tau = \frac{L}{v_d}$, over the gate length L, where v_d is the saturation drift velocity in the channel
- *RC* time constant of the circuit

The cut-off frequency of operation is defined as the frequency fc at which the current through C_{GS} is equal to that of the equivalent current generator $g_m V_{ds}$ in the intrinsic FET, given by

$$f_c = \frac{g_m}{2\pi C_{GS}} = \frac{1}{2\pi\tau} = \frac{v_d}{2\pi L}$$

The maximum frequency of oscillation for maximum power gain with matched terminated circuit is given by

$$f_{\max} = \frac{f_c}{2}(R_{DS}) \quad (10.59)$$

◆ **Advantage** The basic advantage of GaAs MOSFETs are (i) high power, (ii) low noise, (iii) broadband performance, and (iv) compatibility to MICs. The carrier (electron) flow from source to drain is controlled by a Schottky metal gate.

◆ **Disadvantage** The disadvantage of the MESFET structure is the presence of the Schottky metal gate. It limits the forward bias voltage on the gate to the turn-on voltage of the Schottky diode. This turn-on voltage is typically 0.7 V for GaAs Schottky diodes. The threshold voltage, therefore, must be lower than this turn-on voltage. As a result, it is more difficult to fabricate circuits containing a large number of enhancement-mode MESFET.

High-Electron-Mobility Transistor (HEMT)

A **High-Electron-Mobility Transistor (HEMT)** is a field-effect transistor incorporating a junction between two materials with different band gaps (i.e. a heterojunction) as the channel instead of a doped region (as is generally the case for MOSFET). A commonly used combination of materials is GaAs with AlGaAs or more recent is GaN with AlGaN. Devices incorporating gallium generally show better high-frequency performance, and, gallium nitride HEMTs have high-power performance. At room temperature (300 K) electron mobility in HEMT is about 8000 cm^2/V.s and with decreasing temperature, the mobility increases to a very high value of the order of 200,000 cm^2 /V.s at 77 K.

A schematic diagram of HEMT structure is shown in Fig. 10.48. To allow conduction, semiconductors are doped with impurities which donate mobile electrons. However, these electrons are slowed down through collisions with the impurities (dopants) used to generate them in the first place. HEMTs avoid this through the use of high mobility electrons generated using the heterojunction of a highly is doped wide-bandgap *n*-type donor-supply layer AlGaN and an un-doped narrow-bandgap GaN channel layer. The source and drain ohmic contacts are formed by an AuGe alloy and an Au overlay. Schottky contact gate metallization is achieved by depositing Al.

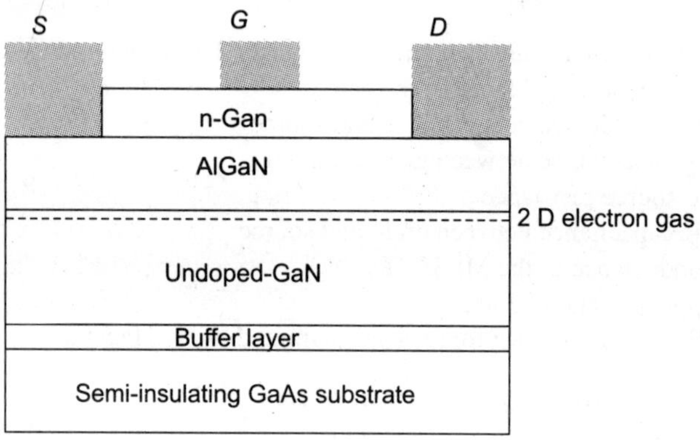

Fig. 10.48 GaN HEMT basic configuration

The electrons generated in the thin *n*-type AlGaN layer drop completely into the GaN layer to form a depleted AlGaN layer, because the heterojunction created by different band-gap materials forms a **quantum well** in the conduction band on the GaN side, where the electrons can move quickly without colliding with any impurities because the GaN layer is undoped, and from which they cannot escape. The effect of this is to create a very thin layer of highly mobile conducting electrons with very high concentration, giving the channel very high electron mobility. This layer is called a 2D electron gas within 100 ångström of the interface. Similar to MESFETs, a voltage applied to the gate alters the conductivity of this layer.

The *n*-type AlGaN layer of the HEMT is depleted completely through two depletion mechanisms:

- Trapping of free electrons by surface states causes the surface depletion.
- Transfer of electrons into the undoped GaN layer brings about the interface depletion.

❑ *Application* HEMTs are used where high gain and low noise at high frequencies are required. HEMT transistors are able to operate up to millimeter wave frequencies, and are used in high-frequency products such as cell phones, satellite television receivers, voltage converters, and radar equipment. HEMTs have shown high current gain upto frequencies greater than 600 GHz and power gain up to frequencies greater than 1 THz. HEMTs can be discrete transistors but are more usually made in the form of a 'Monolithic Microwave Integrated Circuit' (MMIC). Furthermore, GaN HEMTs on silicon substrates are used as power switching transistors for voltage converter applications for their low on-state resistances, and low switching losses due to the wide bandgap properties. Gallium nitride power HEMTs are commercially available up to voltages of 200 V–600 V.

Microwave Transistor Amplifiers 10.8.3

Microwave Transistor Amplifiers are analyzed and designed based on the *S*-parameters of the device and certain performance requirements such as stability, power gain, bandwidth, noise, and dc biasing. Accordingly, the amplifiers can be classified as (i) narrow band amplifiers (ii) low-noise amplifiers, (ii) broadband amplifiers, and (iv) power amplifiers. In a system, a low-noise transistor is used in the first stage since this is the major determinant of the overall system noise. Driver stage transistors are used for power amplification and power transistors are used for final power output.

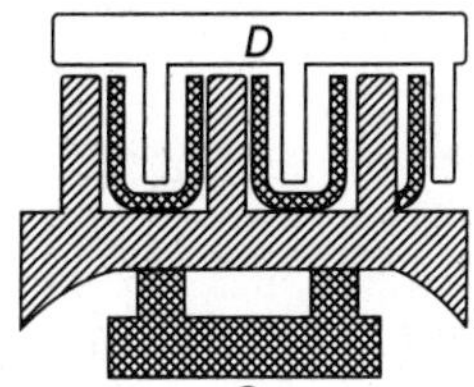

Fig. 10.49 High-power FET

Impedance-matching circuits are also designed at the input and output ports of the transistor for maximum power transfer from the source to the transmitter input and from the transistor output to the load. The following sections describe the design features and the systematic procedure for the design of a microwave transistor amplifier.

Impedance Matching Conditions

In an amplifier circuit, power is transferred from input to outside port efficiently when there is impedance match between the ports. Therefore, a single-stage microwave transistor amplifier can be modeled by the circuit shown in Fig. 10.50. Here a matching network is introduced on both sides of the transistor to transform the input and output impedances of the transistor amplifier to the generator and load impedances Z_g and Z_L, respectively. Each of the impedance matching network input and output ports, therefore, must have conjugate pair impedances looking on both sides, i.e., $Z_{sin} = Z_{in}^*$, $Z_{L\,out} = Z_{out}^*$ as shown in Fig. 10.50. Since perfect matching cannot be achieved, practical input and output reflection coefficients, looking towards the indicated directions at input/output ports, can be defined as follows for the given *S*-parameters of the transistor with reference to Z_0 = 50 ohms:

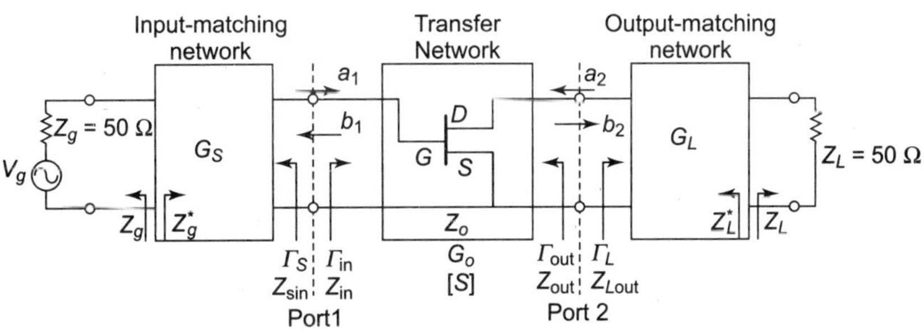

Fig. 10.50 Transistor amplifier circuit

($\Gamma_s = \Gamma_{in}^* = \Gamma_{sm}$, $\Gamma_L = \Gamma_{out}^* = \Gamma_{Lm}$ for conjugate match)

$$\Gamma_L = \frac{Z_{L\text{out}} - Z_{\text{out}}^*}{Z_{L\text{out}} + Z_{\text{out}}^*} = \text{Reflection coefficient looking towards load at the output port} \quad (10.60)$$

$$\Gamma_s = \frac{Z_{s\text{in}} - Z_{\text{in}}^*}{Z_{s\text{in}} + Z_{\text{in}}^*} = \text{Reflection coefficient looking towards source at the input port.} \quad (10.61)$$

$$\Gamma_{out} = S_{22} + \frac{S_{12}S_{21}\Gamma_s}{1-S_{11}\Gamma_s} = \frac{Z_{out}-Z^*_{L\,out}}{Z_{out}+Z^*_{L\,out}} = \text{Reflection coefficient looking into Port 2,}$$

$$= \frac{S_{22}-\Delta\Gamma_s}{1-S_{11}\Gamma_s} \qquad (10.62)$$

$$\Gamma_{in} = S_{11} + \frac{S_{12}S_{21}\Gamma_L}{1-S_{22}\Gamma_L} = \frac{Z_{in}-Z^*_{sin}}{Z_{in}+Z^*_{sin}}$$

$$= \frac{S_{11}-\Delta\Gamma_L}{1-S_{22}\Gamma_L} = \text{Reflection coefficient seen looking into Port 1.} \qquad (10.63)$$

where $\Delta = S_{11}S_{22} - S_{12}S_{21}$ (10.64)

Hence, perfect matching conditions are : $\Gamma_s = \Gamma_{in}^* = \Gamma_{sm}$ (say) and $\Gamma_L = \Gamma_{out}^* = \Gamma_{Lm}$ (say).

Amplifier Gain Equations

In terms of power when the matching networks are lossless (reactive)

P_L = Power absorbed in the load $Z_L = \frac{1}{2}(|b_2|^2 - |a_2|^2)$ (10.65)

P_{in} = Power delivered to the input of the transistor $= \frac{1}{2}(|a_1|^2 - |b_1|^2)$ (10.66)

P_{avn} = Power available from the transistor network

$= P_L|_{\Gamma L = \Gamma out^*}$ (10.67)

P_{avs} = Power available from the source to the network $= P_{in}|_{\Gamma in = \Gamma s^*}$ (10.68)

Under this set-up, three types of power gain equations are derived:

1. Power gain $G_p = \frac{P_L}{P_{in}}$, independent of Z_{sin} but depends of $Z_{L\,ou}$ (10.69)

2. Available power gain $G_A = \frac{P_{avn}}{P_{avs}}$, independent of Z_L but depends of Z_{sin} (10.70)

3. Transducer power gain $G_T = \frac{P_L}{P_{avs}}$, depends on both Z_{sin} and $Z_{L\,out}$ (10.71)

If both the input and output are conjugately matched to the two-port, then $G = G_A = G_T$ and G is maximum.

To express the gain equations in terms of S-parameters of the transistor network and Γ's, we write

$$P_L = \frac{1}{2}(|b_2|^2 - |a_2|^2) = \frac{1}{2}|b_2|^2(1 - |\Gamma_L|^2) \qquad (10.72)$$

$$P_{in} = \frac{1}{2}|a_1|^2(1 - |\Gamma_{in}|^2) \qquad (10.73)$$

S-parameter equations

$$b_1 = S_{11}a_1 + S_{12}a_2 = S_{11}a_1 + S_{12}\Gamma_L b_2$$

$$b_2 = S_{21}a_1 + S_{22}a_2 = S_{21}a_1 + S_{22}\Gamma_L b_2$$

Therefore, $\frac{b_2}{a_1} = \frac{S_{21}}{S_{22}\Gamma_L}$ (10.74)

∴ *power gain*

$$G_P = \left|\frac{b_2}{a_1}\right|^2 \frac{1-|\Gamma_L|^2}{1-|\Gamma_{in}|^2} = \frac{|S_{21}|^2(1-|\Gamma_L|^2)}{|1-S_{22}\Gamma_L|^2(1-|\Gamma_{in}|^2)} \tag{10.75}$$

❑ *Available Power Gain*

Let V_s = Open circuit voltage at Port 1 looking towards the source

Z_s = Impedance at Port 1 looking towards the source.

At Port 1, the total input

$$a_1 + b_1 = a_1(1+\Gamma_{in}) = \frac{V_s}{\sqrt{Z_0}}\frac{Z_{in}}{Z_{in}+Z_s}$$

where
$$Z_{in} = Z_0\frac{1+\Gamma_{in}}{1-\Gamma_{in}} \tag{10.76}$$

Solving for a_1 in terms of equivalent source voltage V_s

$$a_1 = \frac{V_s}{\sqrt{Z_0}}\frac{1}{(1+\Gamma_{in})}\frac{Z_{in}}{(Z_{in}+Z_s)} = \frac{V_s}{2\sqrt{Z_0}}\frac{(1-\Gamma_s)}{(1-\Gamma_s\Gamma_{in})} \tag{10.77}$$

The average power available from the source side is the maximum power delivered to the network:

$$P_{avs} = P_{in}|_{\Gamma in = \Gamma s^*} = \frac{|a_1|^2}{2}(1-|\Gamma_{in}|^2)$$

$$= \frac{|V_s|^2|1-\Gamma_s|^2(1-|\Gamma_{in}|^2)}{8Z_0|1-\Gamma_s\Gamma_{in}|^2} = \frac{|V_s|^2}{8Z_0}\frac{|1-\Gamma_s|^2}{(1-|\Gamma_s|^2)} \tag{10.78}$$

Similarly,
$$P_{avn} = P_L\Big|_{\Gamma_L=\Gamma_{out}^*} = \frac{|b_2|^2}{2}(1-|\Gamma_L|^2) \tag{10.79}$$

Now,
$$a_2 + b_2 = b_2(1+\Gamma_{out})$$

$$b_2 = S_{21}a_1 + S_{22}a_2 = S_{21}a_1 + S_{22}\Gamma_L b_2$$

or,
$$b_2 = \frac{S_{21}a_1}{1-S_{22}\Gamma_L} \tag{10.80}$$

Substituting b_2 and a_1 in expression for P_{avn},

$$P_{avn} = \frac{|V_s|^2}{8Z_0}\left.\frac{|S_{21}|^2(1-|\Gamma_{out}|^2)|1-\Gamma_s|^2}{\left|1-S_{22}\Gamma_{out}^*\right|^2\left|1-\Gamma_s\Gamma_{in}\right|^2}\right|_{\Gamma_L=\Gamma_{out}^*} \tag{10.81}$$

Evaluating Γ_{in} for $\Gamma_L = \Gamma_{out}^*$,

$$|1-\Gamma_s\Gamma_{in}|^2_{\Gamma_L=\Gamma_{out}^*} = \frac{|1-S_{11}\Gamma_s|^2(1-|\Gamma_{out}4|^2)}{|1-S_{22}\Gamma_{out}^*|^2}$$

so that
$$P_{avn} = \frac{|V_s|^2}{8Z_0}\frac{|S_{21}|^2|1-\Gamma_s|^2}{|1-S_{11}\Gamma_s|^2(1-|\Gamma_{out}|^2)}$$

∴
$$G_A = \frac{P_{avn}}{P_{avs}} = \frac{|S_{21}|^2(1-|\Gamma_s|^2)}{|1-S_{11}\Gamma_s|^2(1-|\Gamma_{out}|^2)} \tag{10.82}$$

❑ Transducer Power Gain

$$G_T = \frac{P_L}{P_{avs}} = \frac{|S_{21}|^2(1-|\Gamma_s|^2)(1-|\Gamma_L|^2)}{|1-\Gamma_s\Gamma_{in}|^2|1-S_{22}\Gamma_L|^2} \qquad (10.83)$$

When both input and output are matched

$$\Gamma_L = \Gamma_s = 0,$$

$$\therefore \quad G_T = |S_{21}|^2 \qquad (10.84)$$

Transducer power gain could have three cases:

(a) Unilateral transducer power gain (non-reciprocal), where $S_{12} = 0$

$$G_{TU} = \frac{|S_{21}|^2(1-|\Gamma_s|^2)(1-|\Gamma_L|^2)}{|1-S_{11}\Gamma_s|^2|1-S_{22}\Gamma_L|^2} \qquad (10.85)$$

(b) Maximum unilateral transducer gain, where $\Gamma_s = S_{11}^*$, $\Gamma_L = S_{22}^*$:

$$G_{TU\max} = \frac{|S_{21}|^2}{(|1-|S_{11}|^2)(|1-|S_{22}|^2)} \qquad (10.86)$$

(c) Matched transducer gain ($\Gamma_s = \Gamma_L = 0$)

$$G_{TM} = |S_{21}|^2 \qquad (10.87)$$

Example 10.10 The S-parameters of a transistor at 5 GHz for a conjugate matched transistor amplifier are given by $S_{11} = 0.9\angle{-100°}$, $S_{21} = 2.4\angle 90°$, $S_{12} = 0$, $S_{22} = 0.8\angle 40°$. Determine the maximum gain.

Solution The maximum gain of the conjugate matched transistor amplifier is

$$G_{\max} = \frac{(2.4)^2}{\left[\{1-(0.9)^2\}\{(1-(0.8)^2\}\right]} = 84.21 = 19.3 \text{ dB}$$

Example 10.11 A GaAs MESFET has the following S-parameters at 5 GHz with a 50 ohm reference

$$S_{11} = 0.5\angle 160°,\ S_{12} = 0.1\angle 10°,\ S_{21} = 2.0\angle 10°,$$

$$S_{22} = 0.4\angle 150°$$

The source and load impedances are $Z_{sin} = 30$ ohms, $Z_{L\,out} = 40\ \Omega$. Calculate G, G_A and G_T.

Solution Since S-parameters are measured with 50 ohm reference $Z^*_{in} = Z^*_{out} = 50$ ohms

$$\therefore \quad \Gamma_s = \frac{30-50}{30+50} = -0.25$$

$$\Gamma_L = \frac{40-50}{40+50} = -0.11$$

$$\Gamma_{in} = S_{11} + \frac{S_{12}S_{21}\Gamma_L}{1-S_{22}\Gamma_L} = 0.5\angle 160° + \frac{0.1\angle 10° * 2.0\angle 10° * (-0.11)}{1-0.4\angle 150° * (-0.11)}$$

$$= 0.5182\angle{-161.5970°}$$

$$\Gamma_{\text{out}} = S_{22} + \frac{S_{12}S_{21}\Gamma_s}{1-S_{11}\Gamma_s} = 0.4\angle 150^\circ + \frac{0.1\angle 10^\circ * 2.0\angle 10^\circ * (-0.25)}{1-0.5\angle 160^\circ * (-0.25)}$$

$$= 0.4404 \angle -155.4148^\circ$$

Therefore,

$$G = \frac{|S_{21}|^2(1-|\Gamma_L|^2)}{(1-|\Gamma_{\text{in}}|^2)|1-S_{22}\Gamma_L|^2} = 5.8392$$

$$G_A = \frac{|S_{21}|^2(1-|\Gamma_s|^2)}{|1-S_{11}\Gamma_s|^2(1-|\Gamma_{\text{out}}|^2)} = 5.9591$$

$$G_T = \frac{|S_{21}|^2(1-|\Gamma_s|^2)(1-|\Gamma_L|^2)}{|1-\Gamma_s\Gamma_{\text{in}}|^2|1-S_{22}\Gamma_2|^2} = 5.1398$$

In the amplifier design, the most useful gain is the transducer power gain which accounts for both source and load mismatch and is defined by

$$G_T = G_S \, . \, G_0 . \, G_L \tag{10.88}$$

where ,

$$G_s = \frac{1-|\Gamma_s|^2}{|1-\Gamma_{\text{in}}\Gamma_s|^2},$$ The source-matching network gain due to the impedance matching of the transistor to the impedance Z_0, (10.89)

$G_0 = |S_{21}|^2$, The transistor gain (10.90)

$$G_L = \frac{1-|\Gamma_L|^2}{|1-S_{22}\Gamma_L|^2},$$ The load-matching network gain due to the impedance matching of the transistor to Z_0 (10.91)

Amplifier Stability 10.8.4

In general, the passive matching networks at the input and output produces $Re\ (Z_s) > 0$ and $Re\ (Z_L) > 0$ such that $|\Gamma_s| < 1$ and $|\Gamma_L| < 1$. From expressions of Γ_{in} and Γ_{out} it can be seen that for certain values of S-parameters, even if $|\Gamma_s| < 1$ and $|\Gamma_L| < 1$, $|\Gamma_{\text{in}}| > 1$ and/or $|\Gamma_{\text{out}}| > 1$ representing presence of negative resistance at the input and/or output and oscillations can occur. Therefore, the stability of an amplifier, or its resistance to oscillate is a very important design consideration. The stability is determined from the S-parameters of the transistor, the matching network and the terminations. There are two types of stability defined:

- ***Unconditional Stability*** Achieved when $Re\ (Z_{\text{in}}) > 0$ and $Re\ (Z_{\text{out}}) > 0$

or, $$\left|S_{11} + \frac{S_{12}S_{21}\Gamma_L}{1-S_{22}\Gamma_L}\right| = |\Gamma_{\text{in}}| < 1 \text{ and } |\Gamma_{\text{out}}| = \left|S_{22} + \frac{S_{12}S_{21}\Gamma_s}{1-S_{11}\Gamma_s}\right| < 1.$$

for all passive source and load impedances, i.e.,

$$|\Gamma_s| < 1 \text{ and } |\Gamma_L| < 1 \tag{10.92}$$

- ***Conditional Stability*** Achieved when

$$|\Gamma_{\text{in}}| < 1 \text{ and } |\Gamma_{\text{out}}| < 1 \tag{10.93}$$

only for a certain range of passive source and load impedances. This is potentially unstable.

If the device is unilateral ($S_{12} = 0$), the condition for unconditional stability becomes

$$|S_{11}| < 1 \text{ and } |S_{22}| < 1. \tag{10.94}$$

With the inequalities of Eq. (10.94) and using Smith chart, it can be shown that the condition for unconditional stability can be expressed in terms of S-parameters only, i.e., by a factor

$$K = \frac{1 + |\Delta|^2 - |S_{11}|^2 - |S_{22}|^2}{2|S_{12}S_{21}|^2} > 1 \tag{10.95}$$

and $$|\Delta| = |S_{11}S_{22} - S_{12}S_{21}| < 1 \tag{10.96}$$

If $K > 1$ and $|\Delta| > 1$ or, $K < 1$ and $|\Delta| < 1$, the network becomes unstable. For unilateral device ($S_{12} = 0$), $K = \infty$, $|\Delta| < 1$ for stability.

Example 10.12 A GaAs FET has the following S-parameters at 6 GHz with $V_{ds} = 4\text{V}$ and $I_{ds} = 30$ mA and reference impedance $Z_0 = 50$ ohms.

$$S_{11} = 0.9\ \angle{-60°}$$

$$S_{21} = 3.1\ \angle 124°$$

$$S_{12} = 0.02\ \angle 62°$$

$$S_{22} = 0.8\ \angle{-28°}$$

Determine the stability of this transistor.

Solution

Delta factor $|\Delta| = S_{11}S_{22} - S_{12}S_{21} = 0.7183$

Stability factor $K = \dfrac{1 + |\Delta|^2 - |S_{11}|^2 - |S_{22}|^2}{2|S_{12}S_{21}|^2} = 0.5324$

Since $\Delta = 0.72 < 1$ and $K = 0.53 < 1$, the device is potentially unstable.

Stability Circle

When the two ports of the circuit shown in Fig. 10.51 are potentially unstable, there may be values of Γ_s and Γ_L, i.e., impedances of input and output circuits, for which $R_e\,(Z_{\text{in}})$ and $R_e\,(Z_{\text{out}})$ are positive. These values of Γ_s and Γ_L can be determined using the graphical methods with stability circle on Smith chart.

The input and output stability circles on a Smith chart can be drawn from the equations derived as follows. The stability circles are defined as the loci in the Γ_s plane (input stability circle) or Γ_L plane (output stability circle) for which $|\Gamma_{\text{out}}| = 1$ or $|\Gamma_{\text{in}}| = 1$, respectively. From the condition that

$$|\Gamma_{\text{in}}| = \left| S_{11} + \frac{S_{12}S_{21}\Gamma_L}{1 - S_{22}\Gamma_L} \right| = 1 \tag{10.97}$$

and defining $\Delta = S_{11}S_{22} - S_{12}S_{21}$, we obtain

$$|S_{11} - \Delta\Gamma_L| = |1 - S_{22}\Gamma_L| \tag{10.98}$$

Now, squaring both sides and simplifying after mathematical manipulation we obtain stability for $|S_{11}| < 1$ and $|S_{11}| > 1$, respectively. Under load match condition $Z_L = Z_0$, $\Gamma_L = 0$ (centre of the Smith chart). If $|S_{11}| < 1$, then $|\Gamma_{\text{in}}| < 1$, the centre of the Smith chart ($\Gamma_L = 0$) is in the stable region and all shaded portion of the Smith chart that is exterior to the stability circle defines the stable region for Γ_L. If $|S_{11}| > 1$, then $|\Gamma_{\text{in}}| > 1$ for $\Gamma_L = 0$ and the centre of the Smith chart is in the unstable region and all the shaded portion of the Smith chart that is interior to the stability circle defines the stable region for Γ_L. Similar methods are applicable to the input stability circle.

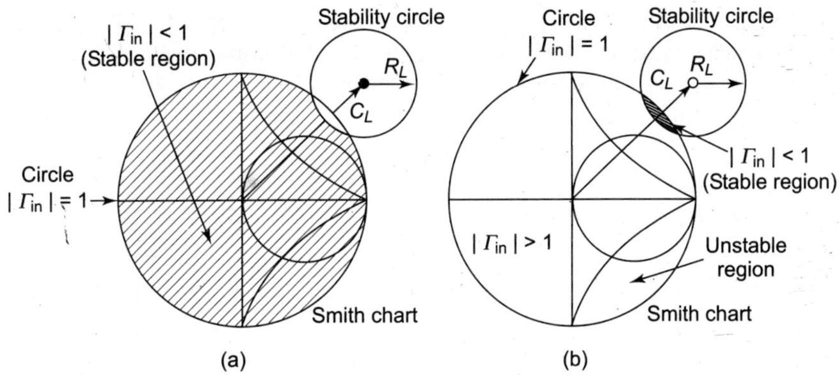

Fig. 10.51 Output stability circles for a conditionally stable device (a) $|S_{11}| < 1$, (b) $|S_{11}| > 1$

For unconditional stability for all passive source and load impedances ($|\Gamma_s| < 1$, and $|\Gamma_L| < 1$), the stability circles lie completely outside or inside the Smith chart where

$$|\Gamma_L - C_L| = R_L \tag{10.99}$$

$$C_L = \frac{(S_{22} - \Delta S_{11}^*)^*}{|S_{22}|^2 - |\Delta|^2} \tag{10.100}$$

$$R_L = \frac{(S_{12} - S_{21})}{|S_{22}|^2 - |\Delta|^2} \tag{10.101}$$

The above equation represents a circle in the complex Γ plane with centre at C_L (complex) and a radius R_L (real) and is defined as the output stability circle. Similarly, there is the input stability circle represented by the equation

$$|\Gamma_s - C_s| = R_s \tag{10.102}$$

where

$$C_s = \frac{(S_{11} - \Delta S_{22}^*)^*}{|S_{11}|^2 - |\Delta|^2} \tag{10.103}$$

$$R_s = \left|\frac{S_{12}\, S_{21}}{|S_{11}|^2 - |\Delta|^2}\right| \tag{10.104}$$

Therefore, from the knowledge of S-parameters of the transistor, we can plot the input and output stability circles to define where $|\Gamma_{in}| = 1$ and $|\Gamma_{out}| = 1$, the boundaries between stable regions $|\Gamma_{in}| < 1$ and $|\Gamma_{out}| < 1$, and potentially unstable regions $|\Gamma_{in}| > 1$ and $|\Gamma_{out}| > 1$. Figure 10.51 shows the shaded stable regions for two cases of conditional output:

$$|(|C_L| - R_L)| > 1, \text{ for } |S_{11}| < 1, |\Gamma_{in}| < 1 \tag{10.105}$$

$$|(|C_s| - R_s)| > 1, \text{ for } |S_{22}| < 1, |\Gamma_{out}| < 1. \tag{10.106}$$

Transistor Amplifier Noise 10.8.5

In a microwave amplifier a small amount of ac output power appears without any ac signal input. This undesired output noise is due to amplified input noise plus the noise power generated within the amplifier circuit on biasing. The noise performance is measured in terms of 'Noise Figure F' defined by

$$F = \frac{\text{Signal-to-noise ratio at input}}{\text{Signal-to-noise ratio at output}} \tag{10.107}$$

The noise input power can be modeled by a noisy resistor R_n at temperature T Kelvin in which noise is thermal noise or Johnson noise caused by random motion of electrons in the conductor due to thermal agitation. The noise power associated with R_n is expressed by

$$N_R = kTB_n \tag{10.108}$$

where $k = 1.38 \times 10^{-23}$ J/K, Boltzman's constant

B_n = Noise bandwidth

Typically, $N_R \simeq -84$ dBm at $B = 1000$ MHz and $T = 290$ K. If G is the gain of the amplifier over the noise band B_n, the noise figure can be expressed as

$$F = \frac{\text{Available noise power at output}}{\text{Gain} * \text{Available noise power at input}} = \frac{N_0}{GkTB_n} \tag{10.109}$$

If the noise generated within the amplifier is represented by noise power N_a, $N_0 = N_a + GkTB_n$ and then

$$F = 1 + \frac{N_a}{GkTB_n}; \tag{10.110}$$

If an amplifier is noiseless, $F = 1$. It can be shown that for a cascaded (m-number) amplifier, the total noise figure

$$F = F_1 + \frac{F_2 - 1}{G_1} + \frac{F_3 - 1}{G_1 G_2} + \cdots + \frac{F_m - 1}{G_1 G_2 \cdots G_{m-1}}; \tag{10.111}$$

Since thermal noise is independent of resistance and proportional to absolute temperature, noise power is expressed in units of temperature in the following manner.

$$N_0 = N_a + GkTB_n = Gk(T_o + T_e)B_n \tag{10.112}$$

Here, $T_o = 290$ K (room temperature), T_e = the equivalent temperature of a noisy resistor at the input that would generate the same noise N_a at the output of a noiseless amplifier as added by the actual amplifiers. Therefore,

$$F = 1 + \frac{T_e}{T_o} \tag{10.113}$$

Example 10.13 Two microwave amplifiers are cascaded. If the gain and noise figures of the first and second amplifiers are $G_1 = 10$ dB, $F_1 = 2$dB, and $G_2 = 20$ dB and $F_2 = 3$dB, respectively, determine the total noise figure of the system. Explain importance of gain of two amplifiers.

Solution

$$F = F_1 + \frac{F_2 - 1}{G_1}$$

$F_1 = 2$dB or, $F_1 = 10^{2/10} = 1.58$

$F_2 = 3$dB or, $F_2 = 10^{3/10} = 2$

$G_1 = 10$dB or, $G_1 = 10^{10/10} = 10$

$G_2 = 20$ dB or, $G_2 = 10^{20/10} = 100$

$$\therefore \quad F = 1.58 + \frac{2-1}{10} = 1.68 = 2.25 \text{ dB}$$

It is seen that the impact on overall noise figure, F, by the second stage is diluted by the gain of the first stage. This second stage impact dilutes even further as gain of the first stage increases.

Microwave Amplifier Design 10.8.6

Microwave amplifier design falls into two categories: small-signal and large-signal design. In small signal operation of the GaAs MESFET, input signal level controls the gate voltage V_{gs} to remain in linear region and the optimum load admittance is equal to the drain conductance $\left(\frac{\partial I_{ds}}{\partial V_{ds}}\right)$. In large-signal operation the gate voltage V_{gs} enters the non-linear region to increase the output current and powers as shown in Fig. 10.52.

There are two basic steps involved in the design of microwave transistor amplifiers—(1) selection of dc biasing depending on the operating mode and (2) design of input/output impedance matching networks. These are described below.

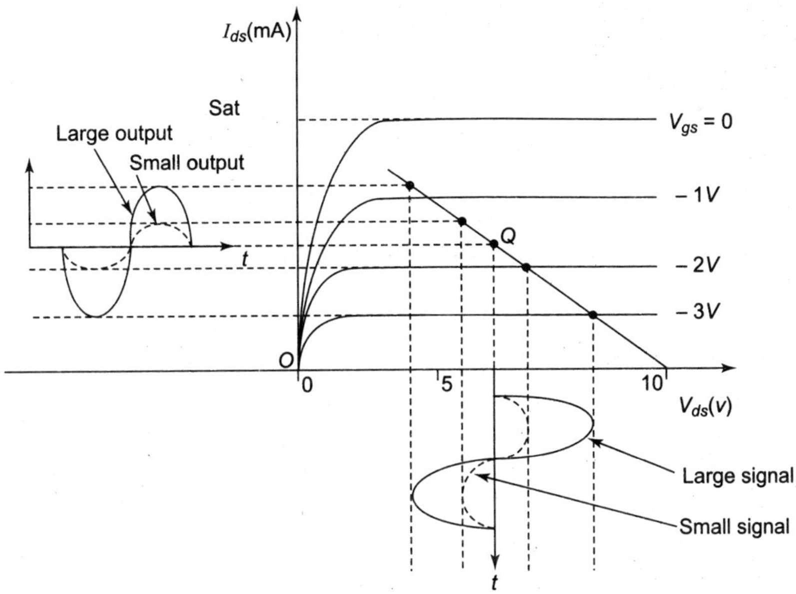

Fig. 10.52 Small-and large-signal operations of microwave transistor amplifier

dc Biasing

Two basic transistors, GaAs MESFET and *Si*-transistors, are considered in the design of amplifier. For safe operating point for a GaAs MESFET following considerations are made in dc biasing.

a. Maximum V_{ds}
b. Maximum I_{ds}
c. Maximum input signal to the gate
d. Maximum power dissipation at junction temperature 175°C.

For the Si-transistors, these are

a. Maximum V_{cE}
b. Maximum I_{cE}
c. Maximum input signal
d. Maximum power dissipation at junction temperature 200°C

The dc biasing circuits for Si-transistor and GaAa FET are shown in Fig. 10.46 and 10.47 and discussed for its importance. In all dc-biasing circuits, the element values can be selected as under:

1. All inductors used are RF chokes to offer high reactance to *rf* signal but very low (≈0) resistance to dc. They are made of 2–3 turns of No. 36 enameled wire on 2.54 mm air core.

2. All source bypass capacitors, C_s, are 0.01 μF disk at 100V.
3. All capacitors in G and D are 0.01 μF disk at 100 V.
4. R_s is selected to set V_s at proper Q-point and transient protection is achieved.

The operating point Q is selected for GaAs FET amplifier based on the following considerations:

1. Low-power and low-noise operation—Relatively low V_{ds} and I_{ds} for class A mode
2. High-gain and low-noise operation—Low V_{ds} as in 1 but higher I_{ds} below saturation value
3. High-power operation—Both V_{ds} and I_{ds} are high but for linear operation in class A, I_{ds} is kept below that in 2.
4. High-power and high-efficiency operation—For high efficiency, class AB or B operation is required. The drain current I_{ds} is lowered as compared to that in 3 with V_{ds} same as in 3.

These four steps are indicated in I_{ds} vs V_{ds} curves of Fig. 10.53.

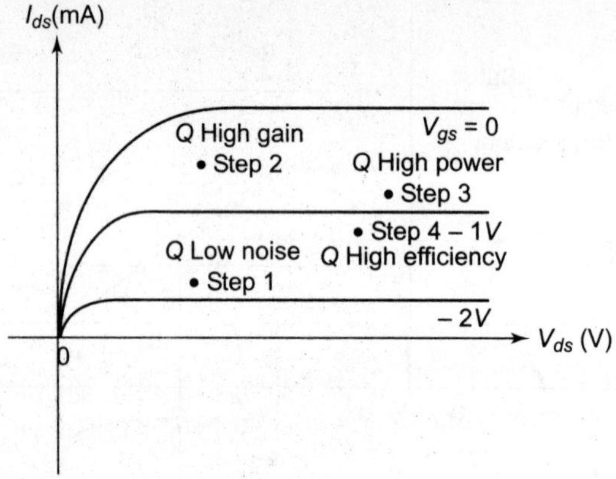

Fig. 10.53 Transistor operating points

Design of Input/Output Matching Networks

Since the transducer gain is the most important gain factor in amplifier design and G_o is fixed for a given transistor, the overall G_T will be controlled by G_s and G_L of the matching sections. To maximize this gain, the matching sections are designed to provide a conjugate match between the source or load impedances and the transistor impedance.

The input and output matching networks are designed for different applications. A very important application of microwave amplifiers is its narrow band operation for maximum-gain and minimum-noise figure. In microwave receiver, the maximum gain and minimum noise figure of an amplifier are equally important. Since from design practices, both cannot be realized simultaneously in most cases, the amplifier circuit can be biased for either maximum power gain or minimum noise figure for given S-parameters. Some cases like preamplifier, the minimum-noise figure is more important then the power gain because, otherwise, the input noise would amplify at the output. The following steps are involved for the matching network design.

Low-Noise Narrowband Design

1. The source and load reflection coefficients Γ_s and Γ_L are converted to the source and load equivalent impedances $Z_{s\,\text{in}}$ and $Z_{L\,\text{out}}$, respectively:

$$Z_{\text{sin}} = Z_0 \frac{1+\Gamma_s}{1-\Gamma_s} \tag{10.114}$$

$$Z_{Lout} = Z_0 \frac{1+\Gamma_L}{1-\Gamma_L} \tag{10.115}$$

where Z_0 = characteristic impedance for reference

and $\Gamma_s = |\Gamma_s| \cos \angle\Gamma_s + j|\Gamma_s| \sin \angle\Gamma_s$; (10.116)

$\Gamma_L = |\Gamma_L| \cos \angle\Gamma_L + j|\Gamma_L| \sin \angle\Gamma_L$; (10.117)

are the source and load reflection coefficients for minimum noise figure. Substituting Γ_s and Γ_L in impedance equations:

$$Z_{sin} = Z_0 \frac{1-|\Gamma_s|^2 + j2|\Gamma_s| \sin \angle\Gamma_s}{1+|\Gamma_s|^2 - 2|\Gamma_s| \cos \angle\Gamma_s} \tag{10.118}$$

$$Z_{Lout} = Z_0 \frac{1-|\Gamma_L|^2 + j2|\Gamma_L| \sin \angle\Gamma_L}{1+|\Gamma_L|^2 - 2|\Gamma_L| \cos \angle\Gamma_L} \tag{10.119}$$

2. The source and load equivalent admittance are determined from step 1.
3. The susceptance components of admittances of step 2 are realized with either short or open-circuited $3\lambda_g/8$ long stubs.
4. The conductance components of admittances of step 2 are realized with $\lambda_g/4$ impedance transformers.

Amplifier Design for Maximum Transducer Power Gain

As shown in Fig. 10.50, the input and output reflection coefficients are given by respectively,

$$\Gamma_{in} = S_{11} + \frac{S_{12}S_{21}\Gamma_L}{1-S_{22}\Gamma_L} \tag{10.120}$$

$$\Gamma_{out} = S_{22} + \frac{S_{12}S_{21}\Gamma_S}{1-S_{11}\Gamma_S} \tag{10.121}$$

For the maximum power gain of the transducers, simultaneous conjugate match conditions to be satisfied:

$$\Gamma_s = \Gamma_{in}^* \tag{10.122}$$

and $$\Gamma_L = \Gamma_{out}^* \tag{10.123}$$

so that $(VSWR)_{in} = (VSWR)_{out} \cong 1.0$. Therefore, under simultaneous conjugate match conditions:

$$\Gamma_s^* = S_{11} + \frac{S_{12}S_{21}\Gamma_L}{1-S_{22}\Gamma_L} \tag{10.124}$$

and $$\Gamma_L^* = S_{22} + \frac{S_{12}S_{21}\Gamma_S}{1-S_{11}\Gamma_S} \tag{10.125}$$

Solving the above two equations, the values of Γ_s and Γ_L for a simultaneous conjugate match can be written as

$$\Gamma_s = \Gamma_{sm} = \frac{A_1 \pm \sqrt{A_1^2 - 4|B_1|^2}}{2B_1} \tag{10.126}$$

$$\Gamma_L = \Gamma_{Lm} = \frac{A_2 \pm \sqrt{A_2^2 - 4|B_2|^2}}{2B_2} \tag{10.127}$$

where $A_1 = 1 + |S_{11}|^2 - |S_{22}|^2 - |\Delta|^2$ (10.128)

$$A_2 = 1 + |S_{22}|^2 - |S_{11}|^2 - |\Delta|^2 \quad (10.129)$$

$$B_1 = S_{11} - \Delta S_{22}^* \quad (10.130)$$

$$B_2 = S_{22} - \Delta S_{11}^* \quad (10.131)$$

It can be shown that [Gonzalez] a simultaneous conjugate match having unconditional stability is obtained for $K > 1$ and $|\Delta| < 1$, which implies that $A_1 > 0$ and $A_2 > 0$. Therefore, for the unconditionally stable two-port network, solutions for Γ_{sm} and Γ_{Lm} with a minus sign in the numerators is to be used. Therefore, under maximum transducer power gain with simultaneous conjugate match:

$$G_{T\max} = \frac{1}{1 - |\Gamma_{sm}|^2} |S_{21}|^2 \frac{1 - |\Gamma_{Lm}|^2}{|1 - S_{22}\Gamma_{Lm}|^2}$$

$$= \frac{|S_{21}|}{|S_{12}|}(K - \sqrt{K^2 - 1}) = G_{A\max} = G_{P\max} \quad (10.132)$$

When $K = 1$, $G_{T\max}$ = maximum stable gain. Under unstable operation, G_T attains a minimum value for Γ_{sm} and Γ_{Lm} with a +ve sign in the numerator of Eq. (10.127):

$$G_{T\min} = \frac{|S_{21}|}{|S_{12}|}(K + \sqrt{K^2 - 1}) \quad (10.133)$$

and the $(\text{VSWR})_{\text{in}} = (\text{VSWR})_{\text{out}} = 1$

The steps for impedance match network for a maximum power gain narrowband amplifier are similar to those for a minimum noise narrowband amplifier. However, for, quick design, Smith chart can be used for input/output matching network design as described in Chapter 5.

Example 10.14 A GaAs MESFET has the following S-parameters measured with a 50 ohm resistance at 10 GHz: $S_{11} = 0.55 \angle -150°$, $S_{12} = 0.04 \angle 20°$, $S_{21} = 2.82 \angle 180°$, $S_{22} = 0.45 \angle -30°$. The noise parameters measured at centre frequency 10 GHz for minimum noise figure $F_{\min}$ are

$$F_{\min} = 3 \text{ dB}, R_n = 10 \text{ ohm}, \Gamma_s = 0.7 \angle 160°, \Gamma_L = 0.5 \angle 100°.$$

The transistor is to be used for a low-noise narrowband amplifier.

Determine the (a) stability, (b) maximum available gain, and (c) input and output matching networks with a 50-ohm reference for a minimum-noise figure.

Solution

(a) ***Stability***

$$\Delta = S_{11} S_{22} - S_{12} S_{21}$$

$$= 0.55 \angle -150° * 0.45 \angle -30° - 0.04 \angle 20° * 2.82 \angle 180°$$

$$= 0.16 \angle 165°$$

$$|\Delta| = 0.16 < 1$$

$$K = \frac{1 + |\Delta|^2 - |S_{11}|^2 - |S_{22}|^2}{2|S_{12} S_{21}|} = \frac{1 + 0.16^2 - 0.55^2 - 0.45^2}{2 * 0.04 * 2.82} = 2.28 > 1.$$

Since $|\Delta| < 1$ and $K > 1$, The device is unconditionally stable.

(b) ***Gain***

Maximum available power gain

$$G_{A\,\max} = \frac{|S_{21}|}{|S_{12}|}\left|K - (k^2 - 1)^{1/2}\right| = \frac{2.82}{0.04}\left|2.28 - \sqrt{(2.28^2 - 1)}\right|$$

$$= 70.5\,|2.28 - 2.0489| = 16.286 = 12.11 \text{ dB.}$$

(c) ***Input and Output matching networks***

(i) *Input matching network*

- Source equivalent impedance for $F_{\min}$:

$$Z_{s\text{in}} = 50\,\frac{1 - |0.7|^2 + j2|0.7|\sin\angle 160°}{1 + |0.7|^2 - 2|0.7|\cos\angle 160°} = 9.0891 + j8.5335 \text{ ohm}$$

- Corresponding admittance

$$Y_{s\text{in}} = \frac{1}{Z_{s\text{in}}} = 0.0585 - j0.0549 \text{ mho}$$

The *Im* ($Y_{s\text{in}}$) is simulated by *OC* stub and *Re* ($Y_{s\text{in}}$) by $\lambda_g/4$ transformer.

- An *OC* stub of $l = 3\,\lambda_g/8$ offers a shunt inductor of susceptance

$$Y = jy_o \tan \beta l = -jY_o$$

and has characteristic impedance

$$Z_{01} = Z_{o\,\text{stub}} = \frac{1}{Y_o} = \frac{1}{\text{Im}\,(Y_{s\text{in}})} = 18.2143 \text{ ohm}$$

- Since the source resistance is 50 ohms, this is transformed to the source equivalent conductance *Re* ($Y_{s\text{in}}$) by a $\lambda_g/4$ transformer with characteristic impedance

$$Z_{02} = \sqrt{\frac{50}{Re\,(Y_{s\text{in}})}} = 29.2412 \text{ ohm}$$

(ii) *Output matching network*

In a similar manner, the output side

- $$Z_{L\,\text{out}} = Z_0\,\frac{1 - |\Gamma_L|^2 + j2|\Gamma_L|\sin\angle\Gamma_L}{1 + |\Gamma_L|^2 - 2|\Gamma_L|\cos\angle\Gamma_L}$$

$$= 50\,\frac{1 - |0.5|^2 + j2|0.5|\sin\angle 100°}{1 + |0.5|^2 - 2|0.5|\cos\angle 100°} = 26.3408 + j34.5875 \text{ ohm}$$

- $$Y_{L\,\text{out}} = \frac{1}{Z_{L\,\text{out}}} = 0.0139 - j0.0183 \text{ mho}$$

- An *OC* stub of $l = 3\lambda_g/4$ simulate shunt inductor of susceptance

$$Y = -jY_0$$

and has characteristic impedance

$$Z_{03} = Z_{0\,\text{stub}} = \frac{1}{\text{Im}\,(Y_{L\,\text{out}})} = 54.6478 \text{ ohm}$$

- Since the load resistance is 50 ohm, this is transformed to the load equivalent conductance *Re* ($Y_{L\,\text{out}}$) by a $\lambda_g/4$ transformer with characteristic impedance

$$Z_{04} = \sqrt{\frac{50}{Re(Y_{L\,out})}} = 59.8986 \text{ ohm}$$

The designed matching networks for the amplifier is shown in Fig. 10.54:

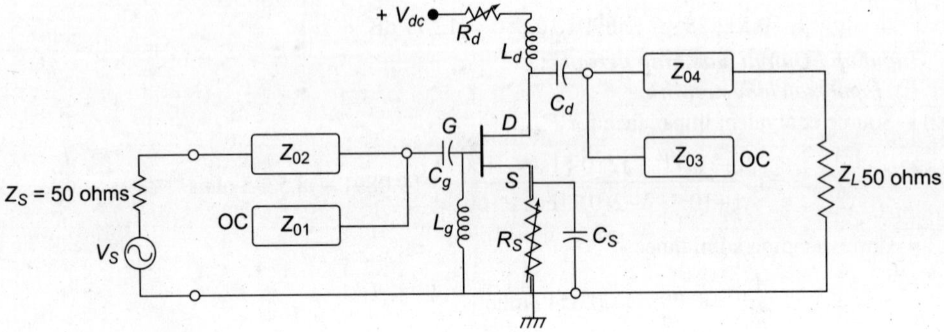

Fig. 10.54

The biasing circuit elements are given below:

RF chokes: L_g, $L_d \rightarrow$ 2–3 turns of No. 36 enameled wire on a 2.54 mm air core

Current control: R_s, $R_d \rightarrow$ 5 *K*-ohm, 1W potentiometers

dc blocks: C_g, C_s, $C_d \rightarrow$ 1 *nF* high-*Q* capacitors

Example 10.15 A GaAs-MESFET amplifier is to be designed at 5 GHz with 400 MHz bandwidth for maximum power gain. The measured parameters at 5 GHz with a 50-ohm reference are

$$S_{11} = 0.52\ \angle{-145°},\ S_{12} = 0.03\ \angle 20°$$
$$S_{21} = 2.56\ \angle 17°,\ S_{22} = 0.48\ \angle{-20°}$$
$$\Gamma_{sin} = 0.75\ \angle 170°,\ \Gamma_{L\,out} = 0.72\ \angle 105°$$

Determine $G_{A\max}$ and design input and output matching networks with a 50-ohm reference for maximum power gain.

Solution

$$\Delta = 0.52\ \angle{-145°} * 0.48\ \angle{-20°} - 0.03\ \angle 20° * 2.56\ \angle 170°$$
$$= 0.168\ \angle 197°$$

$$K = \frac{1 + |0.168|^2 - |0.52|^2 - |0.48|^2}{2|0.03 \times 2.56|} = 3.53$$

Since $|\Delta| < 1$ and $K > 1$, the device is unconditionally stable.

$$G_{A\max} = \left|\frac{S_{21}}{S_{12}}\right| \left|K - \sqrt{K^2 - 1}\right| = \frac{2.56}{0.03}\left|3.53 - \sqrt{3.53^2 - 1}\right|$$
$$= 12.3346$$
$$= 10.913 \text{ dB}$$

Input/Output matching Networks

The procedure for design of matching networks here is same as in previous example and are given below:

Input	*Output*
• $Y_{sin} = \dfrac{1}{Z_{sin}}$	• $Y_{L\text{out}} = \dfrac{1}{50}\left[\dfrac{1+|0.72|^2-2|0.72|\cos\angle 105°}{1-|0.72|^2+j2|0.72|\sin\angle 105°}\right]$
$= \dfrac{1}{50}\left[\dfrac{1+|0.75|^2-2|0.75|\cos\angle 170°}{1-|0.75|^2+j2|0.75|\sin\angle 170°}\right]$	$= 0.0084 - j0.0243$ ohm
$= 0.1026 - j0.0611$ ohm	
• $3\lambda_g/8$ open-circuit stub impedance	• $3\lambda_g/8$ *OC* stub impedance
$Z_{01} = \dfrac{1}{I_m(Y_{sin})} = 16.3719$ ohm	$Z_{03} = \dfrac{1}{I_m(Y_{L\text{out}})} = 41.1846$ ohm
• $\lambda_g/4$ transformer impedance	• $\lambda_g/4$ transformer impedance
$Z_{02} = \sqrt{\dfrac{50}{\text{Re}(Y_{sin})}} = 22.0763$ ohm	$Z_{04} = \sqrt{\dfrac{50}{\text{Re}(Y_{L\,\text{out}})}} = 77.1192$ ohm

The designed circuit is shown without biasing in Fig. 10.55. Biasing design is same as given in Example 10.11.

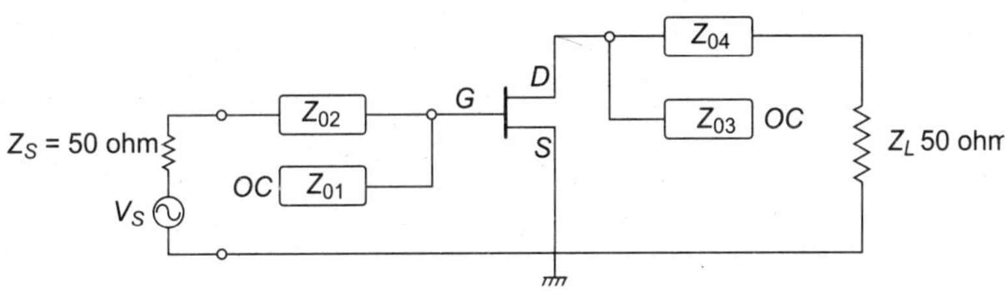

Fig. 10.55

Example 10.16 Design a microwave amplifier using GaAs FET operating at 5 GHz with maximum transducer power gain. The *S*-parameters at the linear bias point are $S_{11} = 0.6\ \angle{-170°}$, $S_{12} = 0.05\ \angle 16°$, $S_{21} = 1.5\ \angle 30°$, $S_{22} = 0.6\ \angle{-96°}$. Use Smith chart for matching network design.

Solution

(i) ***Stability check***

$$\Delta = S_{11}S_{22} - S_{12}S_{21} - 0.6\ \angle{-170°} * 0.6\ \angle\ 96° - 0.1\ \angle 16° * 2.0\ \angle 30°$$

$$= 0.36\ \angle{-266°} - 0.2\ \angle 46°$$

or, $$|\Delta| = 0.3148 < 1$$

$$K = \frac{1+|\Delta|^2-|S_{11}|^2-|S_{22}|^2}{2|S_{12}S_{21}|} = \frac{1+0.270-0.36-0.36}{0.075\times 2} = 2.5273 > 1$$

Since $K > 1$ and $|\Delta| < 1$, the GaAs FET is unconditionally stable.

(ii) ***Reflection Coefficient Calculation for Conjugate Match***

$$A_1 = 1 + |S_{11}|^2 - |S_{22}|^2 - |\Delta|^2 = 1 + 0.36 - 0.36 - (0.31)^2$$
$$= 0.9628$$
$$A_2 = 1 + |S_{22}|^2 - |S_{11}|^2 - |\Delta|^2 = 1 + 0.36 - 0.36 - (0.31)^2$$
$$= 0.9628$$
$$B_1 = S_{11} - \Delta S_{22}^* = 0.6\angle{-170°} - 0.3148 * 0.6\angle 96°$$
$$= -0.4727 - j\,0.0072$$
$$|B_1| = 0.4278$$
$$B_2 = S_{22} - \Delta S_{11}^* = 0.6\angle 96° - 0.3148 * 0.6\angle 170°$$
$$= -0.1372 + j\,0.4524$$
$$|B_2| = 0.4727$$
$$\therefore \quad \Gamma_{sm} = 1.485\angle 179.13° \quad \text{or} \quad 0.5514\angle 179.13°$$
$$\Gamma_{Lm} = 1.211\angle{-106.88°} \quad \text{or} \quad 0.8256\angle{-106.88°}$$

(iii) ***Gain***

$$G_{T\max} = \frac{|S_{21}|}{|S_{12}|}(K - \sqrt{K^2 - 1}) = \frac{1.5}{0.05}(2.5273 - \sqrt{2.5273^2 - 1}) = 6.1877$$
$$= 8.3740 \text{ dB}$$

(iv) ***Design of Matching Networks*** Admittances associated with Γ_{sm} and Γ_{Lm} are read in a Smith chart from Fig. 10.56 as

$$Y_{sm} = (6.5 - j\,0.4) * \frac{1}{50} = 0.13 - j\,0.008 \text{ mho}$$

$$Y_{Lm} = (0.05 + j\,1.3) * \frac{1}{50} = 0.001 + j\,0.026 \text{ mho}$$

The matching networks are designed with an open microstrip shunt stub using smith chart and a series microstrip line on each side as follows:

Input matching network	*Output matching network*
$3\lambda_g/8$ Stub:	$\lambda_g/8$ *OC* stub:
$Z_{01} = \frac{1}{0.008} = 125\ \Omega$	$Z_{01} = \frac{1}{0.026} = 38.46\ \Omega$
$OCl_1 = 3\lambda_g/8 = 2.25$ cm	$l_1 = \frac{\lambda_g}{8} = 0.75$ cm
$\lambda/4$ Transformer:	$\lambda/4$ Transformer:
$Z_{02} = \sqrt{\frac{50}{0.13}} = 19.61\ \Omega$	$Z_{02} = \sqrt{\frac{50}{0.001}} = 223.6\ \Omega$
$l_2 = \lambda_g/4 = 1.5$ cm	$l_2 = \lambda_g/4 = 1.5$ cm

The amplifier circuit is shown below in Fig. 10.57.

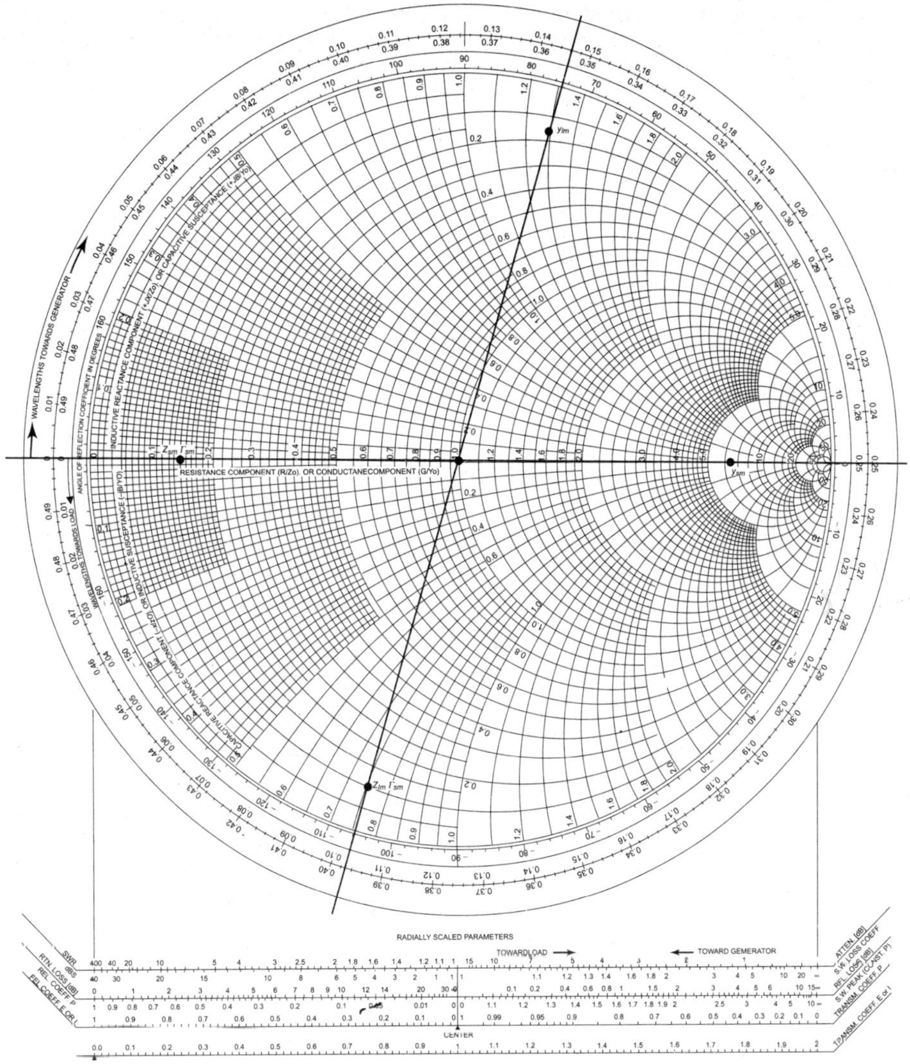

Fig. 10.56

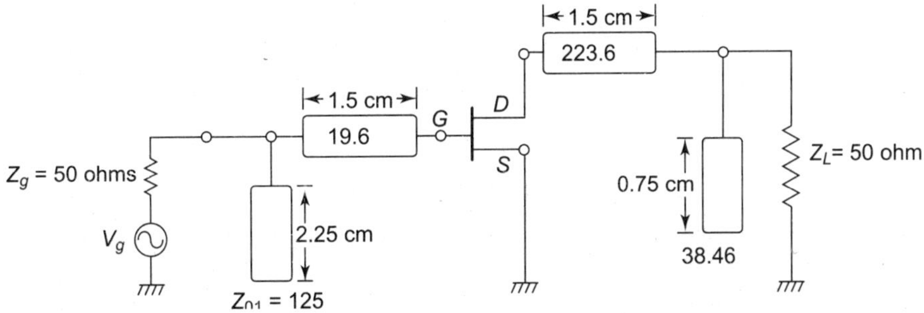

Fig 10.57

Microwave Transistor Oscillator 10.8.7

Microwave transistor oscillators are designed in a similar way as amplifiers using similar dc biasing circuits and devices. The difference is that oscillators do not have *rf* input as shown in Fig. 10.58 and oscillation conditions are different, i.e.,

Stability factor

$$K = \frac{1+|\Delta|^2 - |S_{11}|^2 - |S_{22}|^2}{2|S_{12}\,S_{21}|^2} < 1 \tag{10.134}$$

$$\Gamma_{\text{in}}\,\Gamma_g = 1 \tag{10.135}$$

$$\Gamma_{\text{out}}\,\Gamma_L = 1 \tag{10.136}$$

where $\Delta = S_{11}\,S_{22} - S_{12}\,S_{21}$ (10.137)

Γ_g = Generator reflection coefficient of the input resonant circuit = $\dfrac{Z_g - Z_0}{Z_g + Z_0}$

Γ_{in} = Input reflection coefficient of active device = $\dfrac{Z_{\text{in}} - Z_0}{Z_{\text{in}} + Z_0}$

Γ_{out} = Output reflection coefficient of the active device

Γ_L = Load reflection coefficient of the load matching network.

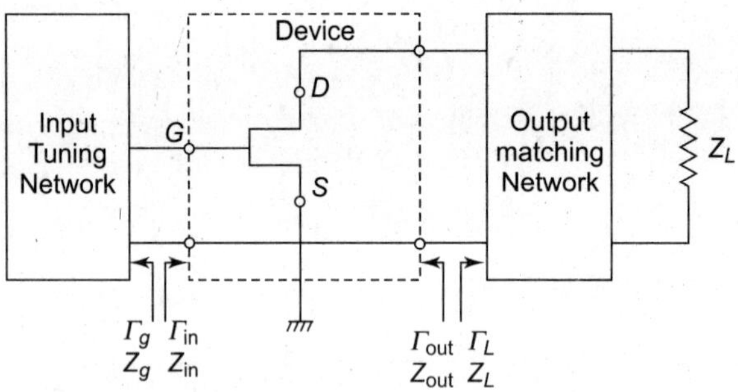

Fig. 10.58 Microwave transistor oscillator circuit

Therefore, oscillation condition becomes

$$\left(\frac{Z_g - Z_0}{Z_g + Z_0}\right)\left(\frac{Z_{\text{in}} - Z_0}{Z_{\text{in}} + Z_0}\right) = 1 \quad \text{or,} \quad Z_{\text{in}} + Z_g = 0$$

or, $R_{\text{in}} + R_g = 0$ (10.138)

and $X_{\text{in}} + X_g = 0$ (10.139)

The convenient value of R_L which maximizes the oscillator power is

$$R_L = \frac{R_0}{3} \quad \text{or} \quad \frac{R_{\text{in}}}{3} \tag{10.140}$$

The detailed description of oscillation conditions has been given by Guillermo Gonzalez.

There are several oscillator circuit configurations. Two important types are shown below.

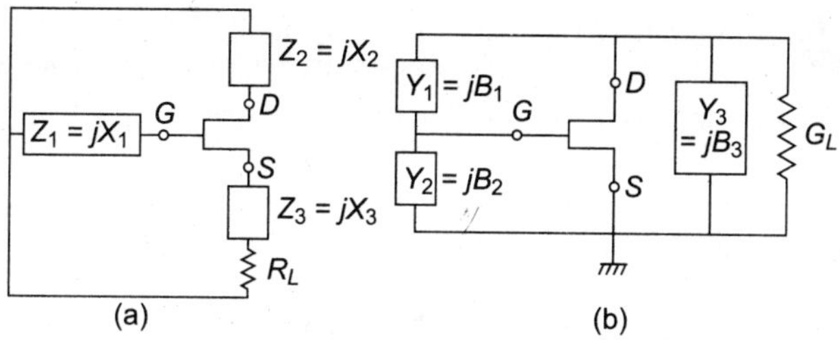

Fig. 10.59 Oscillator circuits: (a) Series-circuit type – Z_1, Z_2, Z_3, are reactive jX (b) Parallel-circuit type – Y_1, Y_2, Y_3 are suceptance jB

A resistive load may be connected in any of these reactive elements—in series with X's; in shunt with B's.

The design procedure for an oscillator is given below:

1. Choose a potentially unstable transistor at oscillation frequency f_0.
2. Design the terminating network to make $|\Gamma_{in}| > 1$. Use series or shunt feedback to increase $|\Gamma_{in}|$.
3. Design the load network to resonate Z_{in} and achieve oscillation condition $R_L = R_0/3$.

Example 10.17 Design an oscillator using a GaAs FET with S-parameters at 10 GHz are

$$S_{11} = 0.9\angle 180°,\ S_{12} = 0.8\angle{-100°},\ S_{21} = 0.9\angle{-160°},$$
$$S_{22} = 0.2\angle 180°.$$

Solution

Stability check

$$\Delta = S_{11}S_{22} - S_{12}S_{21} = 0.9\angle 180° * 0.2\angle 180° - 0.8\angle{-100°} *$$
$$0.9\angle{-160°} = 0.18\angle 360° - 0.72\angle{-260°}$$

$$\therefore \qquad |\Delta| = 0.7719 < 1 \quad \text{and} \quad K = \frac{1 + |\Delta|^2 - |S_{11}| - |S_{22}|^2}{2|S_{12}S_{21}|} = 0.5179 < 1$$

The transistor is potentially unstable at 10 GHz.

The Maximum Output Power

For a transistor oscillator circuit with input-tuning network and load-matching network, the maximum output power of a two-port oscillator with common-source can be approximated by

$$P_{0\,max} = P_{sat}\left(1 - \frac{1}{G} - \frac{\ln G}{G}\right) \qquad (10.141)$$

where P_{sat} = Saturated output power of the transistor in watts

G = Small-signal common source transducer power gain = $|S_{21}|^2$

and the maximum efficient power gain is given by

$$G_{max} = \frac{P_{out}}{P_{in}} = \frac{G-1}{\ln G} \qquad (10.142)$$

Example 10.18 A microwave transistor common-source oscillator is designed with small-signal transducer gain at 10 GHz as 5 dB. The P_{sat} = 1.0 watt: Calculate the maximum efficient oscillator power gain and maximum oscillator power output.

Solution

Given $G = 5 \text{ dB} = 10^{5/10} = 3.16$

$$\therefore \quad G_{max} = \frac{G-1}{\ln G} = \frac{3.16-1}{\ln(3.16)} = \frac{2.16}{1.15} = 1.88 = 2.7 \text{ dB}$$

$$P_{0\,max} = P_{sat}\left(1 - \frac{1}{G} - \frac{\ln G}{G}\right) = 1.0\left(1 - \frac{1}{3.16} - \frac{1.15}{3.16}\right)$$

$$= 0.319 \approx 0.32 \text{ W}$$

Microwave Transistor Oscillator Design

Microwave transistor oscillators are designed by choosing the input and output port terminations in the unstable regions where both the input and output impedances of the transistor circuits will have a negative resistance. The oscillations will occur at a frequency at which the total reactance in the input and output circuits becomes zero. Even if the transistor is stable, it is made unstable by using feedback from the output to the input of the circuit. Any of the standard low frequency oscillation circuits such as the Hertley, Colpitts, or Clapp circuits, are used in which the frequency stability is achieved by using a resonator in either the input or output circuits as part of the feedback loop. Figure 10.60 shows a typical FET oscillator which is stabilized by using a resonator in the input circuit. The magnitude and the phase angle of the input reflection coefficient is controlled by adjusting the coupling spacing *d* between the resonator and the microstrip line and by the length *l* of the input line, respectively. The output is matched by using a stub. Unstability of oscillations is met by using a series capacitance reactance in the common source lead. *RF* chokes are inserted in series with the dc biasing resistances and in shunt with feedback reactance. Si bipolar transistors are used for oscillators at frequencies up to 5 GHz, MESFETs are used from 5 GHz to 40 GHz, and HEMTs are used from 30 GHz to 100 GHz.

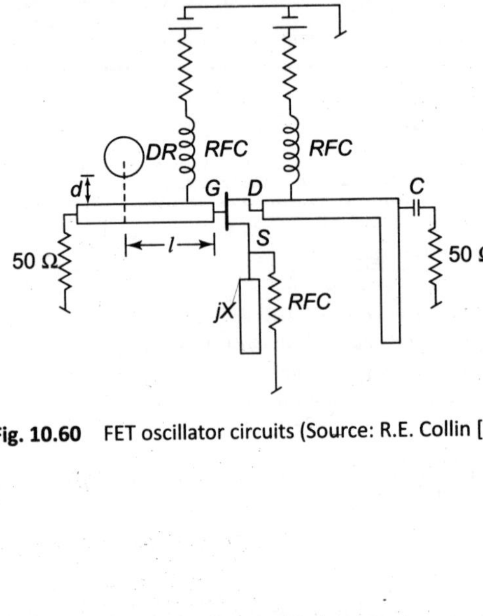

Fig. 10.60 FET oscillator circuits (Source: R.E. Collin [12])

REVIEW QUESTIONS

10.1 The dc current through a point contact diode is 1.5 mA at room temperature. Determine the slope conductance and slope resistance, assuming that $n = 1.4$ and $V_T = 26$ mV.

10.2 A sinusoidal microwave signal with a peak voltage of 50 mV is applied to a microwave diode having $n = 1.4$, $V_T = 26$ mV. $I_0 = 1\mu$A. Calculate the resulting dc current.

10.3 A microwave point contact diode detector has a reverse saturation current of 1μA. Find the detector current for a microwave signal amplitude of 1V at room temperature. $n = 1.4$ and $V_T = 26$ mV.

10.4 (a) A 0.8 V *LO* signal at frequency 10 GHz and a 0.5 V input signal at frequency 9.9 GHz are applied to the above diode when used as a mixer. Calculate the frequency and magnitude of the *IF* signal current.

(b) In a diode mixer circuit the signal frequency is 4 GHz and the local oscillator frequency is 3.93 GHz. Determine the sum frequency, the image frequency, and the *IF*.

10.5 Define and explain the terms conversion loss and noise figure of diode mixer circuits. The conversion loss of a mixer is 5 dB. Find the approximate value of noise figure.

10.6 Explain what is meant by the notation 1-2-3 tripler. When the frequency input applied to a varactor is 3 GHz, determine the idler frequency and the output frequency.

10.7 For a forward bias PIN diode equivalent circuit with $C_p = 0.3$ pF, $C_j = 0.2$ pF, $R_s = 0.3$, $R_j = 0.1$ ohm and $L_p = 0$, show that $(\omega C_j R_s)2 << 1$, and calculate the admittance of the diode at 5 GHz.

10.8 A GaAs Gunn diode has an active region length of 10 μm. If the electron drift velocity is 10^5 m/s, find the natural frequency and the critical voltage. The critical electric field is 3 kV/cm.

10.9 A GaAs Gunn diode oscillator operates at 10 GHz with the drift velocity of electrons = 10^5 m/s. Determine the effective length of the active region. What is the required dc voltage for oscillation? ($E_c = 3.2$ kV/cm)

10.10 A GaAs Gunn diode has a drift length of 10 μm. Determine the intrinsic frequency of the diode oscillations. ($V_d = 10^7$ cm/s).

10.11 A transferred electron diode operates in transit time mode at 10 GHz with the velocity of propagation of 10^7 cm/s. Calculate the approximate thickness of the device.

10.12 The active impedance of a suitably biased Gunn diode is $-10 - j7.5$ ohm at 10 GHz. The diode is to be used in a coaxial line oscillator circuit. Calculate the cavity length required when the characteristic impedance of the coaxial line used is 50 ohms.

10.13 The equivalent circuit parameters for an IMPATT diode oscillator are $R_d = -2$ ohm, $C_j = 0.5$ pF at a breakdown voltage of 80 V with $L_p = 0.55$ nH, and $C_p = 0.3$ pF. The oscillator operates into a resistive load of 2 ohms connected across C_p under a bias current of 80 mA. Determine (a) the resonant frequency, (b) the average power output, (c) the average power input, and (d) the conversion efficiency. *RF* peak current = 1.0A

10.14 A small-signal IMPATT diode amplifier has an R_d of -3 ohm and is operated into a load of 5 ohm using a 3-port circulator. What is the power gain in dB at the resonant frequency?

10.15 A tunnel diode has $R_j = 60$ ohms, $R_s = 9$ ohms, $C_j = 0.6$ pF, and $L_s = 1$ nH. Find the resistive cut-off frequency f_r and the self-resonant frequency f_o.

10.16 A varactor diode has the junction capacitance 0.5 pF with $v = 0$, the barrier potential 1.1 V and $n = 0.3$. Calculate the junction capacitance and the cut-off frequency for a reverse voltage of 0.8 V, if the substrate resistance $R_s = 0.7$ ohm.

10.17 A parametric diode amplifier has $R_s = 4$ ohms, $R_L = R_1 = R_2 = 0$. The characteristic impedance of the amplifier is 50 ohms. Calculate (a) the reflection coefficient, and (b) the power gain.

10.18 A transistor operates at $f = 8.0$ GHz, with $S_{11} = 0.9\ \angle 110°$, $S_{21} = 1.1\ \angle{-30°}$, $S_{22} = 0.55\ \angle{-122°}$. For unilateral operation calculate source and load reflection coefficients for a conjugate matched amplifier.

10.19 Calculate the gain factors G_A, G_T, and G_o for the above amplifier in dB.

10.20 In a bipolar transistor, the emitter charging time is 40 per cent of the total transit time. Calculate the percentage change in the cut-off frequency when the emitter current is increased from 100 mA to 200 mA. All other parameters may be assumed to remain constant.

10.21 *S*-parameters of a microwave transistor are given by $S_{11} = 0.6 \angle -160°$, $S_{12} = 0.045 \angle 16°$, $S_{21} = 2.5 \angle 30°$, $S_{22} = 0.5 \angle -90°$ with 50 ohms reference impedance at source and load ends. The input and output matching networks are designed to produce $\Gamma_s = 0.5 \angle 120°$ and $\Gamma_L = 0.4 \angle 90°$. Determine the transducer gain, available gain and power gain.

10.22 The *S*-parameters of a GaAs FET at 1 GHz in a 50 ohms system are

$$S_{11} = 0.70 \angle -120°, S_{12} = 0, S_{21} = 4 \angle 50°, S_{22} = 0.6 \angle -80°.$$

(a) Find input impedance and G_{TU}

(b) Optimum output termination

(c) Design the input matching network for its max. power gain = 3 dB

REFERENCES

1. Davis, W. A., *Microwave Semiconductor Circuit Design*, Van Nostrand Reinhold. New York, 1984.
2. Liao, S. Y., *Solid-State Microwave Amplifier Design*, John Wiley, New York. 1981.
3. Watson, H. A., *Microwave Semiconductor Devices and Their Circuit Applications*, McGraw-Hill. New York, 1969.
4. Ha, T. T., *Solid-State Microwave Amplifier Design*, John Wiley. New York, 1981.
5. Vendelin, G. D., *Design of Amplifiers and Oscillators by the S-Parameter Method*, New York, John Wiley, 1982.
6. *G.E. Application Note TPD-6104*, "Bulk effect diodes and solid state microwave circuit modules."
7. *Hewlett-Packard Application Note 957-1*, "Broadbanding the shunt PIN diode SPDT switch."
8. *Hewlett-Packard Application Note 962*, "Silicon double-drift IMPATTs for high power CW applications".
9. Pengelly, R. S. and John Wiley and Sons: *Microwave Field-effect Transistor-theory, Design and Applications*, Research Studies Press, Hertfordshire, England, Inc., New York, 1986.
10. Gonzalez, G., *Microwave Transistor Amplifiers, Analysis and Design*. Prentice-Hall, Inc., Englewood Cliffs, N.J., 1984.
11. Gentile, C., *Microwave Amplifiers and Oscillators*, McGraw-Hill Book Company, New York, 1987.
12. Collin, R. E., *Foundation for Microwave Engineering*, Second Edition. McGraw-Hill, Inc., 1982.
13. John, A. Seeger, *Microwave Theory, Components, and Devices*, Prentice-Hall, N.J., 1986.
14. Dennis, R., *Microwave Technology*, Prentice-Hall, N.J., 1986.
15. Bahl, I. J. and P. Bhatia, *Microwave Solid-State Circuit Design*, John Wiley and Sons. 1988.

11 APPLICATIONS OF MICROWAVES

INTRODUCTION 11.1

The microwave devices and circuits described in the previous chapter have widespread practical applications in the development of microwave antennas, radar systems and communication systems. During World War II and shortly afterwards, microwave engineering played a very important role in the development of microwave antennas and high-resolution radar systems capable of detecting and locating enemy planes and ships. Even today, radars of many varied forms, such as missile-tracking radars, fire-control radars, weather-detecting radars, air-traffic-control radars, etc., represent a major use of microwave frequencies. In recent years, microwave frequencies have also come into widespread use in communication links, generally referred to as microwave links between two ground stations directly or via satellites for communication of voice, picture, data and other information. In the modern age, microwaves are increasingly used in domestic, industrial, scientific and medical applications. Some of these applications are described below.

MICROWAVE RADAR SYSTEMS 11.2

Radio Detection and Ranging, called RADAR, is an electronic system which uses an antenna for signal transmission and signal reception and a Trans-receiver (T_X/R_X) for detecting the presence and location of objects and also for tracking its path. The presence of objects and their coordinates, i.e. range R (distance) and direction (angular coordinates θ, φ), are determined from the transmitted signal and the received return signal after the incident signal is scattered from the object, called **target**, as explained in Fig.11.1. The advantages of the use of microwaves in radar are that, very compact, high gain and highly directional antennas can be used. The radar is classified in two ways—**primary radar** and **secondary radar**. In the primary radar, a microwave signal is transmitted and is scattered by the target to be detected.

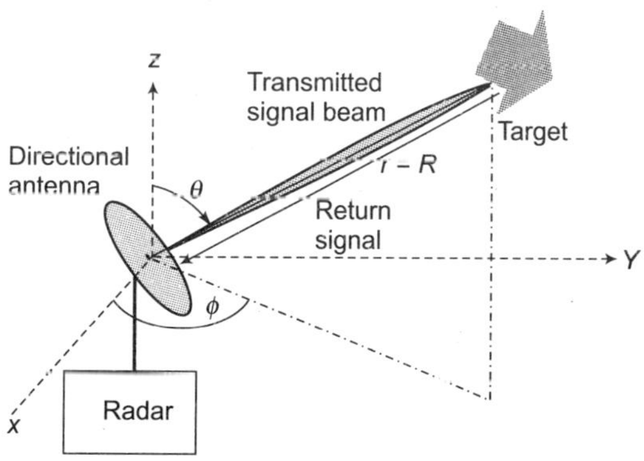

Fig. 11.1 Basic radar system

The reflected signal is received by the radar. From the knowledge of the time between the transmission and reception and the velocity of light in space, the range of the target can be calculated. The application of a primary radar is the detection of position and flight path of aircrafts, ships and missiles.

In the secondary radar, the target in space receives the radar signal and transmits a coded signal for its identification as a friend. The various frequency bands used in radars are given in Table 11.1.

Table 11.1 *Radar frequency bands*

Band Name	*Nominal Frequency Range*	
HF	3–30	MHz
VHF	30–300	MHz
UHF	300–1000	MHz
L	1–2	GHz
S	2–4	GHz
C	4–8	GHz
X	8–12	GHz
Ku	12–18	GHz
K	18– 27	GHz
Ka	27–40	GHz
mm	40–3000	GHz

The Radar Equation 11.2.1

In general, the radar transmitter sends a sinusoidal *RF* wave from a directional antenna. After hitting the target, the signal is scattered back by the target towards the common T_X/R_X antenna (Fig.11.1). The scattered signal is received by the T_X/R_X antenna and processed in the receiver to read the coordinates of the target. If P_t is the transmitted power (W), $G_t = G_r = G$ is the T_X/R_X common antenna power gain relative to an isotropic radiator, and R is the distance (m) of the target from the radar antenna, the power density at the target is given by

$$P_d = \frac{P_t G}{4\pi R^2} \text{ W/m}^2 \tag{11.1}$$

The amount of power that will be scattered back to the radar is given by

$$P_{\text{ref}} = \sigma P_d \text{ W} \tag{11.2}$$

where σ represents the effective echoing area, called the **Radar Cross Section (RCS)** of the target which converts the incident power density into scattered power. If A_{eff} represents the effective area of the receiving antenna, the received power from the scattered component is

$$P_r = \frac{P_{\text{ref}} A_{\text{eff}}}{4\pi R^2} \text{ W} \tag{11.3}$$

Since according to antenna theory, $A_{\text{eff}} = \dfrac{G\lambda_0^2}{4\pi}$, where λ_0 is the free-space wavelength of the signal at the frequency f_0, from (11.1)–(11.3), the equation for the received power becomes

$$P_r = \frac{P_t G^2 \lambda_0^2 \sigma}{(4\pi)^3 R^4} = \frac{p_t G^2}{\left(\dfrac{4\pi R}{\lambda_0}\right)^4} \frac{4\pi\sigma}{\lambda_0^2} \text{ W} \tag{11.4}$$

which means that the received power drops off very rapidly with increase in target range as R^{-4}, increases with the antenna gain as G^2 and wavelength as λ_0^2. From the last equation, the target range is found as

$$R = \left[\frac{P_t G^2 \lambda_0^2 \sigma}{P_r (4\pi)^3}\right]^{1/4} \text{ m} \tag{11.5}$$

Since the minimum possible received power corresponding to the maximum range detection is limited by the receiver noise, the receiver noise power N is important in the above equation:

$$N = FkT_0 B \text{ W} \tag{11.6}$$

where F is the noise factor of the receiver, k is Boltzmann's constant, T_0 is the room temperature in Kelvin, and B is the bandwidth of the receiver. The maximum detectable range R_{max} of the radar corresponds to the minimum detectable signal power S_{min} of the receiver. The signal-to-noise ratio at the input to the radar receiver for the maximum range R_{max} is, therefore,

$$\frac{S_{min}}{N} = \frac{P_{r\min}}{N} = \frac{P_t G^2 \lambda_0^2 \sigma}{(4\pi)^3 R_{max}^4 FkT_0 B} \approx 1 \tag{11.7}$$

The maximum range is obtained from Eq. (11.7) as

$$R_{max} = \left[\frac{P_t G^2 \lambda_0^2 \sigma}{S_{min} (4\pi)^3}\right]^{1/4} = \left[\frac{P_t G^2 c^2 \sigma}{S_{min} f_0^2 (4\pi)^3}\right]^{1/4} \text{ m} \tag{11.8}$$

Navigational distance R_{max} is often measured in international **nautical miles** where 1 nautical mile = 1852 m, c = velocity of light = 3×10^8 m/s.

Equation (11.8) is the basic radar equation describing all the parameters of the radar system which affect performance. However, no account is taken of the various losses, such as atmospheric attenuation and scattering of the radio waves by rain or precipitation.

Example 11.1 A 1 kW, 3 GHz radar uses a single antenna with a gain of 30 dB. The receiver has noise bandwidth of 1 kHz and a noise factor of 5 dB. A target of echoing area of 10 m^2 at a range of 10 nautical miles is to be detected. Calculate the minimum S/N.

Solution

$$\lambda_0 = 30/3 = 10 \text{ cm} = 0.1 \text{ m}$$

$$G = 30 \text{ dB} = 10^3 = 1000$$

$$P_t = 1000 \text{ W}, F = 5 \text{ dB} \rightarrow 10^{5/10} = 3.16$$

$$(4\pi)^3 kT_0 = 7.9433 \times 10^{-18}$$

$$R = 10\ n \text{ mile} = 18520 \text{ m}$$

$$B_n = 1 \text{ kHz} = 1000 \text{ Hz}$$

$$S/N = \frac{P_T G^2 \lambda_0^2 \sigma}{(4\pi)^3 R^4 FkT_0 B_n} = \frac{10^3 \times 10^6 \times 0.01 \times 10}{7.9433 \times 10^{-18} \times (18520)^4 \times 3.16 \times 10^3}$$

$$= \frac{10^5 \times 10^{18}}{7.9433 \times (18520)^4 \times 3.16} = \frac{10^5}{2.953} = 45.3 \text{ dB}$$

Example 11.2 A radar system operates at 10 GHz with a common antenna with a gain of 30 dB. The receiver has a bandwidth of 1 kHz and the noise factor is 5 dB. The transmitted power is 1 kW and the target echoing area is 10 m^2. Calculate its range for $S/N = 10$.

Solution

$$\lambda_0 = c/f = 30/10 = 3 \text{ cm} = 0.03 \text{ m} = 10 \log (0.03) = -15.23 \text{ dB meter}$$
$$B_n = 1 \text{ kHz} = 1000 \text{ Hz} = 10 \log 1000 = 30 \text{ dB Hz}$$
$$P_t = 1 \text{ kW} = 1000 \text{ W} = 10 \log 1000 = 30 \text{ dB W}$$
$$G(\text{dB}) = 30 \text{ dB}$$
$$\sigma = 10 \text{ m}^2 = 10 \log 10 = 10 \text{ dB Sq.m}$$
$$F = 5 \text{ dB}$$
$$S/N = 10 \log 10 = 10 \text{ dB}$$

Therefore,

$$R_0 (\text{dB}) = \frac{1}{4} [30 + 2 \times 30 - 2 \times 15.23 + 10 + 171 - 5 - 30 - 10]$$

$$= \frac{1}{4} [271 - 75.46] = 48.89 \text{ dB meter}$$

or, $$R_0 = 10^{(48.89/10)} \text{ m} = 77.45 \text{ km}$$

Duplexer 11.2.2

A duplexer is a circuit which couples the transmitter and receiver to the antenna while producing isolation between the transmitter and receiver. There are two types of duplexers possible, one using PIN switches and the other using circulators as shown in Fig. 11.2 (a)–(c).

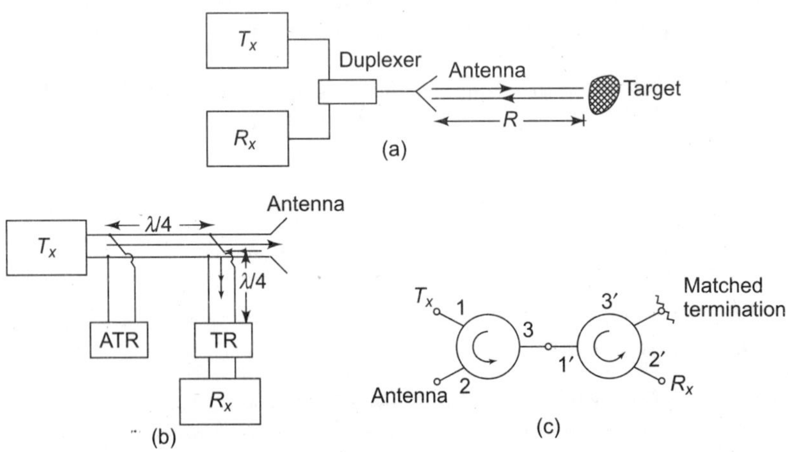

Fig. 11.2 (a) Basic radar system with duplexer (b) PIN switch duplexer (c) Ferrite circulator duplexer

In Fig. 11.2(b), the duplexer uses Transmit-Receive (TR) and anti-TR PIN diode switches. Under transmit condition, the TR and ATR switches are closed or short-circuited. The l/4 sections transform these short circuits to open circuits across the main line to allow power flow from the transmitter to the antenna, while the receiver is protected from high-power output of the transmitter by the TR short circuit. In the receive mode, both the switches are open-circuited. In the ATR branch, this is transformed to short circuit across the main line. This prevents any received signal from entering the transmitter. The short circuit on the main line at the ATR connection point is transformed to an open circuit on the main line at the TR connection point and a receive line appears through the connection to the receiver to receive the signal. Since the operations are controlled by l/4 lines, this arrangement results in a narrowband system. In Fig. 11.2(c), two ferrite circulators do the function of the duplexer over a broader bandwidth by a circulating action.

Pulsed Radar 11.2.3

When a target is stationary, the CW sinusoidal microwave signal returning from the target cannot provide precise radar information since the time of return is difficult to measure with respect to time of transmission of analog signals. Moreover, there would not be any frequency changes to give some information. Hence, short RF pulses are used for the determination of the range of a stationary target as explained in Fig. 11.3.

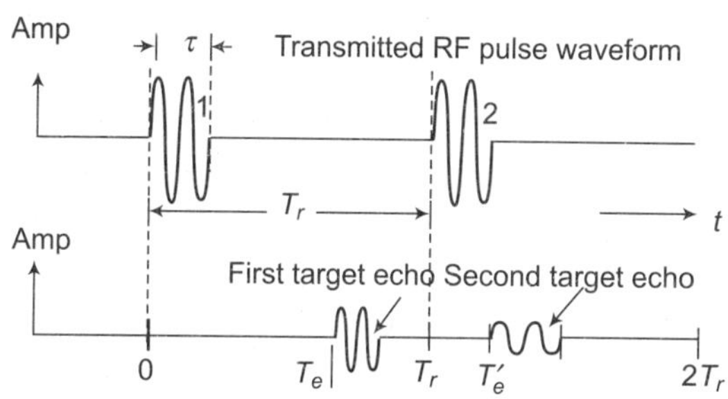

Fig. 11.3 Pulsed radar waveform and detection principle

Here, an RF carrier at the frequency f_0 is pulsed amplitude modulated with modulating pulse repetition frequency $f_r = 1/T_r$ and pulse width τ before transmission.

Let us consider two successive transmitted pulses 1 and 2; the time of receiving echos for the pulse 1 is T_e and that for the pulse 2 is T_e'. For unambiguous detection of the target, the echo time T_e must be less than the pulse repetition period T_r. If $T_e > T_r$, echo would be confused with the reference transmitted pulse number 1 or 2 and the range will be ambiguous.

Range Computation

The **maximum unambiguous range** R_{max} is determined in term of maximum echo time ($T_e = T_r$):

$$R_{max} = \frac{cT_r}{2} \tag{11.9}$$

where c is the velocity of light and the factor ½ comes due to distance travel by the signal for reaching the target and returning back to the radar.

The **minimum range R_{min}** is determined by the pulse width τ because if an echo arrived within time $t < \tau$, it cannot be compared with the time of the transmitted pulse. Hence,

$$R_{max} = \frac{c\tau}{2} \tag{11.10}$$

When two targets are present side by side, they can be resolved only if the second echo arrived after a time τ from the arrival time of the first echo. Therefore, for target resolution, the separation between the pulses should be greater than or equal to τ.

Power Consideration

For the RF pulse of width τ, with period T_r and current maximum I_m, the peak power $P_m = I_m^2 R_r$, where R_r is the radiation resistance of the antenna. The peak average power over a pulse width τ is

$$P = \frac{P_m}{2} = \frac{I_m^2 R_r}{2} \tag{11.11}$$

The pulse radar waveform and the instantaneous power wave are shown in Fig. 11.4. Here, the spectrum of RF pulses contains infinite side frequencies $f_0 \pm nf_r$, $n = 1, 2, 3, \ldots$

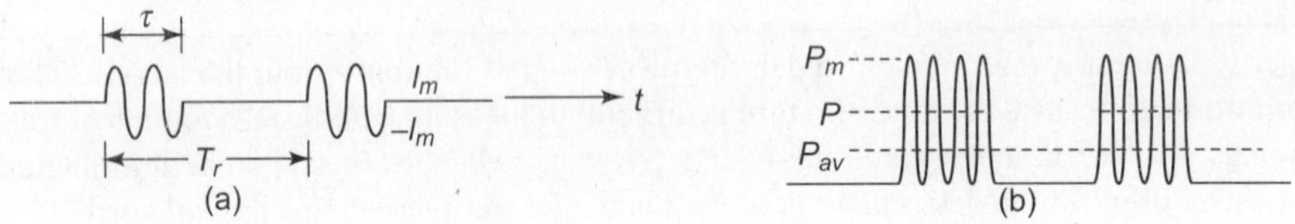

Fig. 11.4 (a) Pulsed radar waveform (b) Corresponding instantaneous power waveform

If $f_r = 1/T_r$ is low, most of the side frequencies will be close to f_0 and the complete spectrum can be received by the receiver bandwidth and the **peak average power *P*** will be used in the radar equation.

If f_r is high, the receiver may receive only the carrier and closer side frequencies within its bandwidth. The **average power in the carrier component** of the modulated wave is

$$P_0 = \frac{P_{av}\tau}{T} \tag{11.12}$$

Here,

$$P_{av} = \frac{P\tau}{T} = \text{True average power} \tag{11.13}$$

Here, τ/T is called the duty cycle of the *RF* pulse. This average power P_0 must be used in the radar equation when PRF is f_r and is high.

CW Radar 11.2.4

CW radar is an analog system in which radar signals are continuous microwaves and frequency-modulated CW. In the radar equation, it is assumed that the target is stationary. This the case for targets such as clouds, buildings and hilltops, etc. In many applications, targets are moving with some velocity. In addition to the position of the target, its velocity and the direction of movement can be determined from the well-known Doppler effect.

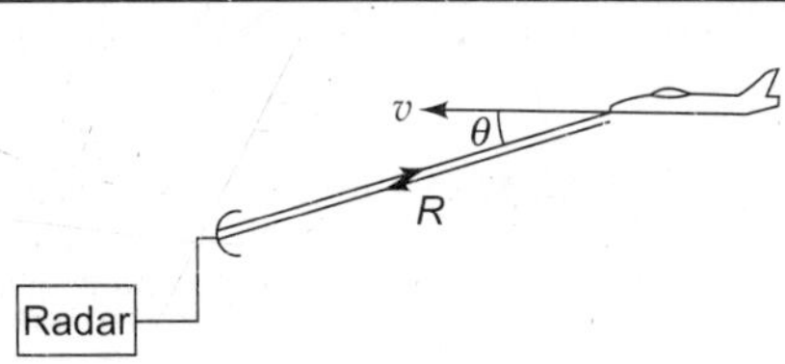

Fig. 11.5 Doppler effect

CW Doppler Radar

When a moving target is detected by a continuous-wave radar signal at the frequency f_0, frequency shift occurs in the return signal given by the Doppler frequency shift f_d

$$f_d = \frac{2\frac{dR}{dt}}{\lambda_o} = \frac{2v\cos\theta}{\lambda_o} \tag{11.14}$$

where v is the relative velocity of the target w.r.t. radar antenna and making an angle θ with the position vector R of the target with respect to the radar antenna, and dR/dt representing the range rate. When the target is moving towards the radar ($0 \le \theta \le \pi/2$) receiver, the signal frequency f_0 increases by f_d and when it is moving away from the receiver $\pi/2 < \theta \le \pi$, the frequency decreases by f_d. If the velocity is perpendicular ($\theta = \pi/2$) to the target, there is no change in the frequency, or the Doppler frequency is zero. By measuring the Doppler frequency shift and noting the polarity of f_d, the range rate for the target can be determined (Fig. 11.5).

The Doppler frequency shift can be utilized to design low-cost radars for intruder alarms, speed detectors and traffic-signal control. Figure 11.6 shows a basic block diagram of a CW Doppler radar. A circulator is used as the duplexer with a single antenna for transmission and reception. A Doppler mixer produces the signal output at the Doppler frequency by mixing the received signal with any signal that leaked through the circulator due to poor isolation.

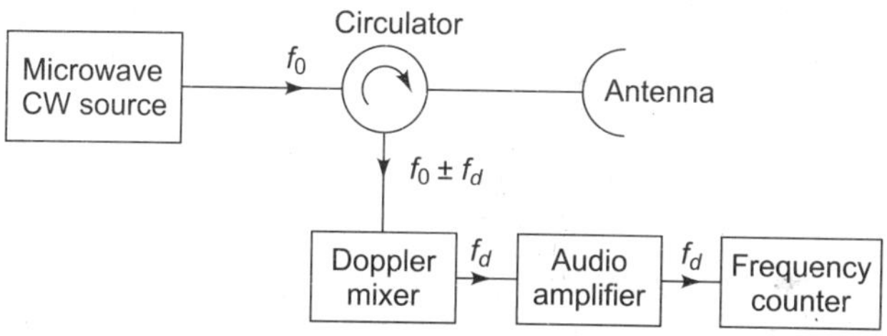

Fig. 11.6 CW Doppler radar

FMCW Radar

❑ ***FM Radar*** The Frequency Modulated Continuous Wave (FMCW) radar uses Doppler frequency information for the detection of a moving target. In this radar, the carrier signal at a frequency f_0 is frequency modulated by the frequency $f_m(t)$ such that the instantaneous FM signal frequency of the transmitted signal is given by

$$f_t = f_0 + f_m(t) \tag{11.15}$$

A basic block diagram of the FMCW radar is shown in Fig. 11.7. If T_e is the time of the echo signal after transmission, the received signal frequency is given by

$$f_r = f_0 + f_m(t - T_e) \tag{11.16}$$

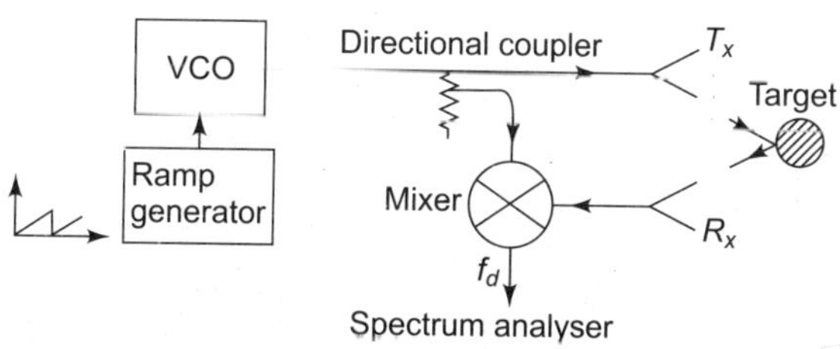

Fig. 11.7 Basic block diagram of an FMCW radar

Therefore, the difference between the transmitted and the received signal frequencies are

$$f_t - f_r = f_d(t) = f_m(t) - f_m(t - T_e) \tag{11.17}$$

The transmitted and received signal waveforms are shown in Fig.11.8. The commonly used frequency modulation wave shape is triangular. The slope of the line in any given time interval is

$$m = f_d/T_e \tag{11.18}$$

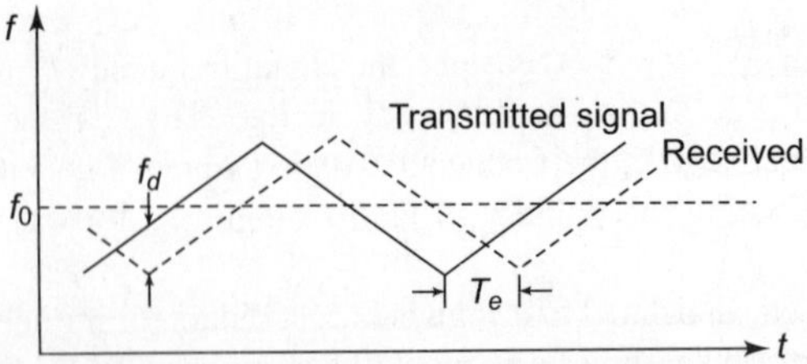

Fig. 11.8 FMCW radar waveforms

The range of the target is, therefore, expressed by

$$R = \frac{cT_e}{2} = \frac{cf_d}{2m} \tag{11.19}$$

The advantages of the FMCW radar are the following:

1. It operates at a continuous power level that is relatively low as compared to pulse radar.
2. It is used mainly for short ranges because its minimum range is not limited.

❑ ***FM Doppler Radar*** The FMCW radar can also be used for the determination of the velocity and range of a moving target. When a triangular frequency modulation waveform returns from a moving target with a velocity v, the frequency difference between the transmitted and the received signals contain Doppler frequency shift

$$f_{\text{dop}} = \frac{2v\cos\theta}{\lambda_o} \tag{11.20}$$

where θ is the angle between the direction of velocity of the target and the radial vector of the target with respect to the radar antenna as the origin. Thus, the T_X and R_X signal waveforms are as shown in Fig. 11.9.

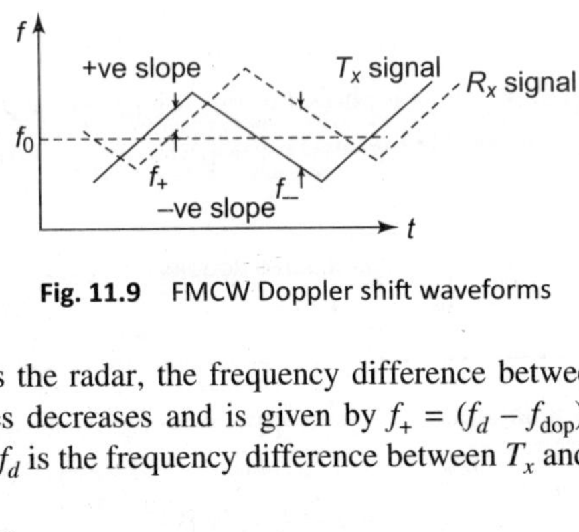

Fig. 11.9 FMCW Doppler shift waveforms

If the target moves towards the radar, the frequency difference between the transmitted and received signals over the positive slopes decreases and is given by $f_+ = (f_d - f_{\text{dop}})$ and over the negative slope it increases $(f_- = f_d + f_{\text{dop}})$ where f_d is the frequency difference between T_x and R_x signals for stationary target. Hence

$$f_d = (f_+ + f_-)/2 \tag{11.21}$$

$$f_{\text{dop}} = (f_- - f_+)/2 \tag{11.22}$$

The range is given by $R = cf_d/2a$

$$= c(f_+ + f_-)/4a \tag{11.23}$$

where a is the slope of the line forming the triangular waves.

Now $$f_{\text{dop}} = \frac{2(dR/dt)}{\lambda_0} \tag{11.24}$$

or, $$v_r = \frac{dR}{dt} = \frac{\lambda_0 f_{dop}}{2} = \frac{\lambda_0}{4}(f_- - f_+) \quad (11.25)$$

Thus, by measuring the frequency shifts over positive and negative slopes, the range and range rate can be found.

An FMCW Doppler radar is used to find the altitude of aircraft, speed of vehicles, and by using a transmitter power of a few watts, to detect road vehicles ahead and give a warning if the close target speed or distance is potentially dangerous.

Tracking Radars 11.2.5

The function of a tracking radar is to keep the radar antenna beam towards the target all the time while the target is moving and process the return signal for finding the target coordinates for taking necessary action automatically.

For scanning, detection, finding range and tracking a target, radar antennas require different beam shapes. For scanning the space above ground, a vertical fan beam is used which will have a narrow beam in the azimuthal plane and a broad beam width in the elevation plane as shown in Fig. 11.10(a).This pattern will be achieved by a truncated paraboloid reflector or a circular paraboloid reflector which is narrow in the vertical plane and wide in the horizontal plane as shown in Fig. 11.10(b). Such antennas are seen in the airports for surveillance of aircraft.

This type of antenna system is also used in height-finding equipment if the reflector is rotated 90 degrees such that the reflector is narrow in the horizontal plane and wide in the vertical plane, producing a beam that is wide in the horizontal plane and narrow in the vertical plane. In this application, the beam of the height-finding radar is a horizontal fan-beam pattern.

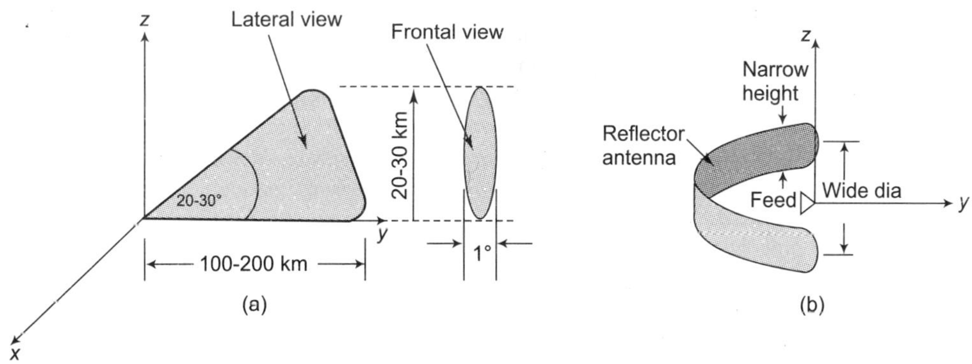

Fig. 11.10 A fan beam: (a) Generated by a truncated paraboloid reflector (b) Generated by a circular paraboloid reflector

For **detecting a target** (aircraft or other objects) and finding its range in the sky, normal pencil-beam antennas are desirable. However, for accurately tracking the target, special antenna arrangements are used, called **monopulse antenna** and the associated radar is called **monopulse radar** as described below.

Monopulse Radar

Monopulse implies that with a single pulse, the antenna produces angle information of an object in the sky. Monopulse radars are used for tracking targets in the sky. A target will be seen by the radar from the moment it enters the main antenna beam. In transmission (T_X) mode, the antenna system produces radiation pattern represented by the black centre beam, which is called the Σ or **sum pattern**. In reception (R_X) mode, the received signals calculate Σ and the **difference Δ signals** as shown in Fig. 11.11(a). Both signals are then compared and their difference is used to estimate the angle of the target accurately.

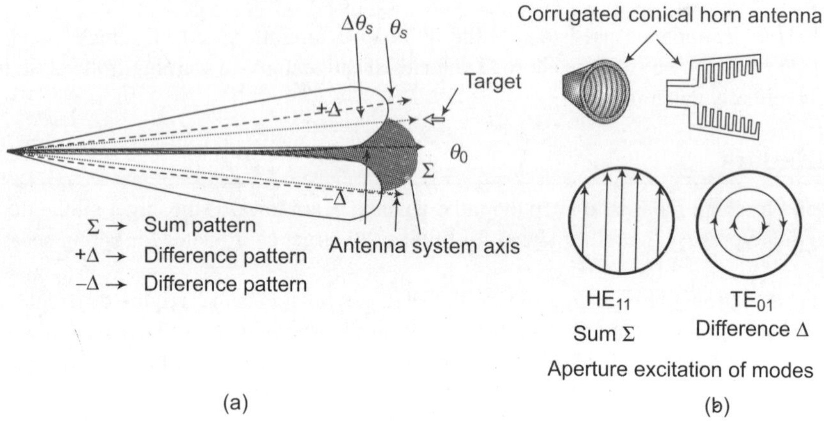

Fig. 11.11 Monopulse antenna radiation: patterns and excitations

The sum signal can be generated by exciting a circular conical corrugated horn antenna in hybrid HE_{11} mode and the difference signal with bore site null can be generated by exciting the same antenna in TE_{01} mode as explained in Fig. 11.11(b). Monopulse radar systems are complex, high-cost equipment. However, a simplified basic monopulse technique is described in this section to explain how the RF part of the radar system is operating and how the angle-measurement accuracy can be improved.

The main advantage of a monopulse radar system in comparison to standard angle-measurement methods is that it is not influenced by amplitude fluctuations of the target echo. This is because the angle information is acquired by comparing signals received by multiple simultaneous beams at multiple channels and produced by a single echo pulse. If the echo amplitude changes, it changes in the same way in all receiver channels.

There are three main monopulse techniques: 1) amplitude-comparison monopulse, 2) phase-comparison monopulse, and 3) the combination of the amplitude and phase-comparison monopulse. In this book, amplitude-comparison monopulse system with sum-and-difference angle-detection system is described in Fig. 11.12.

In this system, the sum-beam direction is θ_0 which is the null or intersection of the difference beams. The difference beams are separated by an angle θ_s as shown in Fig. 11.11(a). If $\Delta\theta_s$ is the angle information of a target with respect to the system reference angle θ_0, $\Delta\theta_s$ is called the error signal for angle measurements. The sum pattern is used in transmission, while both sum and difference patterns are used in reception. Three identical circulators act as T/R switches to maintain the same phase relationship in both receiver channels. The received sum signal is used for target detection, range measurement, and as phase reference for determining the sign of angle-error measurement. The magnitude of the angle error, $\Delta\theta_s$, is determined from the difference signal. For simultaneous angle measurement in azimuth and elevation, the system has two pairs of antennas. It is important that the two antennas are identical, and are positioned symmetrically with respect to the antenna system axis. The antennas must also have reasonably high gain (>~ 10 dB) to improve the overall system performance.

The signals at the antenna outputs are coherently added and subtracted by using a hybrid junction such as a Magic-T or hybrid ring. The hybrid junction is the main component for simultaneously obtaining two radiation patterns. A hybrid ring is shown in Fig.11.12(b) where ports 1 and 2 are the two inputs and ports 3 and 4 are two outputs. Input at the port 1 reaches the port 4 in two paths with same phase ($3\lambda/4$) clockwise

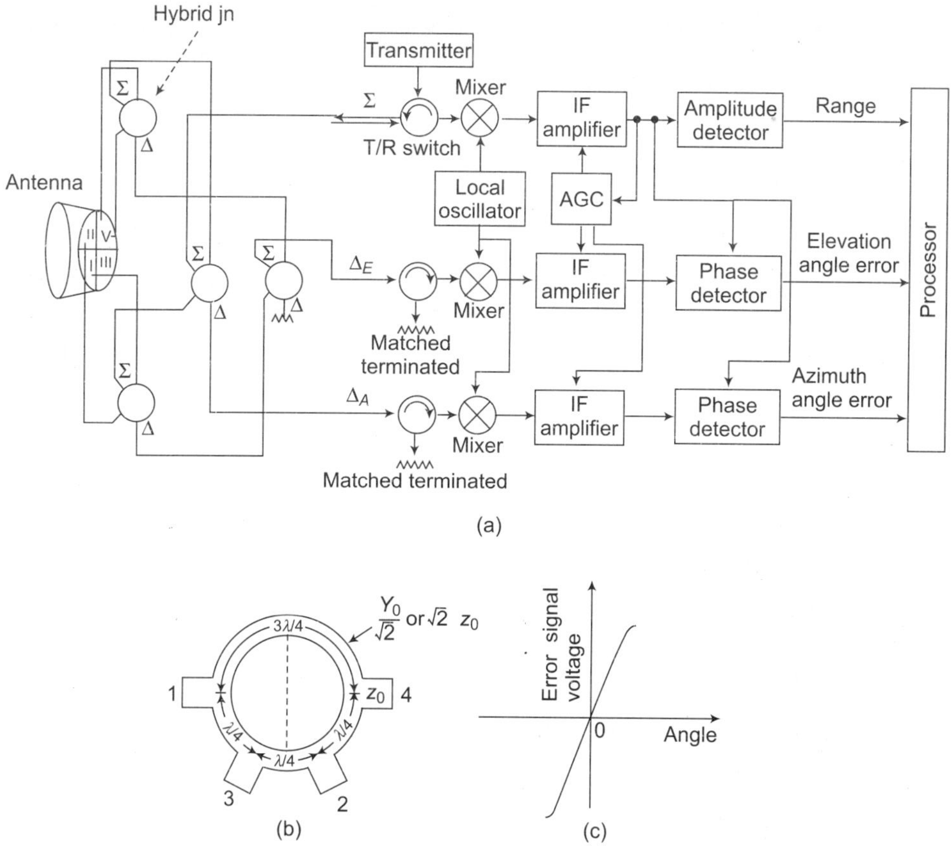

Fig. 11.12 (a) Block diagram of the monopulse radar system (b) Hybrid Junction (c) Error signal vs angle

and counterclockwise and they are added up at this port. Same signal input at 1 reaches the other output port 3 in two paths with same phase and is also added up at 3 (one $5\lambda/4$ and other $\lambda/4$). On the other hand, input at 1 reaches the port 2 in two paths with opposite phase and is cancelled at 2. Similarly, input at the port 2 will appear at 3 and 4 but not at the port 1. Therefore, ports 1 and 2 are isolated. Two different inputs at 1 and 2 will be out of phase at 4 so that the output at 4 is the difference of input signals, but in phase at 3 yields sum signals of inputs 1 and 2. This can be expressed in terms of S-matrix given by

$$S = \frac{1}{\sqrt{2}}\begin{vmatrix} 0 & 0 & 1 & 1 \\ 0 & 0 & 1 & -1 \\ 1 & 1 & 0 & 0 \\ 1 & -1 & 0 & 0 \end{vmatrix} \tag{11.26}$$

With reference to Fig. 11.12, the sum pattern is obtained from I + II + III + IV; the azimuth difference pattern is found from (I + II)–(III + IV); and the elevation difference pattern is from (II + IV)–(I + III) [Skolnik].

For tracking the target, the radar antenna is automatically steered to make the error signal zero by means of signal processing and feedback mechanism to antenna positioner. The nature of variation of error signal with the angle of the target is shown in Fig. 11.12(c).

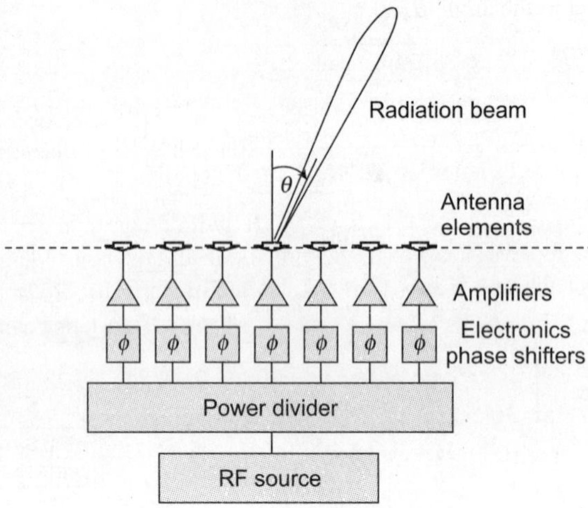

Fig. 11.13 Phased-array radar

The monopulse patterns can also be generated from a two-dimensional planar antenna array by dividing it into four quadrants. The elements are all steered together electrically as shown in the one-dimensional Fig. 11.13. The target is illuminated by all four quadrants equally. These sectorized antenna arrays are placed symmetrically in the focal plane on each side of the bore-sight axis of the radar antenna. In transmission (T_X) mode, both sides will be fed in phase and the Σ radiation pattern is generated. In reception (R_X) mode, the received signals calculate Σ and the Δ signals. Both signals are then compared and their difference is used to estimate the angle of the target very accurately. Both azimuth and elevation angles can be determined.

Radar Cross Section 11.2.6

Radar cross section of a target is defined by

$$A_\sigma = 4\pi \frac{P_{rr}}{P_{id}} \tag{11.27}$$

Here, P_{rr} = Power re-radiated by the target per unit solid angle

P_{id} = Power incident on the target per unit volume

Target scattered power received by the radar can be expressed by

$$P_r(\theta) = \frac{P_t G A_e A_\sigma(\theta)}{(4\pi R^2)^2} \tag{11.28}$$

where, P_t = Transmitted power

G = Gain of antenna relative to an isotropic radiator

A_e = Effective area of the receiving antenna

R = Distance of the target

θ = Aspect angle of the target which is the angle of direction of re-radiated power with respect to the line joining the T_X and the centre of the target

When all the factors remain constant in above equation

$$A_\sigma(\theta) = KP_r(\theta) \tag{11.29}$$

where K is a constant.

Normalizing with respect to the angle $\theta = 0$,

$$\frac{A_\sigma(\theta)}{A_\sigma(0)} = \frac{P_r(\theta)}{P_r(0)} \tag{11.30}$$

$A_\sigma(\theta)$ is a function of the aspect angle θ and hence, P_r is a function of θ. By measuring P_r for different aspect angles, the normalized radar cross section of the target can be obtained for different aspects.

Example 11.3 A radar system operates at 10 GHz with a common antenna with a gain of 30 dB. The receiver has a bandwidth of 1 kHz and the noise factor is 5 dB. The transmitted power is 1 kW and the target echoing area is 10 m^2. Calculate its range for $S/N = 10$.

Solution

$$\lambda_0 = c/f = 30/10 = 3 \text{ cm} = 0.03 \text{ m} = 10 \log (0.03) = -15.23 \text{ dB metre}$$
$$B_n = 1 \text{ kHz} = 1000 \text{ Hz} = 10 \log 1000 = 30 \text{ dB Hz}$$
$$P_t = 1 \text{ kW} = 1000 \text{ W} = 10 \log 1000 = 30 \text{ dB W}$$
$$G(\text{dB}) = 30 \text{ dB}$$
$$s = 10 \text{ m}^2 = 10 \log 10 = 10 \text{ dB sq. m}$$
$$F = 5 \text{ dB}$$
$$S/N = 10 \log 10 = 10 \text{ dB}$$

Therefore,

$$R_0 \text{ (dB)} = 48.89 \text{ dB metre}$$

or,

$$R_0 = 10^{(48.89/10)} \text{ m} = 77.45 \text{ km}$$

MICROWAVE COMMUNICATION SYSTEMS 11.3

Microwaves are used in wireless communication to accommodate an increased number of channels with a high signal-to-noise power ratio. Since the bandwidth can be a small fraction of the microwave carrier frequency f_c, by using a higher carrier frequency, the overall system bandwidth can be increased, and more distinct channels can be transmitted at microwave frequencies than at much lower frequencies. Moreover, high-resolution radiation beams (narrow) can be generated with relatively smaller size antennas and unlike UHF/VHF systems, atmospheric noise and human-made interferences caused by electric transient, automobile ignition system, etc., are negligibly small. Since microwaves penetrate the ionosphere and any possibility of ground-reflected waves and surface waves suffer heavy attenuation, the microwave communication is Line-Of-Sight (LOS) communication. In general, a microwave communication system requires a transmit-receive equipment at each end for the purpose of transmitting the base-band signal to the microwave carrier and receiving the base-band signal from the microwave carrier. The base-band signal is usually a multiplexed signal, carrying a number of individual telephone signals, video signals and data. Microwave communication systems can be broadly classified as terrestrial systems and satellite systems.

Terrestrial Systems 11.3.1

Since ground-based LOS distance is limited due to the attenuation of microwaves caused by the geographical profile, trees, buildings, etc., long-distance microwave communication is achieved through a multisection radio relay system. It consists of two terminal stations and a number of repeater stations spaced at intervals of about 32–80 km, each of which amplifies the microwave signals to make up for transmission losses in space and retransmits again for the next stage. A basic block diagram of the terrestrial microwave communication system is shown in Fig. 11.14.

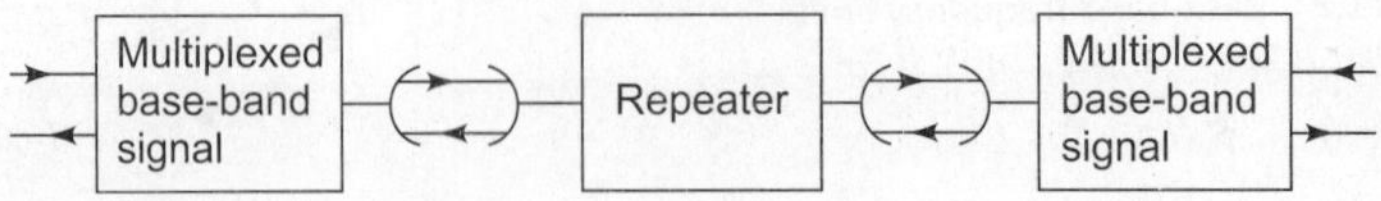

Fig. 11.14 Basic block diagram of a terrestrial microwave communication system

The systems may employ analog or digital modulation. In analog systems (FDM/FM), the information signals are Frequency-Division Multiplexed (FDM) to form the base-band signal which is then Frequency Modulated (FM) onto the microwave carrier for transmission through an antenna. In digital systems (TDM/PSK), the information signals are Time-Division Multiplexed (TDM) to form the base-band signal which is then Phase Modulated (PM) by Phase Shift Keying (PSK) onto the microwave carrier.

Terminal Systems

FDM is more commonly used in that a number of telephone channels (300 Hz–3.4 kHz) are transmitted on the same microwave carrier by FM. The telephone channels are arranged side by side in a base band at intervals of 4 kHz as shown in Fig. 11.15. The base-band frequency limits are given in Table 11.2.

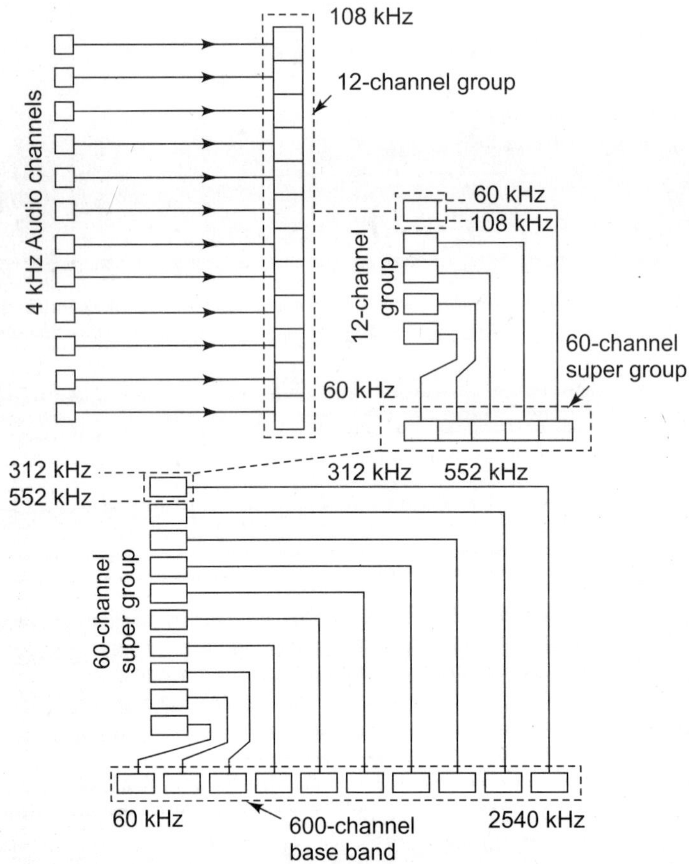

Fig. 11.15 FDM multichannel telephone signals

Table 11.2 Base-band frequency limits

Number of channels	*Base-band frequency limit (kHz)*
12/1 group	12–60 or 60–108
24/2 group	12–108
60/1 super group	12–252 or 60–300
120/2 super group	12–552 or 60–552
240/4 super group	60–1052
600/10 super group	60–2540

In the terminal system, each of the multichannel base-band inputs first modulates a 70 MHz IF carrier and then this IF signal is upconverted to a microwave carrier. All microwave carriers f_1, f_2, ... are passed to a branching filter to form an FDM signal for transmission through a polarization filter and antenna as shown in Fig. 11.16.

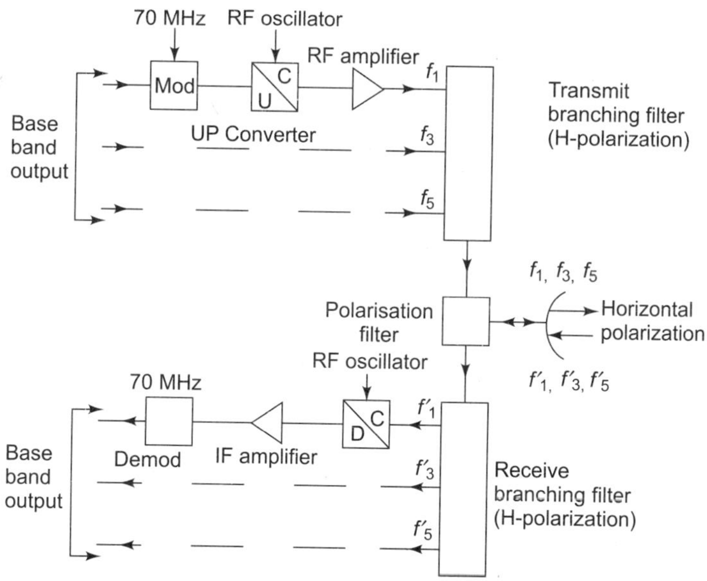

Fig. 11.16 Block diagram of a terminal system

The multicarrier received signal is received by the common antenna and passed through the polarization filter to receive the branching filter. The output of the branching filter consists of a multichannel carrier f_1', f_2', ..., which is downconverted at each separate channel to 70 MHz IF and then demodulated to the base-band signal.

Repeaters

The block diagram of a repeater system is shown in Fig. 11.17. Incoming carriers f_1, f_2, ...,polarized horizontally, are received from the left side by the antenna and converted to 70 MHz IF. The IF signals are upconverted to different microwave carriers f_1', f_2', ..., and retransmitted on the next hop after passing through the branching and polarization filter as vertically polarized signals from the right-side antenna.

Similarly, the incoming vertically polarized signal $f_1, f_2, ...,$ from the right-side is converted to a horizontally polarized signal carrier $f_1', f_2', ...,$ and retransmitted from the left-side antenna for two-way communications. The signals at two sides are made oppositively polarized in order to minimize direct coupling of the transmitted signals on the other side.

Example 11.4 A 30 km long microwave terrestrial link is operating at 4 GHz with a radiated power of 100 W through a parabolic disk having a maximum gain of 50 dB. The receiver uses a similar antenna. Find the free space loss and the received power.

Solution

$$P_t(\text{dBm}) = 10 \log P_t(\text{mW}) = 10 \log 100 \times 10^3$$

$$= 50 \text{ L(dB)} = 32.5 + 20 \log 30 + 20 \log 4 \times 10^3 = 134.1$$

$$P_r(\text{dBm}) = P_t(\text{dBm}) + G_t(\text{dB}) + G_r(\text{dB}) - \text{L(dB)}$$

$$= 50 + 50 + 50 - 134.1 = 15.9$$

Radio-path Planning

In order to plan for obstacle-free communication, the first Fresnel zone ellipsoid around the direct-ray path must be determined. The following factors are considered for radio-path planning.

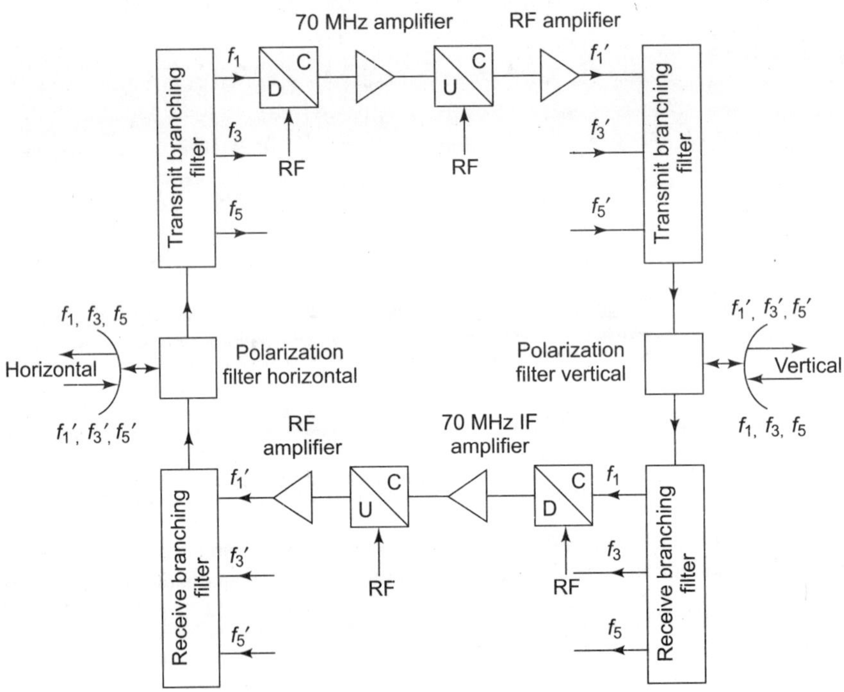

Fig. 11.17 Block diagram of a heterodyne repeater

Atmospheric Refraction

Since the dielectric constant of the atmosphere changes, the refractive index of the atmosphere decreases with increase in height above the earth's surface. The propagation path of microwaves, therefore, bends towards the earth as shown in Fig. 11.18. Since the change in refractive index is very small, such a curved ray path can be modelled by an equivalent straight-ray path over a spherical earth of radius a' greater than radius a of the original earth producing same result, for the received signal, where $a' \gg (4/3)a$. With this model, the maximum distance R_{max} between a transmitter and a receiver, is a function of antenna heights above ground when the ray path just grazes the equivalent earth's surface. This distance R_{max} is called radio horizon. From Fig. 11.18,

$$(a')^2 + R_1^2 = (a' + h_t)^2 \quad (11.31)$$

Therefore, $R_1^2 \approx 2a' h_t$, since $a' \gg h_t$ (11.32)

Similarly, $R_2^2 = 2a' h_r$ (11.33)

The radio horizon

$$R_{max} \approx R_1 + R_2$$

$$= \sqrt{(2a')}(\sqrt{h_t} + \sqrt{h_r}) \quad (11.34)$$

Taking $a = 6376$ km,

$$R_{max}(\text{km}) = \sqrt{17}(\sqrt{h_t}\ (\text{m}) + \sqrt{h_r}\ (\text{m})) \quad (11.35)$$

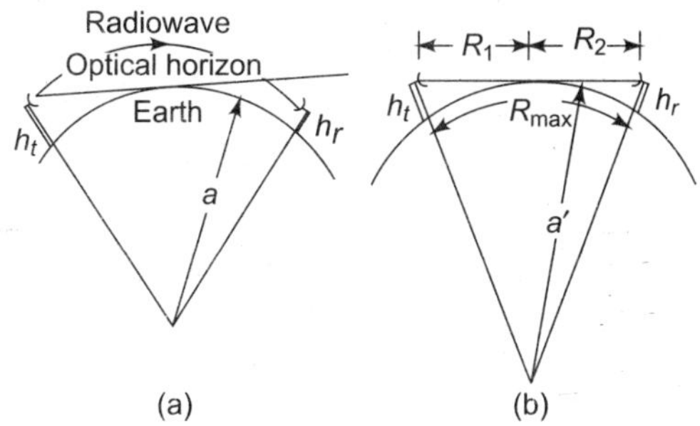

Fig. 11.18 Effect of atmospheric refraction

Contour Maps

If the distance $R = \sqrt{17h}$ of one antenna from the grazing point P is plotted with R(km) as abscissa and h(m) as ordinates, a contour is obtained as shown in Fig. 11.19(a) where the vertical distance h of a point on the curve from the tangent drawn at any point P determines the distance R(km) by the equation

$$R(\text{km}) = \sqrt{17h\,(\text{m})} \quad (11.36)$$

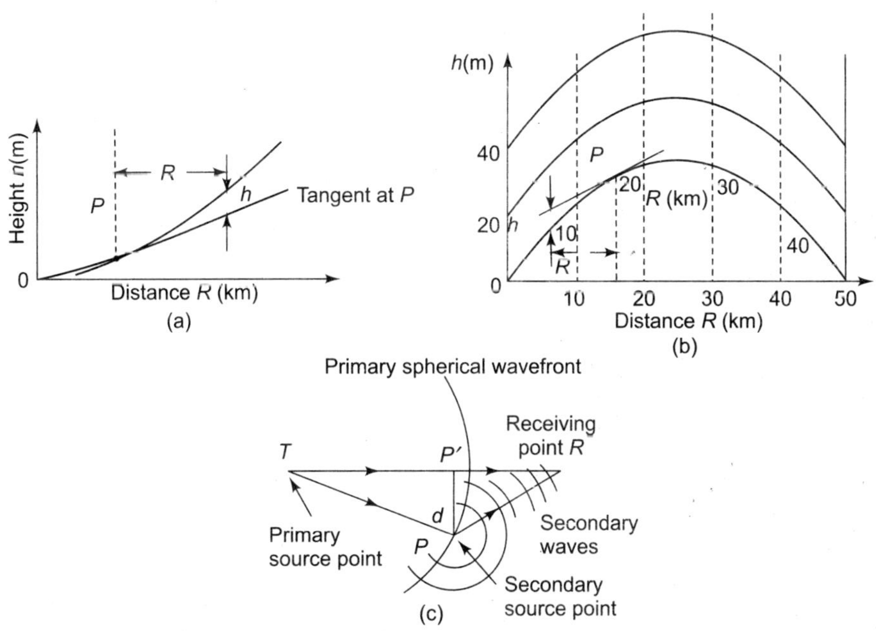

Fig. 11.19 (a) Contour of h vs R (b) Contour maps (c) Determination of Fresnel zone

When this curve is inverted, a baseline is formed by it as shown in Fig. 11.19(b). Then the line-of-sight distance R(km) from the antenna to the horizon is obtained corresponding to the antenna height h(m) measured from the baseline to the tangent drawn on the baseline at P. The profile of the terrain between the transmitter and receiver when plotted on this graph, is obtained and is known as the contour map. This map helps achieve obstacle-free line-of-sight transmission.

Fresnel Ellipse

The above line-of-sight concept is developed assuming that the propagating wavefront is plane. However, in practice, the wavefront is not plane for a finite practical distance between T_x and R_x. If we model the transmitting source as a point source, the wavefront will be spherical. Each spherical wavefront can be assumed to consist of a large number of secondary source points P's which in turn generate secondary waves. These secondary waves reach the receiver with a phase lag in comparison to the direct path. Therefore, even if an obstacle does not block the direct ray path, it may fall on the secondary wavelet generation point P and reduces the signal at the receiver point due to interference with the direct-ray path. It can be shown by the diffraction theory that the indirect path $TP + PR$ = direct path $TR + l_0/2$ must be cleared for obstruction-free transmission. The locus of P to satisfy the equation $TP + PR = TR + \lambda_0/2$ describes an ellipsoid around the direct-ray path TR. This ellipsoid surface encloses a volume called the first Fresnel zone, which must be cleared of any obstacle to obtain free-space propagation condition. The length of the perpendicular drawn from a point P on the ellipsoid to the direct-ray path (axis) is given by

$$d = \sqrt{\left[\frac{TP' \times P'R}{TR}\lambda_0\right]} \tag{11.37}$$

The maximum clearance distance is obtained at the midpoint ($TP' = P'R = TR/2$) of the range and is given by

$$d_{\max} = \sqrt{\left[\frac{TP' \times TP'}{TR}\lambda_0\right]} = \sqrt{\left(\frac{TR}{4}\lambda_0\right)} \tag{11.38}$$

Ground Reflections

The above radio-horizon calculation assumes a perfect smooth reflecting surface. But since the magnitude and phase of reflection coefficient depends on the nature of the earth's surface, frequency and the polarization of the waves, the contour map clearance will be applicable only when the Rayleigh's criterion for surface roughness is met. A reasonable approximation is that in the terrestrial links, microwave reflection coefficients have magnitudes less than unity but close to it and the phase angle is 180° for both horizontally and vertically polarized waves. According to the Rayleigh criterion, the height of surface irregularities should be less than $3.6\ \lambda_0/\theta$, where θ is the grazing angle of the ray path with the earth surface in radians.

Fading

Fading is a phenomenon in which the received signal strength is reduced or lost due to atmospheric effects and ground reflections. Accordingly, four types of fading—absorption fading, reflection multipath fading, atmospheric multipath fading and subrefraction fading—exist.

There are several elements in the atmosphere that cause attenuation of microwaves due to absorption of energy. These are (i) rain, (ii) cloud, (iii) fog and airgas molecules, such as water vapour (H_2O) and oxygen (O_2). Heavy rain seriously attenuates microwaves at frequencies above 10 GHz. Moderate rain, clouds and fog seriously attenuate microwaves at frequencies above 30 GHz. Snow does not have a remarkable effect on microwaves. The airgas molecules attenuate microwaves by absorbing energy due to vibrational resonances. The peak attenuations occur for H_2O at frequencies of 22.22 GHz and 176.47 GHz and for O_2 they are 60 GHz and 120 GHz. Thus, microwave communications must be established excluding these frequencies.

After the radio-path planning for satisfactory signal strength there is possibility of diurnal and seasonal changes in the refractive index of air. These cause changes in the relative path length between the direct and ground-reflected rays as shown in Fig. 11.20(a). The corresponding phase change may reduce the resultant signal strength or even cancellation of the signal. This is called **reflection multipath fading**. Again, due to decrease in refractive index with height, a second atmospheric ray path occurs in addition to the direct-ray path as shown in Fig. 11.20(b). The corresponding phase difference between these two rays produces a cancellation of the signal, or **atmospheric multipath fading**. Another effect in the atmosphere, which reduces signal strength by bending the ray away from the receiving point as shown in Fig. 11.20(c), is known as **subrefraction fading**.

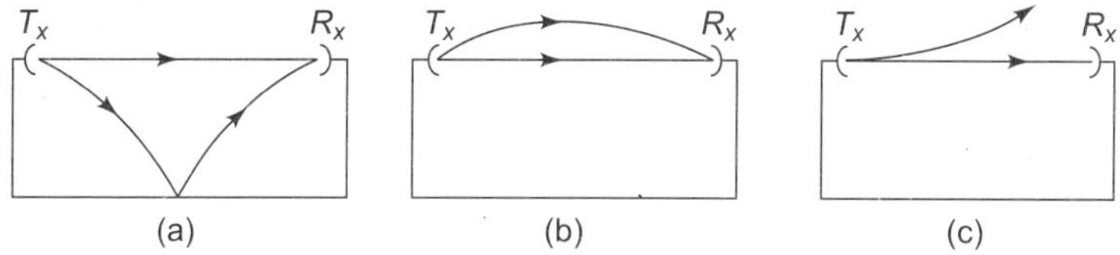

Fig. 11.20 Fading mechanism: (a) Reflection multipath fading (b) Atmospheric multipath fading (c) Subrefraction fading

The fading caused by change of refractive index occurs at frequencies below 8 GHz, whereas that caused by rain, cloud, fog and gas molecules (H_2O and O_2) occurs at frequencies above 10 GHz.

The fading is minimized by two techniques called **frequency diversity** and **space diversity** as explained in Fig. 11.21.

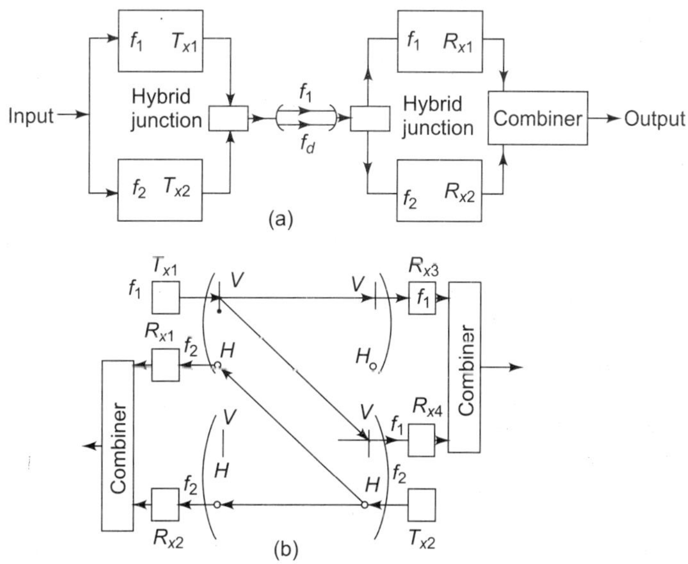

Fig. 11.21 (a) Frequency diversity technique (b) Space diversity technique

In the frequency diversity technique, the signal is sent through the same path over two different frequencies f_1 and f_2, where $f_1 \sim f_2 \geq 100$ MHz so that statistically it is highly unlikely to simultaneously fade both the signal frequencies. Since this method requires two transmitters and two receivers for two frequencies, the system cost is very high. Moreover, licensing restriction on the allocation of frequencies in the microwave band prevents easy use of this technique.

In the space diversity technique, the signals are transmitted from left to right at least through two different paths in space at the same *frequency* f_1 from a single antenna (vertically polarized) and received by two different antennas spaced vertically at a distance equal to a minimum of 50 wavelengths. Outputs of the two receivers are combined to give a single desired output. Similarly, right to left transmission is made at a different frequency f_2 with horizontal polarization.

In case fading cannot be eliminated, a fading margin of 30 dB is given in the system design for a 99.99% availability of the system reliably over the time during the worst fading month.

In communication-link calculations, the free space propagation equation is no longer valid due to practical difficulties in the link as described above. Hence, a modified equation is required to be formulated as follows.

Since the noise level in the receiver would degrade the *S/N* ratio in the analog system and produces errors in the digital system, the output power P_0 of the final amplifier stage of the transmitter and minimum detectable receiving carrier power $S_{\min}$ should be considered in link calculations which results in the equation

$$P_0(\text{dBm}) \geq S_{\min}(\text{dBm}) + FM(\text{dB}) + L(\text{dB}) + L_b(\text{dB}) + L_f(\text{dB}) - G_t(\text{dB}) - G_r(\text{dB}) \tag{11.39}$$

where, FM(dB) is the fading margin, L_b(dB) is the loss in the branching circuits of the transmitter and receiver, such as, filters and circulators, L_f(dB) is the loss in the transmitter and receiver antenna feeder lines. The overall system gain is defined as

$$G_s(\text{dB}) = P_0(\text{dBm}) - S_{\min}(\text{dBm}) \tag{11.40}$$

A typical value of G_s is 100 dB for a well-designed system.

Satellite Communication Systems 11.3.2

In satellite communication systems, the information is transmitted at microwave frequency through a highly directional antenna from a ground station to the geo-synchronous satellite, which receives the signal through an on-board antenna, shifts the frequency, and amplifies it by means of a low-noise wide-band amplifier (transponder). Then this received signal is retransmitted towards the earth at 2 GHz frequency lower than the uplink frequency. The usual uplink frequency is 6 GHz and downlink frequency is 4 GHz covering the C-band. This frequency conversion is necessary to avoid interference between uplink and downlink. Satellites are used to handle the long-distance telephone traffic, to relay TV signals across oceans and to provide national TV cables directly to the home.

The satellite frequencies are chosen such that the effect of the ionosphere is negligible and absorption by atmospheric gases and water vapour is very small. These satellites orbit around the earth in a circle at an approximate speed of 11,000 km/hour but are geostationary at approximately 35,800 km above the earth's surface. Thus, no tracking is required and a fixed antenna in the ground station establishes the link at all times.

The stage between the reception of the uplink signal and the transmission of the downlink signal is called a transponder. A satellite commonly contains 12 transponders having a bandwidth of 36 MHz each. The whole satellite has a bandwidth of 500 MHz. There are satellites that operate at other frequency bands such as 14 GHz/11 GHz and 17 GHz/12 GHz. Figure 11.22 shows the block diagram.

Table 11.3 Satellite bands and frequencies

Satellite Band	*Frequency Range*
6 GHz/4 GHz	5.93 – 6.42 GHz/3.705 – 4.195 GHz
14 GHz/11 GHz	14.0–14.5 GHz/11.45 – 11.7 GHz
18 GHz/14 GHz	

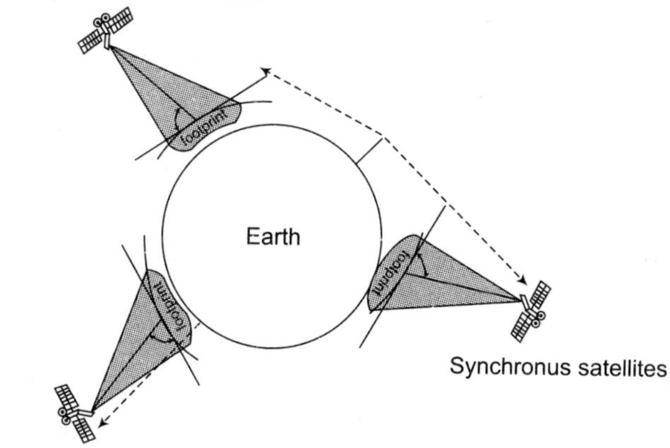

Fig. 11.22 Synchronous satellites around the earth

Path Loss/ Link Budget

A **link budget** is the accounting of all of the gains and losses from the transmitter, through the communication channel (free space, transmission lines, etc.) to the receiver in a communication system. A simple link budget equation is

$$\text{Received Power (dBm)} = \text{Transmitted Power (dBm)} + \text{Gains (dB)} - \text{Losses (dB)} \tag{11.41}$$

The primary source of loss is the decrease of the signal power in a direction proportional to the inverse square of the distance. Due to directional transmitting antennas, the signal power concentrates in the direction of the receiving antenna. The transmitted power is thus effectively increased in the receiver direction and is expressed by including the antenna gain in the link budget. The receiving antenna is also directional and its gain is included similarly to the received power.

Link Margin

It is desired that the received power should be sufficiently large relative to the receiver sensitivity for the link to be useful for sending data. The amount by which the received power exceeds receiver sensitivity is called the link margin. In the link budget, this margin is introduced in the system considering expected fading, polarization mismatch between T_X and R_X antennas, and other losses.

Free-space Path Loss (FSL)

The free-space path loss (FSL) is the loss due to propagation between the transmitting and receiving antennas in LOS communications, and can be written in dimensionless form by normalizing the distance to the wavelength. This is proportional to the square of the distance between the transmitter and receiver antennas in the far-field ranges.

$$\text{FSL} = \left(\frac{4\pi R}{\lambda}\right)^2 = \left(\frac{4\pi Rf}{c}\right)^2 \tag{11.42}$$

where $\lambda = \frac{c}{f}$

Taking logarithm on both sides

$$\text{FSL(dB)} = 20\log_{10}(R) + 20\log_{10}(f) + 32.45 \tag{11.43}$$

where f = Signal frequency in MHz
d = Distance from the transmitter in km
c = Speed of light in a vacuum.

Example 11.5 A 30 km long microwave terrestrial link is operating at 4 GHz, with a radiated power of 100 W through a parabolic disk having maximum gain of 50 dB. The receiver uses a similar antenna. Find the free-space loss and the received power.

Solution

$$P_t(\text{dBm}) = 10 \log P_t(\text{mW}) = 10 \log 100 \times 10^3 = 50$$

$$L(\text{dB}) = 32.5 + 20 \log 30 + 20 \log 4 \times 10^3 = 134.1$$

$$\begin{aligned} P_r(\text{dBm}) &= P_t(\text{dBm}) + G_t(\text{dB}) - L(\text{dB}) \\ &= 50 + 50 + 50 - 134.1 \\ &= 15.9 \end{aligned}$$

Example 11.6 In terrestrial communication, the plane wave signal at 5 GHz covers a distance of 50 km. Determine the free-space path loss.

Solution

$$\begin{aligned} L &= 32.4 + 20 \log f_{\text{MHz}} + 20 \log R_{\text{km}} \\ &= 32.4 + 20 \log (5 \times 10^3) + 20 \log 50 \\ &= 32.4 + 60 + 20 \log 5 + 20 \log 50 \\ &= 92.4 + 13.98 + 33.98 \\ &= 140.36 \text{ dB} \end{aligned}$$

Basic Friis Transmission Equation

The Friis transmission equation is a relation between the power received by the receiving antenna and the power input to the transmitting antenna at a given frequency and a given separation distance between these two antennas as follows. Let

P_r = Power received by the receiving antenna
P_t = Power input to the transmitting antenna
G_t and G_r = Gain of the transmitting and receiving antennas with respect to isotropic, respectively
λ = Wavelength of transmitted signal
R = Distance between the antennas

The Friis transmission equation gives the ratio

$$\frac{P_r}{P_t} = G_t G_r \left(\frac{\lambda}{4\pi R} \right)^2 \tag{11.44}$$

The factor within the parentheses is the FSL, where the wavelength and distance units are the same. This simple form applies only under the following ideal conditions:

- The antennas are in LOS range.
- P_r is the available power at the receive-antenna terminals when match terminated.
- P_t is the power delivered to the transmit-antenna terminals.
- The antennas are perfectly aligned and polarized.

In practical terrestrial communications, the ideal conditions are never achieved, due to obstructions, reflections from buildings, the ground, the trees, etc. However, the equation is reasonably accurate in well-designed satellite communications when there is no reflection and negligible atmospheric absorption.

Modifications to the Friis Transmission Equation

A practical model of LOS communication is shown in Fig. 11.23, where there are cables and connectors are used between antennas to T_X/R_X. Here, the effects of impedance mismatch between antenna and the cables/connectors, misalignment of the antenna polarization, and the absorption in the transmitting medium can be included such that the modified Friis transmission equation becomes

$$\frac{P_r}{P_t} = G_t(\theta_t,\phi_t)G_r(\theta_r,\phi_r)\left(\frac{\lambda}{4\pi R}\right)^2 (1-|\Gamma_t|^2)(1-|\Gamma_r|^2)\left|a_t \cdot a_r^*\right|^2 e^{-\alpha r} \quad (11.45)$$

where

- $G_t(\theta_t, \phi_t)$ is the gain of the transmit antenna in the direction (θ_t, ϕ_t) in which it faces the receive antenna
- $G_r(\theta_r, \phi_r)$ is the gain of the receive antenna in the direction (θ_r, ϕ_r) in which it faces the transmit antenna
- Γ_t and Γ_r are the reflection coefficients of the transmit and receive antennas, respectively
- a_t and a_r are the polarization vectors of the transmit and receive antennas, respectively.
- α is the absorption coefficient of the transmitting medium

When the entire link is a lossless and matched system, $\alpha = 0$, Γ_t and $\Gamma_r = 0$, and $|a_t \cdot a^*_r| = 1$, so that

$$\frac{P_r}{P_t} = G_t(\theta_t,\phi_t)\, G_r\,(\theta_r,\phi_r)\left(\frac{\lambda}{4\pi R}\right)^2 \quad (11.46)$$

which is the basic form of the Friis transmission equation.

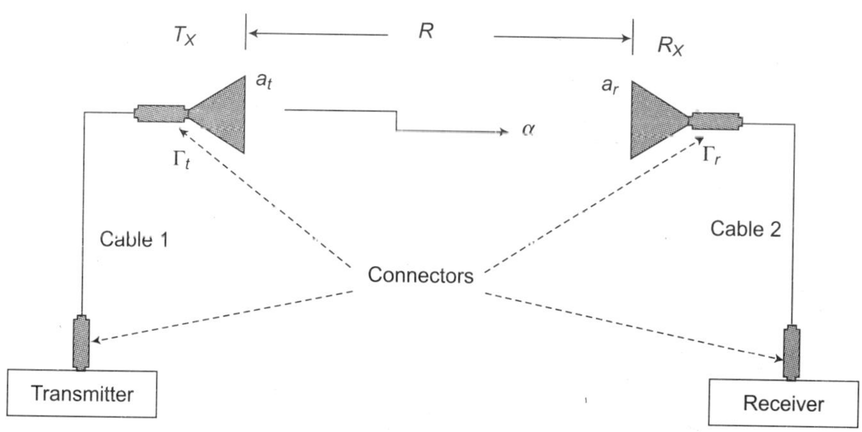

Fig. 11.23 LOS link

Fade Margin

In wireless communication, the term **fade margin** or **fading margin** has the following meanings:

- A fading margin is a design allowance given to produce sufficient system gain or sensitivity to accommodate expected fading. This ensures that the required quality of service is maintained.
- The fade margin is the amount by which a received signal level may be reduced without causing

system performance to fall below a specified threshold value. It is used in satellite communication. As an example, a system like global star operates at 25–35 dB fade margin.

In case fading cannot be eliminated, a fading margin of 30 dB is given in the system design for a 99.99% availability of the system reliably over the time during the worst fading month. For an unprotected and without diversity system, the fade margin is given by

$$\text{FM (dB)} = 30 \log R + 10 \log (6\,\text{ABf}) - 10 \log (1 - R_0) - 70 \tag{11.47}$$

where R = Distance, km

f = Frequency, GHz

R_0 = Reliability objective

A = surface roughness factor = 4 over smooth surface

= 1 over average terrain

= 0.25 over very rough terrain/mountain

B = Factor to convert a worst-month probability to an annual probability

= 1 to convert a worst-month probability to an annual probability

= 0.5 for humid areas

= 0.25 over average land

= 0.125 for very dry or mountain areas

Constant term = 70 dB

In communication-link calculations, the free-space propagation equation is no longer valid due to practical difficulties in the link as described above. Hence, a further modified equation is required to be formulated as follows.

Since the noise level in the receiver would degrade the S/N ratio in the analog system and produces error in the digital system, the output power P_0 of the final amplifier stage of the transmitter and minimum detectable receiving carrier power $S_{\min}$ should be considered in link calculations which results in the equation as follows

$$P_0(\text{dBm}) \geq S_{\min}(\text{dBm}) + FM(\text{dB}) + L(\text{dB}) + L_b(\text{dB}) + L_f(\text{dB}) - G_t(\text{dB}) - G_r(\text{dB}) \tag{11.48}$$

where, FM(dB) is the fading margin, L_b(dB) is the loss in the branching circuits of the transmitter and the receiver, such as, filters and combiner/splitter, L_f(dB) is the loss in the transmitter and receiver antenna feeder lines. The overall system gain is defined as

$$\begin{aligned} G_s(\text{dB}) &= P_0(\text{dBm}) - S_{\min}(\text{dBm}) \\ &\geq FM(\text{dB}) + L(\text{dB}) + L_b(\text{dB}) + L_f(\text{dB}) - G_t(\text{dB}) - G_r(\text{dB}) \end{aligned} \tag{11.49}$$

A typical value of G_s is 100 dB for a well-designed system.

Example 11.7 Determine the fade margin for a 40 km site distance at a frequency of 2 GHz. Reliability is given by 99.99 % in the region of smooth terrain and humid climate.

Solution For an unprotected and without-diversity system, the fade margin is given by

$$\text{FM (dB)} = 30 \log R + 10 \log (6\,\text{ABf}) - 10 \log (1 - R_0) - 70$$

where R = distance, km

f = Frequency, GHz

R_0 = Reliability objective

A = Surface roughness factor = 4 over smooth surface

= 1 over average terrain

= 0.25 over very rough terrain/mountain

B = Factor to convert a worst-month probability to an annual probability

= 1 to convert a worst month probability to an annual probability

= 0.5 for humid areas

= 0.25 over average land

= 0.125 for very dry or mountain areas.

Constant term = 70 dB

For given smooth terrain, $A = 4$ and humid climate $B = 0.5, f = 2$ GHz, $R = 40$ km, $R_0 = 99.99\%$
Therefore,

$$FM(\text{dB}) = 30 \log 40 + 10 \log (6 \times 4 \times 0.5 \times 2) - 10 \log (1 - 0.9999) - 70 = 31.86$$

From the above equation, a link-budget equation including all practical effects, is expressed logarithmically as

$$P_{RX} = P_{TX} + G_{TX} + G_{RX} - L_{TX} - L_{RX} - L_{FS} - L_M \quad (11.50)$$

where P_{RX} = Received power, dBm

P_{TX} = Transmitter output power, dBm

G_{TX} = Transmitter antenna gain, dBi

L_{TX} = Transmitter losses (coax, connectors...), dB

L_{FS} = Free space loss or path loss, dB

L_M = Link margin losses (fading margin, body loss, polarization mismatch, other losses), dB

G_{RX} = Receiver antenna gain, dBi

L_{RX} = Receiver losses (coax, connectors...), dB

Example 11.8 Radio link has a 100-watt transmitted power connected to an antenna of 2 m^2 effective aperture at 10GHz. The receiving antenna has an effective aperture of 0.5 m^2 and is located at a 30 km LOS distance from the transmitting antenna. Assuming lossless, matched antennas, find the power delivered to the receiver.

Solution

Here,
$$\lambda = \frac{c}{f} = \frac{30}{10} \text{ cm}$$
$$= 3 \text{ cm} = 3 \times 10^{-2} \text{ m}$$

From the Friis transmission formula,

$$P_r = P_t G_t G_r \left(\frac{\lambda}{4\pi R} \right)^2$$

Antenna gain $G = \dfrac{4\pi A_e}{\lambda^2}$, A_e = Effective area of antenna.

Therefore,

$$G_t G_r = \left(\frac{4\pi}{\lambda^2} \right)^2 A_{et} A_{er}$$

Substituting in the Friis equation

$$P_r = P_t\left(\frac{4\pi}{\lambda^2}\right)^2 A_{et}A_{er}\left(\frac{\lambda}{4\pi R}\right)^2 = \frac{P_t A_{et} A_{er}}{R^2\lambda^2}$$

$$= \frac{100\times 2\times 0.5}{(30\times 10^3\times 3\times 10^{-12})^2} = 1.235\times 10^{-4}$$

Modulation of Link Signal

For TV transmission, the video broadcast signal bandwidth is 4.5 MHz including the synchronizing, blanking and equalizing pulses, which frequency modulates the miocrowave carrier at 6 GHz band. The sound signal has a bandwidth of 15 kHz which frequency modulates a 6.8 MHz subcarrier. The sound information is frequency division multiplexed with the video information and relayed by a single transponder. The frequency deviation of the video signal and sound signal are 10.5 MHz and 2 MHz, respectively. Therefore, the bandwidth of 2(10.5 + 2 + 6.8) = 36.5 MHz » 36 MHz should be assigned to the transponders. The total bandwidth for 12 transponders plus guard bands is accommodated within 500 MHz.

For data and telephone signal relay, the time division multiplexing (TDM) or frequency division multiplexing (FDM) is used. In TDM, the data is converted to TDM signals which modulate a microwave carrier digitally for transmission. In FDM, modems are used to convert the data into signals which are compatible with voice-frequency bandwidth. The microwave carrier is analog modulated by these signals. The voice frequency signal is also pulse-code modulated (PCM) and then time-division multiplexed and finally merged with the digital signal.

Satellite Antennas

Spacecraft antennas are mounted on the body of a satellite to provide coverage of a certain zone on the earth's surface. Monopole, dipole horn antennas, reflector antennas, and microstrip array antennas are used. Earth-station antennas are large parabolic reflector antennas having high gain to provide a narrow (pencil) beam pointing towards the satellite. The monopole and dipole are used primarily at VHF/UHF frequencies for omni-directional coverage in TTC systems where the aperture antenna would have a very large size to increase the gain at very large wavelengths. Horn antennas are used when relatively wide beams are required for global coverage. It is difficult to obtain a gain much greater than 23 dB. The reflector antenna with feed placed at its focus produces plane waves giving maximum gain and a narrow beam for coverage of a particular zone on the earth. In order to reduce the complexity in the mechanical design for the reflector, an array of pointed antennas is the alternative to paraboloid antennas for high gain or narrow beam width. In an ideal spacecraft, there would be one antenna beam for each earth station, completely isolated from all other beams, for transmission and reception. For multiple earth stations, a separate beam should be provided for each station. This would also require one antenna feed per earth station with a single reflector.

The INTELSAT V spacecraft has four reflector antennas. Each is illuminated by a complex feed that provides the required beam shape to cover all earth stations within a given coverage zone. The largest reflector should be used to transmit at 4 GHz for the zone beams (peanut) to concentrate onto densely populated areas where much telecommunication traffic is generated. The smaller antennas are used to provide hemisphere transmit and receive beams, and the 14/11 GHz spot beams.

Most domestic satellites do not have complex antenna systems, but use orthogonal polarization frequency reuse to double the effective bandwidth at 6/4 GHz for separate channels with cross-polar isolation of 25 dB. Table 11.4 shows various type of antennas used in Indian satellites.

Table 11.4 Indian satellite antennae

Satellite Name	Antenna Type	Transponders Frequency Band	Application
INSAT–1 B (1983)		C BAND K_U BAND	Communication
INSAT 2 A INSAT 2 B ALREADY GIVEN			Communication
INSAT 2E		17 C BAND	
INSAT 3A		12 C Band 6 EXTENDED C BAND 6 K_U BAND	A Satellite Aided Search and Rescue (SAS&R)
INSAT 3C(2002)		24 C BAND 6 EXTENDED C 2 S BAND (BSS)	Communication
INSAT 3E (2003)		24C BAND 12 EXTENDED C BAND	Communication
KALPANA I (MetSat-I)(2002)			Meteorological service
GSAT 2(2003)		C BAND K_U BAND	Communication
EduSat (2004)		1 K_U BAND 5 K_U BAND spotbeam 6 EXTENDED C	Configured for audio-visual medium employing digital interactive classroom lessons and multimedia content
INSAT 4A ('05) (Tata Sky – Tata Group And Star joint venture)		12 C BAND 12 K_U BAND	Communication
INSAT 4 B ('07)		12 C BAND 12 K_U BAND	Communication
INSAT- 4C R (replacement of INSAT 4C after failure)		12 K_U BAND K_U BAND bacon	Communication
Oceansat-2 (IRS-P4)			Gathers data for oceanographic, coastal and atmospheric applications
INSAT 4G (GSAT 8)('11)		24 K_U BAND L_1 & l_5 BAND (GPS navigation system)	Advanced communication satellite
GSAT 12('11)		12 extended C band	for various communication services like tele-education, telemedicine and for Village Resource Centres (VRC), mission
RISAT-1			First indigenous all-weather radar imaging satellite

Example 11.9 Let an illumination zone subtend an angle of 6° × 3° when viewed from a geostationary orbit. What dimension must a reflector antenna have to illuminate half this area with a 3°-diameter circular beam at 11 GHz? Can a reflector be used to produce a 6° × 3° beam? What gain would the antenna have?

Solution

(i) For a 3° circular beam,

$$\theta_3 = \frac{75\lambda_0}{D}; \text{ at 11 GHz, } \rightarrow \lambda_0 = 0.0272 \text{ m}$$

Therefore, $D/\lambda_0 = 75/3 = 25$ or, $D = 25\lambda_0 = 0.68 \text{ m} \approx 2 \text{ ft.}$

(ii) To generate a beam of 6° × 3°, we need an aperture with dimensions in two orthogonal planes as

$$D_{1,2} = \frac{74\lambda_0}{6}, \frac{75\lambda_0}{3} = 12.5\ \lambda_0, 25\lambda_0$$

Therefore, $G = \dfrac{30{,}000}{6° \times 3°} \approx 32 \text{ dB}$

A feed horn of unequal beam width is required to illuminate such a reflector with elliptical aperture. But this results in poor polarization characteristics.

Thus, when orthogonal polarization is to be transmitted or received, it is better to use a circular reflector.

(iii) $G = 30{,}000/3^2 \approx 35 \text{ dB}$

Ground Stations

The ground station (Fig. 11.24) has a need to achieve a low-system noise temperature in the receiving mode or during down link, due to very weak signals being received. The antenna characteristics of the earth stations are very important in their design. In a large station, a 30 m dia dish antenna is used, particularly in Intelsat network. There are small earth stations which use a 0.7 m dia dish antenna for reception of Direct Broadcast Satellite (DBS) TV.

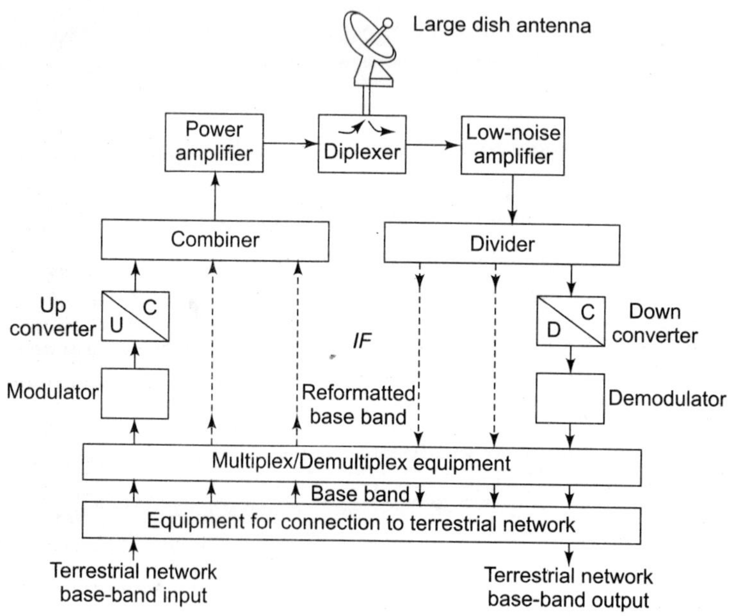

Fig. 11.24 Ground stations

The carrier-to-noise power ratio in a ground station is proportional to G/T (dB/K), where G is the gain of the antenna > 25 dB to satisfy link requirements, T is the system noise temperature in Kelvin. Typically,

$$GT = G(\text{dB}) - 10 \log T$$
$$= -10 \text{ to } + 46 \text{ dB/K} \tag{11.51}$$

G/T = 46 dB/K (max.) corresponds to G = 65 dB, T = 70 K. The optimum G/T for a given application is a compromise between the cost of a large antenna to increase G and the cost of a lower system noise to decrease T. Since the beam width is very small, most large antennas are equipped with automatic tracking facilities so that the motion of the satellite can be followed.

Important Parameters for Satellite Antennas

❑ ***EIRP*** EIRP is the most useful parameter of an antenna which is called the effective isotropically radiated power of the satellite antenna.

$$\text{EIRP} = \text{Antenna gain} * \text{Transmitter power output} = G_t P_t.$$
$$= 10 \log_{10} [P_t G_t] \text{ dBW} \tag{11.52}$$

❑ ***Satellite Received Power*** It is expressed in terms of EIRP

$$P_r = [P_t G_t]\, G_r \left(\frac{\lambda_0}{4\pi R}\right)^2 = \text{EIRP}^* G_r / L_p \tag{11.53}$$

$$\text{Path loss} = L_p = \left[\frac{4\pi R}{\lambda_0}\right]^2 \tag{11.54}$$

assuming no other losses in the antenna.

❑ ***G/T Ratio for Earth Station*** In satellite communication, the noise in R_x is reduced as far as possible because the signals are weak due to large distances involved. The carrier-to-noise ratio at the demodulator is given by

$$C/N = \frac{P_r G}{[kT_s B]G} = \frac{P_r}{kT_s B} \tag{11.55}$$

Here, T_s = system noise temperature = noise temperature of a noise source located at the input of a noiseless R_x which gives the same noise power as the original R_x at the output. $T_s = T_1$ (thermal noise) + T_2 (sky noise). Therefore,

$$C/N = \frac{P_t G_t G_r}{kT_s B}\left[\frac{\lambda_0}{4\pi R}\right]^2 \alpha \frac{G_r}{T_s} \tag{11.56}$$

for a given satellite system. Therefore, G/T should be large to increase C/N.

Example 11.10 In a satellite communication link, the earth station G/T ratio is 10 dBk, the satellite EIRP is 50 dBW and the propagation loss is 200 dB. If the sum of the fading margin, antenna pointing margin and equipment margin is 3 dB, calculate the received C/N ratio.

Solution

$$k = 1.38 \times 10^{-23} \text{ J/K}$$
$$C/N = \text{EIRP} - \text{margin} - \text{propagation loss} + G/T - 10 \log \text{K}$$
$$= 50 - 3 - 200 + 10 + 228.6 = 288.6 - 203$$
$$= 85.6 \text{ dB Hz}$$

INDUSTRIAL APPLICATION OF MICROWAVES 11.4

Microwaves have been used for industrial, scientific, and medical applications at frequencies of 896, 950, 2450, 3300, 5800 and 10,525 MHz. Most industrial applications of microwaves utilized the heating effect of microwaves in industrial processes like cooking, baking, puffing, drying, curing, evaporating, sterilizing, moulding, etc. Many industries such as food, chemicals, rubber, textiles, plastics, paper, ceramic, cosmetics, etc., use microwave energy.

Microwave Heating 11.4.1

In 1946, Dr Parcy Spencer, an engineer with the Raytheon Corporation, USA, while operating a magnetron, discovered that the candy bar in his pocket melted and also the popcorn placed in front of the magnetron sputtered, crackled, and popped all over the lab. He also observed that an egg placed in front of the magnetron exploded and splattered hot yolk all over. In this process, Spencer discovered that microwaves could be used for heating and cooking.

Mechanism of Microwave Heating

Any dielectric material exhibits loss due to non-zero conductivity as well as permittivity. Microwaves penetrate into such materials and dissipate heat due to ohmic losses. The normal conductive effects are made up of free electron conduction, rotational and vibrational losses in the molecules. As a result, an equivalent conductivity σ (S/m) is defined such that the dielectric constant becomes complex

$$\varepsilon_r = \varepsilon_r' - j\varepsilon_r'' \tag{11.57}$$

where ε_r' is the relative permittivity and ε_r'' is the loss factor given by

$$\varepsilon_r'' = \sigma/\omega\varepsilon_0 \tag{11.58}$$

The losses in the dielectric are measured in terms of loss tangent

$$\tan\delta = \varepsilon_r''/\varepsilon_r' - \sigma/(\omega\varepsilon_r'\varepsilon_0) \tag{11.59}$$

The amount of microwave radiation is measured in terms of power density. For incident microwaves, radiation is measured in terms of power density. For incident microwaves, the power density inside the dielectric material is given by

$$P_d = \sigma|\mathbf{E}_i|^2 = \omega\varepsilon_r'\,\varepsilon_0 \tan\delta\,|\mathbf{E}_i|^2 \text{ W/m}^3 \tag{11.60}$$

where $\mathbf{E}_i$ is the internal electric field in V/m. $\mathbf{E}_i$ will not necessarily be uniform and this produces nonuniform heating, resultin in 'hot spots'. The 'hot spots' are generated from the focusing action of high dielectric constant materials and also from the slope of any cavity formed inside the materials.

Microwave Ovens

A microwave oven is a metallic cooking chamber or cavity excited in multiple modes at a frequency of 2450 MHz by means of a magnetron oscillator as shown in Fig. 11.25. The frequency of 2450 MHz is chosen to achieve a desired balance between the degree of heat generated within the food and the degree of penetration of the energy.

The microwave energy is generated by a magnetron oscillator at a frequency of 2450 MHz. The energy is extracted from the resonant cavities by a magnetic loop formed by the centre conductor of output coaxial line. The other end of the coaxial line forms a field probe which feeds a rectangular waveguide of dimensions a = 6.83 cm and b = 3.81 cm. This waveguide is terminated with a short, at a distance of a quarter guide wavelength from the field probe. Thus, the waveguide is excited in the dominant TE_{10} mode. The other end of the waveguide is physically open to launch the microwave energy into the stainless-steel cavity in which the food is placed. A rotating metallic stirrer (blade) reflects the microwave energy and produces multimode fields in the cavity. Due to the rotation of the stirrer, the boundary conditions of the oven change with time and a statistically uniform field is produced throughout the cavity resulting in uniform heating of the food material. A plastic ceiling shield is placed between the stirrer chamber and the oven chamber for environmental protection.

The typical input power to the oven is normally 1 to 1.5 kW, which in turn produces a maximum microwave power of 600–700 W at 2450 MHz. Since a microwave can cook from the inside to the outside of the material, an electricheater is used to brown the food to give a conventional cooked appearance on the outside. The molecules associated with the moisture contained in the food are set in vibration at microwave frequency and the resulting friction produces heat necessary for the cooking process. Usually, microwaves can penetrate food materials up to a depth of 2–3 cm. Therefore, large pieces are still cooked by the conduction of heat to the centre of the food. Since the cooking time is less and the process is even, the food retains its natural flavour and nutritional value.

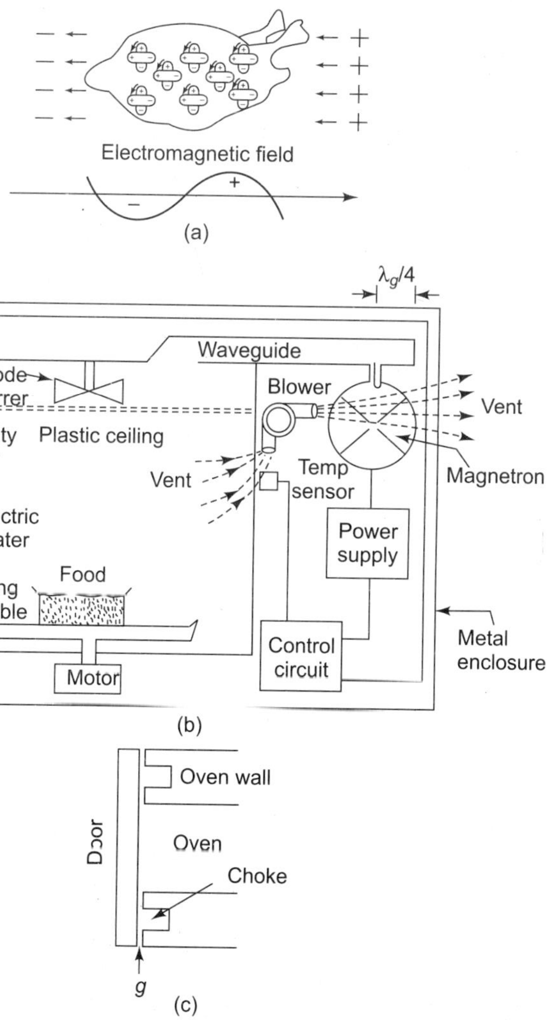

Fig. 11.25 (a) Rotational and vibrational losses due to polarization change in food molecules in microwaves (b) Microwave oven (c) Choke in oven door

The cooking time and temperature are controlled by using a microprocessor-based controller circuit. The temperature is sensed from the exhaust air outside the oven cavity. The time of cooking is predetermined for certain ranges of food. The time and degree of heating can be set in the control circuit and once the required parameter is detected, the magnetron may be automatically switched off.

The microwave oven must be designed such that the door maintains shielding integrity with the oven cavity walls, to prevent leakage of microwave energy above the maximum permissible limit, as per international regulations. The shielding integrity cannot be high for direct mechanical contact between the door and oven cavity. A microwave quarter-wave choke is designed in series with the door to oven gap, so that the gap sees an open circuit towards the oven cavity which prevents leakage. Moreover, the choke is filled with a special ferrite material inside. The quarter-wave section must be cut accordingly. The door must have an additional safety lock which automatically switches off the microwave power if the door is opened.

Industrial Control and Measurements 11.4.2

This application includes precision thickness measurement of metal sheets in rolling machines and determination of moisture content in a substance.

Thickness Measurements

Sheet-thickness measurements involve determination of complex reflection or transmission coefficient at microwave frequencies. The metal sheet in the rolling process is passed between two horns placed face to face as shown in Fig. 11.26. The circulator, connecting waveguides, phase shifter and the horns constitute a resonant cavity whose resonant frequency is made equal to that of the signal source by adjusting the phase shifter. For a given thickness, the reflection coefficient at the plane PP' is made a small value. Any change in thickness will cause changes in reflection coefficient which can be calibrated by using a known thickness of a given metal.

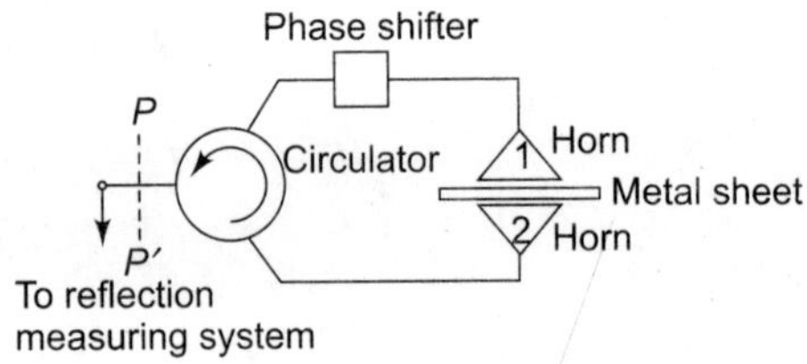

Fig. 11.26 Thickness measurements

Moisture Content Measurements

The moisture content in paper and textile is measured using microwave bridges. In this, the attenuation and phase shift of the electromagnetic wave passing through the material or the complex reflection coefficient of the reflected wave from the material sheet are measured. The principles of measurement set up are explained in Fig. 11.27. The sensitivity of these methods can be increased by suitable selection of spacing between the horn radiator and the material sheet.

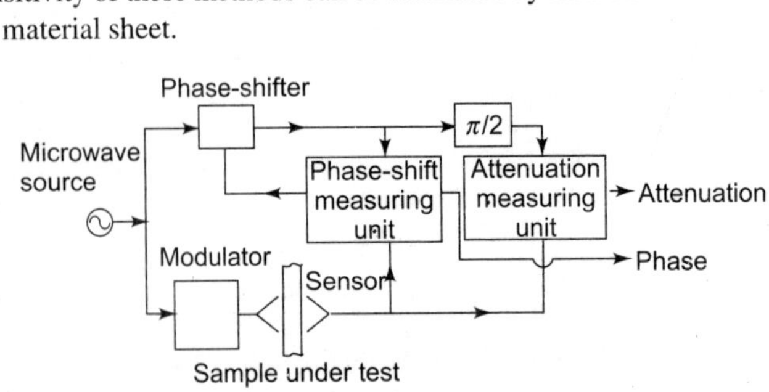

Fig. 11.27 Moisture-content measurement in solid material

MEDICAL APPLICATIONS 11.5

Microwaves are now extensively used in medical applications due to their non-ionizing nature compared with more harmful radioactive or X-ray exposure and because microwave power can be focused selectively for heat-specific tissue regions. Two therapeutic applicators—**diathermy** and **hyperthermia**—are very common. Since microwaves can penetrate the body to reach bones and deep muscles, localized heating can be achieved. Microwave energy is launched into a patient using a focusing device such as dielectric lens and the temperature is noted in a thermometer (Fig. 11.28). In diathermy, microwave heating is used to relieve pain and to treat various inflammatory diseases and ailments. Microwave heating is also combined with radioactive exposure to destroy cancerous cells.

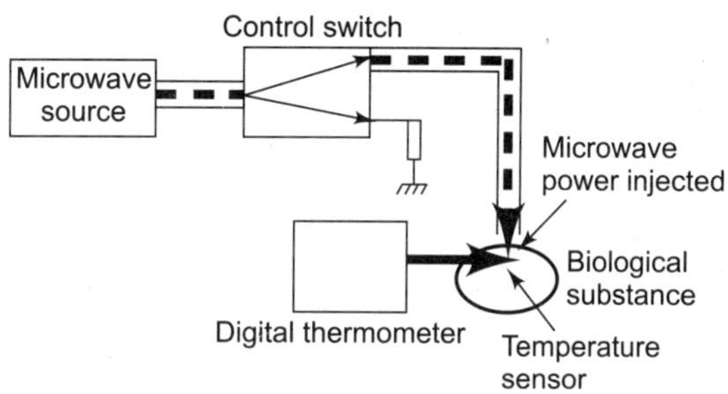

Fig. 11.28 Microwave diathermy

REVIEW QUESTIONS

11.1 An antenna has a gain of 55 dB at a frequency of 6 GHz. Calculate its effective area.

11.2 A 73-ohm antenna is fed by a 50-ohm lossless line with 5 kW input power. The total power radiated by the antenna is 4 kW. Calculate (a) the antenna efficiency, and (b) the power loss in the antenna.

11.3 The complex relative permittivity of a substance at 3 GHz is $60 - j20$. Calculate (a) the loss tangent, (b) the equivalent conductivity, and (c) the power density for an internal field of 100 mV/m.

11.4 A radar transmits 100 W power through an antenna of 40 dB gain. The echoing area of a target placed at a 5 km distance is 1 m^2. Find (a) the power density at the target, and (b) the reflected power from the target.

11.5 An FMCW radar operates at 10 GHz and employs triangular-wave modulation for which the rate of change of frequency is 1 GHz/s. The beat frequency from a target return is 5 kHz over the negative slope and 4.8 kHz over the positive slope. Find the target range, range rate, and the relative direction of motion of the target with respect to radar.

11.6 A 1 kW, 3 GHz radar uses a single antenna with a gain of 30 dB. The receiver has a noise bandwidth of 1 kHz and a noise factor of 5 dB. A target of echoing area of 10 m^2 at a range of 10 nautical miles is to be detected. Calculate the minimum *S/N*.

REFERENCES

1. Balanis, C A: *Antenna Theory—Analysis and Design*, Harper & Row Publishers, New York, 1982.
2. Brown, W C: "Satellite Power Stations: A New Source of Energy" IEEE spectrum, March, 1973, pp. 38–47.
3. Dennis Roddy: *Microwave Technology*, Prentice-Hall, Inc., Englewood Cliffs, New Jersey, 1986.
4. Freedman, G: "The Future of Microwave Heating Equipment in the Food Industries", K, Microwave Power, Vol. 7, No. 4.

5. Gupta, C, *Microwaves in Medicine*, Electron. Power, Vol. 27, No. 5, pp. 403– 406, 1981.
6. Gupta, K C, Bhal, I J: "RF Applicators Find Jobs on Farms, in Factories", Microwaves, 17 June, 1978, pp. 52–54.
7. Gupta, K C: *Microwaves*, New Age International (P) Limited, Publishers, Formerly Wiley Eastern Ltd., New Delhi, 1983.
8. Kraszewski, A: "Microwave Aquametry—A Review", J Microwave Power, Vol. 15, No. 4, 1980.
9. Kraszewski, A and Stanislaw Kulinski: "An Improved Microwave Method of Moisture Content Measurement and Control", IEEE Trans, on Ind. Electron and Control Inst. Vol. IECI-23, No. 4 Nov. 1976.
10. "Special Issue on the Industrial, Scientific and Medical Applications of Microwaves", Proc, IEEE, 62, No. 1. Jan. 1974.
11. Stuchly, M A; Stuchly, S S: "Dielectric Properties of Biological Substances", J Microwave Power, Vol. 15, No. 1, 1980.
12. Stuchly, S; Kraszowski, A; Rzepecka, M: "Microwaves for Continuous Control of Industrial Process", *Microwave Journal*, 12, No. 9, Aug. 1969, p. 51.
13. Veley, V F: *Modern Microwave Technology*, Prentice-Hall, Inc., Englewood Cliffs, New Jersey, 1987.

12 MICROWAVE RADIATION HAZARDS

INTRODUCTION 12.1

Microwaves are used in scientific and industrial applications as well as in the military and civilian world. But there are some adverse effects of high-power microwave radiations. These are Hazards of Electromagnetic Radiation to Personnel (HERP), Hazards of Electromagnetic Radiation to Ordinance (HERO) and Hazards of Electromagnetic Radiation to Fuel (HERF). HERP is the potential of electromagnetic radiation to produce harmful biological effects in humans, HERO is the potential of electro-explosive devices to be adversely effected by electromagnetic radiation, and HERF is the potential of electromagnetic radiation to cause spark ignition of volatile combustibles such as vehicle fuels. In this book, some details on HERP are described.

HAZARDS OF ELECTROMAGNETIC RADIATION 12.2

Hazards of Electromagnetic Radiation to Personnel (HERP) 12.2.1

Since applications of microwaves extend to the human body in medical treatments and there are possible exposures of the bodies of human and animals to high microwave power in domestic, public and military applications, HERP is described in detail in this book for increased awareness and education.

A large amount of literature has been published on the biological effects of microwave radiation. Independent studies have revealed that microwave energy can be hazardous to personnel to cause:

1. Cataracts
2. Skin cancer
3. Headaches and dizziness
4. Blood disorders/leukemia
5. Birth defects in pregnant women
6. Central nervous system damage
7. Temporary sterility in men
8. Cardiovascular problems
9. Interference with some pacemakers
10. Increased stress
11. Decrease in immune system competency

There is also the possibility that weak electric and magnetic fields from high-power transmission lines may affect biological cellular processes at the cell nucleus. HERP is caused by the thermal effect of radiated energy. Since biological substances, such as blood, brain, bone, muscle, and fat, behave as conductive or lossy dielectrics, the microwave energy directed onto the body may be scattered, reflected and absorbed depending on the field strength, the frequency, the dimension of the body and the electrical properties of the tissue. The absorbed microwave energy produces molecular vibration and converts this energy into heat. If the organism cannot dissipate this heat energy as fast as heat is produced, the internal temperature of the body will rise. This heat may damage these biological substances permanently. For example, if the lens of the eye is exposed to microwaves, its circulatory system would be unable to provide sufficient flow of blood for cooling and may cause cataracts. Similarly, the stomach, intestines and bladder are especially sensitive to thermal damage from high-power microwaves.

Microwave frequencies for which the wavelengths are of the same order of magnitude as the dimensions of the human body produce close coupling between the body and the microwave field and large amount of heat can be generated to cause severe damage in the body. Significant energy absorption will occur even when the body size is at least 1/10 of a wavelength. Although the biological damage occurs mostly due to electric field coupling, low-frequency magnetic field coupling can also produce damage when exposure time is large. Some biological effects cannot be explained by a temperature rise in the body. Persons working in microwave fields have reported headaches, eyestrain, overall fatigue and disturbance of sleep. These effects have been associated with the interaction of the microwave fields with the central nervous system of the body. Such effects have been labeled as ***non-thermal interactions***. These may be responsible for some of the long-term effects from prolonged exposure to low levels of electromagnetic fields. These effects usually occur with pulsed or pulse-modulated fields and not with the continuous wave fields.

The common sources of electromagnetic radiations which may cause hazards to biological elements are: radars, communication transmitters, mobile phones, microwave leakage from ovens, high power transmission lines, etc. The effects are reported specially for long-time exposures. A symbolic presentation of the EM fields radiated from a mobile phone to a human head is shown in Fig. 12.1.

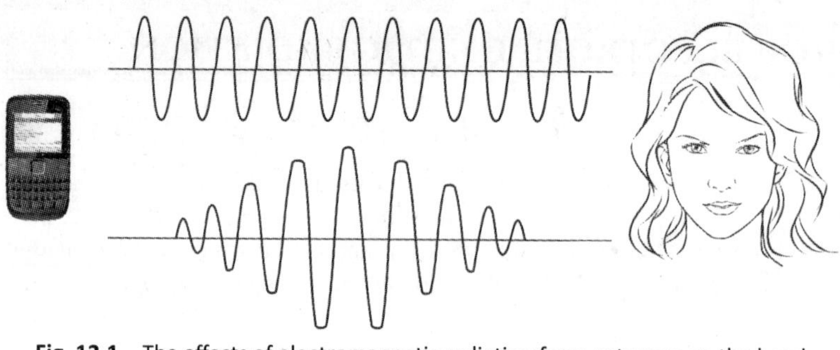

Fig. 12.1 The effects of electromagnetic radiation from antennas on the head

Some of the material properties of biological elements are given in Table 12.1 based on the report published by M A Jenesen and Y Rahmat-Samii. These data can be used to analyze the biological effects of EM fields.

Table 12.1 Material Properties at 900 MHz [Ref M A Jenesen, Y Rahmat-Samii]

Material	ε_r	σ (S/m)	ρ (g/cm^3)
Brain	55.0	1.23	1.03
Muscle	58.5	1.21	1.04
Eye	73.0	1.97	1.01
Skin/fat	34.5	0.60	1.10
Blood	59	1.26	1.00
Bone	8.0	0.105	1.85

Hazards of Electromagnetic Radiation to Ordinance (HERO) 12.2.2

Microwave energy is also dangerous to ordinance, like weapon systems, safety and emergency devices and other equipment containing sensitive electro-explosive devices (EEDs), in addition to the attending personnel and associated equipment. Radiated fields can cause unintentional triggering of EEDs.

Ordinance is more sensitive than humans partially because they do not have a circulatory system to dissipate internal heat. Ordinance reacts to peak power whereas, humans react to average power over some time. However, EEDs can more easily be protected from the effects of RF energy than humans by enclosing them with metallic enclosures which reflect back the incident microwave energy.

Hazards of Electromagnetic Radiation to Fuel (HERF) 12.2.3

HERF occurs due to possibility of accidentally igniting fuel vapours by RF-induced arcs during fuel-handling operations in proximity to high-level RF fields. The probability of ignition may be significant for more than 50 volt-ampere arcs.

RADIATION HAZARD LEVELS FOR PERSONNEL 12.3

The most widely used parameter for the measure of microwave radiation level is average power density for a plane wave in free space

$$P_d = EH = E^2/377 = 377\,H^2 \tag{12.1}$$

Unfortunately, the majority of hazardous fields in practice are not simple plane waves but have complicated amplitude, phase and polarization distributions due to their standing wave, or near field or modulation characteristics. Therefore, some standards consider Specific Absorption Rate (SAR) as the unit of measure to determine the radiation exposure limits. SAR is the rate of energy absorption for unit mass of substance measured in W/kg. SAR depends on the density of the biological substance, the substance conductivity, and the magnitude of the field. Mathematically

$$\text{SAR} = \sigma E^2/m_d \text{ W/kg} \tag{12.2}$$

where E(V/m) is the rms electric field within the material, σ (S/m) is the conductivity of the material and m_d (kg/m^3) is the mass density of the material.

Energy absorption takes place due to penetration of microwaves inside the body. The microwave penetration depth inside a body is a function of frequency. Since both dielectric constant and conductivity

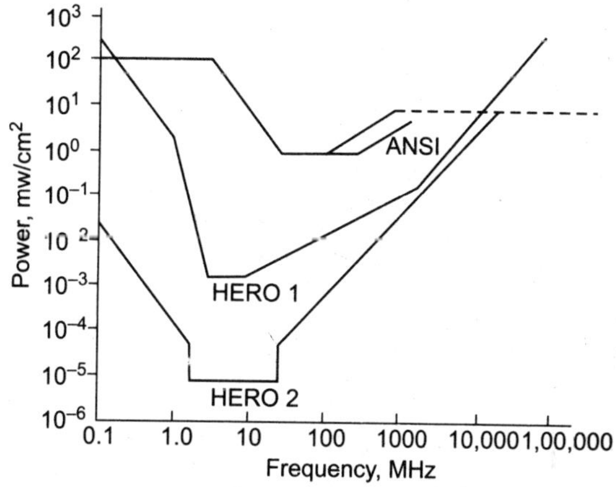

Fig. 12.2 A comparison of HERP and HERO limits

of fat are less than those for other substances in the body the penetration in fat is much greater than that for either the muscle or the blood over most of the microwave frequency range. Since the conductivities of the muscle and blood are higher, microwaves dissipate more in these substances resulting in a greater rate of heating.

Within the body, wherever the dielectric constant is higher, microwave penetration generates hot spots due to focusing action and shape of the cavities formed by bones. It is theoretically predicted that hot spots occur inside the human skull in the frequency range of 918–2450 MHz, whereas within the eyeball, it occurs at frequencies of nearly 1500 MHz. It is also found that the microwave absorption depends on the variation in body size and orientation with respect to wavelength. The typical average SAR value for a human being is about 0.03 W/kg for an incident power density of 1 mW/cm^2 at 700 MHz. This value peaks at 0.25 W/kg at 70 MHz, when the average height of a person is approximately half the wavelength.

Metabolic Heat for Humans 12.3.1

Human requires energy to perform work and produces heat to maintain an internal body temperature of around 37°C. More heat is produced with higher activity level, and the body will sweat causing discomfort. For very less activities, too little heat is produced and blood will be withdrawn from the hands and feet. Correspondingly, skin temperature will fall and the person will feel cold and uncomfortable.

The metabolic rate (M) is defined as the rate at which energy is produced when the body utilizes oxygen and food. Most of the energy produced is measurable as heat but some will be used by the body in performing mechanical work (W). The metabolic heat production (H) is then

$$H = M - W \tag{12.3}$$

The range of W usually is from 0 to 20% of M. For low activity, such as office work, W can be regarded as 0. Since the metabolic heat production is a rate of production of energy with time, hence it has the units of power (watt). The value of metabolic heat production is usually related to surface area of the body or body mass. Thus, units of W/sq. m or kcal/min/kg are used. Usually, values of 1.8 sq.m are assumed for the surface area and 70 kg for the mass of an average man, and 60 kg and 1.6 sq.m for an average woman. The body surface area is considered because this is directly related to the heat exchange between the body and the ambient or environment. However, body weight (mass) is used for persons who walk upstairs against gravitational force which will contribute to the heat produced.

The rate of temperature rise due to microwave heating can be expressed by

$$\frac{dT}{dt} = \frac{Q}{S_p}\ °\text{C/s} \tag{12.4}$$

where Q = (SAR + metabolic rate of heat production per unit mass – the rate of heat loss per unit mass in W/kg) and S_p is the specific heat of the substance in kcal°C/kg. Due to microwave irradiation, the temperature initially increases rapidly for a few minutes and then the thermoregulatory system of the body tends to stabilize the temperature. But the body temperature will start to rise again if the system cannot remove the excess heat at the same rate and, therefore, a hazard occurs.

RADIATION HAZARD LIMITS 12.4

Various international agencies have defined safety limits for radiation exposure. International Radiation Protection Association (IRPA) sets the Permissible Exposure Levels (PEL) for the general public. IRPA guidelines are divided into two categories—Occupational Permissible Exposure Levels, which applies to personnel who work in the vicinity of RF for 8 hours a day, and General Public Permissible Exposure Levels, which applies to residents, who are exposed to the RF electromagnetic environment all throughout the year.

The limits take into consideration the skin depth and Specific Absorption Rate (SAR). Table 12.1 gives the exposure standards for different countries published in earlier days.

The US Navy's Bureau of Medicine and Surgery established the biological hazard level that personnel should not be exposed to a power density above 10 mW/cm^2, when averaged over any 0.1 hour period, in the frequency range of 10 MHz to 100 GHz. This limit is 100 mW/cm^2 where the exposure is not continuous.

The American National Standards Institute Inc. (ANSI) developed ANSI C 95.1–1982 electromagnetic hazard limit of HERP and ordinance, as shown in Fig. 12.1. The ANSI standard uses the idea of a constant average SAR limit of 4 W/kg, which is 1/10 the value of the threshold for adverse effects.

Table 12.2 USSR, US, Canada, and Sweden exposure standards

Standard	*Type*	*Frequency*	*Exposure limit*	*CW/ duration*	*Pulsed*
USSR Govt. 1977	Occupational	10–30 MHz	20 V/m	Working day	Both
		30–50 MHz	10 V/m	Working day	Both
			0.3 A/m	Working day	Both
		50–300 MHz	5 V/m	Working day	Both
		0.3–300 MHz	10 mW/cm^2	Working day	Both
			100 mW/cm^2	Working day	Both
			100 mW/cm^2	2 hours	Both
			1 mW/cm^2	2 hours	Both
			1 mW/cm^2	20 min	Both
USSR Govt. 1970	General public	0.3–300 GHz	1 mW/cm^2	24 hours	Both
US ANSI 1974	Occupational	10 MHz–100 GHz	10 mW/cm^2	No limit	CW
			200 V/m		
			0.5 A/m		
			1 mWhr/cm^2	0.1 hour	Pulsed
US Army and Air Force 1965	Occupational	10 MHz–300 GHz	10 mW/cm^2	No limit	Both
US Industrial Hygientist 1971	Occupational	100 MHz–1000 GHz	10 mW/cm^2	8 hours	Both
			25 mW/cm^2	10 min	Both
Canada Can. Stds Asso. 1966	Occupational	10 MHz–100 GHz	10 mW/cm^2	No limit	CW
			1 mWhr/cm^2	0.1 hour	Pulsed
Sweden worker protection authority	Occupational	0.3–300 GHz	1 mW/cm^2	8 hours	both
		10–300 MHz	5 mW/cm^2	8 hours	Both
		10 MHz–300 GHz	25 mW/cm^2	Any	CW, Pulsed averaged over 1 second

Source: Health and Welfare, Canada, 1978.

The curve "HERO 2 Limit" represents the maximum safe fields for bare EEDs with lead wires arranged in optimum receiving orientation. The curve "HERO 1 limit" represents the safe field strength for fully assembled ordinance undergoing normal handling and loading operations. The limits for ordinance are generally lower than the limit for humans because ordinance is more sensitive than humans as they do not have natural systems to dissipate heat.

The modified military standard for military operations in ships at sea, is based on the ANSI limit but specified as 5 W/cm^2 above 1 GHz. Restricted areas have reduced safety levels for children and pregnant women. The limit for the military in non-operations condition is the same as the ANSI limit.

The RF limits for HERP in the US is based on the body-heating effects. However, the Soviet limit for HERP is 0.01 mW/cm^2, which is 30 dB lower than the US limit because they studied that there were biological effects other than thermal. The Chinese limit is 0.05 mW/cm^2.

RF wireless devices used by the general public operate at frequencies ranging from 300 kHz to 2.45 GHz. The thermal effects due to absorption of radiofrequency energy by soft tissue are the main basis for regulation in various countries for Safety Code. The exposure limits have been set to prevent excessive tissue heating. The rate at which RF energy is absorbed by the body is described in terms of the Specific Absorption Rate (SAR) in units of W/Kg. RF exposure limits for the general public in different countries, are given in Table 12.3 for specific frequencies of mobile-phone usages.

Table 12.3 RF exposure limits for the general public in different countries (Equivalent plane wave power density, W/m^2)

Frequency	*900 MHz*	*1.8 GHz*	*2.1 GHz*	*2.45 GHz*
Country				
1. ICNIRP	4.5	9	10	10
2. Canada, USA, Japan	6	10	10	10
3. Australia, Austria, Belgium, Finland, France, Germany, UK, Spain, Ireland	4.5	9	10	10
4. Bulgaria, Poland, Russia	0.10	0.10	0.10	0.10
5. Switzerland	0.045	0.095	0.095	0.095

Standards for HERP 12.4.1

Many countries have some recent influence on the radiation limits. Some of these are given below:

1. International Commission on Non-ionising Radiation Protection (ICNIRP), 1998.
2. ICNIRP is linked to IRPA and WHO. It sets standard for exposure to time varying electromagnetic fields (0–300 GHz).
3. National Radiological Protection Board (NRPB), 1993.
4. NRPB operates in UK as statutory Adviser to the Health and Safety Commission on human exposure to static and time-varying EM fields and radiations (0–300 GHz).
5. ANSI/IEEE, 1999
6. IEEE C 95.1–1999 standard sets for safety level of human exposure to RF fields in the 3 kHz–300 GHz range.

Figures 12.3–12.9 show these standards for power density, E-field and H-field vs. frequency for different categories. Figures 12.3 and 12.4 show the power density limits of above three standards for occupational and general public. It is seen that IEEE (1999) and ICNIRP (1998) are almost harmonized at lower frequency range (0.01–2.0 GHz). In the entire band, ICNIRP (1998) is most stringent. IEEE and NRPB, however, are harmonized at higher frequency range (> 3 GHz). Figure 12.5 and 12.6 show E-field and H-field

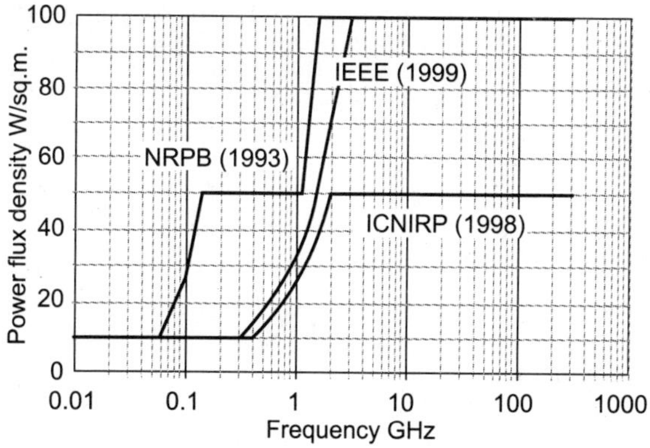

Fig. 12.3 Comparison of power density limits for three standards (occupational)

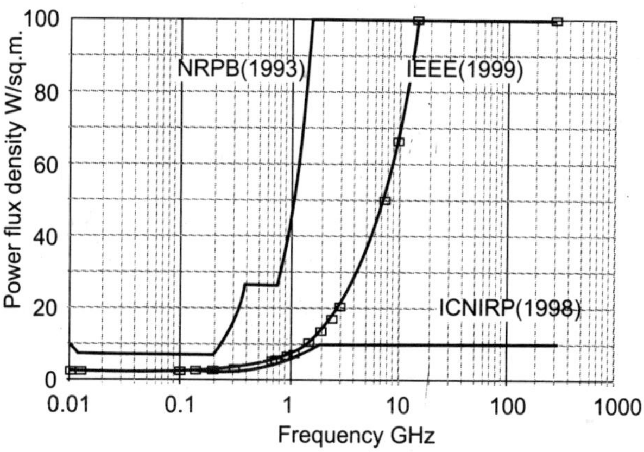

Fig. 12.4 Comparison of power density limits for three standards (public)

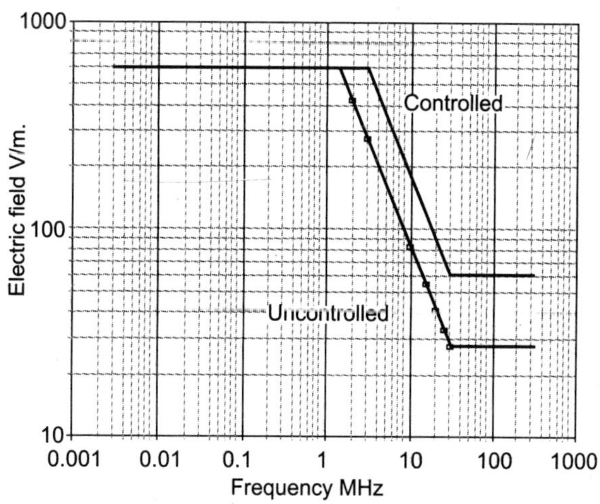

Fig. 12.5 IEEE (1999) electric field (controlled and uncontrolled areas)

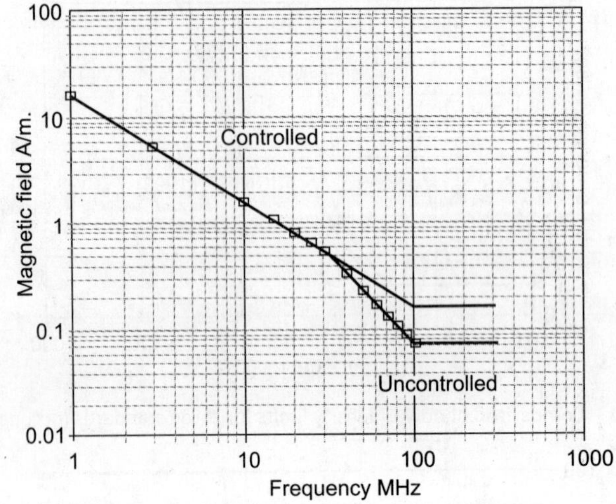

Fig. 12.6 IEEE (1999) magnetic field (controlled and uncontrolled areas)

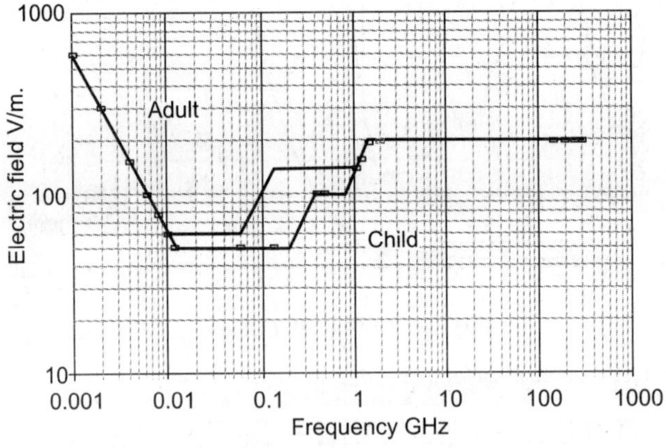

Fig. 12.7 NRPB (1993) electric field (adult and child)

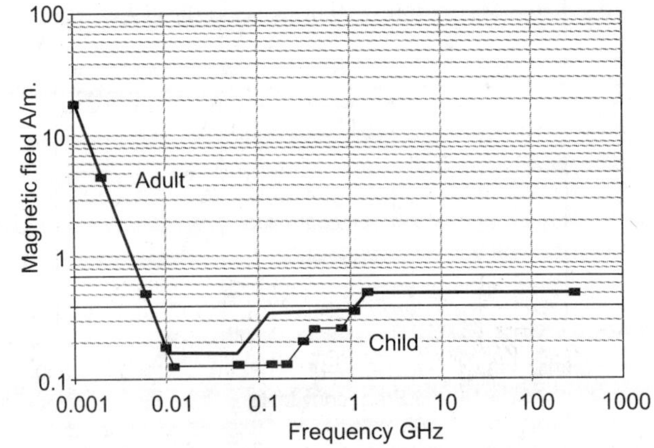

Fig. 12.8 NRPB (1993) magnetic field (adult and child)

limits, respectively, of IEEE (1999) standard for controlled and uncontrolled areas. Figures 12.7 and 12.8 show, respectively E-field and H-field limits for adult and child as per NRPB (1993) standards. Figure 12.9 shows the limits of-field for occupational and General Public according to ICNIRP (1998) standard.

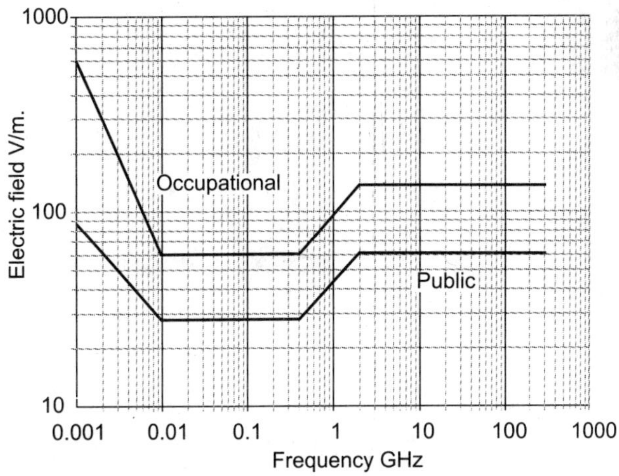

Fig. 12.9 ICNIRP (1998) electric field (occupational and public)

RADIATION PROTECTION 12.5

Radiation protection can be practiced by preventing personnel from entering into the beam of the transmit antenna or from coming close to any microwave generators or propagating medium. In areas where high-power radars are used, the service and maintenance personnel must wear microwave absorptive suits made out of stainless steel woven into a fire-retardant synthetic fibre. The suit is lightweight, comfortable, and easy to put on. The attenuation produced by such a suit is above 20 dB at 2450 MHz, 20–35 dB in the range 650 1150 MHz, and 35–40 dB in the range 1–11 GHz.

REVIEW QUESTIONS

12.1 The microwave exposure limit for the 50–300 MHz frequency range is 5 V/m. If the biological substance has a conductivity of 150 S/m and a mass density of 1.3×10^3 kg/m^3, calculate the incident power density corresponding to this limit and the specific absorption rate.

12.2 A biological substance has conductivity of 140 S/m, mass density of 1.2×10^3 kg/m^3 and 1.2 kcal. °C/kg specific heat. Find the rate of temperature rise in °C/s.

REFERENCES

1. "American National Standard Safety Levels with Respect to Human Exposure to Radio Frequency Electromagnetic Fields", 300 kHz to 100 GHz, ANSI C95.1–1982, July 1982.
2. Bowman, R R: "Quantifying Hazardous Microwave Fields", Trans. Int., Microwave Power Inst., Vol. 8
3. "Electromagnetic Compatibility by Design McDonnel Aircraft Company", McDonnel Douglas Corporation, St. Louis, Missouri.
4. "Health Aspects of Radio Frequency and Microwave Exposure", Health and Welfare, Canada, 1978.

5. "IRPA Guidelines on Limits of Exposure to RF Electromagnetic Fields in the Range from 100 MHz to 300 GHz", 1988.
6. Lin, James, "Microwave Biophysics", Trans. Int. Microwave Power Inst., Canada, 1978.
7. Jansen MA; Rahmat-Samii, Y: "EM Interaction of Handset antenna and a human head in personal communications," Proc. IEEE, Vol. 83, No. 1, 1995.
8. "Military Standard Definitions and Systems of Units, Electromagnetic Interference and Electromagnetic Compatibility Technology", MIL-STD-473A, June, 1977.
9. Michael Barge, J: "Warning: Radhaz", *EMC Technology Magazine*, pp. 21–27, May, June, 1989.
10. Kachmar, Michael: "Radiation Protection, in Small to Extra-large", Microwaves and RF, pp. 41–42, July, 1986.
11. Michaelson, S M: "Biologic and Pathophysiologic Effects of Exposure to Microwaves", Trans. Int. Microwave Power Inst., Vol. 8, 1978.
12. Michaelson, S M; Thomson, R A E and Howland, R J: "Biologic Effects of Microwave Exposure", RADC-TR-67-461, 138, pp. 1967.
13. Repacholi, M H: "Control of Microwave Exposure in Canada", Trans. Int. Microwave Power Inst., Vol. 8.
14. "Radiation Hazards", IEC Tech. Com. TC 77.
15. Ronald Kitchen, RF and Microwave Radiation Safety Handbook, Newnes, 2001. "Standards for Safety Levels with Respect to Human Exposure to Radio Frequency Electromagnetic Fields", 3 kHz to 300 GHz. IEEE C 95.1–1991/ ANSI C95.1–1982.
16. "Preclusion of Ordnance Hazards in Electromagnetic Fields", General Requirement for, MIL-STD 1385 B, August, 1986.
17. Stuchly, M A and Repacholi, M H: "Microwave and Radio Frequency Protection Standards", Trans. Int. Microwave Power Inst., Vol. 8, 1978.
18. World Health Organization. Model legislation for electromagnetic fields protection. Geneva, Switzerland: WHO; 2006.
19. Federal Communications Commission. Specific Absorption Rate (SAR) for cellular telephones. Washington, DC: FCC; 2011.

13 MICROWAVE MEASUREMENTS

INTRODUCTION 13.1

The basic measurement parameters in low-frequency ac circuits containing lumped elements are voltage, current, frequency and true power. From these measurements, the values of the impedance, the power factor, and the phase angle can be calculated. At microwave frequencies, the amplitudes of the voltages and currents on a transmission line are functions of distance and are not easily measurable. However, in a lossless line, the power transmitted is independent of the location along the line. Therefore, it is more convenient to measure power instead of voltage and current. Much of the properties of devices and circuits at microwave frequencies are obtained from the measurement of *S*-parameters, power, frequency, phase shift, VSWR and the noise figure.

Due to the complications and high cost of direct microwave measuring devices and instrumentations, such as, vector network analyzers, spectrum analyzers, power meters, etc., microwave measurements in the laboratory are often carried out using 1 kHz square-wave modulating signal which modulates the microwave test signal. The transmitted and reflected signals are then demodulated and measured using low frequency instruments such as an oscilloscope and a low frequency (1 kHz) tuned receiver, called VSWR meter. The amplitude and phase information of the microwave test signals are available in the detected low frequency signal for calculating the desired parameters. These are described in the following sections. It s found appropriate to include the descriptions of some important measurement devices and instrumentations in brief, in this chapter.

TUNABLE DETECTOR 13.2

The low frequency square-wave modulated microwave signal is detected using non-reciprocal detector diode mounted in the microwave transmission line. These diodes are specially designed point contact or metal-semi conductor Schottky barrier diodes. A detailed description of these diodes are given in Chapter 10. To match the detector to the transmission system a tunable stub is used as shown in Fig. 13.1. Broadband detectors are also manufactured in coaxial form. In order to pick up propagating fields, a coaxial line tunable probe detector is used.

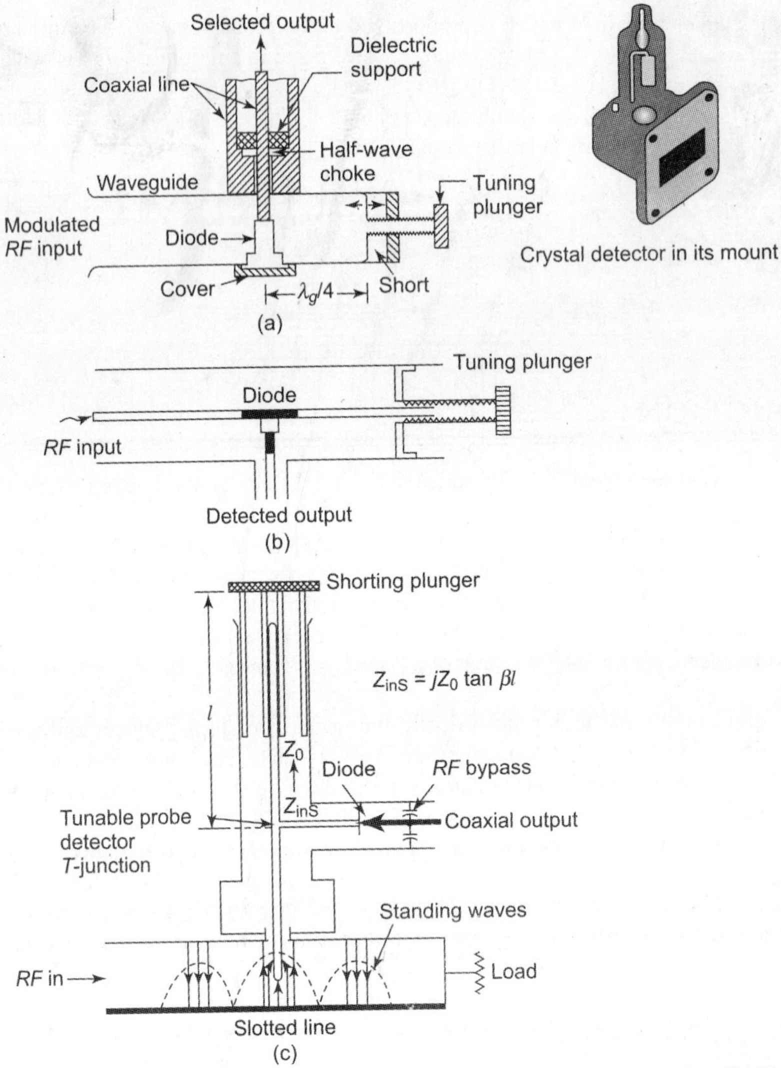

Fig. 13.1 (a) Tunable waveguide detector (b) Coaxial detector (c) Tunable probe detector

SLOTTED-LINE CARRIAGE 13.3

A slotted-line carriage (Fig. 13.2) contains a coaxial **E** field probe which penetrates inside a rectangular waveguide slotted section or a coaxial slotted line section from the outer wall and is able to traverse a longitudinal narrow slot. The longitudinal slot is cut along the centre of the waveguides broad wall or along the outer conductor of the coaxial line over a length of 2–3 wavelengths where the electric current on the wall does not have any transverse component. The slot should be narrow enough to avoid any distortion in the original field inside the waveguide. The two ends of the slot is tapered to zero width for reducing the effect of discontinuity. The probe is made to move longitudinally at a constant small depth to achieve a uniform coupling coefficient between the electric field inside the line and the probe current at all positions.

The probe samples the electric field which is proportional to the probe voltage. This unit is primarily used for the determination of locations of voltage standing wave maxima and minima along the line. The probe carriage contains a stub tunable coaxial probe detector to obtain a low frequency modulating signal output to a scope or VSWR meter. The probe should be very thin compared to the wavelength and the depth also should be small enough to avoid any field distortion.

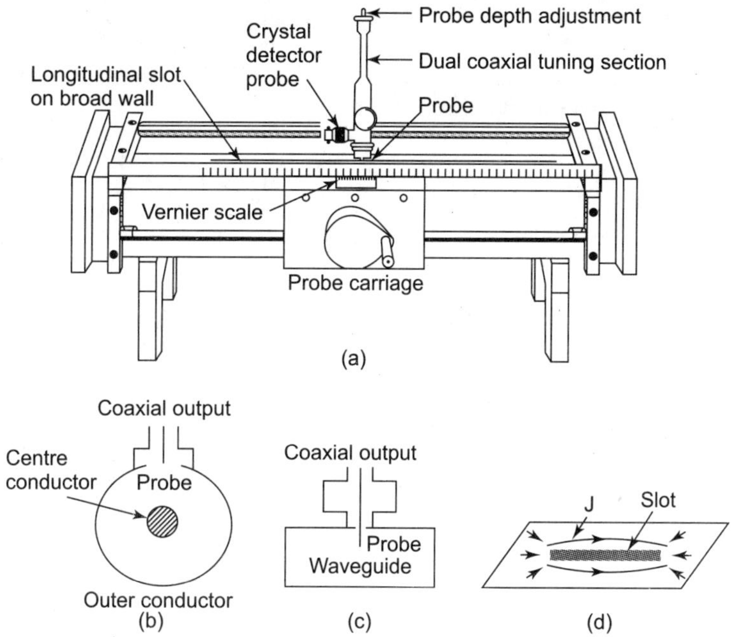

Fig. 13.2 (a) Slotted-line carriage and schematic diagram (b) Cross section of a coaxial slotted line (c) Cross section of rectangular waveguide slotted line (d) Longitudinal slot and electric wall currents

The slotted line with tunable probe detector is used to measure

1. VSWR and standing wave pattern
2. Wavelength
3. Impedance, reflection coefficient and return loss measurements by the minima shift method.

VSWR METER 13.4

A VSWR meter is a sensitive high-gain, high Q, low-noise voltage amplifier tuned normally at a fixed frequency of 1 kHz at which the microwave signal is modulated. The input to the VSWR meter is the detected signal output of the microwave detector and the output of the amplifier is measured with a square-law-calibrated voltmeter which directly gives the VSWR reading V_{max}/V_{min} for an input of V_{min}, after the meter is adjusted to unity VSWR for an input corresponding to V_{max} as shown in Fig. 13.3. A gain control can be used to adjust the reading to the desired value. The over all gain is nearly 125 dB which can be altered in steps of 10 dB.

There are three scales on the VSWR meter. When the VSWR is between 1 and 4, reading can be taken from the top SWR NORMAL scale. For VSWR between 3.2 and 10, bottom of SWR NORMAL scale is used. When the VSWR is less than 1.3, a more accurate reading can be taken by selecting the EXPANDED scale, graduated from 1 to 1.3. The third scale at the bottom is graduated in dB.

SPECTRUM ANALYZER 13.5

A spectrum analyzer is a broadband superhet receiver which provides a plot of amplitude versus frequency of the received signal, i.e., the signal spectrum as explained in Fig. 13.4.

The local oscillator is electronically swept back and forth between two frequency limits at a linear rate. The sweep voltage waveform is saw tooth type with zero flyback time to move the spot on the CRT horizontally in synchronism with the frequency sweep so that the horizontal position is a function of the frequency of the local oscillator. The amplitude of the input *RF* signal is obtained from the vertical deflection of the spot.

The basic design considerations for proper operation are

1. Frequency sweep rate
2. Frequency sweep range
3. Bandwidth of *IF* amplifier
4. Centre frequency of *IF* amplifier

For highest resolution, the bandwidth should be kept minimum and consequently sweep speed should be very low in order to allow time to build up the voltage in the receiver circuit. The range of frequencies to be covered should be as small as possible. The *IF* frequency should be chosen high enough to avoid the image response. If f_i is image frequency, f_0 is local oscillator frequency, f_{if} is *IF* frequency and f_s is signal frequency, then

$$f_i = f_0 \pm f_{if} = f_s \pm 2f_{if} \tag{13.1}$$

is the frequency that beats with the *LO* frequency and produces a frequency difference equal to the *IF*. Thus,

$$f_{if} = f_s - f_0; f_s > f_0 \tag{13.2}$$

$$= f_0 - f_s; f_s < f_0 \tag{13.3}$$

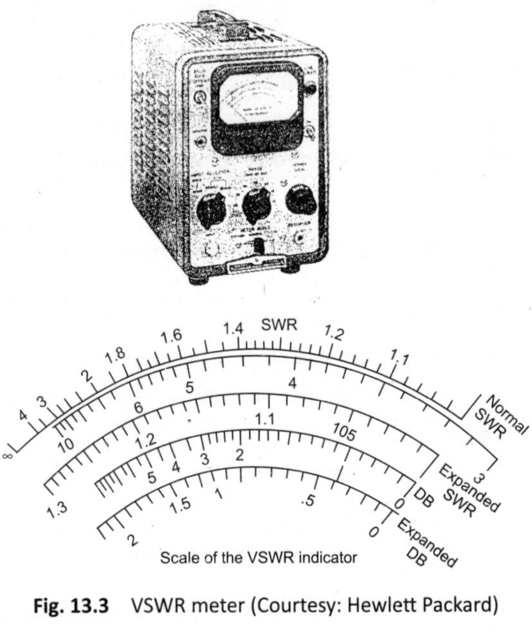

Fig. 13.3 VSWR meter (Courtesy: Hewlett Packard)

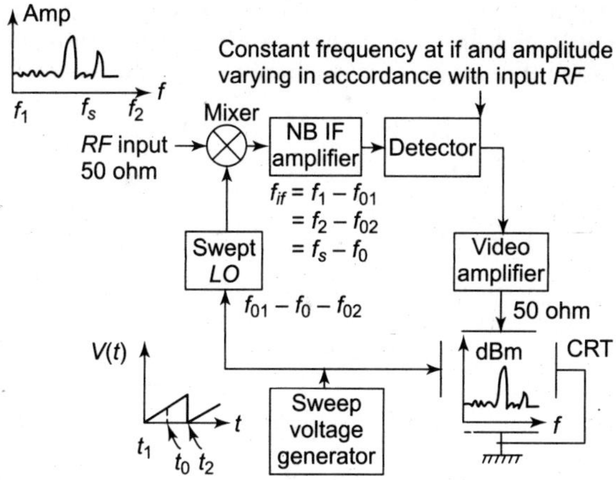

Fig. 13.4 Basic block diagram of a spectrum analyzer

For example, when f_{if} = 450 kHz and f_0 = 8000 kHz, signal on either 8450 kHz or 7550 kHz will beat against f_0 and produce f_{if}. For f_{if} = 450 kHz, image frequency $f_i = 450 \times 2 = 900$ kHz off the signal frequency. For f_{if} = 2000 kHz, $f_i = 2000 \times 2 = 4$ MHz off the signal and can be tuned out easily.

The bandwidth and hence resolution of the spectrum analyzer is determined by the bandwidth of *IF* amplifier.

NETWORK ANALYZER 13.6

The use of the slotted line for microwave measurements has the disadvantage that the amplitude and phase measurements are limited to single frequencies. Therefore, broadband testing is very time consuming and manpower cost is very high. A network analyzer measures both amplitude and phase of a signal over a wide frequency range within a reasonable time. The basic measurements involve an accurate reference signal which must be generated with respect to which the test signal amplitude and phase are measured. A schematic block diagram of a complex network analyzer is shown in Fig. 13.5.

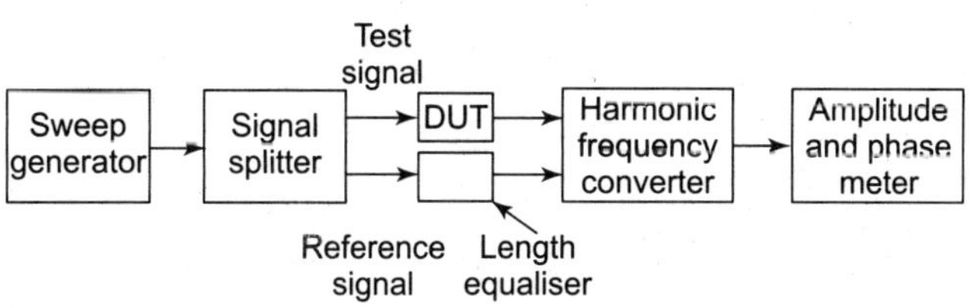

Fig. 13.5 Schematic block diagram of a complex network analyzer

The microwave signal from a sweep oscillator is first divided by means of a power divider into test signal and a reference signal channel. The test signal is transmitted through the device under test, while the reference signal passes through a phase equalizing length of line. Since processing of the microwave frequencies is not practical, both the test and reference signals are converted to a fixed intermediate frequency by means of a harmonic frequency converter. The output signals from the harmonic frequency converter are compared to determine the amplitude and phase of the test signal. The harmonic frequency converter uses a phase locked loop which helps the local oscillator to track the reference channel frequency as shown in Fig. 13.6.

This allows swept frequency measurements. The frequency conversion takes place in two steps. The first mixer converts *RF* to a fixed *IF* in the MHz range and then after amplification they are further converted to another fixed *IF* in the kHz range by means of second mixer for the final amplitude and phase comparison.

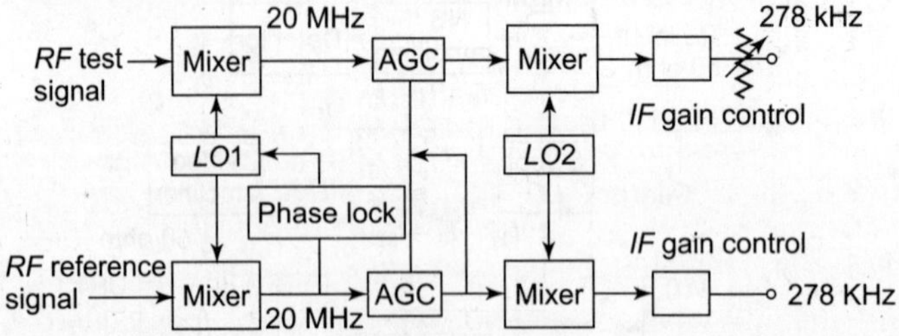

Fig. 13.6 Schematic block diagram of the harmonic frequency converter

The reflection and transmission measurements are carried out by using the reflection-transmission test unit as shown schematically in Fig. 13.7.

The reference-line length can be balanced for transmission measurement, and the device under test is compared to the sliding short for reflection measurements. The direction couplers used in the bridge are accurately matched to ensure a good balance between the two channels.

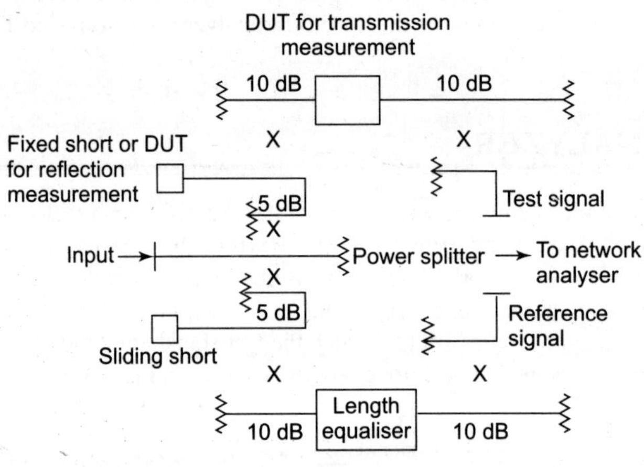

Fig. 13.7 Reflection-transmission test unit

For a two port-network, Fig. 13.8 shows the test set up for *S*-parameters S_{11} and S_{21} measurements using a network analyzer S_{ii} and S_{ij} are computed from the measured output of the dual directional couplers as follows:

$$S_{11} = \frac{V_2}{V_1}\underline{|\phi_2 - \phi_1} \tag{13.4}$$

$$S_{21} = \frac{V_3}{V_1}\underline{|\phi_3 - \phi_1} \tag{13.5}$$

For S_{22} and S_{12}, the signal source and the load position are interchanged, so that

$$S_{22} = \frac{V_3}{V_4}\underline{|\phi_3 - \phi_4} \tag{13.6}$$

$$S_{12} = \frac{V_2}{V_4}\underline{|\phi_2 - \phi_4} \tag{13.7}$$

Therefore, from the measurements of amplitude and phase from the ports of the dual directional couplers, S-parameters of a two-port network can be determined.

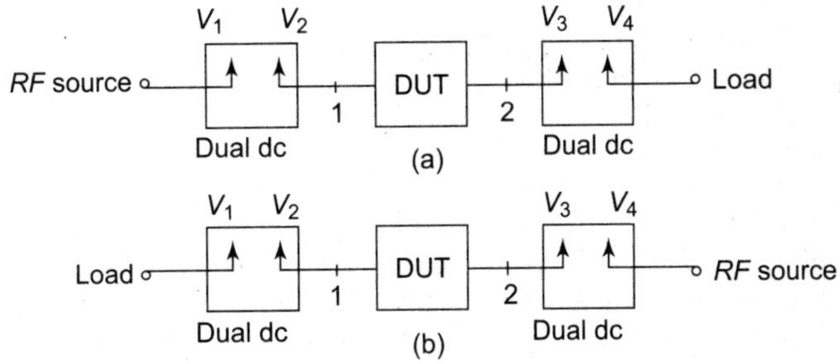

Fig. 13.8 S-parameter test set: (a) S_{11} and S_{21} (b) S_{22} and S_{12}

POWER MEASUREMENTS 13.7

CW power is defined as the quantity of energy dissipated or stored per unit time. The range of microwave power is divided into three categories—low power (less than 10 mW), medium power (from 10 mW to 10 W) and high power (greater than 10 W). The average power is measured while propagation in a transmission medium and is defined by.

$$P_{av} = \frac{1}{T}\int_0^T v(t)i(t)\,dt; \quad \begin{aligned} v(t) &= \text{Instantaneous } rf \text{ voltage} \\ i(t) &= \text{Instantaneous } rf \text{ current} \end{aligned} \tag{13.8}$$

where T is the time period of the signal. For a pulsed signal,

$$P_{peak} = \frac{1}{\tau}\int_0^\tau v(t)i(t)\,dt\,, \tag{13.9}$$

$$P_{av} = P_{peak} * \text{Duty cycle} \tag{13.10}$$

$$\text{Duty cycle} = \text{Pulse width} * p.r.f.d = \tau f_r = \tau/T < 1$$

where τ is the pulse width, T is the period and f_r is the pulse repetition frequency. The most convenient unit of power at microwaves is dBm where

$$P(\text{dBm}) = 10 \log \frac{P(mW)}{1\,mW} \tag{13.11}$$

viz. 30 dBm = 1 W and –30 dBm = 1 μW.

The microwave power meter consists of a power sensor, which converts the microwave power to heat energy. The corresponding temperature rise provides a change in the electrical parameters resulting in an output current in the low frequency circuitry and indicates the power. High power is often measured, especially for standards and calibration purposes, using microwave calorimeters in which the temperature rise of the load provides a direct measure of the power absorbed by the load.

The sensors used for power measurements are the Schottky barrier diode, bolometer and the thermocouple.

Schottky Barrier Diode Sensor 13.7.1

A zero-biased Schottky barrier diode is used as a square-law detector whose output is proportional to the input power. Since diode resistance is a strong function of temperature, the circuit is designed such that the input matching is not affected by diode resistance as shown by the equivalent circuit in Fig. 13.9. The diode detectors can be used to measure power levels as low as –70 dBm.

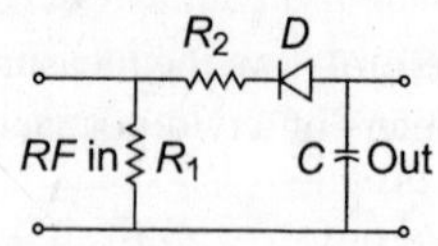

Fig. 13.9 Schottky barrier diode sensor

Bolometer Sensor 13.7.2

A bolometer is a power sensor whose resistance changes with temperature as it absorbs microwave power. The two most common types of bolometer are, the *barretter* and the *thermistor.* The barretter is a short thin metallic (platinum) wire sensor which has a positive temperature coefficient of resistance. The thermistor is a semiconductor sensor which has a negative temperature coefficient of resistance and can be easily mounted in microwave lines as shown in Fig. 13.10 due to its smaller and more compact size. The impedances of these bolometers are in the range 100–200 ohms. However, barretters are more delicate than thermistors, hence they are used only for very low power (< few mW). Medium and high power are measured with a low-power thermistor sensor, after precisely attenuating the signal. The sensitivity level of a thermistor is limited to about –20 dBm. The thermistor mount provides good impedance match, low loss, good isolation from thermal and physical shock and good shielding against energy leakage.

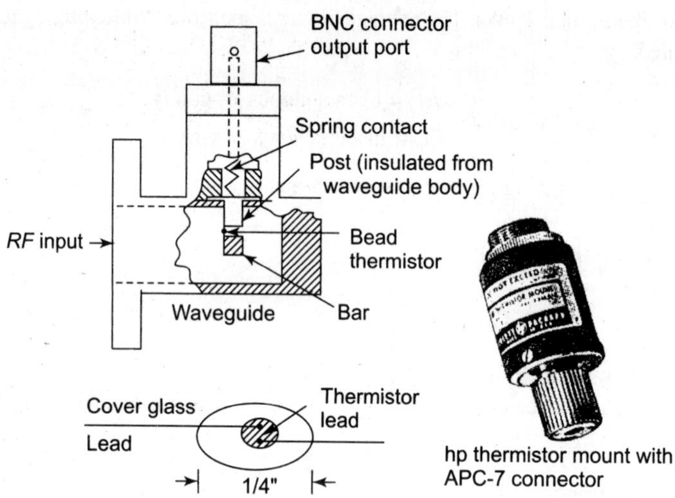

Fig. 13.10 Thermistor mount

Power Meter 13.7.3

The power meter is basically constructed from a balanced bridge circuit in which one of the arms is the bolometer as shown in Fig. 13.11. The microwave power applied to this arm will change the bolometer's resistance causing an unbalance in the bridge from its initial balance condition under zero incident power. The non-zero output is recorded on a voltmeter which is calibrated to read the level of the input microwave power.

The main disadvantages with using a single bridge are that (1) the change of resistance due to a mismatch at the microwave input port results in incorrect reading, and (2) the thermistor is sensitive to changes in the ambient temperature resulting in false reading.

These problems are eliminated by using double identical bridges—the upper bridge circuit measures the microwave power, and the lower bridge circuit compensates the effect of ambient temperature variation ($V_1 = V_2$). The added microwave power due to mismatch is compensated automatically through a self-balancing circuit by decreasing the dc power V_2 carried by the *RF* sensing thermistor until bridge balance is restored or net change in the thermistor resistance is zero due to negative dc feedback.

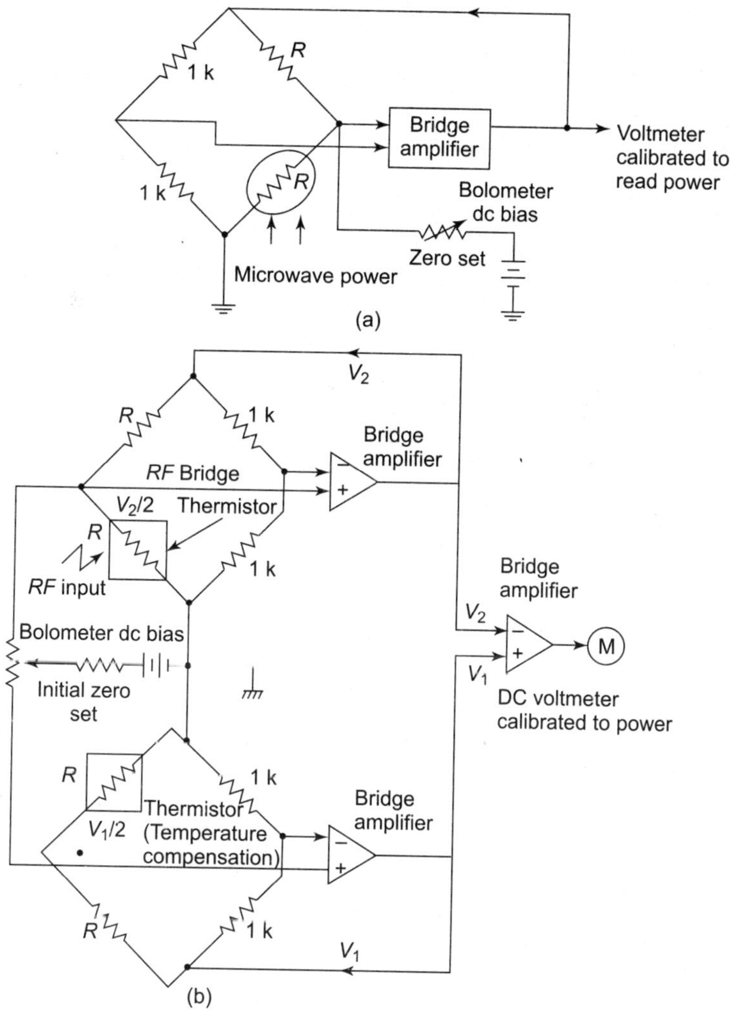

Fig. 13.11 Principle of the power meter bridge circuit: (a) Single bridge (b) Double bridge for compensation

The initial zero setting of the bridge is done by adjusting $V_2 = V_1 = V_0$ with no microwave input signal applied, when R is the resistance of the thermistor at balance. Without and with microwave present, the dc voltages across the sensor at balance are $\frac{V_1}{2}$ and $V_2/2$, respectively. The average input power P_{av} is equal to the change in dc power:

$$P_{av} = \frac{V_1^2}{4R} - \frac{V_2^2}{4R} = \frac{(V_1 - V_2)(V_1 + V_2)}{4R} \tag{13.12}$$

For any change in temperature if the voltage changes by ΔV, the change in RF power is $P_{av} + \Delta P = (V_a + \Delta V)^2/4R - (V_2 + \Delta V)^2/4R$

or $$P_{av} + \Delta P = \frac{(V_1 - V_2)(V_1 - V_2 + 2\Delta V)}{4R} \tag{13.13}$$

Since $V_1 + V_2 >> \Delta V$ in practice, $\Delta P \approx$ o. The meter responds to Eq. 13.13 to read microwave power P_{av}.

Thermocouple Sensor 13.7.4

A thermocouple is a junction of two dissimilar metals or semiconductors (n-type Si). It generates an emf when two ends are heated up differently by absorption of microwaves in a thin film tantalum-nitride resistive load deposited on a Si substrate which forms one electrode of the thermocouple as shown in Fig. 13.12. This emf is proportional to the incident microwave power to be measured.

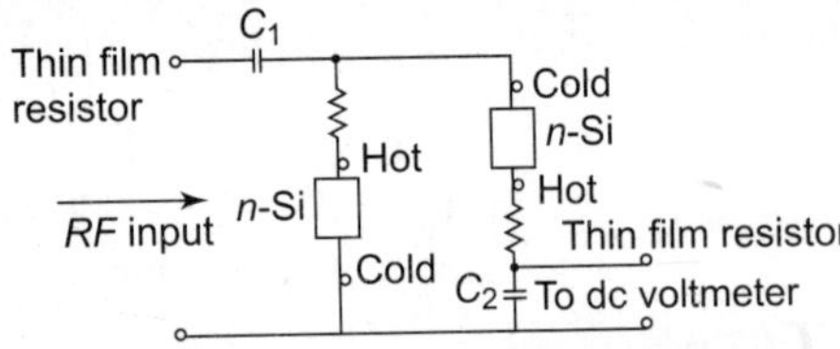

Fig. 13.12 Thermocouple power sensor

Here, C_2 is the RF by-pass capacitor and C_1 is the input coupling capacitor or dc block. The emf generated in the parallel thermocouples are added to appear across C_2. The output leads going to the dc voltmeter are at RF ground so that the output meter reads pure dc voltage proportional to the input microwave power. For a square-wave modulated microwave signal the peak power can be calculated from the average power measured as

$$P_{peak} = P_{av} \times T/\tau \tag{13.14}$$

where T is the time period and τ is the pulse width.

High-Power Measurements by the Calorimetric Method 13.7.5

High-power microwave measurements can be conveniently done by the calorimetric method which involves conversion of the microwave energy into heat, absorbing this heat in a fluid (usually water) and then measuring the temperature rise of the fluid as shown in Fig. 13.13. There are two types: one is the direct heating method and another is the indirect heating method. In the direct heating method, the rate of production of heat can be measured by observing the rise in the temperature of the dissipating medium. In indirect heating method, heat is transferred to another medium before measurement. In both the methods static calorimeter and circulating calorimeter are used.

Static Calorimeters

It consists of a 50-ohm coaxial cable which is filled by a lossy dielectric load. The load has sufficient thermal isolation from its surrounding. The microwave power is dissipated in the load. The average power input is given by

$$P = \frac{4.187\, mC_p T}{t} \text{ watts} \tag{13.15}$$

A schematic diagram of static calorimeter is shown in Fig. 13.13(a).

Here, m = Mass of the thermometric medium in gms

C_p = Specific heat in cal/g

T = Temperature rise in °C

t = Time in seconds

By noting the temperature rise, microwave power can be determined.

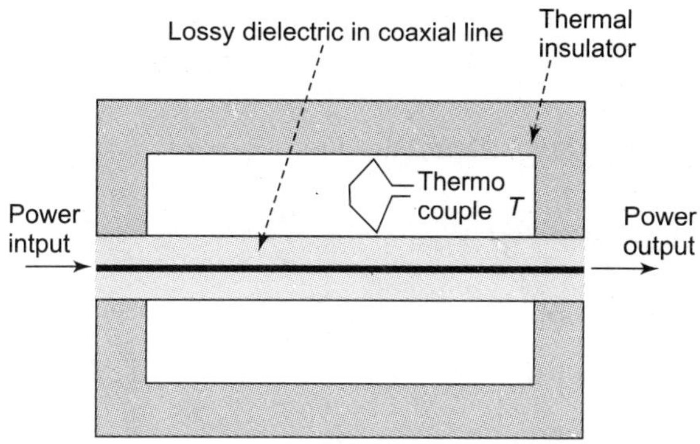

Fig. 13.13(a) Static Calorimeter

Circulating Calorimeters

Here, the calorimeter fluid (water) is constantly flowing through a water load. The heat introduced into the the fluid makes exit temperature higher than the input temperature. Here, average power

$$P = 4.187\, v\, d\, C_p\, T \text{ Watts} \tag{13.16}$$

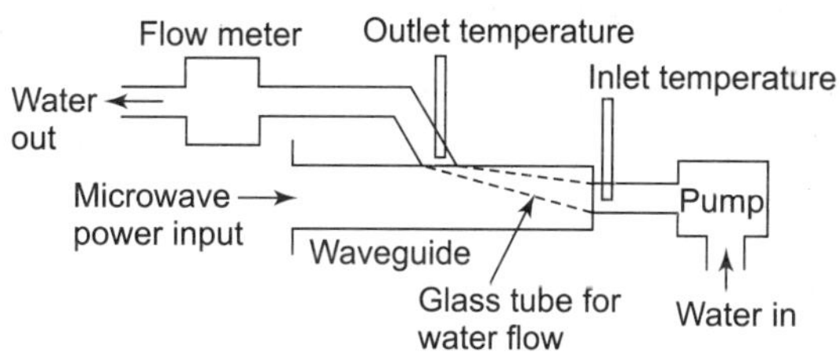

Fig. 13.13(b) Microwave calorimeter

where, v = Rate of flow of calorimeter fluid in cc/sec

d = Specific gravity of the fluid in gm/cc.

T = Temperature rise in °C

C_p = Specific heat in cal/gm

Microwave power is measured from the reading of the temperature rise.

A disadvantage of calorimeter measurements is the thermal inertia caused by the lag between the application of microwave power and the parameter readings.

INSERTION LOSS AND ATTENUATION MEASUREMENTS 13.8

When a device or network is inserted in the transmission line, part P_r of the input signal power P_i is reflected from the input terminal and the remaining *part* $P_i - P_r$ which actually enters the network is attenuated due to the non-zero loss of the network. The output signal power P_0 is therefore less than P_i. Therefore, insertion loss is defined by the difference in the power arriving at the terminating load with and without the network in the circuit.

Since, $$\frac{P_0}{P_i} = \frac{P_i - P_r}{P_i} * \frac{P_0}{P_i - P_r} \quad (13.17)$$

or, $$10\log\frac{P_0}{P_i} = 10\log\left(1-\frac{P_r}{P_i}\right) + 10\log\left(\frac{P_0}{P_i - P_r}\right) \quad (13.18)$$

Insertion loss = Reflection loss + Attenuation loss

where, by definition,

$$\text{Insertion loss (dB)} = 10\log(P_0/P_i) \quad (13.19)$$

$$\text{Reflection loss (dB)} = 10\log\left(1-\frac{P_r}{P_i}\right) = 10\log(1-|\Gamma|^2)$$

$$= 10\log\frac{4S}{(1+S)^2};\; S = \frac{1-|\Gamma|}{1+|\Gamma|} \quad (13.20)$$

$$\text{Attenuation loss (dB)} = 10\log\left(\frac{P_0}{P_i - P_r}\right) \quad (13.21)$$

$$\text{Return loss (dB)} = 10\log P_r/P_i = 20\log|\Gamma| \quad (13.22)$$

For perfect matching, $P_r = 0$, and the insertion loss and the attenuation loss become the same. The experimental set up for insertion and the attenuation measurements are shown Fig. 13.14. The relative power levels are measured by using detectors and a VSWR meter. DC_1 and DC_2 are two identical directional couplers.

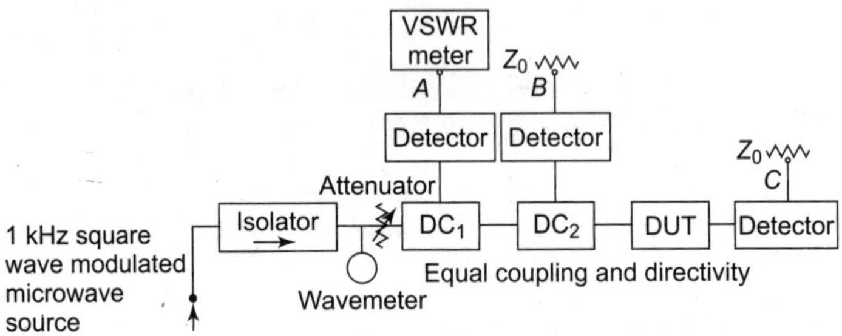

Fig. 13.14 Insertion loss and attenuation measurements

The following steps are involved for the insertion loss and attenuation measurements:

1. The microwave source is set to a suitable frequency and the 1 kHz square wave modulation level is adjusted for a peak reading on the VSWR meter at *A* with minimum input attenuation.
2. For a crystal detector to work in the square-law region the power level is adjusted to get a reading in the 30 dB range of the VSWR meter. The input power from port *A* is set to zero dB or 1.0 using gain control.
3. Frequency is read from the cavity frequency meter when a dip is observed in the VSWR meter.
4. Connecting matched load Z_0 to ports *A* and *C* and VSWR meter to port *B*, without disturbing any other set-up, the reading in the VSWR meter gives the ratio P_r/P_i, the return loss. The reflection loss $1-(P_r/P_i)$ is calculated.

5. Now the input attenuator is adjusted to give an attenuation equal to the dB coupling of the directional coupler. The matched load is connected to ports *A* and *B*, and the VSWR meter to port *C* without disturbing any other setup. The reading in the VSWR meter gives the ratio P_0/P_i, the insertion loss. Attenuation of the network under test can be determined by subtracting the dB reflection loss from the dB insertion loss.

The main errors in this measurement are the following:

1. P_i, P_0 and P_r may not all be capable of operating the crystal detector within its square-law region.
2. Both the directional couplers may not have the same characteristics.
3. There is some degree of mismatch between the various components in the set-up.

VSWR MEASUREMENTS 13.9

VSWR and the magnitude of voltage reflection coefficient Γ are very important parameters which determine the degree of impedance matching. These parameters are also used for the measurement of load impedance by the slotted line method as shown in Fig. 13.15. When a load $Z_L \neq Z_0$ is connected to a transmission line, standing waves are produced. By inserting a slotted line system in the line, standing waves can be traced by moving the carriage with a tunable probe detector along the line. VSWR can be measured by detecting V_{max} and V_{min} in the VSWR meter: $S = V_{max}/V_{min}$.

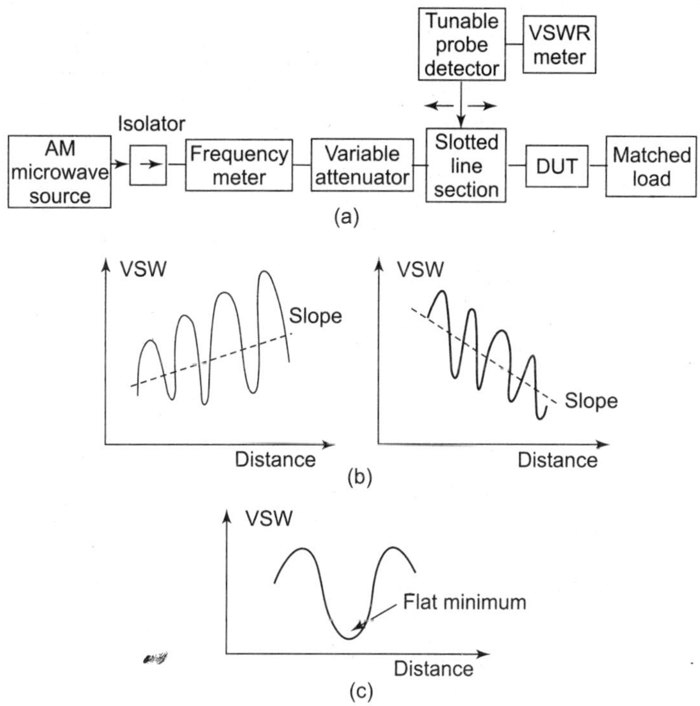

Fig. 13.15 Slotted-line method of VSWR measurement: (a) Basic experimental setup (b) Slant pattern due to mechanical slope error (c) Flat minimum position due to imperfect square wave modulating signal

Low VSWR ($S < 20$) 13.9.1

Low values of VSWR can be measured directly from the VSWR meter using the experimental set-up shown in Fig. 13.15. as follows.

1. The variable attenuator is adjusted to 10 dB. The microwave source is set to the required frequency. The 1 kHz modulation is adjusted for maximum reading on the VSWR meter in a 30 dB scale. The probe carriage stub is tuned for maximum detected signal in VSWR meter.
2. Slide the probe carriage along the non-radiating slot from the load end until a peak reading is obtained in VSWR meter. The meter's gain control is adjusted to get the meter reading at 1.0 or 0 dB corresponding to the position of voltage maximum.
3. The probe is moved towards the generator to an adjacent voltage minimum. The corresponding reading in VSWR meter directly gives the VSWR = V_{max}/V_{min} on the top of SWR NORMAL scale for $1 \le S \le 4$ or on the EXPANDED scale for $1 \le S \le 1.33$.
4. The experiment is repeated for other frequencies as required to obtain a set of values of S vs f.
5. For VSWR between 3.2 and 10, a 10 dB lower RANGE should be selected and reading corresponding to V_{min} position should be taken from the second VSWR NORMAL scale from the top.
6. For VSWR between 10 and 40, a 20 dB RANGE sensitivity increase is required and reading is taken from the top of VSWR NORMAL scale (1 to 4) at the voltage minimum and should be multiplied by 10 to obtain actual VSWR.
7. For VSWR between 32 and 100, a 30 dB lower RANGE must be selected and reading is taken from the second VSWR NORMAL scale (3.2 to 10) from the top at the voltage minimum. The reading should be multiplied by 10 to obtain actual VSWR.

The possible sources of error in this measurements are the following:

1. V_{max} and V_{min} may not be measured in the square-law region of the crystal detector.
2. The probe thickness and depth of penetration may produce reflections in the line and also distortion in the field to be measured. Depth of penetration should be kept as small as possible otherwise values of VSWR measured would be lower than actual.
3. Mechanical slope between the slot geometry and probe movement may cause different values of VSWR for measurement at different locations along the slot (Fig. 13.15 (b)).
4. When VSWR < 1.05, the associated VSWR of connector produces significant error in VSWR measurement. Very good low VSWR (< 1.01) connectors should be used for very low VSWR measurements.
5. If the modulating 1 kHz signal is not a perfect square-wave, the microwaves will be frequency modulated and at each frequency there will be a different set of standing waves. This causes reduction in the sharpness of voltage minima and there may be error in the reading of minimum position as shown in Fig. 13.15(c).
6. Any harmonics and spurious signals from the source may be tuned by the probe to cause measurement error.
7. A residual VSWR of slotted line arises due to mismatch impedance between the slotted line and the main line as explained in Fig. 13.16. Let

   ρ_L = Actual load reflection coefficient
   ρ_s = Slotted line reflection coefficient on main line
   E_i = Incident electric field at any point on the main line
   E_L = Reflected electric field from the load
   E_s = Reflected electric field on the main line because of slotted line

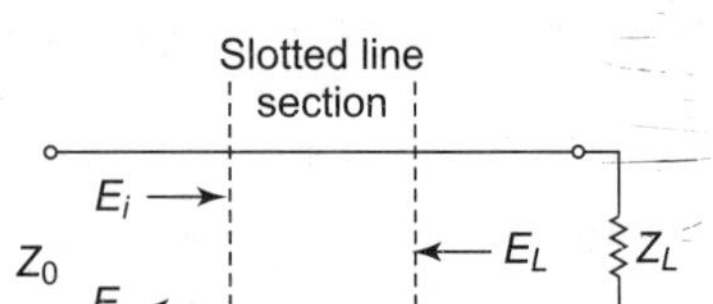

Fig. 13.16 Residual VSWR of slotted line

Then, the total reflected field at a point = $|E_s \pm E_L|$. The maximum and minimum VSWR and reflection coefficients on the main line are

$$S_{max} = \frac{E_i + (E_s + E_L)}{E_i - (E_s + E_L)} \quad (13.23)$$

$$S_{min} = \frac{E_i + (E_s - E_L)}{E_i - (E_s - E_L)} \quad (13.24)$$

$$\rho_{max} = \frac{S_{max} - 1}{S_{max} + 1} = |\rho_L| + |\rho_s| \quad (13.25)$$

$$\rho_{min} = \frac{S_{min} - 1}{S_{min} + 1} = |\rho_L| - |\rho_s| \quad (13.26)$$

The above equations can be solved for ρ_L and ρ_s from the measurements of S_{max} and S_{min} on the line. Then the residual VSWR

$$S_s = \frac{1 + |\rho_s|}{1 - |\rho_s|} \quad (13.27)$$

High VSWR ($S > 20$) 13.9.2

For high VSWR, the difference of power at voltage maximum and voltage minimum is large, so it would be difficult to remain on the detector's square-law region at maximum positions when the diode current may exceed 20 μA. Therefore, VSWR measurement with a VSWR meter calibrated on a square-law basis ($I = kV^2$) will be inaccurate. Hence double minimum method as shown in Fig. 13.17. is used where measurements are carried out at two positions around a voltage minimum point. The theory of this method can be established as follows. Let the ratio of line voltage near a minimum and the voltage at the minimum be

$$r_n = \frac{|V(x)|}{|V_{min}|} \quad (13.28)$$

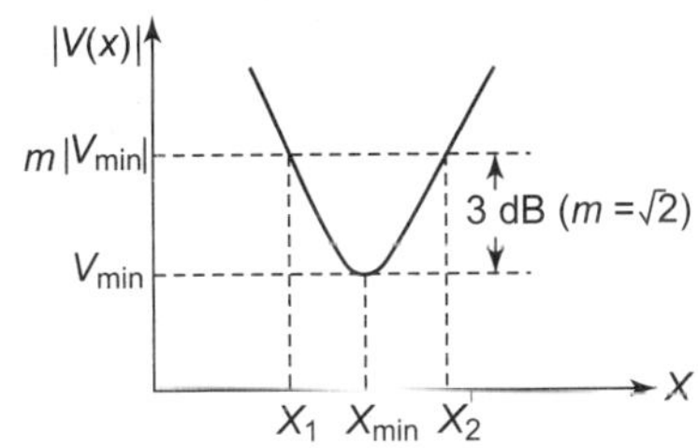

Fig. 13.17 Double minima method

For a load, reflection coefficient $\Gamma = \rho e^{j\phi}$. The line voltage at a distance x from the load end is

$$|V(x)| = |V_{inc}|\,|1 + \rho e^{j(\phi - 2\beta x)}|$$

or, $$|V(x)| = |V_{inc}|\,[1 + 2\rho \cos(\phi - 2\beta x) + \rho^2]^{1/2} \quad (13.29)$$

The voltage minimum

$$|V_{min}| = |V_{inc}|\,(1 - \rho) \text{ at } x = x_{min} \quad (13.30)$$

If x_1 and x_2 are two points around x_{min} where $|V(x_1)| = |V(x_2)| = m|V_{min}|$,

$$m = \frac{|V(x_1)|}{|V_{min}|} = \frac{[1 + 2\rho\cos(\phi - 2\beta x_1) + \rho^2]^{1/2}}{1 - \rho} \quad (13.31)$$

By substituting $\rho = (S - 1)/(S + 1)$, VSWR can be expressed as

$$S = \frac{\left[m^2 - \cos^2\left(\dfrac{2\pi(x_1 - x_{min})}{\lambda_g}\right)\right]^{1/2}}{\sin\left[\dfrac{2\pi(x_1 - x_{min})}{\lambda_g}\right]} \quad (13.32)$$

where $\beta = 2\pi/\lambda_g$ and λ_g is the guide wavelength.

If x_1 is the point in the vicinity of x_{min},

$$\Delta x = 2\,(x_1 - x_{min}) \tag{13.33}$$

and

$$S = \sqrt{\left[\frac{m^2 - \cos^2\left(\dfrac{\pi\,\Delta x}{\lambda_g}\right)}{\sin^2\left(\dfrac{\pi\,\Delta x}{\lambda_g}\right)}\right]} = \left[\frac{m^2 - 1}{\sin^2\left(\dfrac{\pi\,\Delta x}{\lambda_g}\right)} + 1\right]^{1/2} \tag{13.34}$$

For convenience of measurement with a square-law detector, if $m = \sqrt{2}$ is selected, where x_1 is 3 dB above the x_{min} point, then

$$S = \sqrt{\left[\frac{2-1}{\sin^2\left(\dfrac{\pi\,\Delta x}{\lambda_g}\right)} + 1\right]} = \sqrt{\left[1 + \operatorname{cosec}^2\left(\dfrac{\pi\,\Delta x}{\lambda_g}\right)\right]} \tag{13.35}$$

If $\pi\,\Delta x << \lambda_g$

$$S \approx \operatorname{cosec}\left(\frac{\pi\,\Delta x}{\lambda_g}\right) = \frac{1}{\sin\left(\dfrac{\pi\,\Delta x}{\lambda_g}\right)} \cong \frac{\lambda_g}{\pi\,\Delta x} \tag{13.36}$$

where $\Delta x = x_2 - x_1$. Thus, high VSWR can be measured by observing the distance between two successive minima to find λ_g and distance Δx between two 3 dB points on both sides of V_{min}.

The method follows the steps given below.

1. The probe is moved to a voltage minimum and the probe depth and gain control is adjusted to read 3 dB in the VSWR meter.
2. The probe is moved slightly on either side of the minimum to read 0 dB in the meter. This position x_1 is noted. The probe is then moved to the other side of the minimum to read 0 dB again at x_2.
3. By moving the probe between two successive minima a distance equal to $\lambda_g/2$ is found to determine the guide wavelength λ_g,
4. High VSWR is calculated from

$$S \doteq \frac{\lambda_g}{\pi(x_1 \sim x_2)} \tag{13.37}$$

RETURN-LOSS MEASUREMENT BY A REFLECTOMETER 13.10

The return loss and VSWR of a load can be determined by measuring the magnitude of the reflection coefficient with a reflectometer, a set-up in which two identical directional couplers are connected opposite to each other as shown in Fig. 13.18. One coupler couples to the forward wave and the other to the reverse wave.

Let us assume that the directional couplers have infinite directivity, a voltage coupling coefficient C, main line VSWR 1 and the detectors have constant impedance and perfect matching to the line. When a unit input amplitude is fed to Port 1; voltages at Ports 4 and 2 are, respectively,

$$b_4 = C \tag{13.38}$$

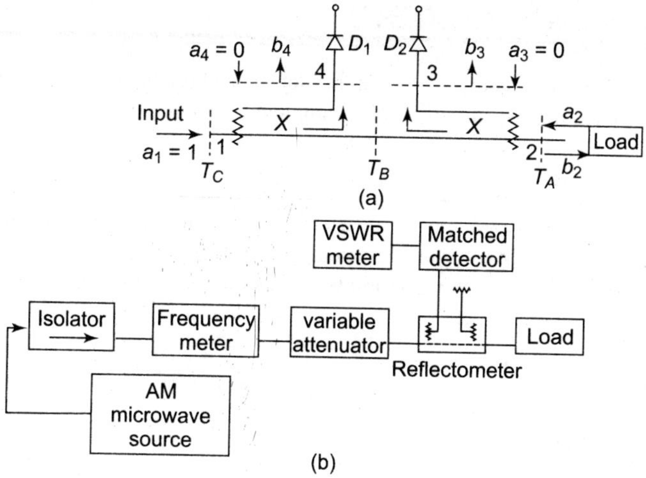

Fig. 13.18 (a) Reflectometer (b) Experimental set-up for a reflectometer

$$b_2 = (1 - C^2)^{1/2} \tag{13.39}$$

Incident voltage at Port 2 is reflected by the load under test. If Γ_L is the reflection coefficient, the reflected wave amplitude at Port 2 is

$$a_2 = (1 - C^2)^{1/2}\, |\Gamma_L| \tag{13.40}$$

This will be coupled to Port 3 to produce a voltage of

$$b_3 = (1 - C^2)^{1/2} C\, |\Gamma_L| \tag{13.41}$$

Then $$|b_3/b_4| = (1 - C^2)^{1/2}\, |\Gamma_L| = K|\Gamma_L| \tag{13.42}$$

If coupling is extremely small *i.e.* $C << 1$, $K \approx 1$. Therefore,

$$|b_3/b_4| = |\Gamma_L|. \tag{13.43}$$

Thus, knowing the voltage ratio between ports 3 and 4 , the reflection coefficient and hence VSWR and return loss can be determined from the following relations.

$$\text{VSWR} = \frac{1 + |\Gamma_L|}{1 - |\Gamma_L|} \tag{13.44}$$

$$\text{Return loss} = -20 \log |\Gamma_L| \tag{13.45}$$

The experiment is conducted first by terminating port 2 with a short and adjusting the output of the detector D_1 at Port 4 to unity in VSWR meter while detector D_2 at Port 3 is matched terminated. VSWR meter and match load at D_1 and D_2 are now interchanged. The output of Port 3 is noted which should ideally be equal to the output from Port 4. Without disturbing the VSWR meter adjustment, the unknown load is connected at Port 2 by replacing the short and the output at Port 3 is noted to obtain $1/|\Gamma_L|$ directly from the VSWR meter.

This method is well suited for loads having low VSWR. The major source of error in this method is that the unstability of the signal source causes a change of signal power level during measurement of input and reflected signal levels at different instants of time. Non-ideal directional couplers and detectors are also sources of error.

IMPEDANCE MEASUREMENT 13.11

Since impedance is a complex quantity, both amplitude and phase of the test signal are required to be measured. The following techniques are commonly employed for such measurements.

Slotted-Line Method 13.11.1

The complex impedance Z_L of a load can be measured by measuring the phase angle ϕ_L of the complex reflection coefficient Γ_L from the distance of first voltage standing wave minimum $d_{\min}$ and the magnitude of the same from the VSWR, S. The following relations are important for the computation of Z_L.

$$Z_L = Z_0 \frac{1+\Gamma_L}{1-\Gamma_L} \tag{13.46}$$

$$\Gamma_L = \rho_L\, e^{j\phi L} \tag{13.47}$$

$$S = (1 + \rho_L)\,(1 - \rho_L) \tag{13.48}$$

$$\phi_L = 2\beta d_{\min} - \pi \tag{13.49}$$

$$\beta = 2\pi/\lambda_g \tag{13.50}$$

$\lambda_g = 2 \times$ Distance between two successive minima.

The method of using slotted line to determine an unknown impedance is explained in Fig. 13.19. The steps for measurement are summarized as follows.

1. Measure the load VSWR to find ρ_L from Eq. 13.48.
2. Measure the distance d between two successive voltage minima to find $g = 2\,d$ and $\beta = 2\pi/\lambda_g$.
3. Measure the distance $d_{\min}$ of the first voltage minimum from the load plane towards generator in the following manner.

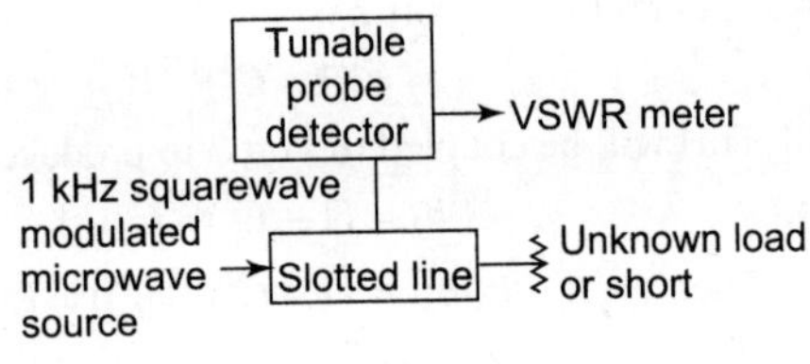

Fig. 13.19 Determination of load impedance using slotted line

Since it may not be possible to reach the first $d_{\min}$ by the probe close to the load directly using slotted line, an equivalent load reference plane on the slotted line is established by means of a short circuit at the load reference plane where a voltage minimum now occurs. Since a series of minima are produced on the slotted line at intervals of $\lambda_g/2$, the load reference plane can be shifted to a convenient minimum position near the centre of the slotted line as shown in Fig. 13.20. The $d_{\min}$ can then he measured by observing the first minimum from this shifted reference plane when the load replaces the reference short.

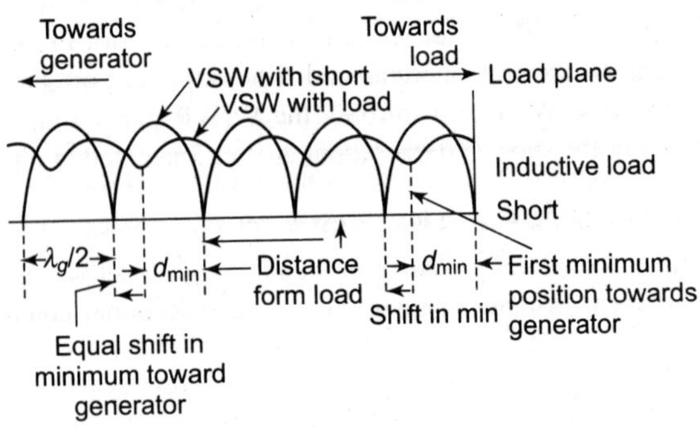

Fig. 13.20 Determination of $d_{\min}$

4. Phase angle ϕ_L of the load is calculated from Eq. 13.49 and hence $\Gamma_L = \rho_L e^{j\phi L}$ is found.
5. The unknown impedance Z_L is then calculated from Eq. 13.46.

To ease the calculation, Smith chart (Fig. 13.21) can be used to determine Z_L from the measurements of S and d_{min} as follows, where load VSWR $S = 2$, and $d_{min}/\lambda_g = 0.2$, say.

1. Draw the VSWR circle centred at 0 ($r = 1$) with radius cutting the *r-axis* at $S = 2$.

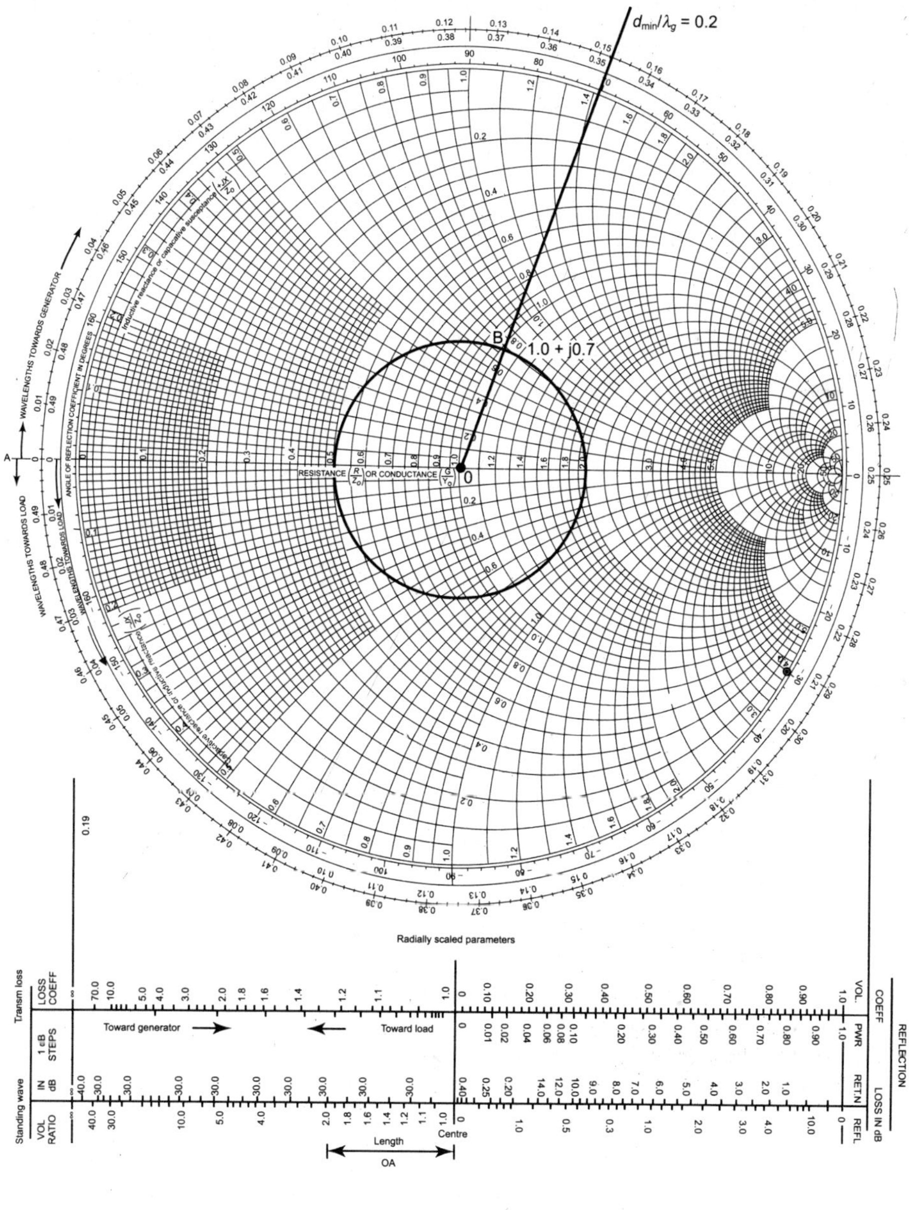

Fig. 13.21 Smith chart for Z_L measurement

2. Move from the short circuit load point A on the chart along the wavelengths toward load scale by distance d_{min}/λ_g to B and join OB.
3. The point of intersection between the line OB and the VSWR circle gives the normalized load $z_L = Z_L/Z_0$ and hence the complex load $Z_L = Z_0\,(1.0 + j0.7)$.

Impedance Measurement of Reactive Discontinuity 13.11.2

The impedance of a shunt reactive discontinuity, such as a post or windows or a step in a microwave transmission line, can be measured using the slotted line method from the measurement of line VSWR and the distance of first voltage minimum from the discontinuity plane as follows.

Let jX be the reactance of the discontinuity at load distance point $d = 0$ The line is terminated by a matched load R_0 at $d = 0$. The total impedance of the combination is

$$Z_L = \frac{R_0 \cdot jX}{R_0 + jX} \tag{13.51}$$

The normalized value

$$\frac{Z_L}{R_0} = \frac{jX}{R_0 + jX} = \frac{X^2}{R_0^2 + X^2} + j\frac{X R_0}{R_0^2 + X^2} = x + jy, \text{ say} \tag{13.52}$$

where,

$$x = \frac{X^2}{R_0^2 + X^2},\; y = \frac{X R_0}{R_0^2 + X^2} \tag{13.53}$$

with X unknown. Here, we see that

$$X/R_0 = x/y \tag{13.54}$$

or,

$$B/G_0 = y/x \tag{13.55}$$

where $B = 1/X$ and $G_0 = 1/R_0$. The values of x and y are obtained from the Smith chart to find the value of the normalized susceptance of the discontinuity.

The procedure of measurement follows the following steps with reference to the experimental set-up shown in Fig. 13.22.

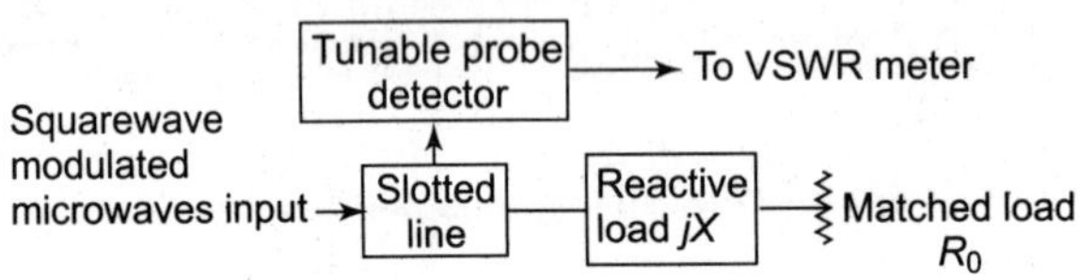

Fig. 13.22 Experimental set-up for the measurement of impedance of a discontinuity

Discontinuity such as a tuning screw or windows in the line is connected to the slotted line and the output port is terminated by a matched load R_0. By adjusting the tuning stub, a voltage minimum position, x_1 is noted along the slotted line and the corresponding VSWR and λ_g are measured.

Next, the discontinuity and matched load are replaced by a fixed short at the point of the discontinuity position. The first voltage minimum position x_2 is noted towards the load from x_1. Thus, x_2 translates the plane of discontinuity on the slotted line scale so that the distance of first voltage minimum with discontinuity and matched load termination is $d_{min} = (x_1 \sim x_2)$.

By locating d_{min}/λ_g, S on the Smith chart, $Z_L/R_0 = x + jy$ can be read and $B/G_0 = y/x$ is calculated.

Impedance Measurement by Reflectometer 13.11.3

The reflectometer arrangement shown in Fig. 13.18(a) cannot have ideal conditions of infinite directivity, constant impedance detectors and perfect impedance matching. When the unknown impedance is connected to the output Port 2, the ratio of the signal amplitudes at ports 3 and 4 is, in general

$$\frac{b_3}{b_4} = \frac{A\Gamma_L + B}{C\Gamma_L + D} \tag{13.56}$$

where A, B, C and D are functions of the S-parameters of the four port 2 formed by the reflectometer. By using two tuners T_A and T_B, a movable short, and a sliding load of low VSWR (<1.02), the above ideal conditions could be achieved as shown in Fig. 13.23.

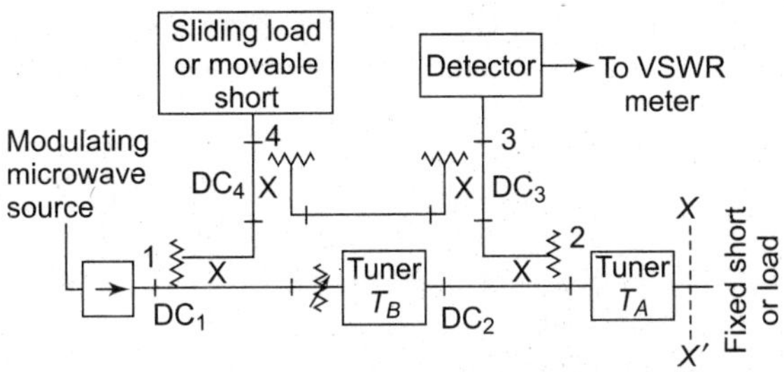

Fig. 13.23 Reflectometer with tuners for amplitude and phase measurements

Tuner T_A is adjusted to make $|b_3/b_4|$ constant while the phase of Γ_L is varied by changing the position of a sliding load at Port 4. This makes $B = 0$. The tuner T_B is adjusted to make $|b_3/b_4|$ constant as the phase of Γ_L is varied by changing the position of a sliding short at Port 4, then $C = 0$. Hence,

$$\left|\frac{b_3}{b_4}\right| = \left|\frac{A}{D}\right| |\Gamma_L| = K|\Gamma_L| \tag{13.57}$$

or,

$$\Gamma_L = \frac{1}{K}\left|\frac{b_3}{b_4}\right| \tag{13.58}$$

For a given bridge, K is determined by noting $|b_3/b_4$ using a fixed short of known reflection coefficient of -1 at Port 4 and, therefore, by measuring $|b_3/b_4|$, the magnitude of reflection coefficient of any load at Port 2 can be determined.

In order to measure the phase of the load reflection coefficient, four identical directional coupler reflectometer can be used as shown in Fig. 13.23. The procedure of phase measurements is as follows:

1. First a fixed short is placed at XX' plane of the precession waveguide section at Port 2. The movable short at Port 4 and the attenuator are adjusted to obtain null in the detector output at Port 3.
2. The fixed short at Port 2 is replaced by the test load and a shift x of the movable short at Port 4 is measured to obtain null in the detector output at Port 3.

The phase of Γ_L is then given by

$$\phi_L = \frac{2\pi\Delta x}{\lambda_g} \tag{13.59}$$

Thus,

$$\Gamma_L = K\left|\frac{b_3}{b_4}\right| e^{j\phi_L} \quad \text{and} \quad Z_L = Z_0\frac{1-\Gamma_L}{1+\Gamma_L} \tag{13.60}$$

The accuracy of phase measurement depends on the sensitivity of the detector for null reading and the vernier scale reading of the movable short.

FREQUENCY MEASUREMENT 13.12

Microwave frequency is measured using a commercially available frequency counter and cavity wavemeter. The frequency also can be computed from measured guide wavelength in a voltage standing wave pattern along a short circuited line by using a slotted line.

Wavemeter Method 13.12.1

A typical wavemeter is a cylindrical cavity with a variable short circuit termination which changes the resonance frequency of the cavity by changing the cavity length. As discussed in Section 7.5.5 TE_{011} mode is most suitable for wave meter because of its higher Q and absence of axial current. Since this is higher order mode, possibility of generation of lower order modes exits. Hence for practical purposes dominant TM_{010} mode is used in wavemeter applications. Wavemeter axis is placed perpendicular to the broad wall of the waveguide and coupled by means of a hole in the narrow wall as shown in Fig. 13.24. This excites TM_{010} mode in the cavity due to the magnetic field coupling. A block of absorbing material (polytron) placed at the back of the tuning plunger prevents oscillation on top of it. Thus the cavity resonates at different frequencies for different plunger positions. The tuning can be calibrated in terms of frequency by known frequency input signals and observing the dip in the display unit (power meter) connected at the output side of waveguide. The accuracy of such a wavemeter is in the range of 1 % to 0.005 % for available Q of 1,000–50,000, respectively. Since the power is absorbed in the wavemeter at resonance this is called *absorption-type wavemeter.*

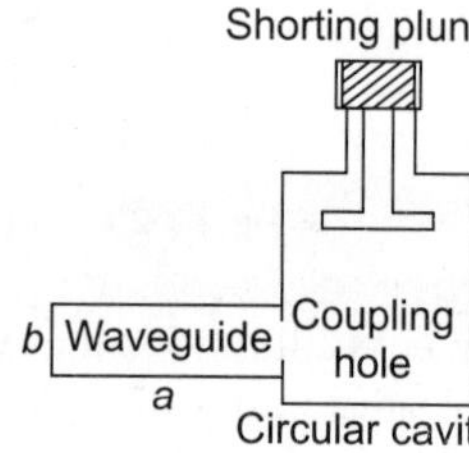

Fig. 13.24 Wavemeter method of frequency measurements

Slotted-Line Method 13.12.2

Since the distance d_{min} between two successive minima of voltage standing wave pattern in a short circuited line is half wavelength $\lambda_g/2$. frequency can be determined from the relations

$$f\,(\text{GHz}) = 30/\lambda_{0\,(\text{cm})} \tag{13.61}$$

$$\lambda_g = 2d_{min} \tag{13.62}$$

$$= \frac{\lambda_0}{\sqrt{[1-(\lambda_0/2a)^2]}} \quad \text{for waveguide} \tag{13.63}$$

$$= \frac{\lambda_0}{\sqrt{\varepsilon_r}} \quad \text{for coaxial line} \tag{13.64}$$

and measuring the by d_{min} by the slotted-line probe carriage.

Down-Conversion Method 13.12.3

An accurate measurement of microwave frequency can be done by means of a heterodyne converter. A heterodyne converter (Fig. 13.25) down converts the unknown frequency f_x by mixing with an accurately known frequency f_a, such that the difference $f_x - f_a = f_{IF}$ is amplified and measured by the counter.

The frequency f_a is selected by first multiplying a local oscillator frequency (known) to a convenient frequency f_1 and then passing it through a harmonic generator that produces a series of harmonics of f_1. The appropriate harmonic $nf_1 = f_a$ is selected by the tuning cavity such that f_a can be added with f_{IF} and display f_x (counter reading + f_a), the unknown frequency. In practice, the system starts with n = 1 and the filter frequency is selected by a feed back mechanism from IF stage until an *IF* frequency in the proper range is present. Typically, f_1 = 100 to 500 MHz for a range of f_x up to 20 GHz. For better accuracy, a low-noise oscillator and noiseless multiplier are to be selected.

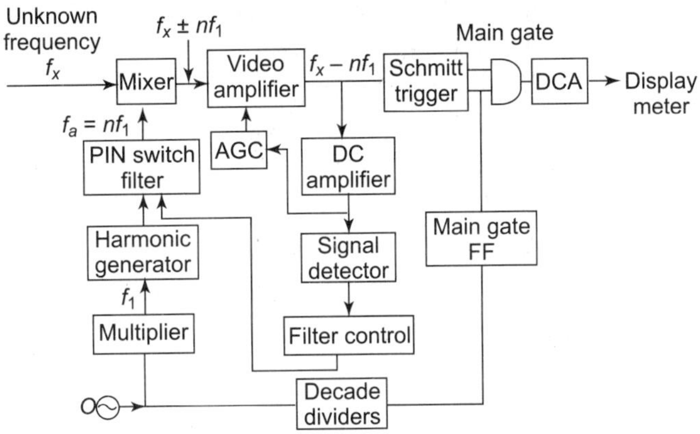

Fig. 13.25 Down-conversion method

MEASUREMENT OF CAVITY 13.13

A difficult measurement at microwave frequencies is the accurate measurement of Q of a high Q cavity. This is due to the fact that the 3dB bandwidth of the cavity response curve is a very small fraction of the resonance frequency. Moreover, the cavity has to be loaded during such measurements and Q becomes lower. There are three definitions of Q, connected to the associated circuit which are summarized below.

❑ ***Unloaded Q_0***

$$Q_0 = 2\pi \frac{\text{Energy stored in the cavity}}{\text{Energy lost per cycle in the cavity}} \tag{13.65}$$

Q_0 is selectivity factor of the cavity, dependent on the geometrical portion of the cavity. The expressions for Q_0 of several cavities are given in Section 7.5 in terms of their dimensions and mode numbers.

❑ ***Loaded Q_l***

$$Q_L = 2\pi \frac{\text{Energy stored in the cavity}}{\text{Energy lost per cycle in the cavity + Energy lost per cycle in the external system}} \tag{13.66}$$

Q_L is the Q of the entire system, including all sources of energy loss.

Q_L can be determined by the formula.

$$Q_L = f_0/\Delta f \tag{13.67}$$

where f_0 = Resonance frequency

Δf = 3 dB bandwidth

❑ *External Q_E*

Q_E is the Q of the external system.

$$Q_E = 2\pi \frac{\text{Energy stored in the cavity}}{\text{Energy lost per cycle in the external system}} \tag{13.68}$$

From above definition, $1/Q_L = 1/Q_0 + 1/Q_E$ and thus $Q_L < Q_0$. For an aperture coupled transmission type cavity, the input and output coupling factors β_1 and β_2 are a measure of the extent to which the power is coupled to the cavity and from the cavity, respectively, where

$$\beta_1 = \frac{4}{4S_0 - (S_0 + 1)^2\, T(f_0)} \tag{13.69}$$

$$= 1 \text{ (Critical coupling)}$$

$$< 1 \text{ (Under coupling)}$$

$$> 1 \text{ (Over coupling)}$$

and

$$\beta_2 = \beta_1 S_0 - 1 \tag{13.70}$$

$$Q_0 = Q_L(1 + \beta_1 + \beta_2) \tag{13.71}$$

Here S_0 = VSWR at the resonance frequency f_0.

$$T(f_0) = P_{out}/P_{in} \tag{13.72}$$

= Transmission loss at the resonance frequency f_0.

Measurement of both the transmission loss $T(f_0)$ and VSWR S_0 at resonance gives the data needed for calculating β_1 and β_2 and determining Q_0. A brief description of several methods of measurement of Q is given below.

Slotted-Line Measurement of Q 13.13.1

A slotted line may be used to measure the Q of a reflection type cavity which is normally used in a microwave tube, through pure VSWR measurements or through measurement of the shift in position of a standing wave minimum as the generator frequency is varied. Here the VSWR in the line that feeds the cavity is uniquely related to the variation in amplitude of the cavity input reflection coefficient and the shift of minimum is related to the variation of phase angle of the complex voltage reflection coefficient. The measurement set-up is shown in Fig. 13.26. The half-power frequency is found directly from the VSWR measurement, where the equivalent resonator reactance is assumed to be equal in magnitude to the equivalent resonator resistance. If $Z_{in} = R + jX$ is the input impedance in the vicinity of resonance of the cavity, VSWR

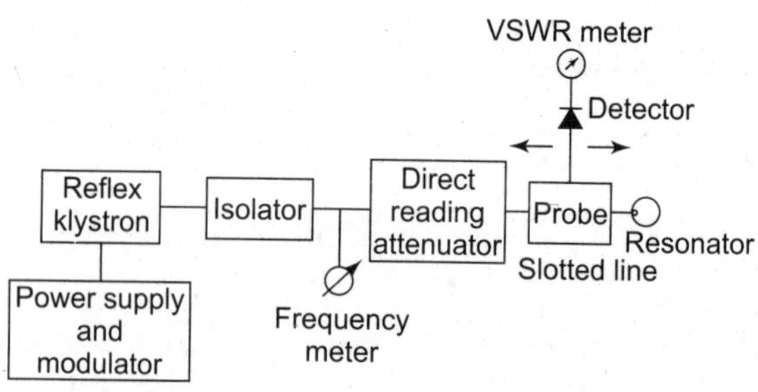

Fig. 13.26 Slotted-line measurement of Q

$$S = \frac{|Z_{in} + Z_0| + |Z_{in} - Z_0|}{|Z_{in} + Z_0| - |Z_{in} - Z_0|} \tag{13.73}$$

At resonance frequency f_0, $X = 0$, so that minimum VSWR S_0 is

$$S_0 = R/Z_0, \quad \text{if} \quad R > Z_0$$

$$= Z_0/R, \quad \text{if} \quad R < Z_0 \tag{13.74}$$

At half-power frequencies f_1 and f_2 of the unloaded cavity, $X = R$, so that

$$S_1 = \frac{\sqrt{[(R+Z_0)^2 + R^2]} + \sqrt{[(R-Z_0)^2 + R^2]}}{\sqrt{[(R+Z_0)^2 + R^2]} - \sqrt{[(R-Z_0)^2 + R^2]}} \tag{13.75}$$

or,

$$S_1 = S_0 + \frac{1}{2S_0} + \sqrt{\left(S_0^2 + \frac{1}{4S_0^2}\right)}; R > Z_0$$

$$= 1/S_0 + S_0/2 + \sqrt{(1/S_0^2 + S_0^2/4)}; R < Z_0 \tag{13.76}$$

The unloaded $Q_0 = f_0/(f_1 \sim f_2)$ can be determined from the above measurements. For a loaded cavity, minimum value S_0 as well as $\Delta f = f_1 \sim f_2$ increase and this results in a lower value of Q. The accuracy of measurement lies on the half- power VSWR and half-power bandwidth. In this method the measurement errors include the departure from square-law behaviours of the probe detector, frequency instability of the source, generator mismatch, probe and generator interaction at high VSWR.

Reflectometer Method of Measurement of Q 13.13.2

This method is suited for a reflection cavity and provides oscilloscope presentation by a swept microwave source. The method determines the magnitude of voltage reflection coefficient Γ at resonance, at half-power points and at a point far away from resonance $\Gamma \approx 1$. A schematic of the experimental set-up is shown in Fig. 13.27. The total errors in the measurement depends essentially on the accuracy of measuring the bandwidth and in setting half-power level for the reflection coefficient. The errors in half-power point for the reflection coefficient depend on the imperfect directivity of directional couplers and instability of the source frequency.

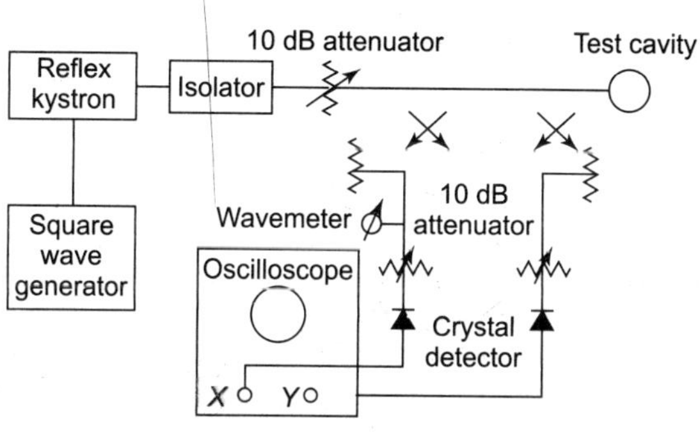

Fig. 13.27 Reflectometer method of measurement of Q

Q From Transmitted Power Measurement 13.13.3

This method uses the power transmitted through a cavity as a function of frequency, for measuring loaded Q. This method can be used for both transmission and the reflection type cavities. The transmission method has the advantage that the loaded Q can be measured directly regardless of the existence of coupling losses. However, this method cannot yield the unloaded Q without a number of additional measurements. The basic procedure can be carried out in several different ways as discussed below.

There are three main sources of error in the transmission method of Q measurement. The first relates to a possible mismatch of the generator and load between which the cavity is inserted, The second relates to the most important error that arises in measuring the bandwidth of the cavity response curve. Third is that due to the inaccuracies in relative power measurements, caused either by imperfect calibration of attenuators or power meters, or imperfect square-law response of detectors, errors in the readings of meters and attenuator dials, or by both calibration and reading errors.

CW Measurement

The CW measurement set-up is illustrated in Fig. 13.28. The transmitted CW power can be monitored with a power meter. Alternatively, the *RF* signal may be square-wave modulated and a tuned amplifier or a VSWR meter may be used at the output of the square-law crystal detector to indicate the transmitted powers at resonance frequency f_0 and half power points, $f_0 \pm \Delta f/2$. Q is calculated from $f_0/\Delta f$. To avoid errors due to non square-law response from the crystal at different power levels, a calibrated attenuator may be used for determining the half- power frequencies by keeping input power level same in all measurements.

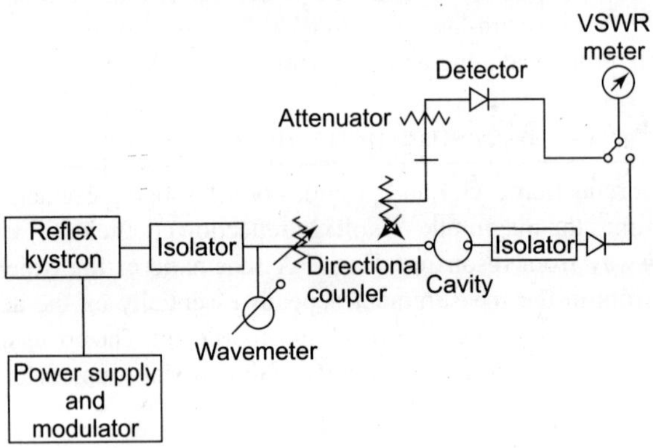

Fig. 13.28 Q from CW transmitted power measurement

Swept Frequency Measurement of Q

A swept frequency technique (Fig. 13.29) requiring less frequency stability of the *RF* source than the above procedures, provides simultaneous display of two oscilloscope traces, one proportional to the incident power, the other to the power transmitted through the cavity. The method uses a pair of matched crystal detectors of the same response law over the power and frequency range of the measurements. The wave meter measures the resonance frequency f_0 and the half-power frequency from the dip in the input response curve at the output peak and half-power points, respectively, to yield loaded Q_L.

The procedure of measurement is as follows:

1. A linear sweep (sawtooth wave) voltage is applied to the repeller of the reflex klystron so that a FM microwave signal is produced. Since the expected cavity bandwidth is much smaller than the frequency swing of FM signal, we can assume constant input voltage to the cavity at different frequencies around the resonance frequency of the cavity. Applying voltage to the triggering input of a double beam CRO from the same sweep generator, a response of the cavity is obtained in Y_2 beam.

   To determine the resonant frequency and 3 dB bandwidth, a marker trace is generated by applying a detected signal from the auxiliary arm of the directional coupler to Y_1 beam on the CRO. This represents the klystron power output mode characteristics.

By tuning the klystron frequency, cavity response is maximized and cavity peak response and peak of the klystron mode is made coincident. The frequency meter is adjusted to obtain a dip at the peak of cavity response. This indicates resonant frequency of the cavity. The flat top of the klystron mode curve is now adjusted to 3 dB point of the cavity response and the frequency bandwidth is noted by observing frequency meter readings corresponding to the dip in the response at 3 dB points.

2. VSWR of the cavity, terminated by match load, is determined at f_0 with a slotted line and VSWR meter.
3. Transmission loss α_0 at f_0 is determined by observing input and output powers of the cavity.

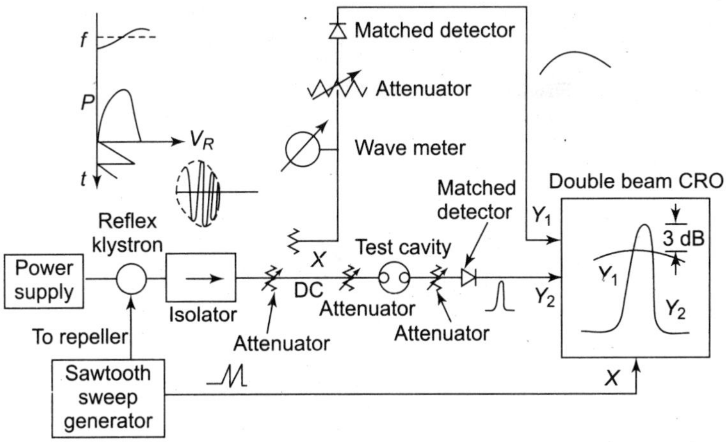

Fig. 13.29 Swept frequency measurement of Q

The Q_L of the cavity is calculated from the above measured data and using Eqs 13.68 – 13.73. This method is a rapid method of Q measurement. Obviously the Q_L of the frequency meter used should be considerably higher than that of the cavity under test to set the marker positions accurately. Reflex klystrons are widely used as sources of swept frequency microwave power. The electronic tuning range of reflex klystron within which frequency modulation is possible, is comparatively narrow. Thus, at 10 GHz, this range is ordinarily about 10–15 MHz and the resonance curve should be completely accommodated within this frequency range.

A further limitation of measurement is that the input power to the cavity must be maintained constant within the cavity band. This limits very low Q measurement by this technique.

Accuracy of the measurement depends on the measurement of very low bandwidth. Therefore, resolution of the frequency meters should be very high. Eye estimation in measuring bandwidth also introduces some error. Errors can be reduced by taking the average of several observations.

Swept Frequency Measurement of Q using Electronic Frequency Marker

Bandwidth measurement accuracy is the most important factor in high Q measurement. Ordinarily, when the test cavity Q is not very high, a high Q cavity wavemeter may be used to measure half-power frequencies. A cavity operating at 10 GHz with a high Q of 10,000 has a bandwidth of 1 MHz. It is difficult to adjust the microwave oscillator for such a small frequency difference. Moreover, it is difficult to measure this frequency difference since standard cavity wavemeters have smallest scale division of 1 to 5 MHz.

For very accurate measurement of high Q in a laboratory set-up, the swept frequency method using electronic frequency marker can be employed as shown in Fig. 13.30.

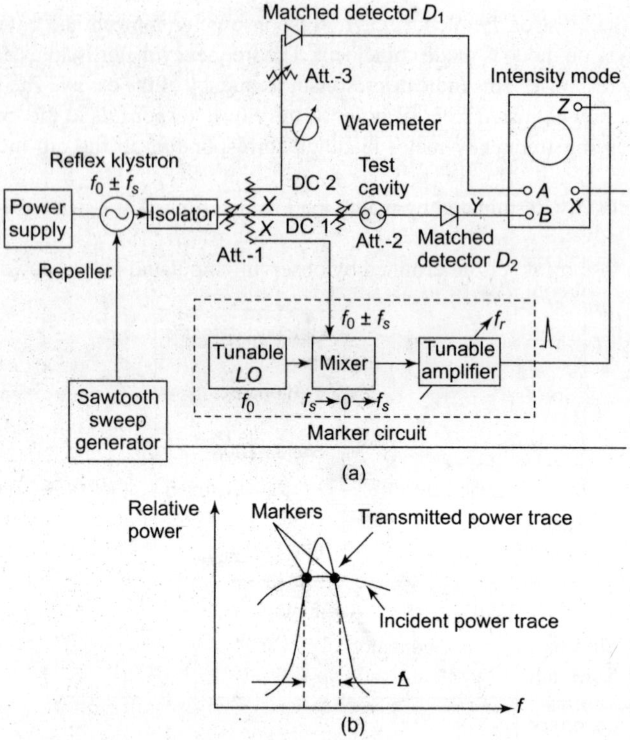

Fig. 13.30 (a) Swept frequency measurement of Q using electronic frequency markers (b) Input and output power traces

The swept source is a klystron and is shown to be sawtooth modulated. The same sawtooth sweep is simultaneously applied to the horizontal input of a dual trace oscilloscope to give displays of incident and transmitted power versus frequency on the oscilloscope screen synchronously. Sawtooth generates FM signal from the klystron with envelope variation in accordance with the klystron mode curve and the cavity response curve. Both the input and output signal of the cavity are detected by means of crystal detectors before being fed to the CRO. The markers are generated with the aid of an auxiliary frequency stable low frequency oscillator and the klystron in the main line which is frequency swept through the cavity resonance, varying in frequency from $f_0 - f_s$ to f_0 to $f_0 + f_s$. A portion of the swept signal from the directional coupler DC-1 and the output of the fixed frequency oscillator (~ 800 MHz) are combined in a harmonic crystal mixer which is connected to the input of a superheterodyne receiver. Harmonics of low frequency signal are mixed with the swept frequencies. Sum and difference frequencies are produced in the mixer. The superhet receiver can respond only to the time varying difference frequency component, $f_s(t)$ and produces a sharp audio output pip whenever the frequency of the input signal to the receiver equals f_r, the frequency to which the receiver is tuned. This occurs twice during a sawtooth sweep, once below resonance, when $[f_0 - f_s(t) - f_0] = f_r$ and again above resonance, when $f_0 + f_s(t) - f_0 = f_r$, that is, whenever the swept signal frequency applied to the cavity differs from the resonant frequency by $\pm f_r$. If the output pip is applied to the Z-axis (intensity grid) of the oscilloscope, two bright spots will appear on the resonance curve, as shown, marking these two frequencies. The frequency separation of the markers is obviously $2f_r$, or twice the frequency reading of the tuning dial of the receiver.

The crystal detectors D_1 and D_2 must be well matched to the line and selected to give identical responses over the required frequency and power range and at the power level used. It is desirable that the *RF* power levels at both crystals be nearly the same. This makes it necessary that the sum of the attenuation in Att.-2 and the cavity insertion loss (in dB) in the transmitted power channel be approximately equal to the (dB) coupling loss and sum of the attenuations in the directional coupler arm.

The directional coupler should have high directivity in the frequency range of interest. The isolator should have at least 20 dB of isolation and VSWR better than 1.2. All the attenuators should have a good VSWR (<1.10). Broadband matching of the crystal detectors can be obtained by using well matched 10 dB fixed pads in tandem with the crystal mounts. The calibrated variable attenuator Att.-3 should be of the precision type. For good matching, isolators may be used before the cavity and the detector.

❑ *Checking Crystals*

1. The klystron is tuned to f_0 by observing maximum transmitted signal through the cavity.
2. The cavity transmission trace *B* from D_2 should be centred with respect to the incident power trace *A* from D_1.
3. The dual trace oscilloscope vertical centering controls are adjusted so that the base lines of the two traces *A* and *B* coincide. During this adjustment, the *RF* power is completely removed.
4. Attenuator Att.-3 is set to zero. Att.-1 and 2 are adjusted so that the peak of the trace from D_2 just touches the trace from D_1.
5. The cavity is removed and Att.-2 is adjusted so that the traces of the two crystals coincide.
6. The coincident traces are checked by increasing the attenuation of Att.-1 by approximately 6 dB. The two traces should remain coincident over this range. Otherwise, discrepancies between the two crystals are noted and later applied as a correction. The amount of the discrepancy can be determined by adjusting Att.-3 starting with some initially inserted attenuation and observing the change in setting required to align both traces exactly.
7. Steps 1–6 are repeated for a few frequencies in the neighbourhood of f_0.

❑ *Measurement Procedure*

1. When the cavity is removed from the test position, the *RF* frequency sweep is adjusted so as to cover a sufficiently large portion of the cavity resonance curve. Calibrated attenuators 2 and 3 are adjusted so that the incident and transmitted power traces coincide. Coincidency of baselines of the two traces for zero *RF* power signal is done previously.
2. The test cavity is then inserted and the klystron repeller voltage is adjusted and it is tuned at resonance f_0 so that both the transmitted and incident power traces on the scope are centred. Att.-2 is adjusted so that the incident power trace touches the maximum power point of the transmitted power trace, corresponding to f_0 as shown in Fig. 13.31 (b). The change in attenuation of Att.-2 is the cavity transmission loss at resonance $T(f_0)$. Att.-3 is kept at 0 dB. Resonance frequency f_0 is observed by using of reaction wavemeter.
3. The incident power trace is lowered down by adding 3 dB attenuation with Att.-3.
4. The markers are placed at the points of intersection of the incident and transmitted power traces at the above setting of Att.-3 by tuning the superheat receiver. The tuning frequency f_r is noted to determine half- power bandwidth $\Delta f = 2f_r$.
5. Q_L is calculated from $f_0/2f_r$.
6. For the VSWR measurement, the klystron is square-wave modulated. A slotted line with a tunable probe detector is inserted at the input of the cavity and output is matched terminated. Detected signal from the slotted line is fed to a VSWR meter. VSWR S_1 are measured at frequencies around f_0. Also S_0 at f_0 is determined.

The Q_0 is calculated from Eqs (13.67–13.72).

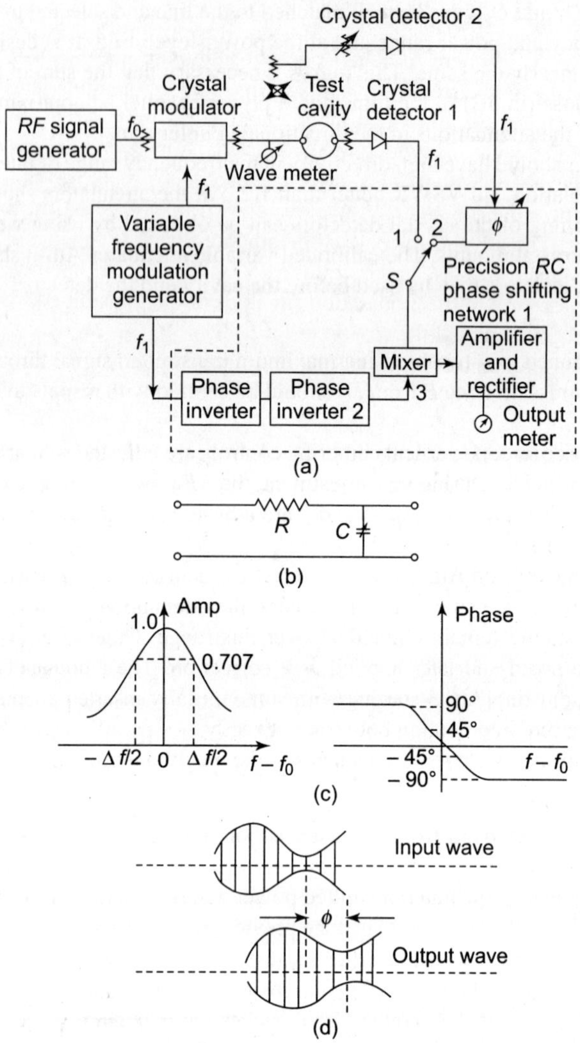

Fig. 13.31 (a) Phase-shift method of measurement of Q (b) Precision RC phase shifting network (c) Transmission characteristics of single tuned resonator (d) Phase change of transmitted signal

❑ *Accuracy of Measurement* Advantages of swept frequency measurements of Q with frequency markers are

- Requires less frequency stability of the *RF* source
- Accurate measurement of bandwidth
- Measurement time is small.

Q Measurement by Phase Shift of Modulation Envelope of Transmitted Signal [13,22]

When a microwave signal of frequency f_0 is sine wave amplitude modulated at a frequency f_1 and applied to a transmission cavity having a resonant frequency f_0, the modulation envelope of the transmitted signal will be delayed in phase with respect to that of the impressed signal by an angle

$$\phi = \tan^{-1}(2Q_L f_1/f_0) \tag{13.77}$$

If f_1 is made equal to $\Delta f/2$, where Δf = half-power bandwidth, and $\theta = 45°$, then $Q_L = f_0/2f_1$. Since f_1 can be accurately known from the frequency calibration of the modulating generator, Δf and hence Q_L can be determined. When a sine wave amplitude modulated carrier wave is applied to a resonant circuit at its resonant frequency, its sidebands are symmetrically shifted in phase on transmission through the resonator, one sideband being advanced, the other delayed with respect to the input wave, as shown in Fig. 13.31. This results in the phase shift in the modulation envelope shown in Fig. 13.31(c), which may be measured by means of a phase meter. The advantages of this method are that it is capable of bandwidth measurements and the measurement is much less affected by frequency stability of the microwave generator. This follows from the fact that the phase shift of the modulation envelope of the transmitted wave is a maximum when the carrier is at the resonant frequency of the cavity.

The accuracy of measurement depends ultimately on the accuracy of the precision phase measuring instrument.

Q Measurement from Inflection Points of the Resonance Curve

The tuned circuit behaves most nearly like a linear network in the neighbourhood of the two frequencies on the resonance curve and therefore, the second derivative, is zero at the inflection points of the curve. To a very good approximation, it can be shown that $Q_L = f_0/(\sqrt{3}\Delta f)$ for the response curve, where Δf is frequency difference between the inflection points of the curve concerned. If the inflection point frequencies are denoted by f_1 and f_2, then f_0 is given very closely by $(f_1 + f_2)/2$, so that the above equation can be expressed by

$$Q_L = \frac{f_1 + f_2}{2\sqrt{(3\,|f_1 \sim f_2|)}} \tag{13.78}$$

The equation applies when a square-law detector is used in monitoring the resonator frequency response.

A general method to determine the two inflection point frequencies is illustrated in the power transmission curve shown in Fig. 13.32(b), for a cavity resonant at f_0. The *RF* signal frequency is modulated about some centre value f at an audio frequency f_1 rate. The envelope of the *RF* (or detected) output of the cavity will be a distorted reproduction of the modulating wave form. The distortion will be a minimum if f is one of the inflection points. The linearity of behaviour of the tuned circuit at the inflection point frequencies provides the basis to determine these frequencies with fair precision. If a pure sinusoid of frequency f_1 is used to frequency modulate the *RF* oscillator, the second harmonic content should be a minimum if the applied *RF* frequency to the cavity corresponds to one of the inflection points. Figure 13.32(a) shows a block diagram for this method.

Decrement Method of Measurement of Q 13.13.4

Since it is difficult to measure VSWR or transmission in a very high Q systems a method is adopted based on measuring the time rate of decay of microwave energy stored in the very high Q resonator:

$$U = U_0 e^{-\omega 0 t/Q} \tag{13.79}$$

This method uses a pulsed excitation for the cavity system and observation of the rate of decay of the oscillations from U_1 to U_2 between exciting pulses at times t_1 and t_2.

Thus, $$10\log_{10}(U_1/U_2) = 4.343\omega_0(t_2 - t_1)/Q \tag{13.80}$$

or, $$Q = 8.686\pi f_0\,\Delta t/A = 27.3\,f_0\,\Delta t/A \tag{13.81}$$

where, $$10\log_{10}(U_1/U_2) = A,\; t_2 - t_1 = \Delta t \tag{13.82}$$

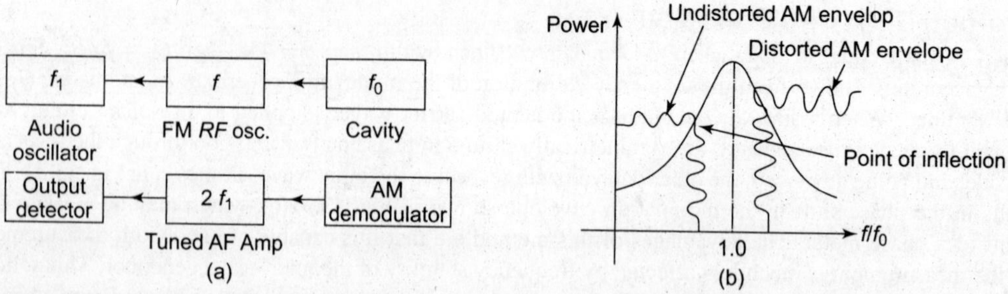

Fig. 13.32 (a) Inflection point method of measurement of 0 (b) Inflection point of resonance curve of cavity

This set-up is shown in Fig. 13.32(c and d). An oscilloscope technique is used in conjunction with a pick-up probe and power indicating equipment. Since the fluctuations of the source frequency do not affect the rate of energy decay in the resonator, the decrement method does not require a very high degree of frequency stability of the generator. However, frequency variations do affect the level of energy build up in the cavity resulting in corresponding fluctuations in the decay curve trace on the oscilloscope. This problem can be overcome by photographing a trace due to a single pulse and then measuring the decay time on the photograph.

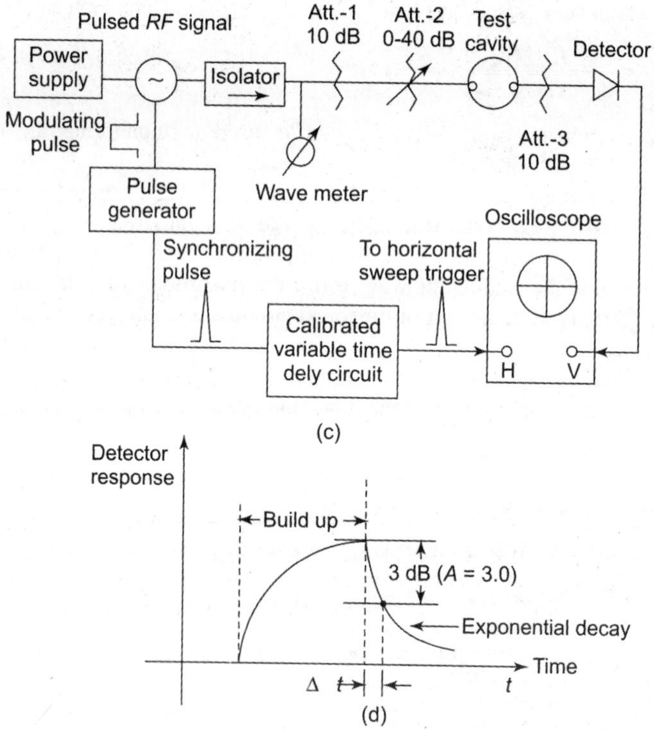

Fig. 13.32 (c) Decrement method of measuring *Q* (d) Transient response of a cavity

Two-port Transmission Cavity Q Measurements using Network Analyzer 13.13.5

In Chapter 7, the transmission loss at resonance of a two-port transmission cavity (7.142) is given as

$$T(\omega_r) = \frac{4k_1k_2}{(1+k_1k_2)^2} = S_{21} \tag{13.83a}$$

If the cavity ports are match terminated with the coupling-line impedance Z_0, and the coupling apertures are identical, $k_1 = k_2 = k_e$.

$$\therefore \quad T(\omega_r) = \frac{4k_e^{\,2}}{(1+k_e^{\,2})^2} = \left[\frac{4k_e}{1+k_e^{\,2}}\right]^2 = S_{21} \tag{13.83b}$$

Under this condition, the input reflection coefficient and VSWR at resonance are

$$\Gamma_i = S_{11} \tag{13.83c}$$

$$\text{VSWR} \quad S_r = \frac{1+k_2}{k_1} = \frac{1+k_e}{k_e} = \frac{1+|S_{11}|}{1-|S_{11}|} \tag{13.83d}$$

The coupling coefficient k_e can be determined by measuring S_{11} or S_{21}.

For a high Q microwave cavity, the loaded Q_L can be determined from the measurement of resonant frequency f_r and half-power bandwidth Δf from the $T(\omega)$ response curve:

$$Q_L = \frac{f_r}{\Delta f} \tag{13.83e}$$

Then unloaded $Q = Q_o$ and external $Q = Q_e$ can be determined from the relations

$$Q_o = (1+k_1+k_2)Q_L = (1+2k_e)\,Q_L \quad \text{and} \quad Q_e = \frac{Q_L Q_O}{Q_O - Q_L} \tag{13.83f}$$

The schematic diagram of the principle of network analyzer measurement set-up for cavity measurements is shown in Fig. 13.32(c). The procedures of measurements are given below.

Measurement Procedures

1. The network analyzer is calibrated for reflection measurements S-test set up (i) and S_{11} of the cavity is measured at resonant frequency (–ve peak frequency of return loss) as follows:
   (a) Initially, the cavity (test) is replaced by a fixed short and in other branch (reference) a sliding short is connected.
   (b) The sweep frequency range is set for the desirable design frequency range and power level is kept at reasonable level below the maximum limit.
   (c) The sliding short is adjusted to obtain zero response in the NA. This ensures that amplitudes and phases in reference and the test channels are equal.
   (d) Without changing anything in the set up the fixed short in the test channel is replaced by the cavity under test with its output and terminated in matched load.
   (e) A response curve of reflection (S_{11}) vs frequency in the NA is recorded to read the resonant frequency corresponding to the peak of the response curve. The –3 dB bandwidth also recorded. S_{11} is measured at resonance.
2. The analyzer is now calibrated for transmission set-up (ii) and S_{21} is measured as follows:
   (a) Initially, the cavity is replaced by an equal length of similar line as cavity and in the reference channel a variable length equalizer in connected.
   (b) Procedure 1(b) is repeated.

(c) Length of the equalizer is adjusted to read zero amplitude and phase responses in the NA.

(d) Without changing anything in the set-up, the cavity in placed in the test channel replacing the fixed length of line.

(e) A response curve of transmission (S_{21}) vs frequency in the NA is recorded to read the resonant frequency corresponding to the peak of the response curve. The –3 dB bandwidth also recorded. S_{21} is measured at resonance.

3. From steps 1 and 2, average value of resonant frequency f_r and –3 dB bandwidth are noted.
4. Q_0, Q_L and Q_e are determined from above relations (13.83a–13.83f) using the measured S-parameters, f_r, and –3 dB bandwidth.

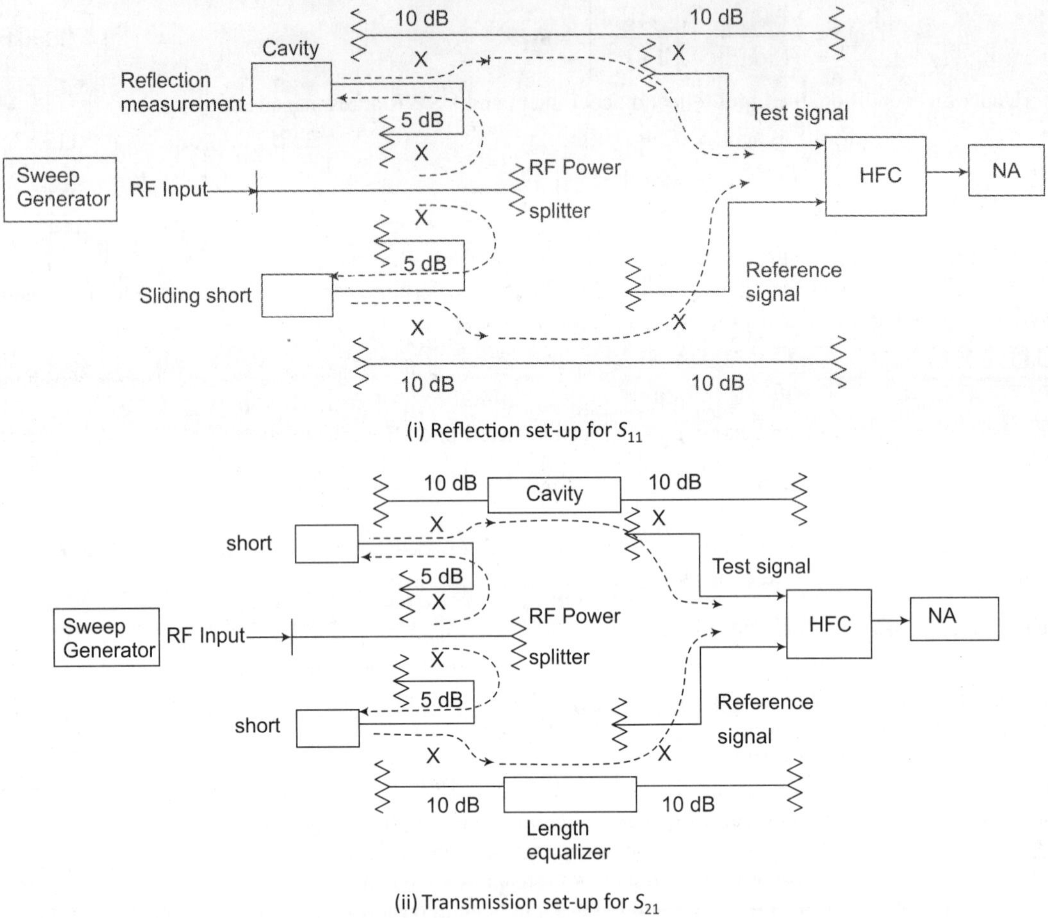

HFC: Harmonic Frequency Converter NA: Network Analyzer

Fig. 13.33(a) Schematic diagram of the network analyzer set-up for cavity measurement

The above procedures describe only the principle of measurements of cavity Q. Modern network analyzers are very sophisticated having automated test set-ups and mostly computer controlled as shown in Fig. 13.33(b). The detail description of measurement procedures is beyond the scope of this book.

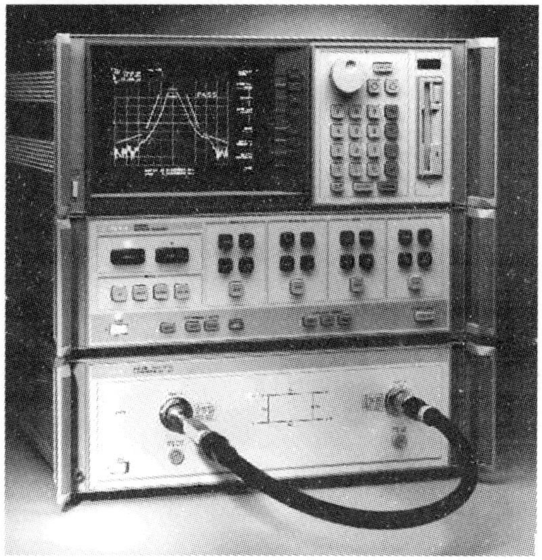

Fig. 13.33(b) Photograph of a modern network analyzer (Courtesy: M/S Agilent Technologies)

DIELECTRIC CONSTANT MEASUREMENT OF A SOLID 13.14

The dielectric constant ε_r is defined by the permittivity ε of the material with respect to that ε_0 of air or free space

$$\varepsilon_r = \varepsilon/\varepsilon_0,\ \varepsilon_0 = (10^{-9}/36\pi)\ \text{farad/m} \tag{13.84}$$

Due to presence of non-zero conductivity, dielectric material exhibits loss, resulting in complex value represented by

$$\varepsilon_r = \varepsilon'_r + j\varepsilon''_r \tag{13.85}$$

The loss tangent $\tan\delta = \varepsilon''_r/\varepsilon'_r$ (13.86)

The measurement of the complex dielectric constant is required not only in scientific application but also for industrial applications such as microwave heating or ovens and to study the biological effects of microwaves.

The dielectric constant is not independent of frequency. Generally, the variation of ε_r with frequency is sufficiently gradual so that it can be considered to be constant over a fairly wide frequency hand for most common microwave applications. On the other hand, the per cent variation in ε''_r is almost always greater than that of ε'_r so that ε''_r should be measured near the frequency or frequencies of interest.

There are several methods available for dielectric constant measurement. The following sections describe two commonly used methods: the waveguide method and cavity perturbation method.

Waveguide Method 13.14.1

In this method it is assumed that the material is lossless. A dielectric sample *AB* completely fills a length of the waveguide and the end is terminated in a short as shown in Fig. 13.34. A voltage standing wave minimum is observed in the slotted line at *C* (say).

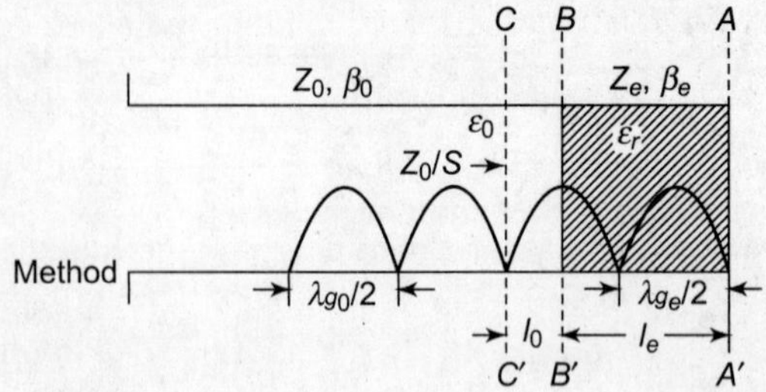

Fig. 13.34 Waveguide method

Let $\qquad l_e = AB = \text{the dielectric sample length} \qquad (13.87)$

$$l_0 = BC \qquad (13.88)$$

Then the distance of V_{min} from short circuit $= l_e + l_0 = AC$. For a dielectric filled guide of characteristic impedance Z_e, input impedance at BB' is purely reactive,

$$Z_{in}' = jZ_e \tan \beta_e l_e \qquad (13.89)$$

where β_e is the propagation constant.

Using this Z_{in}' as termination at BB', input impedance at CC' for the empty guide is

$$Z_{inc} = \frac{Z_{in}' + jZ_0 \tan \beta_0 l_0}{Z_0 + jZ_{in}' \tan \beta_0 l_0} = 0, \text{ at } V_{min} \text{ point} \qquad (13.90)$$

Therefore, $\quad Z_{in}' + jZ_0 \tan \beta_0 l_0 = 0 \qquad (13.91)$

or, $\quad jZ_e \tan \beta_e l_e + jZ_0 \tan \beta_0 l_0 = 0 \qquad (13.92)$

or, $\quad Z_0 \tan \beta_0 l_0 = -Z_e \tan \beta_e l_e \qquad (13.93)$

Assuming nonmagnetic dielectric in the waveguide,

$$\frac{Z_0}{Z_e} = \frac{\beta_e}{\beta_0} \qquad (13.94)$$

or $$Z_0 = \frac{\beta_e Z_e}{\beta_0} \qquad (13.95)$$

Substituting this value in Eq. 13.93,

$$\frac{\beta_e Z_e}{\beta_0} \tan \beta_0 l_0 = -Z_e \tan \beta_e l_e \qquad (13.96)$$

or $$\frac{l_0 \tan \beta_0 l_0}{\beta_0 l_0} = -\frac{l_e \tan \beta_e l_e}{\beta_e l_e} \qquad (13.97)$$

or $$\frac{l_0 \tan Y}{l_e Y} = -\frac{\tan X}{X}; \text{ where } X = \beta_e l_e\ Y = \beta_0 l_0 \qquad (13.98)$$

For dominant mode, $\beta_0 = 2\pi/\lambda_{g0}$ and $\lambda_{g0} = 2$ * distance between two successive V_{min} which can be measured by the slotted line. l_0 and l_e are also measured in the slotted line. Therefore, the left hand side (α *say*) of the above transcendental Eq. 13.98 is known and it can be written as

$$(\tan X)/X = -\alpha \qquad (13.99)$$

The above equation can be solved for determining $X = \beta_e l_e$.

Now, $$\beta_e = \frac{2\pi}{\lambda_{ge}} = \frac{2\pi}{\lambda_0}\sqrt{\left[\varepsilon_r - \left(\frac{\lambda_0}{\lambda_c}\right)^2\right]}, \lambda_c = 2a \tag{13.100}$$

where a is the waveguide broadwall dimension.

Since β_e is known, ε_r can be determined from above equation. This equation has infinite number of solutions for ε_r. Hence, two different lengths of sample are taken for two sets of solutions.

For length l_e: $X = X_1, X_2, \ldots;\quad \varepsilon_r = \varepsilon_{r1}, \varepsilon_{r2}, \ldots$

For length l'_e: $X = X'_1, X'_2, \ldots;\quad \varepsilon'_r = \varepsilon'_{r1}\ \varepsilon'_{r2}, \ldots$

The desired value of ε_r is the solution which is common for the two samples.

Cavity-Perturbation Method 13.14.2

Cavity-perturbation techniques are highly sensitive and accurate and therefore, particularly advantageous in the determination of the dielectric constant and small loss tangents.

In the cavity-perturbation technique, a small volume of the test sample is introduced in a cavity resonator at the position of maximum **E** field for the measurement of ε (and at maximum **H** field for the measurement of μ). Since the volume of the sample is small, the field in the cavity is assumed to remain undisturbed.

But introduction of dielectric sample changes the resonance frequency and the RF loss. These changes are reflected in the change of Q of the cavity from its unperturbed value. All these changes are related to the dielectric constant and the loss tangent.

Let the original cavity containing air having ε_0, μ_0 and resonance frequency f_0 is perturbed by some material ε, μ_0 placed inside the cavity at E_{max} and $H = 0$ position. For a sufficiently small sample the fields in the cavity outside the dielectric material are not greatly perturbed from those for the empty cavity case. Then the well established perturbation formulae for change in frequency and Q are given by

$$\omega - \omega_0 = \frac{\omega(\varepsilon'_e - 1)\int_{V_s} |E_0|^2\, dv}{\int_{V_c} |E_0|^2\, dv} \tag{13.101}$$

$$\left(\frac{1}{Q}\right) = \frac{\varepsilon''_r |E_0|^2 V_s}{\int_{V_c} |E_0|^2\, dv} \tag{13.102}$$

where E_0, ω_0 = Field and resonant frequency of the original cavity
E, ω = Corresponding quantities of the perturbed cavity
V_c, V_s = Volumes of the empty cavity and the sample filling the cavity, respectively

Assuming small perturbation, ω can be replaced by ω_0 and the simplified relations are

$$\frac{\Delta\omega_0}{\omega_0} = \frac{(\varepsilon'_r\ 1)|E_0|^2_{V_s} V_s}{2\int_{V_c} |E_0|^2\, dv} \tag{13.103}$$

$$\left(\frac{1}{Q}\right) = \frac{\varepsilon''_r |E_0|^2_{V_s} V_s}{\int_{V_c} |E_0|^2\, dv} \tag{13.104}$$

where $|E_0|_{V_s}$ has been assumed constant over the region of integration throughout the small sample and

$$\left(\frac{1}{Q}\right) = \left(\frac{1}{Q_c}\right) \sim \left(\frac{1}{Q_s}\right) \tag{13. 105}$$

Here, $Q_c = Q$ of the unperturbed cavity
$Q_s = Q$ of the perturbed cavity

In principle, any type of cavity either rectangular or circular can be used with excitation by suitable mode. The following criteria are important.

(a) The cavity Q should be as high as possible to enhance the accuracy of the measurement and this is very stringent, particularly in low loss materials.

(b) The mode of operation should be such that the dielectric sample is conveniently placed at a uniform field region E_{max} where $H = 0$ for the measurement of ε' and ε''. Otherwise, if $H \neq 0$ in the perturbed region, complex permeability of the magnetically lossy material will add to frequency shift and RF loss in addition to the contribution offered by complex ε. It will be extremely troublesome to mathematically relate the individual contributions (reverse is the case for the measurement of μ).

(c) The accuracy of the method depends on the smoothness of the sample, and fitness of the sample in the cavity. The smaller the sample, the more critically the accuracy will depend upon the sample irregularities. It is, therefore, necessary to machine the sample very carefully for smoothness and size.

In view of the above, rectangular cavity in TE_{103} mode and circular cavity in TM_{010} mode is found suitable in this technique. Both the cavities thus produce E_{max} and $H = 0$ at the centre where the dielectric samples in the form of a thin rod are placed inside the cavity transversely as post parallel to **E** field as shown in Fig. 13.35. With this combination, the perturbation relations are described as follows.

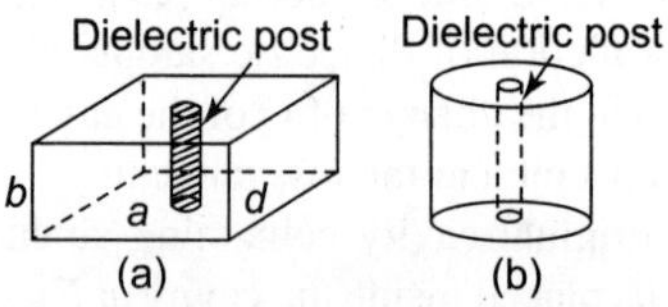

Fig. 13.35 Cavity perturbation method: (a) Rectangular cavity TE_{103} (b) Circular cavity TM_{01}

TE_{103} Rectangular Cavity

The non-zero field components for TE_{103} mode in the rectangular cavity $a \times b \times d$ are E_y, H_x and H_z only. Since the sample length is parallel to y, only E_y is required for finding ε_r.

$$E_y = \frac{\lambda}{2a} \sin(\pi x/a) \sin(3\pi z/d) \tag{13.106}$$

If the sample is placed at $x = a/2$, $z = d/2$

$$|E_0|^2 = |E_y|^2 = |E_{103}|^2 \left(\frac{\lambda}{2a}\right)^2 \tag{13.107}$$

and

$$\int_{V_c} |E_0|^2 dv = \left(\frac{\lambda}{2a}\right)^2 \int_0^a \int_0^b \int_0^d \sin^2\left(\frac{\pi x}{a}\right) \sin^2\left(\frac{3\pi x}{d}\right) dx\, dy\, dz$$

$$= \left(\frac{\lambda}{2a}\right)^2 \frac{abd}{4} \tag{13.108}$$

Substituting Eq. (13.108) and Eq. (13.109) in Eqs (13.104) and (13.105),

$$\frac{\Delta f_0}{f_0} = -2(\varepsilon_r' - 1)\frac{V_s}{V_c} \tag{13.109}$$

$$\left(\frac{1}{Q}\right) = 4\,\varepsilon_r''\frac{V_s}{V_c} \tag{13.110}$$

or,
$$\varepsilon_r' = 1 + 0.5\left(\frac{V_c}{V_s}\right)\frac{f_0 - f}{f_0} \tag{13.111}$$

$$\varepsilon_r'' = 0.25\left(\frac{V_c}{V_s}\right)\left(\frac{1}{Q_c} \sim \frac{1}{Q_s}\right) \tag{13.112}$$

TM_{010} Circular Cavity

For the circular in TM_{010} mode $E_\rho = 0 = E_\phi$ and E_z is parallel to the length of the sample rod which is introduced axially. It can be shown in a similar way:

$$\varepsilon_r' = 0.539\left(\frac{V_c}{V_s}\right)\left(\frac{\Delta f_0}{f_0}\right) + 1 \tag{13.113}$$

$$\varepsilon_r'' = 0.269\left(\frac{V_c}{V_s}\right)\Delta\left(\frac{1}{Q}\right) \tag{13.114}$$

The unique feature of the TM_{010} mode is that its resonant frequency is independent of the length and therefore design is very simple.

MEASUREMENTS OF SCATTERING PARAMETERS OF A NETWORK 13.15

S-parameters can be conveniently measured following the Deschamps method which utilises measured values of complex input reflection coefficients under a number of reactive terminations. In this section measurements of *S*-parameters of a general two-port network and a four-port network called the magic-*T* are described.

S-Parameters of a Two-Port Network 13.15.1

The output end of the two-port network is terminated with a short-circuit plunger to vary the reactive termination by moving the plunger in steps of at least 1/8th of the guide wavelength and the corresponding input reflection coefficients are measured. When attenuation coefficient α in the line is small and the change of total length of the line is less than one guided wavelength, the points *P*′s of the corresponding measured reflection coefficients Γ_1 describe an average circle on the polar chart with centre *C* and radius *CP* as shown in Fig. 13.36. However, for a lossless line $\alpha = 0$ and the reflection coefficients will describe a perfect circle. The reflection coefficients can be measured using a network analyser and reflection bridge set-up. Each pair of reflection coefficient points *P*′s, diametrically opposite, are joined and the lines of intersection *O*′ is marked. *CO*′ joined. Two perpendicular lines at *C* and *O*′ are drawn to intersect the circle at *Q* and *P*, respectively. *PQ* is joined to intersect *CO*′ at S_{11}. A line $S_{11}E$ parallel to *O*′*P* is drawn to meet the circle at *E*.

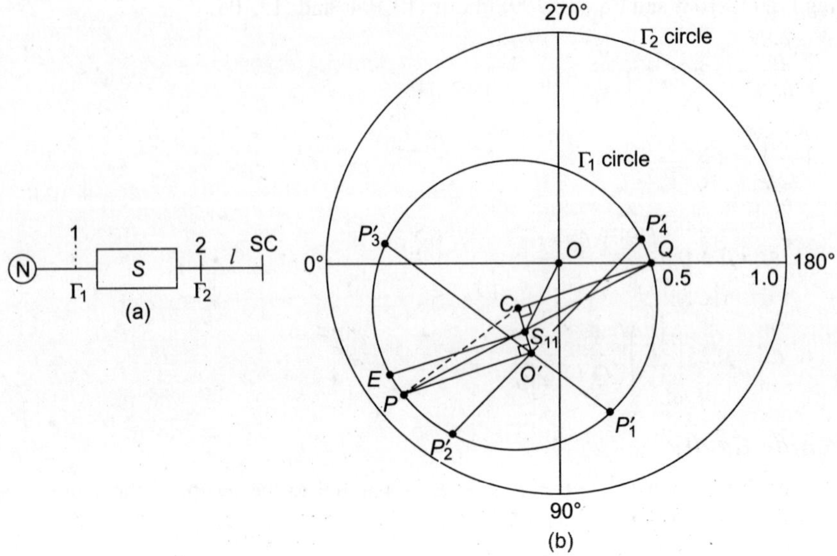

Fig. 13.36 Deschamp's method

It can be shown that

$$|S_{11}| = |OS_{11}| \tag{13.115}$$

and

$$|S_{12}| = \frac{|S_{11}E|}{\sqrt{\rho}\cdot R} \tag{13.116}$$

where $R = CP$, the radius of the circle and $\rho = \exp(-2\alpha l)$, the input reflection coefficient with output end of the line are short-circuited. The attenuation coefficient can be determined from the following equation when experiments are performed using two widely different lengths l and l' of the line. From (13.115) and (13.116),

$$\alpha = \frac{1}{2(l-l')}\ln\left(\frac{R|S_{11}E|^2}{R'|S_{11}E'|^2}\right) \tag{13.117}$$

For $|S_{22}|$ and $|S_{21}|$, measurements are carried out by interchanging the ports.

S-Parameters of a Magic-*T* 13.15.2

The *S*-parameters of a matched magic-*T* can be determined using Deschamp's method by measuring the reflection coefficients at different ports under specific termination of other ports as explained in Chapter 6. The schematic diagram of measurement set up is shown in Fig. 13.37.

Measurement of S_{ii}

The diagonal elements of [*S*], i.e., S_{ii} are determined from the slotted line measurement of the VSWR, S_i at the corresponding ports i with all other ports matched terminated.

$$[S_{ii}] = (S_i - 1)/(S_i + 1) \tag{13.118}$$

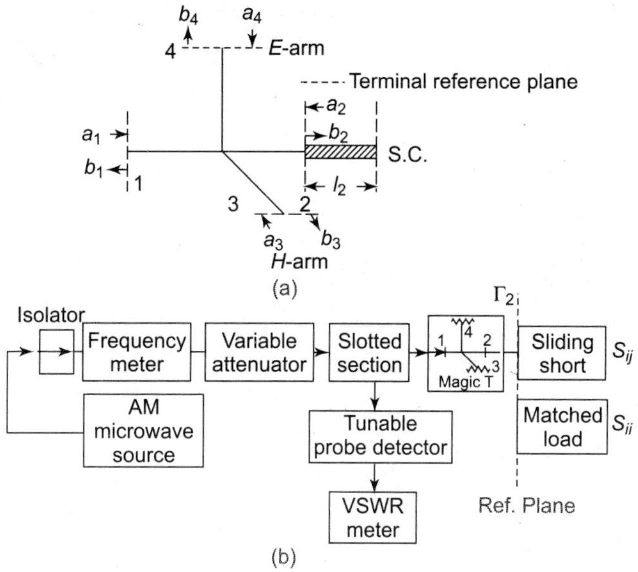

Fig. 13.37 Experimental set-up for *S*-parameters measurement of Magic-*T*

Measurement of S_{ij} ($i \neq j$)

To measure S_{12}, Ports 3 and 4 are match terminated and Port 2 is terminated in a short circuit plunger when the input is fed at Port 1. At the reference plane of Port 2, the reflection coefficient will always be equal to 1 in magnitude but its phase will vary continuously with the distance of the short position from the same reference plane, leading to a reflection coefficient circle $\Gamma_2 = -e^{j2\theta_2}$ with centre O. Since $S_{12} = S_{21}$ for an isotropic medium, as described in Chapter 6, the input reflection coefficient Γ_1 describes a circle with centre C for different short positions of Port 2 termination:

$$\Gamma_1 = S_{11} - \frac{S_{12}^2\, e^{j2\theta_2}}{1 + S_{22}\, e^{j2\theta_2}} \tag{13.119}$$

Equation (13.119) can be rewritten after some manipulation using few steps [1] as

$$\Gamma_1 = S_{11} + \frac{S_{12}^2\, S_{22}^*}{1 - |S_{22}|^2} + \frac{S_{12}^2\, e^{j(2\phi - \phi_{22})}}{1 - |S_{22}|^2} \tag{13.120}$$

where $$S_{22}^* = |S_{22}| e^{-j\phi_{22}} \tag{13.121}$$

Here, 2ϕ is a function of $2\theta_2 + \theta_{22}$ and $|S_{22}|$. As the length of the circuit line increases, ϕ increases and Γ_1 describes a circle (Fig 13.38) with centre C so that

$$OC = S_{11} + \frac{S_{12}^2\, S_{22}^*}{1 - |S_{22}|^2} \tag{13.122}$$

and radius $$CP = \frac{S_{12}^2\, e^{j(2\phi - \theta_{22})}}{1 - |S_{22}|^2} \tag{13.123}$$

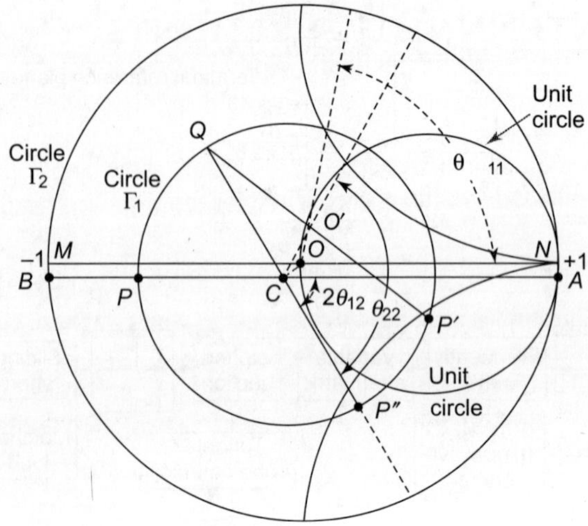

Fig. 13.38 Deschamps's circle for magic-*T* on Smith chart

The experimental measurement procedure is as follows:

1. Ports 2, 3 and 4 are matched terminated and S_{11} is measured directly at Port 1. $S_{11} = OO'$ is located on the Smith chart where O' the 'iconocentre'.
2. Placing a sliding short at Port 2, $\theta_2 = \beta l_2$ is varied to obtain Γ_1 circle on the Smith chart. The center C is determined on the diagram.
3. Placing open circuit at the reference plane of port 2 ($\theta_2 = \pi/2$, $l_2 = \lambda_g/4$) the corresponding $\Gamma_1 = \Gamma_1'$ is noted and indicated on Γ_1 circle as P'.

Construction of Deschamp's Circle is Done in the Following Manner

A straight line is drawn from P' to a point Q on the G_1 circle through O' and another line from Q to a point P'' on the Γ_1 circle through C. The angle of the phase or CP'' is equal to $2\theta_{12}$, where θ_{12} = angle of S_{12}.

Now it can be shown that [1]

$$= \overrightarrow{O'C} - \overrightarrow{OC} - \overrightarrow{OO'} = S_{12}^2 S_{22}^* /(1 - |S_{22}|^2) \tag{13.124}$$

Therefore, $\angle\overrightarrow{O'C} = 2\theta_{12} - \theta_{22}$. Since $\angle\overrightarrow{CP''} = 2\theta_{12} = \angle BCQ$, therefore $\angle\overrightarrow{CP''} - \angle\overrightarrow{O'C} = \theta_{22} = \angle P''CO'$.

Therefore, The S-parameters are obtained as

$$|S_{11}| = OO', \angle S_{11} = \theta_{11} = \angle NOO' \tag{13.125}$$

$$|S_{22}^*| = O'C/CP''; \angle S_{22}^* = \theta_{22} = \angle P''CO' \tag{13.126}$$

$$|S_{12}| = \sqrt{[\text{Radius of the } \Gamma_1 \text{ circle} \times (1 - |S_{22}|^2)]} \tag{13.127}$$

$$\angle S_{12} = \theta_{12} = \frac{1}{2} \angle BCQ \tag{13.128}$$

A similar procedure is followed for other ports.

Ordinarily, a magic-*T* is not matched because of discontinuity at the junction. Hence, *H*-arm is matched by a tuning screw from the bottom wall and *E*-arm is matched by an inductive iris by trial and error method. With this arrangement, the other two arms are automatically matched.

MICROWAVE ANTENNA MEASUREMENTS 13.16

The most important parameters required to be measured to determine the performance characteristics of microwave antennas are radiation amplitude patterns, radiation phase patterns, absolute gain, directivity, radiation efficiency, beam width, input impedance, bandwidth and polarizations. The accurate measurement methods for these parameters require standard antenna test ranges.

Antenna Test Ranges 13.16.1

There are two basic antenna test ranges used for antenna measurements. These are indoor and outdoor test ranges. Usual indoor test range is an anechoic chamber which consists of a rectangular volume enclosed by microwave absorber walls. These walls reduce reflections from the boundary walls and increases the measurement accuracy. Microwave absorbers are carbon impregnated polyurethane foam in the shape of pyramids. The materials are expensive for lower frequency ranges because the typical the size of pyramid is nearly 5′– 6′ for 100 MHz. Most of the antenna placed at a far field distance from a transmitting antenna. Consequently the requirement of a large space limits the use of the costly indoor facility. Special indoor ranges such as compact range and near field range could be used where the former produces a plane wave field in a smaller distance by means of an offset fed reflector antenna having a special edge geometry. The latter one uses mathematical computations of the near field measurement data to obtain the far field information. Both these methods are very costly and have several limitations. In this section, far field outdoor test range is discussed for antenna parameter measurements.

Outdoor Antenna Test Range

The most popular microwave antenna test range is the free space outdoor range in which the antennas are mounted on tall towers as shown Fig. 13.39. The reflections from the surrounding environment are reduced by

1. Selecting the directivity and side lobe level of the transmitting antenna.

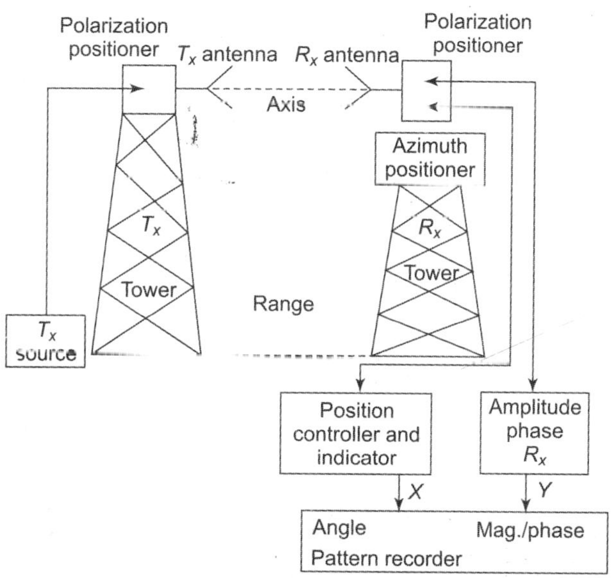

Fig. 13.39 Outdoor antenna test range

2. Making line-of-sight between the antennas obstacle free.
3. Absorbing or redirecting the energy that is reflected from the range surface or from any obstacle.

Radiation Pattern Measurements 13.16.2

The radiation pattern is a representation of the radiation characteristics of the antenna as a function of elevation angle θ and azimuthal angle ϕ for a constant radial distance and frequency. The three-dimensional pattern is decomposed into two orthogonal two-dimensional patterns in **E** and **H** field planes where the *Z-axis* is the line joining the transmitting and receiving antennas and perpendicular to the radiating apertures. Due to the reciprocal characteristics of antennas, the measurements are performed with the test antenna placed in the receiving mode. The source antenna is fed by a stable source and the received signal is measured using a receiver. The output of the receiver is fed to *Y*-axis input of an *XY* recorder. The receiving antenna positioner controller plane and the angle information is fed to *X*-axis input of the *XY* recorder. Thus the amplitude vs angle plot is obtained from the recorder output.

Initially, two antennas are aligned in the line of their maximum radiation direction by adjusting the angle and height by the controller and antenna mast. Effects of all surroundings are removed or suppressed through increased directivity and low side lobes of the source antenna, clearance of LOS, and absorption of energy reaching the range surface.

The following precautions are taken for better accuracy in the measurements:

1. Effects of coupling between antennas–inductive or capacitive–causes error in measurement. The former exists at lower microwave frequencies and negligible if range $R \geq 10\ \lambda$. Mutual coupling due to scattering and reradiation of energy by test and source antenna causes error in measurement.
2. Effect of curvature of the incident phase front produces phase variation over the aperture of test antenna and this restricts the range R. For a phase deviation at the edge $\leq \pi/8$ radians, $R \leq 2D^2\lambda$, where D is the maximum size of the aperture.
3. Effect of amplitude taper over the test aperture will give deviation of the measured pattern from the actual. This occurs if the illuminating field is not constant over the region of the test aperture. Tolerable limit of amplitude taper is 0.25 dB, for which decrease in gain is 0.1 dB.
4. Interference from spurious radiating sources should he avoided.

Phase Measurement 13.16.3

The phase of the radiated field is a relative quantity and is measured with respect to a reference as shown in Fig. 13.40. This reference is provided either by coupling a fraction of the transmitted signal to the reference channel of the receiver or by receiving the transmitted signal with a fixed antenna placed near the test antenna. The fixed antenna output is fed to the reference channel of the receiver and the phase pattern is recorded as the antenna under test is rotated in the horizontal plane.

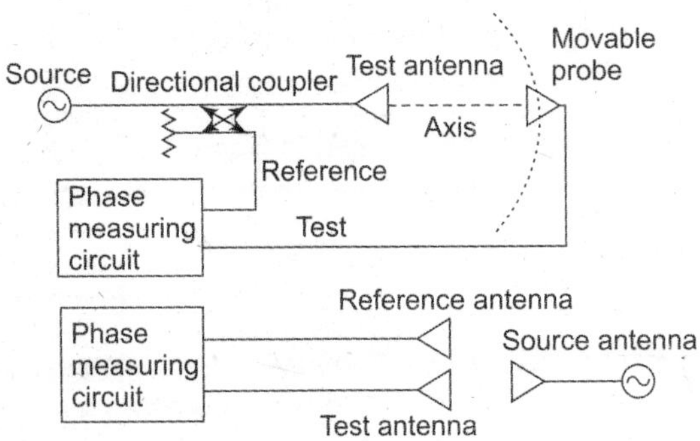

Fig. 13.40 Phase pattern measurement set-up

Phase Centre Measurement 13.16.4

When an antenna radiates, there is an equivalent point in the antenna geometry which represents the radiation centre. At the far field region the phase pattern of this antenna remains constant with angle when measured with respect to this point. Therefore, the phase centre of the test antenna is determined by positioning the rotational axis of the test antenna mast such that the phase pattern within the main beam remains constant.

Beamwidth 13.16.5

The beamwidth of the antenna is calculated from the angle subtended by the 3 dB or 10 dB points on the both sides of radiation maximum in the main beam.

Gain Measurements 13.16.6

The gain is the most important parameter to be measured for microwave antennas because it is used directly in the link calculations. There are three basic methods that can be used to measure the gain; standard antenna method, two antenna method and three antenna method.

Standard Antenna Method

This method uses two sets of measurements with the test and standard gain antennas. Using the test antenna of gain G_r in receiving mode, the received power P_r is recorded in a matched recorder. The test antenna is then replaced by a standard gain antenna of gain G_s and the received power P_s is again recorded without changing the transmitted power and geometrical configuration. Then

$$P_r/P_s = G_r/G_s$$

or, $$G_r(\text{dB}) = G_s(\text{dB}) + 10 \log \left(\frac{P_r}{P_s}\right) \tag{13.129}$$

Thus by measuring the received power with test and standard gain antennas and knowing gain G_s of the standard gain antenna, the gain of the test antenna can be found.

Two-Antenna Method

In this method, the signal is transmitted from a transmitting antenna of gain G_t, and the signal is received by the test antenna of gain G_r placed at far-field distance R. The received power is expressed by

$$P_r = \frac{P_t G_t G_r \lambda^2}{(4\pi R)^2}$$

or $$G_r(\text{dB}) + G_t(\text{dB}) = 20 \log \left(\frac{4\pi R}{\lambda}\right) + 10 \log \left(\frac{P_r}{P_t}\right); \tag{13.130}$$

where P_r is the received power and P_t is the transmitted power. When the two antennas are selected identical, $G_r = G_t$ so that

$$G_r(\text{dB}) = G_t(\text{dB}) = 10 \log \left(\frac{4\pi R}{\lambda}\right) + 5 \log \left(\frac{P_r}{P_t}\right); \tag{13.131}$$

By measuring R, λ and P_r/P_t, the gain G_r can be determined.

Three-Antenna Method

In the two, antenna method, if the measuring systems are not exactly identical, error will be introduced. Hence, the three antenna method is the most general method to find gain of all the three antennas. Any two antennas are used at a time i.e. 1 and 2, 2 and 3, and 3 and 1, respectively. The following equations can be developed for the received and transmitted powers.

$$G_1\,(\text{dB}) + G_2\,(\text{dB}) = 20\log\left(\frac{4\pi R}{\lambda}\right) + 10\log\left(\frac{P_{r2}}{P_{t1}}\right); \quad (13.132)$$

$$G_2\,(\text{dB}) + G_3\,(\text{dB}) = 20\log\left(\frac{4\pi R}{\lambda}\right) + 10\log\left(\frac{P_{r3}}{P_{t2}}\right); \quad (13.133)$$

$$G_3\,(\text{dB}) + G_1\,(\text{dB}) = 20\log\left(\frac{4\pi R}{\lambda}\right) + 10\log\left(\frac{P_{r1}}{P_{t3}}\right); \quad (13.134)$$

Since R and λ are known and (P_r/P_t)'s are measured, the right hand side of the above equations are known. Then three unknown quantities G_1, G_2 and G_3 can be determined from the above three equations.

A typical block diagram of the measurement set-up for two and three antenna methods are shown in Fig. 13.41.

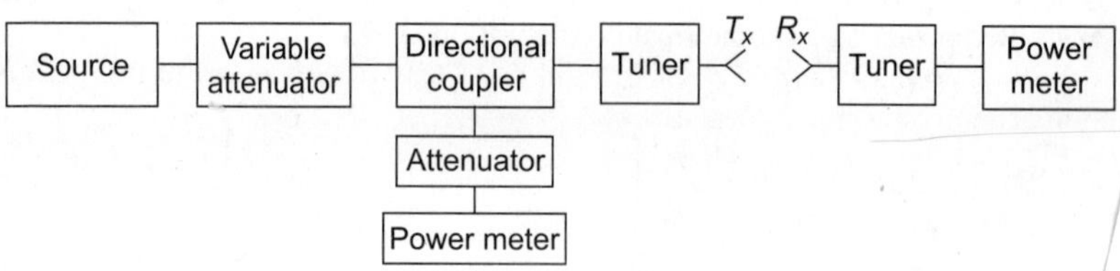

Fig. 13.41 Block diagram of antenna gain measurements

For accuracy of the measurements, care must be taken so that

1. All antennas meet the far field criteria: $R \geq 2D^2/\lambda$.
2. The antennas are aligned for bore-sight radiation face-to-face.
3. The measuring system is frequency stable.
4. Impedance mismatched in the system components is minimum.
5. Polarization mismatch is minimum.
6. Reflection from various background and support structure is minimum.

Directivity Measurements 13.16.7

The directivity of an antenna can be determined from the measurements of its radiation pattern in two principal planes, E and H planes and finding the half- power beam widths θ_E and θ_H degree in these planes, respectively.

$$D_0 = \frac{41{,}253}{\theta_E\,\theta_H} \quad \text{or,} \quad \frac{72{,}815}{\theta_E^2 + \theta_H^2} \quad (13.135)$$

This method is accurate for the antennas having negligible side lobes.

Radiation Efficiency 13.16.8

$$\text{Radiation efficiency} = \frac{\text{Total power radiated, } P_{\text{rad}}}{\text{Total power accepted at its input}}$$

$$= \frac{P_{\text{rad}}}{P_{\text{in}} - P_{\text{ref}}} = \frac{\text{Gain}}{\text{Directivity}} \tag{13.136}$$

where P_{in} is the input power and P_{ref} is the reflected power at the input. Therefore, the radiation efficiency can be determined from the measurement of gain and directivity.

Polarization Measurements 13.16.9

The polarization of an antenna is conveniently measured by using it in the transmitting mode and probing the polarization by a dipole antenna in the plane that contains the direction of the electric field as shown in Fig. 13.42. The dipole is rotated in the plane of polarization and the received voltage pattern is recorded and analyzed as follows.

❑ *Linear Polarization* For linear polarization, the output voltage pattern will be a figure of eight.

❑ *Circular Polarization* For circular polarization, the output voltage pattern will be a circle.

❑ *Elliptical Polarization* For an elliptical polarization, the nulls of the figure of eight are filled and a dumb-bell polarization curve is obtained which will be tilted and a polarization ellipse can be drawn as shown by dashed curve in Fig. 13.43. The sense of rotation of the circular and elliptical polarizations can be determined by comparing the responses of two circularly polarized antenna, one left and the other rightwise rotation. The polarization of the test antenna will be the same as that of one of these two directions for which the response is larger.

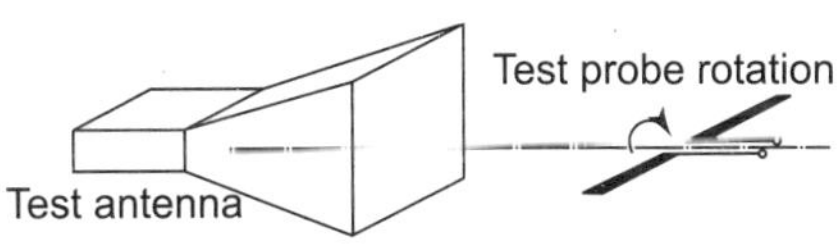

Fig. 13.42 Polarization measurements

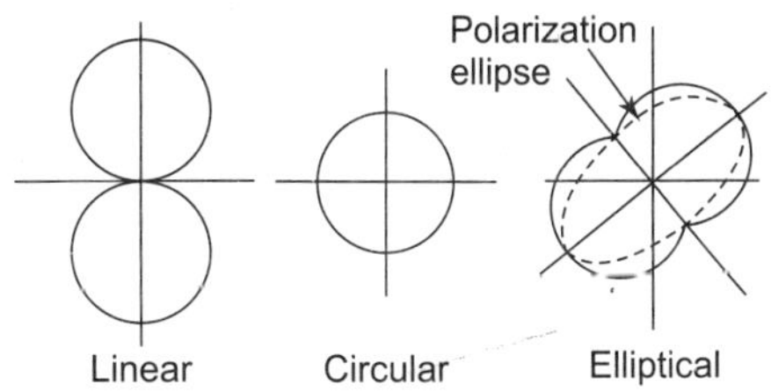

Fig. 13.43 Polarization patterns

RADAR CROSS SECTION (RCS) MEASUREMENTS 13.17

The radar cross section of a target is defined by

$$\sigma = \frac{4\pi \times \text{Power re-radiated per unit solid angle}}{\text{Incident power density}} \tag{13.137}$$

and is expressed in terms of received power

$$P_r(\theta) = \frac{P_t\, G_t\, A_e\, \sigma(\theta)}{(4\pi R^2)^2} \tag{13.138}$$

where, P_t = Transmitted power

G_t = Gain of transmit antenna relative to an isotropic radiator

A_e = Effective area of the receiving antenna
R = Distance of the target
θ = Aspect angle of the target which is the angle of direction of re-radiated power with respect to the line joining the T_x and the centre of the target

When all the factors remain constant in the above equation,

$$\sigma(\theta) = KP_r(\theta) \tag{13.139}$$

where K is a constant. Normalizing with respect to the angle $\theta = 0$,

$$\frac{\sigma(\theta)}{\sigma(0)} = \frac{P_r(\theta)}{P_r(0)} \tag{13.140}$$

Thus, by measuring the received power, the radar cross-section of a target placed at far field location, can be determined.

There are two basic radar cross section terms, i.e., monostatic or back scattering cross section and bistatic cross-section. The former is defined for the reflected signal received by the common transmit/receive antenna. The other one is defined when the scattered signal is measured in the forward direction other than towards the source end. We will be interested in the back scattered radar cross-section in this chapter.

The basic laboratory experimental set up for radar cross-section measurement is shown in Fig. 13.44.

Since signals reflected from the background cause error in the measurement, the tuner is adjusted to obtain minimum reading in VSWR meter-1 ($S < 1.05$) without the target present. Without disturbing the tuner, the target is now positioned and received signal strength is adjusted to read $S = 1$ in the VSWR meter-2 when $\theta = 0°$. The target is low rotated in azimuthal angular directions and corresponding readings in the VSWR meter-2 are noted to obtain normalized RCS vs angle information.

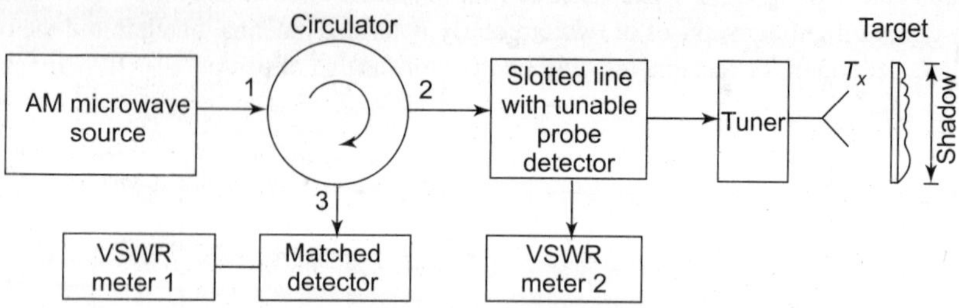

Fig. 13.44 Basic radar cross-section measurement set-up

The possible sources of error in the above method are the following:

1. When background cancellation was done by using the tuner, the target shadow region at its back contributed to background echo which is not present when the target was positioned. This produces imperfect background cancellation.
2. Finite isolations from isolator Port 1 to 3 and from Port 2 to 1 produce error in the measurement.

In order to improve the isolation between transmit and receive signal, the circulator can be replaced by arranging a separate receiving antenna system by the side of the transmitting antenna. Under these circumstances, the RCS is called quasi-monostatic. However, transmit and receive antennas are considered to be located at almost the same origin when the target is at far away from the antennas.

For better cancellation of actual background echo, RCS of a target is often defined in terms of an equivalent sphere whose RCS = σ_{sphere} is precisely known and is independent of the aspect angle. The following measurements are recommended inside an anechoic chamber of low reflection coefficient as shown in Fig. 13.45.

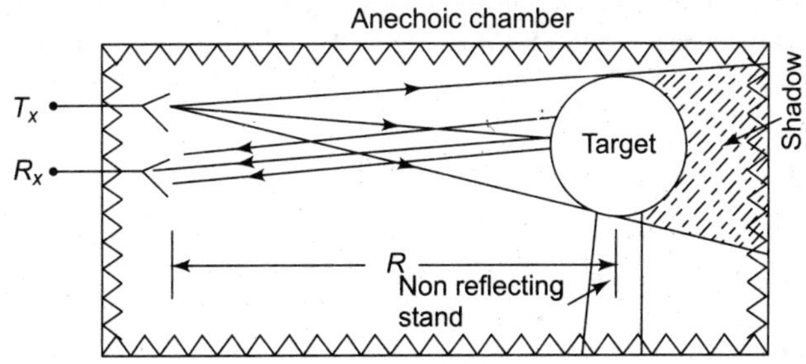

Fig. 13.45 Quasi-monostatic radar cross-section measurements

1. With an equivalent sphere of almost same vertical cross-section of the target the return signal strength is measured as P_{rs}.
2. Without disturbing anything in the chamber the sphere is replaced by the target under test and the received power P_{rt} is measured.

   Then the RCS of the target is obtained as

$$\sigma_{\text{target}}(\theta) = \frac{P_{rt}(\theta)}{P_{rs}}\,\sigma_{\text{sphere}} \tag{13.141}$$

REVIEW QUESTIONS

13.1 The signal power at the input of a device is 10 mW. The signal power at the output of the same device is 0.20 mW. Calculate the insertion loss in dB of this component.

13.2 A crystal detector generates a signal of 10 mV for an incident microwave power of –25 dBm. What is the detector sensitivity in mV/mW? Why is the microwave signal amplitude modulated by a 1 kHz square wave before it is applied to a crystal detector?

13.3 Calculate the VSWR in dB in a waveguide when the load is a 3 dB attenuator terminated by a short circuit.

13.4 The reflection coefficient of a load is 0.5 ∠–30°. Using the Smith chart determine the normalized admittance of the load.

13.5 A waveguide with a load is matched by a tuning screw located at position 1. What impedance will be presented by the waveguide to the generator if the screw is moved by half guide wavelength towards the load from position 1?

13.6 A microwave signal is modulated by a rectangular pulse of width 1 μs. If the average power is 200 W and the pulse repetition rate is 500 pulses per second, calculate the value of the peak power.

13.7 In attenuation measurement of a matched attenuator the microwave source is modulated by 1 kHz square wave. The VSWR meter is peaked to 0 dB with the 30 dB range without attenuator in the circuit. When the attenuator is inserted, the VSWR meter reads a value of 2 dB on the VSWR scale in 40 dB range. Find the attenuation provided by the attenuator.

13.8 A coaxial slotted line is used to measure VSWR of the load at 2 GHz by double minima method. If the distance between the positions of twice minimum power is 0.5 cm, find the value of VSWR on the line and the magnitude of the voltage reflection coefficient.

13.9 In a reflectometer set-up, two identical directional couplers are used to measure the incident and the reflected power. If the power level of the reverse coupler is 12 dB down from the level of the forward coupler, what is the VSWR on the line?

13.10 A rectangular waveguide is terminated with an unknown load. By using a slotted line, load VSWR is found to be equal to 2 and distance between two successive minima is 1.5 cm. When the load is replaced by a short, the position of a given minimum shifts 0.3 cm towards the generator. Find the normalized value of the load.

13.11 A slotted line with a short-circuit termination measures two successive minima at 25.3 cm apart. When an unknown load is connected, the VSWR is 3.1 and the minimum occurs at 16.5 cm form the load . Find (a) voltage reflection coefficient at the load, (b) the load impedance when the characteristic impedance of the line is 50 ohms.

REFERENCES

1. Altman, J. L., Microwave Circuits, D. Van Nostrand CO. Inc., 1964.

2. Balanis, C. A., *Antenna Theory Analysis and Design,* Harper and Row, Publishers, New York, 1982.

3. Barrington, A. E., and J.R. Rees, "A simple 3 cm *Q*-meter", *Proc. I.E.E.* Vol. 105 B, pp. 511–512, November, 1958.

4. Birnbaum, G. and J. Franeau, "Measurement of the Dielectric Constant and Loss of Solids and Liquids by a Cavity Perturbation Method", *J. Appl. Physics,* Vol. 20 No, 8, August, 1949.

5. Boverly, "Inflection Point Method of Measuring *Q* at Very High Frequencies", *Sperry Engineering Rev.,* Vol. 4, No. 3, pp. 16–19 (May-June, 1951). Sperry Gyroscope Co., Creat Neck, New York.

6. Collin, R. E., *Field theory of guided waves,* McGraw-Hill Book Company, p. 482, 1960.

7. Collin, R. E., *Foundation of Microwave Engineering,* McGraw-Hill, 1966.

8. Dakin T. W., and C. Works, "Microwave Dielectrics Measurements", *J. Appl. Physics,* Vol. 18, No. 9, September, 1947.

9. Deschamp, G. A., "Determination of reflection coefficient and insertion loss of a waveguide junction", *A. Appl. Physics,* 1953, 28, pp. 1046–1050.

10. "Electronics Letters", Vol. 10, No.8. 18th April 1966.

11. El-Ibiary, M. Y., "Q of resonant cavities, measurement by phase shift method", *Electronic Tech.,* p. 284 July, 1960,.

12. Ginzton, E. L., "Microwave *Q* Measurement in the Presence of Coupling Losses", *IRE Trans,* on Microwave Theory and Techniques, Vol. MTT-6, pp. 383–389, October, 1958.

13. Harrington, J. F., *Time-Harmonic Electromagnetic Fields,* McGraw-Hill, 1961.

14. Kinzer, J. P. and I. G. Wilson, "Some Results on Cylindrical Cavity Resonators", BSTJ, Vol. 26 pp 410, July, 1947.

15. Lebedev, I., *Microwave Engineering,* MIR Publishers, p. 313.

16. LeCaine, H., "The *Q* of Microwave Cavity by Comparison with a Calibrated High Frequency Circuit", *Proc. IRE,* Vol. 40, pp. 155–157, February 1952.

17. Lerner, D. S., and H. A. Wheeler, Measurement of bandwidth of microwave resonator by phase shift of signals-modulation", *IRE, Trans. on MIT,* Vol. MTT-8. p. 343, May, 1960.

18. Montgomery, C. G., "Technique of microwave measurements", *RLS,* Vol. 11.

19. *Proc IEEE,* 1966, February, 1966 p. 311.

20. *Proceeding IEEE Trans.* Vol. 62, No. 1, January 1974.

21. Ragon, G. L., "Microwave Transmission Circuits", *RLS*; Vol. 9.

22. Reich, Skolnik, Ordung and Krauss, *Microwave Principles,* East-West Press, 1972.

23. Shaw T. M., and J. J. Windle, "Microwave Techniques for the Measurement of the Dielectric Constant of Fibers and Films of High Polymers", *J. Appl. Physics,* Vol. 21, No. 10, Oct. 1950.

24. Singh, A., "An improved method for the determination of *Q* of cavity resonators", *IRE Trans,* on Microwave Theory and Techniques, Vol. MTT-6, pp 155–160, April, 1958.

25. Storer, J. E., et. al., "A simple graphical analysis of a two-port junction", *Proc. I.R.E.,* 1953, 41, pp. 1004–1013.

26. Sucher, M., and J. Fox, *Handbook of microwave measurement,* 3rd Edn. Vol. II, p. 431.

27. Surber, Jr. W. H., and G. E. Jr. Crouch, "Dielectric Measurement Methods for Solid at Microwave Frequencies", *J. Appl. Physics,* Vol. 19, No. 12, December, 1948.

28. The Narda Microwave Corporation Catalog No. 12.

29. Wenger, N. C., "The launching of surface waves on an axial-cylindrical reactive surface", *IEEE Trans.* Antennas Propagate, Jan. 1965, pp. 126–134.

30. Wilson, I. G., C. W. Schramm, and J. P. Kinzer, "High *Q* Resonannt Cavities for Microwave Testing", *BSTJ,* Vol. 25 p. 408, July, 1946.

53. Why are waveguide twists and bends constructed so that the direction of the propagated energy is gradually changed?
54. Describe the action of the phase shifter which is used as a duplexer.
55. In a rectangular waveguide, a bend is followed by a 90° twist. If both the bend and the twist introduce an SWR of 1.1, what is their combined SWR value?
56. Design a lossless T-junction divider with a 30-ohm source impedance to give a 3:1 power split.
57. What is a re-entrant cavity?
58. How does a shorted, quarter-wavelength long line act as a resonator? Sketch a practical coaxial line resonator.
59. Describe how a cylindrical cavity is excited by loops and rectangular waveguides into the TM_{010} mode.
60. Explain how Q varies with frequency. What are the dominant modes of a cylindrical cavity for diameter/length > 1 and diameter/length < 1.
61. An electron is accelerated from rest over a distance of 1.5 cm. If the accelerating voltage is 125 V, calculate the time interval during which the electron was accelerated.
62. An electron is accelerated from rest through a potential difference of 1 kV. What is the electron's terminal velocity in metres per second?
63. If the cavity grid spacing in a klystron tube is large, what harmful effects will be observed?
64. Explain the phenomenon of back heating in a magnetron. Can this phenomenon be put to any advantageous use?
65. What is meant by saturation power in a TWT? How does one improve the efficiency of a TWT?
66. "A TWT is a broadband device whereas a klystron is essentially a narrow-band device." Justify the statement.
67. What are the advantages of the TWT amplifier over the klystron amplifier? What is the purpose of the external magnetic field?
68. Transferred electron bulk effect occurs in —— (pick the right material—Ge, Si or GaAs).
69. The PIN diode is not used as a microwave detector. True or false? Justify your choice.
70. Briefly explain the PIN diode operation. Sketch the construction and explain its applications.
71. What are the high-frequency limitations observed in conventional tubes?
72. What is electronic and mechanical tuning of a reflex klystron?
73. What is the slow-wave structure? What is its significance in TWT?
74. Explain with a diagram why choke flanges are preferred in waveguide.
75. Write S-matrix of an E-plane Tee having straight arm ports number 1 and 2.
76. What is the phase-shift property of a two-port network?
77. A waveguide termination having VSWR = 1.1 is used to dissipate 100 W of power. Find the reflected power.
78. Explain why the output of a microwave source is amplitude modulated in a laboratory measurement set-up.
79. Define phase velocity and group velocity of plane-wave propagation in space and, hence, obtain the expression for it.
80. What is the significance of propagation constant being imaginary, zero, and real?
81. How do you account for the finite conductivity of a conductor forming waveguide walls? Explain.
82. What do you understand by the dominant mode in a waveguide? What is its importance?

83. Explain why TM_{01} and TM_{10} modes in a rectangular waveguide do not exist.

84. Explain the impedance characteristics of fundamental mode in a microstrip line when the line width changes.

85. Define and state the relationship between loaded, unloaded and external Q of a cavity resonator.

86. What do you understand by degenerate modes in a waveguide? Give examples of two such modes in a rectangular waveguide.

87. Explain the difference between the intrinsic impedance and wave impedance in free space and in a rectangular waveguide.

88. Explain with a neat diagram how R, L and C are realized in MIC form.

89. Describe three special considerations in a microwave transistor in comparison with a low-frequency one.

90. Does a non-reciprocal lossless network always have a purely imaginary impedance matrix? Explain

91. Explain dominant and higher order modes in a microstrip line. Describe how microwave signals are launched into a microstrip line from a signal source.

92. Describe with diagram the steps of fabrication of sandwich-type capacitance in microstrip circuits.

93. Explain what you understand by high-impedance and low-impedance filters.

94. Describe the factors which influence the stability of a microwave amplifier.

95. Describe the difficulties encountered in the design of a broadband microwave transistor amplifier.

96. State the conditions of oscillation of a two-port microwave transistor oscillator.

97. Explain the characteristics of uniform plane travelling waves with mathematical representation.

98. Show that in good conductors, the wave is attenuated greatly as it progresses through the conductor.

99. Determine the reflection and transmission coefficients for an electromagnetic plane wave, incident normally from air on a sheet of copper at 1 GHz.

100. Explain why TM_{01} and TM_{10} modes in a rectangular waveguide do not exist.

101. Explain the impedance characteristics of fundamental mode in a microstrip line when the line width changes.

102. Draw electric current lines on the broad and narrow walls of a rectangular waveguide excited in dominant mode.

103. What do you understand by degenerate modes in a waveguide? Give examples of two such modes in a rectangular waveguide.

104. Explain the difference between the intrinsic impedance and wave impedance in free space and in a rectangular waveguide.

PART B: LONG-ANSWER QUESTIONS

1. A helix TWT operates at 3 GHz under a beam voltage 8 kV and beam current 300 mA. If the helix impedance is 30 ohms and the rf interaction length is 30 cm, find the output power gain in Db.

2. Why is a magnetron called a crossed-field device? What is meant by π mode operation in a magnetron containing eight cavity resonators?

3. Using a neat diagram, show the construction of a MESFET and its biasing circuit at microwave frequencies.

4. An Si microwave transistor has reactance of 1 ohm, transit-time cut-off frequency of 4 GHz, maximum E-field of 1.6×10^5 V/m and saturation drift velocity of 4×10^5 cm/s. Determine the maximum allowable power.

5. Explain how a T-section low-pass filter is simulated in microstrip form. How are the component values determined?
6. Distinguish between hybrid MIC and MMIC fabrications. What are their advantages and disadvantages?
7. Determine the operating frequency range for a rectangular waveguide whose internal dimensions are 2.85 cm and 1.06 cm.
8. An EM wave in free space has an electric field intensity of 500 mV/m. Calculate the value of the wave's magnetic field intensity and the value of its Poynting vector.
9. What is the maximum frequency that can be propagated through a rectangular waveguide whose inner dimensions are 16.0 cm and 8.0 cm?
10. What is a double-stub tuner? How does it achieve impedance match?
11. What are standing waves? Define VSWR. Express VSWR in terms of reflection coefficient (ρ) and reflection coefficient (ρ) in terms of VSWR. What do VSWR = 1 and VSWR = ∞ signify with reference to the matching of the transmission line with the load?
12. A transmission line has characteristic impedance of 500. It has been terminated in a 200 Ω load. If the load is dissipating a continuous sinusoidal power of 100 watts, calculate the following.
    (a) Reflection coefficient, ρ
    (b) VSWR on the line
    (c) Magnitude of the reflected voltage
13. Design a two-section binomial impedance transformer to match the 100-ohm load with the 50-ohm line. Calculate the bandwidth for maximum allowable reflection coefficient of 0.1.
14. Design a two-section Chebyshev transformer to match a 75-ohm load to a 50-ohm line. The maximum SWR allowed is 1.2. Find the corresponding bandwidth.
15. Consider a lossless two-port network. If the network is reciprocal, show that $S_{21} = S_{12}$.
16. A certain three-port network is lossless and reciprocal, and has $S_{13} = S_{23}$ and $S_{11} = S_{22}$. Show that if Port 2 is terminated with a matched load then Port 1 can be matched by placing an appropriate reactance at Port 3.
17. A rectangular waveguide is 1 cm × 2 cm in dimensions. Calculate λ_c for TE_{10} and TM_{11} modes and *S*-matrix for matched condition.
18. For a 1 cm × 2 cm rectangular waveguide operating at f_0 = 10 GHz. Calculate the following:
    (a) Group velocity and phase velocity
    (b) Cut-off frequency
    (c) Characteristic impedance
    (d) Possible modes
19. Find the *S*-matrix of *H*-plane T.
20. Why is TE_{10} mode considered to the dominant mode in a rectangular waveguide? What is the dominant mode in a circular waveguide? Why is this mode not preferred in circular guides? What are then the preferred modes in circular waveguides?
21. Describe the features of waveguide series and shunt tee sections. Explain the operation of the magic-tee hybrid junction.
22. Explain the operation of the flap and the precision-vane attenuators.
23. What is an isolator? Describe the action of the ferrite isolator.
24. A 4 W power source is connected to the input of a directional coupler with C = 20 dB, D = 35 dB, and an insertion loss of 0.5 dB. Find the output powers (in dBm) at the through, coupled and isolated ports. Assume all ports to be matched.

25. An input signal V_1 is applied to the sum port of a 180° hybrid, and another signal V_4 is applied to the difference port. What are the output signals?

26. Design a five-hole directional coupler in the Ku-band waveguide with a binomial response. The centre frequency is 17.5 GHz, and the required coupling is 20 dB. Use round apertures centred across the broad wall of the waveguide.

27. Calculate the even- and odd-mode characteristic impedances for a tapered coupled-line 180° hybrid coupler with a 3 dB coupling ratio and 50-ohm characteristic impedance.

28. A coupled microstrip line on a substrate with $\varepsilon_r = 10$ and $h = 0.16$ cm has strip widths of 0.16 cm and a strip spacing of 0.064 cm. Find the even- and odd-mode characteristic impedances.

29. Design a quarter-wave matching transformer to convert the 50-ohm impedance of the output lines to 30 ohms. Determine the magnitude of the *S*-parameters for this circuit, using a 30-ohm terminating impedance.

30. Describe in detail the operation of a directional coupler. Calculate the coupling value if the power in the main waveguide is 60 mW and power delivered to the coupled port is 0.4mW.

31. A waveguide is terminated by a load with a reflection coefficient of 0.3. The incident power is 100 mW and a directional coupler pickup the reflected power to read 1.0 mW. What is the value of the coupling factor?

32. What is the purpose of a circulator? Describe the action of a circulator that employs the Faraday rotation effect.

33. Find the resonant frequencies of the three lowest modes of an air-filled cylindrical cavity of 2 cm radius and 2.5 cm length. List them in ascending order.

34. Find the resonant frequencies of the three lowest modes of an air-filled rectangular cavity having dimensions of 5 cm × 4 cm × 2.5 cm.

35. Define unloaded, loaded and external Q of a cavity resonator. Establish a relation between them.

36. Determine the unloaded Q factor of a cylindrical cavity for TE_{011} mode.

37. Show that Q of a circular cavity in TM_{010} mode is greater than that of square-based rectangular cavity with TE_{101} mode when the circular cavity is circumscribed by the square-based rectangular cavity.

38. Describe how a microstrip disk resonator can be realized. Explain its operation in terms of modes for rectangular and circular geometry.

39. What is meant by velocity modulation of an electron beam? Explain how velocity modulation create density modulation in a klystron amplifier.

40. How does drift space help convert velocity modulation to current density modulation in an electron beam?

41. Why is a magnetron called a crossed-field device? What is meant by π mode operation in a magnetron containing eight cavity resonators. Describe how strapping separates the π mode from other possible modes.

42. What is mode jumping in a magnetron? How is it avoided? What are the typical values of magnetron efficiency?

43. Explain the term *gain* as applied to a TWT. How does a helix slow down the circuit wave? Sketch the behaviour of the circuit wave and the beam wave in a TWT.

44. Explain why the Gunn diode is not a high-power and/or high efficiency mode. Sketch the diagram of the Gunn diode oscillator with its resonant cavity. List the performance parameters of such an oscillator.

45. What is meant by domain formation in GaAs? Explain how a Gunn diode can produce microwave oscillation.

46. "IMPATT diodes may fail to operate if proper heat sinks are not provided". Explain. What are its distinguished features compared to Gunn diodes?

47. Describe the operating principles of IMPATT diodes. Describe its applications.

48. Discuss the differences between transferred electron devices and avalanche transit-time devices.

49. Define varactor diode and its equivalent circuit. Describe the construction of the diode. Sketch the typical characteristics of varactor-diode capacitance versus reverse voltage.

50. Describe the construction of a Schottky diode and explain its operation. What are the applications of such a diode?

51. Describe the construction and equivalent circuit of a crystal diode. Give two applications of this diode.

52. Explain how the PIN diode is used as a microwave switch.

53. Determine the reflection and transmission coefficients for an electromagnetic plane wave, incident normally from air on a sheet of copper at 1 GHz.

54. Explain dominant and higher order modes in a microstrip line. Describe how microwave signals are launched into a microstrip circuit from a signal source.

55. Explain how a T-section low-pass filter is simulated in microstrip form. How are the component values determined?

56. Draw a neat schematic diagram of a SPDT PIN switch in MIC form and explain its operation.

57. With a neat diagram, determine the S-matrix of a three port microstrip circulator.

58. The magnitude of electric field of a linearly polarized plane wave traveling in free space in a direction 60° w.r.t. $Z = 0$ plane is 10 mV/m. Determine (a) the magnitude of the magnetic field, (b) the power flow through a surface of 1 m^2 area that is perpendicular to the Z-axis, (c) wave impedance, and (d) dielectric constant of the medium.

59. Establish the laws of reflection for a linearly polarized plane wave of parallel polarization incident on a plane interface between two lossless dielectric regions 1 and 2 at an angle θi with respect to the normal. Obtain the expression for Brewster angle as well as the angle of incidence for which total internal reflection occurs.

60. A uniform plane wave at 2.5 GHz is transmitted through a medium having $s = 2.0$ S/m, $\varepsilon = 36\ \varepsilon_0$, $\mu = \mu_0$. Find the complex propagation constant, phase velocity and the wave impedance of the medium. If the electric field magnitude is 10 V/m, find the time-average power flow per unit area.

61. (a) (i) Draw the physical configuration of cross section of strip and microstrip lines. What are the dominant modes in these lines? Sketch the electric and magnetic field lines in these structures when excited in dominant modes.

(ii) A microstrip line is composed of zero thickness copper conductors on the lossless substrate having $\varepsilon_r = 9.0$ and 3 mm thickness. If the line width is 1 mm and operated at 9 GHz, calculate the effective dielectric constant of the medium, the characteristic impedance of the line, and guide wavelength.

(b) (i) Show schematically methods of excitation of dominant and TE_{01} and TM_{01} modes in a circular waveguide.

(ii) An air-filled circular waveguide having inner radius of 1 cm is excited at 10 GHz in the dominant mode. Find the guide wavelength, wave impedance, and the frequency bandwidth for operation in dominant mode only.

62. (a) An air-filled rectangular waveguide made of Cu sheet has internal cross section of 0.9" × 0.4" and is operated at 9375 MHz in dominant mode. Calculate the attenuation loss for a 10 cm length (conductivity of Cu = 5.8×10^7 mho/m). Derive the formula used.

(b) Prove that the frequency of minimum attenuation due to conductor loss in a parallel-plate waveguide for TM mode is $\sqrt{3}\, f_c$, where f_c is the cut-off frequency. Find the expression for this minimum attenuation.

63. **(a)** Show that the expression for unloaded Q of a circular cylindrical cavity excited in TM_{010} mode is $Q = 1.202\ \eta/[Rs(1 + a/d)]$, where $\eta = \sqrt{(\mu_0/\varepsilon_0)}$, Rs = surface resistance of cavity walls, a = radius, and d = the length of the cavity.

(b) **(i)** Explain how a cavity is formed from a rectangular waveguide ($a > b$) excited in dominant mode. Draw the electric and magnetic field configurations in the cross section and on the planes parallel to broad walls for dominant TE mode.

(ii) Explain how a circular cylindrical cavity is excited in TM_{010} mode using a waveguide slot and coaxial loop. Describe the method of tuning of this cavity.

64. A reflex klyston is operated with a dc beam voltage of 300 V and a beam current of 20 mA for generating a 10 GHz signal at 1 3/4 mode. If the repeller space is 5 mm and beam coupling coefficient is unity, calculate

(a) Maximum RF power output

(b) Corresponding repeller voltage

(c) Maximum electronic efficiency

(d) Bandwidth over a repeller voltage range $\Delta V_R = 2$ V

65. **(a)** **(i)** Derive an expression of cut-off magnetic field and voltage in a cylindrical magnetron oscillator considering the equation of electron trajectory.

(ii) Explain how a pure π mode can be excited for maximum efficiency. How is this oscillatory tuned to the desired frequency?

(b) Explain using a neat diagram how a TWT amplifier circuit is constructed and operated for amplification of a microwave signal. Explain the gain characteristic of TWTA and compare TWTA with a klystron amplifier.

66. **(a)** Describe the construction and operation of a GaAs Gunn diode. Explain the Gunn mode of operation. Using a neat diagram, show how the Gunn diode is mounted and biased in a rectangular waveguide.

(b) Using a neat diagram, explain the operation of a $n\text{-}^{+}p - i - p^{+}$ IMPATT diode with dc biasing and for microwave oscillation. Draw an electrical equivalent circuit of the device and power amplifier using a three-port circulator. Express the gain formula in terms of the circuit parameters.

67. **(a)** Describe with the help of a neat diagram how insertion loss and attenuation of a two-port device are measured using a laboratory-model microwave test bench. What are the main errors in these measurements? Explain how high VSWR > 20 can be measured by double minima method.

(b) **(i)** Explain using a neat diagram of a set-up, the measurement of an unknown complex impedance using a slotted line.

(ii) Describe how dielectric constant ε_r of a solid dielectric is measured using the waveguide method.

68. **(a)** Draw the stripline form of a branch-line direction coupler according to the schematic shown below. Write down its S-matrix when all ports are matched.

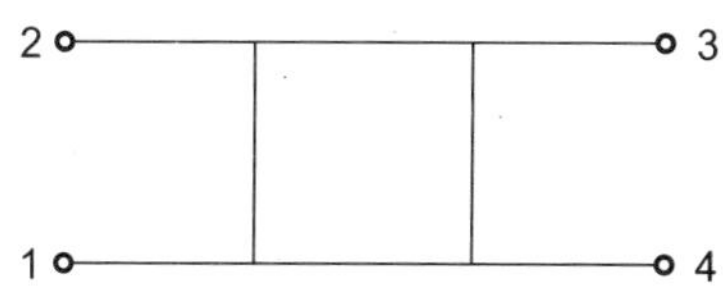

Find the coupling to Port 2 when the signal is fed to Port 1 with ports 3 and 4 connected to $\lambda/4$ short-circuited line section.

(b) **(i)** Prove that the output of a lossless microwave passive filter can be higher than the input when the terminating resistance value is less than the input generator resistance.

(ii) Using a neat diagram, explain the operation of a rotary precision phase shifter.

69. A 50-ohm Cu strip line is to be fabricated using a dielectric substrate of $\varepsilon_r = 4.0$, tan $\delta = 0.001$ and thickness = 0.4 cm between ground planes for operation at 10 GHz. If the Cu thickness is 0.01 mm, calculate (a) line width, and (b) attenuation constant in dB.

70. (a) Describe with a neat diagram the construction and operation of a microstrip circulator. Write its *S*-matrix. What determines the insertion loss and isolation performance of this device? Explain how such a device can be used as an isolator.

(b) Design a 50-ohm microstrip low-pass filter having a cut-off frequency of 2 GHz and 30 dB attenuation at 3.5 GHz for 0.1 dB ripple. Use alumina substrate of 2 mm thickness and dielectric constant 9.8. Outline the realization of the element values with a diagram. The prototype element values are $g_0 = g_6 = 1$, $g_1 = g_5 = 1.15$, $g_2 = g_4 = 1.37$, $g_3 = 1.98$.

71. **(a)** Derive the expression of coupling and directivity of an edge-coupled parallel stripline coupler using even- and odd-mode analysis.

(b) Describe the operation of a branch-line coupler using even- and odd-mode analysis. Determine the *S*-parameters of this coupler

72. **(a)** Describe the equivalent circuit of IMPATT diode using a neat diagram. Explain the impedance matching with load and the condition of stable oscillation. Describe the operation of IMPATT diode power amplifier circuit using a three-port circulator.

(b) Using a neat diagram, give the equivalent circuit of a Schottky diode and MIC mixer circuit using hybrid ring. Explain its operation.

73. **(a)** Describe how a varactor diode frequency quadrupler circuit is realized in microstrip form and explain its operation.

(b) How are amplifier stability at microwave frequencies stated in terms of *S*-parameters?

74. Design input and output matching networks for minimum noise figure of a narrow-band GaAs MESFET amplifier at 10 GHz with 50-ohm line impedance for

F min =3 dB

Optimum source reflection coefficient $\Gamma_{\text{sin}} = 0.541\angle{-140°}$

Load reflection coefficient $\Gamma_{\text{Lo}} = 0.543\angle{-105°}$

75. Derive an expression for the power output of a two-cavity klystron amplifier.

76. **(a)** Describe the principle of operation of a parametric amplifier using a varactor diode. Mention its uses.

(b) Explain the process of transfer electron effect and the working of a Gunn diode oscillator.

77. **(a)** How are the performance characteristics of a directional coupler determined? Derive the *S*-matrix of a 3 dB directional coupler.

(b) (i) State the basic principle of a phase shifter. (ii) Explain the construction and working principle of a precision waveguide phase shifter.

78. **(a)** Explain with proper experimental set-up how high VSWR can be measured. Derive the working formula used.

(b) (i) Explain antenna test ranges and their design criteria. (ii) Explain with experimental set-up how gain of an antenna can be measured with the three-antenna method.

79. An air-filled rectangular copper waveguide (a = 0.9") × (b = 0.4") is operated at 9.2 GHz in dominant mode. Find the expression for the attenuation constant due to conductor loss. Calculate the attention loss for a 20 cm length. Conductivity of copper is 5.8×10^7 mho/m.

80. **(a)** Establish the laws of reflection for a linearly polarized plane wave of parallel polarization incidents on a plane interface between two lossless dielectric regions 1 and 2 at an angle θ_i with respect to the normal interstate. Obtain the expression for Brewster angle as well as the angle of incidence for which total internal reflection occurs.

(b) A uniform plane wave at 2.45 GHz is transmitted through a medium having $\sigma = 2.17$ S/m, $\varepsilon = 47\varepsilon_0$, $\mu = \mu_0$. Find the complex propagation constant, phase velocity and the wave impedance of the medium. If the electric field magnitude is 10 V/m, find the time-average power flow per unit area.

81. (a) (i) Draw the physical configuration of cross section of strip and microstrip lines. What are the dominant modes in these lines? Sketch the electric and magnetic field lines in these structures when excited in dominant modes. (ii) A microstrip line is composed of zero-thickness copper conductors on the lossless substrate having $\varepsilon_r = 9.0$ and 3 mm thickness. If the line width is 1 mm and is operated at 9 GHz, calculate the effective dielectric constant of the medium, the characteristic impedance of the line and guide wavelength.

(b) Derive the expression for minimum attenuation for the TM mode in parallel plate waveguide filled with air dielectric. Calculate its value for copper conductors placed at 5 cm apart for TM_{30} mode.

82. (a) A 10 GHz signal is to be transmitted through a circular waveguide filled with air. Determine the inner diameter of the waveguide such that its lowest cut-off frequency is 20% below this original frequency. Find the bandwidth for operation of dominant mode only and the corresponding wave impedance.

(b) Derive the general field components of TM_{mn} waves in a circular waveguide. Describe how using coaxial line probes excite TE_{11} and TM_{01} modes in these waveguides.

83. The electric field of a uniform plane wave is given by $\mathbf{\acute{E}} = \hat{\mathbf{x}}\,20\sin(3\times10^8 t - \beta z) + \hat{\mathbf{y}}\,20\,(3\times10^8 t - \beta z)$. Find (a) phase velocity, (b) magnetic field, (c) wave impedance, and (d) the polarization of the wave.

84. An infinite parallel-plate waveguide has a lossless dielectric medium within two perfectly conducting plates separated by a distance b in the y-direction. The guide is excited in TM_n mode for propagation in the z-direction. Determine (a) the expression of time-average power flow, and (ii) the velocity at which energy propagates along the guide.

85. Find the displacement current associated with a copper wire ($s = 5.8 \times 10^7$ mho/m) in which the conduction current at 1 MHz is 10 mA. Find the associated electric field.

86. (a) An air-filled rectangular Cu waveguide has cross-sectional dimensions in the xy-plane as 4 cm × 2 cm. It transports 2 mW of energy in the dominant mode at 10 GHz. Determine the peak value of the electric field and the attenuation constant. Surface resistance of Cu is 1.8/100 ohm.

(b) The electric field strength of a uniform plane wave in free space is 1 V/m at 300 MHz. If a very large thick flat copper plate is placed normal to the direction of wave propagation, determine

(i) **E** and **H** fields at the surface of the plate
(ii) Skin depth
(iii) Conduction current density at the surface
(iv) Power loss per square metre of the surface area

87. (a) Explain with diagram the dc biasing circuit of a GaAs MESFET microwave transistor amplifier in CE configuration.

(b) How are low- and high-power MESFET structures realized in MIC form?

88. (a) Explain how ferrite exhibits nonreciprocal behaviour to a linearly polarized plane wave. Using neat diagrams, explain the construction and operation of a reciprocal ferrite phase shifter. What changes have to be incorporated to convert this into a nonreciprocal phase shifter?

(b) Describe the configuration and operation of a microstrip branch-line coupler. Using even-and odd-mode analysis, determine the S-parameters of this coupler.

89. Describe with a neat diagram the construction and operation of a microstrip circulator. Write its S-matrix, What determines the insertion loss and isolation performance of this device? Explain how such device can be converted into an isolator.

90. State different conditions of oscillation of a two-port microwave transistor oscillator. Show the schematic of a bipolar transistor oscillator in microstrip form together with its corresponding lumped equivalent circuit.

91. **(a)** Describe the equivalent circuit of an IMPATT diode using a neat diagram. Explain impedance matching with load and the condition of stable oscillation. Describe the operation of IMPATT diode power amplifier circuit using a three port circulator.

(b) Describe the thin-film hybrid technology for the production of MICs. How does it differs from thick-film hybrid technology? What are the advantages of monolithic MICs over hybrid ones?

92. Prove that the output of a lossless microwave passive filter can be higher than the input when the terminating resistance value is less than the input generator resistance.

93. Calculate the even- and odd-mode characteristic impedance for a coupled-line 180° hybrid coupler with 3 dB coupling and 50-ohm characteristic impedance. The given substrate has a dielectric constant of 10 and a thickness of 0.16 cm. The spacing between the lines is 0.064 cm.

94. A GaAs FET oscillator is to be designed at 10 GHz with transistor *S*-parameters:
$S_{11} = 0.9\angle 180°$, $S_{12} = 0.79\angle -98°$, $S_{21} = 0.89\angle -163°$, $S_{22} = \angle 0.2180°$. Outline the design steps using large signal measurements.

95. **(a)** An Si microwave transistor has a reactance = 1 ohm, transit-time cut-off frequency = 4 GHz, maximum E-field = 1.6×10^5 V/m and saturation drift velocity = 4×10^5 cm/s. Determine maximum allowable power.

(b) State the necessary and sufficient conditions for unconditional stability of a microwave transistor amplifier in terms of S- parameters.

96. The *S*-parameters of a microwave transistor are

$S_{11} = \angle 0.54165°$, $S_{12} = \angle 0.0920°$, $S_{21} = \angle 2.030°$, $S_{22} = 0.5\angle -80°$. The input and output matching networks are designed to produce reflection coefficients towards input and output $0.49\angle -150°$ and $0.5\angle 690°$, respectively. Determine transducer power gain, available power gain, operating power gains and power delivered to the 50-ohm load when a 50-ohm 10 $\angle 0°$ signal source feeds the amplifier.

97. Explain with a diagram the dc biasing circuit of a GaAs MESFET microwave transistor amplifier using a source-coupled bypass resistance.

98. **(a)** Explain the properties of a magic-T. Derive the *S*-matrix of a matched magic-T with collinear arms 1 and 2, and difference and sum arms 3 and 4, respectively.

(b) What are the relations between *ABCD* parameters with *S*-parameters? What is the advantage of *ABCD* parameters over those of *S*-parameters?

99. Derive the *S*-matrix and find the *S*-parameters of the 3 dB attenuator circuit shown below:

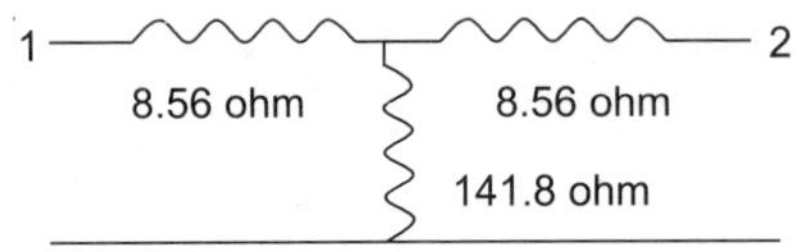

100. **(a)** What are the advantages of monolithic MICs over hybrid ones? Describe the method of fabrication of a lumped capacitor, inductor and resistor in microstrip form. How is a three-port circulator realized in microstrip form?

(b) Describe with a neat sketch the operation and design considerations of a parallel-coupled bandpass filter using J-*K* inverters. Give the design steps of this type of filter.

101. **(a)** A microwave transistor has the following *S*-parameters at 10 GHz, with a 50-ohm reference impedance:
$S_{11} = 0.45{+}150°$, $S_{12} = 0.01\ {-10°}$, $S_{21} = 2.05{+}10°$, $S_{22} = 0.40{-}150°$
The source impedance Z_s = 20 ohms and the load impedance Z_L = 30 ohms. Compute the power gain, the available gain, and the transducer gain.

(b) Define two types of stability of microwave amplifiers. Express the conditions of stability in terms of S-parameters of GaAs FET. How are these conditions simplified for a unilateral device?

102. The S-parameters of a GaAs FET at 2 GHz with bias voltage V_{gs} = O are given as (Z_0 = 50 ohms), $S_{11} = 0.894\angle{-60.6°}$, $S_{21} = 3.122\angle 123.6°$, $S_{12} = 0.020\angle 62.4°$, $S_{22} = 0.781\angle{-27.6°}$. Determine the stability of this transistor from the necessary and sufficient conditions.

103. **(a)** Establish the condition for oscillation of a microwave transistor circuit in terms of terminating and load network reflection coefficients.

(b) Design a transistor oscillator at 4 GHz using a GaAs FET in a common gate configuration, with a 5 nH inductor in series with the gate to increase the instability. Choose a terminating network to match to a 50-ohm load and an appropriate tuning network. The S-parameters of the transistor in a common source mode are (Z_0 = 50 ohms): $S_{11} = 0.72\angle{-116°}$, $S_{21} = 60\angle 76°$, $S_{12} = 0.03\angle 57°$, $S_{22} = 0.73\angle{-54°}$.

104. A telephone line has R = 6 ohms/km, L = 2.2 mH/km, C = 0.005 F/km and G = 0.05 μmho/km. Determine Z_0, α, β at 1 kHz.

105. Derive the expression for

(a) The input impedance of a high-frequency line of length L and terminated by a load Z_L

(b) A 50-ohm lossless line connects a signal of 100 kHz to a load of 100 ohms.

106. The load power is 100 mW. If the generator end is matched, calculate the

(a) Voltage reflection coefficient and VSWR

(b) Position of first V_{min} and V_{max}

(c) Impedance at V_{min} and V_{max}

107. **(a)** A 300-ohm lossless line is fed by a matched generator of voltage = $10\angle 0°$ volts. The line is 100 m long and is terminated by a resistance of 200 ohms. Calculate the return loss and transmission loss.

(b) Distinguish between the characteristic impedance and the input impedance of a line. Under what conditions will they be the same? In a lossless line, how can the input impedance be made purely inductive capacitive, infinite and zero by adjusting only line length?

108. **(a)** With the help of a schematic cross-sectional diagram, explain the electric and magnetic field distributions in strip and microstrip lines. How does their characteristic impedance vary with line width?

(b) Explain how a lossless mismatched line is matched by a single shunt stub using a Smith Chart.

109. **(a)** Explain the use of delay equalizers. Give the general method of designing a lattice delay equalizer.

(b) Design an m-derived low-pass T-section filter to meet the specifications.

(i) Cut-off frequency = 2000 c/s

(ii) Frequency of infinite attenuation = 2050 c/s

(iii) Characteristic impedance R_0 = 500 ohms

C

STANDARD RELATION BETWEEN CARTESIAN (x, y, z), CYLINDRICAL (ρ, ϕ, z), AND SPHERICAL (r, θ, ϕ) COORDINATES

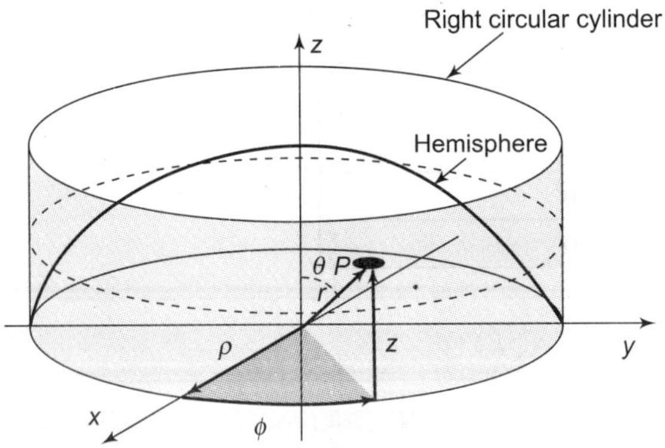

Cartesian and cylindrical	*Cylindrical and spherical*	*Cartesian and spherical*
$x = \rho \cos \phi$	$\rho = r \text{ sine } \theta$	$x = \text{r} \cos \phi \sin \theta$
$r = \sqrt{x^2 + y^2}$	$\phi = \phi$	$r = \sqrt{x^2 + y^2 + z^2}$
$y = \rho \sin \phi$	$z = r \cos \theta$	$y = r \sin \phi \sin \theta$
$\phi = \tan^{-1}\left(\frac{y}{x}\right)$	$r = \sqrt{\rho^2 + z^2}$	$\phi = \tan^{-1}\left(\frac{y}{x}\right)$
$z = z$	$\theta = \tan^{-1}\left(\frac{\rho}{z}\right)$	$z = r \cos \theta$
		$\theta = \tan^{-1} \frac{\sqrt{x^2 + y^2}}{z} = \cos^{-1} \frac{z}{r}$

D VECTORS

UNIT VECTORS D.1

Rectangular in terms of Spherical Coordinates

$$\hat{\mathbf{x}} = \hat{\mathbf{r}}\sin\theta\cos\phi + \hat{\boldsymbol{\theta}}\cos\theta\cos\phi - \hat{\boldsymbol{\phi}}\sin\phi$$

$$\hat{\mathbf{y}} = \hat{\mathbf{r}}\sin\theta\sin\phi + \hat{\boldsymbol{\theta}}\cos\theta\sin\phi - \hat{\boldsymbol{\phi}}\cos\phi$$

$$\hat{\mathbf{z}} = \hat{\mathbf{r}}\cos\theta - \hat{\boldsymbol{\theta}}\sin\theta$$

Rectangular in terms of Cylindrical Coordinates

$$\hat{\mathbf{x}} = \hat{\boldsymbol{\rho}}\cos\phi - \hat{\boldsymbol{\phi}}\sin\phi$$

$$\hat{\mathbf{y}} = \hat{\boldsymbol{\rho}}\sin\phi + \hat{\boldsymbol{\phi}}\cos\phi$$

$$\hat{\mathbf{z}} = \hat{\mathbf{z}}$$

Spherical in terms of Rectangular Coordinates

$$\hat{\mathbf{r}} = \hat{\mathbf{x}}\sin\theta\cos\phi + \hat{\mathbf{y}}\sin\theta\sin\phi + \hat{\mathbf{z}}\cos\theta$$

$$\hat{\boldsymbol{\theta}} = \hat{\mathbf{x}}\cos\theta\cos\phi + \hat{\mathbf{y}}\cos\theta\sin\phi - \hat{\mathbf{z}}\sin\theta$$

$$\hat{\boldsymbol{\phi}} = -\hat{\mathbf{x}}\sin\phi + \hat{\mathbf{y}}\cos\phi$$

Spherical in terms of Cylindrical Coordinates

$$\hat{\mathbf{r}} = \hat{\boldsymbol{\rho}}\sin\theta + \hat{\mathbf{z}}\cos\theta$$

$$\hat{\boldsymbol{\theta}} = \hat{\boldsymbol{\rho}}\cos\theta - \hat{\mathbf{z}}\sin\theta$$

$$\hat{\boldsymbol{\phi}} = \hat{\boldsymbol{\phi}}$$

Cylindrical in terms of Rectangular Coordinates

$$\hat{\boldsymbol{\rho}} = \hat{\mathbf{x}}\cos\phi + \hat{\mathbf{y}}\sin\phi$$

$$\hat{\boldsymbol{\phi}} = -\hat{\mathbf{x}}\sin\phi + \hat{\mathbf{y}}\cos\phi$$

$$\hat{\mathbf{z}} = \hat{\mathbf{z}}$$

Cylindrical in terms of Spherical Coordinates

$$\hat{\boldsymbol{\rho}} = \hat{\mathbf{r}}\sin\theta + \hat{\boldsymbol{\theta}}\cos\theta$$

$$\hat{\boldsymbol{\phi}} = \hat{\boldsymbol{\phi}}$$

$$\hat{\mathbf{z}} = \hat{\mathbf{r}}\cos\theta - \hat{\boldsymbol{\theta}}\sin\theta$$

VECTOR IDENTITIES D.2

$$\mathbf{A} \times (\mathbf{B} \times \mathbf{C}) = (\mathbf{A}\cdot\mathbf{C})\mathbf{B} - (\mathbf{A}\cdot\mathbf{B})\mathbf{C}$$

$$(\mathbf{A} \times \mathbf{B}) \times \mathbf{C} = (\mathbf{C}\cdot\mathbf{A})\mathbf{B} - (\mathbf{C}\cdot\mathbf{B})\mathbf{A}$$

$$\nabla\cdot(\nabla \times \mathbf{A}) = 0$$

$$\nabla\times\nabla V = 0$$

$$\nabla\cdot\nabla V = \nabla^2 V$$

$$\nabla(U + V) = \nabla U + \nabla V$$

$$\nabla(\mathbf{A} + \mathbf{B}) = \nabla\cdot\mathbf{A} + \nabla\cdot\mathbf{B}$$

$$\nabla(UV) = V\nabla U + U\nabla V$$

$$\nabla\cdot(U\mathbf{A}) = \mathbf{A}\cdot(\nabla U) + U(\nabla\cdot\mathbf{A})$$

$$\nabla\times(U\mathbf{A}) = (\nabla U)\times\mathbf{A} + U(\nabla\times\mathbf{A})$$

$$\nabla\times(\nabla\times\mathbf{A}) = \nabla(\nabla\cdot\mathbf{A}) - \nabla^2\mathbf{A}$$

$$\nabla^2\mathbf{A} = \hat{x}\nabla^2 A_x + \hat{y}\nabla^2 A_y + \hat{z}\nabla^2 A_z$$

$$\nabla\cdot(\mathbf{A}\times\mathbf{B}) = \mathbf{B}\cdot(\nabla\times\mathbf{A}) + \mathbf{A}\cdot(\nabla\times\mathbf{B})$$

$$\mathbf{A}\cdot(\mathbf{B}\times\mathbf{C}) = \mathbf{B}(\mathbf{C}\times\mathbf{A}) = \mathbf{C}(\mathbf{A}\times\mathbf{B})$$

$$\nabla\cdot(\mathbf{A}\times\mathbf{B}) = \mathbf{A}(\nabla\cdot\mathbf{B}) - \mathbf{B}(\nabla\cdot\mathbf{A}) + (\mathbf{B}\cdot\nabla)\mathbf{A} - (\mathbf{A}\cdot\nabla)\mathbf{B}$$

$$\nabla(\mathbf{A}\cdot\mathbf{B}) = (\mathbf{A}\cdot\nabla)\mathbf{B} + (\mathbf{B}\cdot\nabla)\mathbf{A} + (\mathbf{B}\cdot\nabla)\mathbf{A} + \mathbf{A}\times(\nabla\times\mathbf{B}) + \mathbf{B}\times(\nabla\times\mathbf{A})$$

$$\iiint_V \nabla\cdot\mathbf{A}\partial v = \oiint_s \mathbf{A}\cdot d\mathbf{S} \quad \text{Divergence theorem}$$

$$\iint_s (\nabla\times\mathbf{A})\cdot d\mathbf{S} = \oint_l \mathbf{A}\cdot dl \quad \text{Stoke's theorem}$$

VECTOR DIFFERENTIAL OPERATORS D.3

Rectangular Coordinates

$$\nabla V = \hat{\mathbf{x}}\frac{\partial V}{\partial x} + \hat{\mathbf{y}}\frac{\partial V}{\partial y} + \hat{\mathbf{z}}\frac{\partial V}{\partial z}$$

$$\mathbf{A} = \hat{\mathbf{x}}\frac{\partial A_x}{\partial x} + \hat{\mathbf{y}}\frac{\partial A_Y}{\partial y} + \hat{\mathbf{z}}\frac{\partial A_Z}{\partial z}$$

$$\nabla \times \mathbf{A} = \hat{\mathbf{x}}\left(\frac{\partial A_Z}{\partial y} - \frac{\partial A_Y}{\partial z}\right) + \hat{\mathbf{y}}\left(\frac{\partial A_x}{\partial z} - \frac{\partial A_Z}{\partial x}\right) + \hat{\mathbf{z}}\left(\frac{\partial A_Y}{\partial x} - \frac{\partial A_x}{\partial y}\right)$$

$$\nabla^2 V = \frac{\partial^2 V}{\partial x^2} + \frac{\partial^2 V}{\partial y^2} + \frac{\partial^2 V}{\partial z^2}$$

Cylindrical Coordinates

$$\nabla V = \hat{\mathbf{r}}\frac{\partial V}{\partial x} + \hat{\boldsymbol{\phi}}\frac{1}{r}\frac{\partial V}{\partial y} + \hat{\mathbf{z}}\frac{\partial V}{\partial z}$$

$$\nabla \cdot \mathbf{A} = \frac{1}{r}\frac{\partial}{\partial r}(rA_r) + \frac{1}{r}\frac{\partial A_\phi}{\partial \phi} + \frac{\partial A_Z}{\partial z}$$

$$\nabla \times \mathbf{A} = \hat{\mathbf{r}}\left(\frac{1}{r}\frac{\partial A_Z}{\partial \phi} - \frac{\partial A_\phi}{\partial z}\right) + \hat{\phi}\left(\frac{\partial A_r}{\partial z} - \frac{\partial A_Z}{\partial r}\right) + \hat{\mathbf{z}}\frac{1}{r}\left(\frac{\partial}{\partial r}(rA_\phi) - \frac{\partial A_r}{\partial \phi}\right)$$

$$\nabla^2 V = \frac{1}{r}\frac{\partial}{\partial r^2}\left(r\frac{\partial V}{\partial r}\right) + \frac{1}{r^2}\frac{\partial^2 V}{\partial \varphi^2} + \frac{\partial^2 V}{\partial z^2}$$

Spherical Coordinates

$$\nabla V = \hat{\mathbf{r}}\frac{\partial V}{\partial r} + \hat{\boldsymbol{\theta}}\frac{1}{r}\frac{\partial V}{\partial \theta} + \hat{\boldsymbol{\phi}}\frac{1}{r\sin\theta}\frac{\partial V}{\partial \phi}$$

$$\nabla \cdot \mathbf{A} = \frac{1}{r^2}\frac{\partial}{\partial r}(r^2 A_r) + \frac{1}{r\sin\theta}\frac{\partial}{\partial \theta}(A_\theta \sin\theta) + \frac{1}{r\sin\theta}\frac{\partial A_\phi}{\partial \phi}$$

$$\nabla \times \mathbf{A} = \hat{\mathbf{r}}\frac{1}{r\sin\theta}\left(\frac{\partial}{\partial \phi}(A_\phi \sin\theta) - \frac{\partial A_\theta}{\partial \phi}\right) + \hat{\boldsymbol{\theta}}\frac{1}{r}\left(\frac{1}{\sin\theta}\frac{\partial A_r}{\partial \phi} - \frac{\partial}{\partial r}(rA_\phi)\right) + \hat{\boldsymbol{\phi}}\frac{1}{r}\left(\frac{\partial}{\partial r}(rA_\theta) - \frac{\partial A_r}{\partial \theta}\right)$$

$$\nabla^2 V = \frac{1}{r^2}\frac{\partial}{\partial r}\left(r^2\frac{\partial V}{\partial r}\right) + \frac{1}{r^2\sin\theta}\frac{\partial}{\partial \theta}\left(\sin\theta\frac{\partial V}{\partial \theta}\right) + \frac{1}{r^2\sin^2\theta}\frac{\partial^2 V}{\partial \phi^2}$$

E

Table E.1(a) *S* parameters in terms of *ABCD*, *Y*, and *Z* parameters

	ABCD	*Y*	*Z*
S_{11}	$\frac{A + B/Z_0 - CZ_0 - D}{A + B/Z_0 + CZ_0 + D}$	$\frac{(Y_0 - Y_{11})(Y_0 + Y_{22}) + Y_{12}Y_{21}}{(Y_0 + Y_{11})(Y_0 + Y_{22}) - Y_{12}Y_{21}}$	$\frac{(Z_{11} - Z_0)(Z_{22} + Z_0) - Z_{12}Z_{21}}{(Z_{11} + Z_0)(Z_{22} + Z_0) - Z_{12}Z_{21}}$
S_{12}	$\frac{2(AD - BC)}{A + B/Z_0 + CZ_0 + D}$	$\frac{-2Y_{12}\,Y_0}{(Y_0 + Y_{11})(Y_0 + Y_{22}) - Y_{12}Y_{21}}$	$\frac{2Z_{12}Z_0}{(Z_{11} + Z_0)(Z_{22} + Z_0) - Z_{12}Z_{21}}$
S_{21}	$\frac{2}{A + B/Z_0 + CZ_0 + D}$	$\frac{-2Y_{21}\,Y_0}{(Y_0 + Y_{11})(Y_0 + Y_{22}) - Y_{12}Y_{21}}$	$\frac{2Z_{21}Z_0}{(Z_{11} + Z_0)(Z_{22} + Z_0) - Z_{12}Z_{21}}$
S_{22}	$\frac{-A + B/Z_0 - CZ_0 + D}{A + B/Z_0 + CZ_0 + D}$	$\frac{(Y_0 + Y_{11})(Y_0 - Y_{22}) + Y_{12}Y_{21}}{(Y_0 + Y_{11})(Y_0 + Y_{22}) - Y_{12}Y_{21}}$	$\frac{(Z_{11} + Z_0)(Z_{22} - Z_0) - Z_{12}Z_{21}}{(Z_{11} + Z_0)(Z_{22} + Z_0) - Z_{12}Z_{21}}$

Table E.1(b) *ABCD* parameters in terms of *S*, *Y*, and *Z* parameters

	S	*Y*	*Z*
A	$\frac{(1 + S_{11})(1 - S_{22}) + S_{12}S_{21}}{2S_{21}}$	$\frac{-Y_{22}}{Y_{21}}$	$\frac{Z_{11}}{Z_{21}}$
B	$Z_0\frac{(1 + S_{11})(1 + S_{22}) - S_{12}S_{21}}{2S_{21}}$	$\frac{-1}{Y21}$	$\frac{Z_{11}Z_{22} - Z_{12}Z_{21}}{Z_{21}}$
C	$\frac{1}{Z_0}\frac{(1 - S_{11})(1 - S_{22}) - S_{12}S_{21}}{2S_{21}}$	$\frac{-(Y_{11}Y_{22} - Y_{12}Y_{21})}{Y_{21}}$	$\frac{1}{Z_{21}}$
D	$\frac{(1 - S_{11})(1 + S_{22}) + S_{12}S_{21}}{2S_{21}}$	$\frac{-Y_{11}}{Y_{21}}$	$\frac{Z_{22}}{Z_{21}}$

Table E.1(c) *Y* parameters in terms of S, *ABCD*, and *Z* parameters

	S	*ABCD*	*Z*
Y_{11}	$Y_0 \dfrac{(1-S_{11})(1+S_{22})+S_{12}S_{21}}{(1+S_{11})(1+S_{22})-S_{12}S_{21}}$	$\dfrac{D}{B}$	$\dfrac{-Z_{22}}{Z_{11}Z_{22}-Z_{12}Z_{21}}$
Y_{12}	$Y_0 \dfrac{-2S_{12}}{(1+S_{11})(1+S_{22})-S_{12}S_{21}}$	$\dfrac{-(AD-BC)}{B}$	$\dfrac{-Z_{12}}{Z_{11}Z_{22}-Z_{12}Z_{21}}$
Y_{21}	$Y_0 \dfrac{-2S_{21}}{(1+S_{11})(1+S_{22})-S_{12}S_{21}}$	$\dfrac{-1}{B}$	$\dfrac{-Z_{21}}{Z_{11}Z_{22}-Z_{12}Z_{21}}$
Y_{22}	$Y_0 \dfrac{(1+S_{11})(1-S_{22})+S_{12}S_{21}}{(1+S_{11})(1+S_{22})-S_{12}S_{21}}$	$\dfrac{A}{B}$	$\dfrac{+Z_{11}}{Z_{11}Z_{22}-Z_{12}Z_{21}}$

Table E.1(d) *Z* parameters in terms of *S*, *ABCD*, and *Y* parameters

	S	*ABCD*	*Y*
Z_{11}	$Z_0 \dfrac{(1+S_{11})(1-S_{22})+S_{12}S_{21}}{(1-S_{11})(1-S_{22})-S_{12}S_{21}}$	$\dfrac{A}{C}$	$\dfrac{Y_{22}}{Y_{11}Y_{22}-Y_{12}Y_{21}}$
Z_{12}	$Z_0 \dfrac{2S_{12}}{(1-S_{11})(1-S_{22})-S_{12}S_{21}}$	$\dfrac{(AD-BD)}{C}$	$\dfrac{-Y_{12}}{Y_{11}Y_{22}-Y_{12}Y_{21}}$
Z_{21}	$Z_0 \dfrac{2S_{21}}{(1-S_{11})(1-S_{22})-S_{12}S_{21}}$	$\dfrac{1}{C}$	$\dfrac{-Y_{21}}{Y_{11}Y_{22}-Y_{12}Y_{21}}$
Z_{22}	$Z_0 \dfrac{(1-S_{11})(1+S_{22})+S_{12}S_{21}}{(1-S_{11})(1-S_{22})-S_{12}S_{21}}$	$\dfrac{D}{C}$	$\dfrac{Y_{11}}{Y_{11}Y_{22}-Y_{12}Y_{21}}$

F

SOME INTEGRALS F.1

$$Si(x) = \int_0^x \frac{\sin\tau}{\tau} d\tau \quad \text{sine integral}$$

$$Ci(x) = -\int_x^\infty \frac{\cos\tau}{\tau} d\tau \quad \text{cosine integral}$$

$$Cin(x) = \int_0^x \frac{1-\cos\tau}{\tau} d\tau = 0.552 + \ln(x) - Ci(x)$$

$$\left.\begin{aligned} C(x) &= \int_0^x \cos\left(\frac{\pi}{2}\tau^2\right)d\tau; \quad C(-x) = -C(x) \\ S(x) &= \int_0^x \sin\left(\frac{\pi}{2}\tau^2\right)d\tau; \quad S(-x) = -S(x) \end{aligned}\right\} \text{Fresnel integrals}$$

PLOTS OF SPECIAL FUNCTIONS F.2

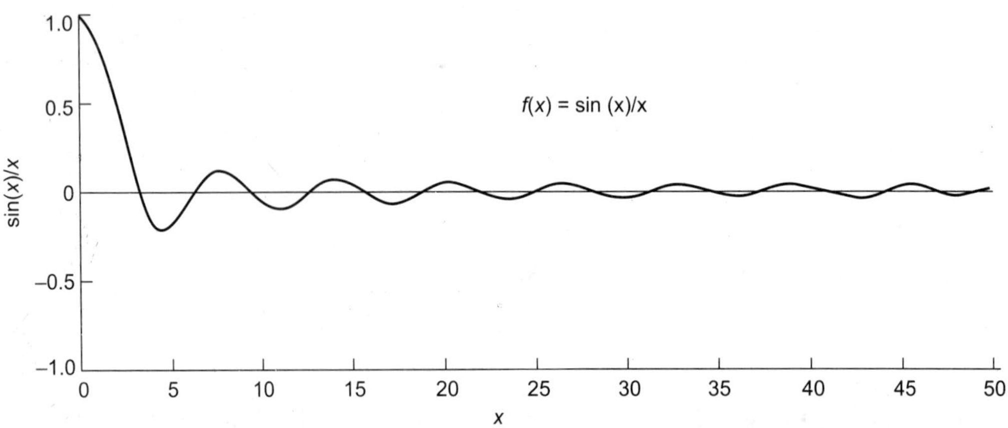

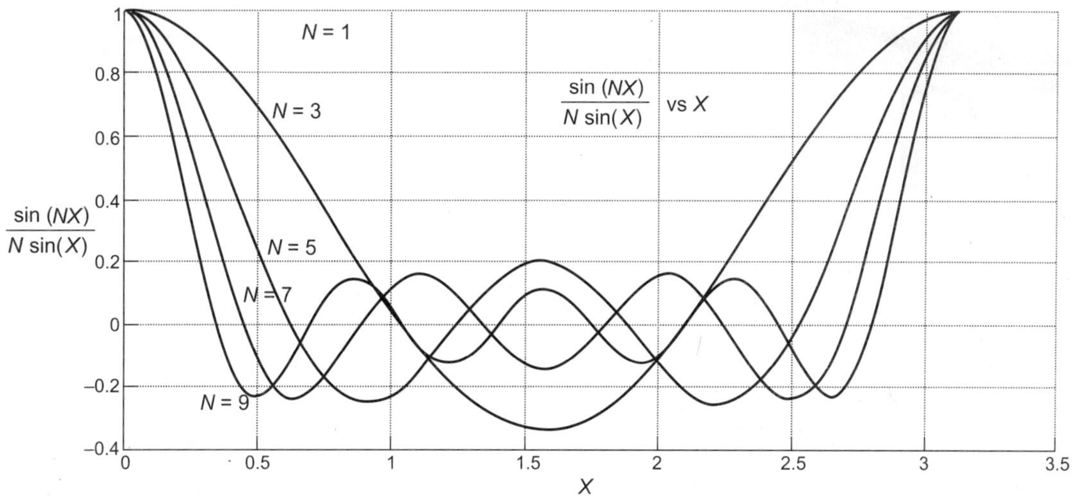

N = 1
N = 3
N = 5
N = 7
N = 9
sin (NX) / N sin(X) vs X
sin (NX) / N sin(X)
1
0.8
0.6
0.4
0.2
0
−0.2
−0.4
0
0.5
1
1.5
2
2.5
3
3.5
X

G

CYLINDRICAL BESSEL FUNCTION

$J_n(x)$ = Bessel function of first kind of order n

$Y_n(x)$ = Bessel function of second kind of order n

$H_n^{\,1}(x)$ = Hankle function of first kind of order n = Bessel function of third kind of order n

$H_n^{\,2}(x)$ = Hankle function of second kind of order n

$$H_n^{\,1}(x) = J_n(x) + j\,Y_n(x)$$

$$H_n^{\,2}(x) = J_n(x) - j\,Y_n(x)$$

$$J_n(x) = \frac{2(n-1)}{x} J_{n-1}(x) - J_{n-2}(x)$$

$$\frac{d}{dx}[x^m J_m(x)] = x^m J_{m-1}(x)$$

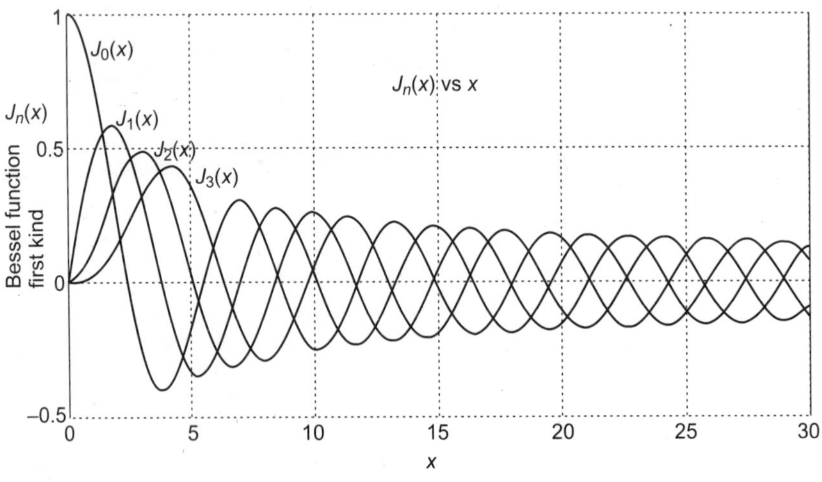

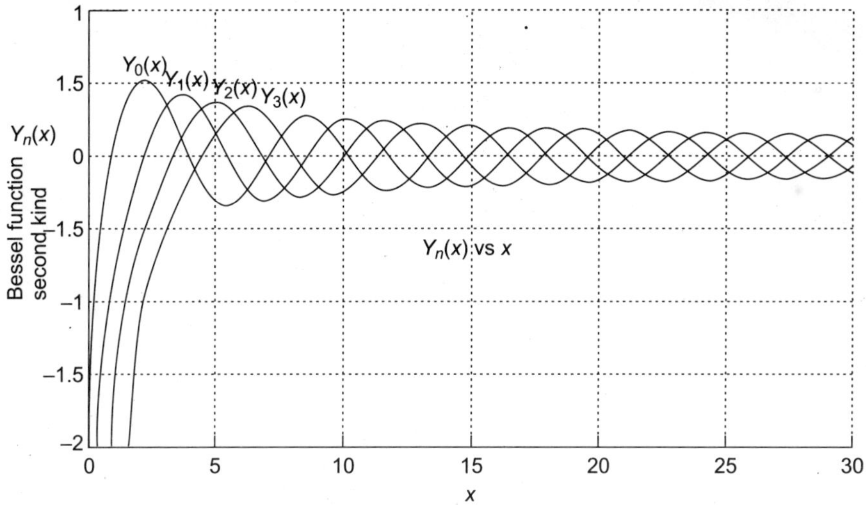
$Y_0(x)$
$Y_1(x)$
$Y_2(x)$
$Y_3(x)$
Bessel function second kind $Y_n(x)$
$Y_n(x)$ vs x
x

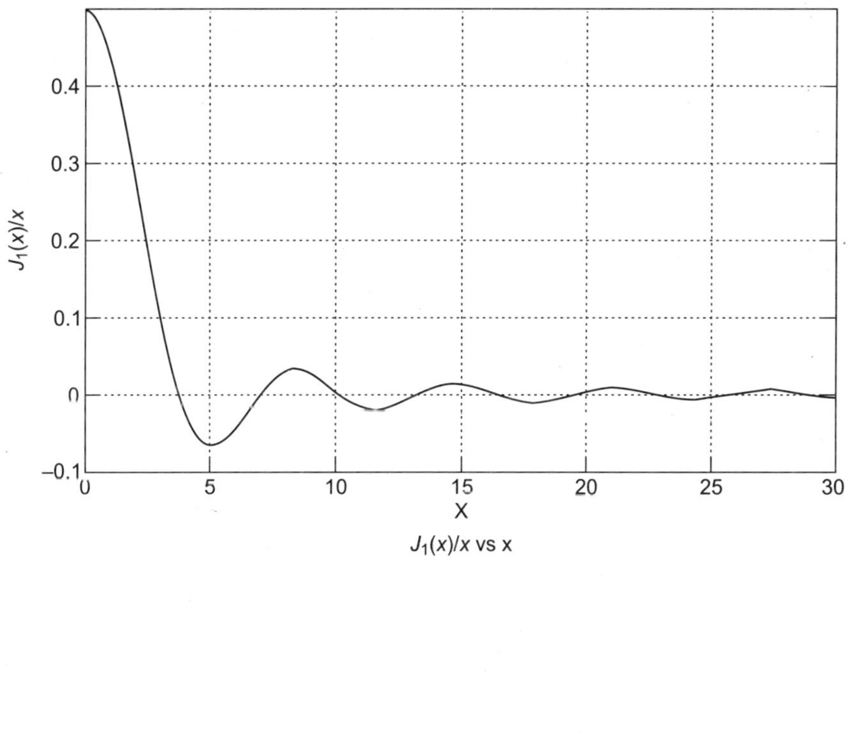
$J_1(x)/x$
X

$J_1(x)/x$ vs x

INDEX

P

Q

R

S